AFRICA
South Africa
Mpumalanga
Gauteng
North West Province
Free State
KwaZulu-Natal
Northern Cape
Eastern Cape
Western Cape
Villa Nora
Marken
1022m
Lephalala
Goud
1453m
Kloof Pass
Melkrivier
Melk
Lephalala
Mogalakwena
24°S
Dorset
Sukses
G
R
E
B
R
kolo
Hanglip
1768m
Mokopane
(Potgietersrus)
24 Rivers
Geelhoutkop
1739m
alwater
1304m
Bokpoort Pass
Doorndraai Dam
Dwars
Sand
Sandrivierspoort
Alma
Mookgophong
(Naboomspruit)
Nyl
1086m
1469m
Nylsvley
Modimolle
(Nylstroom)
SPRINGBOK FLATS
Bela-Bela
(Warmbad)
1080m
Settlers
29°E
N1

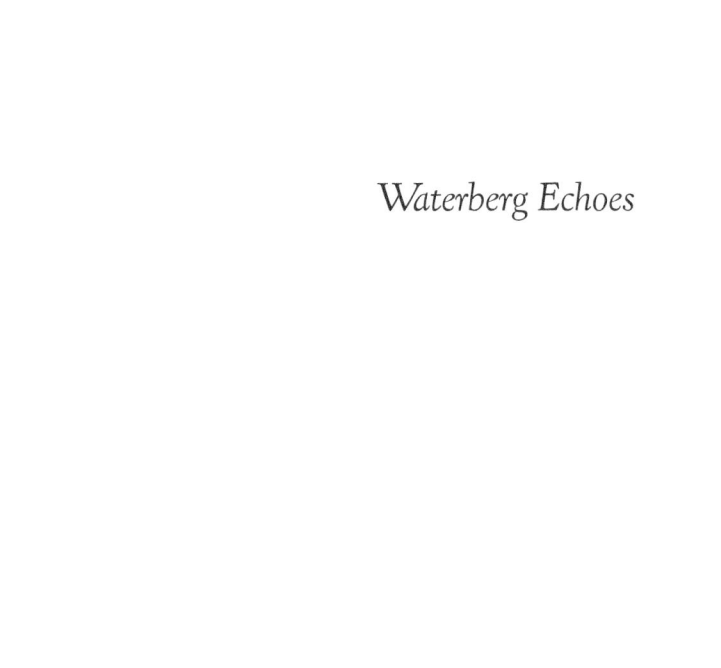

Waterberg Echoes

Waterberg Echoes

RICHARD WADLEY

Richard Wadley
September 2019

Protea Book House
PRETORIA
2019

WATERBERG ECHOES
Richard Wadley

First edition, first impression in 2019 by Protea Book House

PO Box 35110, Menlopark, 0102
1067 Burnett Street, Hatfield, Pretoria
8 Minni Street, Clydesdale, Pretoria
protea@intekom.co.za
www.proteaboekhuis.com

EDITOR: Danél Hanekom
PROOFREADER: Carmen Hansen-Kruger
COVER AND BOOK DESIGN: Hanli Deysel
FRONT COVER IMAGE: "Geelhoutkop" by Richard Wadley
MAPS AND SKETCHES: Colin Bleach
SET IN 10,5 on 15 pt ZapfCalligraphy
PRINTED IN CHINA THROUGH COLORCRAFT LTD., HK

ISBN: 978-1-4853-0935-2 (printed book)
ISBN: 978-1-4853-0936-9 (e-book)
ISBN: 978-1-4853-0937-6 (ePub)

Original text © 2019 Richard Wadley
Published edition © 2019 Protea Book House

No part of this book may be reproduced or transmitted in any form or by any electronic or mechanical means, including photocopying and recording or by any other information storage or retrieval system, without written permission from the publisher.

TO EWAN, ALASDAIR AND BEN:
You are also of this land.

EDITORIAL NOTE

There are many farms in the Waterberg with the same name, which can cause a great deal of confusion. However, each farm has a unique registration number. In the interest of simplicity, these have been excluded from the text, but are shown after names on accompanying maps and also in the index.

Table of contents

FOREWORD 9
LIST OF MAPS 11

CHAPTER 1:	The chasing of tales	13
CHAPTER 2:	Digging into the past (by Lyn Wadley)	19
CHAPTER 3:	Homeland or a place of refuge?	45
CHAPTER 4:	The Wild West: Swaershoek and Rankin's Pass	85
CHAPTER 5:	Baltic connections	119
CHAPTER 6:	The gateway towns	148
CHAPTER 7:	Vaalwater and the plateau settlements	210
CHAPTER 8:	The gap through the hills	252
CHAPTER 9:	The New Belgium story	278
CHAPTER 10:	Sondagsloop and Tarentaalstraat: a Sunday stroll among the guinea fowl?	342
CHAPTER 11:	Making tracks to the plateau	376
CHAPTER 12:	Pests and pestilence	395
CHAPTER 13:	Fitting the land to its purpose	428
CHAPTER 14:	Matters of faith	468
CHAPTER 15:	A futile broedertwis	529
CHAPTER 16:	Domain of the Northern Lions	545
CHAPTER 17: :	Gaining the right to read	576
CHAPTER 18:	Rocks, dynamite and dreams	613
CHAPTER 19:	On the periphery	645
CHAPTER 20:	From Waterberg to Waterberg: the plight of the Herero	676

CONCLUSION 712
ENDNOTES 723
GLOSSARY AND ABBREVIATIONS 785
SELECTED BIBLIOGRAPHY 790
ACKNOWLEDGEMENTS 795
INDEX 801

FOREWORD

This book brings together an amazing part of South Africa, the Waterberg, and an amazing person, Richard Wadley. The result is not only a lively read but also a special contribution to recording the history of our country.

The Waterberg, situated in Limpopo Province, is different. It is not like other parts of South Africa where changes in topography blend into one another. It is like a cake rising abruptly out of extensive plains which extend to the north, east, south and west – though the side of the cake to the north has collapsed over time. The sides of the cake are towering sandstone buttresses visible from afar. Their weathered shapes and colours tell you that you are approaching a very special part of our country.

The cake is surrounded by many towns: Modimolle, Mookgophong, Mokopane, Lephalale, Thabazimbi and Bela-Bela. To ascend to the top of the cake – the Waterberg plateau – it is necessary to travel through one of many beautiful passes – Sandrivierspoort, Bakker's Pass, Bokpoort, the Kloof Pass, the Beauty Road and many less well-known access routes from villages such as Jakkalskuil, Skrikfontein and others down the escarpment. Once on the plateau, a sense of isolation from the rest of South Africa starts to dominate one's senses. In the centre of the plateau lies the only town genuinely within the Waterberg, Vaalwater, a disorderly village where planning played no role.

The plateau is not endowed with agricultural potential – the soils are poor – and it has frustrated mining prospectors because of the absence of meaningful mineral deposits. It is, however, endowed with natural beauty and significant biodiversity – many crystal-clear streams and rivers, wetlands, beautifully shaped mountain-bushveld trees, a wide variety of flowering plants and wildlife, intriguing rock formations and deep kloofs as you approach the edges of the plateau, all offering a very different sense of place.

It has been my privilege to get to know Richard since his arrival in the Waterberg in 2002. Born in London, spending his childhood years in India, high school education in Southern Rhodesia at the time of the Central African Federation, obtaining his tertiary education at UCT and Wits resulting in an M.Sc. in Mining Engineering specialising in Geology and, finally, completing an advanced business management course at Harvard all played their part in shaping his wonderful personality. His very successful career in the mining business world was disrupted by corporate fragmentation and takeovers. To escape any more "corporate punishment" (his words), he and his brilliant wife Lyn moved to the Waterberg. They immediately set about engaging and building bridges with the three language groups – Afrikaans, English and Sepedi – that made up what is a typically divided population.

Richard, refreshingly free of prejudice, clearly enjoys interacting with people of all walks of life and, because of his obvious interest in others and very attentive manner, people equally enjoy engaging in deep conversation with him. He is blessed with a brilliantly sharp mind, a lovely sense of humour backed up by twinkling eyes and an enviable level of energy. He has embedded himself in the Waterberg by giving generously of his time and skills to community activities – chairman of the Waterberg Nature Conservancy, chairman of the Waterberg Academy – a local private school – member of the Vaalwater Community Forum, the Boerevereniging and the Community Police Forum, lecturing at the Lapalala Wilderness School, managing a very effective fire warning and call-out system which has significantly reduced the impact of fires in the dry season, organising successful meetings between disgruntled farmers and Eskom.

The networks he has established combined with his natural talents and fascination for the Waterberg and all its aspects have resulted in an important contribution to South Africa's historical literature. History books telling the national story make a very limited contribution unless underpinned by substantial recordings of local history. Local history always intersects with national history and, in so doing, makes our national history real. This book greatly enriches the large body of literature in South Africa focusing on our local history. It goes further than many other such publications by covering a much greater time span and also by embracing so many different aspects of local history.

This is an important work. At a time in our history where there is great uncertainty about the road that our country will take, it is especially worthwhile having a longer-term perspective on a very unusual and special part of our country such as is offered by this book. A longer-term view can be so helpful in trying to understand what has brought us to where we are and what progress can be made through human struggle and determination to move forward.

SAM VAN COLLER
"Lindani"

LIST OF MAPS

Location of major sites in the northwestern Limpopo Province	20
Location of major archaeological sites in the Waterberg	25
Subcontinental migrations in the fifteenth to eighteenth centuries	51
The difaqane and Boer trekkers clash in the early nineteenth century	57
Principal chiefdoms surrounding Waterberg in the mid-nineteenth century	60
Location of major historical settlements and sites in the Waterberg region	65
Important sites and settlements in the Mogalakwena River valley	67
Swaershoek and the Alma valley	95
The Waterberg journey of Carl Mauch in 1869	128
The gateway towns of Modimolle, Mookgophong & Bela-Bela	148
Modimolle and surrounds	162
Mookgophong and surrounds	185
Bela-Bela and surrounds	207
Vaalwater and surrounds	214
Bulgerivier and surrounds	231
The Witfontein area	234
The Zanddrift and Twenty-Four Rivers area	242
Zandrivierspoort and surrounds	264
Location of the New Belgium Block of farms in the northern Waterberg	281
Location of the 60 farms comprising the New Belgium Block	286
New Belgium Estate during the government settlement scheme era	329
New Belgium in the early twenty-first century	338
The Dorset area	349
Tarentaalstraat	350
Loubad and surrounds	371
The Nylstroom–Vaalwater railway line as completed in 1925	386
Tsetse fly in sub-Saharan Africa	397
Early travellers' journeys around the Waterberg and Jeppe's tsetse zone	403
Historical and current distribution of tsetse fly in South Africa	407
Distribution of malaria risk in northern South Africa, 1938	418
Distribution of bilharzia risk in South Africa	422
Company landownership across the Waterberg, 1899	435

LIST OF MAPS

Roads and infrastructure in the Waterberg, 1910	441
The area around Zion City Moria	501
The 1914 Rebellion graves of Union Forces burghers on Zandfontein	533
The Zandfontein skirmish on 8 November 1914	539
The Waterberg District Municipality and outline of the Waterberg Biosphere Reserve	573
Main rural schools in the Waterberg	578
Geology and mining history of the Waterberg	627
The Grootegeluk complex and power stations today	641
Thabazimbi and surrounds	647
Matlabas and surrounds	652
Ellisras/Lephalale and surrounds	663
Marken and surrounds	665
Marken and Kloof Pass area	670
German South West Africa at the close of nineteenth century	683
The Waterberg plateau of north-central Namibia	687
The route followed by Samuel Maharero and his clan	693
Location of Waterberg farms to which Herero refugees dispersed, 1907–1923	700

CHAPTER ONE
The chasing of tales

"The Waterberg? Where's that?" is still a surprisingly common response from South Africans south of Pretoria when told of the Elysian land my wife and I have called home for the last seventeen years. On the internet, websites promoting the Limpopo Waterberg are thoroughly intermixed with those about a smaller plateau of the same name in northern Namibia – and as we shall see towards the end of this book, there is indeed a little-known relationship between them. Many confuse the town of Nylstroom (Modimolle) on the edge of the Waterberg with Nelspruit in Mpumalanga, which they know from fishing trips to Dullstroom or visits to the Kruger Park. Only those who've travelled by road to northern Botswana are likely to have heard of (and briefly passed through) Vaalwater, the only sizeable settlement on the Waterberg plateau. Indeed, for residents of the Waterberg, one of its attractions is that its marvellous biodiversity, vistas and scenic roads enjoy little attention from hurrying tour buses or the quadbike/4x4 fraternity, for whom there are nearer and easier places to explore for a quick outing.

What is the Waterberg anyway? In its broadest modern context, it could be taken to mean the area contained within the District Municipality of that name – an area of almost 45 000 km^2, making it by far the largest district in Limpopo Province and one which, incidentally, corresponds quite closely to the Waterberg magisterial district of more than a century ago (see map on endpapers). The municipality includes the towns of Bela-Bela, Lephalale, Modimolle (the seat), Mokopane, Mookgophong, Thabazimbi and Vaalwater.[1] Within it lies the "Waterberg" coalfield west of Lephalale and the "Waterberg" limb of the platinum-bearing Bushveld Complex north of Mokopane. News associated with mining and industrial developments of these mineral-rich areas has done much to harm the promotion of tourism in the pristine areas of the District Municipality. Careless journalists, bureaucrats and corporate public affairs spinners tend to use the name Waterberg to imbue their environmentally challenging industrial projects with an aura of mystery and romance.

For those of us fortunate to live on the Waterberg plateau – that elevated region in the centre of the District Municipality and now largely protected in a UNESCO-accredited Biosphere Reserve – this is the *real* Waterberg: an upland of magnificent hill ranges, hidden valleys, scenic vistas, virgin wetlands, perennial streams, sparse human population, limited infrastructural, agricultural or commercial development; and a burgeoning diversity and population of wildlife, both faunal and floral.

The ignorance of the general public about the Waterberg plateau is matched by the scant knowledge amongst most of the locally resident community of the region's fascinating and complex history. This is hardly surprising because, apart from a few generalised coffee-table publications,[2] which contained selective historical summaries interspersed with visual imagery, the story of the Waterberg has barely been told in any detail. There have been, it is true, several, mainly privately published (or unpublished) memoirs about specific individuals, extended families and communities.[3] Some works have focused on topics like the South African (Anglo-Boer) War, or on leading military and political figures whose careers included the Waterberg.[4] Historical notes have been compiled about individual farmers' associations and co-operatives or for publications to commemorate various anniversaries of churches, organisations and towns.[5] But with few exceptions, these were written in Afrikaans (or High Dutch), were unashamedly partisan and were of limited distribution. Almost all are now out of print and increasingly hard to find.

The *natural* history of the region – by which I mean its flora and fauna – has been well covered in particular by the work of the accomplished field guide Lee Gutteridge, whose book is still readily available.[6] Most recently, an outstanding website is being developed by Michèle and Warwick Tarboton to document the flora and fauna of the Waterberg.[7] And there are other writers too who have provided accessible accounts of the plateau's rich biodiversity.[8] But the *human* history of the Waterberg is far less documented, accessible or known.

My own interest in this field evolved out of a desire to learn about the lives of four people whose graves lie on our property in a private cemetery beneath a large fig tree. That project (which almost fifteen years later has not yet been fully concluded) led me to develop some familiarity with genealogy and the intricacies of the National Archives in Pretoria and introduced me to characters and events of a century and more ago.

In particular, a serendipitous encounter with the unpublished (but fortunately archived), meticulous journal of William Caine, a land manager on the New Belgium Estate during the South African War, lured me into an investigation of his

CHAPTER 1: THE CHASING OF TALES 15

background as well as the context of his solitary life on a remote cattle station in the middle of potentially life-threatening hostilities.[9] The story of New Belgium became the subject around which most of the other historical topics in this book have pivoted.

As my stumbling, inexpert research into Waterberg's history became known within the community, I found myself inundated with snippets of information, contacts and names of people to interview, sometimes with introductions, occasionally with a word of caution. Over the last decade and more, I have conducted interviews with scores of residents, present and past, many in far-flung retirement homes, some in the Cape or KwaZulu-Natal, some overseas, others still in

FIGURE 1.1: The splendour of the Waterberg plateau.
Images courtesy of Mich Veldman (LEFT) and Warwick Tarboton (RIGHT).

their original houses. Too many have passed on since our discussions, leaving me saddened by my memory of them, but also glad that at least I'd had the opportunity to record some of their recollections. Others died before I could have this privilege – and sometimes their unique mental archives were lost with their passing.

I was continually struck by the willingness – often excitement – expressed by people whose recollections I sought. People were invariably keen to share their versions of the past; perhaps their own families had grown tired of hearing them! Quite often, these did not entirely accord with what others had said, or with what was formally recorded, but this divergence merely served to remind me of how nuanced every historical recollection can be and of the caution that must be exercised when drawing on uncorroborated oral testimonies alone.

Storytellers abound in the community, but there are few more knowledgeable and entertaining than Louis Nel, born and bred in the Waterberg, a cattle farmer (breeding Huguenot) in the Melkrivier area where his father was once a headmaster, and a longstanding and prominent participant in agricultural and church affairs. Louis's integrity, intuitive sense of right and wrong and his practical compassion for others (he and his wife Ansie established a school for black children on their farm long before it was customary to do so) belie his unfashionable political conservatism. And his sense of humour is legion. Many are the marvellous anecdotes that I could have borrowed for inclusion in this book, were it not for the fervent hope that Louis will soon write one of his own. There are numerous other remarkable dyed-in-the-Waterberg characters whose contributions to the region have also added essential value to this book. They include the indefatigable Baber brothers, Charles and Colin, who emulated the energy and community spirit of their maternal grandfather, E.A. (Ted) Davidson, one of the early English-speaking settlers. Dynamic farmers both (Bonsmara cattle, tobacco and maize), they were active for decades in local agricultural associations and were staunch supporters of the church (St John's), founded by their great aunts. Their determination and ability to bridge the deep ravine separating English- from Afrikaans-speakers was so successful, at least in their lifetimes, that they are spoken of with reverence to this day. Sam Mafora, the teacher-turned inspector, who still lives in retirement in Vaalwater, was celebrated and respected for the maturity he brought to the transformation of education in the town in the 1990s.

Louis Trichardt, a descendant of the famous Boer Trekker and, equally proudly, of Johan Enslin, who led the Jerusalemgangers to the Nyl in the 1840s, still lives in the Swaershoek, where he was born. Another highly successful, prize-winning cattle farmer (Brahmins), he also spent many years as a political representative of his community, culminating in his role as the Member for Waterberg in the last Parliament before the elections of 1994. Dr Peter Farrant, the fourth generation of his family to live on the farm Hartebeestpoort outside Vaalwater, runs a successful diversified farming operation (including Bonsmara stud) and paediatric practice, but he was also instrumental in the establishment of a renowned (NGO) HIV/Aids-treatment facility in the town. With his wife Janet, he set up a thriving, highly successful secondary school on a portion of his property made available for the purpose.

The list would be incomplete without mention of Clive Walker, a name which is almost synonymous with Waterberg in conservation circles. Clive and his wife Conita were critical to the establishment of the Lapalala Wilderness, a huge private nature reserve founded in the 1980s through the passion and vision of Dale Parker, a Cape businessman. He used his authority to establish the Lapalala Wilderness School and then to co-found both the Waterberg Nature Conservancy and the Waterberg Biosphere Reserve, all of which endure many years later. Now in his eighties, Clive is still busy writing books, at least two of which are about his abiding passion for rhino conservation. Conita has recently published an account of her own life in the Waterberg.[10] With his family, Clive also recently opened an absorbing, multifaceted living museum on his property between Vaalwater and Melkrivier.

It is as well that there have been, and still are, such outstanding personalities to provide guidance and strength in the Waterberg. For despite its superlative scenery, this is, and has always been, a tough place in which to live and earn a living. Its history is replete with stories of prejudice, suffering, struggle and misfortune, exploitation of the weak by the powerful, of the ignorant by the cunning and of the oppressed by their rulers. Yet amongst them are accounts of outstanding selflessness, courage in adversity, endurance, faith, perseverance and hope. Together they weave a tapestry rich in the colours that comprise the spectrum of human nature. The history of the Waterberg is a microcosm of the subcontinent in which it is situated, albeit tinted by its own particularly harsh and challenging environment.

The story of the Waterberg is a history that simply had to be written before many of the sources essential to its compilation disappear, are lost or die. This is not to pretend that the present work is a *complete* history of the Waterberg and its peoples. There are many stories and communities in the region that have not been adequately addressed and many more that are already lost forever. However, it is hoped that this publication will at the very least have served to record, accurately and in a reasonably objective manner, the principal features of our past, in a way that is accessible to the interested reader, yet bears sufficient credibility that it can be referred to with confidence by future researchers.

CHAPTER TWO
Digging into the past

Limpopo Province has a long prehistory. Everything that is known about it is derived from the detailed analyses of painstaking excavations conducted at numerous sites across the region over the past hundred years or so. Interpretations of these analyses are made in the context of other similar information gleaned from excavations elsewhere on the subcontinent south of the Zambezi River as far as the Cape coast. In the broader Waterberg region, several archaeologists have been active over the years; the best-known of these include Prof. Revil Mason, Jan Aukema, Jan Boeyens and Maria van der Ryst.

As more sites are identified and excavated (or re-excavated), so the body of information grows; the explanatory models adapt to the new data and become more robust. Before the observations and reports made by early western travellers into the African interior from the seventeenth century onwards, there is no written record of the early history of the region. Even the prehistory of the last thousand years is based largely on archaeological excavation, contextualised and augmented where possible by oral narratives. Whilst inevitably limited in its ability to visualise many aspects of human social behaviour, archaeological data are able to provide powerful evidence of changes in environmental conditions in the development of cognitive thought amongst hominid and human societies and in the evolution of technologies and social organisation. Age-dating methods have become more diversified and reliable in the last few decades and the continuing accumulation of dating records is helping to build a credible model of the times in which our ancestors lived and died. Yet the record remains patchy in several areas. There are many known sites that still require excavation and others that merit reinterpretation, but the rate at which work proceeds is very largely dependent upon available funding. What follows is a consensual summary of the knowledge accumulated to date.

The earliest archaeological remains are from the Makapan Valley near the town of Mokopane (formerly Potgietersrus) (Figure 2.1). Here the record began more than three million years ago when Australopithecines (*Australopithecus afri-*

canus)¹ lived in the region of the modern Makapan Limeworks. The end of the long sequence of events in the Makapan Valley was the 1854 fatal siege of Chief Mokopane and over 1500 of his people in Historic Cave (today known as Gwaša Cave) (see Chapter 3). Not surprisingly, the Valley has been declared a World Heritage site, along with a suite of early hominid (human and evolving human) sites that includes Sterkfontein and Swartkrans within the Cradle of Humankind near Krugersdorp.

Limpopo boasts a second World Heritage site at Mapungubwe on the Limpopo River. The Mapungubwe Iron Age site represents South Africa's earliest example (AD 1220–1300) of a state headed by a king.² The hilltop site housed the royal palace and was where the iconic golden rhino, symbol of kingship, was found. While best known for its Iron Age golden treasures, exotic glass beads and hilltop rain-making sites,³ the Mapungubwe area also has a wealth of Stone Age sites, for example at Kudu Koppie.⁴

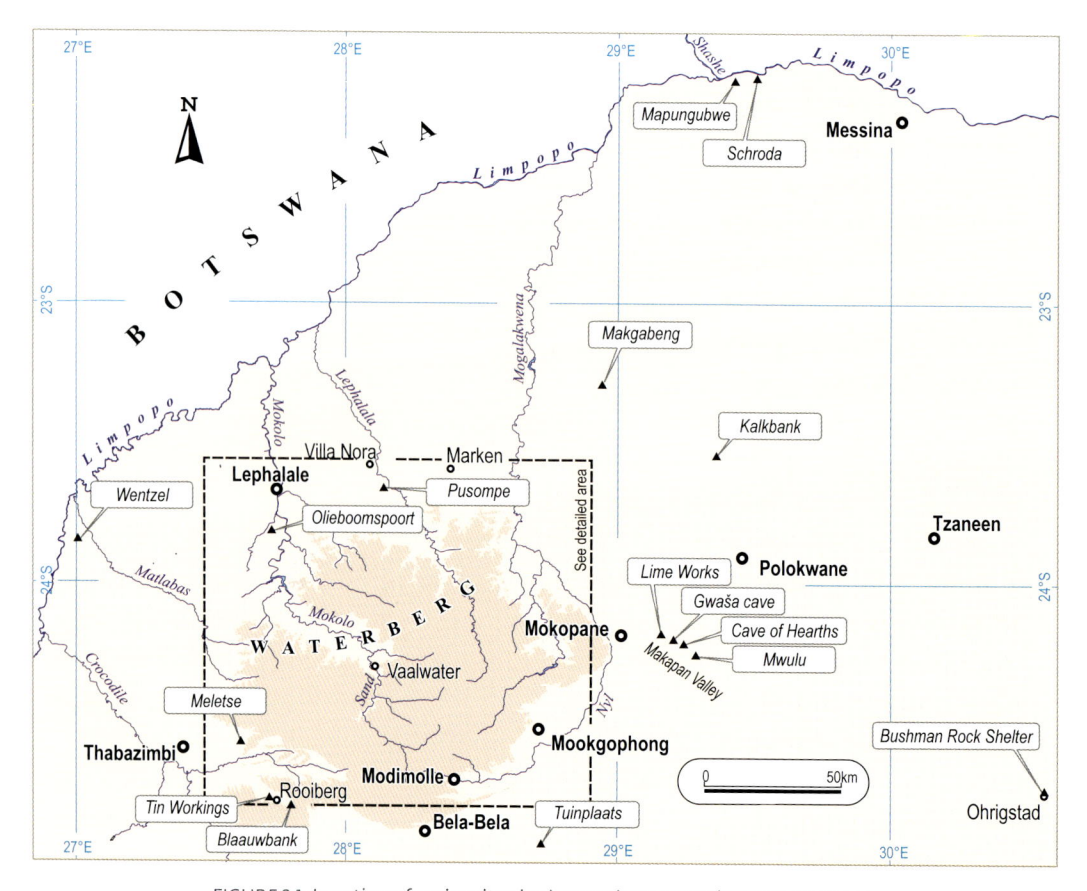

FIGURE 2.1: Location of major sites in the northwestern Limpopo Province.

No early hominid bones have been found in the high Waterberg, but it is possible that it was sometimes visited by early humans, such as *Homo ergaster* (the earliest known ancestor of *Homo sapiens* in southern Africa), who made and used cutting and butchery stone tools somewhere between about 1,5 million and 300 000 years ago. Such tools have been found in the Makapan Valley at Cave of Hearths,[5] which was studied from the 1940s to the 1980s. Notwithstanding its suggestive name, no evidence for controlled fire has been found in the cave, which is thought to have been occupied more recently than 780 000 years ago.[6] Yet, fires were probably lit there. For, prior to fire use, rock shelters and caves would have been too dangerous for human habitation; they would have been predator lairs. Fire had been used in the huge Wonderwerk cave near Kuruman, Northern Cape, more than 1 million years ago[7], so pyrotechnology is known to have been exploited well before the end of the Earlier Stone Age (ESA).

The Cave of Hearths sediments were heavily cemented (brecciated) by lime and were dynamited from the cliff face by Revil Mason (who was also the first archaeologist to hold a blasting licence). The deep sequence was thought to comprise eleven "beds" (members), with the three oldest ones (beds 1–3) containing stone tools and discarded animal bones from an Acheulean phase of the ESA.[8] Acheulean stone tools are characterised by large, multipurpose cutting tools like oval or pointed hand-axes and straight-edged cleavers. Bed 3, the youngest of the ESA occupations, produced the mandible of a hominid thought to be a precursor of *Homo sapiens*.[9] Such fossils are rare in South Africa perhaps because, unlike Australopithecines, early *Homo* communities processed or buried their dead away from their living quarters. Although no dates are available for Cave of Hearths, Bed 3 is thought to be about 300 000 years old. Ancient *Homo sapiens* skeletons are absent from Cave of Hearths and are infrequently found in Limpopo, though one recovered from Tuinplaats on the Springbok Flats was dated to less than 20 000 years ago and possibly as recently as 11 000 years ago.[10]

In the ESA, people were unlikely to have been *active* hunters of large game; instead, they would have scavenged opportunistically, retrieving carcasses trapped in mud or chasing predators from their prey. The ESA beds at Cave of Hearths contained remains of many antelope and zebra and the bones were hammered open to extract marrow.[11]

Beds 4–9 at Cave of Hearths yielded Middle Stone Age (MSA) (which from other sites is given an age range of 300 000 to 30 000 years ago) stone tools that include many blades and unifacial points (for examples, see Figure 2.2). Blades are long, sharp, ribbon-like cutting tools whereas unifacial points are triangular flakes

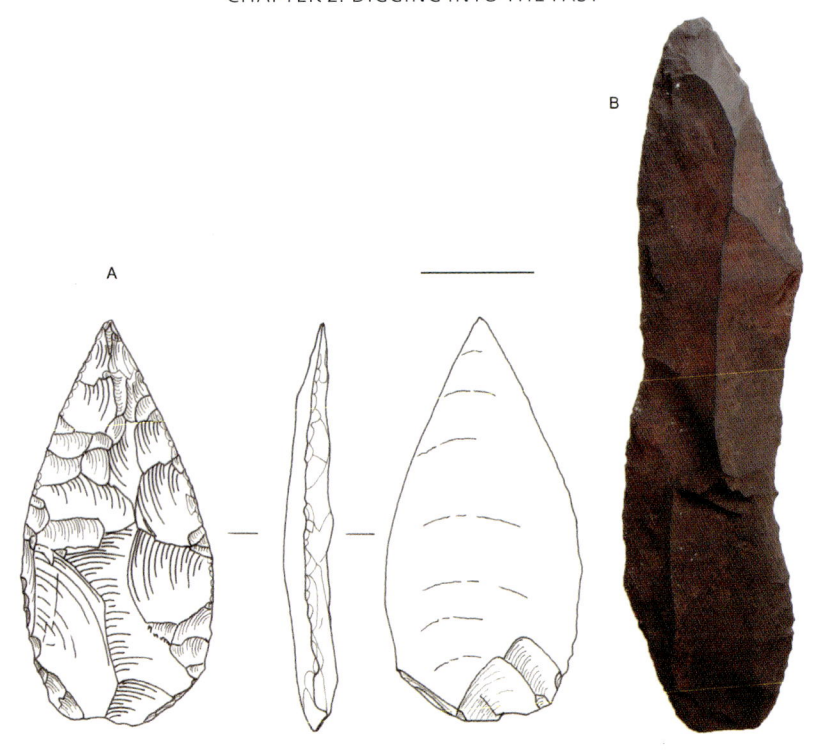

FIGURE 2.2: Middle Stone Age stone tools. A: Three views of a unifacial point. Note that the secondary shaping is on one face only. B: Blade with long, sharp cutting edges. The scale bar is 2 cm.

sharpened on one face to form pointed tips thought to be suitable for hafting as the tips of spears. Unlike their ESA predecessors, people in the MSA were effective and daring hunters. They targeted large and medium-sized animals including buffalo, eland, zebra, wildebeest and hartebeest. We know that they actively hunted dangerous animals because at Klasies River Mouth (near Storms River in the southern Cape) a buffalo vertebra had a stone spear-tip embedded in it.[12] In addition, both Cape buffalo and long-horned (extinct) buffalo were hunted at Cave of Hearths.[13] Elsewhere, researchers have found microscopic traces of blood and animal remains on the tips of stone points.[14] Such points are known to have been hafted onto handles because residues of adhesives are sometimes preserved on their bases, as well as microchipping, where twine would have been used to attach the stones to shafts.[15] The manufacture of MSA weaponry relied on the creation of adhesives by combining and heating ingredients such as powdered iron oxide and plant gum or resin.[16]

From the 1950s onwards, archaeologists excavating MSA sites in the interior of South Africa named stone-tool industries with long blades and unifacial points after the town of Pietersburg (now Polokwane). Pietersburg industries are generally found in Limpopo Province, and have not yet been recognised north of the Limpopo River. Most Pietersburg industries have been found in caves or rock shelters, perhaps because such places attract explorers. While the best-known Pietersburg site is Cave of Hearths, there are other important ones like Mwulu's Cave not far from the Makapan Valley,[17] Bushman Rock Shelter near Ohrigstad[18] (Figure 2.1), and Olieboomspoort.[19]

Olieboomspoort Rock Shelter, first described by Mason in 1962 and excavated by Maria van der Ryst from 1997 onwards, overlooks the Rietspruit, which cuts through Waterberg strata on the farm Fancy, about 26 km south of Lephalale. The shelter is named for the poort through which the original road ran, next to the river, before this section was relocated to higher ground to the west. The shelter was conveniently located no more than 20 metres from the old road, making its access in Mason's time very easy. It has a Pietersburg Industry made from a variety of fine-grained rocks; this overlies an ESA assemblage and underlies a long Later Stone Age (LSA) sequence.[20] Very few Stone Age sites in South Africa have long cultural sequences that include ESA, MSA and LSA industries. This makes Olieboomspoort a site of exceptional importance.

FIGURE 2.3: Olieboomspoort Shelter on the farm Fancy. The original road is visible between the shelter and the river. Image taken by a drone on 9 February 2018, courtesy of Siegwalt Küsel, the son of the excavator, Dr Maria van der Ryst.

Open-air sites that have been excavated in the Limpopo Province below the Waterberg escarpment include Blaauwbank, Kalkbank and Wonderkrater (Figures 2.1 and 2.4). The site on Blaauwbank, in a gravel donga near Rooiberg, has a rich assemblage of Pietersburg tools overlying ESA ones[21] while Kalkbank (an ancient pan on the farm De Loskop, about 45 km north-northwest of Polokwane) also with a Pietersburg industry,[22] generated many broken animal bones that are now known to have been accumulated predominantly by nonhuman agents.[23] Wonderkrater, which is located on the farm Driefontein, north of Mookgophong (formerly Naboomspruit) (Figure 2.4), is an MSA spring and peat site. It is dated (by optically stimulated luminescence dating of quartz grains) to between about 100 000 and 30 000 years ago.[24] Unfortunately, there are few stone tools around the spring and they cannot be securely attributed to anything other than a generic MSA industry. Most seem to be cutting tools made on rhyolite, a local volcanic rock type.[25]

THE WATERBERG PLATEAU
The Middle Stone Age (MSA)

The earliest firm evidence so far of human habitation on the Waterberg plateau is from the MSA, in other words, in the last 300 000 years. There are extensive remains of MSA occupations in the Waterberg, mostly in the form of scattered stone tools, but no MSA site on the plateau has yet been dated, so at present we can only say that the occupations would have been somewhere between 300 000 and 25 000 years ago because this is the span of MSA ages elsewhere in South Africa.

On the plateau, people lived either in rock shelters or open camps, sometimes near pans or rivers, though they were not as dependent on close sources of water as their ancestral ESA counterparts. This independence from water suggests that they had water containers made from products like skin or ostrich egg. A small rock shelter on North Brabant (now a portion of New Belgium called Goudrivier) (Figure 2.4) was excavated by Schoonraad and Beaumont in 1968 and here the MSA component of the site was attributed to a "Middle Pietersburg" industry.[26] LSA artefacts were also recovered from younger occupations with radiocarbon ages of AD 900 and AD 1100. Van der Ryst subsequently excavated the same shelter and retrieved MSA unifacial points from the oldest sediments.[27] A few MSA blades and a broken point were later found in a rock shelter at Schurfpoort

CHAPTER 2: DIGGING INTO THE PAST

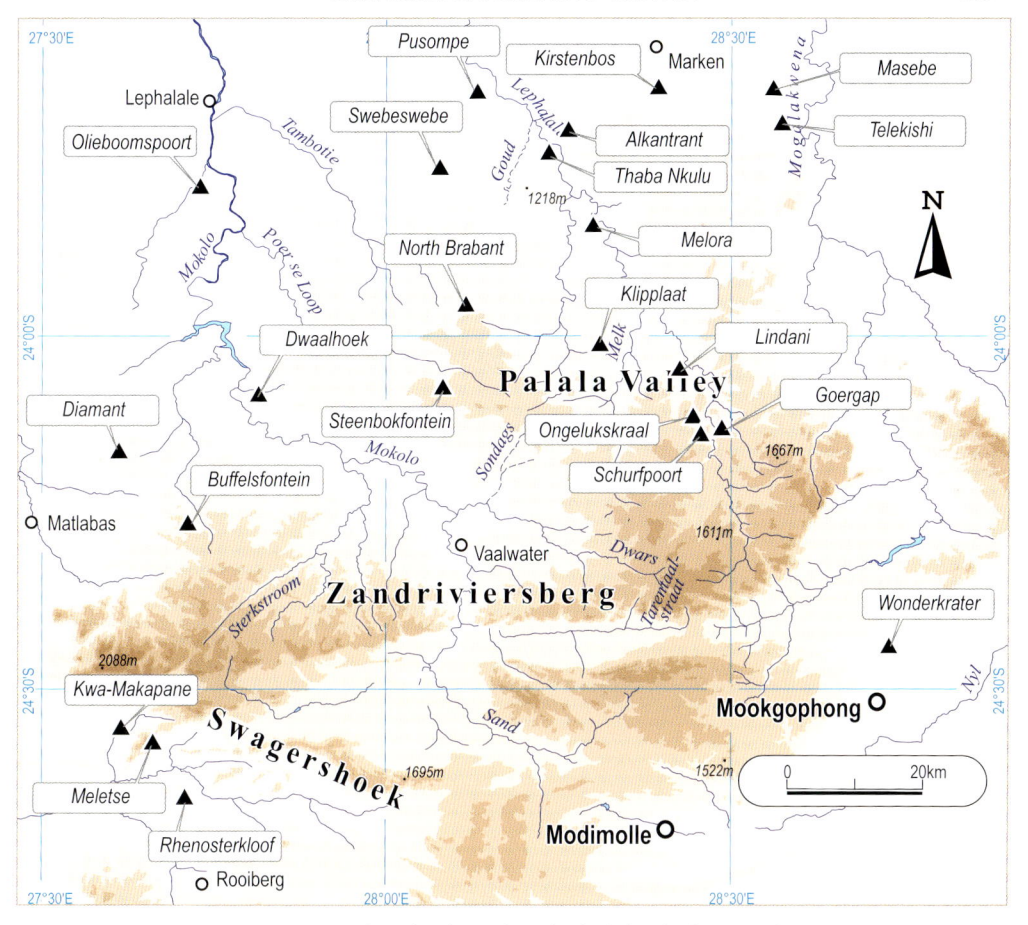

FIGURE 2.4: Location of major archaeological sites in the Waterberg.

FIGURE 2.5: Steenbokfontein spring site and some of its tools. A: The spring eye. B: Blade made from silicified siltstone. C: Partly reshaped, pointed flake made from silicified siltstone. The scale bar is 1 cm.

on the Waterberg plateau, just south of the hill known as Tafelkop.[28] Nearby, in a rock shelter on Goergap, close to the Lephalala River, Van der Ryst excavated a substantial MSA stone tool collection. At the base, the tools were generally longer than 50 mm and the length of some quartzite blades exceeded 150 mm. Since the Pietersburg industry is characterised by long products, this assemblage probably belongs to it. Points and knifelike tools were also common. In younger MSA layers at Goergap, the tools were smaller and there was a change in rock type from quartzite to felsite.[29]

The spring site on Steenbokfontein west of Dorset is the only MSA open site that has been excavated on the plateau (Figures 2.4 and 2.5).[30] The surrounds have abundant outcrops of fine-grained, jointed siltstones exposed around the eyes of springs. This rock is particularly suitable for tool-making so the area is littered with MSA tools. The density of blades and flakes lying on the ground surface points to low soil sedimentation rates and repeated visits by people. Animal and human traffic would have been heavy around the spring so tools have been repeatedly trampled and damaged. This site was probably a tool-making workshop, which makes it unusual among the Limpopo sites discussed here. Not only is it an open site with rock types that do not occur elsewhere, but it seems to be a special purpose destination. Sometimes rocks used for MSA tools were transported considerable distances, presumably in bags or other containers. When this happened, the stone tool "knappers" generally carried out part of the manufacturing process at the rock source. Steenbokfontein seems to be one such "quarry" site. The end result of having specialist quarry sites is that tool assemblages from some MSA home-base sites tend to lack some of the preliminary tool-making products and instead contain predominantly finished products like points and blades.

The Later Stone Age (LSA)

The LSA has been dated to between 30 000 and 100 years ago elsewhere in South Africa. However, radiocarbon dates from sites excavated so far on the Waterberg plateau have revealed that, with one known exception, LSA occupations there were limited to the last thousand years. This is an important finding: on the basis of known archaeological sites on the Waterberg plateau, there is a noticeable gap between sites with MSA tools (the youngest of which would have been about 30 000 years old) and LSA ones of recent origin (not older than about 1000 years). This suggests that the Waterberg may have been a wilderness almost entirely free of human habitation for tens of thousands of years. Currently, we do not know whether or not avoidance of the plateau was caused by environmental conditions hostile to human habitation.

Certainly the Waterberg seems to have become attractive for LSA settlement only after about AD 1000. One hypothesis, which remains to be tested with palaeo-environmental studies in the Waterberg, is that an unusually moist period occurred at and before AD 1000. Cooler conditions rather than high precipitation may have been the cause, as such changes are recorded at Mapungubwe on the Limpopo River where fields may have become waterlogged at the time that the capital was abandoned around AD 1300.[31] The subsequent drying out of the environment would have made the Waterberg more suitable for occupation because it would have reduced the amount of sourveld and encouraged the growth of palatable grasses. In turn, pasture improvement would have provided better winter feed for wild and domestic animals. This hypothesis is not entirely convincing though, because farther south in the Magaliesberg and adjacent Highveld, people used sourveld seasonally in the LSA.[32]

Several of these recent LSA sites have been discovered and excavated on the plateau, most of them in shelters overlooking or at least close to the Lephalala River. Some were excavated by Van der Ryst in the 1990s. She concluded that, after a hiatus following the MSA, LSA occupation in the northwestern portion of the Waterberg commenced only during the late eleventh and early twelfth centuries AD and that the settlement of the plateau by hunter-gatherers coincided with the arrival of Iron Age agropastoralists.[33]

This joint immigration of farmers and hunter-gatherers seems counterintuitive, but it is likely that the hunter-gatherers followed the farmers because they hoped to benefit from the exchange of services for food. Perhaps the hunter-gatherers obtained regular supplies of grain and milk in exchange for products like hunted meat and animal skins. Novel and useful items such as clay pots and metal tools may also have been desired. LSA communities traditionally subsisted by gathering plant foods and hunting, trapping or collecting small and medium-sized animals such as steenbok, duiker, hare and tortoise, although the reports of early European visitors to the region refer to hunter-gatherer communities also following the movement of larger migratory game.[34]

The hunter-gatherers of the Later Stone Age were almost certainly ancestors of modern San (Bushmen). They used bows and arrows and made tiny, sharp stone tools and polished bone points to use as arrowheads. They knapped small stone scrapers for working animal leather (Figure 2.6), and scrapers made from

FIGURE 2.6: Later Stone Age stone scrapers that would have been hafted to handles and used for scraping leather. The scale bar is 1 cm.

FIGURE 2.7: Beads made from ostrich eggshell. A: Beads strung as a necklace, but Later Stone Age beads may also have been sewn on leather bags or clothing. B: Perforated bead blank prior to polishing. It is 0,5 cm wide.

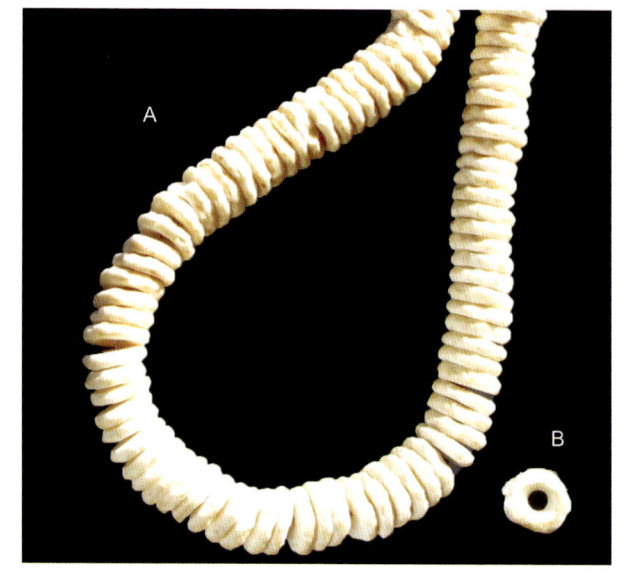

quartz crystals and felsite were the most common tool types at the rock shelter, Goergap, on the Lephalala River.[35] Also at Goergap were polished bone points and linkshafts that would have been arrow components, polished bone needles for basketry or stitching leather clothing, and ostrich eggshell beads (Figure 2.7). Grooved stones were used for polishing the eggshell beads that were clearly made at the site because unfinished beads and bead blanks were excavated. Objects that would have come from Iron Age settlements were also excavated from Goer-

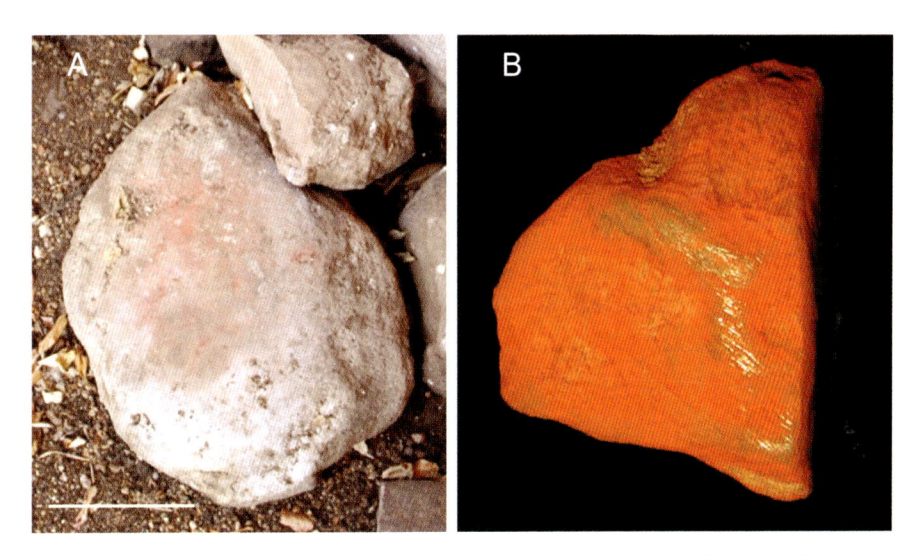

FIGURE 2.8: Ochre-stained grindstone and specular hematite. A: The large grindstone with a heavy red ochre stain from grinding hematite was found at the rock art site on Swebeswebe (on Maria, now part of Swebeswebe along the northern Waterberg escarpment). The scale bar is 15 cm.
B: A nodule of specular hematite from Weltevreden. Such nodules were ground for their red powder, but in the Iron Age the iron-rich pieces would also have been collected for smelting.

gap. These include glass and metal beads, and potsherds of Eiland affiliation (to be explained later, see Figure 2.13) which were also found at the rock shelter on Schurfpoort.[36] On the surface sediment, grindstones with red ochre stains and specular haematite nodules were found. These suggest that Goergap's occupants made paint for the creation of rock art. This was also the case at North Brabant, a rock shelter with a well-preserved rock art panel. Here, many hematite nodules and ground stone fragments were excavated from the LSA occupations dating between AD 900 and 1100.[37] Specular hematite nodules and grindstones with traces of red hematite (ochre) powder are common at Waterberg sites (Figure 2.8).

Rock art

Rock art in the Waterberg seems, like the rock shelter settlements of the LSA, to date to the last thousand years, although surviving panels may well be only a few hundred years old. The painted sites in the Waterberg were important places for expressing the religious beliefs held by the ancestors of the San.[38] Consequently, some of the art is in sacred places where people did not live, for example on the edge of cliffs. Other art is on rocks or walls of rock shelters where people may have lived, worked and practised rituals. Soft iron-rich rocks, like the hematite piece illustrated in Figure 2.8, were ground to make powder for red paint. Charcoal was used for black paint and ground ostrich eggshell or white clay for white paint. These powdered products were mixed with liquid such as water, egg, milk, blood or fat.

Certain curious motifs may represent medicine men (shamans) in trance and these sometimes seem to depict the hallucinations that were experienced during trance. On occasion the paintings may have been intended to teach people about aspects of San religion, including the experiences that medicine men had during trance. Sometimes they were part of rain-making ceremonies. Some figures and animals are represented unrealistically because the artist was giving a message to viewers and therefore emphasised human or animal attributes most appropriate for that purpose. Often strange-looking people were painted, for example paintings of humans with exaggeratedly stretched trunks are widespread in the Waterberg and elsewhere in South Africa. Occasionally medicine men "became" animals during their out-of-body travels and peculiar half-animal, half-human creatures were painted to express this hallucinatory experience.

One of the rock art motifs that seems endemic to the Waterberg is the "Waterberg Posture".[39] This is a highly stylised male human profile with an upright, drawn-out body, a blob of a head, single leg, pronounced buttocks, a single short, pointing arm and an exaggerated, protruding penis. The figures frequently occur as a line of men facing the same direction (Figure 2.9A) as shown at Klipplaat. The arms-forward posture is one sometimes seen in shamanistic dances when men form a single line.[40] Stylised male figures can also be seen at Masebe Nature Reserve as part of friezes that also depict elephant, giraffe and kudu. The Masebe art is open to the public and some of it has consequently been vandalised.

Stylized female figurines (Figure 2.9B) found at the Iron Age site on Schroda, on the south bank of the Limpopo River, have similarly elongated bodies, truncated arms, prominent buttocks and inconspicuous heads.[41] The two art traditions articulate different belief systems, yet use similar methods for their expression.

Another motif commonly painted in the Waterberg is the hartebeest. In some instances the animal is not naturalistically depicted: the back leg may be long and thickset, whereas the foreleg is unnaturally short, the body may be stretched out, and the rump pronounced (Figure 2.10).[42] This unnatural depiction is reminiscent

FIGURE 2.10: Rock art panel from Klipplaat showing superimposed paintings of animals and humans. Note the "Waterberg hartebeest" with arrows superimposed on its neck (top, right).

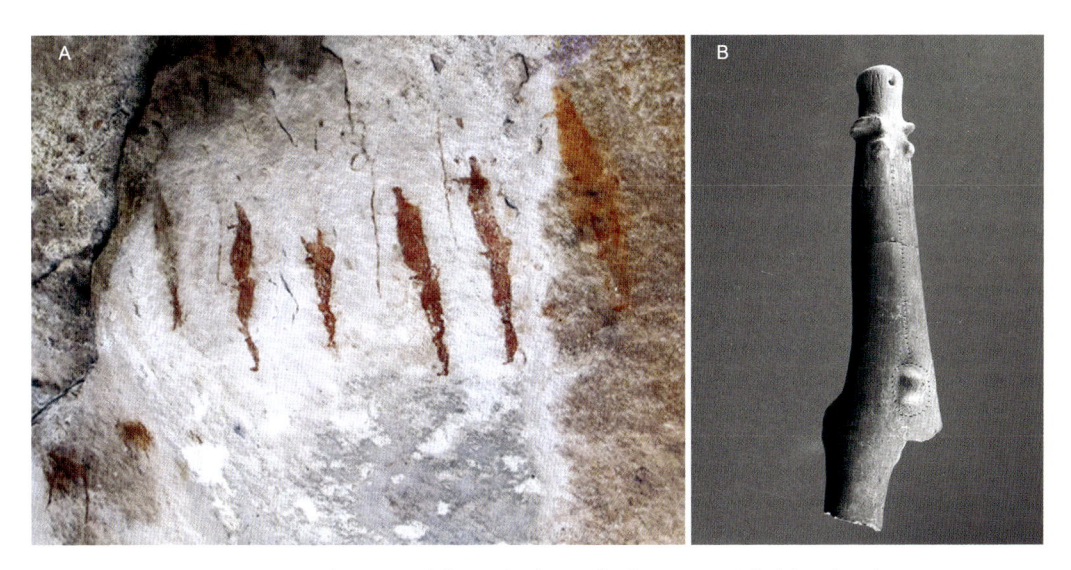

FIGURE 2.9: A: Rock art panel from the farm Klipplaat near Melkrivier showing a line of men in the "Waterberg Posture". The central figure is incomplete. A figure at bottom left appears to be partly human and partly hartebeest and it may represent a transformed shaman in trance. B: Broken female figurine from the Iron Age settlement at Schroda (reproduced by permission of Ditsong National Museum of Cultural History, Pretoria). Note the knoblike head, arm stumps and elongated trunk.

of the figures in the Waterberg Posture; undeniably the animal motifs have some features that are humanlike rather than animal-like and they can therefore be considered therianthropes (creatures that are part human, part animal).[43] Perhaps medicine men hallucinated about them and believed that they transformed into hartebeest whilst in trance. Kudu, giraffe and eland are also commonly painted animals in the Waterberg and sometimes motifs are painted on top of older ones. Expressions of religious beliefs may have been strengthened by laying one meaningful motif on top of another.

Rock paintings are often associated with sites for holding rain-making ceremonies for both hunter-gatherers and farmers. The livelihoods of both were dependent on rain. Indeed, rock art sites were important rain-making places for the first farmers in the area.[44] The association of Iron Age pottery and paintings implies a relationship between the rain-making rituals of the San and those of the Iron Age farmers of the area. The initial stages of contact between San and black Iron Age farmers appear to have been cooperative and San made rain and prepared animal skins for the farmers in exchange for a variety of items, including livestock, milk and carbohydrates.[45] Throughout South Africa there are historic records of San used as rain makers by both black and white farmers,[46] and supernatural powers have long been attributed to San. All over southern Africa the San were used as shamans, even in recent times. Their services were also used for the provision of meat, skins, berries, firewood and ostrich eggshell beads and there are records of them receiving food or livestock in return for their labour. Bushmen were "clients" rather than servants of the people they worked for. They would seasonally or irregularly "melt" back into the veld, disappearing from the view of the settled farmers. On their return they would generally camp on the outskirts of the settlements that they planned to associate with.

In 1881 the Berlin missionary Hermann Schlömann was sent to staff the new mission at Malokong, among the people of the Langa chief Masebe. In 1894 he was sent by Masebe's son Bakenberg to a mission outpost called Pusompe on the

FIGURE 2.11: A painted panel on Thaba Nkulu (Gansieskraal, New Belgium). Three painting styles are evident: finger paintings of animals at the top, a fine-line brush outline of a kudu bull and fine-line kudu females where the figures were first drawn in white outline then filled with colour. The order of the paintings shows that the finger paintings were made before the kudu bull. This demonstrates that the shelter was used more or less contemporaneously by black farmers and San hunter-gatherers.

Lephalala River to minister to the Seleka (Tswana-speakers), who were then subject to the Langa.[47] While with them, he visited a rock shelter of ritual importance close to the river. The rock shelter was in use by the Masele, also known as *vaalpense* (grey paunches). The vaalpense seem to have been people of mixed San and black descent who were observed by nineteenth-century white travellers in the Waterberg, Soutpansberg and Dwarsberg.[48] Vaalpense were hunter-gatherers who formed a marginalised, subordinate class living in abject poverty. Some were enslaved by black communities, but others were seasonally employed to herd cattle or as labourers. Vaalpense were reported to have used the rock shelter near Pusompe Mission for various rituals, including rain-making. When the chief's rain-making magic failed, he would take his people to the painted shelter where they prayed for rain.[49] A local resident said that 50 years earlier when he was a child, people had prayed for rain at the rock painting site North Brabant on the farm New Belgium, west of Lapalala. People apparently called the shelter the *reënkerk* (rain church).

The arrival of substantial Late Iron Age communities in the eighteenth century resulted in land and resources becoming scarce, so tensions arose between the hunter-gatherers and farmers. The hunting and gathering way of life was essentially displaced; some San fled north or west, but there was also considerable intermarriage with black farmers. From this time onward, Iron Age farmers may have conducted most of their own rain-making ceremonies, even though they may have continued to revere and use shelters painted by San. Rather crude depictions of animals and people in red, black or white paint, painted directly with fingers (Figure 2.11), are often found together with fine-line San paintings.[50] Similar finger paintings occur further north on the Makgabeng plateau and also in the Limpopo Valley.[51] Not all of the paintings involve rain-making; some seem to have involved rituals relating to initiation ceremonies for both young men and young women.[52]

Herders

Before Iron Age communities arrived in the Waterberg, Khoekhoe herders may have travelled through the Waterberg or close to it, just below the plateau. Their movements and the date of their arrival are not known. Some archaeologists think that herders brought with them the first pottery to appear in the area, a thin, comb-stamped, highly decorated ware known as Bambata (after the type site in the Matobo Hills, south of Bulawayo, Zimbabwe).[53] Van der Ryst found some pieces of this pottery in the Olieboomspoort sediments. Early Bambata pottery seems to have appeared in Limpopo around AD 220, when Khoekhoe herders journeyed southwards through the area before their arrival in the Western Cape.

Some Waterberg rock art may have been made by these herders. Their art is thought to depict simple geometric forms and is finger-painted, unlike the fine-line brushwork of San art. At some sites handprints were part of the art repertoire because the prints are sometimes linked to the finger-painted designs. Where handprints occur together with fine-line San art it is not possible to be sure that they were done by herders. The herders owned fat-tailed sheep and these are often depicted in Waterberg rock art, for example at Dwaalhoek, near Bulgerivier.[54]

Farmers of the Iron Age

The archaeologist Jan Aukema, who conducted the first comprehensive archaeological survey of the Waterberg in the 1980s, distinguished several phases of Iron Age occupation below and on the Waterberg plateau and recognised that more might exist. During his Motlhabatsi (Matlabas) survey in the western region, Aukema excavated evidence for the first phase of farmer activity at Diamant west of Bulgerivier, in the sweetveld at the base of the western Waterberg. The Diamant site has two occupation phases, represented by Happy Rest facies and Diamant facies ceramics. The earliest decorated ceramics of Happy Rest style at this site date back to about AD 570.[55] The Iron Age pottery specialist Prof. Tom Huffman suggests AD 500–750 and AD 750–1000 as the most likely time ranges for Happy Rest and Diamant facies ceramics, respectively.

The early farmers brought to the Waterberg Indonesian glass beads (probably obtained from Schroda near Mapungubwe on the Limpopo, which was on an early international ivory trade route), decorated pottery, cattle, sheep and crops such as sorghum and millet, although they still hunted wild animals and collected wild plant foods. At Diamant, remains of dog, cattle, sheep/goat were found alongside wild animals such as zebra, wildebeest, hartebeest, impala and ostrich.[56]

They brought iron-working skills, too, and metal production took place at Diamant amongst other sites.[57][58] Smelting sites are fairly common in the Waterberg and its foothills. Several iron-working sites occur at Thaba Nkulu on the farm Gansieskraal, now part of New Belgium, where the Lephalala River emerges from

FIGURE 2.12: White finger paintings at Swebeswebe. They are either herder or farmer paintings. Although some of the paintings are of animals it is not possible to decipher their identity.

the Waterberg plateau onto the Bushveld. One site is isolated and screened from the remains of villages by a low cliff. Here, vitrified slag and *tuyères* (clay pipes) (Figure 2.13A, B, C) are scattered across an area of about 6 metres x 4 metres. Tuyères and bellows were used to pump air into the furnace to heat the wood charcoal there so that the correct smelting temperature could be reached and reduction could take place.[59] At another site on Thaba Nkulu, a team excavated iron slag, tuyères, furnace lining, iron ore, copper artefacts and iron artefacts, and later identified chemical signatures for metal artefacts, and metal smelting and smithing material.[60] Early iron mining took place at Phalaborwa in the seventh century, but there would certainly have been other sources of iron closer to the various dispersed Iron Age settlements. In the Waterberg, the iron-rich sandstones frequently contain nodules of magnetite, especially in proximity to later diabase intrusions. Slag, associated with smelting sites, has been discovered in the vicinity of areas where these nodules occur in high concentrations.

FIGURE 2.13: Evidence of iron smelting and rain-making ritual in the foothills of the northern escarpment of the Waterberg. A: Smelting site at the base of a cliff, hidden from the closest Iron Age settlement on Thaba Nkulu (Gansieskraal, part of New Belgium). B: Fragment of a clay tuyère used with bellows for pumping air into the furnace. C: Iron slag, a solid waste product of smelting. D: Small Eiland pot found in a hilltop crevice at Alkantrant (Uitkomst), a few kilometres east of Thaba Nkulu. It is too small to be a pot for domestic purposes and it would have been used for storing herbs or similar products as gifts for the ancestors. Ancestors control rain and must be persuaded to part with it. Iron spearheads were also left in crevices on the same hill. The scale bar is 5 cm.

The next of Aukema's phases incorporated an Eiland ceramic facies characterised by herringbone decoration on vessels (Figure 2.13D). The Eiland ceramic facies is part of what Huffman calls Middle Iron Age (roughly dating AD 900–1300) and it overlaps with the more northerly traditions at Toutswe and K2-Mapungubwe.[61] People who made the Eiland facies ceramics also created clay crucibles for working tin and bronze at Rhenosterkloof in the Rooiberg area at the beginning of the second millennium AD.[62] This is the earliest known working of tin and bronze in southern Africa. One Eiland site, Wentzel, about 80 km southwest of Lephalale, is near the Limpopo/Matlabas confluence in an area of good agricultural potential, where soil is deep. However, Eiland settlements are usually open sourveld sites in high river valleys on the Waterberg plateau. Eiland is not associated with stone-walled settlements, nor hilltops (except for rain-making sites that will be discussed later).

Iron Age farmers appear to have settled on the high Waterberg plateau in considerable numbers by the eleventh century. The reason for the shift of farmers from the sweetveld of the lowlands to the sourveld of the Waterberg plateau is not understood. Aukema thought that the animals herded in the sweetveld of the lowlands by the early farmers overgrazed this veld which might, as a result, have encouraged tsetse-fly infestation.[63] The Waterberg plateau is situated too high for tsetse flies and its proximity to the lowlands would have provided a useful escape route from these dangerous pests. The immigrant farmers made the distinctive Eiland pottery with herringbone motifs (Figure 2.13D) that was popular between the eleventh and sixteenth centuries. It was found, amongst other sites, at Kirstenbos, south of Marken, where both Late Stone and Late Iron Age settlement are evident.[64] Eiland pottery and grindstones are frequently found in crevices or small rock shelters on hilltops or cliffs of the Waterberg. As Aukema pointed out, the rock shelters are generally unsuitable for habitation, even of a defensive nature, and it is far more likely that the remains of clay pots, stone cairns and grindstones represent tokens left for rain-making ceremonies.[65] As part of these ceremonies there was an appeal to the ancestors of the chief. The ritual was performed by the chief and assistants called "rain doctors".[66] These rain doctors were probably San in the early days of farmer/San contact. The grindstones were for grinding medicines that were then mixed and stored in clay pots. At times the lower grindstone was the bedrock of the cliff face on which the medicines would be offered. At Thaba Nkulu, the small medicine cupules are sometimes so close to the edge of a cliff that we can be sure that they do not represent receptacles for normal domestic grinding activities (Figure 2.14).

FIGURE 2.14. Two rain-making hills on Thaba Nkulu (Gansieskraal, part of New Belgium) where herbal rain-making medicines were probably ground as offerings to ancestors. A: Small grinding hollows known as "cupules" are ground into bedrock close to the cliff edge of a hill in the foreground. B: Lower grindstone pecked and ground into bedrock on the edge of a cliff, on the conical hill shown in A.

FIGURE 2.15: Baobabs around an Iron Age settlement on Thaba Nkulu (Gansieskraal, part of New Belgium). A: Circle of "young" baobabs around the eroded ground once housing unwalled homesteads. B: The village looked on to the rain-making hill that features in Figure 2.14.

Thaba Nkulu was favoured by Iron Age farmers from the time of their first settlement in the area and certainly well before their permanent settlement on the plateau. The area marks the southern boundary for baobab (*Adansonia digitata*) and some low altitude Iron Age villages have baobabs associated with them (Figure 2.15), either because they were deliberately planted or because discarded seeds germinated around homesteads.

Another rain-making place that was used for a long time is a small tunnel cave on a hilltop at Telekishi on the eastern escarpment on the farm Schrikfontein. Here there are many Eiland potsherds as well as historically made pots, so the cave was a repository for gifts for the ancestors for hundreds of years.[67]

In contrast to settlement in the Early and Middle Iron Age, Late Iron Age settlement is found on hilltops with stone-walled settlements and undecorated pottery. These settlements may be linked to the arrival of Nguni-speakers (Ndebele people) in the region, between the sixteenth and seventeenth centuries.

In the seventeenth century, the ancestors of the Northern Sotho and Tswana peoples seem to have settled in sweetveld near the Matlabas River, as the earlier Iron Age farmers had done before them.[68] These Late Iron Age farmers used multichrome Moloko pottery with decorations comprising multiple comb-stamped bands and bands burnished with red ochre and black graphite. At the same time, Sotho-Tswana people controlled the Rooiberg tin mines,[69] Sotho-Tswana settlements were built on defensive hilltops,[70] and stone-walled settlements appear on the Waterberg plateau (though not all of them made by Sotho-Tswana communities). These plateau settlements are sometimes on cliffs, surrounded by perimeter walls, implying that they were designed to defend their human occupants as well as their cattle and other stock. Some Moloko pottery was found at Kirstenbos and a large sample was excavated from Ongelukskraal,[71] but Moloko pottery was absent from the Buffelsfontein and Melora Hilltop sites where undecorated ceramics from another tradition were found. Ongelukskraal is below the vertical cliffs of Tafelkop and it was excavated by both Aukema (who called it Skeurkrans) in 1989 and Jan Boeyens in 1991. In the centre of the stone-walled complex are steep cliffs and shelters and the area was clearly occupied by different groups of people before the Late Iron Age. For example, there are some fine-line rock paintings of antelope that would have been made by San. Furthermore, a single date of about 1800 years ago was obtained for a layer containing two Bambata potsherds, grindstones and a few quartz flakes.[72] The date is an outlier in the

Waterberg, but Bambata ware at Jubilee Shelter in the Magaliesberg was dated to a similar age,[73] so Ongelukskraal is not an isolated occurrence of this early pottery. The Ongelukskraal early date and the Bambata potsherds suggest pre-Eiland occupation either by LSA communities that made or obtained these wares through trade, or by herders or Iron Age people with links to settlements like Toteng in Botswana where Bambata ware is abundant.[74] Huffman favours the interpretation of hunter-gatherers obtaining the pottery through trade. A few pieces of Eiland ceramics were also recovered from this site, but they are also not associated with the stone walling. Instead, the Late Iron Age builders of the stone walls at Ongelukskraal made Moloko-style pottery attributed to Sotho-Tswana speakers. A radiocarbon date for the stone-walled component at Ongelukskraal is 310 ± 45 BP (Pta-5153), that is, from the late seventeenth century.[75] This date implies that Sotho-Tswana speakers (for example at Ongelukskraal) and Nguni speakers (for example on Melora Hilltop) may have coexisted on the Waterberg landscape during the Late Iron Age.

Aukema excavated a defensive hilltop site on Buffelsfontein south of Bulgerivier in the western Waterberg, dated AD 1550 ± 70.[76] The site has perimeter walls that occasionally abut cliff edges to provide extra security. The walls meander to form lanes and enclosures for cattle and domestic spaces for women. Upright monoliths within the walls provide striking markers (Figure 2.16A). Whilst the monoliths may simply have acted as "road signs", they may equally have embodied symbolic meaning. For example, Boeyens and Van der Ryst[77] suggest that monoliths marking the central courts of nineteenth-century Tswana towns and Venda capitals probably signify rhino horns. Architectural structures may therefore express attributes of strength and power associated with traditional leadership. Buffelsfontein is not, however, a Tswana town, so the iconography may not have been the same. While excavating Buffelsfontein, Aukema uncovered remains of beehive-shaped huts on small terraces behind kraals.[78] These huts were placed towards the back of the dwelling units, rather than in the middle as is characteristic amongst Sotho-Tswana speakers. The huts were therefore most likely produced by a Nguni-speaking group. Aukema found circles of carbonised wooden stumps and thin slabs of stone slanting towards the centre of each hut.[79] The flexible branches and stone "pillars" would have acted as supports for the beehive huts. The undecorated globular pots at Buffelsfontein are not like the multichrome Moloko ware and this is a further reason for recognising that this is not a Sotho-Tswana settlement.

Melora Hilltop, in the Lapalala Reserve (on Landmans Lust), is northeast of

FIGURE 2.16: Late Iron Age stone-walled settlements. A: Buffelsfontein, south of Bulgerivier: monolith set in stone walling on the perimeter of the site. B: In the Telekishi settlement a small stone-walled kraal was possibly used for calves or small stock like goats.

Buffelsfontein and is younger, with its midden dating to AD1700 ± 50.[80] It has a similar spatial pattern and house form, though there is only one known monolith at Melora. The evidence for beehive huts at Melora is not as secure as that at Buffelsfontein, but the lack of *daga* (burnt clay) hut floors implies that beehive huts made of portable reed mats were erected in domestic areas. The Melora Hilltop dry stone walling encloses a village area of about six hectares to form what is interpreted as a defensive position, although there are also remains of some hut dwellings outside the enclosure. The types of undecorated pottery at Melora and elsewhere on the Waterberg plateau suggest that Nguni speakers lived in these hilltop villages. Indeed, all the evidence from Aukema's excavations on top of Melora Hill, and subsequent ones there by a Unisa team led by Boeyens, supports its Nguni affiliations. These Nguni were presumably predecessors of the Northern Ndebele chieftaincies (the Langa and Kekana) now living in the Mogalakwena River valley area. At its peak, the Hilltop site may have accommodated up to a thousand people. Large stone grain-bin platforms suggest that crops were grown successfully on the dolerite-rich soil below the hill. The presence of an iron adze[81] demonstrates the use of iron technology.

The Pedi (a group of Northern Sotho) grew to dominate the area east of the Waterberg in the eighteenth century, perhaps as a result of their control of key trade networks. In 1822 the "Zimbabwean" Ndebele (the Matebele), led by Mzilikazi, moved into Pedi land and defeated the Pedi, also causing shortages of food and domestic animals in the area. In a period of turmoil known as the difaqane (sometimes called mfecane) (see Chapter 3), the Matebele caused havoc in Northern Sotho and Tswana areas for fifteen years. A joint Tswana and Boer army eventually drove the Matebele from the Marico in present-day North West Province into Zimbabwe. Raiding from other groups continued though and many people hid in the mountains.

A nineteenth-century refuge village was erected below the Melora Hilltop on the Melora Saddle. It contains Moloko pottery, believed to have been made by Sotho-Tswana people. An infant pot burial was uncovered from next to the foundations of one of the huts.[82] Considerable soil erosion also exposed stone-built granary platforms, hut rubble, ceramics, and lower and upper grindstones. In addition, many artefacts have been found, including copper bangles, a copper earring, iron slag and tools, glass beads, a clay spindle whorl and clay figurines.[83] One of the figurines was that of a rhinoceros (Figure 2.17A). The radiocarbon date of AD 1830 ± 45 supports oral tradition about the area as a refugium during the difaqane.

Late Iron Age sites are common in the Waterberg and one that is open by prior arrangement to the public lies on the Telekishi Community Walking Trail (Figure 2.4). The beginning of the trail traverses a deeply wooded kloof sealed at its entrance by a stone wall about 20 metres wide. A second boundary wall at a higher altitude is close to an eighteenth- or nineteenth-century walled settlement. A free-standing stone circle built with two lines of large rocks filled with smaller rubble has a back courtyard wall, which is waist-high to knee-high. In the courtyard are lower and upper grindstones and the area would have been used by women for processing grain. Further into the kloof, above a stone-walled kraal,

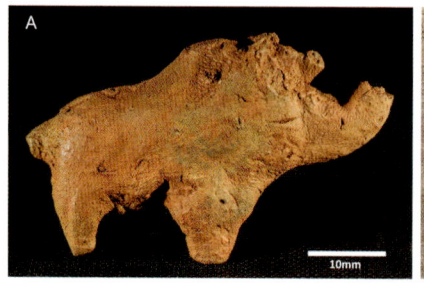

FIGURE 2.17: Rhino symbolism in the Waterberg. A: Melora Saddle clay figurine of a rhino (image courtesy of Boeyens and Van der Ryst). In addition to being symbolic of leadership among farming communities (Boeyens, J. & van der Ryst, M. (2014), the rhino features prominently in the rain-making rock art of San hunter-gatherers (Eastwood & Eastwood (2006). B: A black rhino mother and calf are depicted in this Thaba Nkulu motif. Black rhino mothers walk ahead of their calves.

is a terrace that may represent a high-status area because altitude correlates with status. Here, a communal wall seems to separate space possibly intended for men and women. Semi-circular courtyard walls are women's household space; they might have benches at the back, but these cannot be seen and might be revealed by excavation. On the other side of the communal wall is men's space. Here are stone circles, some with entrances linked to others without. The kraals without entrances may be small-stock enclosures (Figure 2.16B).

Summary of Waterberg plateau archaeology

Human occupations on the Waterberg plateau are abridged compared with the impressively long cultural sequence in Limpopo Province as a whole. The province has a World Heritage site with hominid fossils that are three million years old as well as a second World Heritage site at Mapungubwe. In contrast, the earliest archaeological sites currently known on the plateau are from the Middle Stone Age. These sites, although undated, are unlikely to be older than 300 000 or younger than 25 000 years because this is the span of Middle Stone Age ages elsewhere in South Africa. The plateau sites occur in rock shelters and in the open, and their stone tool kits include long blades for a variety of cutting tasks, and triangular points that were probably weapon tips hafted to wooden shafts. Assuming that some of the plateau Middle Stone Age sites were as young as 30 000 years, there is a curious 29 000 year gap in occupation of the Waterberg plateau between the end of the MSA and the arrival of the first farmers. We can only speculate that it was abandoned for environmental reasons that remain obscure. The demographic movement into the Waterberg that occurred at about 1000 years ago is unprecedented elsewhere in South Africa. Hunter-gatherers and farmers appear to have arrived on the plateau simultaneously as part of a symbiotic relationship that lasted several hundred years. The evidence from various rock shelter occupations and painted panels suggests that hunter-gatherers were rain-makers for the earliest farmers in the area. In exchange they may have received clay pots, milk and possibly carbohydrates in the form of grain.

Nutrient-poor soil and unreliable water resources (notwithstanding the name Waterberg, water resources are scarce in the area) may ultimately have led to food shortages and tensions between hunter-gatherers and farmers. Finger-painted rock art as well as hilltop sites with Iron Age artefacts hidden in rock crevices implies that farmers began to conduct their own rain-making rituals, at least by the Late Iron Age. Late Iron Age-walled settlements are ubiquitous on the plateau, suggesting that it was, for a time, well populated. However, the difaqane had a devastating effect on settlements, and the environmental constraints of the plateau must also be taken into account. Crop growing would not have been successful for more than a few years on the nutrient-poor sandstone-derived soils and the richer, dolerite-derived soils are of limited extent on the plateau. Therefore, it is not surprising that Iron Age populations were scarce in the Waterberg by the time of Chief Makapan's short-lived defensive Mabotse settlement (1872–1880) on Kwa-Makapane Hill below the southwestern edge of the plateau (Figure 2.4).[84] At this stage, the archaeological record merges with the historical one and from this point onwards, the Waterberg story becomes entirely historical.

CHAPTER THREE
Homeland or a place of refuge?

Lebowa – an artificial homeland

Scenic Kloof Pass, opened in 1988, descends the northeastern escarpment of the Waterberg plateau via a series of steep gorges etched into the earliest sandstones comprising that geological feature (see Chapter 20). At its foot, the pass opens out onto the flattish pre-Waterberg plain, the road passing between spectacular outliers (*inselbergs*) of Waterberg rocks which form pillars or columns seemingly guarding the entrance to the pass.

The biggest of these is *Mmamatlakala* (literally, "mother of the leaves"), which stands some 300 metres above the surrounding plain. A short distance to the north is Phatsane ("splinter"), a tall thin column with its cap-rock precariously balanced. Between them lies the village of Mmamatlakala, its fields running down to the north-flowing Mmetlane River which, having merged with the Mokamole River a little further downstream, flows into the Mogalakwena River (known far upstream to the south as the Nyl) about 12 km later.

These rivers defined the western boundary in this area of one of the two large (and several smaller) blocks of land that comprised Lebowa, the *homeland*, or Bantustan created in the 1960s in terms of the separate development (*apartheid*) ideology of the National Party government. Lebowa was first designated as the homeland for the loosely defined North-Sotho ethnic group in 1962, when a Territorial Assembly was established. In 1971, a Legislative Assembly was formed, leading to self-government in 1972 and elections in 1973. These were won convincingly by the Lebowa People's Party, with former schools inspector Dr Cedric Pathudi being elected as the homeland's first Chief Minister.[1] Pathudi died in 1987 and was replaced by Nelson Ramodike, who led the state until its dissolution after the South African elections of 1994. In later years, Ramodike joined a succession of political parties in the region, gaining the epithet "political chameleon". He died in 2012.[2]

Today, Mmamatlakala is a much bigger settlement than it was in the 1960s; yet

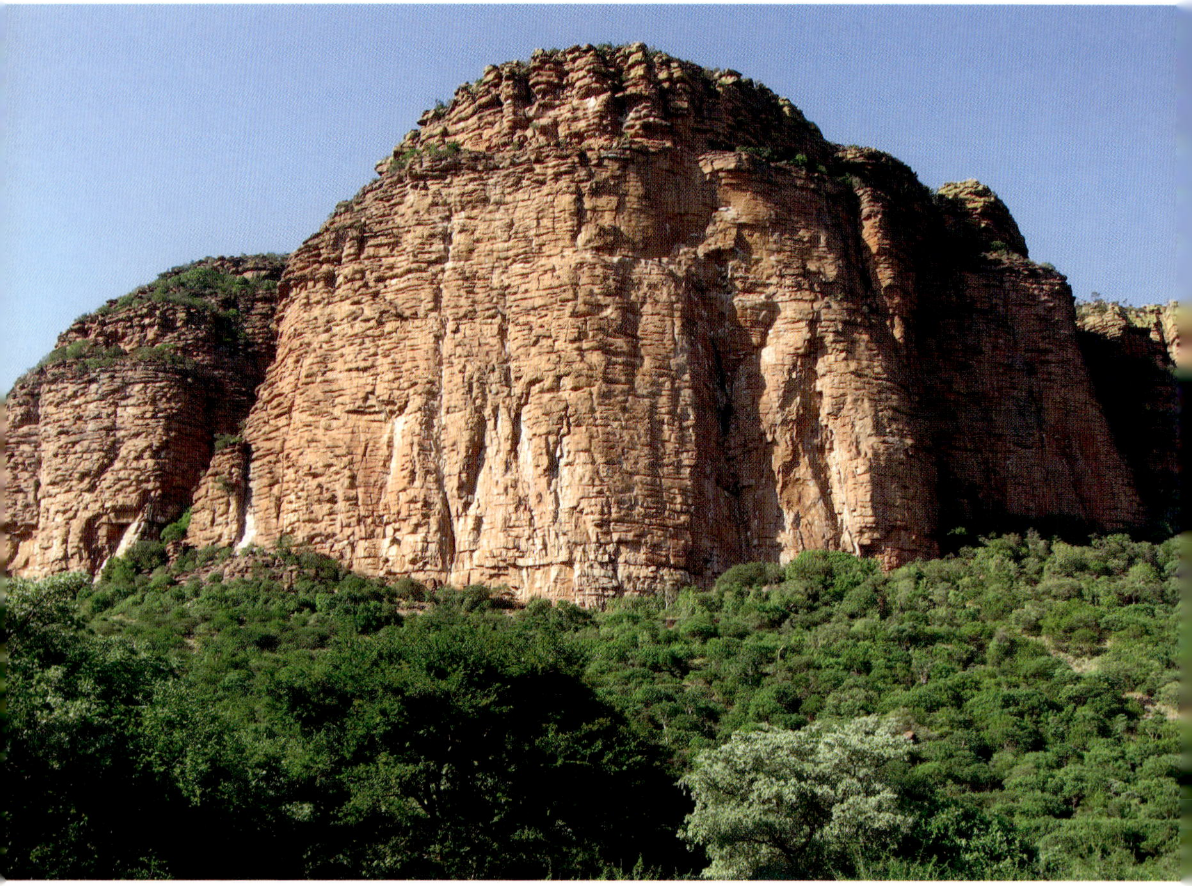

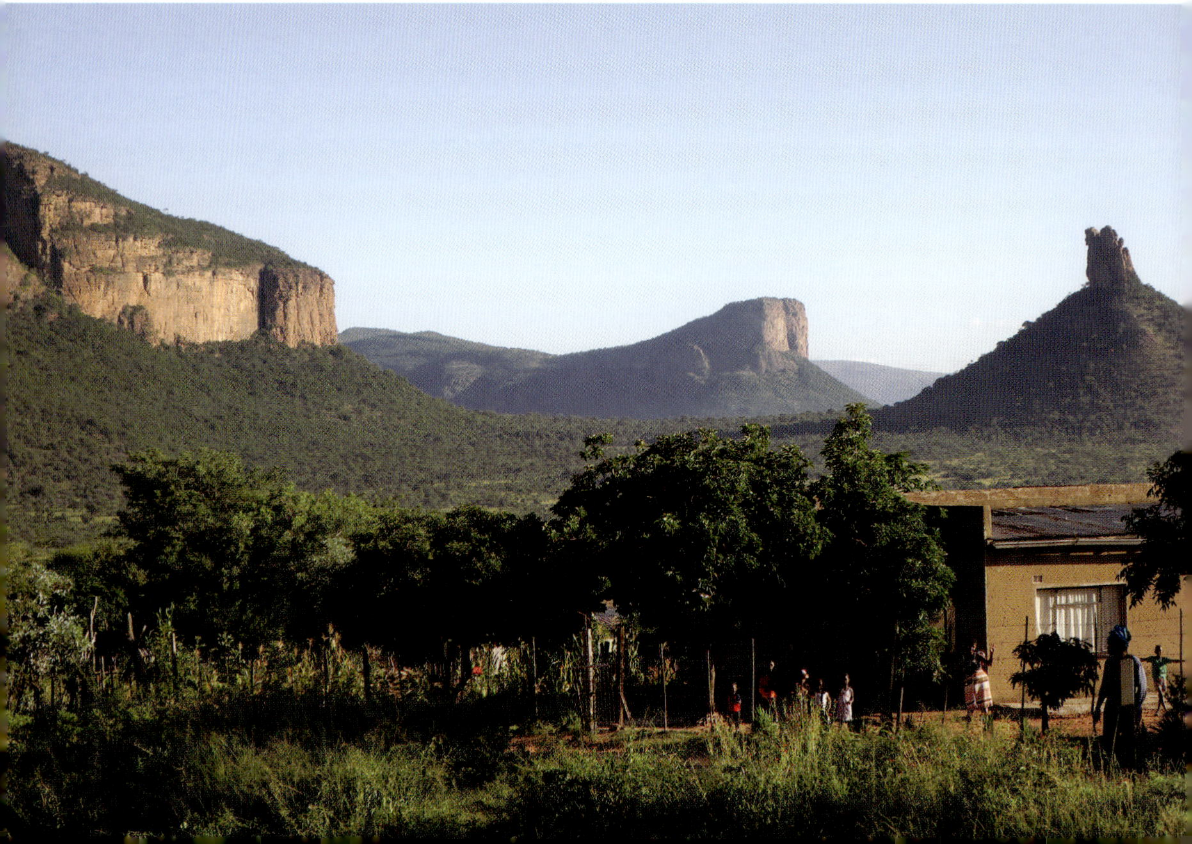

FIGURE 3.1: Mmamatlakala Hill (top), Waterberg escarpment and Phatsane column (bottom), from Mmamatlakala village at the foot of Kloof Pass.

FIGURE 3.2: Dr Cedric Phatudi, Chief Minister of Lebowa, receiving an honorary doctorate from the University of the North in 1985 (SANA SAB 17355, National Archives Photographic Library, Pretoria).

it and the dozens of similar villages that dot the plain to its north and east still closely define the extent of the former Lebowa state that ceased to exist in 1994. This part of the homeland was made up of three magisterial districts: Bochum, Seshego and Mokerong (the latter comprising three separate blocks). In a particularly bizarre piece of social engineering, Mokerong, the southern tip of which touched the municipal boundary of the old Potgietersrus (now Mokopane), was initially separated from Seshego by a narrow corridor which for 70 kilometres north from Potgietersrus ensured that the road to Steilloop and Tom Burke (the R35) ran safely through white farmland and the settlements of Limburg and Gilead. Later, these properties were purchased by the SA government and incorporated into Lebowa.[3]

The road down Kloof Pass (the R518) joins Mokopane with Marken, the pass itself being almost equidistant from these points. This road too was sited outside the homeland boundary. Of historical interest is that at a point about 15 km from Mokopane, just before the R518 crosses the Mogalakwena River, there is a grove of tall trees close to the southern side of the road. These are *Faidherbia albia* (formerly *Acacia albida*), the ana tree or *anaboom,* a deciduous species that favours Lowveld watercourses and alluvial plains. The grove is a national monument (proclaimed in 1949,[4] although the plaque commemorating this has been stolen so many times that the authorities no longer bother to replace it). It is known as "Livingstone's trees" on account of a belief – unsubstantiated – that the missionary explorer David Livingstone camped here on his way to or from the North in 1847. There is little doubt, though, that Thomas Baines passed by the grove of ana trees in November 1871 on his way back to Pretoria from a visit to Lobengula. On the 10th, he'd camped opposite the villages of Chiefs Mokopane and Lekalakala (where the township of Mahwelereng is today), less than 10 km to the south of the trees.[5] These trees are certainly outliers, in that their distribution is confined principally to three areas: western Namibia, the Caprivi-Zambezi valley, and the Limpopo valley around Kruger Park and down to Swaziland.[6] The grove comprises the only specimens of the tree known in the Waterberg region, although some have also been reported from the northeastern Springbok Flats, close to the former Lebowa capital, Lebowakgomo.[7]

FIGURE 3.3: Livingstone's trees (*Faidherbia albida*).

Lebowa was the designated home of the North Sotho, in the same way that Venda, Qwa Qwa, Gazankulu, KwaZulu, KwaNdebele, Bophuthatswana and Transkei and Ciskei were created in the time of apartheid as the homelands of the Venda, South Sotho, Shangaan, Zulu, (southern) Ndebele, Tswana and Xhosa people respectively. However, in this context, unlike the other Bantustans, the term "Northern Sotho" was more of a political than an ethnographic construct – a convenient sack-name to encompass all the black residents of the northern part of the Transvaal who were deemed to be neither Venda nor Tswana.

"Northern Sotho" is now understood to represent all the chiefdoms that occupied the northern parts of the Transvaal other than those specifically living in Venda, KwaNdebele or Bophuthatswana. The Sotho people were long recognised as comprising "an enormously complex group of tribes. Very diverse as to ancestry and lacking the homogeneity of the Venda, they are divided into numerous small independent tribes which consist of individuals of widely differing ancestry, indicated by the tribal origin or foreign totem of the individual."[8]

There are still over a hundred North Sotho chiefdoms; many of them (e.g. Bakone and Bakgaga on the Pietersburg plateau area) were indeed Sotho-speaking, but others, including the Langa, Kekana and Ledwaba, were (and to an extent remain) Ndebele-speaking, as are the Seleka people of the northern Lephalala River, despite their more recent Batswana origin. To a large extent, these non-Sotho speaking communities had over time become what has been rudely termed "Sotho-ised": assimilated into a predominantly Sotho society.[9]

This coalescence of two languages and cultures understandably created considerable confusion, misinterpretation and even prejudice amongst early white chroniclers of black communities in the region, aggravated by the fact that their conclusions were often drawn from translated interviews and shaped by their own Eurocentric sociological models. Whereas the Ndebele communities south of the Springbok Flats (the Manala and Ndzundza, for example) – the so-called Southern Transvaal Ndebele – were seen to be relatively homogeneous and to have retained their language and culture (a viewpoint that led directly to the creation of KwaNdebele in the 1960s), those in the north were found to have lost much of these ancestral attributes in favour of a mixed Sotho-Ndebele culture.[10]

In contrast, Sotho chiefdoms that had not merged with Ndebele seemed to predominate in the eastern part of Lebowa, which included Sekhukhuneland (the Bapedi and Bakwena) and the Pietersburg and Haenertsburg districts.[11] In this context, the reader should be reminded that, in the same way that the prehistoric record of the Waterberg region (Chapter 2) is derived largely from interpretations of the excavated archaeological record, so too is the pre-European history of the area dependent to a great extent on the observations and records of the first European travellers. Later, professional ethnographers, amongst whom Jan and Eileen Krige, N.J. van Warmelo, Isaac Schapera and A.O. Jackson were prominent, conducted structured interviews with many traditional leaders in an era prior to urbanisation and the fragmentation of rural societal fabric due to forced removals. Much of what follows can be attributed to their work and that of subsequent academic research.

THE COMING OF THE NDEBELE
The Langa Ndebele

The inhabitants of the western block of what was called Lebowa (the three magistracies listed earlier) were and remain of predominantly Ndebele rather than Sotho descent. The so-called Langa Ndebele, who still predominate in the area between Mmamatlakala and Mokopane, are known to have been resident in that area for at least two hundred years. (Langa means "sun" in Sindebele.) It seems that they originated from the Hlubi area of northern Nguniland, near the headwaters of the Buffalo River, long before the rise of the Zulu dynasty, and that they migrated northwards from there in the mid-seventeenth century. Having traversed through Swaziland, they settled briefly near where Leydsdorp is today (about 70 km west of Phalaborwa), before moving on to an area southeast of Polokwane under the leadership of Masebe I where they remained for several generations. (Map 3.1). There, their nearest neighbours were the Matlala, a Sotho chiefdom, and the Kekana, an even earlier group of Zulu/Ndebele migrants.[12] Various Langa groups, for example the Molekoa, led by sons of the ruling chiefs, moved away westwards during the late eighteenth century, in some cases subjugating and absorbing existing Sotho chiefdoms in the region. By the 1850s, the Molekoa for example seem to have settled to the west of the Mogalakwena River, in the foothills of the Waterberg, although remaining subjects of the Langa chief.[13]

In about 1795, Mapela, grandson of Masebe II, rose to become the chief of the dominant Langa clan. During his rule, the group increased in size, in part through the absorption of small Sotho chiefdoms, including for example the small Pedi group of Chief Matlou, who occupied fertile land against the Mogalakwena River, and the Phalane. (A subset of the latter clan fled and their descendants can now be found around the Pilanesberg of North-West Province.) Another Sotho

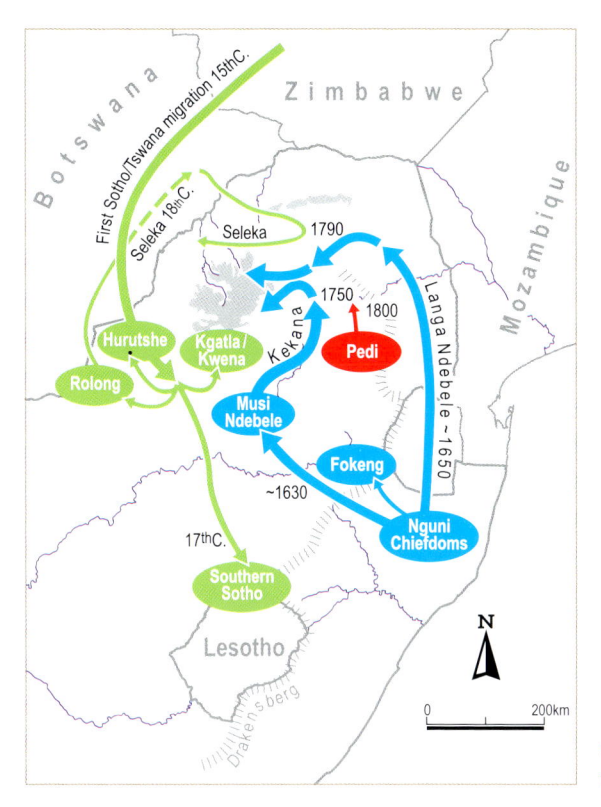

FIGURE 3.4: Subcontinental migrations in the fifteenth to eighteenth centuries.

group attacked by the Langa under Mapela were the Bididi of Songwane (now Shongoane or Setateng) on the Lephalala River, upstream from the Seleka. They took refuge on a hill (Bobididi), on the west bank of the Lephalala River, about 10 km south of the modern settlement of Villa Nora.[14] Many years later, after Mzilikazi and his warriors had been driven north by the Boers (with the help of the Langa), the Bididi became subject to the Langa, with heavy tribute being extracted from them at least until 1890.[15]

However, the Langa did not have everything their own way: at the turn of the nineteenth century, a Pedi expedition under Chief Thulare (a Sotho chief) ventured far northwards from the Pedi homeland in Sekhukhuneland into the Waterberg and Zoutpansberg districts, extracting tributes from all whom they encountered, including the Langa. During this time, the Pedi were dominant in the northern and eastern Transvaal, but their influence waned following infighting between claimants to the throne after Thulare's death in about 1820.[16] It was the dominance of the Pedi that prompted missionaries from the Berlin Missionary Society to develop a written form for the Pedi language – the reason why to this day, Sepedi (or Sesotho) is considered, incorrectly, to be the mother tongue of all people of so-called North Sotho extraction.[17] A wave of Ndebele (Matebele) raiders associated with Mzilikazi arrived in the area in the 1820s, while Mapela was still alive, forcing him to relocate once again, this time to the hill called Fothane (now shown as Moordkoppie), on the east bank of the Mogalakwena, about 30 km northwest of Mokopane. Today, the village lies just to the west of the country's largest platinum mine, Mogalakwena Platinum. Here Mapela died in around 1825, and was succeeded in about 1836, after some internal rivalry, by a grandson, Mankopane, whose own son, Masebe III, would eventually succeed him.[18]

The Kekana or Valtyn Ndebele

The other principal Ndebele group within the area around Mokopane is the Kekana, or Valtyn chiefdom. It seems likely that this community is descended from the Southern Transvaal Ndebele, with a common ancestral chief called Musi (Msi), whose people had originally settled in an area just north of the Magaliesberg range at Pretoria, about where Wonderboompoort is today and whose migration from the Hlubi in Nguniland preceded that of the Langa, having taken place in the period 1630 to 1670.[19] Conflict between Musi's five sons resulted in their dispersal towards the north and east (Figure 3.4).

One of these sons, known as Mathombeni (or Kekana), travelled with his followers to the northeast into the Steelpoort valley. A war with one of his brothers, Ndzundza, forced Kekana's clan to move on, first to the Zebediela area and eventually, in about 1750, to Chidi or Pruissen, a promontory about 13 km to the southeast of present-day Mokopane – by which time the group was led by Kgaba, one of Kekana's sons. Only after the calamitous events of 1854 did the remainder of the chiefdom relocate to a site below the Sefakaola Hill just north of Mokopane (the hill that overlooks the modern township of Mahwelereng). Here they encountered an existing Sotho-speaking community known as the Mashishi, with whom eventually there was conflict.[20]

The Kekana Ndebele people were already settled along the alluvial plain of the Mogalakwena River when the Langa group, under Mankopane, arrived to settle to the north of them, around Fothane. Although sharing a common distant Nguni ancestry, the two Ndebele groups appeared not to build on this relationship immediately, but rather to focus on expanding their individual influence through subjugation of the disparate Sotho clans and chiefdoms.[21] Kgaba's brother, Khuba, remained with his part of the Kekana clan in the Zebediela area to form the fourth main Ndebele chiefdom in the region.

Apart from the Langa and Kekana Ndebele chiefdoms described above, there is archaeological evidence that other Ndebele groups had migrated into the northern Transvaal region and even onto the Waterberg plateau, as interpreted at sites such as Buffelsfontein south of Bulge River and at Melora Hill on the Lephalala River drainage.[22] (See Chapter 2.) These migrations certainly preceded that of the Langa and may even have preceded the Kekana, although it is believed that like the Kekana, they had all owed allegiance to the ancestral chief Musi; this probably occurred in the early seventeenth century. However, there is little oral ethnographic record of these early Ndebele migrations and their history is based largely on interpretation of archaeological excavations. According to this evidence, there may have been an even earlier incursion of Nguni affiliates into the Waterberg. A particular style of pottery and walling, named Moor Park, seems to be associated with the Fokeng (Bafokeng), who built structures in the Waterberg in the mid-fifteenth century, having migrated from their base along the Drakensberg escarpment in the eastern Free State, a staging post on their way from the Natal midlands.[23] Some members of the Hurutshe cluster, a Sotho-Tswana group living in the Marico area of the western Transvaal, were also found to have migrated from their heartland to beyond the northern limits of the Waterberg plateau (Figure 3.4).[24]

The Seleka of Lephalala

One of these groups was the Seleka, whose ancestry was long uncertain amongst ethnographers. They were thought to have been of Tswana descent, because of their cultural characteristics, their use of Setswana and their totem which was a duiker (*phuthi*) – unlike Ndebele clans, who venerate the elephant. Recent research has proved conclusively that the Seleka are in fact descended from the Nguni of northern KwaZulu-Natal and that they migrated onto the southern Highveld in the late seventeenth century, settling close to the Lesotho border.[25] Later the chiefdom moved in stages to eastern Bechuanaland to live with the Bakwena of Molepolole to whom they had become related by marriage. From there, one group moved east across the Limpopo and around the north of the Waterberg plateau to settle north of Potgietersrust, while another group settled near a hill called Ngwape on the Bechuanaland side of the river. Eventually, the eastern group, harried by the arrival of the Langa Ndebele from the south, moved west again and, under Chief Seleka I in around 1840, established itself around a hill called Mmatshwaana, to the south of the later settlement of Beauty on the lower Lephalala River.[26]

However, the Seleka seemed endlessly restless – and also willing to take on hostile neighbours, including the Langa Ndebele, who defeated them and forced them to pay tribute. In 1858, Seleka I's son, Kobe, decided to take his people back across the Limpopo to Bechuanaland, where intermarriage with the Bamangwato (the clan of the Khamas) took place. But when Kobe's son, Seleka II, was found to be plotting against Khama in 1887, the clan was forced to flee back to its former home near Beauty. Kobe died two years later and Seleka II succeeded him, moving the seat of the chiefdom from Mmatshwaana to Thothwane, where it remains to this day. Seleka II also initiated discussions with President Kruger and Commandant Joubert about securing land for his people (the clan's long absence in Bechuanaland had rendered it ineligible for a location like the other Ndebele groups further east). Although relations with the Boers were good – indeed, during the South African War, the Seleka sheltered Boer cattle, returning the stock to their owners afterwards – no suitable land could be made available. In 1903, the clan was able to purchase the farm Beauty for the equivalent of R2000. Later, in terms of the Bantu Authorities Act of 1951, the chiefdom was granted five other farms: Rietfontein, Kafferskraal, Harry Smith, Olifantsdrift and Witfontein.

Seleka II died in 1917 and was succeeded by Mananya, the eldest son of his principal wife. Unusually amongst the chiefdoms of the region, Mananya's long reign of 29 years was peaceful and constructive for the clan. His successor, Seleka III, in 1946 continued this trend (which included the promulgation of the Bantu Authorities legislation, the acquisition of the additional farms and the establishment of the Seleka Tribal Authority in 1954). Seleka III died in 1960 without having married a principal wife and so his brother, Tompi Zachariah, was appointed regent. Tompi ruled for the next 30 years and gained such a reputation for his wise and sound leadership that he was appointed Minister of Justice in the cabinet of Dr Cedric Phatudi, the first Chief Minister of the Lebowa government.[27] The Seleka chiefdom was undoubtedly unique in having enjoyed sound, peaceful leadership, free of fraternal disputes and intraclan rivalry for over 75 years.

The difaqane[28]

From 1750–1825 profound changes took place in the structure of black societies in southeastern Africa, commencing amongst the Nguni-speaking groups living in the region from southern Mozambique south along the Natal coast and interior. Some chiefdoms began to expand their influence through forging alliances with surrounding smaller groups to form loose confederations. Two examples of such confederacies were the relatively benign Mthethwa chiefdom under Dingiswayo, and the warlike Ndwandwe under Zwide, which was more inclined to attack neighbours, seize their assets and cause the rest of the community to flee.[29] The Zulu were then a small chiefdom within the Mthethwa confederacy and the young Shaka (he'd been born in about 1787) was a member of Dingiswayo's army. Shaka rose to head the Zulu chiefdom in 1816 and instilled in his own army some of the military skills he'd learned from the Mtethwa, adding his own ruthlessness and readiness to eliminate his enemies. Soon afterwards, the Mtethwa were attacked and defeated by the Ndwandwe and in the process Dingiswayo was killed. Shaka, who now headed the most powerful chiefdom within the Mtethwa, managed to regroup its remnants around his Zulu chiefdom and in 1819, he exacted revenge on the Ndwandwe, defeating them completely, killing members of the ruling family and forcing large numbers of the chiefdom to flee into the interior as refugees. By the mid-1820s, the Zulu kingdom had become the most formidable power in the subcontinent.[30] However, Shaka's reign of terror gradually bred enemies among his own chiefdom and in 1828 he was assassinated by his half-brother Dingane. These processes of coalescence and growth of some chiefdoms at the expense of others, as described above in respect of the Nguni, were also taking place amongst other communities in the region, such as the southern Sotho under Moshoeshoe. They led to the creation of a few powerful, dominant chiefdoms that were well organised both militarily and politically, at the expense of hundreds of smaller chiefdoms that were destroyed, assimilated or broken up into small groups of people desperate to escape annihilation and to secure new lands and assets for themselves.

One of the groups to flee the Nguni-controlled areas now forming the northern part of KwaZulu-Natal was the Khumalo clan, led by its young chief Mzilikazi. He had been born in about 1795; his group had been under the sway of Zwide and the Ndwandwe until the latter's defeat in 1819. Unwilling to fall under the tyrannical authority of Shaka, the Khumalo suffered many casualties at the hands of the Zulu before Mzilikazi led the survivors up onto the Highveld.[31] Here, his people, whom we shall call the Matebele to differentiate them from the earlier Ndebele chiefdoms that were already long-settled on the Highveld, continued their aggressive subjugation of the Sotho communities and other Nguni refugee groups who lay in their path.

But the Matebele were not the first to cross the Drakensberg: The Hlubi, under their chief Mpangazitha, having fled from Shaka's destruction of the Ndwandwe, crossed the divide and fell upon the Sotho communities living on the Highveld. The first of these were the Tlokwa (1822–1823), whose regent was MaNthatisi, the mother of a boy who was still too young to become chief. Survivors of that slaughter fled across the Highveld, leading others to refer to them as the Mantatees. Another group of invaders from east of the Drakensberg who wrought havoc among the Sotho-Tswana communities was the Kololo, led by Sebetwane (1823–1828).[32]

Some conflicts were characterised by great violence, leading to massive loss of life and stock, to the annihilation of some clans, the dispersal of others and the assimilation of women into invading groups.[33] The whole Highveld region became destabilised, with some communities in flight, others in bitter, deadly conflict either to protect lands and cattle or to acquire them. Thousands of people are estimated to have died and many thousands more forced to flee – mainly northwards – to wherever they could find refuge. "Raiding and pillage spread as far west as the Tswana. Early visitors described scenes of destruction on the Highveld. The landscape in places was littered with human bones, stone walling was destroyed, and refugees were forced to live as hunter-gatherers, in extreme cases resorting to cannibalism." This was the difaqane ("the hammering").[34]

There is no doubt that towards its end, while Mzilikazi and his Matebele wrought havoc across the Highveld in pursuit of their own place to settle, the chaos that was the difaqane must have been aggravated (and exploited) by others: In the west the Korana and Griqua people, armed with guns and on horseback, were able to exploit the vacuum to their own territorial advantage; the earliest Boer trekker groups, who began leaving the eastern Cape in 1836, found little resistance as they advanced northeastwards; black and white entrepreneurs would have seized homeless individuals or groups for sale into the slave trade at Delagoa Bay; missionaries (like Robert Moffat at Kuruman) discovered fertile ground for their proselytising. Whatever the causes, the result was a broad swathe of societal disruption that migrated northwards to and around the Waterberg by the early nineteenth century. Many once-settled clans had abandoned their villages, lost their livelihoods and had sought refuge in inhospitable places.[35]

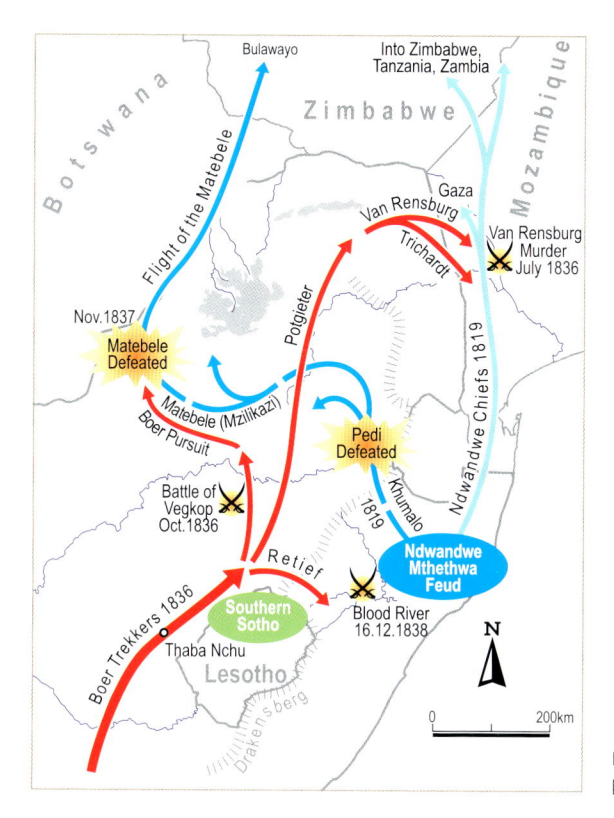

FIGURE 3.5: The difaqane and the Boer trekkers clash in the early nineteenth century.

The Matebele quickly conquered and subjugated the weaker Pedi clans in the Groblersdal area before establishing their headquarters on the Crocodile River near the present town of Brits. From here, they set about gaining control of the entire western Transvaal region. Each defeated clan in turn fled elsewhere, destroying or displacing communities as they went. For some time, the Matebele were able to consolidate their position in the western Transvaal, eventually incorporating (or destroying) the Hurutshe and their vast capital, Kaditshwene, and moving into their former territory in the Marico area. Throughout, they were harried by Zulu impis from the south (now under the command of Dingane, Shaka's successor) as well as by Korana and Griqua groups.

Then, early in 1836, the first of the Voortrekker groups appeared on the scene, driven to leave the eastern part of the Cape Colony by their intolerance of British rule, taxation and the abolition of slavery. Initially, they met little resistance from the scattered and fractured black communities they encountered. On the contrary, their help was often enlisted to defend these communities against the Matebele. In August 1836 a Voortrekker group, which may have strayed unwittingly onto land claimed by Mzilikazi, was attacked by a Matebele impi on the Vaal

River near where Parys is today. Soon afterwards, one of the Voortrekker leaders, Andries Potgieter, returned from an expedition to the north of the country (to the Soutpansberg) and, finding his compatriots in disarray, organised them into a sound defensive position. This was as well, because it was not long before the trekkers were again attacked, this time by a much larger Matebele force.

This engagement, known as the Battle of Vegkop, took place on 20 October 1836. Whilst in one sense the Boers were victorious in that they drove the Matebele away with the loss of only two of their number, it was also a huge loss: the retreating Matebele took with them some 6000 head of cattle and over 40 000 sheep, virtually all the Voortrekkers' stock. Clearly, retribution would follow. In January 1837 a Boer commando attacked the Matebele north of the Vaal River and recovered several thousand head of cattle. By April, the Voortrekker numbers had increased sufficiently (thanks to the arrival of trek groups led by Gert Maritz, Piet Retief and later Piet Uys) that a serious punitive expedition against Mzilikazi could be contemplated.

Eventually, early in November 1837, a large Boer force moved across the Vaal, heading northwest to Mzilikazi's headquarters in the Marico district. In a series of battles spread over nine days, the Matebele were decisively routed and fled north across the Limpopo River into what became Zimbabwe where their descendants live to this day.[36]

The Boer trekkers quickly occupied territories abandoned by communities that had been displaced – although many of those communities still considered the land to be theirs, awaiting their return from temporary exile. The historian Hermann Giliomee summarised the situation well:

> As a result of the [difaqane], many areas were sparsely populated or seemingly devoid of people …The implications … for the Afrikaner trekkers who moved into the deep interior in the late 1830s and the 1840s were profound. It did make the Voortrekkers' task of settling in the deep interior much easier because large areas were temporarily depopulated. Desirable land seemed to be there for the taking.[37]

It should not be thought that the effects of the difaqane were limited to south of the Limpopo River. When the Ndwandwe were defeated by Shaka in 1819, several surviving chiefs fled with their people. One clan ultimately settled in Mozambique and southern Tanzania where they founded the Gaza kingdom;

a second deposed chief moved into southern Zimbabwe where he defeated the powerful Rozwi state of south-central Zimbabwe, which had held sway since the late seventeenth century (under its legendary king, Changamire). Nguni descendants whose ancestral migrations were part of the difaqane have also been identified in Mozambique, Zambia and Malawi.[38]

The Sotho-Tswana people of the Waterberg

From the above it is apparent that successive waves (at least four) of Nguni (Ndebele and Matebele) invaders/migrants into the northern part of South Africa between the fifteenth and early nineteenth centuries had encountered communities already resident there. In the area surrounding the Waterberg, these were representatives of the Sotho-Tswana group of chiefdoms, of which five separate lineages have been identified. These were the Hurutshe, who lived near the confluence of the Marico and Crocodile Rivers at the turn of the fifteenth century and developed a tradition as workers and traders in iron;[39] the Kgatla and Kwena, who occupied the central Highveld around where Brits lies today; the Rolong, who lived south of Zeerust (and who may have been displaced by the nearby Hurutshe); and the Fokeng of the eastern Transvaal and Free State (who were originally of Nguni origin).[40] The Fokeng are often stated as having been the most senior of these clans, reaching the height of their influence in the early eighteenth century under their chief, Sekete IV. When he was killed by a conspiracy of the other clans, the Fokeng dispersed, to be replaced as the leading clan by the Hurutshe. The early Hurutshe were said to have been ancestral to the Kgatla and Kwena clans, representatives of which migrated both south (where they gave rise to the Moshoeshoe lineage), and north to form the ancestral groups of the Ngwato (or Bamangwato) of Botswana.[41]

It seems that those chiefdoms most closely associated with the area around the Waterberg (as opposed to those further east) were principally of Tswana descent – the Tswana being the earliest member of the Sotho group to migrate from central Africa into the region later defined as the western Transvaal – and that they were established in their heartland by the end of the sixteenth century.[42] Unfortunately, their subsequent experience of repeated dispersal and flight episodes at the hands of successive waves of Ndebele/Matebele incursions has made a detailed reconstruction from oral history difficult. In general though, the literature suggests that many Sotho-Tswana communities became either subjugated by, or assimilated into the more dominant, militaristic and cohesive Ndebele chiefdoms, which however adopted their language and several of their customs.

Hence by the mid-nineteenth century, the principal societies occupying the plains to the immediate east of the Waterberg plateau were essentially Sindebele- and Sesotho-speaking Sotho-Tswana-Ndebele chiefdoms like those of the Langa and the Kekana/Valtyn. In the south and southwest, between Hammanskraal and the Botswana border, the Tswana ancestry was more predominant, chiefdoms including the Kgatla, Hwaduba, Fokeng Tlhako and Hurutshe.[43] In the northwest, an isolated Tswana group, the Seleka, an offshoot of the Rolong, had settled close to the confluence of the Lephalala and Limpopo rivers, at Beauty (now Ga-Seleka). Whilst there were undoubtedly scattered communities living on the Waterberg plateau itself at this time, they are not referred to in any studies written about the Ndebele, the Tswana or the Sotho. There seems to have been no cohesive chiefdom on the plateau; rather, the inhabitants seem to have been refugees from one of the many conflicts that had been taking place in the surrounding countryside. All the records that have been examined concur that the plateau was sparsely populated by the 1850s.

FIGURE 3.6: Principal chiefdoms surrounding Waterberg in the mid-nineteenth century.

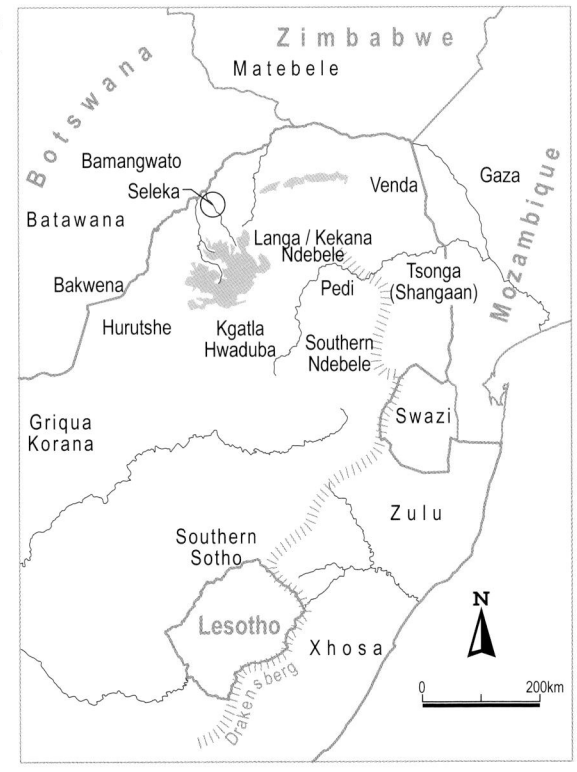

This conclusion is supported by the observation that whereas the various missionary societies were extremely active in attempting to establish operations to the south, east and northeast of the plateau in areas where population density was attractively high and settled, no mission station was ever built or planned on the plateau itself. Early maps of the region, which accurately depict the widespread settlements below the escarpment, show only two black settlements on the plateau: one, unnamed, on Dordrecht (now part of Lapalala) and another, named Adam Matsega's Kraal, on Vogelstruisfontein, south of Dorset.[44] The settlement on Dordrecht first appeared on a diagram produced by the Surveyor General's office in 1894, and the one on Vogelstruisfontein on a cadastral map published in 1911. In both cases, owing to the delays between survey and publication, the entries may have referred to villages that were no longer in existence by the time the maps were compiled.

Meletse Mountain and Kwa-Makapane

Immediately to the south of the Sandriviersberg range, which defines the southern margin of the Waterberg plateau, is a prominent hill called Meletse (or Madimatle) mountain, which rises to 1862 metres above mean sea level (amsl), almost 800 metres above the surrounding countryside. The hill, which is made up of dolomites and ironstones belonging to rocks older than the Waterberg, contains a labyrinth of caves, for which reason it is also known as Gatkop. For centuries, these caves have been considered sacred places by many Sotho-Tswana clans in the region and are still used frequently by members of the traditional healer community as places of prayer.[45] There are numerous accounts of pilgrims who have journeyed to the caves to be cured of one or other condition. In recent years, Meletse mountain has become the focus of intensive exploration for iron ore, containing iron mineralisation that is similar to the ore that was mined for over 80 years at Thabazimbi, about 25 km to the west. In 2015 the South African Heritage Resources Agency suspended planned mining development on the hill for at least two years to enable it to assess the heritage value of the site.[46]

Adjacent to Meletse, on the farm Buffelshoek, is a low hill known as Kwa-Makapane. For eight years, from 1872 until 1880, the hill was occupied by a group of about 5000 Kgatla people (the Mabotse community), under the leadership of their chief, Makapan, who moved them from their traditional home at Mosetlha, just north of Pretoria.[47] It's not clear why Makapan chose this particular site for settlement (although its proximity to the sacred Madimatle caves may have been a factor), but the reasons for the move were fears that the Kgatla might lose control of their land at Masetlha, that members of the community might be indentured for labour and that they might become subject to taxation by the impoverished Zuid-Afrikaansche Republiek (ZAR, or Transvaal government).

Conflict between Boer Voortrekkers and the Ndebele

Once Mzilikazi and his Matebele followers had been put to flight across the Limpopo River early in 1838, the advancing Boer trekkers found little to impede their progress northward. By the early 1840s they had established communities in the Marico district and soon thereafter at Eersbewoond between the modern towns of Bela-Bela and Modimolle. Only once they moved further north into areas settled by powerful chiefdoms less disturbed by the difaqane, such as the northern Pedi in Sekhukhuneland, the Venda in the Soutpansberg and the various Ndebele clans to the northeast of the Waterberg, did they begin to encounter serious resistance.

The only incident in the north up until this point had been the killing of the entire Van Rensburg trek on the lower Limpopo River[48] in July 1836 by an impi of Manukozi Soshangane, a nephew of the defeated Ndwandwe chief Zwide who, having fled the might of Shaka, had established the powerful Gaza chiefdom in the area. The Van Rensburg party had left Louis Tregardt's (Trichardt's) group at Zebediela and had continued north at a faster pace to Soutpan (at the western end of the Soutpansberg) where they camped for some time. By the time Tregardt arrived at Soutpan, the Van Rensburg party had already left in an attempt to find a way to the coast. Van Rensburg had made the mistake of persisting with entering the Gaza region despite warnings from local headmen not to do so. Tregardt, the first Voortrekker leader to have left the eastern Cape, was an intelligent and wealthy man, tactful and patient and at pains to avoid undue conflict with those whose territories he crossed. Having searched in vain for the missing Van Rensburgs, Tregardt prudently decided to take a different route to the coast. After an extremely difficult journey, which included negotiating a route down the Drakensberg with his wagons, the Tregardt group did succeed in reaching the coast at Delagoa Bay (Lourenco Marques, now Maputo) on 13 April 1838, eleven months after setting out from Soutpan. Sadly, Tregardt, his wife Martha and almost half the party of 53 were soon to die there of malaria.[49]

On 6 February 1838 the trekker Piet Retief and a hundred of his colleagues and servants had arranged to meet the Zulu king Dingane (the man who had killed his half-brother, Shaka, ten years before) in the hope of negotiating a land treaty. On arrival at the king's kraal, the Boers were tricked into leaving their weapons outside, and were then clubbed to death – with Retief having to witness the process before being killed too. Retribution was inevitable: In November, Andries Pretorius arrived from the Cape with a force of 60 men and a cannon. Together with his commando leader, Sarel Cilliers, and the remaining trekkers in the area they established a laager in Dingane's territory near the Buffalo River. On 16 December the combined force of about 500 defended the laager against an attack by over 10 000 Zulu. After a two-hour battle in which some trekkers were slightly wounded, but none killed, the Zulu army withdrew, leaving behind over 3000 casualties. This decisive battle (which became known as the Battle of Blood River) brought Dingane's rule speedily to an end; he was killed and his half-brother Mpande, who was proclaimed king, assisted the Boers in recovering about 40 000 head of cattle.[50]

On the (Transvaal) Highveld the Boer settlers had established an informal capital at Potchefstroom under the leadership of the autocratic trekker, Andries Potgieter. In 1845 Potgieter decided to move the capital to the northeast of Transvaal to be closer to a trade corridor to Delagoa Bay he had hoped to negotiate with the Portuguese. But his chosen site, Ohrigstad, north of Lydenburg, was disastrous: Not only was it within the tsetse-fly belt and a malarial area, but it was surrounded by chiefdoms that were becoming increasingly incensed by the inroads being made on their territory by white invaders. Moreover, Ohrigstad had already been settled by a group of Natal Boers, who refused to accept Potgieter's authority. Unlike Tregardt, Potgieter was belligerent, intolerant and confrontational, although he was a popular leader. He was also inclined to secure the support of some chiefdoms in return for protecting them against others, thereby setting up interclan feuds. Before long, his attitude and behaviour would precipitate a long series of battles between black chiefdoms and the white settlers which would characterise the region for the next 40 years. Potgieter was also attracted by the business opportunities offered by trade in ivory and slaves with the Portuguese at Delagoa Bay. In 1844 he and 24 mounted Boers escorted a march by 300 ivory-laden slaves to the coast, where all (ivory and slaves) were sold to the Portuguese.[51]

Frustrated by the lack of support from his Natal compatriots, Potgieter led his people further north in 1848 to find a new capital at Zoutpansbergdorp, about 17 km to the west of the modern town of Louis Trichardt (Makhado). From here he promptly launched an expedition to attack Mzilikazi across the Limpopo – the reason for this is unclear, but when his commando failed to locate the Ndebele, they decided instead to make an unprovoked attack on the Langa Ndebele (allies of the Ohrigstad Boers), killing many men, capturing large numbers of sheep and cattle and seizing many children as slaves. Obviously, there would be repercussions, although Potgieter was not to know that the death of his son would be one of them. The trekkers left behind at Ohrigstad moved south to settle in a somewhat better location at Lydenburg. By 1850 Ohrigstad had become a ghost town.[52]

The Moorddrift murders

On 16 December 1852 Andries Potgieter died at Zoutpansbergdorp after falling ill while leading a commando against the Pedi chief Sekwati. He was succeeded as Commandant General in the Soutpansberg by his son Pieter Johannes Potgieter, aged 30. In August the following year, the local trader/hunter João Albasini learned from Chief Sekwati that the Pedi were resentful of the Boer settlements and commandos and might attack groups of Boer travellers. A few months later, a party was about to set out from Zoutpansbergdorp for the western Transvaal; it ignored Albasini's warnings not to travel through increasingly hostile territory. The route south ran directly through Kekana Ndebele territory, between the villages of two prominent chiefs, one of whom was Mokopane (in Sesotho) or Mughombane (in Sindebele), the other Lekalakala. Mokopane was the grandson of Kgaba.

The party of 12, including two women and six children led by Willem Prinsloo, stopped at a crossing over the Nyl (Mogalakwena) River just below Mokopane's village one day in September 1854. There they were killed cruelly and their bodies dismembered. Prinsloo and a colleague had gone to pay respects to the chief, but were killed on arrival. The crossing, less than 10 km south of the town of Mokopane, became known as Moorddrift. (The site was proclaimed a national monument by the National Party government in 1949 and is still marked by a monument to those who died.) Two other men were also killed at Mokopane's kraal.

On the same day, some distance to the north, another group of 14 Boers, a hunting party led by Hermanus, the notorious and violent-tempered younger brother of the late Andries Potgieter, were allegedly lured onto the property of the Langa Ndebele Chief Mankopane (grandson of Mapela) by the promise of

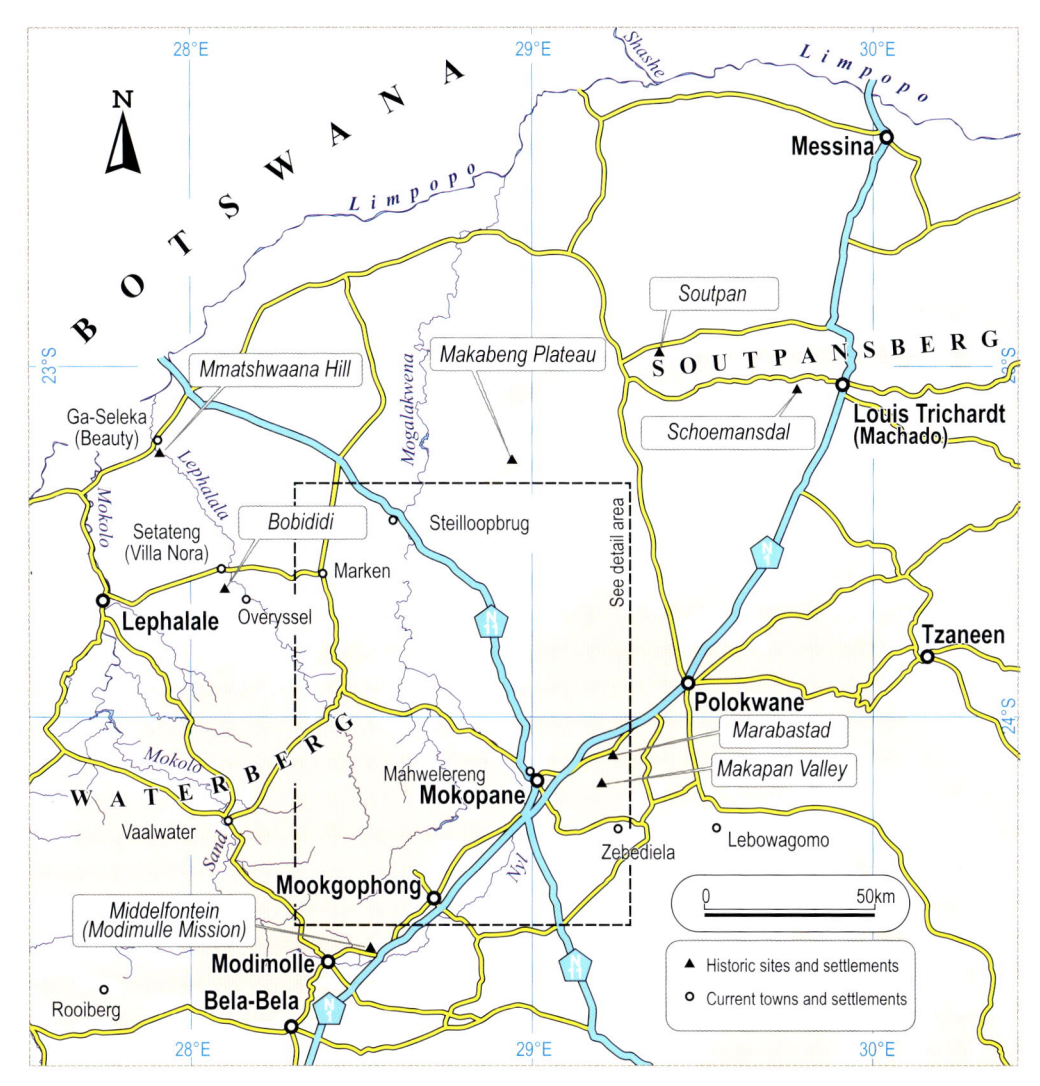

FIGURE 3.7: Location of major historical settlements
and sites in the Waterberg region (more detail in Figure 3.9).

elephant tusks and were also murdered and mutilated. It is not certain to what extent Hermanus Potgieter may have provoked this deadly incident, although there are many stories attesting to his brutal behaviour towards blacks, including the kidnapping of children as slaves.[53]

A total of 28 whites died that day.[54] There is ample evidence that the three separate attacks on Boer groups on the same day were coordinated between the Kekana and the Langa Ndebele, but explanations for the rationale behind them differ according to the perspective of the analyst. Afrikaner republican historians and Boer apologists (for example the journalist Gustav Preller) maintained that the "massacres were but a continuation of the general plan of murder and pillage on the part of the Ndebele".[55] However, what should not be forgotten is the unprovoked attack on the Langa by Andries Potgieter in 1848, his kidnapping and enslavement of Ndebele children and the nefarious actions of his brother Hermanus; all in addition to the invasion and seizure of Ndebele territory by the Boer trekkers. In general, it is concluded that the intention of the two Ndebele chiefdoms was to take advantage of the prevailing drought and horse sickness in the area to launch an attack that they hoped would have the effect of persuading the Boers to leave the region and return to the south.

Makapansgat

Instead of leaving, the infuriated Boers, under the leadership of Commandants General Piet Potgieter and M.W. Pretorius (who came up from Rustenburg) quickly mustered a combined force of about 500 men, and a month after the massacre, laid siege to the Kekana group who, under Mokopane's command, had taken refuge in a huge network of caverns they called Gwaša in some nearby dolomite hills. The Boers initially tried to block the caves by blasting. When that failed, they dropped

CHAPTER 3: HOMELAND OR A PLACE OF REFUGE?

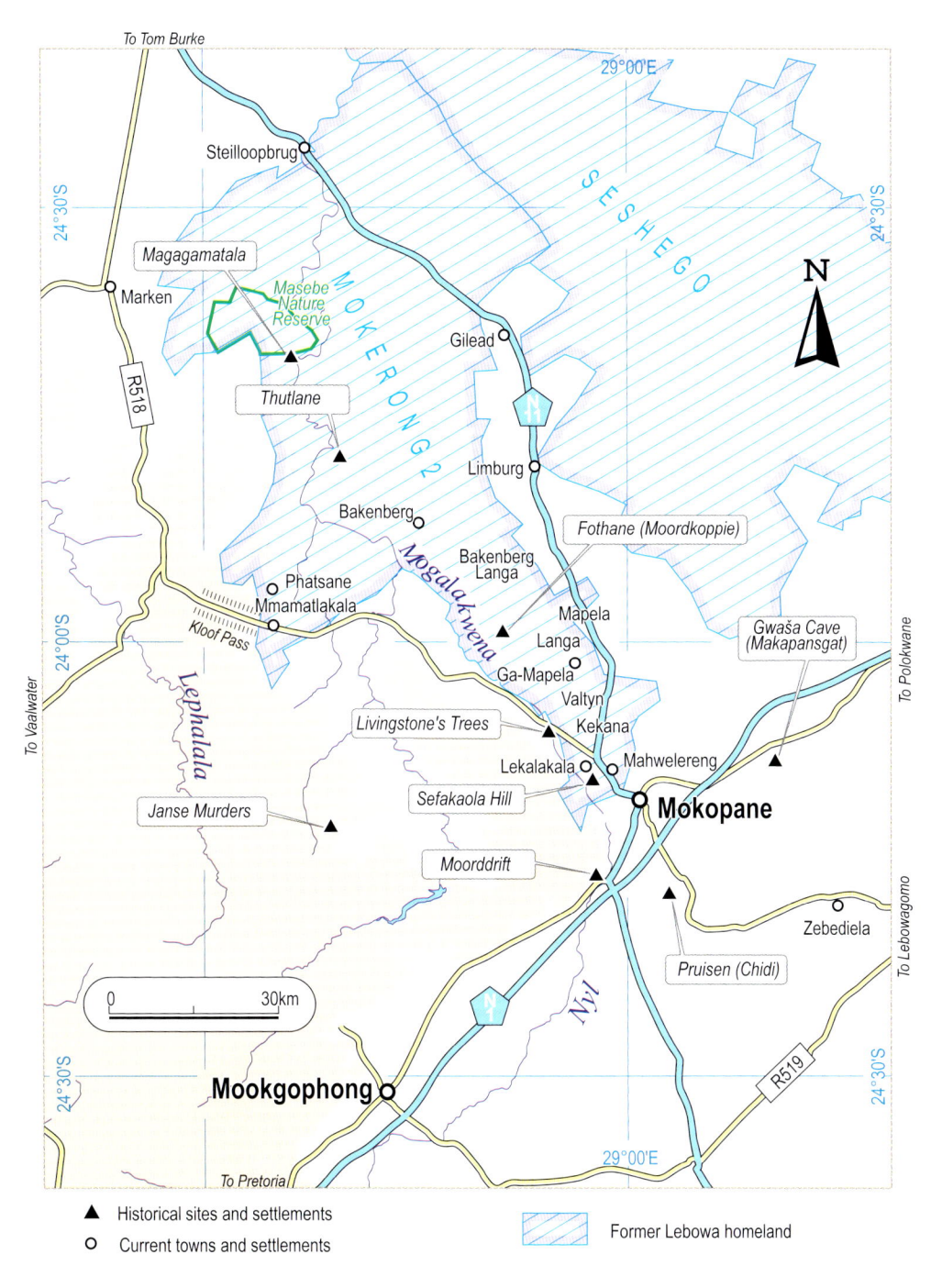

FIGURE 3.8 (OPPOSITE): A wagon crossing the Nyl/Mogalakwena River at Moorddrift. SANA TAB Photographic Library, No. 16316 (by H.F. Gros).
FIGURE 3.9 (ABOVE): Important sites and settlements in the Mogalakwena River valley.

masses of branches in front of the openings from above and lit them in the hope of either asphyxiating those trapped within or driving them out. The Kekana finally surrendered late in November, by which time more than 1500 of their number had died of starvation, thirst or disease. Several hundred others, including many women and children, had surrendered during the period. Chief Mokopane himself is said to have escaped from the cave, but committed suicide later.[56]

On 6 November 1854 during the siege, Piet Potgieter had made the fatal error of standing silhouetted at the mouth of one of the cave entrances and was shot by a sniper from within. Potgieter's body was retrieved and carried to safety in order to prevent its mutilation by the Kekana. Afrikaner mythology relates that the man who singlehandedly performed this daring task was a 29-year-old field cornet called Stephanus Johannes Paulus Kruger. However, the primary source for this story is in Kruger's own memoirs (happily embellished by Gustav Preller). Other reports from the time, including one by Commandant General Pretorius himself (a political opponent of Kruger), were that, at most, Kruger was one of a group of men tasked with the retrieval of Potgieter's body. Another version even suggests that this dangerous task was delegated to a team of black auxiliaries (the Kgatla).[57] Be that as it may, Potgieter was initially buried on the farm Middelfontein (about 16 km north of Nylstroom), which he'd purchased the previous year, but his body was exhumed in 1964 for reburial in Potgietersrus. A Berlin Mission (Modimulle) was established on part of the farm in 1867.

Black ivory

Among the children who escaped from the siege at Makapansgat (as it came to be known) was Mokopane's young son and heir to the chiefdom. As with the other children, he was taken into bondage by the Boers. This practice was widely and extensively adopted, especially in the northern part of the Transvaal. Boers defended it on the basis that it did not constitute slavery, because the children were well fed and looked after and at age 25, male captives (at least) were released without any further obligation. Strictly, they could not be sold but could be bartered, which encouraged a trade in kidnapped or stolen children – known offensively as "black ivory" – that prompted many of the raids Boer commandos made on black villages. Such trading in, or "transferring" of, indentured servants (called *inboekelinge*) continued at least until the annexation of the Transvaal by Britain in 1877.[58] Mokopane's son, who would become Mokopane II, or Chief

FIGURE 3.10: Magagamatala Hill.
The Mogalakwena River flows past its base at the right.

Klaas Mokopane, was later tracked down by his uncle to a Boer farmer in the Brits area who had traded for him, and was released upon payment by the Kekana (assisted by the Langa) of stock (sheep or cattle) and ivory.[59]

Potgieter was succeeded as Commandant General of the Soutpansberg by Stephanus Schoeman, who also not only married Potgieter's widow Elsje (Elsie-Maria), but changed the name of Zoutpansbergdorp, the town founded by Potgieter's father, to Schoemansdal.[60] A fine example of the Yiddish term chutzpah.

Magagamatala

As soon as the siege on the Kekana had been lifted, Pretorius organised a commando to be sent in pursuit of the Langa Ndebele, who had been responsible for the murder of Hermanus Potgieter and his party. In anticipation of this action, the Langa had fled from their seat at Fothane and retreated to a prominent flat-topped hill overlooking the west bank of the Mogalakwena River called Magagamatala ("the cliffs of green" due to the green lichen that grows on the vertical sandstone faces) – now included in the Masebe Provincial Nature Reserve, about 35 km north of the foot of Kloof Pass at Mmamatlakala. This proved to be as unwise a refuge as the caves at Makapansgat had been for the Kekana. The rainy season had commenced, however, and having established the location of the Langa, the Boers withdrew for the time being.

When they returned more than three years later on 14 April 1858, they were led by Commandant General Schoeman. Surprisingly, the Langa under their chief Mankopane were still ensconced on Magagamatala. Schoeman split his force of 350 men into two groups: One, led by Commandants S.J.P. Kruger and Barend Vorster was to scale the mountain from one side on foot, while a mounted group under Schoeman would attack from another side. The southern and eastern edges of the hill dropped precipitously down to the river over 100 metres below. The assault was made at midnight in the middle of a thunderstorm. When Kruger's men summited the hill having surprised and overpowered the guards, the Langa on the plateau panicked; many were caught in the crossfire between the two Boer groups, while hundreds more fell or hurled themselves over the cliffs. Estimates of the numbers killed varied from about 800 to the unlikely figure of 4000. Mankopane himself survived the attack, which became known to the Langa Ndebele as the War of Nterekane (the name by which the loathed Hermanus Potgieter had been known).[61]

Following his defeat at Magagamatala, Chief Mankopane of the Langa Ndebele relocated to a hill complex called Thutlane, 15 km to the south. Here he began to solicit the interest of missionaries in establishing a station at his village, presumably in the hope of forestalling further conflicts between his people and the Boer settlers. Initially, an evangelist of the Paris Mission (from Lesotho) was stationed at Thutlane from 1865, but in 1867 he was replaced by Kühl of the Berlin Missionary Society, which had also been trying to establish stations among the Kekana.

In the meanwhile, some reforms were enacted in respect of the governance of the Boer Republic of Transvaal: A new constitution was adopted and slavery was formally forbidden in a proclamation of 1857; Schoeman was appointed the sole Commandant General in 1858 (succeeded in 1863 by Kruger); "Native Commissioners" were appointed with João Albasini being the first in 1859; and the terms of the Sand River Convention of 1852 requiring custodians of native orphans to become formal guardians were enforced.[62]

Relationships between the neighbouring Langa and Kekana Ndebele chiefdoms were mixed: Cooperation in some respects, but incidents of hostility and intrigue were perpetrated by both sides. Reconciliatory efforts included the late Kekana Chief Mokopane's grandson, Mokopane II (or Klaas Mokopane) taking three of Langa Chief Mankopane's daughters as wives; one of them, Madikana, would be the mother of Valtyn, the eventual heir to the Kekana chiefdom.

The end of Schoemansdal

The principal Boer interest in the north, from the Soutpansberg to the Limpopo, was ivory. "To outsiders, the prodigious herds of elephant in the Limpopo River Basin, coupled with an established market and trade partners, made this economically attractive. To the trekkers … the opportunity to take control of the Limpopo ivory trade and set up profitable trade relations with the Portuguese not only held promise of wealth but also offered total independence from British rule."[63]

Conditions in the northernmost Boer settlement Schoemansdal began to deteriorate, mainly as a result of fraught relationships between the Boers and surrounding Venda chiefdoms whose territories were extensively exploited by the Boers for ivory hunting. A series of convoluted alliances, conflicts, attacks and interpersonal feuds took the surrounding area to the brink of civil war and eventually necessitated the despatch of a commando under Commandant General Kruger from Pretoria in June 1867. It was clear to Kruger that continued occupation of the town was not sustainable and a forced, unpopular evacuation and abandonment of Schoemansdal was undertaken on 15 July.[64] It was later sacked by the Venda.

Piet Potgietersrust

After the siege of Chief Mokopane and his followers at Makapansgat in 1854, when Piet Potgieter was also killed, the Transvaal Republic established a settlement on the Mogalakwena River in the strategically important Makapanspoort immediately to the south of the new headquarters of the defeated Kekana, around Sefakaola Hill. The village had been planned some years before by Andries Potgieter and was to be called Vredenburg, but now it was named Piet Potgietersrust in memory of the slain Commandant General. (In 1904 its name was shortened to Potgietersrust and in 1954 the last "t" was dropped. It became Mokopane in 2002.[65]) From the outset this was a vulnerable site, afflicted with malaria and adjacent to the large resentful Kekana chiefdom. Its only attribute was that it lay squarely on the main wagon and trade route to the Soutpansberg from the south. Until 1865, when Nylstroom was founded, it was also the first settlement north of Pretoria. However, unrest in and around Potgietersrust grew, and in January 1867 the village was attacked and partially sacked by Mankopane in alliance with the Kekana regent Mogemi (Mokopane II being still a youth). The north was certainly keeping Commandant General Kruger busy! After completing the evacuation of Schoemansdal, he returned with a new commando, albeit poorly equipped owing to the now straitened finances of the ZAR government, and attempted, with little success, to chastise the Ndebele attackers. Instead, the Boers suffered several minor defeats.

Marabastad

Eventually, following a serious malaria outbreak and a protracted series of peace negotiations with Mankopane (facilitated by Berlin missionaries) – which concluded favourably for the Ndebele – the decision was made to evacuate Potgietersrust and the town was abandoned in May 1870. Many of its residents joined the evacuees from Schoemansdal at the new settlement of Marabastad, some 40 km to the northeast (now within the Kuschke Game Reserve, 13 km southwest of Polokwane, adjacent to the N1). At the time, Marabastad was little more than a defensive laager, but after Edward Button discovered gold at nearby Eersteling in 1871, it enjoyed a brief life as a mining town.

The German explorer Carl (Karl) Mauch passed through Marabastad on 7 June 1871, two months before Button's discovery:

> ... the junction of two spruits that combine into the Sand River and in whose fork lies the newly founded village of Maraba's Town, which has a very extended row of houses about 1½ miles long ... The country is bare steppe and though the altitude is considerable (I suspect about 4,000 feet) [he was correct], fever carried away about 50 people in 1870 – this year 17.[66]

Given his reputation (and own opinion of himself) as a prospector, especially for gold, Mauch would have been mortified to learn that he'd passed almost across the top of a major gold-bearing reef complex which during the course of the next 130 years would yield almost three tons of the metal! But perhaps he never learned of the discovery. This journey was his last – after visiting Albasini in the Soutpansberg, Mauch carried on northwards across the Limpopo, visited and was the first to describe the ruins of Great Zimbabwe, then continued on to the Zambesi River and the coast, whence he took a steamer back to Europe, suffering repeatedly from the ravages of malaria. He died in hospital in Stuttgart on 4 April 1875 after falling from a window and fracturing his skull. (See Chapter 6).

On 20 November 1869 several Soutpansberg chiefs and headmen concluded an agreement with the Transvaal Republic, represented by Commandant General Kruger, Magistrate R.A. van Nispen and Commandant D.B. Snyman in terms of which the chiefdoms became subject to the ZAR government and were required to pay an annual levy. What they received in return for this sacrifice of their sovereignty is not clear. It seems that whilst the Kekana of Mokopane were amongst the signatories, Mankopane's Langa were not. Instead, Mankopane continued to ferment strife with surrounding chiefdoms throughout the 1870s until his death on 30 May 1877, a month after he'd despatched his son and chosen heir, Masebe (III) to attack one of his enemies, Monyebodi, of the Sotho Matlala chiefdom, at Makgabeng. The attack was unsuccessful, in part because the Langa were said to have been intimidated by the loud ringing of the mission bell at Makgabeng, but Masebe returned a hero nevertheless, because he'd captured a large number of Monyebodi's cattle.[67] This incident took place shortly after the British annexation of the Transvaal (12 April 1877).

The Langa Ndebele in the time of Chief Masebe III

Masebe III ruled the Langa Ndebele for 14 years, from 1877 until 1890. The white population saw him as a refreshing change from his belligerent, cantankerous father. Within months of his accession, he'd acknowledged that he was a subject of the Transvaal Government, he attended church services, wore European dress, built himself a European-style house, prohibited Sunday work and would not authorise rain-making ceremonies. All these actions, while endearing to the missionaries, began to alienate him from his people, providing an opportunity for his brother Tokodi to gain power. In order to safeguard his position, Masebe arranged for his brother to be killed and for Tokodi's family to be expelled (they took refuge with the Kekana). Masebe's actions attracted the ire of some local chiefdoms as well as the attention of the British administration, which sent an officer (a Captain King, Native Commissioner for the Waterberg District) to remind Masebe of his obligations to obey the law. The British should have punished Masebe for the murder of his brother, but at the time were distracted by the military campaign against Sekhukhune (the war that the last Boer president, Burgers, had failed so dismally to conclude and which was largely the cue for British annexation).

Throughout Masebe III's rule, members of the Berlin Mission played an important role in maintaining the peace, in mediating disputes and in providing support for the chief. Several mission stations were established in Langa territory and continued to be staffed until 1925. However, during 1882, by which time the Transvaal had regained its independence, Masebe began to develop a dependency on alcohol, which he obtained from traders he'd allowed to enter Langa territory. His ill-treatment of family members resulted in his principal wife and eldest son Hans fleeing to safety among the Kekana. This in turn led to an acrimonious dispute between Masebe and Mokopane II, which culminated in a decision by the Langa to declare war on the Kekana. Despite their numerical superiority and better military skills, the Langa lost every one of the several campaigns they launched against the Kekana from 1883–1886. In the end the conflict was settled by Kruger, then President of the Transvaal: Summonsing the two chiefs to appear before him in Pretoria in October 1886, he told them that he was not interested in who had been responsible for the war between them, but that their fighting must cease immediately or there would be severe consequences. Whereupon, it is said, the two men (one old, the other young) shook hands and the war was over![68]

During the conflict, a schism had developed between Masebe's two sons – Hans, who'd fled to the Kekana and later lived with Berlin missionaries at Wallmansthal before being protected by Kruger himself in Pretoria, and the younger brother, Bakenberg, who'd remained with his father and gradually gained favour. Towards the end of his life, though, Masebe renewed his relationship with his older son and arranged a marriage for him. Masebe died on 4 May 1890, having shot himself in the head while suffering from the effects of alcohol abuse. It was said that he was beset with guilt about the death of his brother Tokodi.

> He was buried in his cattle-kraal at midnight on the day on which he died. He was clothed in the wet hide of a freshly slaughtered bull. His face, which was not covered by the hide, was made to face southeast. This is the direction of the old home in Natal from which the Ndebele of Langa had emigrated.[69]

The Kekana of Chief Valtyn

After the devastation of the Kekana chiefdom in 1854 and its subsequent move to Sefakaola, the clan spent several years rebuilding its status, at least until the young chief (Mokopane II) who had been rescued from the Boers came of age. In 1865 the Berlin Mission obtained permission to establish an operation at Sefakaola, mainly because Mokopane hoped the missionaries would provide him with political and diplomatic assistance. When the Langa, with help from the Kekana, attacked Potgietersrust in January 1867, the Boers retaliated by commandeering the new mission and using it as a base from which to fire upon the Kekana headquarters. The missionaries had to evacuate, and in their absence the irate Kekana destroyed the building. Undaunted, the missionaries returned, rebuilt the mission and continued their work until 1877 when relationships with Mokopane finally broke down and they had to abandon their church once again. In the 1880s the Kekana had to put up with the numerous attacks made on them by Masebe III's men, as described above. But at least the Boers for the time being posed no threat, having essentially withdrawn from the northern part of their republic after the abandonment of both Schoemansdal and Potgietersrust. The government was also extremely short of money (and would remain so until the discovery of gold on the Witwatersrand in 1886), so it was reluctant, if not unable, to fund commandos to the region.

Mokopane II's second wife was one of Mankopane's daughters (that is, one of Masebe III's sisters) and they had a son called Valtyn Lekgobo, also known as Lesiba Kekana. He acceded to the Kekana chiefdom upon the death of his father in 1890 (the same year in which Masebe III died). Unlike Mokopane, Valtyn was a pragmatist: He allowed the Berlin mission to return and he hired a secretary literate in both English and Dutch so that he could communicate with various officials – and also glean information from letters that passed through his hands, such as those to surveyors.

This was important, because in the same year (1890) the Transvaal Government established a Location Commission, the objective of which was to formally identify and demarcate areas to be set aside for the various chiefdoms. The Kekana, for example, who then numbered about 10 000, were to be confined within an area of 12 726 ha. Worse, the western boundary of their chiefdom – and of the neighbouring Langa – was to be fixed by the Mogalakwena River, thereby denying them access to the good pastures along the narrow alluvial plain on the west bank of the river.

FIGURE 3.11: Chief Mokopane (second from the left) above his village.
SANA TAB Photographic Library, No. 16219 (by H.F. Gros).

Moreover, both chiefdoms had long adopted the practice of holding their annual male initiation schools amongst the hills to the west of the river. Strictly, this would no longer be possible. Unfortunately, Valtyn was also addicted to alcohol and became quite tyrannical, causing his people to seek a replacement.[70]

The cultural historian Isabel Hofmeyr, who had spent some years studying the oral culture of the Kekana, made an interesting point about their education and that of the neighbouring Langa. She found that the two Ndebele chiefdoms shared a view that literacy was symptomatic of the oppression imposed on them by the Boers and as a result they resisted efforts by the Berlin missionaries to provide them with education. This view was aggravated by the fact that the medium of instruction adopted by the missionaries was Sesotho – the language of the commoners they had vanquished over the years.

This resistance to learning would have long-term consequences: The Sesotho-speaking commoners, who had less aversion to education, soon became more skilled and better able to deal with the inroads made by western cultures than the more chiefly aloof Ndebele.[71]

The Langa partitioned

The death of Masebe III in 1890 left the Langa dynasty in the hands of two brothers, Hans and Bakenberg. A dispute arose amongst the Langa over who was Masebe's rightful successor, a problem created in part by Masebe's own indecision in the matter. Hans's mother was a sister of Masebe's principal wife, who'd borne no sons, and according to custom, she ranked higher than Bakenberg's mother, who was a secondary wife. But Masebe had, until close to the end of his life, seemingly favoured Bakenberg over Hans. In the end the dispute was resolved within a month of Masebe's death by the Native Commissioner (who else but Pieter Potgieter, the son of the slain Commandant General?) at a meeting in Potgietersrust, resettlement of which had recently taken place. Potgieter concluded that it was impossible to select one of the sons ahead of the other and decided instead to award each of them part of the former chiefdom. The negotiations took place at precisely the time (June/July 1890) when the Location Commission was busy demarcating the boundaries of land to be assigned to the various chiefdoms. Therefore, it was convenient for the Commission to split the Langa chiefdom in two; it awarded the southern part to Hans and the northern part to Bakenberg.

Hans's portion included Fothane Hill (or Moordkoppie), once the capital of his ancestor Mapela, and so this area became known (rather confusingly) as Mapela's Location. Hans built his capital at the foot of Magope Hill, adjacent to Fothane. Soon, it grew to number over a hundred homesteads, two-thirds of which belonged to the Langa clan. Included among them were the homes of two of Hans's brothers, Marcus and Cornelius, as well as numerous paternal uncles and several widows of Hans's father and grandfather.

The most recent Langa seat, Thutlwane in the north, was for some reason initially excluded from the Location Commission's demarcation and so Bakenberg had to establish a new settlement for himself. Members of the Langa Ndebele were invited to choose which of the two new chiefdoms they preferred. It seems that a majority of subjects of Langa (Ndebele) descent chose to join Bakenberg, while nearly all those of Sotho stock moved to Hans's area.[72] This voluntary division is still reflected in the local demography today.

So came about the three newly demarcated chiefdoms of Bakenberg (also known as Backeberg), Mapela (Hans Masebe's Location) and Valtyn (Makapan's Location), together forming a block 45 km long and about 10 km wide, accommodating at the time (1890) a total of some 30 000 people in an area of about 45 000 ha. Almost no provision was made for the inclusion of pasture and crop-farming land in this area, which was supposed to be the sustainable "homeland" of rural, agrarian communities. This block, with numerous additions and exclusions through the next century, would eventually form the core of the Mokerong (2) magisterial district that comprised the western part of the Lebowa homeland from the 1960s. During the first half of the twentieth century, the Mapela chiefdom managed to raise funds (in part through the sale of mineral rights) to purchase many farms adjacent to its property and these were gradually added to the chiefdom.[73] The Bakenberg chiefdom may have done the same, but it seems that the Kekana of Valtyn were unable materially to increase the size of their property.

Langa involvement in the South African War

In the decade from the partition of the Langa chiefdom until the outbreak of the South African War in October 1899, relations between the two brothers, Hans and Bakenberg, seemed to have been cordial – or at least, any disagreements between them were speedily resolved through the intervention of Native Commissioner Potgieter, who remained in the area throughout this period. At around the time of the outbreak of war, however, it appears that Hans and Bakenberg may have taken advantage of the distraction of the Boers by the British invasion from Natal to re-open issues of dispute. One of these led to a violent clash in the area of Villa Nora. The two chiefs were summoned to a hearing on the matter, and as punishment each was to supply a number of men (levies) to assist the Boer forces under Commandant "Groot Freek" Grobler guarding the northwest frontier against attack from Rhodesia. When they returned from this expedition in March 1900, they immediately began to launch raids on each other's territory, burning villages. Then Hans invaded part of Bakenberg's territory and in suing for peace, the latter was forced to accede to his brother's demand that a number of senior Langa uncles, who had opposed Hans's bid for succession, be sent to Hans – who then had them shot in public, presumably as a demonstration of his ascendancy. This incident led to another hearing in Pretoria, to be followed by a trial, but it was derailed by the British occupation of the capital in June and the effective end of Boer rule.

In the meanwhile, Bakenberg appealed to his western neighbours, the Seleka, for ammunition and, having received it, resumed the fight. Soon though, he found himself surrounded and fled with his people to the old capital at Thutlwane, from where he continued to engage with Hans in sporadic skirmishes – as well as in occasional raids on neighbouring chiefdoms – until they were ordered to stop fighting by the British, whose forces under the command of Colonel (later Brigadier General) Plumer occupied Pietersburg on 8 April 1901.[74]

The Janse murders

As the British advance moved inexorably northwards, many Boers who were neither *bittereinders* (those who continued the guerrilla war under General Beyers) nor *Gamlanders* (those who fled with their herds to the safety of Bechuanaland) sought to hide themselves and their stock in the mountainous regions surrounding and on the Waterberg plateau. Among such fugitives were Abraham Christoffel Janse (or Jansen), his oldest son Johannes Hermanus and his nephew, Johannes Horn, who farmed in the upper Palala region of the Waterberg plateau, not far north of Bokpoort Pass and the farm of Hans "Purekrans" van Rooyen. It seems that they had decided to hide their cattle in a remote, secluded valley overlooking the upper reaches of the Klein Sterk River on the farm Charles Hope (about 7 km west of Hanglip, the well-known escarpment peak). While thus engaged, they were spotted on 8 May 1901 by a party of men sent by Chief Hans Masebe to raid white farms along the plateau edge for cattle. The raiding party, of about 18 men, descended on the trio and killed them, removing various body parts (genitalia, bearded facial skin, and arms) *before* doing so. They then returned to the Mapela capital with their trophies as well as a number of cattle. A young herd-boy, who'd witnessed the gruesome murders, managed to escape and report the incident to the Boers, who returned to the site and buried the remains of the three victims. (A later visit to the site discovered that the bodies had been dug up and removed, presumably by other medicine men seeking body parts.)[75]

One explanation for the attack on the three Boers that became popular amongst Republican Afrikaners in later years was that the Langa raiding party had been sent out at the instigation of the British with the specific intention of murdering resisting Boer farmers,[76] but this has been found to lack substantiation. Early in 1907, a white detective, H.S. Garaty, acting on evidence submitted by General Beyers on behalf of the Palala Boers, visited the Mapela chieftancy in the disguise of a herbalist. In due course he managed to gain the trust of the chief (Hans's successor, Marcus), as well as the senior medicine man and was led to a cave in which parts of the slain men were identified. This in turn culminated in a trial which took place in Nylstroom on 26 September 1907 when nine men appeared before the court (several others had either died or were granted amnesty because of their senior positions in the Mapela chief's court). They were found guilty of murder and sentenced to death.

The sentence was not enforced and six years later they were granted amnesty, released and allowed to return to Mapela, essentially on the basis that their crime had been committed in wartime. This gave strength to the Boer argument that the murders had been orchestrated by the British.

The advent of British administration

In February 1903 the new British Native Commissioner for Waterberg, S.W.J. Scholefield, tabled a report on the status of the native population under his jurisdiction. Scholefield had the advantage of having worked amongst the Batswana in Bechuanaland since 1891. He was quite fluent in both Dutch and Setswana and had developed a good understanding of Tswana customs and culture.[77] He referred to the four established native locations in his district (Valtyn Makapan, Hans Masibi, Hendrik Bakenberg and Zebediela), which he estimated to accommodate about half the total of around 15 000 black males in the district. He also referred to "Seleka, an exile from Bechuanaland, without having a recognised location, lies on the Palala River … on a farm he has hired, making use of adjoining private and Crown land. It will be necessary that he be properly located." [78]

As noted above, Scholefield's comment led to the Seleka being enabled to acquire the farm Beauty shortly afterwards. Scholefield, for all his defence of the conduct of the British towards the black population, at least recognised the severity of the cattle question, which he acknowledged to be

> … a very difficult problem. There is no doubt that many cattle were taken by the Boers. I am personally of the opinion that for every receipt produced, three others are due … Cattle have so appreciated on the market that the price paid for cattle to natives on commandeering notes will not replace those cattle. There is no doubt that the native generally is very short of breeding stock. It is further practically impossible for him to buy, as no matter what number of cattle may be imported, all native are very chary of spending money on cattle which in all probability will not stand the climate.[79]

From 1903 to 1905, the South African Native Affairs Commission considered submissions from numerous parties in an effort to gather accurate information about the status of the black population in the country with the objective of facilitating appropriate policy formulation. Unfortunately, only nine chiefs, none of them from the areas described above, came forward to testify. Land tenure was one of the most critical issues considered by the Commission: Whereas those chiefs who testified were strongly in favour of communal land tenure continuing (not surprisingly, for that was the basis of their power), the Commission favoured individual ownership or tenure of land. But in respect of tribal lands, the chiefs would not yield.[80]

In 1905 the British administration published a comprehensive report, *The Native Tribes of the Transvaal*. This document has attracted a great deal of justifiable criticism from later researchers for its racist content – which did no more than reflect the prevailing views of the white population – but is seldom given credit for the remarkable amount of detail that the administration had been able to gather and collate in the short and frenetic period since the end of the South African War. It reported that from a census conducted the previous year (1904), the total resident "native" population of the Transvaal was 811 753, of whom 61 138 (less than 8%) lived in the magisterial districts of Warmbaths, Nylstroom and Potgietersrust. Included in the latter number were the populations of the three chiefdoms northwest of Potgietersrust: Valtyn Makapan (9047); Hans Masibi (9450); and Hendrik Backeberg (3831). Of Chief Hans Masebe the report commented:

> [He] is a brutal and depraved chief, much addicted to drink, very cruel and greatly feared by his tribe. He is however a strong man and has his tribe well in hand.[81]

The Mapela Langa in the twentieth century

Hans Masebe was also described by one of the Berlin missionaries as a coarse, haughty and violent person, who personally administered many beatings to his tribesmen and sexually abused women within his broader family. These misdemeanours led in due course to his arrest at the end of 1904 and to his certification and commitment to an institution where he died on 29 November 1905 because of wounds sustained during an attempted suicide. He was buried in the cattle kraal of his homestead at the foot of Magope Hill. Following Hans's arrest, the new Native Commissioner, King, camped for over a month in the courtyard of the chief's homestead to maintain stability.

Hans died while his heir was still a child and so his brother Marcus was installed as regent, a role he was to fill with dignity and competence until 1918. This was a period of peace and constructive development for the Mapela community. But in 1918 Marcus was forced into exile with the Bakenberg Langa people to make way for Hans Masebe's son, Alfred Sedibu, who was then old enough to assume the chieftancy. However, Alfred, like his father, turned out to be violent, abusive, undisciplined and determined to obtain whatever he wanted by whatever means necessary. When Alfred died in 1937, he was succeeded by his brother Nkgalabe Johannes, a good man who also continued with the purchase of additional properties in order to expand the size of the Mapela territory.

By the 1980s the Mapela chieftancy had acquired five farms adjoining the eastern boundary of the demarcated Mapela land: Drenthe, Overysel, Zwartfontein, Vaalkop and Blinkwater. Many of the people resident on these farms are descended from Sotho-Tswana ancestors rather than from the Ndebele.[82] The Molokomme people on Drenthe, for example, are descendants of the Hurutshe from the Marico area; the Malebana people on Overysel are of Kgatla origin, from near Rustenburg. In addition, the Mapela had purchased several properties far to the west of the chieftancy, for example the three farms Martinique, Neckar and Abbotspoort, all on the Lephalala River, about 15 km downstream from Villa Nora and 25 km downstream from the Bididi of Shongoane (the people who were once subjects of Chief Mapela). The occupants of the farm Olifantsdrift, immediately to the north of Martinique, were of Tlokwa descent (Sotho/Tswana),[83] but elected to become subject to the Mapela chiefdom.

Unfortunately, neither Johannes nor Alfred had met the requirements of tribal succession by marrying a principal wife and fathering rightful heirs to the chieftancy. The death of Johannes precipitated a succession crisis. After much debate and dissension, the Mapela elders appointed one of Hans Masebe's grandsons by a subordinate wife as the chief of the clan. He was Chief Hendrik Madikwe Langa and his rule commenced in 1958. He was found to be a concerned and consultative chief who communicated readily and frequently with his headmen to resolve issues before they became serious.

In the background, however, heated debate continued over who would be ranked first as Hendrik's successor.[84]

The Valtyn Kekana in the twentieth century

The Valtyn Kekana people appear to have been more compliant and orderly than their northern neighbours during the transition from Boer to British rule and then to that of the Union – notwithstanding that they were far more defiant about the admission of missionaries, the introduction of education, and in their response to restrictive legislation such as the Bantu Authorities Act of 1951.[85]

The Valtyn Kekana seem not to have been able to embark on a programme to acquire additional farms adjacent to their chiefdom in the way that the Mapela Langa had done in the time of Masebe III and later (the latter had had the benefit of being able to finance some of their acquisitions from the proceeds of selling mineral rights). As a result, overcrowding on the small area demarcated for the chiefdom soon became problematic, aggravated by a reduction in this area (later rectified to some degree) and by the fencing off of surrounding properties by their (white) owners. By 1918 almost three quarters of the men in the chiefdom were obliged to seek work elsewhere (principally on Witwatersrand gold mines). The Valtyn chiefdom rapidly became overgrazed due to insufficient land being available for both residential and agricultural use.[86] Valtyn (Lekgobo) died in 1923 and was succeeded by his son, Kgatabedi II (also known as Bernard), who ruled until his death ten years later. Because his own son, Alfred, was too young to succeed him, the Kekana were ruled by Alfred's uncle, Gojela I, as Regent until 1961 when Alfred inherited the chieftaincy.[87]

In 1925 after the discovery of platinum in the area by Dr Hans Merensky (especially on the farms Overysel and Zwartfontein owned by the Mapela chiefdom), the area was visited by the new Prime Minister, General J.B.M. Hertzog. He called for the Ndebele locations to be moved to facilitate the exploitation of the deposits. Despite receiving support from the Potgietersrust Municipality and many members of the local white population, the idea failed to gain the approval of the relevant department.

In the 1930s the government introduced what it termed "betterment" programmes, designed to make more effective use of the limited land. This involved the sacrifice of traditional cluster homesteads in favour of more formal, higher-density grid pattern villages, the internal fencing of the chiefdom into various allotments and the introduction of stringent soil erosion control measures. In 1949 the Chief sold part of the chiefdom's scarce land to the government to enable it to build a township. The Valtyn Kekana chiefdom held out against the imposition of the Bantu Authorities Act of 1951 (which was accepted by the two Langa chiefdoms in 1954) but in 1961, at a time when the royal lineage was fragmented and under dispute, a "simple-minded" heir won the chiefdom and approved both the provisions of the Act as well as the establishment of the township of Mahwelereng. The latter was formally established the following year and incorporated residents relocated from the old Potgietersrus native location. Opposition to the township from within the chiefdom was very strong, and it was referred to disparagingly as *"Anthande lekeišeneng"* (meaning "We don't like the location"). Conversion of traditional cluster homes in the chiefdom to conformance with a regular grid commenced late in the 1960s and was complete by 1970, leading to many deep, irreversible changes in the culture of the community.[88]

CHAPTER FOUR
The Wild West: Swaershoek and Rankin's Pass

The coming of the *swaers*

On a rainy December morning in 1869, a lonely farmer living on a remote property in the far west of the Waterberg was astonished by the arrival on foot of a bedraggled, heavily bearded traveller. The traveller was the German explorer Carl Mauch (about whom more is written in Chapter 5) and the farmer was Johannes Guillame Brink.[1] The farm was Rhenosterpoort. Brink and his wife Susanna Sophia (née Triegaardt [sic], daughter of the Voortrekker leader Louis Tregardt, or Trichardt), had settled on the farm two years earlier. When Brink died a few years later, the property passed into his wife's name and thereafter into the Trichardt family.[2] Today Louis Trichardt, the great, great grandson of his famous namesake, still lives on the farm where he was born over 80 years ago, in a house built in 1893 by his maternal grandfather, Johan Adam Enslin.

Enslin had been born in Rustenburg on 7 December 1857, the son of Christoffel Bernardus Enslin (22 January 1826–5 January 1858) and Jacomina Elizabeth de Beer (11 April 1834–25 July 1898), who had travelled with their parents as members of the Potgieter Trek in 1836–1837. Christoffel was himself the son of an earlier Johan Adam Enslin (I). Christoffel and Jacomina were married in Rustenburg on 16 December 1852. Johan was their fourth child and the only one to live beyond infancy. His father (Christoffel) was killed by a buffalo when the boy was just a month old. He was baptised in the newly formed Gereformeerde Kerk in Rustenburg by the Church's founder, Rev. Dirk Postma, on 20 March 1859.[3]

Initially, several families from the Potgieter Trek, led by Johan Adam Enslin I, settled in the Marico district (whence the local name Enselberg, about 25 km northeast of the modern town of Zeerust). Interestingly, the site they chose for their settlement was very close to the ruins of Kaditshwene – Tswenyane (also referred to in the literature as Karechuenya or Kurrichane), which, until it was sacked by Sotho groups related to the difaqane in the early 1820s, had been the capital of the Tswana polity the Bahurutshe – and which with a population of 14 000–20 000 was possibly the largest urban settlement in the interior of South Africa at the time (See Chapter 2).[4]

Early in 1858 several of the Marico trekker families decided to trek further north to settle at Eersbewoond on the farm Tweefontein between Nylstroom and Warmbaths. This group comprised Stephanus Arnoldus de Beer, his four sisters, their husbands and families. One sister, the recently widowed Jacomina Enslin, travelled with her only surviving child, Johan. The three other De Beer sisters had all married members of the Marico party: S.J.M. Swanepoel; Nicolaas Johannes Grobler (married Johanna Petronella de Beer); and Philippus Eloff.

This group had set out from the Marico with the intention of trekking all the way to The Promised Land, and for this reason was known as the *Jerusalemgangers*. It happened to be an unusually wet summer rainy season. On arriving at the floodplain of the river now known as the Nyl, the Jerusalemgangers found their route to the north blocked: by the Waterberg hills to the left, by the marshy black turf of the Springbok Flats to the right, and by the flooded Nyl plain in front of them. Although anxious to proceed across the Lowveld during the coming winter to avoid the threat of tsetse fly, the group was obliged to settle temporarily at Eersbewoond. As it turned out, they found this experience so pleasant that they soon resolved to travel no further. According to Louis Trichardt, the group chose to name the river before them the Nyl – but contrary to popular legend, were under no illusion that this was the Egyptian river of that name (or that the lone hill, Kranskop, was a ruined pyramid!). They merely wished to signify that they had reached their promised land of freedom.

Shortly afterwards, though, all except De Beer moved north over a low range of hills into the valley of the Sand River (now known as the Alma valley). Eloff stopped along the way, at the source of the river they had called the Nyl, and settled there on a piece of land he named Nyl-Zyn-Oog. Swanepoel settled down in the middle of the valley, at Swanepoelsrus (subsequently changed to Weltevreden), but Grobler, his widowed sister-in-law and their families continued westwards to the head of the valley, to the area around present-day Rankin's Pass. Here the widow Enslin decided to put down roots on the property to be called Rhenosterpoort. This property of over 5000 ha was originally awarded to J.J. Scheepers in 1860; he sold it to the Brink brothers (Guillame and Cornelis) in 1867.[5]

Despite the deeds records, there is a popular story that Jacomina Enslin purchased the farm from Scheepers for the price of a salted horse (a valuable animal in those parts).

Grobler continued up a rough track through the hills until he came to the narrow plain below the ramparts of the western Sandriviersberge, the range with Marikele Point as its highest peak, and there he too finally stopped, in the shadow of the range. He named his piece of land Groothoek – and his descendants live on the property to this day. The settlement of the brothers-in-law along the Alma valley late in 1859 was to give the district the name by which it is still known today: Swaershoek.

The Enslins and Trichardts of Rhenosterpoort

The widow Jacomina Enslin married a widower, Hans Jurgens Steijn (1825–1904), whom the records show as having been awarded the farm Groenfontein 319 to the southwest of Rhenosterpoort in 1869. He was a wealthy big game hunter in

FIGURE 4.1: An early Voortrekker wagon. Redrawn by C. Bleach from a photograph at the old Gereformeerde Kerk, Modimolle, by R. Wadley, 2015.

northern SA and Mozambique. He was apparently related by marriage to the well-known João Albasini, the colourful early white settler in the eastern Lowveld. The Steijns (or Steyns) remained on their northern portion of Rhenosterpoort for the rest of their lives.[6]

Jacomina's only child, Johan Adam Enslin II (1857–1922) married one of the Grobler daughters, Catharina Maria (1861–1946). They had a son (also called Johan Adam Enslin) and four daughters, all of whom married members (that is, cousins) of the same group that had come from Marico via Eersbewoond. The farm was divided into five portions, one for each of their children. Swaershoek was living up to its name.

These second-generation *swaers* were Johan Enslin III, who moved to Nylstroom and settled there; Johanna Petronella, who married Sarel Swanepoel and lived on a southern portion of Rhenosterpoort; Catharina Maria, who married Abraham de Beer and lived on another portion in the south of the farm; Jurgina, who married Phillipus Eloff; and Anna Susanna (Sannie), who in 1916 married Jan Christiaan Trichardt and lived with her parents on the northern portion. They had one child, Louis Trichardt, the current resident and owner of that portion. Jan and Sannie Trichardt also became the foster parents to another 16 children, all offspring of two couples (one being Jan's sister and her husband) who had settled on the farm and died while their families were young.

Johan Enslin II and his son-in-law Jan Trichardt made many improvements to the farm over the years. Amongst these was the excavation of a tunnel into the krans that formed the northern boundary of the farm with Sterkfontein, then occupied by the Bells. A spring emanated from the krans, and Jan Trichardt arranged for a contractor to open up the spring in the hope of increasing its supply. This was successful and the spring has flowed strongly ever since. At the time, however, there were some questions about Enslin's right to the water (at the expense of his northern neighbour), although these seemed to have been resolved.[7]

Jan Trichardt became the chairman of the Swaershoek Farmers' Association, a member of the Ossewa Brandwag and chairman of the local National Party. Through the Farmers' Association, he persuaded Gerrit Bakker, the Nylstroom pharmacist who was Member of the Provincial Council for Waterberg from 1936–1947, to build a proper road down to the new town of Thabazimbi (founded in 1930): the section down below Kransberg became known as Bakkers Pass. He also arranged with Jannie Hofmeyr, then Administrator of the Transvaal, for this road to be linked to Rankin's Post, later Rankin's Pass, by another pass named Jan Trichardt's Pass or Nek. This was first built in 1923 and upgraded in 1956. In 1926

Jan bought a Chevrolet and was one of the first people to own a car in the district. Jan Trichardt Senior died in 1981 at the grand age of 91. His wife Anna Susanna (Sannie) had died in January 1966.[8]

The Rankins and Reads of Rankin's Pass

In the southern part of Rhenosterpoort, between the foot of Trichardt's Pass and Rankin's Post, there was a Transvaal Republic police (ZARP) post before the South African War. Further on, on the farm Rhenosterfontein, the owner, Phillipus A.

FIGURE 4.2: Johan Adam Enslin (II) and his wife Catharina Maria (Grobler).

FIGURE 4.3: Entrance to the Geloftefeesterrein at the foot of Trichardt's Pass.

Eloff (one of the swaers), had established a school with John Rankin as treasurer. After the war, the British built a new school on Rhenosterpoort and later a new police station on Rhenosterfontein, on a piece of land (11 morgen) donated for that purpose by its new owner, Rankin. The southern portion of Rhenosterpoort, owned at the time of writing by the delightful Coen Weilbach ("Waterbergers only speak English in self-defence …") not only hosts the old dressed stone school building and teacher's home, but also the Swaershoek Geloftefeesterrein and the grave of Burgher A.S.A. Opperman, a Union Defence force soldier from Commandant Geyser's Waterberg Commando, who died there on 25 November 1914 during the short-lived rebellion (see Chapter 15).

The origin of the name Rankin's Pass is an enigma, other than that it relates to John Rankin, who in 1909 became the owner of the portion of the farm Rhenosterfontein on which the current settlement is situated, having purchased it from P.A. Eloff. John Rankin was born in Scotland (his nickname was Jock) on 28 September 1859 and came to the Waterberg in the late 1880s or early 1890s looking for a way to earn a living. He married one of Eloff's daughters, Hester Carolina (her uncle was private secretary to President Paul Kruger) and presumably settled on his father-in-law's farm long before purchasing a portion. In due course he and Hester had six children: two sons (James and Phillip) and four daughters (Agnes, Hester, Nora and Bertha).

FIGURE 4.4: Mrs Hester Carolina Rankin, wife of John Rankin. Lewis, Thomas H. (ed.) (1913), p. 217.

One son, probably Phillip, was severely disabled from birth. The other son, James Creighton (Jimmy) Rankin, was presumably born in around 1895, because by 1919 his first wife, Jacoba Wilhelmina Swanepoel, had already died. (He later married a young lady who worked in the Rankin's Pass post office, Nicolina Jacoba Dorothea Strydom.)

In September 1897, in his capacity as treasurer of the Rhenosterfontein school

Committee, John Rankin wrote a letter to the Department of Education in Pretoria requesting permission to employ a kindergarten teacher, Emma Petronella Eloff (possibly his sister-in-law). One of the Rankin daughters, Hester Carolina (named after her mother), was born in 1898. At the outbreak of war in October 1899, Rankin considered himself to be a burgher, although it is unclear whether he went off to join a commando.

On 22 April 1901 John Rankin accepted an ultimatum to surrender to a Major Brereton of the British forces. Subsequently (16 November 1901), he stated that at that stage, he was the owner of the farm Rhenosterfontein and also the lessee of a portion of Rhenosterpoort belonging to J.A. Enslin, who, he claimed, had surrendered to the British at Magalapye [sic] in May.[9] Following his surrender, Rankin joined the British army intelligence units, serving as a guide with Major McMicking and Lt Col Wilson and the mounted columns under their command. After Rankin's departure from his farm to surrender and join the British forces, a column under the command of Lt Col Colenbrander destroyed several of the buildings and equipment on both his property and that of Enslin. In an affidavit submitted for compensation in May 1903, Rankin claimed that he had been instrumental in persuading some 200 Boers with 10 000 cattle to surrender at Palapye. Internal military notes affirmed that Rankin had provided valuable service to the British columns. However, as was so often the case, his application for compensation was unsuccessful.

Once peace had been restored, Rankin returned to his farm with his family and began rebuilding his livelihood. He built a store with postal agency at the northern boundary of the farm, together with a mill – both of which still stand. In due course, he was appointed Resident Justice of the Peace for the district. These developments have provided some alternative explanations for the name Rankin's Pass that became associated with the settlement. According to Louis Trichardt, the postal agency was originally called Swagershoek Pos, but because there was already an agency of this name in the eastern Cape, mail was often sent to the wrong address. To resolve this confusion, the Waterberg address was changed to "Rankin's Pos" and later to Rankin's Pass.

Another explanation for the name of the settlement is that because travel permits for the movement of cattle were required at this point, to limit the spread of disease, especially east coast fever in the period 1908–1912, the place became known as Rankin's Pass because it was John Rankin, the Resident Justice of the Peace (RJP), who dispensed the permits or passes.[10] Yet another explanation for the name has been that it was originally Rankin who pioneered the pass up the escarpment to the Enslin/Trichardt/Grobler properties at the top, and that Jan Trichardt – after whom the pass is named today – merely arranged for it to be properly built and maintained.

In the period between the end of the South African War and World War I, the local police post had been located in the southeast corner of Rhenosterpoort and there had been complaints about the nature and condition of the accommodation there. In 1908, for example, the Attorney General of the Transvaal was advised that the regional inspector had threatened to evacuate the premises, although Rankin, as the RJP asserted that "the quarters of the two police stationed there are not at all so bad and that this step is taken to force the government to build new quarters for them soon".[11]

John Rankin established a highly profitable business by growing maize on the portion of Rhenosterpoort he leased from Enslin (and later Enslin's son-in-law Sarel Swanepoel), milling the crop in a water mill he built and supplying the meal at the rate of 50 bags a week to the tin mine which opened at Rooiberg in 1908. Swanepoel won the contract to transport the maize meal to the mine by ox-wagon. As this wagon could only carry 25 bags at a time and the return trip to the mine from Rankin's mill took three days, Swanepoel was kept busy.[12] Rankin also had a contract to supply the local police post with feed for its horses. Unfortunately for the partners, a sharp dip in the tin price between 1921 and 1923 forced Rooiberg to close temporarily and their business foundered, together with that of a Jewish trader called Nathanson, who had opened a shop and mill at Rankin's Pass as well as a hotel at Rooiberg.[13] In January 1910, Constable John William Chaney – who features elsewhere in this book (Chapter 8) as the owner of the hotel at Zandrivierspoort and later of a store and mill at Vaalwater – was placed in charge of the Rhenosterpoort police post. On 15 January the Acting Attorney General, J.C. Smuts, signed a proclamation appointing Chaney as the Acting Public Prosecutor and Clerk of the court of the RJP at Rhenosterpoort.[14]

In February 1914 John Rankin donated 11 morgen (9,5 ha) from his farm Rhenosterfontein to the government for the purpose of establishing a new police station. This was done in 1916 with the station initially taking the name of the farm, although the name Rankin's Pass police station was adopted soon afterwards. The fine building is still in use. In 1924 the police property was fenced, shortly after the portion of Rhenosterfontein of which it had been part was sold by Rankin to a man called Knobel. The new owner saw the fence being constructed and raised no objection, maintaining friendly relations with the resident officers. During the next five years, additional buildings were constructed for police use within the fenced property. Suddenly, in 1929, Knobel discovered that the enclosure had incorporated significantly more land (18 morgen) than had been donated and lodged a strong objection with the Police Commissioner in Pretoria.[15] The matter was eventually resolved to mutual satisfaction – but it provided ideal material for one of the stories of the local writer C.R. Prance.

FIGURE 4.5 (TOP): The old mill at Rankin's Pass.
FIGURE 4.6 (BOTTOM): Rankin's Pass police station.

In 1923, John Rankin not only sold a portion of his farm (604 morgen less 11 morgen for the police station) to Knobel, but also another 1000 morgen to his son, James Creighton Rankin. Jimmy Rankin, like his father before him, became a loved and respected member of the community, and also a RJP in his father's stead.

The family cemetery next to the old mill at Rankin's Pass contains not only the grave of John Rankin, who died on 12 June 1930, but also that of William Rankin, presumably John's older brother, who lived from 1851 to 1938 and who may have resided near Pretoria before moving to stay with his brother. In 1905 William married Catherine Bell (1885–1944), the daughter of Edward and Mary Bell, and incurred a fine from the Immigration Department in the process.[16] It would seem likely that she was related to the Bells of Sterkfontein (see p. 101), but this has not been established. There are two other graves in the cemetery: those of two children, Gerty Rankin (1923–1938) and John Charles Edwin Read (1929–1931). The latter was the son of Raymond Read and Hester Rankin, John and Hester Rankin's daughter.

Raymond Sydney Read was born near Plettenberg Bay on 30 December 1888, the son of a local farmer Hendrik Petrus Johannes Read and his wife Louisa Mary Morgenrood, and grandson of an Eastern Cape settler, James Oliver Read, one of a later group of settlers to arrive in the Eastern Cape. James Read had been a shipwright and married a French-Huguenot girl called Terblanche. After school, Raymond became a travelling salesman for a Cape company, Fletchers. He

FIGURE 4.7: Headstone of John Rankin's grave at Rankin's Pass

travelled throughout the interior of South Africa and on one occasion visited the hamlet of Rankin's Pass in the southern Waterberg. Here he met a pretty girl called Hester, described as "the daughter of the local farmer, miller, store and bottle-store owner and mampoer distiller Oupa John Rankin." [17]

After the big flu epidemic of 1918, Raymond and Hester Rankin were married. They had five children of whom one (John Read) died in 1931, aged two and is buried at Rankin's Pass. A son, Henry Rankin Read, and three sisters survived.

After the marriage, Raymond Read left his job as a salesman and settled on the farm Kliprivier just west of Rankin's Pass where they established their home. However, these were hard days: This was a sandy farm and in those days there was no fertiliser available. Drought and the Depression took its toll, and in 1933, Read decided to find another means of livelihood. He managed to secure a position as a shop assistant on the farm Tuinplaas east of Settlers.

The Reads moved to a farm called Highlands (a portion of Hamburg) north of Settlers and after several years of hardship were eventually able to purchase the property in 1954.[18]

Before leaving the Waterberg, Raymond Read was an active member of the community. In 1919, as secretary of the Swagershoek Farmers' Association, he was involved in lobbying for the construction of a railway line into the Alma valley. In 1926, in the same capacity and a year after the line to Vaalwater had been built, Read wrote to the railways to request (unsuccessfully) that the siding at Alma be upgraded to a station.

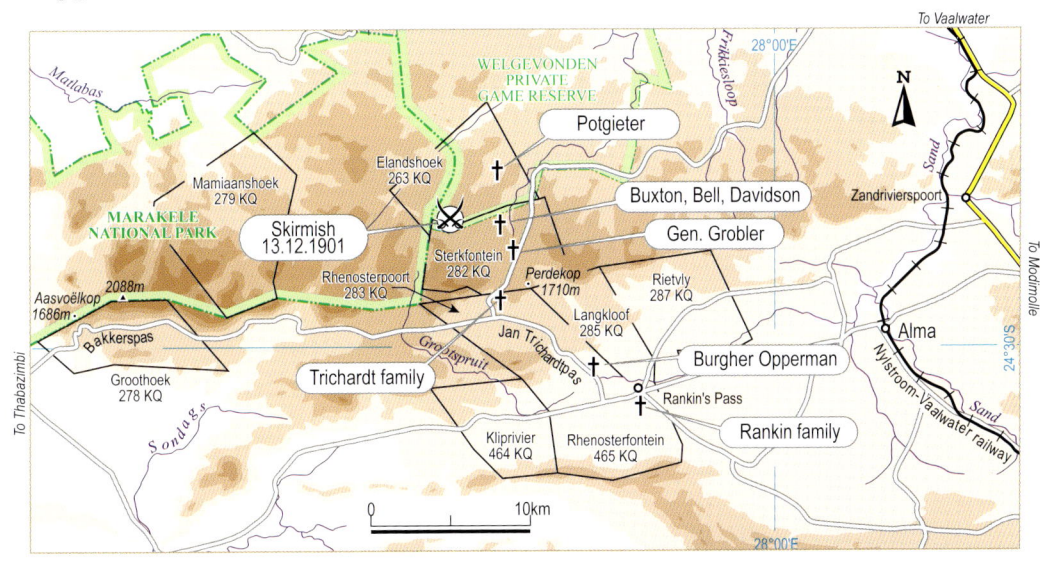

FIGURE 4.8: Swaershoek and the Alma valley.

Kliprivier – a pioneer homestead

The farm Kliprivier on which Raymond and Hester lived for some years had belonged to Johannes Guillame Brink (the owner of Rhenosterpoort who was surprised by the visit of Carl Mauch) until a portion was purchased by John Rankin in 1905 and another by Johan Adam Enslin in 1906. (He in turn gave it to his daughter Miena on the occasion of her marriage to Kerneels Smit.) Research undertaken in the 1980s into early Waterberg homesteads included a detailed description of the ruins of the old homestead on Kliprivier which were still discernible. This provided a good idea of the infrastructure typical of such a rural home:[19]

> The building was constructed on a stone foundation and had a clay floor. The walls were constructed of unbaked bricks, plastered and limewashed. The trusses were made of "birch poplar" and the doors of Oregon pine. The windows were made of pine, poplar and other bushveld wood. The roof was thatched with reeds (*riete* or *ruigte*), tied to the brandering with *rieme*.
>
> A "kook" rondavel was used for food preparation outside the dwelling. The walls were constructed with clay and were about 1,5 metres high, with the void between the walls and the roof being protected by wire mesh, in order to provide ventilation. The remains of an outdoor oven were found behind the rondavel. A water furrow passed directly next to the rondavel on its way to the irrigation dam.
>
> Water flow was regulated with a sluice made from boekenhout. A place where water could be scooped out of the furrow (a *skepplek*) was located next to the building, and was reinforced with timber poles. A flat stone (*trapklip*) was laid on the bank of the furrow to enable scooping of the water. There was also a tobacco shed (a *skuur*), constructed of short thick timber posts with a large thatched roof. The building had no walls; there was a small pen for calves and sheep under the roof. A smithy also existed on the farm, away from the dwelling under the shade of a tree. Bellows, hearth, anvil and vice were found there, with the water furrow passing nearby. The furrow was also used as a place to wash clothes.[20]

The Groblers of Groothoek

To the west of Rhenosterpoort, along the track that would later become the road to Bakkers Pass, lies the farm Groothoek – the first property to be entered into the deeds ledgers for the Waterberg District. It had been settled by Nicolaas Johannes Grobler and his wife Johanna Petronella (née De Beer) in about 1858 or 1859 when Grobler was about 30. A house built on the property in 1897 was continuously inhabited by members of the Grobler family until the death in 2015 of the last resident, Johannes Hendrik (Hansie), one of Nicolaas's sons. A second son, Zacharias Christiaan, known as Grysman, was renowned in the area for his craftsmanship with wood; he was a popular furniture maker.[21] One of Grysman's daughters, Catharina Maria, married her cousin Johan Adam Enslin II of Rhenosterpoort. Several Grobler sons became ministers in the Gereformeerde Kerk.

Grysman's brother Hansie had several sons, one of whom, Andries Nikolaas or Juttie, was renowned for his illicitly distilled *mampoer* (home-made brandy or moonshine). (See a reference to this in the section about the town of Thabazimbi in Chapter 19). The road past his farm to Thabazimbi was very rough and slow and many passers-by (including the travelling *dominee* from Nylstroom) would stop at Juttie's house for refreshments – he became a very popular man! Juttie kept a ready supply of mampoer under the bedstead in his room; depending upon his assessment of the credentials of the visitor, he might retire to his room to fetch some of the precious liquid, mindful of the fact that making mampoer without a licence was a jailable offence and that it was therefore essential to trust the recipient. The still itself was safely hidden away in the bush, downwind and some distance from the house – although it could be found by following a well-worn footpath. Peak time for the manufacture of mampoer was when the peaches ripened in late spring, followed by the fruiting of *stamvrugte* (stemfruit, *Englerophytum magalismontanum*) in December and then, if lucky, by the fruiting of the *mispel* (medlar, or *Vangueria*, of which there are several species, including *Pachystigma triflorum* = *Vangueria triflora*, the Waterberg medlar) in late summer. Almost any fruits were used though: oranges, *naartjies*, *grys- en goorappels* (sand-apples), wild dates or wild pears.[22] One informant said that a bag of medlar fruit could produce 12 bottles of mampoer and that the best liquor had no aroma and would not contain the flavour of the fruit from which it had been distilled.[23]

There are numerous claims that mampoer was effective in treating ailments:

> Take a lekseltjie (literally a lick) of mampoer in a cup of black coffee for palpitations. If your heart now begins to beat normally, take half a bottle of Jamaican ginger and a bottle of witdulsies (white dulcis) and drink a tot of the mixture daily; for a disturbed stomach, cloves soaked in mampoer are recommended; there is nothing that terminates a cold as fast as mampoer mixed with warm sugar-water.

It was also said to be effective against arthritis, varicose veins, ear- and toothache, and as a disinfectant.[24] It is unclear to what extent these assertions were valid, or whether they were simply intended to justify consumption of the liquor.

Groothoek may no longer be famous for its mampoer (or if it is, it is a closely guarded secret), but its location in an unusual micro-environment at the foot of the Kransberg escarpment (aptly named the Kasteelberg by the writer C.R. Prance, who frequently referred to its ramparts in his stories) lends itself to the occurrence of rare plant species. For example, *Crassula cymbiformis* Toelken, a small succulent occurs only on Groothoek; it is on the national red list of flora and is ranked as "critically rare".[25] There are several other plant species, mostly bulbs, which are also confined to the southern slopes of the Sandriviersberg.

A lonely memorial on the hill

One day in April 1886, a young mother went for a walk amongst the rugged hills of the western Waterberg and never returned. Despite an extensive search by family and neighbours under the leadership of her brother, the local field cornet, no trace of her was ever found and her disappearance remains a mystery to this day.

Jacoba Johanna Carolina van Heerden was born at Noodhulp, close to the thermal springs that became known as Warmbaths, on 20 November 1863, the daughter of Carl Sebastiaan van Heerden and his wife Catharina Frederika (née Swanepoel, daughter of one of the Jerusalemgangers). In January 1884 she married Pieter Johannes Potgieter (Jr). It seems that they had two children in quick succession: a son, named after his father (but who must have died as an infant because his name is not reflected on his mother's death certificate) and a daughter, Catherina Frederica, born on 15 September 1885. She was seven months old when her mother disappeared.[26]

The Potgieters lived on the farm Elandshoek, a grassland property on top of the escarpment and included in what is today the Welgevonden Private Game Reserve. According to Field Cornet Carel S. van Heerden's report, Jacoba had gone missing on the neighbouring farm Rietvly (also spelt Rietvlei), which at the time was owned by members of the Van Heerden family.

Her husband, Pieter Potgieter, was born on 19 January 1855 on the farm Middelfontein (soon to be the home of the Modimulle Berlin mission), which then was owned by the estate of his late father Pieter Johannes Potgieter (Sr). [Some records state Pieter Jr's birth date as 19 November 1855 – which would not be possible if Pieter Senior was his father – see below.]

Pieter Senior in turn was the son of the famous Voortrekker leader Commandant General Andries Hendrik Potgieter, who had died on 16 December 1852. General Pieter J. Potgieter (Sr) was born in 1822 and was killed by a sniper during the siege of Makapan's Caves on 6 November 1854 and buried at Middelfontein. (The town of Potgietersrus(t) was named after him.) His wife was Elsie Maria Aletta van Heerden (born in 1823). At the time of her husband's death, Elsie-Maria was seven months' pregnant with their son, Piet Jr.

When General Piet Potgieter (Sr) was killed at Makapan, his comrade-in-arms, Stephanus Schoeman was elected Commandant General for the Zoutpansberg in his stead (on 19 February 1855). He also became engaged to his friend's widow, Elsie-Maria, in June 1855 (however, the marriage was only formalised on 20 January 1856 by the Catholic priest Joaquim de Santa Rita Montanha, probably at Middelfontein).[27]

Schoeman was responsible for the building of the town he called Schoemansdal on the site of the village founded by A.H. Potgieter, originally named Vredeburg or Zoutpansbergdorp. Schoemansdal survived until 15 July 1867 when it was abandoned in the face of repeated attacks by Makhado and others, after Commandant General Paul Kruger failed to receive promised support and ammunition from Pretoria.

Apparently Pieter Potgieter Jr had accidently been shot in the leg as a child. Complications from this wound resulted in the leg being amputated when he was 25, four years before his marriage to Jacoba, but he remained very active.[28] Van Heerden family records suggest that Pieter Jr and his young wife Jacoba may have been cousins. Now bereaved, disabled and the father of a very young daughter, Pieter Jr faced a difficult future. But he overcame these hurdles to become a Commandant and Native Commissioner for the Waterberg, a role he fulfilled with distinction between 1881 and 1889 (see Chapter 3).[29] In 1887, he was remarried to

Catharina Aletta Magdalena Maré, who bore him eight children, the last of whom, Paul Maré Potgieter, was born in Nylstroom in 1901.

At some point after their marriage, the family moved to Pretoria. Here Potgieter was appointed as the first full-time mayor of the town in its last days as capital of the Transvaal Republic,[30] from 29 December 1898 until 3 June 1900. On that date he was forced by Smuts and Botha to step down, immediately prior to the surrender of Pretoria to Lord Roberts.[31] Evidently he then fled to Germany, only returning to the Transvaal after the Peace of Vereeniging was signed in June 1902. He died aged 56 on the farm Oudekloof near Wakkerstroom on 18 February 1911 and was buried in Pretoria.

On a windswept grassy slope in front of the former Potgieter home on Elandshoek, a small marble memorial to Pieter Potgieter Jr's first wife was discovered in 2006 lying in the grass. The lead lettering had long fallen from the plinth, but it was still possible to make out the inscription: "Jacoba Johanna Caro-

FIGURE 4.9: Pieter Johannes Potgieter (1855–1911). Engelbrecht (ed.) (1955): p. 54.

FIGURE 4.10: Memorial to Jacoba Johanna Carolina Potgieter (née Van Heerden) on Elandshoek.

lina Potgieter, Geb. Van Heerden. 20sten Nov. 1863–15den April 1886. Diep Betreurd [Deeply mourned]".

At a public auction in 1906, Elandshoek (which at 4115 ha was a very large property) was acquired by the mining magnate Sir Thomas Cullinan for £1000. The farm, which he visited frequently, was used for breeding cattle and had a dairy that made cheese. During 1912 R.J. de Villiers, presumably Cullinan's manager on the farm, corresponded with the Department of Agriculture about building milk and cooling rooms.

A Scottish colony

In the 1920s, Cullinan decided to lease the farm. Two of the early tenants – on different portions – were the Davidsons and the Bells, Scottish families who may have been related. William Davidson and his wife Mary were from Edinburgh, William having come to South Africa to be an engine driver on Paul Kruger's railway. He was born in 1873 and died on the farm in 1933, and is buried next to his friend Henry Bell on the adjacent farm Sterkfontein. He and Mary had four children: Robert Wilson, William, Douglas Haig and Louise, the last two still minors when their father died.[32] The farm was sold by the Cullinan family trust in the 1960s.[33]

Henry Bell also became a Cullinan tenant on Elandshoek, but prior to this, in 1914, he'd taken out a lease with intent to purchase Sterkfontein where he lived in terms of the 1912 Land Settlement Act. According to the agreement concluded with the Department of Lands, the property (the Remaining Extent of Sterkfontein, amounting to almost 5390 morgen) was valued at £3295.14.6d. Bell had put down a deposit of £646 (a large sum of money) and was required to pay off the balance in forty equal half-yearly instalments (including interest at 4% per annum) of £96.17.3d over the next 20 years.[34] In June 1918 there was an exchange of correspondence between Bell and the Department of Agriculture about dealing with leaf scale.[35]

Bell was an engineer who'd served with the British during the South African War before turning to farming. Born in Scotland in 1865, he married Margaret (Mary) Stewart Galloway in Edinburgh in 1901 – which suggests he had returned to Scotland before coming back to the Waterberg. He and Mary had two children, Henry (Harry) and Margaret (Peggy), who were schooled at home, presumably by their mother, because there were no schools nearby at that time. Unusually for those times, the Bells were well liked by their Afrikaans-speaking neighbours.[36] At the time of his death in 1936, Henry was trying to finalise the purchase of Sterkfontein from the government, a process eventually completed by his children. His widow, Mary, continued to live on the farm with her children, and died at the grand age of 91 in August 1961.[37] It seems that Peggy never married, but Harry married a woman called Elizabeth (also known as Gladys), who had no children and who became blind at a relatively young age.

Harry is remembered fondly by the descendants of his neighbours. Koos le Roux of Rietvly (the farm where Jacoba Potgieter had disappeared years before) recalled Harry riding down to the post office at Rankin's Pass to collect his mail. Louis Trichardt of Rhenosterpoort told several stories about Harry and his exploits. One of these was that Harry was persuaded to sign up to fight in World War II only after two neighbours, both Van Heerdens who then ran the store at Rankin's Pass, pretended that they had already done so by taking the "Red Oath" (*Rooieed*)[38] and teased him about being a Scot who was too afraid to fight for his country. On arrival at Alma station for embarkation, however, Harry found that he was alone. Too late, he had signed his enlistment papers! He was sent to North Africa and when he returned from the war several years later, he immediately sought out the two brothers and beat them both up in retribution.

Another Harry Bell story recounted by Louis Trichardt concerned a lonely neighbour. North of Sterkfontein on the road to Vaalwater was the farm Klippoort, on which lived a widower, Frans Roets, who used to communicate with potential partners via the *Sonskynhoekie* column of *Huisgenoot* magazine. One day Harry Bell drove in his *bakkie* to Vaalwater station to collect some equipment that he had ordered. The stationmaster was pleased to see him, saying that he must have been "sent" – because on the platform, sitting demurely on a bench next to her suitcase, was a neatly dressed middle-aged lady. She said she was waiting to be collected by Mr Roets to whom she'd written a week previously to say she was coming up from the Cape to visit him and would he please collect her from the station. She was not to know that the mail to Vaalwater was only delivered weekly and that she and her letter had arrived in the village on the same train! Obligingly, Harry gave the lady, whose name was Marie, a lift to Frans's farm – and was present to witness the shock on the latter's face when introduced to his unexpected visitor. After tea, which Marie efficiently made to ease the tension, Harry took his leave despite protestations from Frans about how his being left alone with a lady would look to the dominee. That afternoon Harry received a visit from Frans, who'd ridden around on his horse (he had no car). Frans pleaded with Harry to wash

FIGURE 4.11: Grave of Henry Bell at Sterkfontein.

FIGURE 4.12: The old Grobler/Bell homestead at Sterkfontein. Original photograph by R. Wadley, redrawn by C. Bleach.

his bakkie and to pick up him and Marie the following morning to take them to Nylstroom so that they could be married without delay before the magistrate! So began a happy relationship between Marie and Frans that ended only when, after Frans's death several years later, Marie returned reluctantly to the Cape to take over a property that she'd inherited.[39]

Groot Freek Grobler

Henry Bell of course was by no means the first tenant or owner of Sterkfontein farm. It had first been inspected in 1862 by J.C. de Klerk when the original owner was Jacobus J. Scheepers, who'd also been awarded the adjacent farm Rhenosterpoort in about 1859. In 1891 the farm was inspected again, this time with somewhat more rigour, by F.A. Grobler, who was then the local field cornet and who conducted numerous inspections in the area. By 1899 the inspector had become the owner of Sterkfontein and a senior office-bearer for the republic.

F.A. "Groot Freek" Grobler had been born in the Waterberg on 8 June 1851 and grew up in the area. He served with distinction in the First War of Independence (1880–1881) during which he participated in the siege of the British garrison at Marabastad. He also participated in various campaigns in the north of the Transvaal Republic against Chiefs Magato, Makgopa, Mmalaboho and Mphepu. In October 1899, at the outbreak of the South African War, he was appointed Assistant Commandant General for Waterberg and Soutpansberg with the responsibility of safeguarding the northern and western borders of the ZAR. He was tasked with taking the Waterberg Commando across the Bechuanaland border to blow up the railway line to Rhodesia, while the Zoutpansberg Commando was to proceed to Rhodes Drift on the Limpopo to head off any attack from Colonel Plumer's force based at Fort Tuli. The Zoutpansberg Commando succeeded in taking Rhodes Drift, but Grobler was unable to persuade his reluctant Waterbergers to cross the border, and the attempt was abandoned.

In January 1900 he was ordered to go to Colesberg in the Cape as second-in-command to General Lemmer, but did so reluctantly. He was evidently not well regarded by his superior, despite showing courageous leadership and initiative in the defence of Colesberg. Following the fall of Bloemfontein, he returned with Lemmer via the Eastern Free State to the ZAR, where in May 1900 he fought against Lord Roberts at Kliprivier south of Johannesburg and also at Tabaksberg.

When Pretoria fell in June 1900, Grobler returned to the north where he acted as liaison between Generals Louis Botha and Koos de la Rey. But the latter was not impressed with his leadership and he was dismissed from his position on 6 September 1900 to be replaced by General C.F. Beyers. Grobler returned to Sterkfontein from where he continued to fight as a burgher, but contracted fever and died there on 13 May 1901.[40]

This benign account of Grobler and his exploits is not shared by conservative Afrikaner historians. In their view Grobler was deeply distrusted by many in the Boer command, as well as by fellow burghers in the Waterberg region. In particular, he was accused, with some other burghers in the area, of rounding up cattle left behind by neighbours who had joined the commando and herding these animals across the border to Bechuanaland, ostensibly to save them from being taken by the British. However, at the end of the war, the cattle were sold back very profitably to returning, impoverished burghers. These men were considered turncoats and became known as *Gamlanders* or *Gamjanders*, after the area to which the cattle had been driven – Khama's Land. By this time of course, Grobler had already died (May 1901), but this does not seem to have exonerated him from his alleged traitorous behaviour. One writer even suggested that Grobler's early demise might have been due to poisoning and that his wife might even have been responsible.[41]

It seems likely that Grobler together with a brother and possibly a neighbour, Johan Adam Enslin, were amongst a group of Boers who concluded (correctly as it turned out) that the war was lost and that the sensible approach would be to protect and safeguard their assets in order to rebuild their lives. However, this attitude was not shared by those few Boers who continued to defy the British for two years after June 1900 – the so-called *bittereinders*. When the conflict finally ended, albeit still without the endorsement of President Steyn of the Free State

FIGURE 4.13: Commandant General F.A. "Groot Freek" Grobler. SANA TAB Photographic Library, No. 32822.

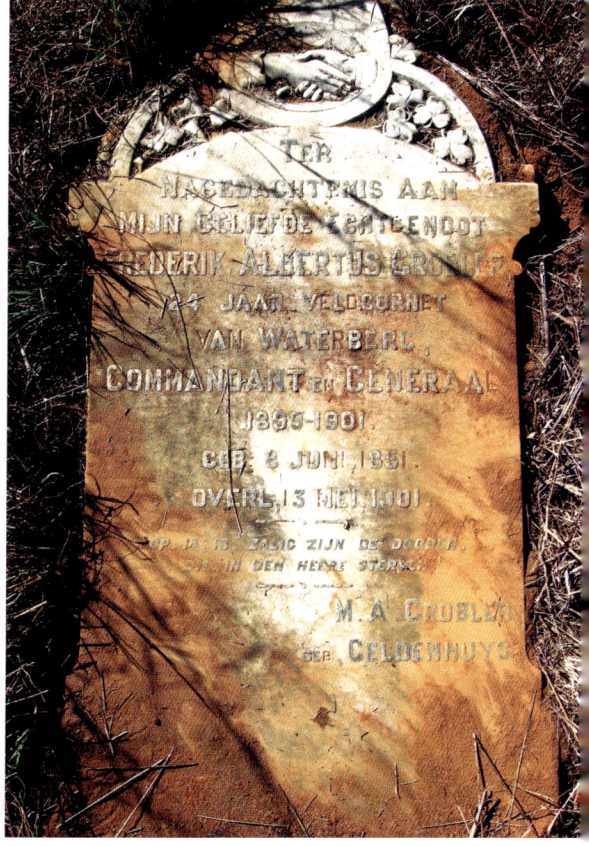

FIGURE 4.14: Grobler's Grave in Sterkfontein.

(who was seriously ill), with the signing of the Treaty of Vereeniging on 31 May 1902, burghers like Grobler and his group who had already made peace, or worse, joined the British forces, were denounced as *"hensoppers en joiners"*, especially by those men who'd been captured and interned overseas, whose families had been interned in camps and whose properties and stock had been destroyed. To this day, families whose ancestors were, or may have been, among the Gamjanders are coy about discussing the matter.

Groot Freek Grobler's grave near the Sterkfontein homestead was discovered almost by chance, the result of a sharp-eyed ranger on a helicopter-borne game count. Game, probably white rhino, had trampled the grave and the chain-link fence surrounding it into the ground, although the headstone, which had fallen face forward into the long grass, had remained intact. Coincidentally, the site was visited and recorded on 8 June 2010, the anniversary of the late general's birth 159 years earlier. Nearby lay the grave of his grandson and namesake who had died just 13 months old in June 1907.

Perdekop

Along the southern boundary of Sterkfontein with Rhenosterpoort and Langkloof is a prominent hill on the edge of the escarpment overlooking Trichardt's Pass. This hill (1710 metres amsl) is called Perdekop. It was so named on account

FIGURE 4.15: Commandant General "Groot Freek" Grobler (right) in the field.

FIGURE 4.16: The grave of Grobler's grandson.

of the practice adopted by the early farmers of corralling their horses on its flat top during the late summer months (February to April) to reduce the risk of fatalities due to horse sickness. It seems that the midges carrying the disease rarely if ever ventured onto this high plateau, perhaps because of the wind coming up from the valley to the south. Hans Jurgens Steijn (Jan Trichardt's stepfather) acquired Arab horses from his friend Albasini and made good use of the hill in midge season.

As a matter of interest, the northwest-facing slope of Perdekop was found in 1980 to host a unique species of butterfly, *Erikssonia edgei* (the Waterberg Copper), the caterpillars of which feed on a local poisonous-leaved plant (*gifbossie*) called *Gnidia kraussiana*. By the 1990s prevention of fires had resulted in denser bush growth, the demise of the gifbossies and the disappearance of the butterfly. A search throughout the southern escarpment was carried out and in 2013 a small colony of the butterfly was found in the hills north of Bela-Bela.

FIGURE 4.17: Perdekop along the southern boundary of Sterkfontein.

The last Waterberg engagement of the South African War

Sterkfontein, or more likely Buffelspoort immediately to the west of it, was the scene of one of the last engagements of the South African War in the Waterberg. For some weeks, British forces based at Warmbaths and Nylstroom had been pursuing groups of Boer guerillas under the overall command of their elusive tactical leader, General C.F. Beyers. By late November 1901, Beyers had left a force of 300–400 Boers at Zandriviersport under Commandant C. Badenhorst, while he moved further north. Lt Cols J.W. Colenbrander and J.W.G. Dawkins began to harass Badenhorst's commando with a series of enclosing movements, taking several hundred prisoners in the process and severely depleting the Boer force.

In early December, when the British units withdrew to Nylstroom for supplies, Badenhorst encamped with the remnant of his men "in fancied security" in an obscure kloof, which was reported as being on the farm Sterkfontein, but which was more likely one of several such valleys on Buffelspoort. Here, the commando was discovered on 12 December by Colenbrander and his Mounted Infantry units, who, with Dawkins in support, then surrounded the Boer position during the night. The next morning, after a brief resistance, Badenhorst surrendered with five officers and seventeen burghers. The wily Beyers, meanwhile, had managed to escape a similar, simultaneous British pincer attack at Geelhoutkop further east and had moved quickly out of reach down the Palala River to the north.[42] During this short battle, one of the British subalterns, Lt Buxton, was killed.

Ronald Henry Buxton was born at Flixton, Suffolk (just south of the River Waveney, which forms the border between Suffolk and Norfolk) on 3 November 1874, the fourth of seven children born to Henry Edmund Buxton, a successful banker, Justice of the Peace and Honorary Colonel of the Volunteer Battalion of the Norfolk Regiment, and Mary Rosalind (Upcher), the daughter of a priest.[43]

Ronald was educated at Harrow, the elite English public school. He entered the Norfolk Regiment in June 1896 as a 2nd Lieutenant and was promoted to Lieutenant in December 1897. He served in West Africa in operations on the Niger River in 1897–98, being mentioned in despatches (23 May 1898), and was employed with the West African Frontier Force from February 1898 to February 1900.

He then proceeded to South Africa to rejoin his regiment, which had arrived by sea at Cape Town on 23 January 1900. The 2nd Battalion, Royal Norfolk Regiment formed part of the 7th Division of the Imperial Army under Lt Gen. Tucker. It

CHAPTER 4: THE WILD WEST

109

FIGURE 4.18: The grave of Lt Ronald Buxton at Sterkfontein.

FIGURE 4.20: Lt Ronald Buxton (Detail of Figure 4.19).

FIGURE 4.19: Lt R.H. Buxton, seated left, with fellow officers outside the Pretoria Electrical Works, sometime after 5 June 1900. Image courtesy of the Royal Norfolk Regimental Museum, Norwich.

participated in the advance from Modder River to Bloemfontein, including playing a perimeter role at the Battle of Paardeberg. The battalion entered Bloemfontein on 14 March (a day after Lord Roberts) and then accompanied him northwards, being involved in numerous skirmishes en route. It entered Brandfort on 3 May 1900 and reached Johannesburg on 31 May where the battalion formed part of the march-past in the town. It then accompanied Lord Roberts to Pretoria, which it entered on 5 June. Thereafter, the battalion was garrisoned in Pretoria (under Gen. Maxwell) and also in Rustenburg.[44]

In August 1901 Lt Buxton was seconded as Adjutant to the 20th Mounted Infantry unit and served with it until his death in action at Sterkfontein on 13 December 1901. Lt Buxton's grave is across the valley from the Sterkfontein farmstead in a small cemetery shared with Henry Bell and William Dawkins. A monument to the men of the various Norfolk regiments that had served in the South African War was erected in Norwich and unveiled by Maj. Gen. Wynne on 17 November 1904. Lt Buxton's name is among those inscribed on the monument. There is also a brass plate in his name on a wall in the south nave of Norwich Cathedral.

In the church of St Edmund's at Fritton, a small Norfolk village and the Buxton family home near Great Yarmouth, about 18 km northeast of Lt Buxton's birthplace, there is a memorial plaque replicating the inscription on the officer's headstone:

> In loving memory of Ronald Henry Buxton, Lieutenant, 2nd Battalion, Norfolk Regiment. Born 1874, Killed in Action at Sterkfontein, Transvaal, 13th December 1901. And Reached By Duty's Path, A Life Beyond The Life He Lost.[45]

Ronald's elder brother, Abbot Redmond Buxton, born in 1868, gained the rank of Major in the Norfolk Yeomanry. He also fought in the South African War as well as in World War I. He died in 1944, aged 76. A younger brother, Edward Hugh Buxton (born 1880), became a Captain with the Suffolk Hussars Yeomanry during World War I, and died in 1955.

The youngest Buxton brother, Knyvet (born 1882), also joined the army. He was a lieutenant with the famous 19th Lancers when he was killed in an accident while playing polo in Rawalpindi, India, on 14 December 1905 (exactly four years after the death of his brother Ronald). None of the Buxton brothers married; of their three sisters Winifred, Violet and Rosalind, Violet never married and the other two had no children. As a result, Ronald's last direct family member was Rosalind, who died in 1968.

Cyril Rooke Prance

To the west of Buffelspoort in the very rugged country that today forms part of the Marakele National Park, lay the farm Mamiaanshoek, a very large (7100 ha) but almost inaccessible property, the only habitable area being in the extreme northeast. The southern corner of the farm rose to a beacon on the escarpment overlooking the road to Bakkers Pass. It is doubtful whether this farm had been inhabited prior to the South African War, but from 1906 it was leased by its then owners, Transvaal Consolidated Lands Co. Ltd (TCL) to a new farmer in the district, Cyril Rooke Prance. Prance was also an author and among the short stories he wrote was one about the remains of a Boer he'd come across in a kloof on Buffelspoort. This man, he was told, was one of Badenhorst's men who'd been wounded in the engagement with Lt Buxton and his troopers on 13 December 1901, left behind and died there. Another story he heard, though, was that the man had guided the British to the Boer hideout, had been caught by his former compatriots and shot.[46]

Cyril Rooke Prance (1872–1955) was born on 3 April 1872 at Annesley, Nottinghamshire, the son of a local vicar. He was educated at Charterhouse School and at Trinity College, Cambridge, qualifying as a civil engineer. At the outbreak of the South African War, he came out to South Africa to join Brabant's Horse, a colonial regiment. Following the occupation of Pretoria in June 1900 (about which he wrote some amusing and irreverent letters to his younger sister Dorothea), Prance joined the newly formed South African Constabulary (SAC), in which he served until 1907, first in the Free State and later in the Waterberg District of the Transvaal. From 1908 until 1921 Prance farmed as a tenant on Mamiaanshoek before moving to a smallholding outside Nylstroom. In 1913 he married Gabrielle Elizabeth Massey; she died in 1942, aged 55 "after an agonizing illness with complications of snake-bite". In 1946 Prance received a letter from a client, who referred to "a snapshot showing an attractive girl, prettily dressed, mounted on a white donkey" which he'd discovered in a book Prance had sent to him. Prance requested that the long-lost photograph of his wife be returned to him.

In about 1930 the Prances moved to Port St Johns where they would live for the rest of their lives. For some time, Prance had been writing stories about his experiences as a policeman and farmer in the rural "backveld" of the Free State and Waterberg. Initially published in a variety of South African and English magazines and newspapers, the stories were gradually collated and published as books, most of them after the move to Port St Johns.

Described by one who knew him as "a classical scholar, an historian, a poet and an author ... who wrote many articles on the district as well as poetry and several novels, all based on actual happenings in the lives of inhabitants", Prance was an acute observer with a keen sense of humour. His stories of the Waterberg, several of which can be tied to specific incidents and personalities, were cryptic, however, making use of false names and places, these also often being amusing. The Sandriviersberg became the "Skranderberg", the Swaershoek was "Schlenterhoek", Nylstroom "Nergensdorp" and "Dooidonkieshoek" was probably Rankin's Pass. "Doodstildorp" appears to have referred to Naboomspruit where Prance was the SAC officer in charge for a period. Principal characters in his stories included Piet van der Blouberg of the farm Bontpootbosluisfontein and his widowed daughter Tante Rebella.

Prance eventually published 13 works during his lifetime, including seven collections of anecdotes, some seafaring histories, a guide to furniture-making and a pamphlet for hikers around Port St Johns, where he died in 1955. He was an important, though underappreciated commentator about the society in which he lived.

FIGURE 4.21: Cyril Rooke Prance with Brabant's Horse (left) and later in life (right). Image (left), as a member of the yeomanry in 1900 (from N.A. Coetzee (1969): The Occupation of Pretoria by the British Forces (part III), taken from "War Letters from the Veld", AD 1900 by C.R. Prance and published in *Pretoriana* 1969 No.059, (April), p. 5 and (right) from SANA TAB Photographic Library, No. 9665.

C.R. Prance had four siblings: an older brother, Ernest, two younger sisters, Edith and Dorothea, and a young brother Hubert. Three of them also had South African – and Waterberg – connections.

Ernest Reginald Prance, a year older than Cyril, was schooled at Wellington College, Berkshire, before being sent up to Clare College, Cambridge, where he completed a BA in 1897 and an MA in 1903. Ernest was ordained as a deacon and by the outbreak of World War I was a parish rector. He joined the war as a chaplain and served in France throughout that terrifying conflict, an experience that evidently left him deeply scarred and disturbed. After the war, he came out to South Africa to become the Rector of the St Michael's Anglican Church in Nylstroom, a role he filled for only two years before retiring at the age of 53. (See Chapter 14 for more details.)

Ernest was a bachelor and lived outside Nylstroom on a small property he shared with his sister Edith until around 1927 when he returned to England to Maldon, Essex, where he was eventually joined by Edith and Hubert. He died in 1947.

Edith Lena Prance, two years younger than C.R., became a chauffeur during the Great War, first in France and later with the Scottish Women's Hospitals in Greece. Having joined her brothers in Nylstroom, Edith, who was unmarried, decided some years later to explore the newly opened Kruger National Park. In 1934, she drove herself through the park and wrote a book about her experiences: *Three Weeks in Wonderland – the Kruger National Park*, the publication of which, in 1935, was credited with providing a boost to the park's tourism industry. Before that, she travelled by sea to India to visit her younger brother Hubert, who was working for the Indian Civil Service as a telegraph engineer. On retiring, Hubert joined Ernest and Edith back in Essex.

Before her death in 1957, Edith became involved with the Boy Scout movement and left her home and property to the local branch of that organisation.

FIGURE 4.22: Edith Prance whilst serving in France as an ambulance driver in World War I. Photograph kindly provided by Ian Valentine, who in 2017 wrote a book about her involvement with the Boy Scout movement in England.

The fifth Prance child, Dorothea, also had a South African connection: In 1908 she married Basil Elmsley Coke, a member of a leading military family. His cousin was Major General John Talbot Coke, who'd commanded the British 10th Brigade under Sir Charles Warren at the Battle of Spioenkop in 1899.

Below the Sandriviersberg

Immediately east of Rhenosterpoort, with the hill Perdekop at its northwestern corner, is the farm Langkloof which, as its name implies, consists principally of a long valley that runs in a northwesterly direction up into the Sandriviersberg range. Before the development of Jan Trichardt's (or Rankin's) Pass, the only access onto the upper plateau in this area was via a rough track along the Langkloofspruit. The farm was originally granted to Hendrik Johannes Swanepoel, one of the original *swaers* of Zwagershoek in February 1862, but as with all other properties in the Waterberg, it was only formally surveyed many years later – in 1904, by the government surveyor F.V. Watermeyer.[47] In 1919 almost half the farm (about 1100 ha) was purchased by Pieter Willem ("Oom Willie") Botes. Oom Willie and his family lived on the property for almost ten years before moving to Moddernek near Nylstroom where he became involved in many community activities: an elder in the Hervormde Kerk (NHK), a school councillor and chairman of the Doornfontein *Geloftefees* committee.[48]

Much of the Botes land on Langkloof was subsequently sold, but in recent years, a great grandson, Anton Botes, a Gauteng businessman, has returned to live on the property and has succeeded in repurchasing all the land originally owned by the family.

FIGURE 4.23: The Botes family homestead on Langkloof in the 1940s. (Photo: Anton Botes)

Adjacent to Langkloof in the east, and lying on the plain below the Sandriviersberg, is Rietvly. For many years, this farm has been the home to members of the Van Heerden clan, who have featured in numerous aspects of the district's history. Today, Aletta le Roux (née Van Heerden) lives on the property with her husband Koos and their two sons. Her grandfather was Charles van Heerden, the uncle of J.C. van Heerden who once owned part of Rhenosterfontein.

Rietvly (also spelled Rietvlei) is the farm made famous (or notorious) by one of the stories of Eugène Marais. Marais visited the Swaershoek valley in about 1910 to treat a kidney condition of Andries van Heerden, and stayed with the family for three months, either on Rietvly itself or close by. (Marais himself was in poor health, due to his growing morphine dependency.) Very soon, Marais learned that Rietvly had shortly before become infested with black mambas and that there had been several fatalities, including that of Andries's youngest brother Fanie, who had died after being bitten while herding sheep. The boy's father, Oom Piet van Heerden, reported having had several narrow escapes himself. It seemed that there was a nest of mambas living in a nearby reed-marsh; they would travel past a quince hedge and across a road to search for *dassies* and other food on a nearby bank. Andries van Heerden apparently decided to set watch for the snakes near the hedge – and was said to have shot 52 of the snakes in a single week![49]

In addition to their experiences with black mambas, the Van Heerdens were active farmers on their property. This is attested to by the correspondence saved in government files. For example, Mrs W.S.J. van Heerden wrote to the government entomologist on 20 November 1916 to enquire about treatment for mealy bugs in her granadillas, and two years later her husband corresponded with the entomologist regarding red scale on his oranges.

The Van Heerdens were active in the community. In the 1930s, J.C. van Heerden was the proprietor of the General Dealership in Rankin's Pass. Two brothers, Carel and Willem, acted as burial officials in the absence of a dominee. Carel was a SAP supporter whilst Willem was a member of the National Party – so their services were selected according to the political allegiance of the recently departed![50] Carel van Heerden and J.J. Prinsloo were amongst a long list of petitioners in 1910 who urged the government to allow the RJP (John Rankin) to issue permits for the movement of cattle. (The list included another four Van Heerdens, six Eloffs, five Swanepoels, L.R. Peacock, two Mortimers – and John Rankin himself!)[51] Gert van Heerden, in his capacity as chairman of the Zwagershoek branch of the South African Party, wrote to Prime Minister Smuts on 28 May 1924 appealing for the new railway line from Nylstroom to Vaalwater, then under construction, to incorporate a detour across the Sand River to Koppie Alleen in order to assist farmers in his district – and effectively threatened to withdraw his members' support for Smuts in the forthcoming election if this was not done. It was, but to no avail, for Smuts lost the election. (See Chapter 11.)[52]

Alma

Aletta le Roux, the Van Heerden descendant currently living on Rietvly, used to teach at the primary school in the settlement of Alma, about 20 km to the east. Alma is situated on the farm Koppie Alleen, separated from its northern neighbour, Knopfontein, by the Sand River (which becomes the Mokolo River once it passes through the Sandriviersberg). Alma began its existence as a siding on the railway line between Nylstroom and Vaalwater, which opened on 1 October 1925. The siding was originally called Koppie Alleen, but within months after the line had opened, the name had changed officially to Alma (according to railways correspondence).

The origin of this name is uncertain. Some believe it to have been the name of the wife of either a local farmer or of a government official. Louis Trichardt says it is derived from *"almal"*, because the siding represented a place where everyone used to gather for various functions and to collect supplies and the mail from the train. John Rankin, for example, used to send a Scotch-cart to Alma to collect the mail.[53] But Alma was also a name made famous by a battle (at the Alma River in the Ukraine) that took place on 20 September 1854 during the Crimean War. The railway line through the siding would have been built almost exactly 70 years after that battle. A Spanish visitor to the Waterberg was quick to point out that the word means "soul" in her language, but was hesitant to imply that Alma is the soul of the Waterberg!

During 1925 Swagershoek farmers appealed to have the siding upgraded to a station. However, a memorandum from the General Manager of the South African Railways and Harbours Administration dated 4 December 1925 said:

> The question of appointing a Caretaker at Alma siding (late Kopje Alleen) has been enquired into, but for the information of the Minister I have to state that according to figures taken out for October last, the total revenue in respect of all traffic received at and despatched from this Siding only amounted to £14-17-9 for the

> month. This ... would not appear to warrant the appointment of a
> Caretaker at present.⁵⁴

The Swaershoek community did not give up easily. In a letter to the Minister of Railways, Mr C. Malan, dated 21 September 1926, the secretary of the Swaershoek Farmers Association, R.S. Read, reminded the Minister of his promise to reconsider the conversion of the Alma siding to a station after October 1926. Evidently, their appeal was again unsuccessful.

Within a short period, a few shops, a postal agency and eventually a school (but no police station) were established around the siding, and in 1945 a co-operative (the WLKV, predecessor of the NTK) established a receiving depot for produce there.

In 1930 a school was established at Alma with the intention of consolidating several existing district schools (there being no school at Vaalwater at that time). Known as the Alma Central School (later to become Laerskool Alma), it offered classes up to Standard 6. Nearly all its pupils were boarders; two *koshuise* (hostels) accommodated up to 350 pupils from far and wide, many of whom arrived by train. For example, there were pupils who travelled all the way from the Lowveld, where Lephalale is today – they would take the Stockport RMS bus as far as Vaal-

FIGURE 4.24: The galvanised-iron silos at Alma.

water, then the evening train to Alma, arriving just in time for supper. The school had two hostels (one each for boys and girls) and eight classrooms. In the mid-1940s Nico Swart was the headmaster.[55]

The school is still in operation today, although the boarding hostels appear derelict, and there are two other schools, one a high school in the burgeoning formal/informal settlement across the Sand (Mokolo) River on Knopfontein.

The NTK had experimented with using galvanised iron and steel as building materials for its silos, and in 1973 one of the first of these was built at Alma. It turned out that the railway line ran over a portion of school ground. This enabled the NTK to run a road next to the line in order to avoid two dangerous rail crossings – but only after the Minister himself was lobbied, at his Pretoria house on a Saturday morning, to allow NTK to purchase the requisite land from the state.[56]

CHAPTER FIVE
Baltic connections

Carl Mauch – an early Waterberg meanderer

The Dwars River is a very pretty spruit but has, due to this year's extraordinary drought, only very little running water. However, it is sufficient to turn the wheel of a miserable little mill. It joins the Pongola [Mokolo] about 4 miles downstream and springs from a yellow wood kopje in the vicinity of Hanglip Mountain ... from which there also emerges the Palala, flowing in a more northerly direction. The region acquires a flat, rolling appearance towards the north after one has passed [over] the not so steep slope of the main range [Sandrivierberge]. The vegetation which, curiously, grows on termite hills is thus nicely distributed. Several spruits, almost all dry however, furrow the region in every direction. Not infrequently one sees horizontal sandstone slabs ... protruding from the sandy ground which is only a few feet deep.[1]

This was written by one of the first adventurers to have visited the Waterberg plateau and certainly the first to have recorded his observations for later generations to enjoy. He was an unlikely visitor: a 32-year-old impoverished carpenter's son and junior teacher from a small village near Stuttgart in Germany.

Carl (or Karl) Mauch had been inspired to explore Africa ever since he saw its blank face in an atlas given to him when he was

FIGURE 5.1: Carl Mauch (1837–1875).

FIGURE 5.2: "The Dwars River is a very pretty spruit ..."

ten. For years, he saved and studied – English, French and Arabic, natural history, astronomy, geography and geology – until in 1865, at the age of 28, he embarked as a crew member on a ship bound for Natal. With meagre resources, he worked his way to Potchefstroom, the former capital of the ZAR, from where he was eventually able to join various expeditions into the northern hinterland – including that of Henry Hartley, the famous hunter.

Mauch is best known for his discovery of the goldfields at Hartley in Zimbabwe and especially for his description of the ruins of Great Zimbabwe, which he visited in 1872, the first European to do so. In the course of his numerous journeys as far north as the Zambezi River, Mauch contracted malaria and other tropical diseases from which he was never to recover. He died in April 1875, aged 38, alone, unrecognised and poor, as a result of injuries sustained in a fall from a window of his room in Stuttgart – to which he had returned for a menial job in a cement factory.

The journals of his travels, which were never published in his native German although they were used as source material for a biography in 1889, were eventually donated to the Linden Museum in Stuttgart. In 1969 they were transcribed from the original Gothic script into modern German and then translated into English for publication by the National Archives of Rhodesia.

For a period of three weeks in November and December 1869, Mauch travelled into and across the Waterberg, both by wagon and on foot, either alone or in the company of two of the earliest ZAR government surveyors, Brooks and Froude. His diligent journal entries make it possible to track his course with surprising accuracy, thanks to his estimates of distance travelled daily, his accurate descriptions of topographical features and names (of rivers and passes) which are still in use, or which can be traced today.

The village of Nylstroom (now Modimolle) was formally registered in February 1866, having been established the previous year on the farm Rietvlei, on land donated by an English businessman, Ernest Olferman Collins.[2] Mauch had spent the month of October 1869 travelling north from Nylstroom as far as Blouberg and the Soutpansberg. He returned to the Berlin Mission at "Modimulle", northeast of Nylstroom, on 21 November. This mission had itself been established only two years earlier by the missionary Koboldt on a piece of land donated by a supportive farmer, Gert Lottering. The mission was built on the farm Middelfontein, just north of the point where today the N1 freeway crosses the railway line and is joined by the R101 between Modimolle and Mookgophong.[3]

Two days later Mauch set off on foot along the low escarpment that lies to the north of what is today the old main road from Mookgophong, examining the rocks and vegetation. After a night on the farm of Mr J. Verdoorn (the first postmaster for the Waterberg), he joined the surveyors' wagon(s) on the track leading northwest from Nylstroom. They passed close to where the first new section of the R33 was rebuilt in 2008 and then along what is now the road (and railway line) to the hamlet of Alma, into the valley of the upper Sand (or Zand) River. Downstream, this river becomes the Mokolo. In Mauch's time, both were referred to as the Pongola.

By 1 December the party had reached the foot of the Sandrivierberge, close to the point where the present R33 to Vaalwater crosses the Klein Sand River on a dangerous curved bridge at the foot of the poort on the farm Zandrivierspoort. The Klein Sand joins the Sand at the entrance to a gorge through the mountains about a kilometre west of the road; and the railway line to Vaalwater (built only in 1925) now follows the river through the gorge. The following description of his "Waterberg meander" is based on his journal:[4]

> A spruit, flowing from the E, joins the Pongola. The road through the poort is 1½ miles long; however, it does not lead along the river, but through a kind of pass in the eastern main range [where Mashudu Lodge is today, on the farm Modderspruit 150KR] ...The river ... has extended sections which, near deep pools, grow thick reeds and rounded swamp grass which affords good cover for buffalo.[5]

Beyond the pass, but before the site of the present village of Vaalwater (which would not be established for another 35 years), the party cut northwards for a while until it reached the Dwars River. Mauch's description of the Dwars was given at the start of this narrative. The "miserable little wheel" referred to in that description was probably located on the farm Sanddrif (Zanddrift) where there were already several early homesteads close to the river. On one of these, a farmer called Manie Henderson would later become renowned for the mampoer he distilled from *mispel* (medlar).[6]

The party turned downstream along the Dwars before crossing into the course of the Sontagsfontein (now the Sondagsloop) which runs into the Dwars on Slypsteendrift. They followed the Sontagsfontein upstream to its source, on the farm Buffelfontein. The route taken by Mauch's group from Zandrivierspoort used an existing rough track, depicted on early maps of the region, for example, that of Jeppe in 1899.[7]

On the way they would have passed very close to the outcrop of a mineralised quartz vein that was to become in 1948 the short-lived Nooitgedacht lead mine on the farm of that name.[8] On Friday, 3 December, having reached the source area of the river, Mauch wrote in his journal:

> During the ascent of a nearby promontory, about 200 feet high, [possibly the high ground on McCabe Zyn Hoek] of the flat range forming the watershed between the Pongola and the Palala, I observed the horizontal layers ... [of] sandstone ... On the height of the promontory there is a large plateau. This is a peculiarity appertaining to all the hill ranges between [the] Palala and [the] Pongola; hence the region is called the "Platten".[9]

This description refers to the high ground running from Martjies se Nek (at Ant's Nest, on Gunfontein) to the Dorset police station and westwards past Visgat in the direction of the R33 tar road to Lephalale. At this point, Mauch refers to crossing "Jan's Nek" on the watershed between the Mokolo and Palala river basins. This has not been reliably identified, but seems to be close to the present Dorset police station, possibly on Trouw Zyn Nek.

Beyond the pass, he encountered:

FIGURE 5.3: The "Seven Sisters" or Sandrivierberge at Zandrivierspoort.

> ... a very pleasing sight to the eye. Euphorbias, thorns and Proteas are predominant. They mainly shed their shadows on anthills. The numerous watercourses form gentle valleys without any groves of trees ... About 2½ hours [by wagon] from Jan's Neck there is the farm "Oversprung", a very pretty situation which leans towards S onto a rather high, wooded mountain range. Two springs join with the perennial spruit on the farm. Several hundred acres of ground are fit for agriculture, while the rest of about 4000 acres serves for cattle grazing.[10]

This is the farm now known as Oversprong, which borders on the Melkrivier near its confluence with the Lephalala. Interestingly, the Deeds Office records that this farm was issued to one Carl Mauch (by State President Pretorius) on 28 May 1868, more than a year *before* this journey. Given that it is unlikely for either date to be incorrect, it seems that Mauch might have acquired the property "blind" and that this trip was his first visit to it. Either way, it is remarkable that his journal makes no mention of his ownership of the farm. According to the Deeds records, Mauch sold Oversprong to Oscar Wilhelm Forssman (a member of a prominent ZAR family, active in land transactions, see later in this chapter) on 15 July 1870 for £50.[11] OWA (Alaric) Forssman was a friend and benefactor to Mauch; also, having been appointed the Consul General for Portugal in 1869, he was able to assist Mauch in preparations for his next (and final) trip into Mozambique in 1872.[12]

From this point Mauch and his group moved westwards.

In the next two days, they crossed the watershed between the Tambotie River to the north and Bar's Kuil or Boer's Loop (now Poer-se-Loop) to the south. The latter is described as:

> ... a natural dam formed in the small river, in which are found barbel and whitefish in great numbers ... there is no question of a carriage-road [here] and only the spoor of a hunting-trap is visible. Game is fairly plentiful and the veld is ideal for cattle-raising.[13]

The next day (6 December), they travelled south again and crossed another pass, Claas's Neck (now Klaas Zyn Nek) on a track leading down to the Mokolo River.

FIGURE 5.4: "The Rhinosterhoek Mountains, with their scraggy peaks ..."

> From the height there appear [to the south] the Rhinosterhoek Mountains with their scraggy peaks and needles; especially prominent is the striking and curiously shaped Marikele Point, the highest peak in the Transvaal to the north of the Witte Watersrand. This, for the moment, is my next goal.[14]

The reference here is clearly to the main southern range of the Waterberg, which forms the southern boundary of Marakele National Park from Kransberg and Aasvoëlkop in the west, past the unnamed highest peak (2088 metres amsl) and the adjacent communications masts, towards Welgevonden and Rankin's Pass in the east. Although these hills are not called the Rhinosterhoek Mountains today – they are called simply the Waterberge – there are two farms by the name of Rhenosterfontein that lie on the northern slopes of the range, and a third, Rhenosterpoort, along its crest.

Then commenced the most arduous – and rash – part of Mauch's journey. Abandoning his colleagues and their wagons, he set out alone and on foot to cross the main Waterberg range and head for Rustenburg, 180 km to the southwest even *"soos die kraai vlieg"*. It is unclear why he chose to undertake this trek, which seems, from his journal entries, to have been more or less spontaneous, but there is later evidence, hinted at by the editor of his journal, that Mauch may have been a difficult character to travel with, that he "sponged" on those who befriended him and that he was by nature something of a loner. So it is possible that he was encouraged or even obliged to leave the wagon train. Mauch speaks only of the northfacing slope of the Rhinosterhoek Mountains appearing in the southwest to be so gentle and overgrown,

> ... that I decided to make the party a very pleased one by hunting and thus, by supplying it with meat, less weakened. But I was soon taught otherwise.[15]

Whatever the reason, there is no further mention of the colleagues he left behind as he struck out across the Mokolo up the tangled and rocky course of the Taaiboschspruit (on Haakdoringdraai and Platbank) towards the crest of the Rhinosterhoek range. His route took him across the heart of what is today the Welgevonden Private Game Reserve. He reached the crest of the range after a seven-hour scramble, only to discover to his disappointment that

> ... I found myself separated from Marikele's Point by several transverse valleys and [I] did not encounter any tree on this watershed between the Taaiboschspruit and the Sterkstroom, which flows towards the E.[16]

Skirting the upper valley of the Sterkstroom, Mauch soon reached

> ... a kind of elongated depression, which proved to be soothing to sore and scorched feet. Water flows from this point in three directions, namely W towards the Matlabas, N towards the Pongola and E to the Sterkstroom.[17]

Figure 5.5: Carl Mauch – lone adventurer. Redrawn from a photograph in the Photographic Library of the National Archives: SANA TAB No. 13396: Carl Mauch and Fernando da Costa Leal (not shown in drawing), 1870.

This feature is clearly visible on a modern topographic map: Located on the farm Buffelspoort (in the southeastern part of the Marakele National Park), the elongated valley forms the source of both the Matlabas and Sterkstroom rivers.[18] In the late afternoon Mauch followed this valley towards the southeast until eventually approaching darkness forced him to stop for the night.

> A few dry twigs from the stunted bushes serve to maintain a modest fire, a pipe of tobacco has to take the place of food and a drink from the nearby brook replaces coffee. I take off my leather jacket, use it as a mat, take my guns and instruments for bedfellows and cover myself with the mackintosh ... The fire had almost died down ... [when] I saw the shape of [a leopard] only a few paces away on the other side of the fire. It observed a movement of mine towards the gun and uttered its dull "Huh! – Huh!" I shout at it with full voice: "Fuzzek!" (a call with which in this country one chases dogs away), and soon it disappeared with big bounds.[19]

At first light Mauch continued his journey, eastwards now, until after four hours he came across an old wagon spoor, which led him in due course to the isolated home of a farmer called Brink, whom Mauch had by chance met over three years earlier while on an expedition to the Pilanesberg.

> He remembered me and was not a little astonished, when he saw me, as if fallen from the sky, and during heavy rain, enter his house. Of course, there was no question of my continuing my march today, for rest, as well as refreshment and the gathering of strength by way of consuming large quantities of milk and buttered bread, were essential. There were ripe figs and also some grapes.[20]

Archival records show that from 1868 to 1874, the farm Rhenosterpoort was owned by Cornelis and Johannes Guillame Brink.[21] The precise location of Brink's homestead on the farm is uncertain, but would appear from the journal entries to have been very close to the junction of the modern Bakkers Pass–Rankin's Pass road with that leading north towards Vaalwater (See Chapter 4).

The next day, 9 December, Mauch trekked on, *"with two half-ripe maize cobs"* to sustain him, in a westerly direction, probably along the narrow plain leading towards Bakkers Pass. This is supported by the entry that:

FIGURE 5.6: "I found myself at the foot of Marikele Point … with its three peaks …"

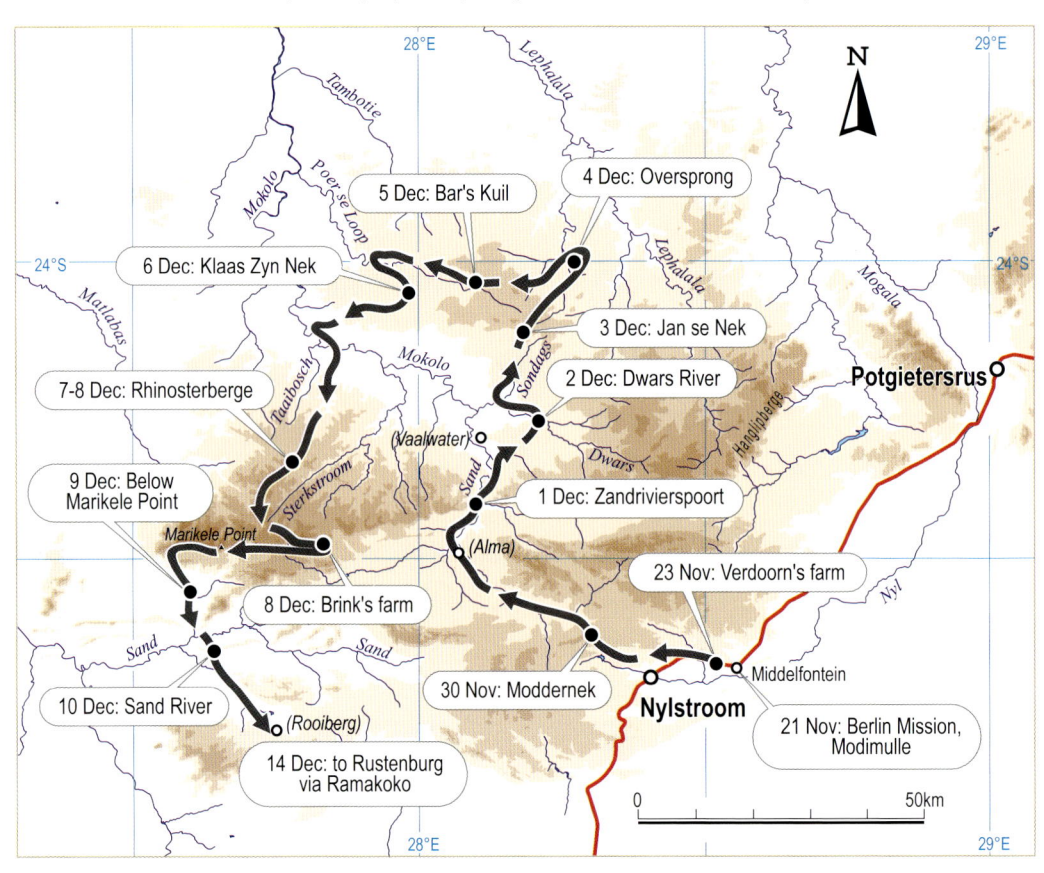

FIGURE 5.7: The Waterberg journey of Carl Mauch in November/December 1869.

> ... after a four hours' walk ... I suddenly found myself at the foot of the Marikele Point, which forms a mighty mountain mass with its three peaks. Its height is, surely, 3000 feet.[22]

In fact, the three peaks concerned – Aasvoëlkop, the peak that today has radio masts; and one further east are, at 1765, 2088 and 2004 metres above sea level respectively, all about 600 metres (2000 feet) above Mauch's path.

Mauch turned south again at this point, descended the Sandriviersberge and found his way down the Sand River in the direction of what he called Roeberg (Rooiberg today). After another five days of wandering generally southwards, passing the village of Ramakoko (east of the Pilanesberg), he eventually came across a wagon that gave him a lift to Rustenburg, which he reached on 14 December, three weeks after leaving Nylstroom. In that time, he'd covered a distance of about 400 km, at least 150 km of which had been on foot – alone, in the middle of a dry, hot summer, over rugged terrain with minimal food supplies. A remarkable achievement by any measure.

Mauch never returned to the Waterberg. His next journey was to take him up to the Soutpansberg and then eventually on into Rhodesia and Mozambique and the discoveries for which he is best remembered (the ruins of the Great Zimbabwe settlement). We are fortunate to have this brief, diligent record of his excursion into the Waterberg almost 150 years ago.

THE FORSSMANS AND THE JEPPES

The names Forssman and Jeppe crop up frequently in the journals of Carl Mauch – and also in the history of the Transvaal Republic between 1850 and the end of the century. Both were large, talented and educated families whose members participated in substantial and diverse ways in the economic and infrastructural development of the country. The Waterberg region was no exception to this involvement.

The Forssman family

Oscar Wilhelm Alarik (Alaric) ("OWA") Forssman was born in Kalmar, Sweden, in 1822, one of a family of seven. Their father was an advocate and judge and the family claimed a noble lineage. At the age of 20, OWA sailed for South Africa, seen by many Europeans as a land of opportunity and adventure.[23] In 1852 he moved from the Cape to Potchefstroom, capital of the new Transvaal Republic (ZAR) and there, in 1853, he married Emelie Henrietta Amalia Landsberg, who at the time was a mere 17 years of age. Her parents, Carl and Catharina Landsberg, farmed in the Rustenburg district. Carl in turn was the elder brother of Otto Landsberg, a prominent Cape Town artist, musician and apothecary (mainly a snuff merchant), about whom more will be said in a while. The brothers had been born in Germany and immigrated to South Africa with their parents in 1818.[24] In 1826, Otto founded a firm of apothecaries called OLCO (Otto Landsberg & Company), which still exists. He died in 1905, aged 102. One of his grandsons, Auguste d 'Astre, who later lived in Potchefstroom with his aunt's family, donated 70 of his grandfather's paintings to the Potchefstroom Museum in 1956.

Emelie and OWA Forssman had ten children, of whom only the second-born, Carl, is relevant to this story.

In the meanwhile, OWA had persuaded *his* elder brother Magnus (born in 1820) to join him in South Africa and to establish a settlement of Swedish immigrants here. Magnus duly sailed with a large party of Swedes on the ship *Octavia* in 1863. They established themselves in an area along the Vaal River to the south of Potchefstroom. However, the initiative failed and most of the families eventually moved elsewhere in the country. Even today, one will find evidence of this short-lived settlement: the bridge across the Vaal on the road (R501) linking Potchefstroom with Viljoenskroon is known as Skandinawiedrif, with the farm Scandinavia on the Free State side just downstream from the crossing.

Magnus had trained as a surveyor in Sweden and was actually employed by the government there; he'd been given extended leave to visit South Africa. The ZAR government was desperately in need of such skills and OWA persuaded his brother to apply for a job. Magnus was duly appointed as the first Surveyor General of the Transvaal in 1866; his wife and children joined him in the country the same year, but returned to Sweden in 1868, having found the conditions and prospects unfavourable. Magnus never saw any of them again – not even a son born after his wife had returned home. In due course, he resigned from his post in Sweden and devoted himself to his new homeland.

Magnus Forssman remained Surveyor General until his death in 1874. During his tenure the government frequently found itself unable to pay his meagre salary (£300 per annum) in cash and instead offered him the occasional farm in recompense. Over the years he acquired numerous properties, most of them having little value at the time. In the course of his work, he had occasion to meet several travellers – especially Mauch – who shared their personal maps with him. These contributions enabled him, together with his friend and colleague Friedrich Jeppe, to produce the first map of the ZAR in 1868.[25]

While his elder brother was making a name for himself in government, OWA (who was sometimes also known by his third name, Alaric) was doing so in business from his Potchefstroom base. An advertisement dated 26 June 1866 in the Potchefstroom paper *Transvaal Argus* announced that A. Forssman was prepared to buy ostriches and wool for cash. In October that year, he chaired a meeting of the town's shopkeepers at which it was decided to accept government notes in lieu of cash (there being a perennial shortage of the latter).[26]

Like his elder brother, OWA accumulated a large number of farms in the Transvaal in payment for various services rendered or goods supplied. In 1872 he published a book with the grand title *A Guide for Agriculturalists and Capitalists, speculators, miners etc wishing to invest money profitably in The Transvaal Republic, South Africa; containing a description of a number of first class farms situated in different districts of the republic, and general information*. A table in the book listed a total of 118 farms in the ZAR said to be the property of OWA Forssman, including 58 in the Waterberg District alone! This was followed by a description of some of the properties, all of which had been surveyed by J. Brooks (one of the two surveyors who led the trip through the Waterberg to which Carl Mauch had attached himself in 1869):

> Malmanies River Drift (19), Bulge River (20), Bulspruit (21), Tanbootiedraai (22), all situated on the road from Nylstroom to the Limpopo and the Tati goldfields, on the Malmanie and Sand [Mokolo] Rivers. Also, Malmanies River (at the junction of Malmanies and Sand Rivers); Houtboschkloof (north bank of Sand, within 500 yards of Tambotiedraai. "Adjoins Mr W Munroe's farm, Zeekoegat". Also, Bankfontein, Roodebokhoek, Creamoftartfontein, Roodepunt, Buffelshoek, (all on the stream "marked as Barkuilsloop") about 11km NE of Tambotiedraai. A waggon road through the corner of the block goes directly to Nylstroom, 45 miles distant. On neighbouring farm, Bulfontein, lives Mr A Smith, while close by are several farms belonging to Mr GJ Verdoorn "the largest speculative landowner in the district" (29,680 English acres). On the top of the hills are the farms Goudfontein, Yzerfontein and Grootvley – "splendid grazing

grounds, the steep hill sides and deep valleys affording excellent cover and runs for cattle. Bulfontein, adjoining, is used by its owner for winter grazing; and on the adjoining farm, a number of Kafirs are located, proving that the ground is to their liking for gardening and grazing purposes.[27]

Some of the other Waterberg plateau farms listed as being owned by Forssman were Klipplaat, Palala, Libanon, Jonghanshoek, Bulgerivier, Krokodilrivier, Boschdraai and Oversprong (which he had purchased from Carl Mauch the previous year). OWA and Mauch seemed to have been good friends, despite Mauch being some 15 years younger than Forssman. Certainly, OWA provided Mauch with many of the provisions for his expeditions, advanced him credit at his store and arranged for goods to be sent to Mauch on request.[28] Also, as the Portuguese Consul General, OWA assisted Mauch's preparations for his expedition into Mozambique. At the same time, Mauch's accurate and detailed map notes were of great value to Magnus Forssman in the compilation of his maps of the region.

One account stated that at his peak, OWA owned more than 250 farms in the ZAR, with a total area of about 300 000 ha, but that after the First South African War most of these were lost.[29] Neither of these statements seems valid: In the period when Forssman was acquiring property, most farms in the region were of around 3000 morgen (2571 ha), hence OWA's own tally of 118 farms would accord well with a total area of about 300 000 ha. Moreover, in 1886, the year in which gold was discovered on the Witwatersrand, OWA sold his farms into a new land company, the Transvaal Lands Company (TLC), for which he went to London to have registered. These properties are reflected in a report on the various Transvaal

FIGURE 5.8: Transvaal Lands Company Limited share certificate.
From a pamphlet written in 1962 by his grandson, Oscar Wiliam Alric Forssman (Jr).

land companies published in a London financial newspaper in 1891, which listed the area held by TLC in October that year as 522 730 acres (211 544 ha).[30]

In 1873 OWA started a passenger coach service between Potchefstroom and the diamond fields around Kimberley, later expanding it to provide a passenger service to the goldfields at Lydenburg via Pretoria. He called the company the Transvaal and Goldfields Extension Transport Company. He used light, four-wheeled buggies that were closed to keep out the rain. The journey from Potchefstroom to Lydenburg took just over 100 hours to complete![31]

OWA, who held the title *Chevalier* on account of his noble lineage, was appointed as the Consul General for Portugal in the Transvaal. (Jeppe's Almanac for 1877 reported that only one other country, Belgium, had consular representation in the ZAR at that time.)[32] During the first South African War in 1881, OWA and his family were besieged in Potchefstroom. When the British garrison left the town after the Armistice in March that year, it is said that the Forssman family

FIGURE 5.9: Chevalier Forssman (in the uniform of the Portuguese Consul General) and his wife (left); and Magnus Forssman, Surveyor General of the Transvaal (above). SANA TAB Photographic Library, No. 3817: OWA Forssman, Portuguese Consul General and his wife; and No. 4630: Magnus Forssman, First Surveyor General of the ZAR.

left with the column of troops returning to Natal and travelled with them as far as Harrismith in the Free State, where they settled.[33] Following the war, OWA lodged a claim against the British government for almost £200 000 in respect of farms and property lost or damaged in the course of the brief war. The matter was discussed in the House of Lords. Prime Minister Lord Gladstone had to state that he thought the claim to be "extraordinarily large" and expressed his doubt that it would secure payment. At least half the amount was in respect of farms, which were restored to OWA by the ZAR. It is not stated how much compensation he was eventually paid by the British.[34]

OWA died in Pretoria on 14 April 1889 and is buried at the Church Street cemetery. His wife Emelie remarried (to Andreas Marthinus Goetz, 1834–1905). With the wealth bequeathed to her by OWA, she visited Sweden where she purchased a large wooden house which she had shipped to the Cape. It was erected on a stand in Kenilworth, where Emelie was to stay until her death in 1898.

OWA and Emelie's son Carl Wilhelm Gottfried Forssman (1855–1934) was born in Grahamstown. Little is known of his upbringing or education. By the outbreak of the South African War he was living in the Nylstroom area, having in 1890 secured the licence to operate a hotel (with liquor licence), store and postal agency at Zandriviers poort (see Chapter 8). He had married Maria Elizabeth Pronk (1861–1941), daughter of Dr I.J. Pronk, a member of the Potchefstroom School Commission and the sister or cousin of Ms A. Pronk, who taught at the Nylstroom government school for many years from 1898 – and who taught at least some of the Forssman children too. A picture of the staff and pupils of the school in 1915 shows Ms Pronk as a teacher and Louis Forssman as a pupil (see Figure 6.12 in the next chapter).[35]

Carl and Maria had ten children, two of whom died before their father. During the South African War, Carl was interned at Irene (possibly first at Nylstroom) from 5 April until 10 November 1901, possibly for collaborating with the Boers.[36] Certainly the diary of William Caine on New Belgium Estate refers on several occasions to Forssman's role on behalf of the Boers, but despite this, he and Carl were clearly good friends and Forssman went out of his way to make Caine's life easier (see Chapter 9). The British evidently didn't take Carl Forssman's wartime misdemeanours too seriously, for in 1904 he was appointed as a Justice of the Peace for the Waterberg area.[37] Carl's occupation in Nylstroom after the war is

FIGURE 5.10 (TOP): Carl Forssman and family, 1912. From O.W.A. Forssman (Jr) (1962).
FIGURE 5.11 (BOTTOM): Forssman's Pass on Naauwpoort.

not recorded. There is a pass on the old coach-road between Nylstroom and Naboomspruit that was known as Forssman's Pass in the 1930s.[38]

At the time of his death on 11 December 1934 (aged 79), Carl and his wife were living on the farm Mooihoek in the Potgietersrust district, just north of Tinmyn.[39] The eldest son of Carl and Maria Forssman was named Oscar William Alaric after his grandfather. He wrote mini-biographies of both OWA and Magnus. In 1938, Oscar married Cythna Lindenberg Letty (1895–1985), the well-known illustrator of botanical works, especially *Wild Flowers of the Transvaal* for which she produced about 730 outstanding illustrations. She also designed the motifs on some of the set of floral stamps of the RSA issued in 1965.

The Jeppe family

For historians the most valuable member of the Jeppe family was Friedrich Heinrich, the author of a series of detailed maps depicting the evolution of the Transvaal Republic. But he was only one of three brothers, all born in Germany, the sons of Carl Wilhelm Friedrich Jeppe, an economic adviser to the Duke of Mecklenburg. The eldest brother, Herman Otto (born in 1819), was the first to immigrate to South Africa in the late 1840s. He was followed by his younger brother Friedrich in 1861 and by a third brother, Julius, in 1870, as well as by their sister and her husband.[40]

Herman Otto Jeppe had trained in medicine, but his poor eyesight had precluded his being able to practice in Germany. On arrival in Cape Town, whom should he begin to work for but Otto Landsberg, the apothecary and artist: the very man whose niece, Emelie, married OWA Forssman!

Otto apparently soon fell in love with Landsberg's daughter Anna Maria (Mimi). However, her parents did not approve of the relationship. The story goes that in January 1849, Mimi, quite a society girl, was invited to a reception at Government House. Herman Otto managed to gain access to the event disguised as a Malay servant, rendezvoused with Mimi and eloped with her to Swellendam where he set himself up as an apothecary.[41,42] Otto and his wife Maria were not pleased, but soon realised that young Herman Otto was not a bad fellow and forgave the young couple for their rash behaviour. Later the couple moved to Rustenburg (where Mimi's uncle lived) and Herman Otto established himself in a medical practice. Before long, however, he was the town's postmaster too and this soon led him to being appointed the first Postmaster General of the ZAR, based in Potchefstroom. Although Pretoria had become the legislative capital of the ZAR in 1860, many government departments continued to be located at the first capital, Potchefstroom, for several years afterwards. Indeed Fred Jeppe (see next page), in his almanac published in 1877 still listed Potchefstroom as the capital of the Transvaal Republic in that year with Pretoria as the "Seat of Government". (The same list shows Nylstroom as being the chief town of the Waterberg District.)[43]

For almost two years (1859–1861) Herman Otto also acted as the State Attorney. He then returned to Rustenburg, picked up his medical practice, moved to Pretoria[44] and was elected to the *Volksraad* in 1866. In 1869 he became the first Master of the Supreme Court. Unfortunately, life fell apart for Herman Otto Jeppe: His wife Mimi died in 1872 (aged 39) and the following year he became embroiled in a financial scandal. This resulted in his dismissal, impoverishment

FIGURE 5.12: Friedrich Jeppe (1834–1898). SANA TAB Photographic Library No.22799: Fred Jeppe, FRGS, Deputy Surveyor General.

and subsequent relocation to a farm near Barkly West. He died in Prieska in 1892.

Herman Otto's younger brother Friedrich Heinrich ("Fred") Jeppe (1834–1898) came out to South Africa in 1861 and like the Forssmans and his brother Otto, moved to Potchefstroom. He married a Rustenburg girl in 1863 and in 1867 followed in his brother's footsteps by being appointed Acting Postmaster General, a position in which he was confirmed the following year. This job persuaded him of the need to produce a map of all the farms, hamlets and villages served by his department.

Jeppe's first map of the Transvaal was published in 1868 together with Alexander Merensky, a Berlin missionary and Magnus Forssman, the Surveyor General, making use of the sketches of the explorer Carl Mauch and others. Like OWA Forssman, Friedrich Jeppe was a friend of Mauch's and the two corresponded frequently when Mauch was away on his travels.[45]

In 1877 Jeppe published a map to accompany his *Transvaal Almanac*. His next major map of the Transvaal came out in 1889, in conjunction with Gustav Troye, a draughtsman. But Friedrich's biggest project, his *magnus opus*, a six-sheet map of the Transvaal Republic, showing each farm accurately in remarkable detail, was only published posthumously in 1899, earning him a fellowship of the Royal Geographic Society before his death.[46]

He designed and produced the first postage stamps for the ZAR. In 1871 he was appointed Treasurer General and became a member of the Executive Committee from then until 1874. After the hiatus caused by the British annexation of the Transvaal (1877–1881), Friedrich became a court translator/interpreter. In 1886, he was asked by the Chief Justice Sir John Kotzé, to record all the laws of the ZAR. Friedrich Jeppe was a supporter of the interests of the Afrikaner: In 1873, he, together with President Burgers and three others, contributed £80 each towards the establishment of "a patriotic Transvaal newspaper". This was launched shortly afterwards as *De Volksstem* (to become *Die Volkstem*), under the editorship of Jan

Celliers, a longstanding Cape editor (and father of the poet of the same name).[47] Friedrich died of a liver complaint in July 1898.

The third Jeppe brother to arrive in South Africa from Germany was Julius (1829–1893), who came out in 1870 with his family and farmed for a while at Morgenzon, near Rustenburg, before moving first to Pretoria and then to Johannesburg. Julius and his wife had two sons, Carl (1858–1933) and Julius Jr (1859–1929). He and a business partner, L.P. Ford, a lawyer and erstwhile Attorney General, missed the gold rush on the Witwatersrand, but instead invested shrewdly in land along the line of the discoveries – the Johannesburg suburbs of Jeppestown and Fordsburg are their legacy. Upon Julius's death, his son of the same name – and a man renowned for his remarkable energy – succeeded him in the family business and went on to be knighted for his services to the development of Johannesburg (he was a town councillor and served on the boards of the Red Cross, the Chamber of Mines, Rand Water Board, Johannesburg Hospital and the Jockey Club amongst others). The elder brother, Carl, went into law and followed an interesting career that included a brief association with the Waterberg:

> In 1876, less than eighteen years old, I had left school, acted as a reporter and entered a lawyer's office. I was then "commandeered" for the Sekukuni Campaign and went through it from first to last. In 1877 I was appointed Clerk of the Peace [Public Prosecutor][48] in Waterberg, then a district inhabited only by natives and a few hundred families of Boers. There, and subsequently in Standerton, where I held a similar post in 1879–80, I often acted as Landdrost – probably one of the youngest ever magistrates in office. In 1880, when the first Boer War broke out, I resigned my appointment and was articled to Mr LP Ford, late Attorney General of the Transvaal.[49]

Whilst serving in the Waterberg, Carl Jeppe (aged 19!) wrote a long letter to the Secretary of the ZAR (April 1877), stating eloquently a range of issues that were proving of concern to the magistrate (C. Moll) in Nylstroom: the problem of collecting taxes, a shortage of police, field cornet and water inspector, and the need for a school were amongst the matters raised.[50]

In 1893 Carl Jeppe, who now had a legal practice in Johannesburg, was elected to the Transvaal Volksraad (later to be known as the Eerste Volksraad) as the Member for Johannesburg and represented that constituency on behalf of the Progressive Party until after the Jameson Raid in 1896.[51] Other prominent members of

FIGURE 5.13: Carl Jeppe in 1893, as a member of the Volksraad. SANA TAB Photographic library, No. 20517: Carl Jeppe and Lukas Meyer illustration as Volksraad members, published in a supplement to *Der Pos*, Wednesday 6 September, 1893.

that party included Gen. Piet Joubert (its leader), Gen. Louis Botha, Lukas Meyer and Gen. Koos de la Rey. There were up to 12 members of the Progressive Party in the 26-seat Volksraad. Despite Jeppe's political allegiance, he was appointed by President Kruger to serve on a Peace Committee appointed to administer Johannesburg in the aftermath of the failed raid. In fact, Jeppe had known "Oom Paul" since he was a boy: The farm Morgenzon to which he and his parents had come in 1870 was next door to Kruger's property. In his memoirs, Jeppe paints an interesting picture of the Boer leader:

> In time of war he was Commandant General of the Republic; during peace, merely a private citizen, though always one who was respected for his shrewdness and force of character, so that his advice was often sought. He was already wealthy, the owner of large herds and many farms. He was no longer young (I doubt whether he ever had been), but had not grown stout as he did in later days; and when you noticed the breadth of his shoulders, his figure akin to that of a prize-fighter, the stories told of his prowess in camp and veld seemed easily believable. On one occasion a friend had fallen under the wheel of a heavily laden waggon, and was pinned down by it in great agony. No screw-jack was at hand. Kruger put his shoulder under a beam and lifted up the enormous weight until his comrade could be relieved. [52]

In 1897 Jeppe was appointed as the ZAR's Consul General in Cape Town. At the outbreak of the South African War, while on his way to Pretoria, he was arrested by the British, but released on parole.[53] In 1908 he became seriously ill and returned to the Highveld where in 1913 he was appointed as an authoritative judge on the Water Court. He was a keen naturalist and farmer, experimenting with the cultivation of tobacco and cotton. He was on the management board of the Pretoria Zoo for 25 years as well as serving on the committee of the Transvaal Museum. He collected plants and brought several collections to the botany division. One of his sons married Barbara Brereton – the famous botanical illustrator Barbara Jeppe (1921–1999).[54] In 1990 Barbara Jeppe received the Cythna Letty gold medal from the Botanical Society of SA – Letty being the late wife of O.W.A. Forssman Jr.

Ludvig (Ludwig) Field

In the 1930s the town of Nylstroom and its citizens benefited greatly from the civic involvement and generosity of Ludvig Johannesen (known in South Africa as Ludwig Field), an immigrant from Norway. Field was born at Rakkestad in the Smaalenene district of Norway on 27 January 1862.[55] One source claimed that he was Jewish,[56] but this was not the case. Both his baptism and funeral were conducted in a Norwegian Lutheran church and he was brought up in a strictly religious protestant family in which his elder brother was a local Sunday school teacher.[57] Field emigrated to Minnesota in the US in 1887 and for the next ten years undertook a variety of tasks, including farming, from which he learned much about new farming equipment technology. He also worked on railroads. In 1897 he was persuaded by his younger brother Einar (1871–1948) to join him in the Transvaal where Einar had arrived to become a farmer. Ludvig was evidently a serial entrepreneur. Having started off working for the De Beers group looking for diamonds, he moved across to the gold mines, but in due course joined his brother on the farm Rietfontein between Nylstroom and Loubad. There he established a blacksmith shop, a wagon factory and a mill.[58] By the outbreak of the South African War, Einar had purchased a portion of the adjacent property, Vaalkop, where he farmed with stock and crops, and Ludvig operated his businesses. After the war Ludvig became well known for undertaking the harvesting of grain for farmers using his own threshing equipment, but in 1899 he was already engaged in producing feedstock for his mill. In September of that year, he took delivery of a set of equipment from the J.I. Case Threshing Machine Company of Wisconsin, which had been sent by sea to Lourenço Marques and from there transported by train to Nylstroom station via Pretoria.⁵⁹

Throughout the war, Ludvig Field remained neutral and continued to operate his threshing equipment. However, for a period, he also had to look after the farm, as brother Einar became caught up in the hostilities. According to Einar's statement to the British after the war, he was ordered in January 1900 by the local acting field cornet (Jacob van Roronter at Moddernek, on the neighbouring farm Donkerpoort) to take his bicycle and deliver mail to a Boer outpost at Basters Nek.

The only place of that name known today lies almost 150 km away to the northeast, north of Hanglip. Field was apparently told to collect mail once a week from a place called Dronthaal and deliver it to Basters Nek. This he did for two months between January and March 1900, whereafter he returned to his farm. On 11 June 1901 he was "captured" on his farm by a column under the command of Major McMicking. He was taken to Pretoria where under military cross-examination, he admitted that during his brief service at Basters Nek he had carried dispatches between Boer outposts in addition to mail, although he was never armed. As a result, despite protests from the Norwegian consulates in Johannesburg and London, Einar Field was sent to a prisoner-of-war camp near Madras, India. He was not permitted by the British colonial administration to return to South Africa after the war.⁶⁰ Instead, he went to Madagascar where he married, had a family and died in 1948.⁶¹

FIGURE 5.14: Threshing equipment supplied to Ludvig Field from Case in Wisconsin, USA. SANA TAB CJC 272 CJC899: Illustrations of equipment included in claim for compensation (1902).

FIGURE 5.15: Ludvig Field. Breedt (1998); p. 34. Field was known locally as Ludwig, but his official records spell his first name as Ludvig. His surname was also given as having originally been Fjeld Johanneson, according to the certificate of Norwegian citizenship mentioned in note 59.

FIGURE 5.16: Ludvig (l) and Einar (r) Field in exile in Madagascar, 1918. Photograph courtesy of Tord Tutturen and family.

FIGURE 5.17: Products of Field's wagon factory in Nylstroom.

Meanwhile, Einar's brother Ludvig had remained on the farm to continue with his milling and no doubt to use his newly arrived threshing equipment. But the British were not going to leave him alone either, despite his neutrality. In June 1902, after the war's end, Ludvig made a statement to the Compensation Commission in Pretoria in which he said that on 1 April 1901, when General Plumer came to Nylstroom, he (Ludvig) was forced to go into the magistrate's office in town. As he left the farm, he saw soldiers breaking into his lounge. When he protested, he was told to "clear off". When he returned later, he found many items of value missing. The next day his machinery (the threshing equipment) was destroyed by dynamite, even though Ludvig had pointed out to the officer in charge that the machines could be disabled simply by removing some parts.

Subsequently, he was obliged by the British to go into the camp, first in Nylstroom, then to Irene. During that absence, additional equipment, as well as part of the actual mill (which belonged to someone else) was damaged beyond repair. His claim for a total of £1164 was disallowed in its entirety on the grounds that a) the documents showing his receipt of the threshing equipment from Lourenço Marques were insufficient to prove his ownership and b) he was alleged to have been involved in aiding the Boers.[62] After the war Ludvig spent some time in Madagascar with his brother Einar.

Eventually, Ludvig returned to Nylstroom, picked up his threshing and milling business and became a prosperous citizen of the town. He also acquired several farms in the area, bred cattle and won a contract to build roads for the government.

Unsurprisingly, given the treatment meted out to him by the British at the end of the war, he aligned himself with Afrikaner interests. He was a bachelor with no immediate family. In 1929 he placed many of his property assets in a Trust, with the objective that the proceeds from their sale or rental would be used to assist the poor and needy. The only condition was that recipients had to be members of one of the three sister churches of the Dutch Reformed Church. So arose the Ludwig Field Arme Fonds (now the Field Fonds), with its curator nominated by the three Afrikaans sister churches in the town. It is still active today after almost a century of dispensing valued assistance to its recipients (see Chapter 14).

For three years (1921–1924) Ludvig chaired the Health Committee of Nylstroom village (effectively its mayor) and in 1930/1931, was elected mayor of the formal *Dorpsraad* (Village Council). During this last session, in February 1931, Ludvig offered to donate £2000 towards the cost of building a town hall for Nylstroom.[63] This was gratefully accepted and of four squares in the town, Market Square was selected as the optimal site (see Chapter 6). A year later Ludvig handed over the completed building (a competition for its design was won by none other than F.H. "Fox" Odendaal, who just happened to be Field's attorney, J.G. Strijdom's legal partner and the future Administrator of the Transvaal).

An interesting insight into Ludvig's character is provided by the condition he attached to his very generous donation:

> In handing over the Town Hall, it is however my express and earnest wish and request that one of the galleries be reserved for the Indian Community for Bioscope only, where the admittance is by tickets sold in the ordinary way and the Indians not be charged more than the Europeans. This wish is a result of my promise to the Indian Community that I would make provision for them in the Town Hall.

The letter was signed by Fox Odendaal on Field's behalf.[64] The road in front of the town hall – and present municipal offices – used to be named Field Street in memory of the town's most generous benefactor. Today it is called Harry Gwala Street.

FIGURE 5.18: Freemasons Hall, Nylstroom. SANA TAB Photographic Library, No. 31739: Masonic Lodge, Nylstroom, by E. Tamsen.

In 1924 Ludvig took a trip to Europe and the Middle East (via Madagascar), writing articles for publication in English-medium South African newspapers, including the *Zoutpansberg Review*. Some of these were later collated and published.[65] He also paid several return visits to his Norwegian home where he donated money to charitable causes. In 1934, he was awarded a Royal Medal and proclaimed Knight of the Order of St Olav in recognition of his support for the sick and poor in his home town of Rakkestad, where he established a foundation similar to the Arme Fonds.[66]

Ludvig Field left South Africa for the last time in 1937 and returned to his home town in Norway. In 1941 he concluded a Will in which he left all his assets to be divided amongst six members of two descendant families, the Tutturens and the Skulleruds. He died at his home in Rakkestad on 17 January 1944 at the age of 82.

Emil Tamsen

Tamsen was born in Hamburg, northern Germany, in 1862, the son of an exiled Danish army officer. The family then moved to America where Emil's uncle, also a former Danish officer, fought in the Civil War as a Staff Officer in General Grant's Union army. Emil came to South Africa in his teens. He fought with the British in the first South African War and participated in the siege of Pretoria, but became a citizen of the ZAR when the Transvaal regained its independence. In 1882 Tamsen opened the first store in Nylstroom. He met his wife (Carla Pauline Richter, the daughter of a German missionary living near Newcastle in Natal) while working in the Woodbush area near Haenertsburg. (His business partner, Carl Natorp, married one of Carla's sisters.) After the South African War, the Tamsen home in Nylstroom became well known for its hospitality and gracious accommodation. Emil and Carla had a son, Adolph (1892–1961) and six other children.[67] Tamsen maintained a remarkably long, seemingly untrimmed beard throughout his adult life, making him easily recognisable in photographs of the time.

Tamsen was very active in local affairs. In 1896 he was elected the first "mayor" of Nylstroom and, after the war, served several terms as chairman of its Health Committee (1910/1911 and 1912–1918). He was still a member of the town council in 1930, aged 68.[68]

He founded the Pyramids Masonic Lodge in Nylstroom in 1903, having joined the fraternity in 1887, a few years after arriving in South Africa. Foundation members of the Nylstroom Pyramids Lodge included Tamsen, Arnold Orsmond (the magistrate) and J.J. McCord from Warmbaths. One of the members of Tamsen's Pyramids Lodge was John William Chaney, who joined in 1908, according to memorabilia in the possession of his granddaughter Anne Howe of Vaalwater. Chaney was then the proprietor of a hotel and postal agency at the entrance to Zandrivierspoort – the very same business founded by Carl Forssman in 1890. Meetings were held in the Masonic Lodge Room, Nylstroom "on the Saturday in every month nearest to full moon, at such hour as the Worshipful Master may direct."[69]

Owing to a loss in numbers caused by the departure of men to fight in World War I in 1914, together with the aftermath of the 1914 Rebellion, the Pyramids Lodge had to close in 1916, but in 1928 Tamsen sponsored a new Pyramid Lodge, which survived into the 1980s. Chaney rejoined the new lodge.

Another prominent Waterberger who was a freemason (until he entered politics) was Jozua Francois (Tom) Naudé, the popular Pietersburg lawyer who became the MP for that area in 1920 (first for the UP, then for the Nationalists), later Speaker of the House of Assembly, then a cabinet minister, President of the Senate and for a little over a year (1967–68) State President.[70] (See Chapter 10.)

FIGURE 5.19: Tamsen's Store, corner Collins and Potgieter Streets. Nel (1966): p. 31.

Tamsen was not only a concerned citizen, successful businessman and dedicated freemason, but also a shrewd and prolific collector of stamps, especially those of the various colonies and republics that existed in southern Africa. He made a fortune from trading in stamps and became well known in international philatelic circles.[71]

FIGURE 5.20: Emil Tamsen (1862–1957). Photograph from the collection of Rosalind King & Elizabeth Hunter (Davidson family).

FIGURE 5.21: Carla and Emil Tamsen, c.1950. Photograph courtesy of Elizabeth Hunter.

CHAPTER SIX
The gateway towns

Within the more than one million hectares that comprise the Waterberg plateau, Vaalwater is today the only settlement of any consequence. However, that was not always the case. There used to be a number of very small settlements, some of which predated Vaalwater, such as Rankin's Pass, Zandriviespoort, Zanddrift, Twenty Four Rivers, and Bulgerivier.

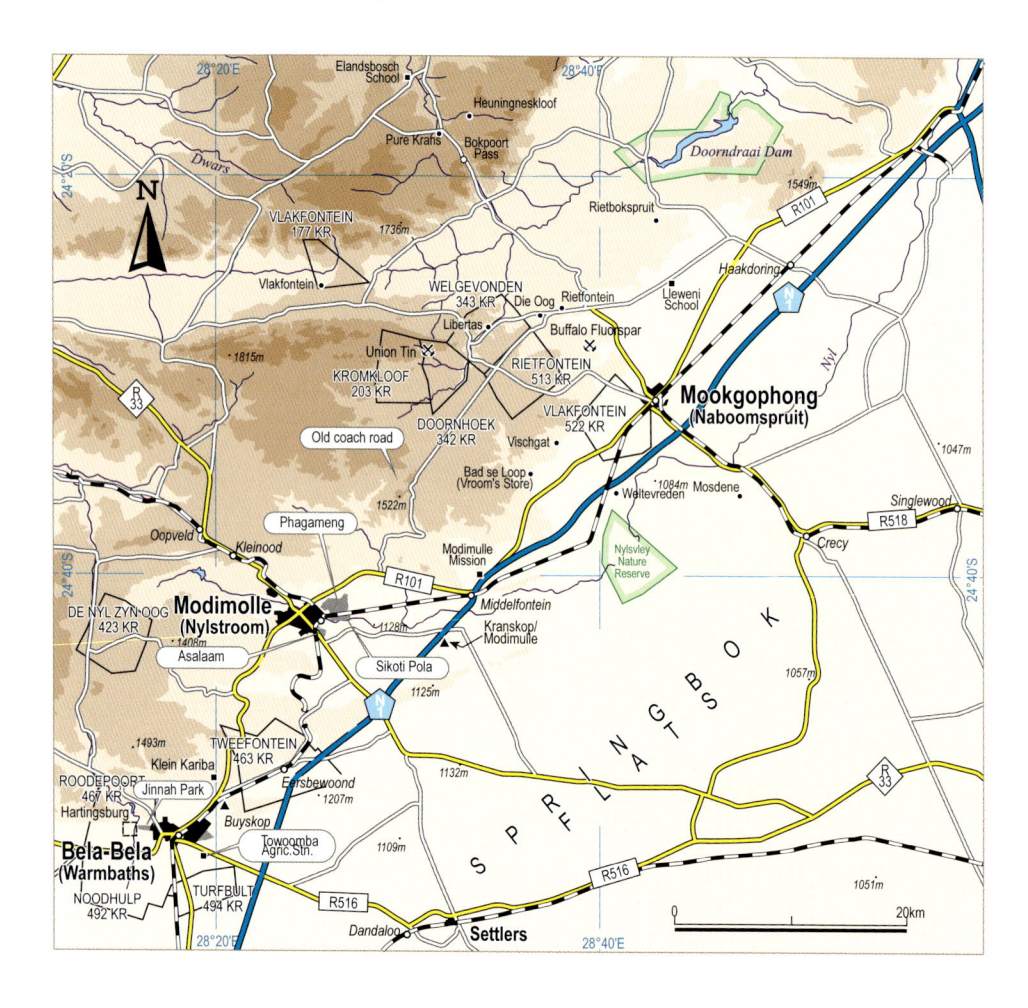

In addition, the Waterberg massif is surrounded by centres much bigger than Vaalwater, including Modimolle (formerly Nylstroom), Lephalale (Ellisras), Thabazimbi, Bela-Bela (Warmbaths) and Mookgophong (Naboomspruit). While Vaalwater clearly earns a place in this history (not least because its story has never been written), a record of the region's past would be incomplete without a discussion about these other towns and how they came into being. Here we look at the so-called Gateway towns – Modimolle, Mookgophong and Bela-Bela – that lie to the south of the plateau and that were established before Vaalwater, and through which lay the only access to the plateau.

NYLSTROOM/MODIMOLLE
Altitude: 1160 metres amsl
Population (2011) including Phagameng: 42 000[1]

Unlike other parts of the Waterberg, a history of Nylstroom was written[2] fifty years ago at the time of its centenary (in Afrikaans and from a partisan perspective). More recently, in 2004, a group of architecture students from the University of the Witwatersrand spent a week in Modimolle, interpreting the architecture of the town in the context of its history and socioeconomic background.[3] The following summarises information from these two sources, augmented by other perspectives and more recent information.

The *Jerusalemgangers*

The first settlement of farmers on the southern margin of the Waterberg was, surprisingly, at neither Nylstroom nor Warmbaths, but rather at the appropriately named site of Eersbewoond on the farm Tweefontein, about halfway between these two towns. First settled in around 1847 by the families of Daniel Janse van Rensburg and Petrus van der Merwe, it seems that in the mid-1850s, several other families arrived in the area from the Marico where they had initially settled for a decade or longer, having parted from the Andries Potgieter Trek in around 1838. They included the Enslins, De Beers, Swanepoels, Groblers and Eloffs. The remainder of the Potgieter Trek had continued into the hills of the eastern Transvaal where

FIGURE 6.1: The gateway towns of Modimolle, Mookgophong and Bela-Bela.

they founded the settlement of Ohrigstad. (This was not a success and Potgieter soon resumed his northward journey until his people finally halted at the foot of the mighty Zoutpansberg to establish the village of Zoutpansbergdorp, later called Schoemansdal.) Commandant Enslin was one of the Jerusalemgangers. On reaching the Nyl floodplain, he decided to settle at what seemed to be the source of the river, at De Nyl Zyn Oog, about 15 km west of Eersbewoond.[4]

The Eersbewoond group had spent several years on the road, escaping the strictures of the Cape Colony while searching for the freedom they dreamt lay ahead. They were deeply religious, yet having been forsaken by their ancestral church in the Cape, the NGK, they lacked a formal church organisation – not for long, however, because in 1859 the Gereformeerde Kerk (GK) or *Dopperkerk* would be founded in Rustenburg and the Eersbewoond families joined it en masse. Most of them soon moved on again, this time westwards to the western end of the formidable rampart that lay to the north (the Zandriviersberge); and intermarrying between the clans, they formed the community of the Zwagershoek (see Chapter 4).[5]

FIGURE 6.2: Trekking near Kranskop. SANA TAB Photographic library No. 16311: Trekking near Kranskop by H.F. Gros.

Ernest Olferman Collins

With the establishment of the settlement of Schoemansdal in the far north of the Transvaal in the early 1850s, a need arose for an intermediate settlement between it and Pretoria, some 350 km to the south, especially in order to render a reasonable postal service. Although a town had been proclaimed at Piet Potgietersrust (Mokopane today) in 1859, this was not considered suitable for the purpose. In 1865 an English-speaking farmer, entrepreneur and owner of the farms Elandspoort and Rietvallei on the Klein-Nyl River, Ernest Olferman Collins, shrewdly offered land from Rietvallei on which to establish a township, including erven for the two existing churches in the region, the Gereformeerde Kerk and the Nederduitse Hervormde Kerk (NHK) (see Chapter 14). The ZAR government readily accepted this offer and the town of Nylstroom was proclaimed on 15 February 1866.[6] Collins himself laid out the erven, including several set aside for government purposes, others for commercial and agricultural use and the rest for

FIGURE 6.3: Ernest Olferman Collins (1821–1868), founder of Nylstroom. Nel, J.J. et al (1966): p. 23.
FIGURE 6.4: Grave of Ernest Olferman Collins in the Modimolle cemetery.

residential occupation. Streets were named after popular English icons, including Victoria and Albert and George Rex, Collins's son-in-law and the grandson of Rex of Knysna.[7]

Legend has it that the name of the town was derived from that of the river, which the Jerusalemgangers (an early group of Voortrekker zealots) had optimistically identified as the Nile (because it was surrounded by reed-beds, it was northflowing and in flood). Having only Biblical descriptions to guide them, they were under the impression that it was possible to travel by ox-wagon from the Cape to the Promised Land via Egypt and were even said to have mistaken the blocky shape of Kranskop, or Modimolle, for a crumbling pyramid.[8] However, as mentioned in Chapter 4, at least one descendant of the founding families refutes this story.

Modimolle, as this isolated hill (inselberg) of Waterberg sandstone was known before the arrival of the first white settlers, means "the spirit has eaten" in Sepedi. The origin of this name and the beliefs behind it are varied, ranging from ancestral spirits (*badimo*) that live on the hill and eat the flesh of children, to a legend that a chief in the past used to throw his prisoners off the top of the sheer cliff on the southern side of the hill. Many black people in the region still consider the hill to be a sacred site and attribute special powers to it, for example that the medicinal herbs growing on its slope are of a greater potency than elsewhere.[9]

Nylstroom's future was secured at the time of its founding when the Volksraad decided to merge the Nylstroom and Schoemansdal/Potgietersrust regions into a single district, to be called Waterberg, and to appoint a magistrate (J.J. Prinsloo) and a Justice of the Peace (G.J. Verdoorn – later also to become the postmaster) for the district, both based at Nylstroom. Understandably, this news did not please the residents of Piet Potgietersrust or Schoemansdal. In 1869 the huge district was split in two; the northern one, its seat at Marabastad, becoming Zoutpansberg. (Two years later, though, Potgietersrust was once again included in Waterberg).

Collins died in 1868, aged 47, having founded a town but not living long enough to see it flourish. In fact, so poor were the farmers in the neighbourhood that the growth of Nylstroom was very slow. The town was proclaimed a municipality in 1881 (delayed in part by the British annexation of the Transvaal from 1877 to 1881). In 1882 the first store opened in Nylstroom. Emil Tamsen opened his store on the corner of the Warmbad and Roedtan roads, having earlier run a store with Carl Natorp at Eersbewoond in the 1880s. The name changes of roads provide an interesting guide to the town's history. For example, the corner of Warmbad & Roedtan became Collins & Main, then Voortrekker & Potgieter and is currently named Thabo Mbeki & Nelson Mandela.[10]

One of the least appropriate sets of changes in street names took place quite recently: Church, or Kerk Street, which, since the beginning of the town's existence

FIGURE 6.5: Kranskop (Modimolle).
FIGURE 6.6: Changing street names.

has hosted all the main churches, was renamed Luthuli Street after the Nobel Prize-winning ANC freedom fighter. The short road in front of the municipal offices, named Field Street in honour of one of the town's major, apolitical benefactors, Ludvig Field (see Chapter 5), was inexplicably given the name of Harry Gwala. Gwala was a particularly violent member of the SA Communist Party whose principal claim to fame in his native KwaZulu-Natal before his death in 1995 was his alleged involvement in arranging the murderous deaths of many supporters of the Inkatha Freedom Party. Neither Luthuli nor Gwala had ever visited the town.

Development of Nylstroom town

In 1882, apart from Tamsen's store there were only two *hartbeeshuisies* and one two-roomed brick building (the magistrate's office) in the town. In addition, there were the two small churches of the GK and NHK. "That was the town – no post office, school, boarding house, doctor or officials."[11] These were hard years in the Transvaal Republic. A serious drought aggravated the already straitened finances of the government. The manager of the Standard Bank in Pretoria wrote to his London superiors in February 1882:

> So large a country as the Transvaal, containing within itself so many elements of disquiet and lawlessness would, even with an honest and strong government, take years to settle down. This state of unrest is scarcely to be wondered at, when we think what the country has lately gone through: that almost all the male inhabitants have been engaged in the active pursuit of war, that its revenues have been crippled in every direction – and that there are no statesmen or men of influence to guide the people.[12]

The following year was described as the worst ever experienced:

> The financial position of the Transvaal government has been temporarily improved by granting various mining and other concessions, but the Boer leaders, with the exception of Commandant Joubert, have no knowledge of finance or of the need for strict economy & much caution is necessary in dealing with them.[13]

The year 1884 was described as

> ... on the whole, the year of the greatest depression within our knowledge.[14]

In this year another new township was proclaimed in the area – at the hot springs of Het Bad some 25 km to the south of Nylstroom. This town, initially named Hartingsburg, would become Warmbaths (Warmbad) and later Bela-Bela. Nylstroom residents were offered an opportunity to exchange their erven for plots in Hartingsburg, but there's no evidence to suggest anyone took up the option. The difference was that there were two established churches and a magistrate in Nylstroom as well as the biggest store in the region, all of which caused more people to visit the town, despite its reputation as an unhealthy place.

Malaria was endemic in the Nyl valley in those days and many local residents were "saturated" with the disease according to doctors who attended to them during the South African War (see Chapter 13). An unattributed source commented in the early 1880s that

> The town itself lies on an open slope surrounded on all sides by dense bush, the river Nyl running on its NE side ... the banks of the river in some places being very thick with weeds and the adjoining ground being swamps. The site of the town is particularly unhealthy, the soil on which [it] stands is sandy and a water furrow between the town and the river, the water percolating through the porous soil collects on the low-lying portions of ground converting it into swamp. (...) the fever is propagated to such an extent that ... out of five families resident here, which is the entire population, not one escaped the fever, and in three families each individual member was suffering from it at one time or another.[15]

A few years later (1891), W.L. Distant, an English naturalist who travelled for a year through southern Africa, wrote of his impressions of Nylstroom in February of that year:

> ... by 7 am we had reached Nylstroom [from Warmbaths], a forlorn spot, where the imposing appearance of a post office and landdrost's court, unsurrounded by any apparent business life, give it the

appearance of a stillborn township. But fever has been the retarding cause of Nylstroom's future, and its character for unhealthiness will long survive, though the natural beauty of the surrounding country and its little disturbed condition, should make it a district beloved of sportsmen. Woods, park-like tracts, undulating country, from which views could be obtained of endless and varied landscape, tall, wooded, isolated hills, and ranges of mountains with forest slopes, alternately meet the eye.[16]

Establishment of local government

Gradually the town developed. In 1888 the telegraph arrived and the first mill in the town was opened by Messrs Forssman, Siemsen & Zoon. The year 1890 was an important one. Following lobbying by Tamsen and other local residents, a proper magistrate's court and post office was built on the Main Street (alias Potgieter Street or Nelson Mandela Drive). That grand old building, which still stands next to the current post office, had four functions: the *Landdrost's* office; an Assistant Magistrate's office; a post office; and a jail at the rear. As the principal representative in the town of the authority of the state, the building needed to create an impression of power, order and stability, and its imposing façade succeeded in achieving this.[17] Interestingly, the government, which had a range of options, chose to locate the building on what was then a secondary road (the road to Roedtan) rather than on Collins Street, the main thoroughfare from Warmbaths to the north and where two of the major churches were situated. The reason for the decision is not recorded but may have been made deliberately to separate the spiritual from the secular and to encourage the lateral development of the town.[18]

In the recent past, the building was home to the local Commando. It also enjoyed fame (or notoriety) in a recent humorous novel about the visit of a revivalist church group to Nylstroom.[19] A visit is rewarding (provided it is made voluntarily, for today, the offices are occupied by the Vehicles and CID branches of the SAPS!). The offices behind the court room are designed around a shady courtyard and protected by a covered veranda, which not only keeps the premises cool, but also serves to protect the fine Oregon pine sash windows and floors from the sun.

In the same year (1890), a Dr J.J. Hoffman took up residence, a coach hotel was built, as well as a gaol.[20] The postal service to Pretoria had advanced to a scotch-cart drawn by four oxen, the latter shortly to be replaced by mules. A second store,

FIGURE 6.7: Nylstroom magistrate's court and post office, 1910 and today.
SANA TAB Photographic library, No. 31742 by E. Tamsen.

Landsberg & Forssman, had opened. Forty erven had been advertised for open sale in the *Staats Courant* in 1888, and the first was purchased in 1890 by Isaac van Alphen. A census carried out in 1890 estimated the total white population of the Waterberg at 2036.

The first legal agents were J.L. de Lange and Jean Daly and by 1894 a land surveyor had also set up his practice. In 1899 De Lange would join Messrs Forssman and Tamsen in constituting the town's first Health Committee, which, after the interregnum caused by the South African War, became the first municipal council in 1902 with Tamsen as its chairperson and additional members including H.J. Kroep (the pre-war magistrate) and C. Hickey.

At some point in the 1890s a local newspaper, *Den Boerenvriend*, began publication in Dutch and English with J.R. Reynolds as its editor. An article in the issue of 21 September 1894 announced that

> Several buildings are now being erected in Nylstroom and we know of others shortly to be commenced. The erven given by the Government for burgher-rights are ready for transfer to their owners and probably several will be built upon, especially if it is decided that the railway is to pass through the town.[21]

In fact, the railway from Pretoria to Pietersburg only reached Nylstroom in 1898 (the first train was on 1 July) and, rather than go through the town, it skirted it, with a station built on the outskirts. The story goes that this strange detour was the result of pressure from the magistrate's wife, who said she didn't want to be disturbed all day and night by the noise of passing and shunting trains in the middle of the town! The original route had been planned further east towards Kranskop, but lobbying by the citizenry succeeded in bringing it (almost) to the town. The section of the line between Pretoria and Warmbaths (it bypassed Hartingsburg) was officially opened by President Paul Kruger on 5 March 1898. The railway historian, the late Boon Boonzaaier, recounted the following story relating to that event:

> ... the President was in a very good mood on the inaugural train, accompanied by friends and dignitaries. At one stage, he challenged his Jewish friend, Sammy Marks, to a race. If the President won, Sammy would convert to Christianity; but if Sammy won, he could become President of the Republic. Evidently Sammy was not interested in either outcome, and declined the challenge.[22]

The line reached Potgietersrust in October 1898 and Pietersburg in May 1899, shortly before the outbreak of the South African War. At that point the trip from Pietersburg to Pretoria took one comfortable, safe day by train for a cost of £1 9s, compared with the alternative of three days by stage coach for a cost of £7 10s. The days of the pioneering Zeederburg coach service drew rapidly to a close.

The South African War and its aftermath

The outbreak of the war in October 1899 created a hiatus in the administration and development of Nylstroom: The departure of many male citizens to join one or other side of the conflict was accompanied by the evacuation of many of their families, mainly to Pretoria, for the duration. The town was occupied by the British military forces in late 1900. They commandeered several buildings, including the premises of all three Dutch Reformed Churches, for use as quarters, a hospital and mortuary, and also as stables – actions that were to leave deep scars in the hearts of the congregations of those churches.

The Nylstroom concentration camp was established by the British on 31 May 1901 and by June there were already 1100 inmates. The camp was not large and initially was somewhat of an improvisation, comprising "a jumble of ancient white bell tents, sail-covered tents and a marquee. A number of people lived in the little town houses, while the gaol was also occupied, as was the church."[23] The camp was sited immediately to the south of the present Gereformeerde Kerk property (on the grounds occupied by Laerskool Eenheid today), the original church building being used as the camp hospital.[24] Unlike some other camps, Nylstroom camp was unprotected and open, allowing inmates to come and go as they wished. The first Camp Superintendent was Henry Cooke and the camp doctor was Dr Pierce Green, who had been seconded from the bigger camp at Irene. Initially, health in the camp seemed to have been good, although the poor condition of many of the inmates – malarial, ill-clothed, of poor hygiene and unused to dwelling in close proximity with others – did not bode well. Several were said to have been *bywoners* (tenant-farmers or share-croppers) and were of limited means.

The Highveld winter of 1901 was unusually severe. Casualties among the human and livestock populations around the Transvaal were inevitably greater than usual. In the camps the cold took its toll, especially amongst the elderly and sick. By August when the camp was visited by Dr Kendall Franks, a medical consultant to the British army, conditions had begun to deteriorate: The accommodation was often unsuitable and dirty; the corridors of the gaol were hung with lines of drying biltong; and several houses and tents were both overcrowded and filthy. One house in the town was found to accommodate 49 people in squalid conditions. An "adequate" hospital had been set up in the *pastorie* of the Gereformeerde Kerk, but the superintendent had difficulty in persuading pregnant women to use it when giving birth; they preferred the services of an untrained midwife among them.[25]

It was in August that the measles epidemic burst upon the camp. Mortality, especially amongst young children, soared. The disease was aggravated by enteric (typhoid) fever and malaria. It afflicted adults as well as children – although the male population of the camp was very small (and mainly elderly) and most of the adult casualties were women. In that month alone, there were nearly 120 deaths, over 90 being of children under 16. A second peak, in November 1901, claimed another 90 inmates, and again, three-quarters of that number were children. The mortality rate seemed to be higher than that at other camps in the Transvaal and Free State, a trend which Dr Green attributed to the fact that most of the residents of Nylstroom camp were already "saturated" with malaria and therefore more susceptible to succumbing from other diseases.

Cooke's replacement by R. Duncan as camp superintendent in September 1901 improved the conditions and administration of the facility. He responded to concerns about scurvy – due to inappropriate diets – by encouraging the inmates to grow vegetables and fruit. Gradually the incidence of this condition declined. On 24 March 1902, however, the military authorities closed the camp and evacuated the inmates to the existing large camp at Irene. During the nine months of its existence, the Nylstroom concentration camp had accommodated a total of some 1850 Boer women and children and a few, mainly elderly, men. According to one record, 544 inmates died during this brief period, of whom about 450 were children under the age of 16.[26] Another, possibly more authoritative source quoting official British Army statistics, lists a death toll of 525, of whom 429 were children under the age of 15.[27] Either way, the number of deaths was very high, particularly if it is borne in mind that the Nylstroom camp existed for only nine months. In total, almost 28 000 Boers died from disease in the approximately 60 camps established during the war; over 22 000 were children under the age of 16, about 4300 were women and 1700 men.

Few sources refer to the number of black people who may have died in either "white" camps or their own camps in the vicinity (there were 36 in the Transvaal). Records were either poorly kept or destroyed, but it is estimated that at least 116 000 black people – mainly those who had lived on or close to Boer farms – were interned across the country, including perhaps 4000 in "white" camps, and that at least 14 000 had died.[28] At the Nylstroom camp there certainly had been some black inmates, because their names had been inscribed in the camp register – but in 1911 all but one of the pages were removed deliberately, for reasons unknown.[29]

Peace returns

The end of the war and declaration of peace in June 1902 enabled Nylstroom's citizenry to return to their homes and the farmers to their devastated properties. The British administration under Lord Milner lost no time in establishing basic governance measures both nationally and locally. Even before peace had been concluded, a detachment of the newly formed South African Constabulary (SAC) had been established in Nylstroom under the command of a Captain Swift, with Lieutenants Willets and Humfrey in support. For the next seven years – until the creation of the Transvaal Police in 1908 – the SAC would be responsible for a wide range of duties, quasi-military initially, changing to more conventional policing as the country returned to normal. Many constabulary members were appointed Justices of the Peace.

The National Bank of SA, which would become Barclays, opened its doors in August 1902 in a building on Tamsen Street (now Chris Hani Street), opposite the

FIGURE 6.8: The Nylstroom concentration camp cemetery in Modimolle.

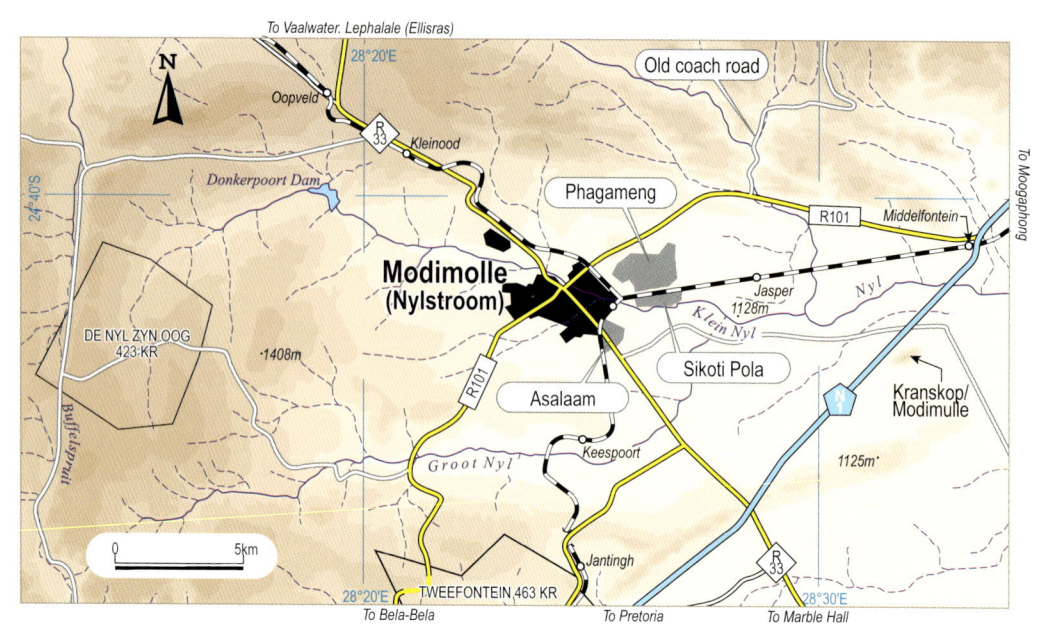

FIGURE 6.9: Modimolle and surrounds.

FIGURE 6.10: Tamsen Street, Nylstroom, with National Bank on the right, c.1920. Museum Africa Photographic Library, No. PH2002-381 (Gabbott Collection).

square where the original town hall and offices were built in that same year. The municipal council, with six members (Tamsen was its chairperson), was established in these premises and began its work. The old gravity-fed water-furrow, or *leiwater* scheme was replaced by pipe-borne reticulation.[30] A new resident magistrate, A.R. Orsmond, was appointed. In 1910 Gerrit Bakker, the first of three generations of important participants in the town's affairs, opened his pharmacy on Main Street.

At the time of the Spanish flu epidemic of 1918, the town's doctor was Dykema (known as Dekeman), a Dutchman.

The development of churches in Nylstroom and elsewhere in the Waterberg is described elsewhere, but with the construction of St Michael's Anglican Church in 1904, there were already four churches in town with the white population given as 361, although there were more in the surrounding area.

In 1905 Tamsen became the first mayor of Nylstroom. Tamsen's store was still flourishing and earned the following description from a traveller in 1906:

> Mr Tamsen established himself in premises overlooking the Market Square and as his enterprise prospered, made numerous additions. The business consists of the store of a general merchant, added to which there is a bottle store – the only one in the township. Mr Tamsen also holds the only licence in the district for supplying firearms. The stock carried is large and extremely comprehensive. The name of the firm is a 'household word' with the farmers for miles around, whose confidence and respect Mr Tamsen has gained and retains. Lumber and building material yards are attached to the premises and the handling of the materials stored there forms an important part of the firm's trade. Agricultural implements form a prominent part of the stocks. Tamsen's supply clothing, furnishing goods, groceries and drapery in large variety. In a word, this old-established house is a general emporium where all the necessities of life may be obtained.[31]

Nylstroom schools

Schooling had made painfully slow progress in Nylstroom since its foundation almost 40 years earlier. The first move towards formal schooling in Nylstroom occurred only in 1870 when President Pretorius approved the formation of a

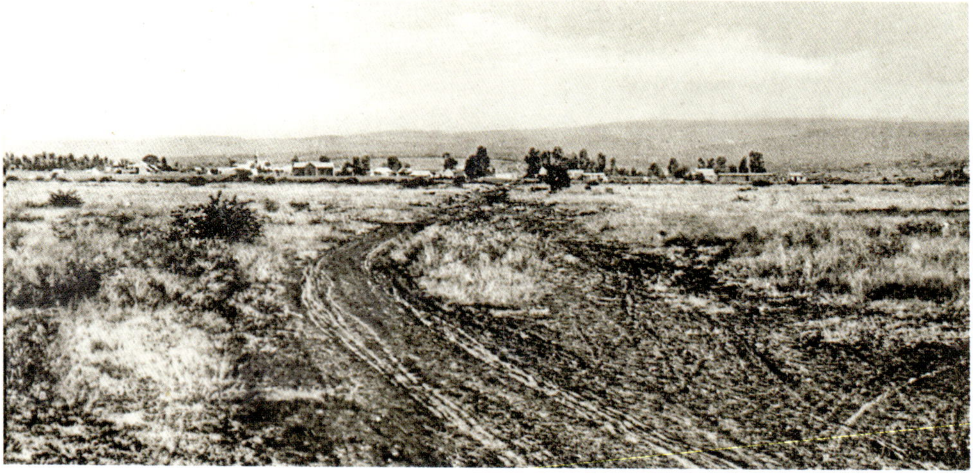

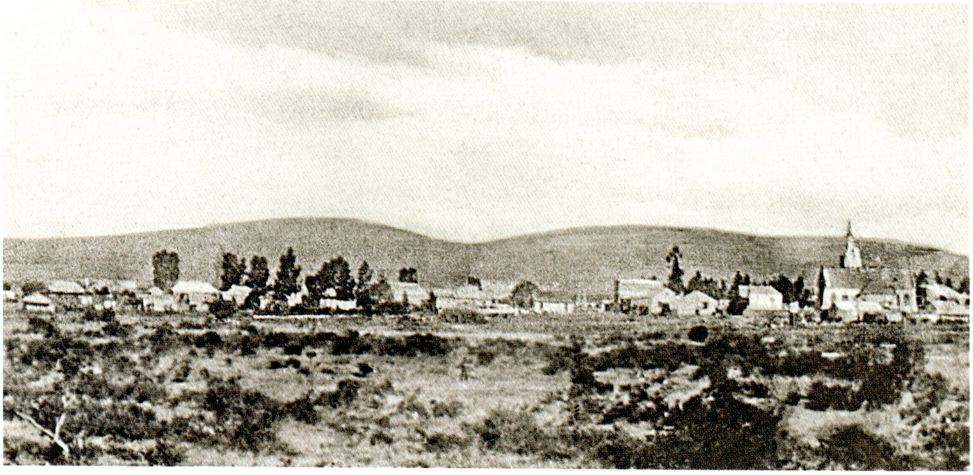

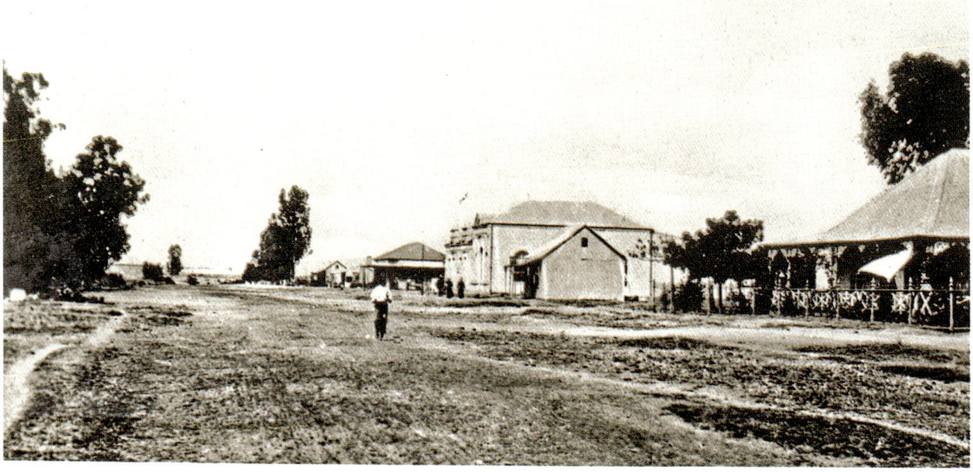

Waterberg school committee under the chairmanship of the then Landdrost, Zacharias de Beer. Unfortunately, little seemed to come of this initiative, until the progressive policies of President Burgers led to the creation of numerous state-assisted schools for white children in the farming areas, including one in Nylstroom in 1876. However, conservative local farmers, of the view that education need not extend further than an understanding of the Bible, were slow to support the new schools. Following representations from a private teacher, C.W. Nelson, and the postmaster, G. Verdoorn, a new school committee was constituted in September 1878 (Carl Jeppe was included among its members). Eventually, a school with 15 pupils opened in the pastorie of the Nederduits Hervormde Kerk on 14 April 1879, with Mrs Maria Impey as teacher. But its existence was brief: Neither the committee nor the teacher enjoyed the support of the Superintendent for Education and it was soon closed. In 1882, after the end of the First Anglo-Boer War, the government established a new inspectorate, a new school committee and new rules for state-assisted education on Christian principles and this found favour with the community. Numerous schools, most of them short-lived, opened in the district. It was no different in Nylstroom but finally, in 1891, it could be stated that there were two schools in the town and eight in the district, with a total of 190 pupils among them. However, the quality of teaching was poor, because the teachers were unqualified – until 1893 when Mr M. Smuts became the first qualified teacher in Nylstroom. By 1898 he had been joined by Ms Maria Emily Pronk (a sister or cousin of Maria Elizabeth Pronk, Carl Forssman's wife). There were now 15 *wykskole* (district schools) in addition to the two in town, and a total of 388 pupils were registered.[32]

After the war, the school in Nylstroom reopened with 66 pupils administered under the new British colonial regime. Because of the subsequent withdrawal of several pupils by their parents, who favoured Christelik-Nasionale Onderwys (CNO)-based education, the number had dropped to 48 by 1905. Officially, the sole medium of instruction was to be English, but the principal, Mr William White, allowed in June 1905 that "a little conversation in Dutch is also being attempted"![33]

New classrooms were added to the existing little building and over the next decade, two headmasters in succession – Messrs Norman Cardwell and C.H. Ackhurst – did much to enlarge and improve the facility and the quality of its edu-

FIGURE 6.11: Scenes of Nylstroom, c.1910. Bottom image is of Main Road viewed towards the east, with the magistrate's court on the right. SANA TAB Photographic Library: Nos. 31741 top, 31740 middle and 31745 bottom, all by E. Tamsen, c.1910.

cation. In 1918 the school closed for the duration of the influenza epidemic, but by January 1920 the number of registered pupils had risen to 160, of whom 40 were boarders. Two issues were pressurising the school administration: the need to extend the school to secondary grades up to matric and the urgent need for proper, formalised boarding facilities to cater for the many children of school-going age who lived too far away from town to make the journey daily.

In 1918 a Standard 7 class was added with the principal, Mr Ackhurst, teaching Dutch, English, accounting, algebra and natural science. Standard 8 followed the next year and so on until 1925 when the secondary phase of the school had 78 pupils with five teachers, while the primary phase had grown to 175 pupils, also with five teachers. The school had quite outgrown its limited facilities and classes were held in the Masonic Hall, in Tamsen's chambers, and even in the charge office.

By 1935 in addition to a couple of additional (but still woefully inadequate) classrooms, there was a laboratory, domestic science centre and wood-working shed. The population of the school was now 260 in the primary phase and 149 in the secondary phase, with a total of 18 staff.[34]

FIGURE 6.12: Nylstroom government school (5 November 1915). Among the pupils are: 11 – W. Nicholson; 12 – D.H. Kirkman; 15 – L. Forssman; 18 – I. Moerdyk; 27 – Nicholson; 30 –Mr Kriel (teacher); 31 – Miss Pronk (teacher); 32 – C. Ackhurst (principal); 34 – Moerdyk; 42 – Moerdyk. Nel et al (1966): p. 66.

In the meanwhile, the three Afrikaans *susterkerke* had joined together to finance and build two *koshuise*, one each for boys and girls. These two fine buildings, similarly built with their double-storeyed, gabled facades and central towers – and which today still grace the town, although they are no longer used as hostels – were built in 1927/8 for what was then the enormous sum equivalent to R28 000. Initially, they were staffed and administered by a committee drawn from the three benefactors. It is notable that whereas the boys' hostel, Ons Toekoms, was built on the main road into the town, almost opposite the NG Church's "*witkerk*", the girls' hostel, Ons Hoop, was located in a quiet suburban street, along the road leading to J.G. Strijdom's home. Architectural students have pondered over the symbolism of these choices, speculating that whereas both the siting and the name of the boys' hostel was intended to symbolise power and growth, those of the girls' home symbolised domesticity and *"ons hoop vir die toekoms"* (our hope for the future) – the role of women in support of their men.[35] Ten years later, the

FIGURE 6.13: Ons Hoop (girls) and Ons Toekoms (boys) hostels in Modimolle today.

buildings were handed over to the Education Department. In 1949 and 1951 two new hostels were added to the accommodation. In 1958 a further two hostels were added when the two original structures were taken out of service. Today Ons Hoop serves as an old-age home and its counterpart houses the municipal Public Works Department.

By the time of the outbreak of World War II, the school had a total of 630 pupils, about half in each phase, and space had become intolerably cramped. Although the two phases were formally separated into two schools that year, they continued to share the existing space until a new high school was finally opened in 1954. Mr Jooste Heystek (later the MP for Waterberg) was then the principal at the primary school and Dr M.J. Coetzee the principal at the high school.

By now, many of the original buildings were in need of replacement and in 1961 the primary school too moved to its own new facility, Laerskool Eenheid, built on the site of the former concentration camp. By 1966 it had an enrolment of over 500 pupils.

The history of black education in the Waterberg is described elsewhere (in Chapter 17) but here it is sufficient to say that until the 1960s there was effectively no public schooling available for black children in this or any other town. Education of black children had since the beginning of the century been delegated by successive governments to churches and missions – and to a few farm schools, which were established and administered by altruistic landowners. Government provided a degree of financial assistance to these operations, without which they could not afford to function.

The nearest mission/church school to Nylstroom was the Berlin Mission school at Modimulle (on Middelfontein) about 15 km from town. There was (and still is) another school at the Catholic church in Warmbaths.

The passing of the Bantu Education Act in 1953 led to the withdrawal of state subsidies to private institutions – which, for over 50 years had shouldered the responsibility of black education – and consequently, to the closure of most of these bodies (which was the governing party's intention). The state itself now had to invest in the building, equipping and staffing of schools for black children. In Nylstroom this process was accompanied by the relocation of the black population of the town in terms of another piece of apartheid era legislation (the Group Areas Act of 1950) to a new, formally laid-out township.

Phagameng

Phagameng (the name was chosen by the authorities, from Sepedi meaning "we have risen") was officially proclaimed in 1967, although it had been conceived over a decade earlier when the area originally designated for black residential habitation, the Sikoti Pola Location, became overcrowded.[36] Sikoti Pola, which was situated along the banks of the Klein-Nyl River immediately east of the town and southeast of the railway line, had been formally established in 1926 to solve a growing squatter problem. It was a poor choice of site: In the rainy season, local flooding would destroy or wash away several homes and the clay soil made access and building difficult. By 1954 the population of the community had risen to over 1000 and no further extensions were possible within the confined area bounded by the river and railway.[37] In 1963 tenders were issued for the construction of 300 formal family houses and a hostel to accommodate nearly 300 single men on a new site north of the river; construction commenced the following year.[38] However, the implementation of further apartheid legislation during the 1960s, designed to curtail further black urbanisation in "white" areas, brought the development of Phagameng to a halt, causing renewed informal settlement around the township. It was only in the 1980s that development of the township was resumed and further extensions added.[39] Today Phagameng has several schools, including two secondary schools.

The Indian, mainly Muslim, population of Nylstroom from before the South African War also became a victim of the Group Areas Act in the late 1950s. The Indian community had traditionally been concentrated at the northern end of the town, around the intersection of the two main streets (Main and Collins, now Mandela and Mbeki). They lived behind their trading stores and built a mosque between the residences and the river. For over half a century, they coexisted peacefully and cooperatively with the white citizens of the town. But from 1950, with the passing of a raft of apartheid legislation, the attitudes of the white local authorities began to harden. Indian shops were confiscated and the owners were forced to move out without any reimbursement or formal monetary compensation. Instead, the government offered to lease them space in several new warehouses it built for the purpose on a designated piece of land on the southern perimeter of the town – a new Indian township to be called Asalaam. By 1957 the move was complete – with the exception of the mosque, removal of which the pious authorities were too nervous to enforce.[40] In 1987 a beautiful new mosque was built on the site of the original one – a poignant reminder of how things used to be and a lasting symbol of defiance against the government of the time and the white population who had allowed their Indian neighbours to be so poorly treated.

Asalaam is mainly where the Indian community of Modimolle lives to this day, although, as with the black population of Phagameng, it is almost 30 years since restrictions as to where they may reside were removed. There is still a distinctive Indian trading centre at Asalaam too, noticeable as one approaches Modimolle from the N1. There's even a primary school: Nylstroom Primary, built in 1957 to accommodate about 20 relocated Indian children in three classrooms, with two teachers. (It now serves a largely black pupil population of over 300 pupils and over 20 teachers.) For secondary education, Indians had to go to Pietersburg, the Witwatersrand or further afield.

Nylstroom after World War II

The development of Nylstroom town proceeded slowly in the period between the two world wars, but sped up in the mid-1940s. The principal farmers' co-operative organisation in the region, the Waterberg Landbouwers Kooperatieve Vereniging (WLKV) had been established in the town in 1909. After many years of considerable difficulty, the farming fraternity embraced groundnuts as a crop well suited to the local conditions. By the late 1940s, the region had become one of the major peanut-producing areas in the world with exports to Europe and the Far East. The coop, now renamed the WKLM (in the 1950s, to become the NTK) built a large factory on the edge of town near the station to process peanuts into peanut butter – and soon the brands of "Apie" and "Nyl" were household names throughout the country.

In 1931 a local businessman, town councillor, mayor (in 1924/25 and 1930/31) and philanthropist, Ludvig Field, donated £2000 for the construction of a new town hall (see Chapter 5). The new hall was opened on 4 March 1932 by the Administrator of the Transvaal, J.S. Smit (unfortunately Field was ill and could not attend the function).[41] Four years later, during the mayoral tenure of Gerrit Bakker, the Donkerpoort Dam was built across the Klein-Nyl and for many years afterwards the town's water supply was assured. More recently, a pipeline delivering water from the Roodeplaat Dam north of Pretoria was built to augment the water supply, which unfortunately soon became inadequate, once again owing to the growth of the town. Electricity supply was always a problem. In the period after World War II, it was only available at night but, eventually, the generators

CHAPTER 6: THE GATEWAY TOWNS   *171*

FIGURE 6.14: Gerrit Bakker, pharmacist and mayor of Nylstroom, 1924–1930; 1935–1943. Nel, J.J. et al (1966): p. 89.

FIGURE 6.15: Ludvig Field's Nylstroom Town Hall in the 1930s. Museum Africa Photographic Library, No. PH2006-10111 (the Gabbott Collection). FIGURE 6.16: Modimolle Town Hall facade in 2018.

were expanded to provide power during the day too. From 1956 until the arrival of Eskom power in the early 1970s, locally generated electricity was augmented by purchases from Warmbaths.

The town's Health Committee had evolved into a Dorpsraad (Village Council) back in 1923. In 1959 its status was elevated further to Stadsraad (Town Council) with Mr A Bakker (Gerrit Bakker's son) as mayor. He filled this position a second time in 1966. In 1957 the first wing of the F.H. Odendaal Hospital was opened, almost 30 years after the first moves had been made to obtain such a facility in the town. The hospital was named after the local attorney and estate agent ("Fox" Odendaal), who at that time was the Waterberg's representative on the Transvaal Provincial Council.

In 1974 the Land Bank opened its regional offices in the town in a fine modern building that occupies a whole block on the corner of Kerk and Chris Hani (Tamsen) Streets. Four storeys high, it is the tallest building in town and its monumental architecture makes it an imposing feature of the CBD.[42] Since 2014 it has become the headquarters of the NTK, whose own former head office, built in 1984, now houses the local offices of several provincial government departments.

For over a century, Nylstroom stood astride the only route linking Pretoria with the north (the R101) and its trade and growth benefitted accordingly. In the 1980s the N1 freeway advanced gradually towards, then past (10 km to the east), and

FIGURE 6.17: The Land Bank Building (now the head office of the NTK).

then beyond Nylstroom en route to Pietersburg. This initially had a major negative impact on traffic and trade in the town, but later it resulted in bringing more people into the district as tourists, weekend farmers and traders. In addition, from the early 1980s to the present day, the dramatic growth of Ellisras (Lephalale), 150 km to the northwest (via Vaalwater), brought a huge increase in through-traffic that more than offset the loss of the original traffic on the old Great North Road.

NABOOMSPRUIT/MOOKGOPHONG
Altitude: 1110m amsl
Population (2011): 25 000[43]

The history of Naboomspruit has been recorded three times: once, in 1952, as a short pamphlet to commemorate the tercentenary of the arrival of Jan van Riebeeck at the Cape;[44] more comprehensively in 1985, on the occasion of the 75th anniversary of the founding of the town;[45] and more definitively in 2010, when it celebrated its centenary.[46] Most of the following material is drawn from these sources.

The town derived both its names from the ubiquitous local occurrence of the striking *Euphorbia ingens*, the giant euphorbia, *naboom* (in Afrikaans) or *mookgopho* (Northern Sotho), "a spiny succulent tree up to 10 metres high, with a short stem and massive, dark green crown".[47]

The first settlers

The area was first settled in about 1859 by the Van Heerden family – Nicolaas and Dorothea – who established themselves on the farm Kromkloof, some 20 km to the west of the present town. Dorothea was a sister of Piet Potgieter, who founded the town of that name some 60 km to the north in 1854. The Potgieter family in turn had settled on the farm Middelfontein where the Berlin Missionary Society would build its Modimulle Mission in 1867. The Van Heerdens had ten children, most of whom continued to live on the farm, with the result that by 1907 there were over a hundred people living on Kromkloof, mostly related to the original family.

Another early settler family was the Van Zyls: The patriarch, Jan Kasper, had trekked from the Eastern Cape and in 1885 purchased the farm Vischgat (also known as Bad se Loop on account of a warm water spring on the property), some 10 km northeast of the Berlin Mission. Unfortunately for them, the fine stone house they built on a hill overlooking the plain of the Nyl River was in clear view of the new railway line. When a train was derailed on this section during the guerrilla phase of the South African War, infuriated British troops assumed the house had been involved in some way, demolished it and imprisoned the family.

About 15 km to the northwest of the modern town of Mookgophong is the scenic Bokpoort Pass, one of only a few routes onto the Waterberg plateau. At the top of this pass, on the catchment of the Lephalala River, is the farm Purekrans, which Petrus Jacobus van Rooyen bought in 1895, becoming one of the first permanent residents on this part of the plateau (the Smit family of Nooitgedacht was another). Its altitude rendered it free of the malaria that plagued the Nyl valley. It was not long before other settlers followed Van Rooyen up the pass to the area that became known as Palala. Hans, one of the Van Rooyen sons, returned from the South African War to his own farm, Heuningneskloof, just north of Purekrans. He became a friend of Eugène Marais. A formal road up Bokpoort Pass was built in 1916 – by hand, using pick and shovel, scotch-carts and dynamite.

FIGURE 6.18: Passenger and Government mail coach.
SANA TAB Photographic Library, No. 16309, by H.F. Gros.

The routes to the north

The original route to the north (the Ou Voortrekkerpad, or Coach Road) carried mail in "Spider" coaches and later in the Zeederberg coaches. The first person to have the postal contract was F.C. Eloff, Paul Kruger's nephew; he sold the contract to the Zeederbergs. Along the way, it was necessary to have staging posts for the change of animals pulling the coaches. At these points, stores, inns and bottle stores were established – initially by Natorp – at places like Middelfontein, Badseloop and Naboomfontein. The latter, about 20 km north of where Naboomspruit would be founded, was the site of possibly the earliest store in the region – that of Kaufmann, who established a store and inn on Naboomfontein, a hamlet that became known as "Kaufmann's Place". The stables at his inn were used to accommodate the horses used on the mail-coach service. After Kaufmann was murdered, his store was taken over by Willy Drummond, who in 1886 had established a store a little further north, where the Drummondlea rail siding is today.

Once the railway line to Pietersburg was commissioned (1898/99), the Zeederberg service fell away and the stores, inns and bottle stores relocated to sidings and stations along the railway line. The store/hotel on Badseloop was originally owned by a Dutchman, A. Vroom, and later by Kuschke.

The area east of Badseloop, on Vlakfontein, was frequented by lions and only travelled through during daylight. In 1888 the *Volksraad* allocated money to build a new road towards the north, lying a short distance east of the old coach road. Only in the 1920s were funds allocated to the building of a road across the Nyl floodplain, which necessitated that a stretch of about 8 km was elevated above the floodplain, with storm water drains every 100 metres.

FIGURE 6.19: A section of Road 600 across the Nyl floodplain in the rainy season. Erasmus & Du Plessis (compilers) (1952): p. 6.

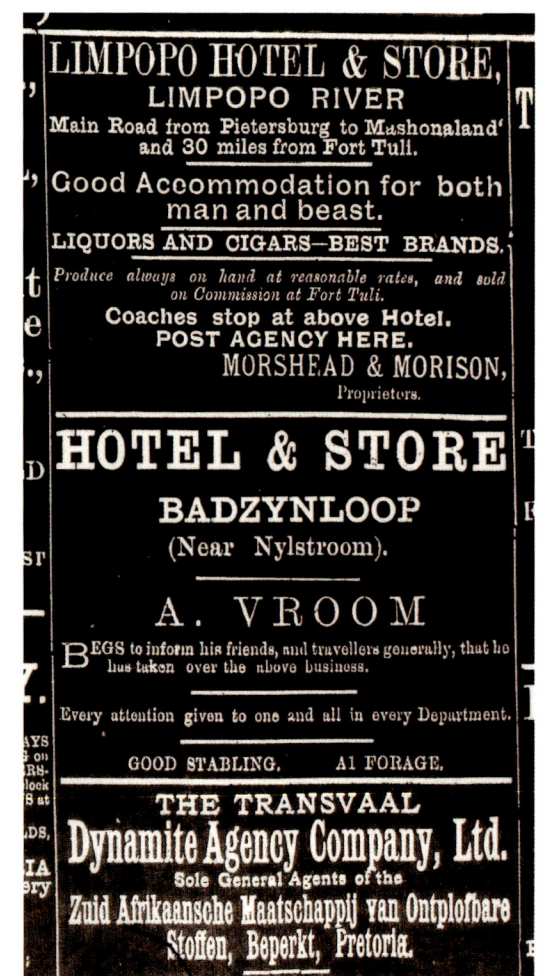

FIGURE 6.20: Vroom's store and inn at Badseloop in the 1890s. Illustration drawn by C. Bleach from original photograph SANA TAB Photographic Library, No. 6615: Vroom's Store, Bad-Zyn-Loop. Photographer unknown. FIGURE 6.21: Advertisement for Vroom's Store in the *Zoutpansberg Review & Mining Journal*, c.1895. Filed at the British Library, Paddington, London, photographed February 2015.

According to a law of 1896, trains were obliged to stop between midnight on Saturday and midnight on Sunday, irrespective of where they found themselves. This was a source of great inconvenience for travellers, who had to find food and water (and somewhere to sleep) for the period. When trains stopped in the middle of nowhere, some held impromptu services, while others (for some reason) borrowed chairs from the toilets to go off into "no-man's land" to see what they could hunt for "the pot".

Two alternative routes led north from Nylstroom: One, developed as a post route before the South African War, ran via Middelfontein (and the Modimulle Mission) and on across the Nyl floodplain past Vroom's store on Badseloop towards the Naboomspruit siding. This was known as Road 600. Another route turned into the hills shortly after leaving Nylstroom and ran up through them – via a cutting called Forssman's Pass – to Doornhoek and the tin mines. Thereafter, it headed in a straight line towards Moorddrift, crossing the road linking the village of Naboomspruit (11 km to the south) and Bokpoort Pass – where the "elephant" garage was later built – and passing Lleweni School. This latter route was for many years the official road to the north, despite being longer than Road 600, which suffered from numerous sandy patches and flooding during the rains. The result though was that Naboomspruit was actually not on the main road between Nylstroom and Potgietersrust.

During the period between the two world wars, many attempts were made by citizen groups in both Nylstroom and Naboomspruit to have Road 600 upgraded and declared as the national road. Nancy Courtney-Acutt, wife of a Nylsvley farmer and the local correspondent for several Witwatersrand newspapers, did much to lobby on their behalf, as did Councillor Bakker in Nylstroom. Eventually, in 1938, the repair and upgrade of Road 600 started, and in 1941 the direct link between Naboomspruit and Nylstroom was finally completed – although tarring of the route was not undertaken until the early 1950s.

Naboomspruit station

In 1898 a rail siding (a train-passing loop) had been built on the new Nylstroom–Potgietersrust line on a piece of flat ground about halfway between the two towns, on the farm Naboomspruit, although no village stood there yet. This location for a siding was chosen despite every effort made by farmers in the Haakdoorn area further along the line (near Kaufmann's Store) to have the siding placed there. In

1903 the little station was manned by the station master, Van Ravenswaay. A few years later, the station burnt down and after a period during which the station master had to live in a railway wagon, a bigger wood-and-iron structure was erected. Communication along the line was by Morse code (and later telephone). The first standalone post office was built next to the station in 1918–1919. A farm telephone line was laid out for 55 km in 1920. Mail was originally carried by coaches and later by train. From the station it was delivered by road to recipients, including the nearby tin mines.

In 1903/4 the Geyser and Engelbrecht families moved onto Vlakfontein (adjacent to the farm Naboomspruit) from their previous farm near Heuningfontein. They and their descendants (notably Frikkie Geyser and Koot Engelbrecht) were to form a large and influential component of future Naboomspruit society.

FIGURE 6.22: Naboomspruit station in the early 1900s. SANA TAB Photographic Library, No. 31746: Naboomspruit station, c.1910.

Actions during the South African War

There may have been several skirmishes in the Naboomspruit area during the war, but only three were of significance. The first involved the murder of members of the Janse family by Masebe clansmen on 20 March 1901 (described in

Chapter 3). Both the other incidents involved attacks on the British-operated railway line by a commando under the leadership of "Captain" Jack Hindon of the Boer reconnaissance corps.

On 4 July 1901 his group derailed a train at Tobias-Zyn-Loop, about 6 km north of present-day Mookgophong. The train was guarded by a detachment of Gordon Highlanders under the command of 22-year-old Lt Alexander Archie Dunlop Best. The derailed train was ambushed by a unit from the commando of General Beyers. Best, at least 15 of his men, three black enlisted men and the driver of the train were killed. Several other men were wounded and/or taken prisoner. There were no Boer casualties. A monument to the fallen stands at the site today; it was unveiled exactly a hundred years after the incident.

Just over a month later, on 10 August 1901, Hindon's men ambushed an armoured train about 7 km south of where Mookgophong is today. Explanations vary (according to the allegiances of the tellers), but the action was a failure for the Boers, who lost seven men that day. It was this incident that caused the British to destroy the Van Zyl homestead on a nearby hill at Badseloop. In 1930 the Afrikaner community erected a monument at Ypres siding in memory of their compatriots who'd died in the incident.

The arrival of the English

One of the first English-speaking settlers in the district was Murray Jackson, who settled on Rietbokspruit north of present-day Naboomspruit in 1901. Another was Jerome Dennison, an Edinburgh lawyer who served in the war and afterwards joined the SA Constabulary for a while. He lived on the flats south of Crecy and grew timber for supply to the mines in the area. Dennison was a founder of the Turf Farmers Association and was very active in agricultural affairs. He was partially responsible for the branch rail line to Zebediela starting from Naboomspruit.

Several English families settled in the area as part of the initiative of the post-war British Administrator, Lord Milner, to establish English-speaking communities in rural Transvaal. Although most put down roots around the hamlet of settlers on the southern Springbok Flats, others moved into the area between Haakdoorn and Hanglip (later partially inundated by the Doorndraai Dam). Names included Finch-Dawson, Robertson, Doyle, Grigson, Cresswell, Quirk, Maddocks, Todd, Turner, Pullinger and Donnisthorpe. During, and especially after World War I, other English families, including several ex-servicemen, arrived, mainly settling on the middle and northern Springbok Flats: the Galpins of Mosdene, the Skirvings (later Whitehouses) of Nylsvley, the Grahams of Weltevreden (Topsy Eschenberg's family), the Courtney-Acutts and others. Jim Galpin later built a magnificent double-storeyed home of Karoo sandstone on Mosdene.

Establishment of Naboomspruit town

The town of Naboomspruit was proclaimed on 8 June 1910 around the rail siding of that name. Its founding was due mainly to the prospecting activities in the area, following the discovery of tin on Doornhoek and Welgevonden by Adolph (Dolf) Erasmus and others (see Chapter 18). At the time, at least a portion of the farm Naboomspruit seems to have belonged to the Waterberg Land Company, its owner being an entrepreneur called Richard Currie. In 1910 Abe Bailey, the mining tycoon, purchased the property and made application for a township through his company SA Townships, having realised the need for a township on the rail line close to the tin deposits his associated businesses were developing in the area. Erven were laid out in a rectangular grid pattern according to the new American style with 232 morgen being set aside for over 800 commercial and residential plots.

The first residents were Cornelius Johannes Lamprecht and his wife Anna Jacoba in 1910, followed by his brother Albertus Lamprecht. The latter had been a transport rider on the route between Beira (in Mozambique) and Rhodesia and on arrival in Naboomspruit, took on the role of delivering goods and equipment to the nearby mines. Next came Gerhardus ("Mossie") Oosthuizen and Willem A. de Klerk and his family in 1912, taking the total white population to eleven. In the following years, they were joined by the families of Hendrik Geyser, Martiens Olckers, Snyman, Griesel and others, until by 1919 there were 130 residents, which was sufficient to justify the formation of a Health Committee.

Eugène Marais

There is probably no name more closely associated with the Waterberg than that of Eugène Marais (1871–1936), the Afrikaans writer, poet, journalist, lawyer, folk doctor – and morphine addict. Yet, in truth, Marais spent almost no time on the Waterberg plateau. For eight years, from late 1907 until early 1916, he lived in its foothills, about 12 km west of the fledgling village of Naboomspruit and 30 km northeast of Nylstroom – well below the ranges of hills that define the plateau edge. Much has been written about Marais, what follows is a mere summary of the events of his life relating to the Waterberg.[48]

Marais first visited the Waterberg in the company of a friend, Dolf Erasmus, who had been associated with the discovery of tin (and would later find platinum) in the hills to the northeast of Nylstroom (see Chapter 18 and the map in Figure 6.25). The principal tin outcrops were on the farm Donkerhoek, soon to become the site of the Union tin mine. Marais had visited the mining camp at Donkerhoek and it was there that he returned late in 1907 for a short-lived partnership to prospect for tin – and probably

FIGURE 6.24: Eugène Marais. SANA TAB Photographic Library, No. 4944. FIGURE 6.23: View from Grootfontein looking across Rietfontein, (the Van Rooyen's farm) towards Hanglip on the Waterberg escarpment.

gold. It was on this farm too that he began to make observations on the behaviour of a troop of baboons and later to conduct a number of arbitrary experiments on them.

Early the next year, after his tin venture had been overtaken by new holders of the mineral rights, Marais moved to the neighbouring farm Rietfontein, where he would remain, on and off, for the next eight years with the owners of the farm, Oom Gysbert and Tant Maria van Rooyen. When Marais came to stay, they were in their fifties and had rebuilt their home and farm lost during the South African War. Gys was multiskilled: a wagon-maker, smith and builder. Tant Maria ("Tamaria") was a well-known nurse and midwife in the district; she recognised that Marais was in need of help to deal with his addiction and agreed to look after him.

Undeterred by his lack of medical qualifications, Marais saw an opportunity to set himself up as a doctor. His life consisted of treating patients with medicine, psychology and hypnosis, studying baboons and termites and making the notes that would form the basis of much of his subsequent written work. What medical skill he did have was based on a combination of common sense, charisma and his experiences from dealing with his own addiction. In a land of ill-educated people where there were no qualified doctors, a persuasive charlatan could easily gain credence and Marais became a local celebrity, *die Wonderwerker*, or Miracle Doctor.

Marais did not stay permanently on Rietfontein, but used it as a convenient base from which to undertake local travels. For a while, he was employed by the tin mine on Doornhoek, in a role that included doctor, engineer and administrator. On another occasion, he stayed for several months in Swaershoek where he nursed a farmer, Andries van Heerden, back to health (see Chapter 4). Then he went to Purekrans, a farm at the top of Bokpoort Pass that belonged to Piet van Rooyen, Gys's brother, who had fallen seriously ill. Unable to assist, Marais called for the help of a proper doctor from Warmbaths, who operated on the old man and prolonged his life for several months.

During this period, Marais made numerous excursions in pursuit of his love of natural history. On one of them he came across an unusual plant, the seeds of which he recognised as ones that had been found to make local children ill when eaten. It is possible that he planted seeds of the plant, for in 1927, long after he had left the Waterberg, Marais sent a sample of the plant to Dr Rudolf Marloth (1855–1931), the eminent Cape botanist and author of the magisterial six-volume series *Flora of South Africa*.[49] Marloth immediately identified the plant as a cycad, but, thinking that cycads, a feature of the Drakensberg range, could not grow so far to the west, he labelled the origin of the sample as Nelspruit (rather than

Nylstroom!). Shortly before his death in 1936, Marais visited his niece Inez Clare Verdoorn (1896–1989), a botanist at the National Herbarium in Pretoria, and told her about his cycad discovery years before. Intrigued, she found the original sheet on which Marloth had entered the origin of the plant and provisionally changed its location to "Waterberg", subject to visual confirmation. Marais undertook to take his niece to see the locality but died before he could do so.

In 1942 Verdoorn – accompanied by two other botanists from the Herbarium, one of whom was working in the area – eventually visited the southeastern Waterberg escarpment (west of Bokpoort Pass on the farm Vlakfontein, which Marais had visited 30 years previously) where they were shown (by a relative of the Van Rooyens) two cycad plants corresponding to the sample Marais had provided. Verdoorn later produced a scientific description of the species and named it *Encephalartos eugene-maraisii* in memory of her late uncle.[50]

Piet van Rooyen's son, Hans "Purekrans" van Rooyen, became a firm friend – one amongst many for Marais could be a charming, sociable man. Marais made several friends also in the neighbouring mining community on Kromkloof, in Naboomspruit and in Nylstroom. In the latter town, Emil Tamsen was a good friend, as was Ebrahim Ravat, a Muslim trader who shared his morphine habit and was able to keep him supplied. (Ravat's grandson still runs a general dealership in Modimolle.)

During 1911 an outbreak of east coast fever in the Waterberg necessitated the appointment of Resident Justices of the Peace (RJP) in each of the areas demarcated for quarantine. From June 1911 Marais was installed in an office (at Rietfontein police station) with two constables, a short distance from the Van Rooyens' home. It was as RJP that Marais travelled to the north of the Waterberg to visit the Herero refugees living there on the farms Groenfontein and Nachtwacht (see Chapter 20). He seemed to empathise with their situation, but was unable to offer them a solution. In September 1914 when the outbreak of World War I precipitated a rebellion amongst a faction of the Boer community (see Chapter 15), Marais supported them, although as RJP he was not allowed to take part in their short-lived revolt. But he did make his horse available to the local rebel leader – and, amazingly, the government later returned it to him.

At the end of 1915, following another failed attempt to break his morphine habit, a weakened and ill Marais was forced to leave the Waterberg for good to be treated at the home of his sister in Pretoria.

The Galpins of Mosdene

Ernest Galpin (1858–1941) bought several farms, totalling 500 ha between Naboomspruit and Crecy, in 1913 and built a house on one of them (Roodepoort) which he called Mosdene. Galpin had been born in Grahamstown and became a bank manager in Port Elizabeth, Barberton and eventually back at Queenstown in the Eastern Cape. From an early age, he developed an interest in collecting plant specimens, a hobby that he would maintain throughout his life. When he retired in 1917, he donated his collection of over 16 000 specimens to the National Herbarium in Pretoria. Galpin's wife, Marie Elizabeth (née De Jongh), shared her husband's botanical interest and was also a useful illustrator.

In 1920 Galpin was commissioned by Dr I.B. Pole Evans, the Chief of the Division of Botany and the Director of the Botanical Survey of the Union of SA, to undertake a botanical survey of the Springbok Flats. This work resulted in the publication of two works: *The Native Timber Trees of the Springbok Flats* (Memoir No. 7 of the SA Botanical Survey, 1924) and *Botanical Survey of the Springbok Flats* (Memoir No. 12, 1926).

Marie Elizabeth died in 1933 of a heart attack during one of their many collecting expeditions together. He continued collecting, sometimes with the assistance of his son E.A. (Jim) Galpin almost until his death. More than 200 plant species are based on his discoveries and many bear his name. Two of the best known are *Acacia galpinii* (the type specimen of which was on Mosdene) and *Bauhinia galpinii*. There is also a genus of small shrubs, *Galpinia*, named after him.

The Galpins' son, Jim, served in East Africa in World War I, but was invalided out in 1917 suffering from malnutrition and malaria. On his return to SA he attended the Potchefstroom Agricultural College, but left before completing the course to return to Mosdene to help his father. In 1928, in Naboomspruit, Jim married Peggy Courtney Acutt (also given as Rankin), the daughter of Renault and Nancy Courtney Acutt (née Lovelock). Peggy was born in Buenos Aires where her grandfather, an engineer, had been responsible for building the harbour. Her father, Renault, became a well-established cattle breeder on the western Springbok Flats, from where her mother Nancy was a longstanding rural correspondent for *The Star*.[51]

During World War II, Jim served as a staff officer with the SAAF, in charge of training pilots about armaments. In his absence, his wife Peggy managed the farm with great success – and afterwards retained responsibility for the herd of 2000 cattle.

The Galpins had three children (two sons and a daughter). Jim shared his father's interest in botany, but was also fascinated by birds (over 400 species on Nylsvley) and archaeology. Jim and Peggy celebrated their 50th wedding anniversary on Mosdene in March 1978. Jim died in Naboomspruit in 1982, aged 84. At the time of his death, Jim was apparently working on a book about the history of the Galpin family in South Africa and that of the Naboomspruit District, to be called *The Naboom Story*.[52]

It was never published.

The farm adjacent to Mosdene, Nylsvley, had remained in the Skirving family since its purchase in 1914. From 1946 to 1974, it was run by a grandson, George Whitehouse. In 1974 the farm was purchased by the state and became a nature reserve managed by the Transvaal Division of Nature Conservation (now LEDET).[53]

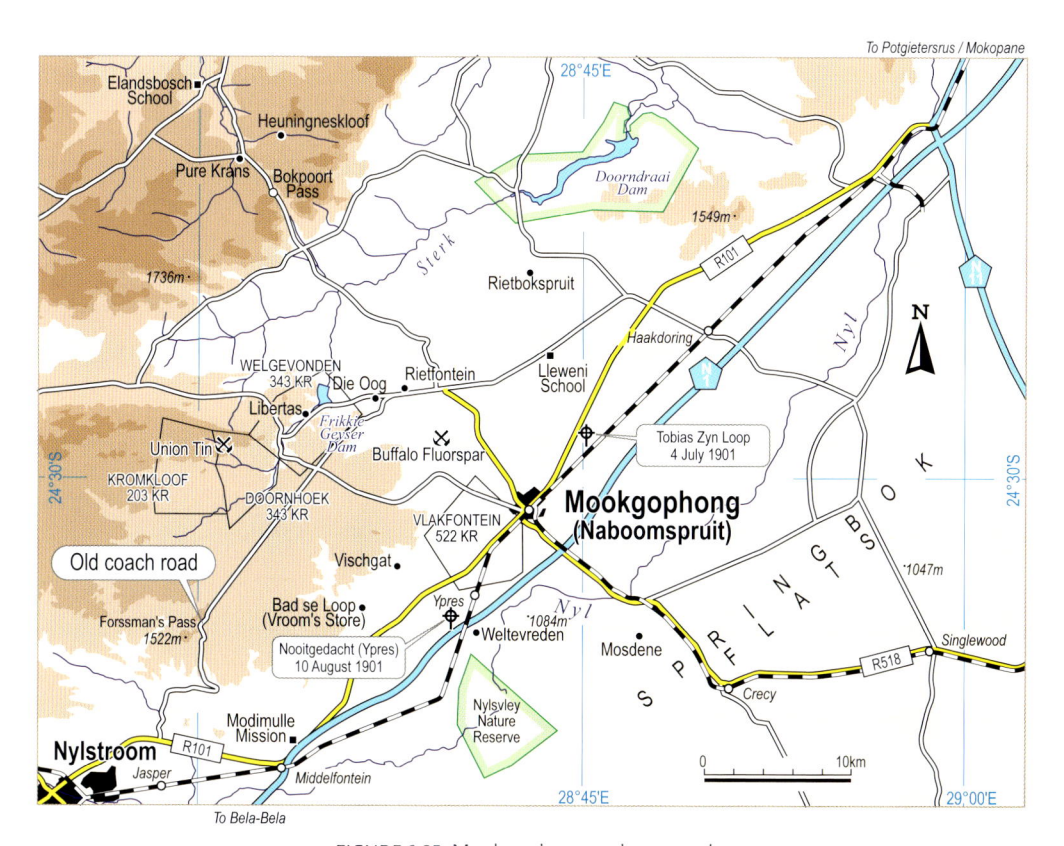

FIGURE 6.25: Mookgophong and surrounds.

Naboomspruit develops

The village's first Health Committee included Gert Smit from Roedtan (chairman), Danie Erasmus and Willem de Klerk (secretary). Its first premises were built in 1921. De Klerk (whose job would become that of town clerk and who would later also be the Waterberg Member of the Provincial Council) was followed by Lewis and later by George Todd. The town grew slowly but steadily through the succeeding years and by 1938 there were 650 residents – leading to the decision to elect the first Dorpsraad or Village Council. Those elected were Dolf Erasmus (mayor), G. Todd (town clerk), and councillors P.J. van Rensburg, J.B.H. Herbst, S. Freeman, H. Devonport, A.S. Lamprecht (later the mayor) and Z. Jaffet.

A Nylstroom trader, Philip Cohen, was quick to see the potential of the new settlement and after his Nylstroom shop burnt down in 1910, he arranged to have a two-roomed wood-and-iron structure built in the new village. Shortly afterwards, he built next to his store the hotel that still stands today. Edgar Armstrong was manager of the hotel in the early days. George Meads, a former policeman, followed him in the 1920s before going off to become a ranger in the Kruger Park. Cohen went on to establish a small butchery and a mill in the village

FIGURE 6.26: Naboomspruit's first town council (1938).
Standing at the back (from left to right): P.J.J. van Rensburg, J.B.H. Herbst, H. Devonport, Z. Jaffet. Sitting in the front (from left to right): S. Freeman, Dolf Erasmus (mayor), A.S. Lamprecht. Front: G. Todd (town clerk). Erasmus & Du Plessis (compilers) (1952): p. 10.

FIGURE 6.27 (TOP): View northwards up Louis Trichardt Avenue (now Nelson Mandela Str.) in Naboomspruit in the 1920s, with the hotel to the left of the blue gum trees and Cohen's store beyond. (BOTTOM) View from hotel veranda. Museum Africa Photographic Library, Nos. PH2006-10054 & PH2002-373C.

in 1918. The mill was demolished in the 1980s. Another prominent businessman was Barney Reichman, who, about a year after Cohen, also opened a store and built a mill. Around 1940, a much bigger mill was built for Busch and his son-in-law Schmedeskamp. An old resident described the scene outside the mill:

> It was a wood and zinc construction, equipped with sophisticated milling technology and could easily fulfil the anticipated needs of the times. With its distinctive roar, turning wheels and clapping drive-belts, it became a wild place [when in action]. Ox-wagons, horse wagons, donkey carts and even occasional motor lorries periodically came to a standstill in front of this throbbing heart of Naboomspruit. Full sacks of grain on the cement slab in front, and white bags of meal in the mill inside, the white dust clinging to everything, the bustle all around, full bags, half bags, empty bags, and above the monotonous roar, the loud shouts of those giving and receiving instructions – that was the mill.[54]

In 1945 the mill was bought by Charlie Willard, who was not only an excellent miller and popular personality, but also a leading boxing coach.

Mining and agriculture formed the basis of Naboomspruit's economy (and even the reasons for its existence) from the outset. Both needed a reliable supply of provisions. Philip Cohen, ever the entrepreneur, opened a hotel and shop on Doornhoek (the site of the Union tin mine) in 1917.[55]

In the same year Walter Cresswell, a farmer and South African Party politician, decided that the village needed a multipurpose hall. He raised funds from local residents and paid David Anderson to build the structure, which had a proper stage, dressing rooms and high windows. The Anderson Hall was used extensively for church services, political meetings, parties, cinema shows and other events.

Barclays Bank opened an agency in the town in February 1924. In 1926 it became a sub-branch under Nylstroom and a full branch in 1927.

Law and order in Naboomspruit

Towards the end of the South African War, Milner's government had established, under a concept attributed to Gen. Baden-Powell, the South African Constabulary (SAC), a police force of about 6000 men for the Transvaal and Free State. In 1906 the SAC was reconfigured as the Transvaal and OFS Police. In 1902 an SAC post had already been established on the farm Rietfontein.

In 1913 the provincial police were incorporated into the South African Police. In 1919 a "C"-class police station was established at Naboomspruit, in a two-roomed thatched structure that belonged to Edgar Armstrong, a local hotelier. The main room was used inter alia as a court for RJP Gert Smit's visits from Roedtan. With the closure of the Rietfontein police station later that year, the personnel were transferred to Naboomspruit, which became a "B"-class station, with Sergeant Keen in charge and Constable Coetzee on the staff. The station was rebuilt in 1921. Meanwhile, Const. Coetzee was transferred to Nylstroom as Public Prosecutor. He moved quickly up the promotional ladder to Lieutenant in 1930, Captain in 1932 and District Commandant at Ixopo in 1937. By 1939 he was a Major in Durban. After service in North Africa during the war, Coetzee returned to service in the SAP, and retired in 1953 with the rank of Major General.

From 1926, Gert Smit became Naboomspruit's first Resident Justice of the Peace. The Cousyn brothers were probably the first lawyers to open a practice there, although Zimmerman & Jaffet also opened a practice in the town; Joe Shandoss (later town clerk) served as the Public Prosecutor, working alongside Constable (later Sergeant) Jan Marais (from 1936). Marais and his horse, Zulu, were well known in the area in the 1930s. They did all the horse patrols, covering over 300 km per month.

In his unpublished history of the area, E.A. (Jim) Galpin recounted the following:

> On 26 March 1914, the postcart en route to Welgevonden mine was held up while crossing the drift at Sandspruit, by a man with his hat pulled down and a scarf over his face and armed with a rifle. Mr Rigby, the Mine Secretary, was made to hand over the payroll (£620). The driver of the wagon was forced to outspan the mules and drive them off. Nobody was to move for an hour – an accomplice watched over them with a rifle from a nearby hill. The same evening, a previously penurious man, Tromp van Dyk, flashed a brand new £5 note in the bar of the Naboom Hotel and stood drinks all round. [Someone] alerted the police and Van Dyk was arrested, handcuffed and locked up in the hotel. He broke away and dashed off into the dark with everyone in hot pursuit. He almost got away by diving through the window of a house across the railway line - but landed on a couple asleep in bed and their outcry gave

his position away. The money, quite a fortune in those days, was pointed out … hidden under a bush behind the hotel. It was intact, except for the amount spent in the pub.

Van Dyk was sentenced to 13 years' imprisonment.[56] The town obtained its first resident magistrate, J.M. Marx, in 1963 only.

Railway extension to Zebediela

After World War I, loans from the Land Bank enabled farmers to settle in the northern part of the Springbok Flats. This area was isolated, farms were far from one another, there were no roads or any other form of communication and the black turf made travel impossible in the rains. Fortunately, among the settlers were several men with vision and organisational skills, people like Jerome Dennison, Tom Cresswell and Toby Berrange. Berrange, a former secretary to the Minister of Railways, had retired to a farm near Naboomspruit. He seized on the development of the Zebediela citrus project as a motivation for the construction of a light, narrow-gauge railway line. This was the brainchild of Major Frank Dutton, an engineer in the railways.

The Stronach-Dutton road-rail system consisted of a tractor which could run on rails, or alternatively, jack itself off the rails, undergo a change of wheels and then run along the hard surface on each side of the rails. Its advantage was speed and low construction costs. Known as the *Songolole* (millipede), it was approved in July 1923 and opened in April 1924, running from Naboomspruit to Singlewood, a distance of 32 km. Within three years, traffic on the line had grown so much that it was replaced with a standard-gauge conventional line and train set, which was taken on as far as Zebediela where by 1926 the first citrus crops were picked – more than 30 000 cases per day. This in turn led to the expansion of the station at Naboomspruit. By 1984, the line was handling 50 000 tons of Zebediela oranges a year (many will still remember the famous "Outspan" brand) and the line was converted to diesel. The subsequent collapse of the whole Zebediela enterprise, following a "successful" land claim on the property at the turn of the century resulted in the closure of the line in 2003.[57]

The traffic along the main line also grew rapidly and by 1984, when it was electrified, 2700 passenger trains passed through Naboomspruit each year (over seven per day) as well as almost half a million tonnes of general goods traffic, plus 10 000 animals and thousands of trucks loaded with Venter trailers (see below). The first electrically powered train passed through Naboomspruit on 4 April 1984.

Naboomspruit's growth after World War II

The completion of the gravel road from Nylstroom in 1941 placed Naboomspruit on the main arterial route to the north and this, together with its railway station, contributed strongly to its growth after the end of the war. A face-brick municipal building opened in 1947, followed by an electricity generator in 1949. Volkskas Bank opened a branch in Naboomspruit in 1951. With increased traffic (rubber-wheeled, as opposed to horse-drawn) came the need for petrol stations – Pegasus, Caltex and Shell, with their hand pumps next to the road: Princess Garage, Van's Garage, Gerbers Garage. The town was allocated the vehicle registration identity TNS (Nylstroom was TAH, Warmbaths TWB and Louis Trichardt TAJ – which people from the south of the country said stood for "Transvaal Anderkant Johannesburg"!).

By 1950 the town's white population had grown to 1200. Water had now become a problem. In 1958 the Frikkie Geyser Dam (named after the mayor of the time) was opened on a tributary of the Sterk River near the Libertas resort, following long negotiations that involved Geyser, Hodges, J.G. Strijdom (the MP for Waterberg) and Percy du Toit, the manager of a nearby fluorspar mine. In 1959 a black township was proclaimed to the east of the town.

Another entrepreneur in the town was Maarten Venter. Born in Germiston, he was apprenticed as a young boy to several local engineering firms, working (for example) as a welder for Dorman Long during World War II. Eventually, he set up his own small welding business in Primrose, later joining forces with a Dutch wagon-maker to start manufacturing trailers. Then he tried farming, but this didn't work, so he began making farming implements. Eventually, he happened to meet Dolf Erasmus, who persuaded him to leave the Free State and move to Naboomspruit to open a sheet metal engineering works there. Soon, he began making light caravans and, from 1961, when his son Jasper took control of the business, baggage trailers for cars. Today the town is famous countrywide for its ubiquitous Venter trailers. (Transvaalers on holiday in the Cape even became known disparagingly as "Venters"!)

In 1965 the impressive municipal offices built in 1947 were almost totally destroyed in a fire. They were replaced ten years later by the current Civic Centre, and in 1980 the present-day post office was opened.

Naboomspruit obtained town council status in 1981, which was also the year in which the town received electricity directly from Eskom and when the wall of the Frikkie Geyser Dam was raised. Most prominent among the various councillors and mayors were J.B.H. Zerbst, who was mayor from 1942–1947 and again from 1960–1962; M.F. (Frikkie) Geyser (1952–1960); and S.E.S. Ferreira, who was mayor from 1972–1974 and 1981–1985. He was also the Waterberg Member of the Provincial Council (LPR) from 1981 and was the chief whip of the Conservative Party. Another prominent official was Joe Shandoss, who succeeded George Todd as town clerk in 1941 and occupied that position for 30 years until 1970. The last mayor under the National Party administration was Koos Opperman (1992–1995); he was succeeded by R.M. Kekana in the transitional administration and by Councillor J. Rekwale after 1995.

FIGURE 6.28: The centenary cenotaph in front of the Mookgophong Civic Centre.

Education and healthcare

The first government school in the district had been on the farm Kromkloof and several early residents, for example Koot Engelbrecht, Bennie Marais and Petrus Geyser, went there from around 1908. When Cornelius Lamprecht came to Naboomspruit in 1910 – when there was only the station and Cohen's shop/inn – he bought a small shop building to use as a home. Lamprecht had children who needed schooling, so he built a zinc room onto the side of the structure to serve as a classroom. The "school" opened on 3 August 1910 with a Dutch teacher called Kromhout – who lasted only two months! When it reopened in October, T.A. Grundlingh was the teacher with 24 children on the roll. The first inspector's report noted that: "The school building is in an iron room, part of a dwelling house, stifling and very unsuitable." A better building was constructed in late 1913 when there were 41 pupils up to Standard 5, with Eduard Langschmidt (from the Cape) as principal and H. Geyser as chairman of the school committee. The next year, with pupil numbers at 47, a second teacher was appointed, both teaching in the same room. (A second classroom was built in 1916.)

In 1927, as the school grew, a third teacher joined the school: J.B.H. Zerbst. He and Langschmidt were to spend their whole careers at the school. By 1944 the Laerskool had 414 pupils. Langschmidt, who also earned a reputation as a great organiser of events in the town, was the superintendent and Zerbst the vice-principal (whilst also fulfilling the role of town mayor for five years). Langschmidt retired in July 1949 after 36 years as principal when there were 486 pupils at the school, which had since developed a good reputation. Zerbst remained, eventually becoming principal in 1966, but retiring in 1969 after 42 years with the school. In 1976 the school was renamed the Laerskool Eugène N. Marais. By the mid-1980s, the school housed 450 pupils, with a teaching staff of 22. In later years, these numbers dropped.

The need for a high school developed as soon as the first children of the Laerskool reached Standard 6 (in 1913), but it was not until 1951 that the Laerskool was permitted to extend its classes to become a junior high school. The first group of 13 matriculants graduated in 1957 when J.B.H. Zerbst was still vice-principal of the whole school and P.W. Botes, who had succeeded Langschmidt in 1950, was head. The following year the two sections of the school were split and a formal high school was created. The school was named the Hoërskool Hans Strijdom ("Hansies") and was opened in 1960 by Strijdom's widow Susan. In 1973 a decision was taken to focus on technical skills to attract learners. By 1974 the school had over

500 pupils. The year 1980 saw the retirement of D.E. Erasmus, who had been head since 1965 and teaching in Naboomspruit for 26 years. In 1986 Mauritz Hansen, a Naboom *seun* and former pupil at Hansies, was appointed head; the school had grown to about 650 pupils and had a staff of 46 (pupil numbers dropped to about 530 by 2000, and to under 450 by 2006). During the succeeding decades, Hansies became well known for its sports prowess and for the school's rugby tournament held there each year. By 2018 it was hosting over 80 teams at the event, making it one of the largest rugby tournaments in the world.

In 2007 the Euphorbia Christelike Onafhanklike Skool (an independent, or private school) opened in the town.

Health was a major concern on the Springbok Flats and around Naboomspruit in the early years. Many lives were lost through exposure to a wide range of illnesses that sometimes assumed epidemic proportions, like influenza and malaria. Bilharzia was also a serious condition. The influenza epidemic of 1918 took the lives of 142 000 people in the country. However, the biggest danger in the northern and eastern parts of the Transvaal was malaria, which claimed thousands of lives (see Chapter 12). An outbreak of epidemic proportions in 1926 forced the Department of Public Health to intervene. The swampy Nylsvlei area was identified as a major breeding ground for mosquitoes. But it was only after a second serious epidemic in 1944/45 that funds were made available to tackle the disease. Quinine was the treatment.

After Eugène Marais's departure in 1917, there was no local medical skill. From 1919, a doctor called McPhearson, with a practice in Potgietersrust, travelled to Naboomspruit by train on a regular schedule, stopping at sidings along the way to treat patients who were waiting for him (train schedules were more relaxed then). He was succeeded in 1920 by a Dr Kennedy, also from Potgietersrust, who followed a similar schedule. The first resident doctor in the town was Dr Joel from 1930. In 1926 Gerrit Bakker, the pharmacist from Nylstroom, established a medicine depot in Naboomspruit. Bakker died in 1947 and was succeeded by his son, Anton, who in 1955 sold the depot to Willem van Heerden. As owner/chemist, he developed the business and delivered a valuable service to the community. From 1925 the first resident nursing sister, Sister Van de Velden, served the town also as a midwife; she was followed in this role by Sister Miemie Marx.

When the Prince of Wales toured South Africa in 1925, his train stopped briefly at Naboomspruit station. He was presented with a platinum lapel button with material from the newly discovered deposit at Welgevonden. There were subsequently numerous other festivals: the Louis Trichardt symbolic ox-wagon trek in 1938; the Van Riebeeck festival in 1952; the celebration of 50 years since becoming a Union in 1960; and in 1961, that of South Africa's conversion to a Republic.

The mineral springs

When Eugène Marais moved in with the Van Rooyens on Rietfontein in 1908, the warm mineral springs nearby were already in use. Around the *"oog"* where the water bubbled out, was a pond or tub where the vegetation had been removed. The farm was eventually bought in 1932 by O.J. Viljoen from Ficksburg, who decided to develop the sulphurous spring as a small resort. He dug a hole about 2 metres square and 1,5 metres deep, channelled the spring water into it and surrounded the pool with a reed screen. It was known as Die Oog. The property changed hands in 1934 when it was purchased by a Potgietersrus teacher, Mr Bergh, who developed three circular cement swimming baths, together with a small café, change rooms and toilets. The place became known as Bergh-se-Bad and operated well until the Rondalia Group bought the property in the 1960s and greatly expanded the springs into a large chalet and caravan park resort.

On 10 August 1965 a big function was held to mark the opening of the Rondalia resort with speeches by the chairman of the Rondalia Group, the Member of the Provincial Executive, the MP for Waterberg (Jooste Heystek), and by the mayor of Naboomspruit, chemist Willem van Heerden. The latter allegedly stated in his speech that "Naboomspruit is 'n kak dorp" [a shitty town]. This caused a furore that quickly precipitated his resignation, although he insisted that what he'd actually said was to thank Rondalia for putting "so 'n ou twakdorpie" [crummy town] like Naboomspruit on the map! This unmayoral comment made national headlines, including the following comment in a Sunday newspaper:

> The announcement threw [the town] into an uproar. There was a run on the bank, mothers hid their children, strong men turned pale and a long line of refugees could be seen fleeing to Bronkhorstspruit. Down the streets, along the alleys, in the pubs, the shocked whispers flew:
>
> Said Dominee Ginsberg
> To Mrs van Rinsberg

> Have you heard? Have you heard?
> The Mayor has used
> A three-letter word.
>
> Replied Mrs van Rinsberg
> To Dominee Ginsberg
> I am shocked, I am rocked,
> In fact I am pained
> Let the Mayor be unfrocked
> Or promptly de-chained. [58]

In 1942 O.J. Viljoen purchased a second property nearby on which there was another hot spring and which he also developed into a resort, known as Viljoen-se-Bad. Several other resorts were also developed in the area around small thermal springs; several of them operate to this day. Close by, on Driefontein, is the spring mound Wonderkrater, used by people 100 000 years ago (see Chapter 2).

WARMBATHS/WARMBAD/BELA-BELA
Altitude: 1130m amsl
Population (2011): 45 000[59]

In 1952 a pamphlet was compiled on the occasion of the Van Riebeeck tercentenary, describing the history of the town of Warmbaths and surrounds. This is the only official publication dealing with the history of the town.[60] Despite the name of the author (J.J. McCord), the pamphlet reflects the predominant ideology that prevailed across the northern Transvaal in the 1950s:

> Warmbaths was born from the craving for liberty, out of the love for a homeland and its people brought here by the Voortrekkers. It must be remembered, that Warmbaths is a milestone on the road of South Africa. Also to be remembered is that South Africanism, South African Nationalism, is not a question of race or language, but something of the heart and spirit! In my own case, I stem from Scotland (McCord) and Holland (Potgieter), but South Africa is my fatherland. Every white who gives whole-heartedly to South Africa is a member of the South African Nation.[61]

And so on, for two pages, before getting down to the history of the town! (In fairness to Mr McCord, he was married to Johanna Hendrina Potgieter, the granddaughter of Boer General P.J. Potgieter.)[62]

In 2003 a group of Wits students who studied the architecture of the town compiled a comprehensive report that included extensive reference to municipal records.[63] This was most fortunate, because only a year later, the municipal offices, including the records room and library, were largely destroyed by fire, causing the loss of invaluable and irreplaceable data. The description of the town is drawn largely from these two scarce sources, augmented by others where relevant.

The legend of Coenraad de Buys

The existence of the thermal springs at the junction between the expanse of the Springbok Flats and the first hills of the Waterberg ranges will have been known for many thousands of years, but the first "white" people to visit them and note their existence are likely to have been members of the early trekker groups. Local legend has it that the springs were first visited by members of the famous (or notorious) Coenraad de Buys clan. However, a recent comprehensive article makes it clear that this was not the case. Members of the Buys family did not arrive in the area until about 1840, by which time the springs were already well-known. Buys, who was born in 1761 near Swellendam, was evidently a restless soul who could not endure authority. In his early twenties, he lived with a woman of mixed race in the Swellendam area, and together they had seven children. Later, after the acquisition of the Cape Colony by the British, he moved to a quit-rent (tenement) farm on the Bushmans River (between present-day Grahamstown and the Addo Elephant National Park). In around 1789 Buys was involved in several frontier disputes between the Boer settlers and various Xhosa clans. His actions were partially responsible for the Second Frontier War. Later, his failure to pay rent resulted in the Governor of the Cape Colony banishing him, with a reward of 100 rix-dollars on his head.

Crossing the border, he had an affair with the mother of a local chief (Geika) before marrying a sister of the Zulu induna Mzilikazi, who bore him five sons. He then returned to the Langkloof for a while, until the animosity of his neighbours forced him to move north with his large family.

The German traveller Henry Lichtenstein met Buys in 1803 and described him as follows:

> His uncommon height, for he measured nearly seven feet, the strength, yet admirable proportion of his limbs, his excellent carriage, his firm countenance, his high forehead, his whole mien, and a certain dignity in his movements, made altogether a most pleasing impression. (...) We found in him ... a certain modesty, a certain retiredness [sic] in his manner and conversation, a mildness and kindness in his looks and mien.⁶⁴

But despite his evident charisma, Buys seems to have been an incorrigible brigand. The resumption of British rule in the Cape in 1806 forced him and his family to leave the Cape again, with an even greater price on his head. They moved further north. Near the confluence of the Orange (Gariep) and Vaal rivers, he sowed more dissent, this time among the Korana communities settled around the mission station at Klaarwater (Griquatown)⁶⁵ before moving away again towards the northeast. He is said to have travelled down the Limpopo valley in about 1821, finally settling in the Soutpansberg area, his faithful first wife dying from malaria along the way.

It seems that one night, well into his sixties, Buys walked off on his own and was not seen again. It is presumed that he too succumbed to malaria, although there were many rumours about his disappearance. Two of his sons worked with Louis Trichardt when the latter arrived in the Soutpansberg in 1837; but when Trichardt left for Delagoa Bay (dying from malaria on arrival), the rest of the clan were forced to flee once more from hostile Venda. They eventually settled on the prominent hill a couple of kilometres to the northeast of the present town of Bela-Bela.

A popular legend is that after arriving at the hill, the clan was besieged by AmaNdebele, who intended to cause their death through thirst. The clan leader, Gabriel, one of Buys's sons, responded by emptying a container of water out in full view of the besiegers, declaring that he could call upon a higher authority to ensure that there would always be plenty of water for his people on the hill. Dispirited, the AmaNdebele abandoned their attempt.[66]

The hill, known appropriately as Buyskop, is built of hard, reddish Karoo (Clarens Formation) sandstone, an unusual occurrence in this region. The main railway line north from Pretoria runs at the foot of the hill and in 1911, following a favourable report from the Geological Survey of South Africa, material from the hill was quarried for use as part of the cladding on the Union Buildings in Pretoria as well as other buildings.[67]

The arrival of the main body of trekkers in the 1840s prompted surviving members of the Buys clan to move back towards the Soutpansberg again, sometimes harried by trekker parties. There is a defile along the eastern escarpment of the Waterberg called Basterspas, down which the clan (disparagingly named Basters because of their mixed race) is alleged to have escaped.[68] They finally settled at Buysdorp, near Mara, at the foot of the Soutpansberg range, after one of Coenraad de Buys's sons, Michael, persuaded President Kruger in 1888 to grant the clan 11 000 morgen in perpetuity for the services various clan members had rendered to the ZAR over the years. The descendants of Coenraad de Buys live on the property to this day.[69]

The attraction of the hot springs

The first Boer settlers in the area included the Janse van Rensburg and Van der Merwe families. Soon the warm springs acquired a reputation for their healing powers, causing others to visit them. McCord stated that in 1852 the trekker leader

Figure 6.29: Buyskop outside Bela-Bela on the road towards Modimolle.

Andries Potgieter, after abandoning Ohrigstad, and founding a new town in the Soutpansberg (Zoutpansbergdorp, later renamed Schoemansdal) fell ill. He was brought to the hot springs, then known as Het Bad, for treatment. While there, he was visited by Ds. Andrew Murray Jr. However, neither the thermal waters nor Murray's ministrations were successful, for Potgieter died that December.[70] Another record suggested that Potgieter had brought his wife Christina to the springs in 1848, and that she succumbed to flu and died there.

Apparently, the area surrounding the springs was first cleared by two early trekker settlers, Grobler and Van Heerden, who made it possible for visitors to access the springs and to dig their own holes in the muddy ground, enabling them to sit in the therapeutic waters. This practice apparently was responsible for the name attributed to the springs by the local Tswana-speaking residents: *bela-bela*, or "boiling, boiling", loosely translated as meaning "he who cooks himself".[71]

The baths were located on the farm Het (De) Bad 1198. It was flanked on the west by Roodepoort, on the southwest by Nood(s) Hulp , and on the southeast by Turfbult. The first registered owner of Roodepoort, from April 1859, was Jan Adriaan Pieter Grobler. Cornelis Jacobus Minnaar was the first to own Het

FIGURE 6.30: The original warm springs at Het Bad, c.1888. SANA TAB Photographic Library, No. 16221: Waterberg Hot Springs (Warmbad), by H.F. Gros, c.1890.

Bad (1861). On Nood(s) Hulp lived the Swanepoels, members of the Jerusalemgangers. Roodepoort was the pantry of the area, according to McCord, whose grandfather once owned the property on which he produced vegetables and fruit and operated a water mill on the Platrivier that ran through the farm.[72]

In 1852 the first political control was introduced, with the appointment of H.J. van Staden as field cornet, followed in 1858 by the appointment of Johannes Stephanus Viljoen as Commandant of the Waterberg region. The springs were visited by President Burgers in 1873. He was sufficiently impressed by the benefits and potential of the resource that he managed to persuade the Volksraad to purchase the farm De Bad for the State (but only after threatening to use his own money for the purpose). The popularity of the springs continued to grow, with visitors camping in the vicinity of the water in their wagons because of the absence of any other facilities.

Establishment of Hartingsburg

On 8 June 1882 the Volksraad passed a resolution to establish a township in the vicinity of the baths and suggested that it be named Hartingsburg in tribute to Professor Pieter Harting, a Dutch academic biologist who had championed the Boer cause in Europe during the recently ended period of British annexation (1877–1881). The township was proclaimed by President Kruger on 14 December 1882, the area to include the farms De Bad, Roodepoort, Nood(s) Hulp and Turfbult. The township itself was laid out in a strict north-south-oriented rectangular block of 150 stands, located about 2 km to the southwest of the springs and just east of the Plat River, on Roodepoort.[73] Apparently most erven were sold, but because of the competition in development from Nylstroom only 28 km away (where the Landdrost had been stationed), Hartingsburg did not take off. By the time the railway line arrived from Pretoria in May 1898, Hartingsburg was already a ghost town, and the station for the warm springs (Warmbad) was built 2 km to the east, where the line, having been laid almost due north from Pienaarsrivier, made a sharp turn northeast towards Buyskop. Today, there is no trace of the embryonic town of Hartingsburg and the area is covered by smallholdings and orchards.

Despite the failure of Hartingsburg, the warm springs continued to grow in popularity during the 1890s, although the conditions remained primitive. Frederick Young described these in 1889 as "miserable and wretched receptacles called baths", which attracted a "motley crowd of individuals".[74] Nevertheless, a visit

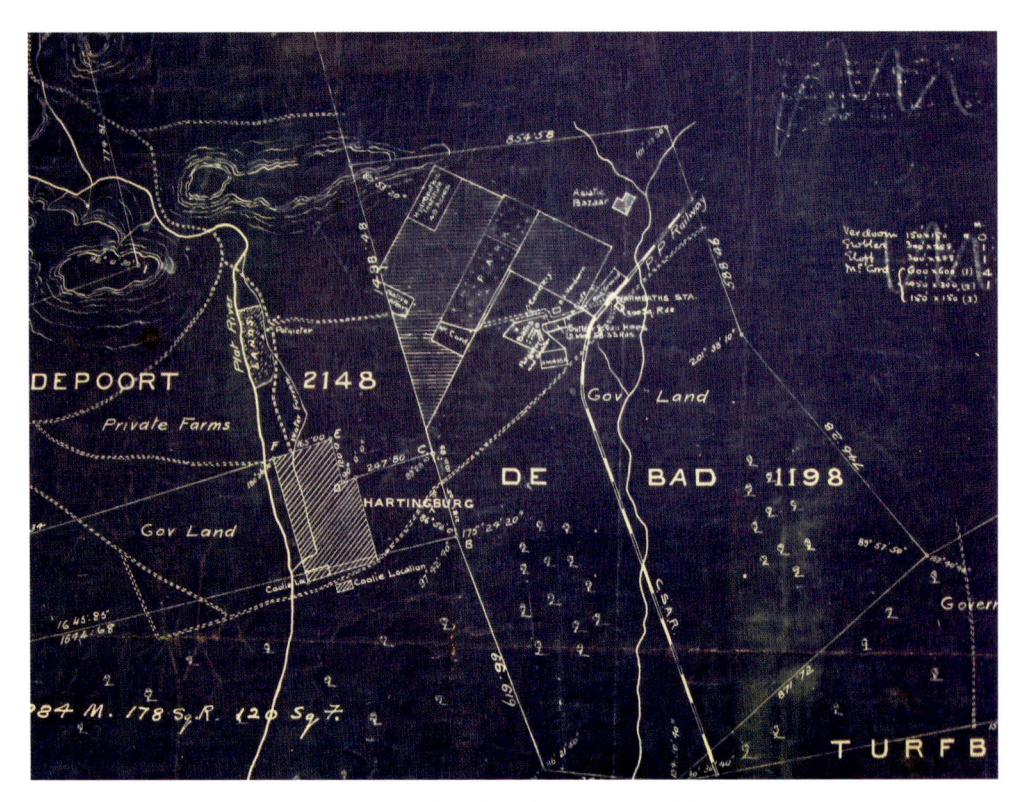

FIGURE 6.31: Town plan of Hartingsburg, probably c.1903.

FIGURE 6.32: The Rotunda, bathhouses and changing rooms at Warmbaths in 1907. Museum Africa Photographic Library, No. MA2006-5112, from a postcard published by E. Tamsen of Nylstroom, 1907.

to the baths was an attractive outing for the well-to-do from Pretoria, including President Kruger and his wife, who were frequent visitors. One Johannesburg paper reported unkindly that Oom Paul had taken his wife for her annual bath! They travelled by ox-wagon, which would be their home for the visit, but others preferred to stay in a hotel built nearby, and to which water was piped from the spring. Eventually, the ZAR Public Works Department built a dome (a rotunda) to cover the spring itself, surrounded by rudimentary bathhouses, making bathing safer, cleaner and a little more private. A bath supervisor was appointed in 1895 and regulations were drawn up to govern the use and administration of the baths. A bathing season, May–September, was established and the baths became a choice winter holiday destination for the residents of Pretoria and Johannesburg.[75]

The intervention of the war

The outbreak of the South African War in October 1899 brought an end to the tourist trade, although for a while Boer burghers made use of the baths between patrols. However, the facilities quickly fell into disrepair because they were not maintained and by the end of the war they were no longer usable. In 1901 the British established a military base near the railway station, and built one of their classic blockhouses (now a national monument)[76] next to the railway line using rocks from Buyskop. They brought "friendly" rural residents (including the Peacock family from the northern Waterberg) into the settlement for their own safety for the duration of the guerrilla phase of the war. It seems that these refugees were accommodated in buildings at the baths. J.J. McCord's father had *winkelregte* (retail shopping rights) at Het Bad. According to his son, all the shops were plundered and burnt during the war, presumably by the British.[77]

In 1903 a post office established at the station just before the war, as well as the original hotel, was given the name Warm Baths (soon to be conflated to Warmbaths) by the British administration. This, or Warmbad, its Dutch equivalent, became the formal name of the settlement for the next hundred years (the current name of Bela-Bela was originally restricted to the black township that was developed on the east side of the railway line, adjacent to the area designated for industrial activity). McCord states that the first erven were laid out in the new settlement in 1903 by the surveyor Watermeyer: 100 erven of 1 morgen each; and 90 of half a morgen, including a plot designated for use as an Asiatic bazaar.[78]

FIGURE 6.33: The South African War blockhouse next to the railway line in Bela-Bela.

FIGURE 6.34: Warmbaths railway station in 1906. Museum Africa Photographic Library No. MA2006-7498, from a postcard published by E. Tamsen of Nylstroom, 1906.

Warmbaths township proclaimed

Warmbaths was formally proclaimed a "township" in 1920, and in the following year the sale by government of erven in the settlement was advertised in a glossy pamphlet. A total of 650 freehold erven were put up for public auction at the Pretoria Town Hall on 2 November 1921. The township was designed by the architect John Abraham Moffat and laid out by W.H. Gilfillan, a surveyor. It seems that an early design, resembling some spa garden-towns in Europe was abandoned and that Gilfillan's adaptation involving a more modest triangular grid, with no obvious civic centre and few green spaces, was eventually adopted.[79] The centre of the base of the triangle was located at the spring resort with its apex pointing north. The layout, always confusing to visitors, resulted in unwieldy shapes of erven. One rumour was that the town's layout mirrored the design of the Union Jack!

The auctioneers were Dely & De Kock. The brochure included photographs of the baths, as well as a diagram of a planned large swimming "pond", describing the site in glowing terms:

> The incomparable hot spring is renowned throughout South Africa – and beyond the seas – for its wonderful curative properties. Undoubtedly the healing powers of the water are assisted by the mild and delightful climate, which prevails during the autumn,

FIGURE 6.35: Sutter Road, Warmbaths, with Nathan Sacks Chemist on the right. Museum Africa Photographic Library, No. PH2006-11938, (Gabbott Collection).

winter and spring months of the year. The fascinating sub-tropical surroundings, combined with the attractions of the marvellous waters, must tend in time to make the Warmbaths Township not only the Baden-Baden of South Africa, but the most popular winter resort of Inland South Africa.[80]

FIGURE 6.36 (TOP): Warmbaths Hotel, c.1910. SANA TAB Photographic Library, No. 34829: Warmbaths Hotel, by E. Tamsen, 1910. FIGURE 6.37 (BOTTOM): Brown's Hotel, Warmbaths. Museum Africa Photographic Library, No. PH2006-11917.

In 1932 Warmbaths was proclaimed a *dorp* and acquired a village council. The following year the Towoomba Agricultural Research Station was established outside the town. The Plat River Dam outside the town was built in 1942. In 1950 the Warmbaths magisterial district was proclaimed with a Landdrost based in the town and in 1960 the town's status was elevated to *stad* with a town council. Its name was officially changed to Bela-Bela on 14 June 2002.

Post-war growth and segregation

During the 1940s and 1950s the town enjoyed a growth spurt with the development of an industrial area to the east of the railway line and the opening of several hotels, catering for people visiting the baths. From July 1958 to July 1959, the town recorded 260 000 visitors. As with other rural towns, apartheid legislation in the 1950s forced the creation of a separate black township (Belabela) to the southeast of the town, and an Indian area, Jinnah Park, in the west, both these suburbs

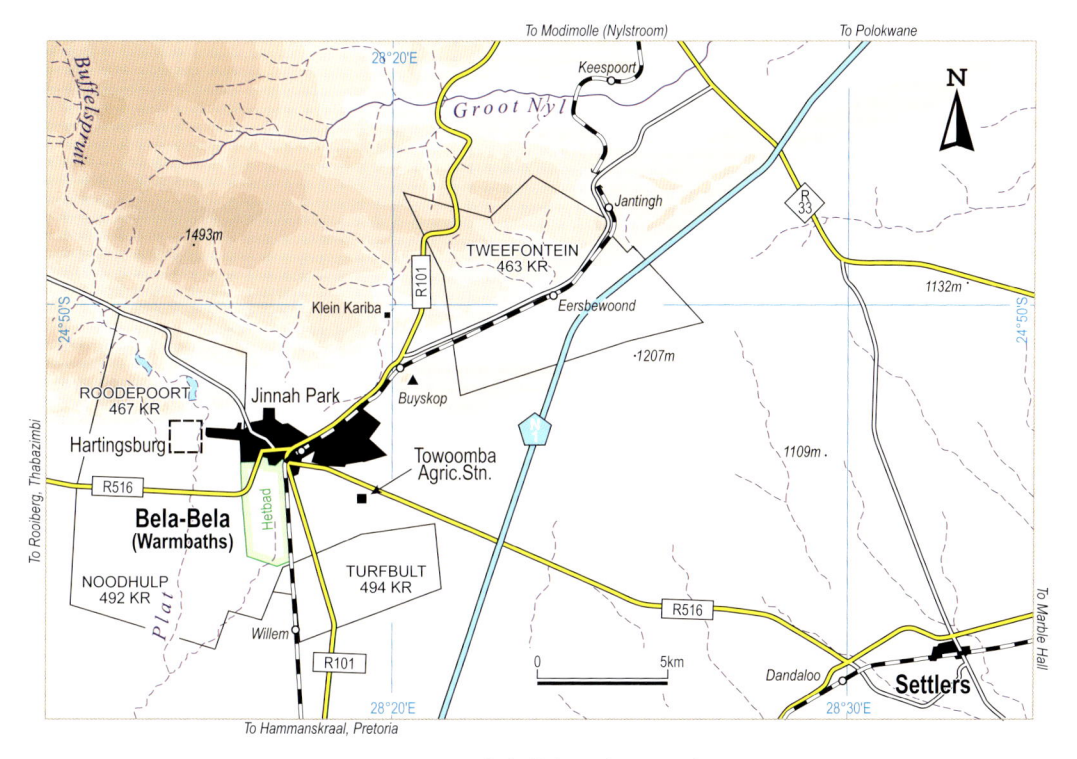

FIGURE 6.38: Bela-Bela and surrounds.

being outside the proclaimed boundary of the town (the farm Het Bad). Indian shops (e.g. Bham's Store) that had been within the triangular town centre were forcibly moved to Jinnah Park, without compensation, just as had happened in Nylstroom. Jinnah Park was "conveniently" separated from the white residential area by the St Vincent's complex of the Roman Catholic Church and Belabela by the railway and industrial area. Further to the west of Jinnah Park was Spa Park, an area designated for the coloured population. The springs too were segregated, with separate baths for the various race groups.

In the 1970s further investment was made to the thermal springs resort (still government-owned). Following several visits to European spas, the state body charged with the administration of the baths, the Board of Public Resorts, decided to introduce a "medical character" (in addition to the existing leisure character) to the facility at Warmbaths, which was by now one of 14 resorts in the Public Works Department's Aventura stable.

This led in 1978 to the opening of the Hydro Spa in which carefully selected building materials, flowing water and subtle control of the movement of patrons was built into the design and architecture, along modern European lines.[81] In 2003 the government concluded the sale of its 14 Aventura resorts to various private companies. Warmbaths was one of eight resorts sold to the Forever Group. The continued improvements and refinements at the Forever Warmbaths resort up until the present day have had a significant impact on the development and prosperity of the town, although they also imposed several burdens on the municipality, not least of which was the provision of an adequate water supply to supplement that from the springs.

On the opposite side of town on the road to Nylstroom, a private resort, Klein Kariba, was developed in the late 1960s. In 1975 the resort was sold to the Afrikaanse Taal- en Kultuurvereniging (ATKV), which was a partially state-controlled and -funded organisation that had been formed in 1930 to provide for the welfare of Afrikaners in the workforce, especially in the railways, but also in other state-owned industries. It was closely aligned with the Afrikaner nationalism movement, promoting the use of Afrikaans in government departments (particularly the railways) and celebrating Afrikaner events. Klein Kariba became one of the ATKV's low-cost (state-subsidised) resorts for the preferential use of its members.[82]

The route of the new N1 freeway linking Pretoria with Pietersburg passed some 8 km to the east of Warmbaths when it opened in 1983 (in the same way that Nylstroom was bypassed). This resulted in a significant loss of through-traffic and business in the town. Later, though, the rapid access provided by the freeway made the development of other weekend resorts – golf estates (e.g. Mabalingwe), game lodges (e.g. Mabula) – in the area surrounding Bela-Bela attractive to residents of Gauteng and these developments have brought plenty of new business.

CHAPTER SEVEN
Vaalwater and the plateau settlements

VAALWATER
Altitude: 1190 metres amsl
Population (2011) including Leseding: 16 500[1]

Vaalwater (also known as Mabatlane, the derivation of which is obscure) is unique among the larger settlements of the Waterberg region in that it has never been proclaimed as a public town. Therefore, it has never had an establishment date to celebrate with a civic function or the publication of a commemorative volume or *gedenkboek*. Since its formation, its welfare has always been at best an administrative afterthought, or latterly, the inconvenient responsibility of a reluctant, tight-fisted and almost bankrupt foster parent (albeit one that is happy to collect its rates). As we shall see, the consequences of this history of neglect, not always benign, contribute to the day-to-day experiences of the current inhabitants of the place.

The farm Vaalwater in the Waterberg Wyk was first awarded by the President of the ZAR, President Pretorius, to John William Fisher on 23 September 1867, having been "inspected" in December 1865 by F.C. de Clerck.[2] Like all other old farms in the Waterberg, its original number (5) was derived from the page number of the huge leather-bound ledger in which the first entries were made. (*Inspektie* was the initial process of farm demarcation adopted by the ZAR, by which properties were roughly outlined according to distances travelled on horseback from a central point. Depending upon the integrity and diligence of the inspector, this process could be of greatly varying accuracy and value.) Within two months, Vaalwater changed hands for the princely sum of £5, the new owner being Henry Austin, who made an enormous profit by selling the property on to Alexander McCorkindale for £120 a mere two months later. The latter was presumably only an agent, for the farm quickly changed hands again, the new owner being David Dale Buchanan.

Vaalwater remained the property of Mr Buchanan for the next 19 years until he sold it back to the government (in April 1887) for £30, presumably because he'd been unable to pay his taxes. It was to remain government property for the next 28 years: that of the ZAR until 1902; of the British administration until 1910; and of the Union government until 1915. During this period, in July 1895, its size was formally measured – at "1484 morgen 515 vierkant rode" (1271,8 ha), the surveyor being M.D. le Vos.[3]

In the meanwhile, with effect from January 1907, the farm was leased to two settlers, William Rufus Kirkman and Robert MacLachlan Armstrong, for a period of five years under the Settlers Ordinance of 1902.

Who were Kirkman and Armstrong?

The Kirkmans – founders of Vaalwater

According to the outstanding genealogical compilation of the Kirkman family made by Peter Kirkman in 2012,[4] William Rufus Kirkman had been born on 29 July 1857 in the Karoo, the sixth child of John Kirkman (1822–1905) and Anne Nash (1828–1885), both of 1820 settler families in the Albany district of the Eastern Cape. Growing up in the Karoo, Rufus (as he was known) initially farmed in the Steytlerville area where he met his second cousin Ida Biggs, a local teacher and also an 1820 settler descendant, whom he married in 1883. In the early 1890s the couple and their young family moved to the Transvaal where Rufus became involved in both farming (south of Pretoria) and a transport business. At the outbreak of the South African War, he left the farm, sent the family back to Graaff-Reinet and joined the Rimington's Guides (subsequently renamed Damant's Horse), a colonial regiment. After his discharge from the unit on medical grounds in August 1900, Rufus attempted to salvage his farm which had been ruined by the Boers, but eventually abandoned it and, with his family and some war reparation payments, set off for the farm Vaalwater in the Waterberg on which he'd secured a lease in terms of the Settlers' Ordinance of 1902, in partnership with the Armstrong family.

Little is known about the Armstrongs. According to notes from another long-time Vaalwater resident, the late Mary Chaney, the family comprised two brothers, Robert and Ben, their sisters Winifred and Annie (the latter married to a Bertie Grewar), and their mother, also Annie.[5] The Armstrongs and Grewars soon established themselves on Zandrivier, a neighbouring property across the Pongola (now Mokolo) River from the Kirkmans.

A government horticulturist, R. Davis, visited the Kirkmans during 1907 and wrote to the Director of Agriculture in the Transvaal saying: "During my recent visit to the Pongola River, I stayed a night at the house of Mr Kirkman, a settler on the farm Vaalwater, no. 5, Waterberg. This gentleman has done good work during the eight months he has been on the place, having made substantial improvements in the way of domestic buildings, kraal etc."[6]

To quote from the Kirkman biography: "Ida [Kirkman, née Biggs] resumed her role as a teacher … and Rufus commenced with general farming, establishing orchards and putting land under irrigation from the river."[7] He erected a grinding mill on the banks of the river and built a trading store with postal agency. Grain was milled for neighbouring farmers and later the water mill was replaced by a steam-operated one which was built within the slowly growing village of Vaalwater, opposite a new Kirkman home and store. Peter Kirkman noted that farmers were often unable to pay in cash for milling, in which case they had to do so with wood for the steam engine. Rufus helped establish the North Waterberg Farmers' Association, which he also chaired for many years.[8]

Enter the Farrants

Amongst his many investments in the community, Rufus built a tennis court. Through this recreational pursuit, his daughter Rhoda met and in due course married (in 1912) Jack (actually John) Farrant, a young settler who lived on the neighbouring farm Hartebeestpoort. Farrant (1880–1966), the son of a surgeon from Somerset, England, came to South Africa to join the Yeomanry at the outbreak of the South African War. Afterwards, he and a friend, Jack Mortimer, began transport riding, which is what brought them to the Waterberg. Farrant was later able to buy Hartebeestpoort (which he renamed Culmpine after his Somerset home), while Mortimer bought a property at 24 Rivers.

Rhoda and Jack Farrant had four children: three daughters and a son. Their son Rupert (1917–2002) trained as a teacher but joined the SA Airforce at the outbreak of World War II, attaining the rank of Major and being extensively decorated (including DFC with Bar) for his services. After the war, he returned to Vaalwater and took over his father's farm when his parents moved to establish a general dealership on a corner of their property closest to the village. He and his wife Margaret (Garry) made a huge contribution to the development of the Vaalwater farming community over the following 50 years and were held in great esteem by all.⁹ Today, Culmpine, the citrus and Bonsmara stud farm, is the home of Rupert and Garry Farrant's son, Dr Peter Farrant and his family.

One of Jack and Rhoda's daughters, Kathleen, born in 1914, married Harvey Zeederberg. Harvey was a grandson of Dolf Zeederberg, whose name was made famous through the transport business he and his brothers had founded in 1884 and which by the mid-1890s was the largest in the Transvaal, running wagons and stagecoaches from Kimberley to the Witwatersrand and then all the way to Rhodesia via Nylstroom and Pietersburg.¹⁰ Kathleen and Harvey's son Arthur eventually took over his grandparents' general dealership. Today the Zeederberg complex (supermarket, filling station, restaurant and other shops) is a thriving shopping precinct – still outside the township of Vaalwater – under the management of Arthur and Shelley Zeederberg's son James. Kathleen later remarried; her second husband, Alex Furman, like other members of this enterprising family, also became a leading figure in the Vaalwater community, best known for his accounting skills and secretarial duties.

On 13 January 1915, ownership of the farm Vaalwater was transferred from the government to Kirkman & Armstrong in terms of a Crown Grant dated 20 July 1914. Three years later ownership changed again when the partnership was evidently dissolved and Rufus took sole ownership. The amount paid was £514.8.9d. Rufus died suddenly of a heart attack in August 1919 and ownership passed to his widow Ida. For many years their sons Gilbert and John took over the management of the mill, store and farms.

In 1923 Ida applied to the Townships Board for the right to establish a small township of 50 erven on the farm Vaalwater in anticipation of the arrival of the railway line from Nylstroom, construction of which had just commenced (see Chapter 11). The proposed township was to be located adjacent to and immediately to the west of where the railway station is located today. Unfortunately at the time, the final site for the rail terminus had not been determined and even as late as January 1925, a decision from the railways was still awaited.¹¹ As a result, Ida never received permission to proceed with the township, although everyone was in favour of both the idea and the site. The Postmaster General had even registered the name of Vaalwater for the post office. Despite this disappointment, Ida generously donated a portion of the farm to the South African Railways for a station. The opening of the railway line from Nylstroom on 1 October 1925 placed the still small settlement of Vaalwater firmly on the map and ensured its survival, notwithstanding its lack of municipal status.

In the absence of a formal township development, growth of the settlement

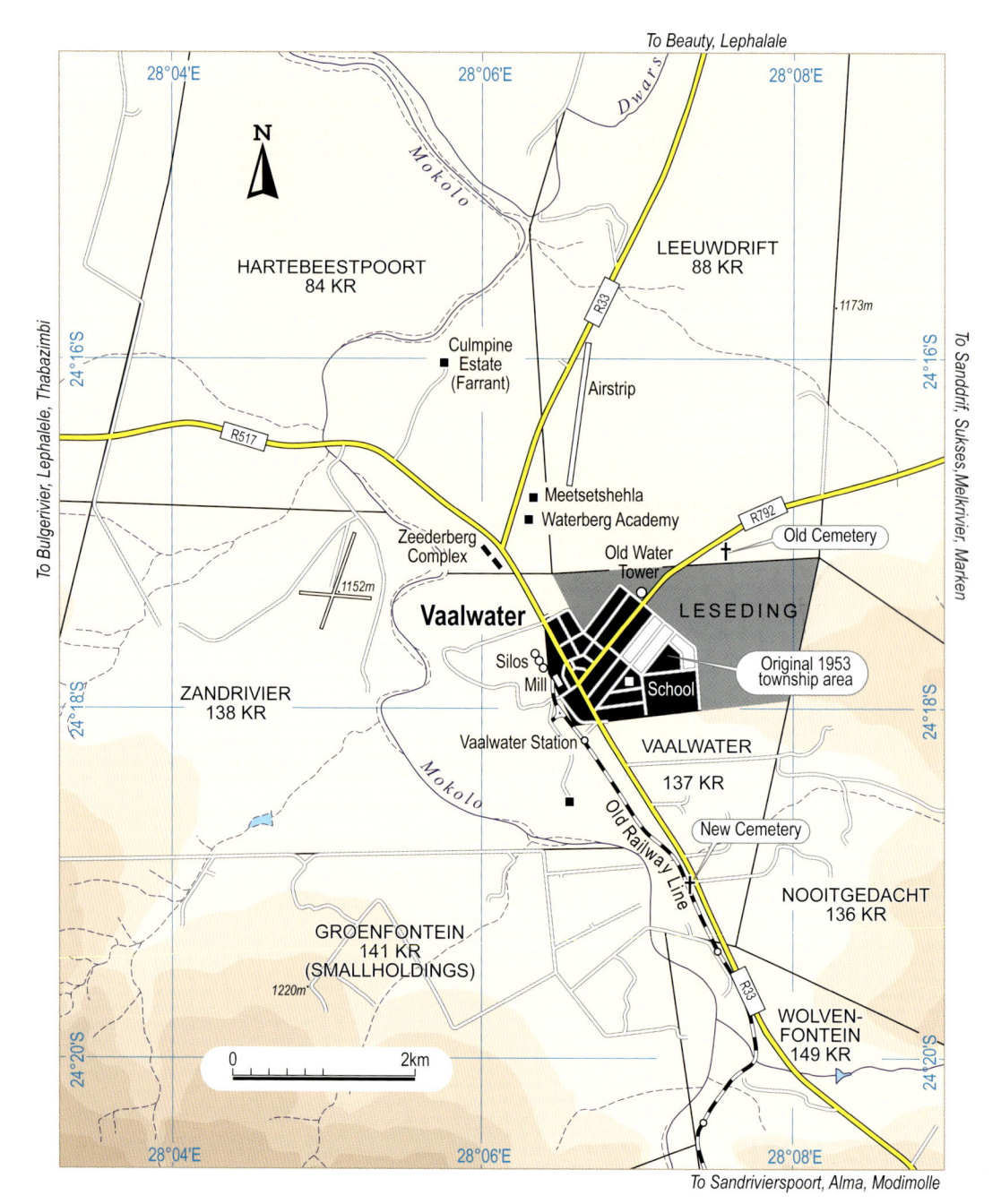

FIGURE 7.1: Vaalwater and surrounds.

FIGURE 7.2: The Kirkman/Chaney mill on the main road in Vaalwater with the NTK silos behind the building.

remained slow. A Road Motor Service (RMS) from the station to Stockpoort on the Bechuanaland border was secured by Ted Davidson of the local farmers' association in an effort to ensure that the rail line was not closed. The road from Nylstroom was surfaced for the first time in 1954 after repeated appeals from the organised farming community. In the meanwhile, the Kirkmans continued to operate their mill in the centre of Vaalwater until the business (Kirkman & Armstrong) was purchased by John William Chaney (previously of Zandriviersjoort) in 1938.

A private township created

In 1947 a second application was made by the Vaalwater Development Company (Pty) Ltd to establish a "private" township on the farm, this time to the north-east of the now extant railway station and its own small residential and technical settlement. The company had just acquired a 500 morgen portion of the farm from Chaney, who in turn had purchased it from the Kirkman family some eight years earlier. The developer was a syndicate of private investors, the most prominent of whom was I.C.H.O. Holtzhausen. Various government departments, including that of Education, dragged their heels in dealing with the application. Finally, on 18 December 1953, the Deputy Administrator of the Transvaal, S.A. Lombard, approved the proclamation of the private township of Vaalwater, as published in the Provincial Gazette the following week.[12] The proclamation made provision for erven reserved for a school and various other government facilities. These included the present location of Laerskool Vaalwater, the post office and municipal offices, although not a police station or clinic, which probably explains why even today the Vaalwater police station and adjacent clinic occupy rented property.

A primary school (for white children) opened in Vaalwater in 1953. (More about this school in Chapter 17.)

In 1954 the Vaalwater post office was upgraded under the administration of Postmaster P. Viljoen (shortly replaced by H.B. Griffiths): it had responsibility for a large region with 29 postal agencies that included those on Ellisras, Stockpoort, Matlabas, Visgat and Beauty.

Notably and not surprisingly, given the times, the proclamation of the township of Vaalwater included a clause that specifically prohibited the sale or lease of erven to any "coloured" person (defined as African, Native, Asiatic, Cape Malay). It also prohibited the residence on erven of such persons other than the servants of (white) owners or occupiers of erven. This restriction was made in the context of the Group Areas Act, a pillar of apartheid-era legislation, which had been promulgated in 1950. The proclamation required that the developer make land available for the development of a "coloured" township – but imposed no obligation on the developer to build one.[13] And indeed, despite several subsequent appeals from civil society, including agricultural organisations, no formal township to accommodate black residents of Vaalwater was built until 1994 when Leseding township was established under the new government's Reconstruction and Development Programme (RDP).

The Vaalwater township plan drawn up by the developers provided for approximately 375 erven (stands). Those to the southwest of the main road from Nylstroom to Stockpoort (later to Ellisras) – in other words, on the side where the grain silos stand today – were designated for light industrial activities linked to a rail siding line that extended beyond the station to a triangular shunting area near the grain silos. A small shopping centre precinct was located across the road from the existing mill. The irregular perimeter of the township and the inconvenient hub-and-spoke arrangement of some streets were necessitated by the shape of the property on which it was developed.[14] The Zeederberg shopping and garage complex was located outside the township on the adjacent farm Hartebeestpoort, as remains the case today, together with the Meetsetshehla Secondary School, the Waterberg Academy independent school, the Farmers' Hall and an Eskom depot.

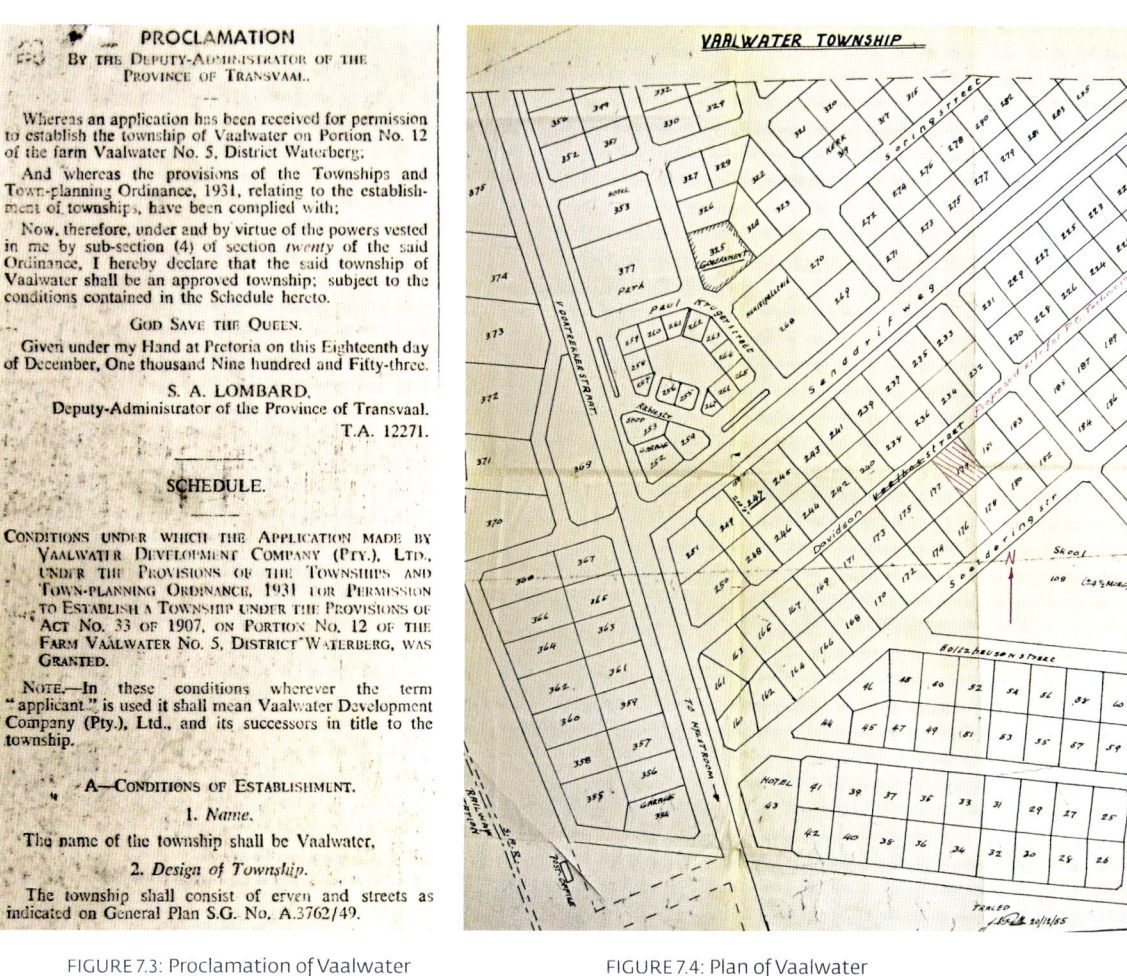

FIGURE 7.3: Proclamation of Vaalwater (private) township, 18 December 1953.

FIGURE 7.4: Plan of Vaalwater township, December 1953.

Despite the belated establishment of a township in 1953, albeit in private ownership, the inhabitants of the settlement of Vaalwater did not receive municipal services of any kind from the provincial or central government. The developer was required to make potable water available, to make provision for refuse removal and to provide rudimentary sewage treatment facilities, as well as to maintain roads within the township.

Within a few years of its introduction, the railway service between Nylstroom and Vaalwater struggled to remain viable. Indeed, it can be said that the existence of the service owed more to the relentless lobbying of the farmers' associations than to any potentially viable business model. In 1956 the railways began exploring the replacement of the thrice-weekly passenger service along the line with

FIGURE 7.5: The original township water tower on the Leseding boundary.

a daily bus service (RMS), similar to the one already in operation from Vaalwater on to Parr's Halt and Surrey in the Lowveld. At that time the route was the gravel road via the small settlement of Bulgerivier, then down the winding course of the Rietspruit, through the picturesque Olieboomspoort to the village of Ellisras, and then onwards to the Limpopo River and Bechuanaland border.[15]

Residents appeal for governmental involvement

In September 1954 the Administrator of the Transvaal, Dr W. Nicol, received a petition from the residents of Vaalwater and surrounds applying for the creation of a Health Committee for the township (the first stage in the formalising of an urban centre with government funding). This application was made in terms of the Peri-Urban Areas Health Board (Ordinance No. 20 of 1943), which had been proclaimed following a Commission established a few years earlier to consider the administration of areas that were becoming urbanised but which were not subject to local governmental control. Typically, very little happened until the report to the Transvaal Provincial Council of the Peri-Urban Areas Health Board Commission of 1952, which then forced the Transvaal provincial government to start dealing with the issues relating to peri-urban settlements.[16]

The 56 signatories to the petition included the names of the township developers (Holtzhausen, Currie and Aulest) as well as several other well-known names in the district: Faber, Hewson, Kirkman, Swanepoel, Lensley, Hoffman (the school principal), Bekker, Vosloo and Schoeman. The nearest magistrate (at Nylstroom) was asked to provide details of the population. He responded that in the township itself there were only a pair of white families, about 50 blacks, a general dealer and the mill. If the immediate surrounds were included, there were approximately 70 white citizens, grouped in 30 families, and about 500 "naturelle". Developments, apart from the school, already included two general dealerships, a butchery, mill, garage, boarding house, post office, railway station and a farmers' hall. A new hotel was under construction on Mr Bekker's property.[17]

The matter was referred back and forth between various government departments. Two major concerns were that a) there were more facilities established *outside* the township than within it – including the railway station and its housing settlement; and b) that there was a growing population of black residents within and outside the township for whom no facilities whatsoever, let alone a formal township, had been established. In May 1956 a comprehensive, powerfully worded report compiled by the assistant secretary for the Peri-Urban Areas Health Board, P. Pretorius, was submitted to the Province, effectively appealing to it to take control of the situation and create a formal jurisdictional area incorporating the township and its immediate surroundings. However, the Provincial Administration remained unconvinced of the urgency of establishing a Local Management Authority for Vaalwater. It was disappointing that this view was endorsed by the Provincial Council representative for Waterberg, Mr "Fox" Odendaal of Nylstroom. (Politics no doubt played a role here. Most of the residents of Vaalwater at the time would have been supporters of the United Party, whereas Odendaal was a dyed-in-the-wool Nationalist.) The Province decided that it would wait to see the extent to which the town continued to develop before reviewing the matter.

So the status quo continued with no decisions taken, while the community – and the township developer – became increasingly dissatisfied with conditions, especially those relating to water supplies, sewage and rubbish disposal, road maintenance and provision for black housing. Unfortunately, the opportunity to express this unhappiness through the ballot box no longer existed, following the repeated trouncing and consequent effective disappearance from the Waterberg of the only political opposition at the time (1950s), the United Party.

In the meanwhile, as if to add insult to injury, the discovery of a vast coalfield in the Lowveld 80 km to the north of Vaalwater and the investment by state-owned Iscor and Sasol in coal-bearing properties there, quickly led to the proclamation and establishment of a formal new township in December 1960 at the little settlement of Ellisras (see Chapter 19). Members of the farmers' associations in that area (the Bosveld Group), excited, but anxious about the rumours that Iscor and Sasol were planning a steel plant and a tar plant in the area respectively, wanted to know about plans to extend both the tarred road and the railway line from Vaalwater down into the Lowveld.[18] In the end the decision was taken to extend the existing heavy-haul rail line northwards from Thabazimbi, always a cheaper option. However, the extension of the tar road to Ellisras did take place – via Bulge River – and was opened in November 1964.[19] From the beginning, this was a better-built road than the one from Nylstroom to Vaalwater – and would remain so for almost the next 50 years!

In 1957 competition arose between the community on the Waterberg plateau and that surrounding the new developments in the Lowveld for the acquisition of a high school. The total numbers of (white) school children attending senior classes in the various primary schools of the "agter Zandriviersspoortberge" region of the Transvaal were stated at the end of 1956 as 116 pupils in Standard 5 (Grade 7) and 81 in Standard 6 (Grade 8).[20] Clearly, there was need for secondary education somewhere in the region, and in 1958 the existing primary school (Waterkloof) at Ellisras was permitted to add further grades (its first 21 matriculants graduated in 1962). The formal separation of the primary and secondary phases took place in 1966 with the completion of the high school buildings at Ellisras.[21] The result of this decision was that Vaalwater would not have its own secondary school for many years into the future.

In 1965 the government announced that it was examining possible sites for the construction of a dam on the Mogol (Mokolo) River in order to meet the anticipated water demand in the growing town of Ellisras. The J.G. Strijdom (now Mokolo) Dam was completed in 1978. Later, a provincial nature reserve was established on its shores. A new post office – the present building – was erected opposite the shopping centre on the corner of Sering and Paul Kruger Streets; it was opened by Postmaster General M.C. Strauss in March 1967.

Vaalwater Local Area Committee established

It was fortunate that assessments of the needs of the Lowveld necessitated the repeated travel of officials through the neglected village of Vaalwater. This forced them to see the plight of the community for themselves – and on 4 December 1969, 14 years after the original application, the Administrator of Transvaal, S.G.J. van Niekerk, proclaimed that the farm Vaalwater and Portion 1 of the neighbouring farm Hartebeestpoort would henceforth be included under the jurisdiction of the Transvaal Board for the Development of Peri-Urban Areas. This was followed by a public meeting, held on 19 February 1970, at the invitation of the chairman of the Transvaal Board for the Development of Peri-Urban Areas "to consider the establishment of a local area committee of the Board in respect of the Vaalwater area, to determine the jurisdiction of such a possible committee, and to nominate persons to serve on a possible Local Area Committee (LAC). [This] is a body consisting of local persons for the promotion of municipal affairs under the general guidance and supervision of the Board."

Six members of the local community were nominated to become members of the proposed LAC: Johannes Cornelius Coetzee; Dr Felix du Plessis; Alex Furman; Theunis Cornelius Keyter; Christoffel Bouwer Nel; and Gert Hendrik Oosthuizen.

A further proclamation, dated 15 July 1970, created the Vaalwater Local Area

Committee (LAC).[22] On 1 September 1970, the first meeting of the LAC was held. It was chaired by Peri-Urban Board member W.T. Lever. Four of the elected members were able to attend, together with four officials from the Board. I.C.H.O. Holtzhausen, the original township developer, attended as a guest.[23] An early Vaalwater LAC meeting was asked to consider an application for the establishment of sport and rugby facilities in the township.

Subsequent meetings of the LAC took place periodically, but it was clear that the authority of the committee was limited to making recommendations to the Peri-Urban Board, which seemed, through the 1970s, to be preoccupied instead with implementing aspects of legislation affecting the residence of blacks. In July 1977 the Deputy Minister of Bantu Affairs, W.A. Cruywagen, proclaimed the existing informal black residential areas of Vaalwater and Ellisras as "emergency camps" in terms of anti-squatting legislation and prescribed several regulations concerning their administration. Central to this issue was the government's determination to prevent any permanent settlement of black people in white areas, and instead, their relocation to new sites in "nearby" homelands. In the case of Vaalwater, the principal site designated for this relocation was Steilloop, a place on the Mogalakwena River some 130 km to the north of the town in the homeland of Lebowa. (Other sites to which black Vaalwater residents were forcibly relocated included Vaalbank, Elaret and Cyferskuil near Radium, on the Springbok Flats). During 1977–78 several meetings were held and memoranda produced to give effect to this policy.[24] From the issues they raised (loss of trade, absence of commercial and domestic staff at weekends, etc.), local white residents seemed more concerned about the impact of these relocations on their businesses and lifestyles than about the consequences for the black families – many of whom had lived in and around Vaalwater since the first white settlement.

To accommodate the needs of businesses and farmers within the jurisdiction of the LAC, a hostel was to be built in the area of the Emergency Camp for single black men. Other black employees, now to be moved to Steilloop, would commute to and from work weekly, using a bus service to be provided by government for this purpose. The hostel was built and served its purpose until the early 1990s, when it was converted for use as a school. The road between Vaalwater and Steilloop was gravel and attention had to be given to tarring it to accommodate the heavy commuter traffic that would result from the application of the separate development policy. The road from Vaalwater to Marken was accordingly rebuilt and tarred, and opened on 12 December 1981[25] – perhaps the only benefit to be derived from all this social engineering.

In the absence of an effective vehicle for local administration, like a municipal council, the farmers' associations in the area were commendably active in debating and lobbying for facilities to benefit the community in and surrounding the township of Vaalwater. Minutes of their meetings are replete with resolutions, appeals and reports of meetings relating to various community services. These issues included the building of high-level bridges over the Mogol River on the Ellisras road and on the access road to the Groenfontein smallholdings; the search for premises for the post office and police station and a house for the doctor; repeated calls for the establishment of a clinic; reticulation of Eskom power to the town and surrounds; the provision of TV and microwave radio reception; Telkom's replacement of the old party line system; and the tarring, or at least, maintenance of key roads; etc. Their persistence, in the face of an astonishing and inexplicable sustained "wall" of governmental disinterest, was quite remarkable – and a tribute to the community interest of all those involved.

Newspapers

Throughout the nineteenth and early twentieth centuries newspapers were the most important medium by which news was communicated to and within communities (for some of us, they remain so!). The initial population of the Waterberg was too small to warrant its own papers, so people were reliant on those from the Witwatersrand (for example, the *Star*, *Pretoria News*, *Die Transvaler*, *Land en Volk*, *Vaderland*, *Volksblad*) or on papers published further north. The first of these was the *Zoutpansberg Review & Mining Journal* first published in the village of Smitsdorp by a group of local businessmen in 1888. Shortly afterwards, the paper, together with the rest of the village, relocated to Pietersburg, which was formally established in 1886 and which the local Landdrost, Dietlof Mare, and Mining Commissioner Munnik wished to make the principal settlement in the region.[26] One of the paper's first editors was Frederic Hamilton, a graduate of Caius College, Cambridge, and the Inner Temple in London, who arrived in South Africa in 1889. He soon found himself to be not only the editor and staff, but also the proprietor of the paper (which he purchased for £50 to forestall its closure). Within a year though, Hamilton was recruited as assistant editor of the *Star* in Johannesburg, escalating to editor in 1895.[27]

Other newspapers, written in Dutch and later Afrikaans, made attempts to capture some of the market. They included *Die Zoutpansberg Wachter* and *Den*

FIGURE 7.6: Offices of the *Zoutpansberg Review and Mining Journal* in Pietersburg, 19 December 1896.
Image courtesy of Polokwane Municipality/Museums, photo by Hugh Exton.

Boerenvriend, but the outbreak of the South African War (and in its aftermath, British retribution) put paid to their efforts. By 1904 the *Review*, then edited by William Brown, was the only survivor in the north. It was subsequently joined by an Afrikaans-medium stablemate, *Die Zoutpansberger*. During the Depression years, the paper's fortunes tumbled and the business was purchased out of bankruptcy in 1930 by Solly Marcus, a Pietersburg businessman. Marcus favoured a liberal editorial policy and the politics of the South African Party, which was not well-received by the growing conservative faction of Afrikaans speakers, and in the mid-1930s a new Afrikaans-medium weekly paper, *Die Noord Transvaler*, was established in Pietersburg.[28]

In April 1943 the resignation of the editor forced Marcus to seek a replacement. He was persuaded to hire a journalist who had recently returned to South Africa from London where he'd lived with his wife since 1932. They had left the country after the writer had completed a jail sentence for the murder of his stepbrother in 1926. The man was Herman Charles Bosman.

Bosman's mandate was "to be the voice of confrontation as editor of a UP-supporting paper in a Nationalist stronghold". Unfortunately, his character caused him to stumble before long: Within a couple of months of arriving, he'd met and impregnated a local teacher and was implicated in her near-death following a clumsy, though "successful" abortion. He was charged, but later acquitted because of insufficient evidence. His successor as editor, a Scot called Hugh Hayes, was speedily appointed and arrived in Pietersburg, staying with the Bosmans while familiarising himself with his new role – which included an affair with Bosman's wife. Bosman left for the Witwatersrand where in April 1944 he married the teacher from Pietersburg (also fired for her part in the scandal).[29]

Solly Marcus sold the *Zoutpansberg Review* after the end of World War II, and under the new owners the paper drifted to the right, becoming an English mouthpiece of the National Party. It lost popularity and circulation, until in 1953 when it was purchased by Harry Cooper who, with his assistant Naomi de Jager, revitalised the paper, selling off the firm's printing business and moving publication to Potgietersrus. Naomi's husband Robert joined the firm and in 1975 they bought it outright. At some point, the title was also changed to the *Northern Review*, the name it has today. In 1997 the De Jagers sold their newspaper business to the national media giant Nasionale Pers (Naspers).[30] The other large Afrikaner media company of the time, Perskor, purchased *Die Noord-Transvaler* and renamed it *Die Monitor*.[31]

Meanwhile, other papers had sprung up to compete for the market, especially amongst Afrikaans speakers. An eccentric Pietersburg farmer had started *Die Morester* after World War II. By the 1950s *Die Morester* was the main weekly regional paper on the Waterberg plateau; it had incorporated the *Noord-Transvaler*, which in turn included the *Zoutpansberger en Kooperateur*.

Within the next 20 years, *Die Morester* was acquired by Perskor and incorporated into *Die Monitor*. After Perskor had been purchased by the Caxton Group in the late 1990s, rationalisation of assets between the two major media groups (Caxton and Naspers) led to Caxton gaining control of the *Northern Review*, which then incorporated *Die Monitor*. New opposition for the *Review* came in 2004 with the establishment of the *Polokwane Observer* by Willie Esterhuyse.

FIGURE 7.7: *Die Morester*, Friday 20 June 1958.

An erstwhile marketing manager for the *Northern Review*, Johann du Plessis, moved to Louis Trichardt (Makhado) where, in 1985, he started his own newspaper named *Die Zoutpansberger* after the former companion to the *Review* (as well as the *Venda Mirror* – now called *Limpopo Mirror*) from his home in the town's caravan park.[32]

FIGURE 7.8: First edition of *Die Pos*, Friday 8 October 1982. Image kindly provided by Keina Swart.

FIGURE 7.9: First issue of the *Mogol Pos*, 16 February 1996.

In Nylstroom, a young writer named Eugene Diener started a community paper with his wife Nicola called *Die Warmbad Nylstroom Pos* in 1982. Reports and advertisements were handwritten, proofs were typeset and material was laid out and manually pasted onto sheets for printing. Lacking the means to set up a public office for the paper, the Dieners distributed it via Ster Radio, a shop on the main road (Nelson Mandela Street). In 1986 they obtained their own typesetting machine in Warmbaths, as well as a darkroom for the photographers. In 1991 the paper was sold to the late Danie and Isabel Stoltz when a full-time office was opened in Nylstroom, but the makeup and printing of the paper was moved entirely to Warmbaths, using modern technology. In 2001 the business was purchased by its present owner/editors, Bea Emslie and Keina Swart. The paper, now titled *Die Pos/The Post*, with an English-medium subsidiary (*The Beat*) is currently published in

Bela-Bela with an office in Modimolle, and has enjoyed great success across the Waterberg region, with both the paper and its editor winning numerous national awards for community-based journalism.[33]

Although too small to warrant a newspaper of its own, Vaalwater is fortunate to enjoy an overlap in distribution from both *Die Pos/The Post* and another, younger bilingual community paper, the *Mogol Pos/Post*. Established by Dries and Elsa van Rooyen in Lephalale,[34] the first issue of *Mogol Pos*, "die hartklop van die Bosveld", appeared on 16 February 1996, in the midst of an unusually heavy rainy season which saw the Mokolo River near the town (Ellisras at the time) burst its banks – and which provided dramatic copy for that first issue.[35] The paper has subsequently benefitted from the substantial growth of the twin towns of Lephalale and Onverwacht due to the expansion of coal mining and electric power generation in the neighbourhood. Under the management of Leoni Kruger, its editor since 2012, the *Mogol Pos* has become essential reading for Waterberg residents and although it is, like *Die Pos*, essentially an Afrikaans-medium paper, it includes an increasing proportion of content written in English.

The community leaders

There were many personalities who participated in lobbying for community services in Vaalwater. In the early years, they included Ted Davidson and the Grewars (father and son), who led the call for a railway line from Nylstroom. Later, Herman Willemse and Rupert Farrant coordinated community efforts; they were followed into the troubled 1970s and 80s by men like Jan Jacobs, Ben Vorster, J.C.F. (Jan) Lourens, Alex Furman, Z. van der Merwe and the two Baber brothers, Charles and Colin. They in turn were succeeded in the 1990s by leaders like Dr Peter Farrant, Louis Trichardt and Ben Mostert. What was remarkable was that despite their differing political and other allegiances, they worked together on a voluntary basis towards the common goal of improving the lot of their agricultural fraternity and the Vaalwater community as a whole.

As the national government's apartheid model began to crumble during the 1980s and realisation dawned that a new, more equitable, nondiscriminatory dispensation was inevitable, enforcement of segregationist legislation became less rigorously enforced, although control still lay in the hands of adherents to the official policies.

From around 1990 Vaalwater fell under the responsibility of the Bosveld Regional Services Council (based in Nylstroom), which replaced the old (and largely dysfunctional) Peri-Urban Board in the area. At that time, the chairman of the BRSC was Eben Cuyler, the CEO was Dr Van Zyl and members of the Bosveld Land Board included Louis Nel and Fanie van Zyl, both local farmers from the Melkrivier area. This arrangement, with some changes, lasted until the incorporation of Vaalwater into the Modimolle Local Municipality in 2000. The location of the seat of municipal governance in Modimolle, 60 km away from Vaalwater, meant that the latter continued to suffer from inadequate allocation of resources and services, as had been the case since its inception.

Vaalwater in the democratic era

Under the new government that took office in 1994, the most significant development in Vaalwater was the formal demarcation of the black township of Leseding, accompanied by the large-scale building of houses under the Reconstruction & Development Programme (RDP). For the first time, this programme afforded homes with full title to the black population of the area – a population whose numbers swelled after the collapse of the labour-intensive tobacco industry in the 1990s and from the influx of mainly illegal (undocumented) immigrants from Zimbabwe, Mozambique and elsewhere. As with many other rural towns across South Africa, there was also a rash of small informal shops, mainly operated by foreign nationals from China, Pakistan, Somalia and other countries. Serious unemployment in the area was aggravated by the arrival of immigrants and led to increased crime and incidents of xenophobia in and around the town.

In the first decade of this century, a new modern clinic (albeit without a resident doctor) was established in Leseding; at least as importantly, a nongovernmental organisation (NGO), the Waterberg Welfare Society (WWS) was founded in 2005 by two concerned Waterberg residents, Dr Tanya Baber and Ms Jane Whitbread, to address the serious and growing incidence of HIV/Aids in Leseding. Mary Stevenson was appointed to run the WWS and to raise funds for its operations. Soon, a separate initiative run by Dr Peter Farrant to educate farm workers and their families about the disease was merged with the WWS. Dr Farrant became the "resident" medical practitioner at the hospice and related facilities that were quickly established by the WWS with mainly foreign donor funding. These initiatives took

FIGURE 7.10: Trader bringing wood for sale in Leseding. Watercolour by Elizabeth Hunter (2008), reproduced with the kind permission of the artist.

place during the era when the SA government, under President Mbeki and his Minister of Health, was adhering to a bizarre (and tragic) philosophy that denied a causative association between Aids and HIV. Hundreds of thousands of South Africans, including many small children, are estimated to have died as a result of being denied medication that would have been able to manage their condition and prolong their normal lives. (The exponential growth in the size of the new Vaalwater cemetery since the beginning of the twenty-first century is testament to the severity of the disease in the local community.)

The Waterberg Welfare Society was established to fill this vacuum in the Vaalwater area; it was and remains enormously successful in this regard. After initial suspicion, related to the fear, stigma and denial associated with the disease, the WWS has come to be adopted and embraced by the community it serves. Eventually, government policy towards HIV/Aids changed and there are now comprehensive state-funded programmes to treat and distribute medication to people living with the condition, but the WWS, at the time of writing under the management of Mr Lesiba Masibe, a member of the Langa royal family, still fulfils a vital role as a hospice and as a counselling and training establishment. It even launched a vibrant radio station, Waterberg Waves, to extend its services to the community.

In 2015, ignoring vocal community opposition, government decided to merge the local municipalities of Modimolle and Mookgophong into a single entity – LIM368 – later imaginatively named the Modimolle-Mookgophong Local Municipality. In the local government elections of 3 August 2016, a surprise outcome was the transfer of political power in the LIM 368 and Thabazimbi local municipalities from the ANC to improbable coalitions of the DA and the EFF. This was mainly due to the prolonged lack of basic service delivery during the first 22 years of ANC governance – nowhere more evident than in Vaalwater and Leseding. For several years, residents' organisations had appealed in vain to the municipality to attend to numerous service deficiencies, including the potholed and patched section of the main road (R33) running through the village. Within three months of the election of the new council, the road was resurfaced. The new mayor, Ms Marlene van Staden (née Vermaak), came from Vaalwater and hopes were raised that at last the village would begin to enjoy some benefits from the rates its citizens had been paying for so long and that the record of municipal neglect it had suffered since its establishment over 60 years earlier was coming to an end. However, the aspirations of the new political leadership were stymied by cadres of the ousted ANC who continued to occupy appointed positions in the municipality, and by enormous levels of debt incurred through their prolonged mismanagement.

Bulgerivier (Bulge River)

Fifty kilometres west of Vaalwater on the R517, the original tarred road that leads to Thabazimbi, Lephalale and Botswana (it is still the preferred route to Lephalale for many), the hamlet of Bulgerivier straddles the road just after the Malmanies (or Mamane) and Bulge River crossings. It had its origin in a school that was built in 1920 on what was then the isolated farm of Bulge River by its tenant/owner A.S.

(Arnoldus, or "Nols") van Emmenis (see Chapter 17). Since 1903, the property had been owned by the government, which had purchased it from the Transvaal Land Company for £1895 – a considerable sum in those days. Its principal advantage, shared with a neighbouring farm to the south, Schoongelegen, was that it had plenty of water, both surface and underground.

Both farms, together with those surrounding them, had been laid out and inspected by C.A. Smit in 1867. Bulge River (apparently named for the meandering course of the stream of that name that runs through the farm; its better-known Afrikaans name Bulgerivier has no meaning of its own) was first issued by President M.W. Pretorius to Petrus Botes and his son in July 1868, and sold on almost immediately to Oscar Wilhelm Forssman when it was valued at £35.

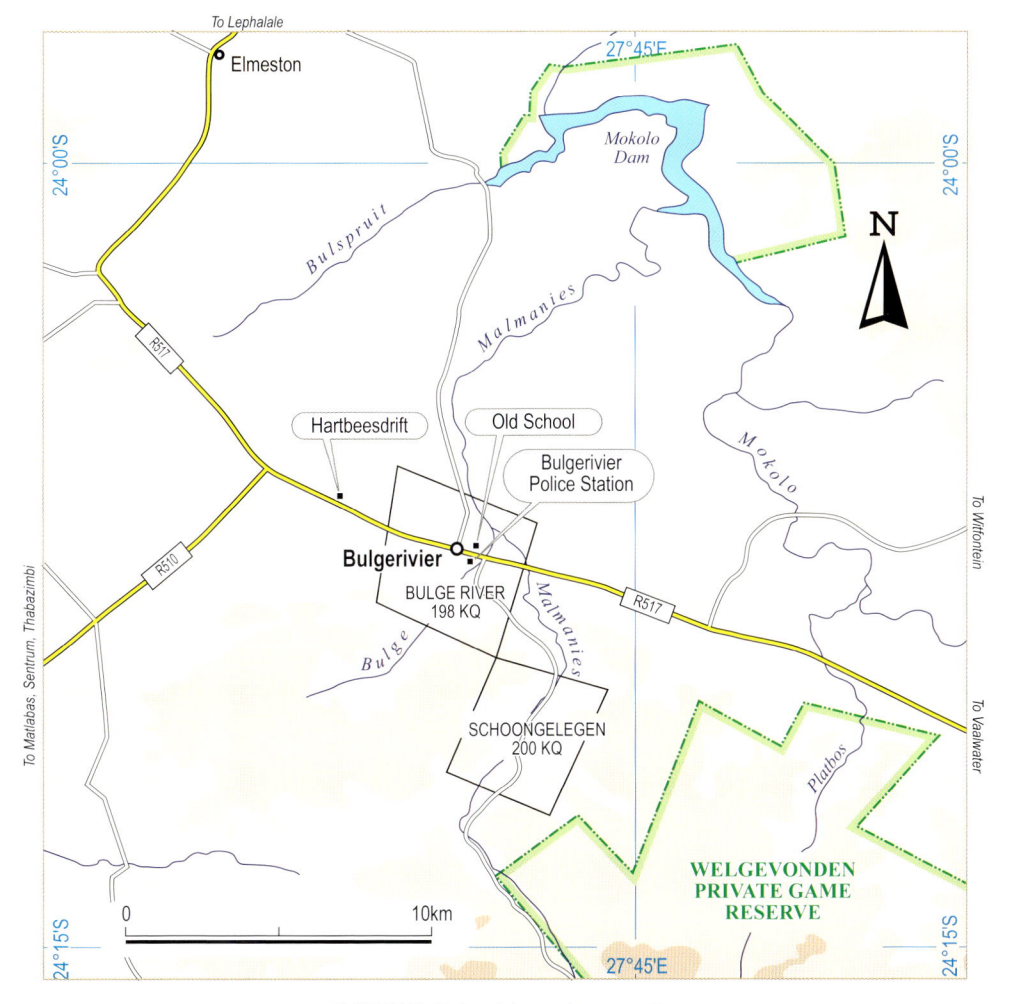

FIGURE 7.11: Bulgerivier and surrounds.

Forssman kept it for 20 years before selling it to Transvaal Land Co in 1889 for £500 – a healthy profit in those distant days of negligible inflation.[36] In about 1919 two Eloff brothers, Freek and Faan, moved onto Schoongelegen and set about making a reasonable road to their property from the main road over a *bult* (ridge) and along the Malmanies River. They allowed a number of *bywoners* to move onto the farm and for a while there was quite a large community in the vicinity – as attested to by the school which operated there briefly and by two large cemeteries next to the river.[37]

The road built by the Eloffs was later extended southwards up the Malmanies River valley and then west through the hills to Matlabas. In recent years, a succession of washaways, lack of maintenance and consolidation of farms has resulted in the closure of this former through-route.

A store was soon built at Bulge River to take advantage of the school traffic. In the late 1960s, the police station at Hartbeesdrift, about five kilometres further west along the road to Ellisras, was relocated to Bulge River. When construction of the Strijdom (Mokolo) Dam commenced in the early 1970s, the access road to the construction site 15 km to the north left the main Vaalwater-Ellisras road at Bulge River. A hotel sprang up alongside the main road (it now offers self-catering accommodation) and directly opposite it, The Fold, an American–funded orphanage, was established early in the current century. Today Bulgerivier is best known for the juicy sweet oranges and *nartjies* that can be purchased along the roadside in winter, grown in the orchards of Lenie and Ben van der Walt.

Incidentally, the "new Ellisras road" (the R33 to Lephalale) which followed part of the old Beauty road from Vaalwater via Witfontein, was only built in the late 1980s, ostensibly to provide strategic alternative and quicker access to the new Matimba power station outside Lephalale. Local opinion is that it was really intended to facilitate the ease with which senior members of government could reach their favourite secret *bosberaad* venue at D'Nyala – now, sadly, a run-down provincial nature reserve bisected by the R33 where the road reaches the plain below the northern Waterberg escarpment. The road was certainly built on the cheap, as evidenced by the frequency with which potholes repeatedly develop along it, and even the provincial roads department concluded that the original road via Bulgerivier was much more soundly constructed.

FIGURE 7.12: The Witfontein store (formerly Gedenk postal agency) with the original school in the background (left) and century-old blue gum. Photograph taken by courtesy of the owner, Len Fletcher, 26 March 2018.

Witfontein

If you take the "new Ellisras road" out of Vaalwater towards Lephalale, you will pass, after about 27 km, a small leafy settlement with a cemetery close to the road on the left, a few old houses and sheds and a charming little store on the right, shortly before a turnoff to Hermanusdoorns. This is Witfontein, which was established on the farm of the same name and which has been in existence for over a hundred years. Frans Petrus Faber and his wife Anna Margaretha (née Nel) purchased a portion of the farm in about 1918 and soon built a simple daub-and-thatch school on the property. They had several children. Amongst them was a son, Rudolph, who was born in 1920 and attended the school, which by then had been rebuilt by his father and a neighbour, John R. "Meester" Oosthuizen, who also happened to be the first headmaster of the school.[38]

Oosthuizen had come from the Cape Colony as a young teacher (he was born in 1903). He was to remain associated with Witfontein school for his entire career, while also developing a farming livelihood on a nearby property, Baverkuil, towards Vaalwater. He married Magdalena (Maggie), sister of the well-known farmer Herman Willemse, who farmed at Zanddrift on the Melkrivier Road. Herman in turn married Meester's sister. When Meester, who initially boarded

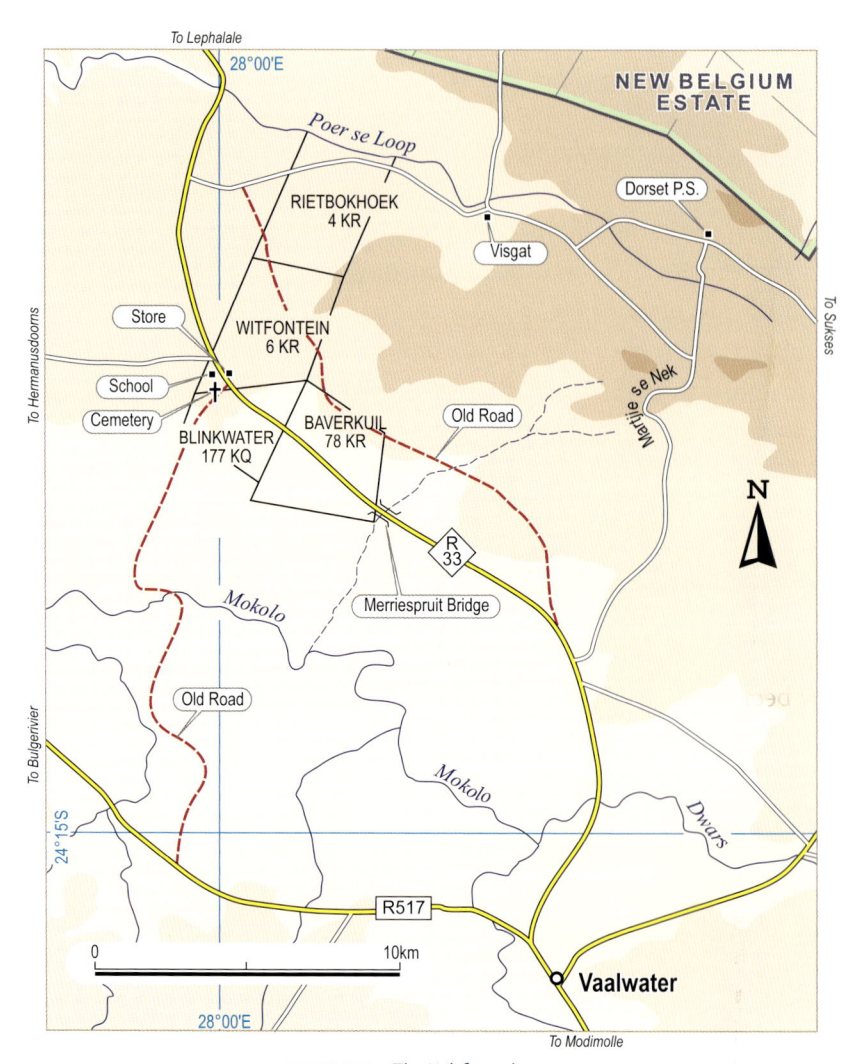

FIGURE 7.13: The Witfontein area.

with the Fabers, built a home for himself next to the new school, Maggie invested great energy and talent in developing a garden around the house. This garden became one of the most spectacular sights in the district, filled with magnificent trees, flower beds, lawns, vegetable beds and an orchard. It was the venue for numerous fêtes, bazaars, weddings and funerals over the years. Unfortunately, it was all destroyed in later years when a son, Sampie, decided to build a furniture factory on the site. The project failed and he moved on, but the derelict shed still stands where the Oosthuizens' beautiful home once stood. Meester died in 1978 and is buried in the old Vaalwater cemetery. Tant Maggie left the area for an old age home, and when she died in 1994, she was buried next to her husband. Hers

FIGURE 7.14: Frans Faber in front of the cemetery at Witfontein.

may have been the last burial in the old Vaalwater cemetery, before it was closed in favour of the present site alongside the road to Modimolle.

Immediately south of Witfontein is Blinkwater, which for many years was owned by the Louw family. W.P. (Willem Petrus) Louw (1899–1981) was not only an industrious and successful farmer, he also established the Witfontein Store, which he used as a depot for the receipt of many items of produce, fertiliser and equipment that local farmers ordered through him. A postal agency, named Gedenk ("Remember"), was granted and operated out of the store, which is still running today (albeit no longer as a post office). One of Louw's daughters married Ben Vorster; the couple also lived on Blinkwater and ran a highly successful dairy farm. Ben became an important leader in the farming community in the 1970s. Louw and his wife Maria were also laid to rest in the old Vaalwater cemetery.[39]

Oupa Frans Faber had taken care to ensure that not only a school but also a graveyard was demarcated on his property. This cemetery, although private, is formally registered and today accommodates almost 30 former family members and selected friends, including Oom Frans (1885–1956) and his wife Anna (1887–1970) and several of their children. The current generation of the family, including residents Frans and Hennie Faber, ensure that the cemetery is diligently maintained. Because of the depth of whitish, unconsolidated sand on the property, it is necessary for each newly excavated grave to be lined with brick walls built on a concrete floor prior to burial to prevent subsequent collapse of the surface and headstone. Frans Faber is among a small group of residents able to respond quickly to complete this unfortunate but necessary process ahead of a funeral. The Fabers are adherents of the Hervormde Kerk (HK); Rudolph and Isabella Faber (Frans's and Hennie's parents) were married in this church in Nylstroom in 1948, where Frans was baptised a year later by Ds. Snyman, who is buried at Witfontein. Rudolph, Hermann Willemse and others were key players in the construction of the church at Vaalwater in 1967 (see Chapter 14).

Bordering Witfontein to the north is Rietbokhoek, traversed by the original *wapad*. On this farm stand the remains of a house built by W.R. Howard, an English-speaking settler and tenant, who eventually took possession of the property via a Crown Grant in 1944.[40] By 2017 the farm was owned by Jan Vermeulen, whose father, Jan Hendrik, had married Anna, one of Rudolph Faber's sisters.

The original road from Zanddrift north to the bushveld, the wapad, lay to the east of the present R33 and circled westwards around the foot of the range of hills between Witfontein and Dorset. It passed to the east of Witfontein en route to a crossing over Poer-se-Loop, where it was joined by another road coming across from Dorset and Visgat. In the 1950s a new shorter gravel road was built from Vaalwater to Beauty, passing through Witfontein; it was later straightened out as cuttings were made through the ridges. In the 1970s this new "Beauty road" (it led to a minor road that went down the escarpment across the Tambotie River and eventually to Beauty, now called Ga-Seleka) was tarred as far as Witfontein, at the urging of Louw and others, particularly through the local farmers' association.[41]

The road was renamed the R33 and tarring was continued to Lephalale in the late 1980s. The rationale for this project is unclear (as mentioned earlier): It was a very quickly constructed road and was probably justified by a combination of objectives, including faster (and alternative) access to the strategic site of Grootegeluk coal mine and the Matimba power station, security considerations – and the government's *bosberaad* hideaway at D'Nyala. Whatever the reasoning, the road has suffered severely from its poor initial construction and the heavy traffic associated with the building of Medupi, the power station outside Lephalale. In the same way, a bridge across the Merriespruit between Witfontein and Vaalwater, washed away in a flash flood in 2014, has still not been replaced. Instead, a hazardous one-way low-level causeway on this national route is left to carry some of the heaviest traffic in the province.

Zanddrift

The third road out of Vaalwater (excluding the R33 from Modimolle), leads to Marken via Sukses and Melkrivier (see Chapter 10). But long before then, less than 10 km from the four-way stop that marks the centre of the village, one crosses the Dwars River – literally, a river that traverses the southern plateau from its source near Bokpoort Pass in the east to its confluence with the Mokolo just north of Vaalwater. This crossing, preceded by a turnoff to the east to Witklip, and followed by one to Twenty-Four Rivers and another (to the west) to Waterval), is easily overlooked today. But perhaps the observation that three roads lead away from the tar at this point should alert one to the fact that in the past this was an important waypoint. Indeed, the farm Zanddrift, on which all these turnoffs as well as the Dwars River bridge are located, existed long before Vaalwater and was the first settlement on the plateau.

Unlike most other farms on the Waterberg plateau, which were initially "inspected" (i.e. roughly laid out, possibly without even a site visit) by Jan Antonie Smit, a local field cornet with an eye for easy money (see Chapter 13), Zanddrift was first defined by J.C. de Klerk in December 1865. It was issued by Deed of Grant on 24 April 1867 to Frans Petrus Johannes du Toit. Another aspect that is unusual about this property is that it effectively remained in Du Toit's name for the next 40 years, by which time most of its neighbours would have changed hands a dozen times and probably ended up being owned by a land company.[42]

Over the intervening years, Du Toit and his descendants followed the usual practice of bequeathing their portions of the property among their children; but in this instance, the shares were *undivided*. That is to say, each descendant received an undivided, increasingly diminutive share of the whole farm. Consequently, by the time the Transvaal colonial administration came to review the status of Zanddrift in 1907, it found that ownership of the property had been divided into 34 775 shares, with many descendant owners, some of whom had claim to no more than a 1/25th stake in the farm. There was no longer a Du Toit listed amongst the many names registered as being stakeholders: these included Van Staden, Snyders, Engelbrecht, Van Rensburg, Erasmus, Visser, Du Preez, Human and Wannenburg.

In 1911 H.J.C. Engelbrecht sold his 831/6955th undivided share in the farm (about 12%) to a newcomer, Hermanus Willemse. Three years later, Willemse was able to purchase another undivided share (a 2493/34775th, or 7% of the total) from the estate of D.P. Wannenburg, in each case for £250. Therefore, for an outlay of £500,

Willemse had acquired a total of almost 20% of the property, albeit not in a way that would enable anyone to point out his particular legally defined piece of land.

This patently unsatisfactory situation was eventually resolved in 1923 when the farm was formally repartitioned according to the shareholdings. These were grouped into 13 composite shareholdings, each of which was allocated a specific surveyed portion of the farm; the largest portion, called the Remaining Extent, was allocated to Willemse to make up his 20%. Fortunately, Zanddrift was a large farm (4292 morgen, or 3676 ha), split conveniently in half by the Dwars River. Hence it was possible to divide the property into 13 blocks, each of which enjoyed some river frontage.[43]

Willemse's Remaining Extent on the eastern side of the farm was by far the largest of them, at about 690 ha. In 1924 he donated one morgen to the government as a site on which to establish a formal school. Not surprisingly, considering the large resident population on the property, a school had already been operating on the farm since 1905. Equally logically, a maternity home had been built with a resident midwife, almost across the 24 Rivers road from the school. And there were stores, at least one of which survived until the 1970s. In 1930 consideration was given to building a new church for the NGK North Waterberg community at Zanddrift – but, eventually, it was built on the Melk River east of Sukses.

Hermanus Willemse was born in the Netherlands in 1868 and was one of a group of teachers who immigrated to the Transvaal in 1898. He fought with the Boers during the South African War, was captured and sent as a prisoner of war to Ceylon where he continued to practise his profession for the benefit of his fellow prisoners. Upon his return to the Transvaal after the war, he joined the Christelik-Nasionale Onderwys (CNO) movement, teaching at a small school on Kareefontein just north of Nylstroom. In 1906 Hermanus became the *skriba* (parish secretary) for the Hervormde Kerk in Nylstroom – a position he would hold for 24 years, a period that included building a new church for the HK in Nylstroom.[44] He moved permanently to Zanddrift upon his retirement in 1930 as both teacher and skriba. But he remained a leading member of the community: He was for many years chairman of the district agricultural union, cofounder of the Vaalwater Hervormde Kerk – and in-law of John "Meester" Oosthuizen, the teacher over at Witfontein.

Even in the years before the outbreak of the South African War, Zanddrift was the centre of quite a large community of farmers, most of whom appeared to have been tenants or bywoners rather than owners, because their names do not appear on the title deed records. One of these was Albrecht Dietrich, who had a store on Zanddrift and another on Koenap in the northeastern corner of the New Belgium Block (see Chapter 9). Albrecht was the son of Captain Moritz Dietrich, formerly an officer in the Royal Saxon Army, who had later married Franziska Carolina Jeppe, a sister of the Jeppe brothers described in Chapter 5. The couple accompanied Friedrich Jeppe when he came out to South Africa in 1861. Their son Albrecht married Angelina Cornelia Mary-Ann (or Marian) Meyer and they had two daughters, Engela and Alberta (actually Cornelis Elizabeth).[45]

During the war, Albrecht Dietrich was imprisoned first (briefly) by the Boers and then by the British, who shipped him off to Bermuda. His wife died from complications after giving birth to Alberta and was buried on one of the nearby farms, probably on Zanddrift itself. (Although she died in the spring of 1901, Angelina's death certificate was only issued in 1903, after the war.[46]) The daughters were taken into the care of a neighbour, Magdalena Adriana (Lenie) Dreyer of the farm Vier-en-Twintig-Rivier. She and several other Boer women and children then fled the area for the Bechuanaland border in several wagons, but they were captured by the British and sent to the Nylstroom concentration camp. There Engela and Alberta Dietrich, aged three years and two months respectively, died in November 1901, together with several Dreyer children, during the darkest days of that camp's existence (see Chapter 6). When Albrecht returned to the Transvaal from his imprisonment, he discovered, in common with many other Boer prisoners of war, that he had no surviving family members and no business, his stores having been ransacked and destroyed in his absence. But he was a survivor. He remarried (Susanna Josina Barnard), settled in the Rustenburg area, and started a new family and a new livelihood. He died in 1941 and his second wife seven years later.[47]

One of Albrecht's grandsons, Keith Dietrich, later a distinguished professor in Visual Arts at Stellenbosch University and an accomplished artist with works in many prestigious collections locally and internationally, spent several years researching the family's history. This prompted him to create an artwork in memory of those residents of the farms around Zanddrift who were imprisoned by the British and who either survived or perished in the concentration camps at Nylstroom, Pietersburg or elsewhere. The work, entitled *Dwarsrivier 1901*, was completed in 2015 and is a poignant memorial to that community and their hardships.

Zanddrift's demise as a centre could be attributed to the decision to make nearby Vaalwater the terminus of the railway line from Nylstroom in 1925, despite lobbying from the Zanddrift community; but equally, the continued diminution in size of properties on the farm (for some portion owners maintained the futile practice of dividing their stakes amongst their children) soon made many of them too small to provide a sustainable livelihood.

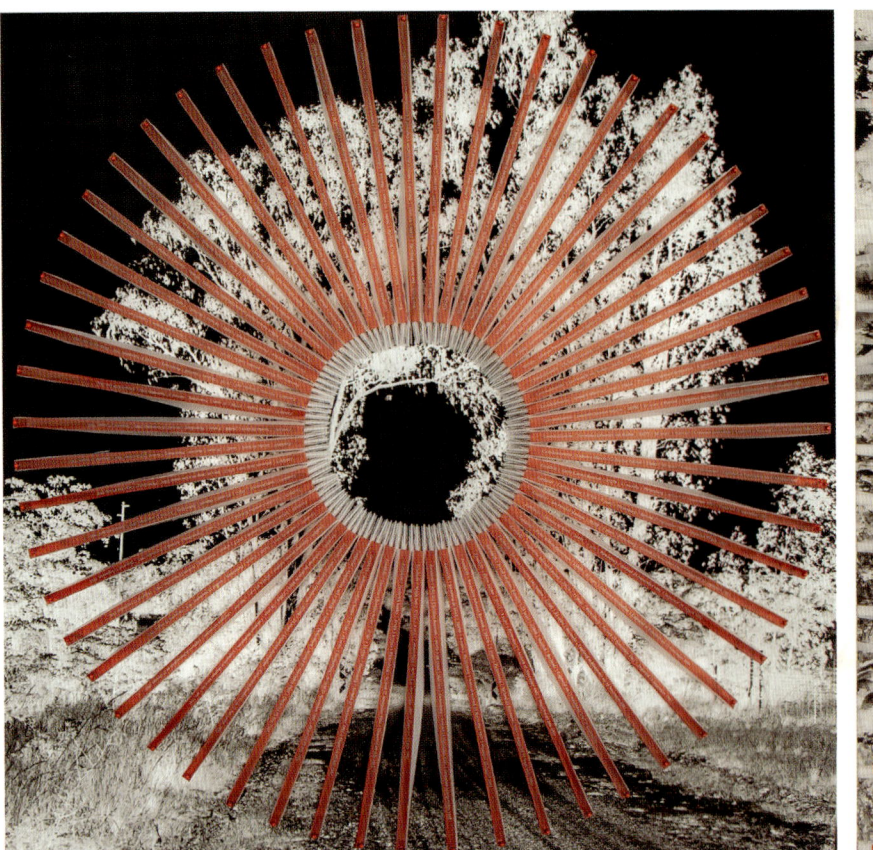

Twenty-Four (24) Rivers

24 Rivers was another small settlement, lying about 9 km east of Zanddrift and centred on the farm Vier-en-Twintig-Rivier. This farm had been demarcated by J.A. Smit in March 1867 – two years after Zanddrift. In the same year, it was awarded by *grondbrief* (deed of grant) to John Johnstone, who six weeks later sold it to Pieter Johannes Marais for £5. Marais, a speculator, found a buyer in William Munro within two weeks, making a 50% profit in the process. Munro owned the property, which amounted to about 1554 ha, for the next 20 years until some event

FIGURE 7.15: *Dwarsrivier 1901* by Keith Dietrich (2015): a diptych from the exhibition project "Between the Folds". It lists the names of 91 people who were captured on several farms along the Dwars River and sent to the Nylstroom and Pietersburg concentration camps. The right-hand panel displays names of 36 women and children who died in the Nylstroom camp (printed strips of tracing paper pinned to a negative image of the Nylstroom concentration camp cemetery). The left-hand panel displays names of 55 people who survived (strips pinned to a negative image of the road across the Dwars River on Zanddrift). Images of the diptych were provided by courtesy of the artist.

The farms along the Dwars River from which the names of the captives were taken are Zanddrift 94KR, Witklip 100KR, Vier-en-Twintig-Rivier 102KR, Goedgedacht 130KR, Boshoek 131KR, Gemsbokfontein 132KR, Klipfontein 157KR and Ratelhoek 158KR. The surnames of the 91 women and children listed on the two panels include Banning (Bannink), Benade (Benadi/Bernardie/Bernardi), Dietrich (Diederick), Dreyer, Engelbrecht, Nel, Olivier, Seegers, Snijders (Snyders) and Van Staden.

triggered a High Court decision to award the farm to the trio of Pieter Johannes Potgieter, Charl Andries Celliers and Johannes Christoffel Minnaar in three equal undivided shares.[48]

Potgieter immediately bought out the other two shareholders (for a total of £209) and was briefly the owner of the whole farm, thereby forestalling the sort of process that had occurred at Zanddrift. He then sold three quarters of the farm to Hermann Ludwig Eckstein for £300 in 1889. Those familiar with the history of the Witwatersrand will recognise Eckstein as one of the Randlords who founded the Cornerhouse Group, forerunner of Rand Mines – and as the philanthropist who bequeathed to the City of Johannesburg the large and enormously valuable piece of land comprising both the Johannesburg Zoo and the Zoo Lake Gardens. Eckstein was acting on behalf of one of his companies, Transvaal Consolidated Land Company (TCL), which took ownership of the property in 1892 and placed a manager there.

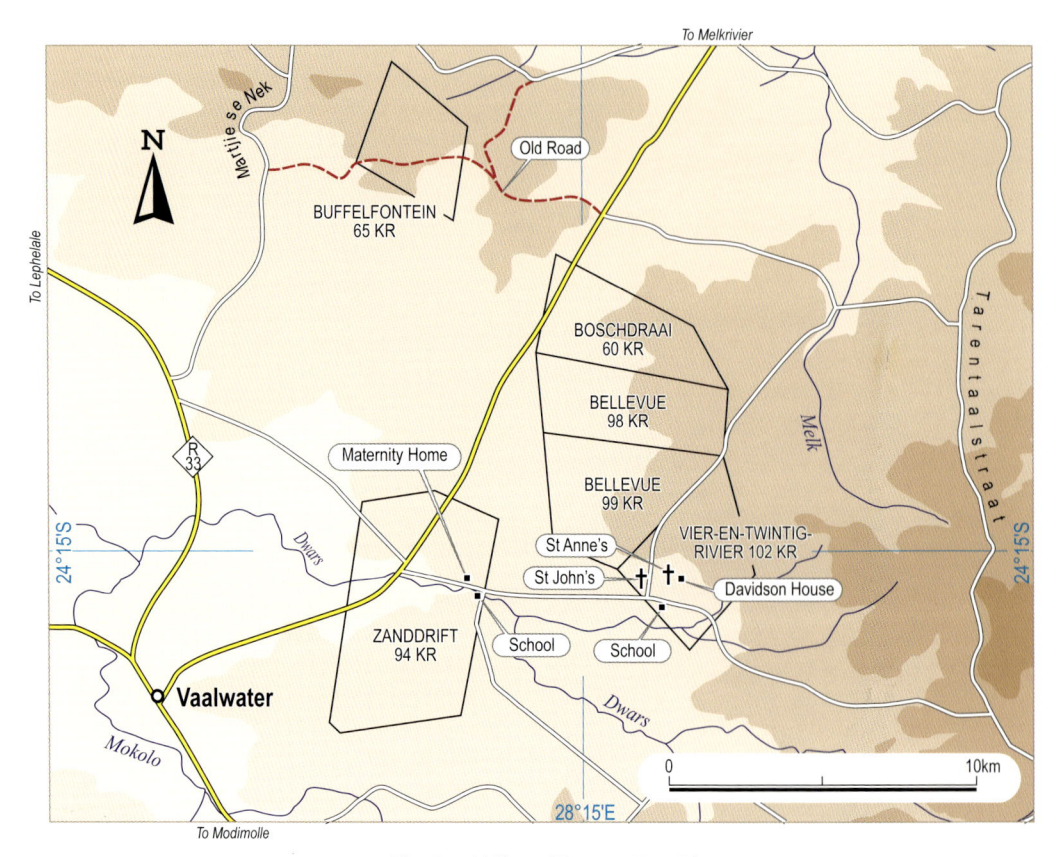

FIGURE 7.16: The Zanddrift and Twenty-Four Rivers area.

Transvaal Consolidated Lands and Arthur Madge

By the turn of the century, 1/8th shares (c.194 ha) of the farm were typically being sold for about £150. One of the buyers was Edith Anne Fawssett, the spinster sister of Katherine Peacock, wife of Arthur Peacock. She purchased the quarter of the farm not owned by TCL in May 1903 for £400.[49] She then moved with her sister, brother-in-law and niece (Molly) onto the property from their temporary accommodation in Nylstroom. Previously, they were housed in the change houses at Warmbaths, to which they'd been taken from their farm on Blaauwbank west of Dorset (also a TCL-owned farm) in October 1901, under escort of a British column. Molly Fawssett had arrived in the Waterberg in about 1898 to live with her aunts Edith and Katherine, the latter being married to Arthur Peacock, then a tenant on Blaauwbank. Until rinderpest wiped out the company's cattle herds in 1896, Peacock had been a manager for TCL on its malarial properties in the Lowveld, near present-day Lephalale.

Transvaal Consolidated Lands owned a third property in the area: Gemsbokfontein on the Vrymansrus road. On all three, it ran cattle (mainly Africander/Hereford cross) and had a total herd of 668 in 1912, according to C.A. Madge, its land manager.[50] Charles Alfred Madge was a dynamic and capable character. He was born in 1874, the son of a London medical doctor and was educated at the prestigious Westminster School, next to the famous abbey of that name. Having served as a Captain with the 6th Battalion of the Royal Warwickshire Regiment during the South African War, he was appointed by TCL soon afterwards. In 1908 TCL owned the surface rights to 720 farms in the Transvaal (half of them in the Waterberg region), amounting to over 1,35 Mha and Madge had to oversee them all, initially from his base at 24 Rivers.[51]

FIGURE 7.17: Lt Col Charles Alfred Madge.

Archival records show him considering the farming of cotton, sheep, ostriches and tobacco as well as cattle on the various properties under his management. Persian-cross-bred sheep were introduced on Blaauwbank, but failed when the flock was infected with wire-worm. Ostriches were introduced on Rondeboschje (south of Mokopane) and by 1912 there was a thriving flock of 169 birds. In 1907 Madge proposed the establishment of a tobacco processing factory at Rustenburg – an idea that may have been the inspiration behind the foundation of the famous, later infamous, Magaliesbergse Koöperatiewe Tabakplantersvereniging (MKTV) which was founded there in 1909.

Following the outbreak of World War I, Madge was appointed to the HQ Company under General Louis Botha for the campaign in German South West Africa. Afterwards, he joined other South African volunteers to support the Allies in Europe. He was serving as a Lieutenant Colonel with the staff of the 33rd Division when he was killed on 10 May 1916 by a *minenwerfer*, a short-range mortar shell, whilst inspecting trenches near the formidable German defensive position of Hohenzollern Redoubt in northwestern France. He was buried at Béthune, a nearby village. A plaque in his memory hangs on the wall of the St John's Church at 24 Rivers.[52] The following January, a meeting of the fledgling North Waterberg Farmers' Society sent a letter of condolence to his widow, who by then had returned to England.[53]

Fawssetts and Davidsons

In 1919 the farm was formally partitioned, with the result that a demarcated 389 ha portion (Portion A) was allocated to Edith Fawssett and the Remaining Extent of 1166 ha was allocated to TCL, which immediately onsold it to John Mortimer, a returned serviceman and early friend of Jack Farrant of Vaalwater, for £952 15/-. Mortimer sold all but 67 ha of his portion of the farm to the Union Government for £1415 in 1921. In the meanwhile, Edith Fawssett and her sister, Katherine Peacock, had managed to raise the funds (which included a substantial contribution from TCL) to build a small church – St John's – on Edith's portion of the farm. In 1941 Edith donated the land on which the church and its graveyard stand to the Trustees of the Pretoria Diocese of the Anglican Church, still the owners today.[54]

This donation was prescient, for Edith died the same year, aged 86 (her sister Katherine had died in 1935, also at the age of 86). In 1942 ownership of her property passed to her niece Mary (Molly) Fawssett, who had married Edward Alexander (Ted) Davidson in 1905 and who had been living on the property ever since. The old Davidson home on 24 Rivers, built in 1910, is still in use and is a veritable museum of memorabilia from the last century. The story of the Davidsons is well told and illustrated in a book produced by Ted's granddaughter, Elizabeth Hunter.[55] Only a brief summary is necessary here.

Ted Davidson was born in England. He came out to South Africa as a teenager and went mining in Rhodesia. At the outbreak of the South African War, he joined up and did service with several units, including the Middlesex Yeomanry, 17th Lancers, Plumer's Column and the West Australians under Grenville. The end of the war found him in Warmbaths, engaged in supervising the repatriation of Boers to their homes. There he met his bride Molly Davidson. After they were married, they settled on 24 Rivers where Ted initially helped Arthur and Katherine Peacock before acquiring his own property.

FIGURE 7.18: Ted Davidson as a young man in the Waterberg. Photograph kindly provided by Joan Baber.

An energetic man with many skills, Davidson not only farmed at 24 Rivers, but also opened a trading store and spent several months in the year travelling by wagon to remote parts of the bushveld, trading goods for cattle and other items. In 1931 he was appointed as a Sworn Appraiser for the Palala Ward of the Waterberg District and in 1942 as a Justice of the Peace for the same ward. As secretary of the North Waterberg Farmers' Association (formerly Society) for over 30 years, he was in the front line of community efforts to secure services and facilities for the Waterberg plateau (see for example Chapters 11 and 13). On his farm he was a pioneer in the provision of a church (St Anne's), a school (the 24 Rivers, now E.A. Davidson Primary School) and library books for his black employees and their families and neighbours.

Ted (1879–1961) and Molly (1881–1975) Davidson had four daughters, three of whom survived to adulthood. The eldest, Lois (1906–1988) married an Australian ex-serviceman called Alfred Baber, thereby founding what became one of the most prominent families on the plateau over the next century. [56]

The Baber dynasty

For much of the second half of the twentieth century, the organised farming community around Vaalwater derived an enormous benefit from the voluntary efforts of two men who were the third generation of their family to live in the district: C.E. (Charles, "Oom Chalz", or "Ouboet") and his younger brother H.C.F. (Colin) Baber. They and their much younger sister, Jennifer, were the children of Alfred and Lois Baber.

Alfred Henry Baber, born in Petersham, New South Wales, Australia, in October 1886, was the seventh of eight children of a couple who'd emigrated there from Somerset, England, in 1876.[57] His father was an Anglican minister. Three of his siblings died in babyhood; his eldest surviving brother, Charles Edward (born in Newton, New South Wales in 1878), a lawyer, came to South Africa with Australian volunteers in 1900 to fight for the British in the South African War. Charles served as a trooper with the Pietersburg Horse (previously the Bushveld Carbineers).[58] After the war, he settled in Pretoria where he practised as an attorney until his death in 1947. He and his wife Anna lived in a house called Towoomba in Hatfield. (The home was named after a Queensland town called Toowoomba. It was also the name given to the agricultural research station outside Bela-Bela, presumably also by an Australian ex-serviceman.)

In 1915 Alfred Baber, aged 29, volunteered to join an Australian battalion sent to augment the Allied forces in France during World War I. In his enlistment papers, he is described as being 6′ 4½″ (1,94 m) tall; his occupation was given as "grazier", and his next of kin was stated to be his Pretoria-based brother Charles (his parents having died several years earlier).[59] In April 1918, during an Allied defence against the German Ludendorff Offensive in northern France, Alfred sustained severe wounds which necessitated his evacuation to England. By the time of the Armistice the following November, he was in convalescence and in March 1919 he was repatriated to Australia.[60]

At some point, Alfred decided to go across to South Africa to join his brother Charles and three surviving sisters, all of whom had followed Charles there. One of the sisters, Ruth Baber, established and ran the Arcadia Nursing Home in Pretoria and it was here that Charles met and became friends with Ted and Molly Davidson from the Waterberg. Charles mentioned to Ted that his brother Alfred was struggling to farm in Australia and that he'd suggested Alfred should try his

FIGURE 7.19 (LEFT): Alfred Baber at the end of World War I. FIGURE 7.20 (TOP): Alfred Baber and his young bride, Lois Davidson, outside the St John's Church at 24 Rivers.

luck in South Africa. When Alfred arrived in 1927, Ted sold him a portion of the farm Boschdraai, close to his own property at 24 Rivers.

Boschdraai, lying to the north of 24 Rivers and separated from it by two farms called Bellevue, was another of the farms originally inspected by J.A. Smit, in March 1867. It was first awarded by President M.W. Pretorius to Johannes Hendrik Venter in October 1868 and sold the same day to J. Vos, a speculator, for £40. Within a few days, Vos passed the property on to Oscar W.A. Forssman for £60 – an outrageous 50% margin seemingly being the norm in those days. Forssman kept the farm (among his many Waterberg properties – see Chapter 5) for 21 years before selling it to Malcolm Thompson, an agent for Transvaal Consolidated Lands, for £500. The farm was eventually purchased for £572 from TCL by Ted Davidson in 1919. Ted later sold a portion of the property, 1000 morgen (857 ha), to Alfred Baber for £1500 – little realising at the time that the purchaser would soon become his son-in-law.

In 1928 when Alfred married Lois[61] he was 42 and Lois 22. On 27 January 1930 a son was born. He was named Charles Edward after his uncle, who was still alive in Pretoria. Brother Colin (in full, Hugh Colin Fawssett) followed in 1935 and Jennifer in 1943. For several years, Lois continued to work for her father in his trading stores, whilst also helping her husband Alfred establish Boschdraai.

Although permanently disabled from his war injuries, which continued to trouble him throughout his life, Alfred participated in farming with Lois and their two sons. A great deal of the responsibility for running the farm and family, as well as ministering to her husband's wounds, fell upon the shoulders of Lois in those early years. Alfred died in 1965, aged 79. A plaque in Alfred's memory in the church at 24 Rivers noted that "Faith and courage in suffering were an inspiration to all those who knew and loved him". Lois lived until August 1988 and is buried in the St John's graveyard, together with Alfred and his sister Eulalia.

Charles and Colin, and later Jennifer Baber too, grew up on Boschdraai. Initially, Charles boarded with his grandparents at 24 Rivers and together with several other local children, was taught by his aunt Jane before being sent off to boarding school in Pretoria at the age of eight. Colin went to boarding school at an even younger age, there being no suitable schools nearby. Both stayed at a St Albans hostel during their primary school years – by all accounts a very difficult time for them. Later, both boys were sent to Pretoria Boys' High (Rissik House) and then to the Potchefstroom Agricultural College, before returning to take up their farming roles on the family properties.[62]

Whilst on Boschdraai, Charles and Colin took loans from the Land Bank to purchase their grandfather's farms Toulon and Sunnyside, east of the village of Ellisras in the Lowveld, mainly in order to assist their grandfather (Davidson) alleviate a cash flow problem. Colin moved down to manage the Lowveld farms while Charles remained on Boschdraai, which was eventually bequeathed to him by his father. Colin met and married Joan Fraser in 1963 and for the next five years, the couple lived in a rustic cottage Colin had built on Sunnyside. The first of their six children was born while they were there. In 1967/68, during a severe drought, Colin and Joan decided to trek with their cattle to the Waterberg plateau and soon they were established in a home they called Summer Place on the farm Buffelsfontein between Vaalwater and Dorset. Their eldest son David still farms there.

Ted Davidson also owned the two farms named Bellevue which lay between 24 Rivers and Boschdraai; these he bequeathed to the daughters of Elizabeth, his second daughter, who had died in a car accident in 1954 when aged only 42. (Charles later purchased Bellevue from his cousin.) Charles and Colin worked hard to develop and improve their properties in order to acquire further farms in the Vaalwater area, using loans from the Land Bank. Charles took great satisfaction from the knowledge that by the time of his retirement from active farming, all his loans had been repaid.

FIGURE 7.21: Colin, Jennifer (Pullinger) and Charles Baber.

When Charles started farming, he experimented with a wide variety of crops in addition to cattle: cotton, maize, tobacco, groundnuts, chillies, wheat and *babala* (*Pennisetum glaucum*, a fast-growing grazing and silage grass). After receiving advice from a visiting group from Rhodesia (now Zimbabwe), he began to specialise in cattle and tobacco, growing maize and groundnuts in rotation with the latter. His wife Nina (née Wynne), whom he married in the early 1960s, was a teacher. She assisted with the farming, and became known as an outstanding cattle breeder, whilst simultaneously managing a large household of children and in-laws as well as much of the administration of the farm. Charles and Nina's three surviving children (Juliet, Anthony and Rupert) and their own families all still live and work on local properties acquired many years ago.

In 1992 Charles was one of five farmers nationally to receive a special award from the Minister of Agriculture for his services to soil and environmental conservation. In October 1993 the provincial branch of the Agricultural Journalists Association named Charles Baber "Transvaal Farmer of the Year". By then, he'd been on the Executive of the local North Waterberg Farmers' Association (NWFA) for over 20 years, on the Executive of the Vaalwater District Agricultural Union for seven years (much of that time as its secretary) being its representative for soil and conservation for 26 years and of tobacco for nine years. In an article written about him at the time, he was quoted as saying:

> Die veld is 'n lewende wese. Ek voel in alle nederigheid ons is maar tydelike opsigters daarvan. Op die plaas is ons 'n biologiese entiteit, ons lewe daar saam. Mense sê ek is 'n grootboer, een van die 20 persent wat 80 persent van die land se landbouproduksie lewer. Hulle vergeet dit is 'n hele span op die plaas wat daarvoor sorg. Ons doen dit saam.[63]

Charles and Colin were very close throughout their lives, despite the difference in their ages. As Charles's daughter Juliet remembers, "they were the best of friends – spoke to each other on the phone every day without fail; they supported each other in a myriad of ways, were tennis partners every Saturday (Colin was instrumental in building the courts in Vaalwater) and shared a deep Christian belief."[64] Charles was devoted to his grandfather ("Gabba" Davidson), and the two of them spent a great deal of time together, driving around the farm(s) while Ted passed on his wisdom, anecdotes and values to his grandson. Colin, although five years younger, also felt the need to "walk in his grandfather's shoes".

Charles, an extrovert with a ready and mischievous sense of humour, loved to engage with people, and meeting him was inspirational. He was always sensitive to the needs of others less fortunate than himself and spared no effort to assist them. He was not driven – or intimidated – by wealth or status; his ready approachability and compassion endeared him to many in the community, across language, cultural and racial barriers:

> On Sunday afternoons, he would phone friends across the country to check on their health and wellbeing. He would always pop in to see a friend if he happened to be in their area, and would often unexpectedly announce that he had come to spend the night, always expecting his friends to be as delighted to see him as he was to see them![65]

Colin, by contrast, was less of an extrovert and more irascible than his brother; he would speak his mind frequently and forcefully. But he was a natural patriarch and commander of his home. He was deeply committed to his family, his Lord and the good of the community. He was widely acknowledged as an outstanding leader and a valuable person to have nearby in times of crisis when quick decisions and organised action were needed. Many were those who found refuge in his home, a place of safety and hospitality under his wise and strong leadership.

Like his elder brother, Colin was extremely active in agricultural affairs. He was a member of the NWFA for 32 years (until his resignation in 2002 when politics overtook the organisation), 26 of them as its secretary. For several years he chaired the Laerskool Vaalwater Board, of which his wife Joan was also a member, until forced to step down by conservative members who objected to the fact that the Baber children attended a private multiracial high school. Although English-speaking, both Colin and Charles were fluent in Afrikaans, as well as in Sepedi. In their capacity as secretaries of the two agricultural bodies, they wrote most of their minutes in flawless Afrikaans.[66] Colin insisted that his family accompany him on 16 December each year to attend the *Geloftefees* celebrations at Melkrivier; he considered this an important social duty.[67]

Colin became not only a successful farmer, but also a successful cattle speculator. He had a business mind and could see the opportunities offered by calculated risks. For many years, he ran a butchery in Vaalwater, supplied from his own abattoir. He was also an officer in the local Commando. In later years, Colin became involved with the Peri-Urban Board and its successors and oversaw the operation of a number of small towns in the northern Transvaal, including Haenertsburg and Roedtan.

Both men lived life to the full, exerting their energies in farming, family life and on community work; they were "larger-than-life, make-it-happen men for whom every problem was simply another challenge to be overcome".

Colin's death in 2003 at the early age of 68, of complications after an operation, was a shock to family and community. Charles lived on to the ripe age of 82 before passing away peacefully in his sleep on Boschdraai in 2012.

CHAPTER EIGHT
The gap through the hills

Zandrivierspoort (now spelt Sandrivierspoort) is the sinuous gap through the seemingly impregnable barrier formed by the Sandriviersberge, which guard the southern limit of the high Waterberg plateau and confronts every traveller on the road from Modimolle to Vaalwater. To the left, or west, as one approaches the *poort* from Modimolle, is the remarkable series of promontories well known to local residents as the "Seven Sisters" (there are actually more) – geomorphological phenomena known as *cuestas* and caused by the erosion of the northwest-dipping Waterberg strata, probably sculpted by the Dwyka glaciation event, 300 million years ago.

The north-flowing Sand River, joined just south of the mountain range by the Klein Sand River coming in from the east, combines to form the Mokolo River and has carved a narrow gorge through the range a little more than a kilometre to the west of the poort. The Mokolo was known to early travellers as the Pongola.

The confluence of the two rivers created a marshy area in front of the *kloof* that proved too difficult for early wagon travellers to traverse on their way to the plateau and the bushveld beyond; so instead, they exploited the somewhat steeper but firmer route offered by the poort. Their original track is still more or less that followed by the present-day multilane highway, the R33, which conveys traffic from the south through Vaalwater to Lephalale and on to Botswana.

When the railway line from Modimolle to Vaalwater was built in 1924/25, the steep gradients in the poort presented a greater engineering concern than the high, piered bridge needed to cross the marshy ground along the Mokolo River, with the result that the line was built next to the river through its kloof (Pongolapoort), only approaching the road outside Vaalwater.

Travellers hurrying through Sandrivierspoort today can be forgiven for failing to recognise the challenges presented by the feature in the past, or even its stra-

FIGURE 8.1 (TOP): The "Seven Sisters", the western part of the Sandriviersberge, on the R33 between Modimolle and Vaalwater.
FIGURE 8.2 (BOTTOM): Sandrivierspoort.

tegic military importance. But the poort has a rich history. It is not known who first "discovered" the route to the plateau via the poort or when a recognised track through it was first made. But the first recorded description of the poort was that of German traveller and naturalist Carl Mauch, who mentioned it in his journal describing his visit to the Waterberg in December 1869, when this was a sparsely populated and little-visited part of the Transvaal (see Chapter 5):

WEDNESDAY, 1 DECEMBER 1869: N to the Sand River Poort, all the way on sandveld in which the waggon wheels plough a furrow of at least 5 inches depth. Near the Poort two (almost 1000 feet high) mountain ranges come together, also of sandstone, of which the one extends from E to W, the other from W to NE. Course NNW to a spruit [the Klein Sand] which, flowing from the E, joins the Pongola. The road through the poort is 1½ miles long; however it does not lead along the river, but through a kind of pass in the eastern main range, as the river has dug its bed at a distance of ½ – 1 mile. As the ground, of course, is still sandy, but now frequently alternating with wet and swampy stretches, there is added to the sand vegetation that of swampy environments, that is to say Acacias.[1]

The first settlement: Carl Forssman

The road could not have been much used at this time: Nylstroom (now Modimolle) itself had only been founded in 1865 and there were no settlements further to the northwest until the formation of a small community around a mill on the farm Vaalwater in 1905 (see Chapter 7). The occasional hunting groups, surveyors and traders would have been the only traffic along the track, apart from quarterly treks made by isolated settlers to attend *nagmaal* services at one of the three newly founded churches in Nylstroom.[2]

However, by 1890 the route was apparently sufficiently well used, and the local farming population along the Sand River valley large enough that appeals were made to the ZAR government for a properly built road to be made from Nylstroom through the poort. On 22 May 1890, Road Inspector J.C. de Beer wrote to the government architect explaining the rationale for such work:

- It was the main transport route used by all burghers for their annual trek from the Waterberg to the bushveld;
- It was a major – and the only – trade corridor linking the ZAR with the Khama and Lobengula chiefdoms and the important trade that took place with their people; and

FIGURE 8.3: The Nek in Sandrivierspoort – early 1900s (TOP) and in 2017 (BOTTOM). Early 1900s photograph courtesy of Museum Africa Photographic Library, No. PH2006-11939.

CHAPTER 8: THE GAP THROUGH THE HILLS

- Occupiers of the land on the far (north) side of the Zandrivierspoort, including private owners and company employees were unable to make regular trips to the nearest town and postal agency because the existing track was in such bad condition.[3]

By this time, Carl Forssman (1855–1934), scion of the entrepreneurial Potchefstroom-based family (see Chapter 5), had acquired a piece of land at the entrance to the poort and built a simple hotel and store there.[4] Soon, Carl applied for a liquor licence[5] and not long afterwards, for the right to open a postal agency on his premises.[6] This application was supported by a petition signed by 70 farmers and land managers living to the north of the poort: They included names such as Messrs Houshold and Desborough of the New Belgium (Transvaal) Land & Development Company; nine members of the Snijders family from Zanddrift on the Dwars River; Purrell Thompson of the Oceana Transvaal Land Co; and Arthur Peacock of Cremartardfontein (a property now part of the D'Nyala Reserve, on which there was – and is – a magnificent baobab tree). Permission to open the postal agency was granted, and Forssman received a salary of £12 a year for his trouble. The agency only closed in 1965.

Petitions were a popular way of lobbying government in those days. The postal agency petition had been preceded by one requesting that the government build and maintain a proper road from Nylstroom to Zandrivierspoort and beyond. De Beer's letter (above) had been written in support of the petition.[7] This request joined a long list of roads the government needed to establish or upgrade, as reflected in the Annual Reports of the Department of Public Works. Eventually, in 1896, an amount of £4500 was awarded to build the road from Zandrivierspoort to Zoutpan (southwest of present-day Lephalale); it was augmented by a further £2000 the following year.[8]

Therefore, by the time of the outbreak of the South African War in October 1899, Zandrivierspoort was served by a reasonably well-maintained, all-seasons wagon road with a small facility at the southern entrance to the poort providing accommodation, refreshments, supplies and post. This was the only all-weather route onto the Waterberg plateau. Another, longer road, from outside Nylstroom via Zwagershoek in the west (where Rankin's Pass would later be established) skirted the southern range of the Sandriviersberge until it reached the Matlabas River in the bushveld to the west of the Waterberg. The track called Tarentaalstraat some 25 km to the east of the poort was extremely steep and rough and, until about 1912, passable only in the dry season (see Chapter 10).

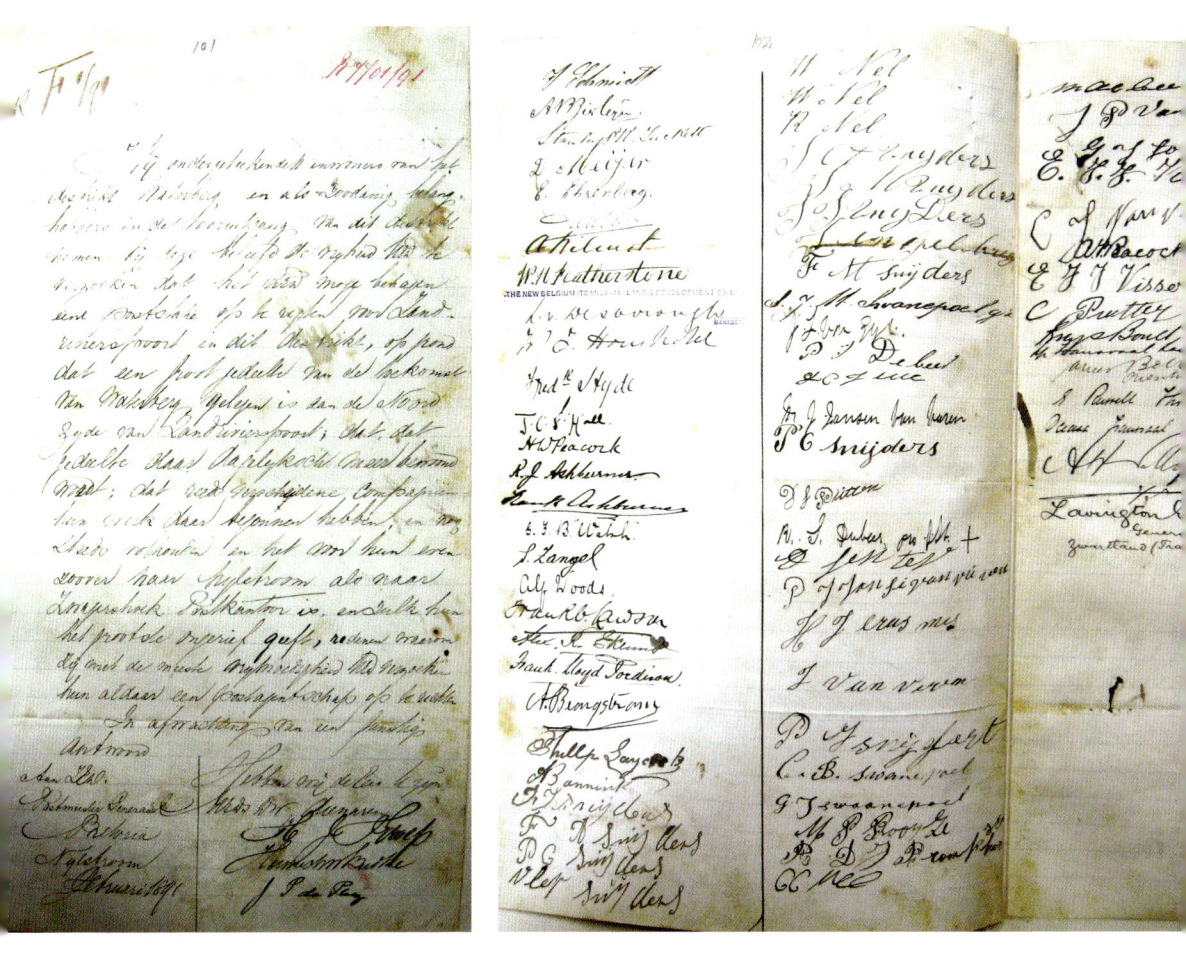

FIGURE 8.4: Petition in support of Forssman's post office application, January 1891.

The role of Zandrivierspoort in the war

With the outbreak of war in October 1899, Boer residents of the Waterberg immediately met to set up commando groups in preparation for the coming conflict – even though the battles themselves at that time were taking place in Natal and Free State, along the line of advance of the British forces.

One of the few accounts of the course of the war in the Waterberg was provided by William Caine, the manager of the New Belgium property on the northern part of the plateau. Caine, who lived alone on the farm Noord Brabant 48 where he managed a herd of over 1000 head of cattle as well as horses and donkeys, maintained a diary throughout the war, until forced by the British to leave the property in November 1901 (see Chapter 9). He recorded being instructed on 5 October 1899 to attend a meeting of all burghers at "Honiefontein" [Heuningfontein] two days later:

> OCTOBER 6TH – Left with Mr Edwards on horseback and we were the first to arrive at Sand River Poort and received our room. Burghers continued arriving until dark and we had to let 3 or 4 lie in the outer room to sleep, for which they paid nothing. Mr Peacock and F Linsley [sic] had the other room which was full and noisy.
>
> OCTOBER 7TH – Rode on in company with 8 other men to the meeting place. We arrived early and old Oosthuizen gave us some breakfast, he and his family being very civil and polite to me. General FA Grobler arrived about 11am and there were about 200 burghers present.[9]

Caine also referred a number of times to Carl Forssman, for example:

> 20TH JANUARY 1900 – I must state here that I have found Genl. [I du Plessis] De Beer very civil and obliging. He is the General in command about here during the absence of FA Grobler at the front. Genl. De Beer gave me a certificate enabling me to travel [from his compulsory temporary camp at Sondagsloop] to Noord Brabant and back, also to Nylstroom any time I wished; and Mr Carl Forssman was a friend to me in these times. He was commissariat at the Head Laager [on Brakfontein, site of the future Dorset police station] and often passed on his way between Nylstroom and the laager.[10]

In May, Caine was told to move his stock well away from the plateau and as he needed to move the herd anyway for winter grazing, he acquiesced. Once again, Carl Forssman came to his assistance, offering the use of one his farms, "Hartebeeste Laaghte" (Hartbeestlaagte), about 25 km northeast of Nylstroom. This trek, which followed a route through Nylstroom and involved a distance of over 120 km, was accomplished in ten days, including rests, without loss. While on this property, Caine learned that some of his black staff and their children on Noord Brabant had been murdered by gangs sent out by Chief Hans Masebe. Concerns about such violence had been one of the reasons the Boer commanders had called settlers in from outlying areas to the security of laagers.

The war did not reach the Waterberg until after the occupation of Pretoria by Lord Roberts in June 1900 and the commencement of the bitter, increasingly desperate and ultimately futile guerrilla phase of the conflict, which was to drag the war out for another two full years. In the meanwhile, of course, commandos from both the Waterberg and Soutpansberg had seen action with other Boer forces in the Free State and Transvaal where their already considerable skills in horsemanship, weaponry and bushcraft had been augmented by combat experience and some degree of military discipline. In other words, they were ideally suited to guerrilla warfare – a style of combat of which the British army at that time had almost no experience.

In August 1900 Caine was returning to Hartbeestlaagte after a visit to Noord Brabant to check that all was in order there. On the evening of the 24th he stopped at Zandrivierspoort where he wrote, "I met a man who said the English had taken the Warm Baths and would be in Nylstroom the next day. At the laager [which he had passed earlier that day], they said the English had been driven back to Pretoria!"[11]

In fact, Nylstroom was occupied briefly on 26 August, but the British withdrew to Warmbaths a few days later, apparently pursued by a Boer force under General F.A. Grobler. Caine stayed on at Hartbeestlaagte for a few more days before setting off with his cattle herd for New Belgium once more. To avoid becoming involved in any fighting around Nylstroom, he followed a back road from Naboomspruit by way of Kromkloof and Doornkom to Rietfontein, near Heuningfontein, where the cattle were rested for several days. Along the way, he was harried by groups of burghers on horseback with varying instructions to commandeer cattle from his herd, all on behalf of a Commandant Barend Vorster. While at Rietfontein, Caine heard the sound of cannon fire from the direction of Nylstroom. He moved on to Zandrivierspoort during the night of 6 September under a full moon, arriving there at sunrise.

> At the poort there is a store and hotel, but the storekeeper left for Germany last May and there are only two [blacks] looking after the place, who are always very attentive to me. One of them told me that they were disturbed the night before by their pigs in the pens making a great to do and rushing out with their guns they saw two men jump out of a pen and make for the steep hill which is close to the back of the store. They fired and one man stopped. Then they saw he was a Boer, so they did not fire any more but caught him

and as they had found one of their pigs lying stunned by a heavy stone, they asked him what he was stealing pigs for. He said he had nothing to eat and was hungry, so they only took his name (of course a fictitious one) and let him go, but the boys tell me the hills are full of Boers that are keeping out of sight so as not to be commandeered.[12]

By the time Caine and his herders reached New Belgium, over a hundred head of cattle had been commandeered by one or other instruction from Boer leaders desperate to keep their men fed. Being the diligent company manager he was, Caine insisted upon, and received, receipts for all the animals so taken.

The Mounted Infantry units

To counter the mobile, elusive Boer commandos, the British had begun to create Mounted Infantry units, formed by volunteers from the many regiments already in the country as well as from units of volunteers or "irregulars" established in England and shipped out to join the conflict. This had originally been the idea of the beleaguered General Buller back in December 1899 after his defeat at Colenso: "Would it be possible for you to raise 8000 irregulars in England?" he asked. "They should be equipped as mounted infantry, be able to shoot as well as possible and ride decently. I would amalgamate them with colonials."

Buller's concept was endorsed by Field Marshall Roberts in October 1900: "Corps of mounted infantry must be raised, say 200 or 300 strong, and each corps given a district and told to live as much as possible on the country. We can go on till Doomsday hunting these Boers with infantry – they only laugh at us."[13]

However, it was found that often infantrymen put on horses turned out to be bad cavalrymen, as they could barely ride. Indeed, it was soon realised that colonial troops (from South Africa, Rhodesia and Australia) proved more successful as mounted infantry, as they were more used to riding horses and were more at home in the open veld. By February 1900 there were already eight Mounted Infantry battalions and a year later more than double that number. Their members were drawn from many different regiments, both colonial and British. From mid-1901, the Waterberg region was primarily the responsibility of the 20th Mounted Infantry Battalion.

The Boer forces in the northern Transvaal came under the overall leadership of General Christiaan Frederik Beyers, a 32-year-old Stellenbosch-trained lawyer with no formal military training prior to the outbreak of the war, but who was a natural leader, had a brilliant mind and great tactical skills.[14] By mid-1901, groups of Boer guerrillas were harassing and "tying up" substantially larger British army units throughout the Transvaal north of Pretoria while repeatedly threatening the security of the vital rail line linking Pretoria with Pietersburg. Frequently, they made their base in the refuge of the mountain fastness that was the Waterberg plateau, which had few – and readily defended – access points, the most important of which was Zandrivierspoort.

In April 1901 William Caine, who had spent the last six months continuously on New Belgium – where successive requisitioning visits from Boer envoys, including one from General Beyers, had reduced his cattle herd to under 200 – sent one of his men to reconnoitre the road to Nylstroom.

The man returned to inform him that the Boers had established positions at Zandrivierspoort and that through-passage was not possible.[15] Many entries in Caine's diary at this time record the movement of Boer families with their wagons and stock northwards towards the Bechuanaland border, often furtively in order to avoid discovery by Gen. Beyers's men. He also recorded (on 24 April) the surrender of Gen. F.A. Grobler with 25 men from Zwagershoek, and subsequently (on 10 May) that "Grobler (ex-General) has crossed the border [into Bechuanaland] with all his livestock, surely now the war must be near its end".[16]

FIGURE 8.5: The young General C.F. Beyers. SANA TAB Photographic Library, No. 32986.

British forces in the Transvaal north of Pretoria by this time comprised two columns under the command of Lt Cols H.M. Grenfell and Henry Hughes Wilson (who was later to become famous as a Field Marshall in World War I). On 18 May 1901, Wilson's column, from Naboomspruit, supported by another from Pretoria under Maj. H. McMicking, encountered Beyers near Boekenhoutskloof northeast of Nylstroom – no doubt en route to disrupt the railway line. Almost surrounding Beyers, and taking several prisoners, they forced him back towards Zandrivierspoort, where "he fell back on a strong main body laagered in the almost inaccessible fastness [of the poort], where he had to be left until more strength could be brought against him". [17]

This laager was located in a hidden valley in the middle of the poort, surrounded by protective hills (hence the name Omsingel, of the rail siding later constructed on the site) yet with access to the ever-flowing Mokolo River. On the hilltops, the Boers built rough forts, or *skanse*, from which they were able to monitor the approach of enemy troops and if necessary, fire upon them. An example of the remains of one of these fortifications can be seen today, high up on the hill above the Mashudu Lodge resort on the farm Modderspruit in the middle of the poort, from where there is a commanding view of its southern entrance.

Caine's diary referred to some of the skirmishes that took place in May 1901:

> SATURDAY 25TH MAY – Plenty of news this week. On Tuesday 21st I heard cannon firing in the direction of Sand River Poort from sunrise up to 11am. Again Wednesday some four or five shots in the morning, Thursday one of my old boys turned up, coming from the Boer laager, he stated that the Boers had all run away from the laager he was in and that the English had captured another laager this side of Sand River and that there were no more Boers between me and Sand River, glorious news if true.[18]

A month later, on 21 June 1901, Grenfell, with a column of 1300 men and three field guns, marched from Potgietersrust (Mokopane) with the intention of making his way through the northern ramparts of the Waterberg in order to approach Zandrivierspoort from behind. At the same time, McMicking, with 550 men and two guns, moved north-westwards out of Nylstroom.[19] Inevitably, Grenfell's column became seriously delayed through having to manoeuvre his transport wagons through the hills, with the result that by the time he eventually reached Zandrivierspoort a week later, he found McMicking's men already in occupation, having driven the Boers away to the west after a 64 km night march and a brief but intense fight. Grenfell immediately set out in pursuit, returning

FIGURE 8.6: Remains of a Boer skans (LEFT) overlooking southern entrance to the poort (RIGHT).

to Zandriviersspoort a few days later with 133 prisoners and 77 wagons, the Boer group having been hindered in its escape by the company of several hundred women and children.[20]

Caine reported the following in his diary:

> WEDNESDAY [3RD JULY] – I rode over to the Peacocks and heard from them that Mr Boshoff [a former member of the Volksraad] had been there last Sunday … with all his livestock, also several other Boers and that numbers had passed flying towards the Crocodile [Limpopo – the border with Bechuanaland]. Boshoff stated that the British troops had come through the poort at Sand River and were pursuing the Boer laager, which fled down the Sand River [Mokolo].[21]

McMicking's column, which included the 20th Mounted Infantry Battalion, now made Zandriviersspoort its base for the next few months, using it as a central point from which to secure the area.

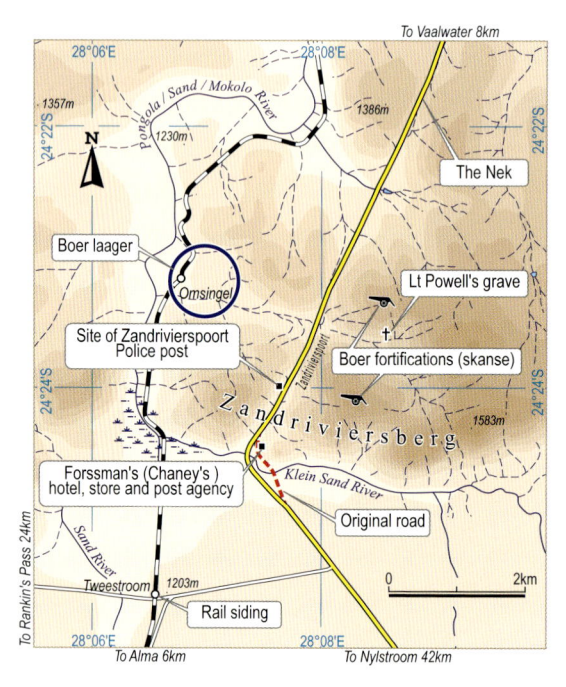

FIGURE 8.7: Zandriviersspoort and surrounds.

FIGURE 8.8 (OPPOSITE): Lt Charles F.B. Powell in ceremonial dress, and his grave in Sandriviersspoort. Photograph of Lt Powell provided courtesy of the Doncaster Museum & Art Gallery (see note 27).

Lieutenant Charles Folliott Borradaile Powell

Charles Folliott Borradaile (misspelt Borrodaile on his headstone) Powell, born 19 December 1879, was the fourth of five children of Sir Richard Douglas Powell, MD, and Juliet Powell (née Bennett). His father was a leading doctor and a Fellow of the Royal College of Physicians with a practice in London.[22] Charles attended Rugby School in Warwickshire from 1894 to 1897 where school records show that he performed well in shooting competitions.[23] In 1898 he was accepted into the Royal Military College, Sandhurst and in August the following year, he joined the 1st Battalion, Kings Own Yorkshire Light Infantry (KOYLI) as a 2nd Lieutenant, being promoted to 1st Lieutenant in April 1900.[24] Lt Powell was selected to command one of four sections of a new Mounted Infantry company, comprising two sections each from the 1st Battalion KOYLI and the Liverpool Regiment, formed at Curragh, Ireland, under the overall command of a Captain Colquhoun. On 18 March 1901 the company embarked for South Africa from Cork, arriving in East London on 7 April. It then travelled by train to Pretoria where it joined the column created under Major McMicking, as part of the newly created 20th Mounted Infantry Battalion.[25] Soon, the battalion was deployed to the Waterberg, based at Zandrivierspoort, under McMicking's command.

On 13 July a small patrol led by Lt Powell came under fire while out on the farm Rietfontein on the Sand River plain south of the poort. Lt Powell was killed and a Lance Corporal Dewsnap wounded in this engagement. Lt Powell was buried near the British encampment in the poort and his grave there can still be seen.[26]

A memorial service for Lt Powell was held in a packed St Mary's Cathedral, Limerick, Ireland, on Sunday 29 July 1901. His obituary, published in the August 1901 edition of the KOYLI regimental journal *The Bugle*, recorded that "Lt Powell was a splendid type of British Officer. His courage, straightforwardness and attractive manner won the affection of all ranks and the Commanding Officer feels sure the whole Battalion will mourn in the loss of a gallant comrade".[27]

During July and August 1901, Grenfell's and Beyers's forces engaged fleetingly several times, but the Boers were always able to make off without serious losses. On three occasions during this period, trains along the line to Pietersburg were attacked and wrecked, twice under the expert leadership of "Captain" Jack Hindon. These repeated attacks forced Grenfell to continue to hunt Beyers and his men in the hills west of Nylstroom, with limited success. By September both Grenfell and Wilson had been transferred to other duties in the Cape, their places being taken by Lt Cols Colenbrander (from Rhodesia) and J.W.G. Dawkins, transferred from the Orange River Colony. Colenbrander set off to secure the region westwards as far as the railway line from Mafeking to Rhodesia, while Dawkins harried Beyers along the Waterberg, including his laager at Geelhoutkop and another on the Palala.[28]

In early November 1901, Colenbrander, remaining in the area, encountered a Boer force of three or four hundred men, established once again at their old laager in Zandriviersport, where a Commandant C. Badenhorst was acting in place of the ever-elusive Beyers. Colenbrander attacked at once and then pursued the Boer commando for over 60 km to the southeast, before turning into Warmbaths on 19 November for supplies. Three days later, Colenbrander resumed his pursuit of Badenhorst, joined now by a force under Dawkins from Nylstroom. They soon routed the Boer commando, which split into many smaller groups. One of these, joined somehow by Beyers himself, would almost certainly have been captured at Geelhoutkop in early December, had not poor weather and miscommunication among the British forces allowed an escape. Two days later Badenhorst and the bulk of the Boer group, encamped on Sterkfontein north of Rankin's Pass, was surrounded by a combined force of Dawkins and Colenbrander and forced to surrender.[29] (See Chapter 4 for details of this action.) This effectively concluded the war along the Waterberg, although Beyers was never captured and continued to harry the British units in the hilly country to the east of Potgietersrust and Pietersburg, until the signing of the Treaty of Vereeniging on 31 May 1902 finally brought the war to an end.

After the war: Policing the district

The British administration ("Crown Colony rule", under the governorship of Lord Milner) that came into effect following the end of the South African War in May 1902, set about the reconstruction of the war-ravaged Transvaal, commencing with the resettlement of Boer prisoners and black refugees on the lands they had been forced to vacate during the conflict. Commissions were established to consider applications for reparations and compensation for property and livestock lost during the war, and police posts were established to keep the peace.[30]

As early as September 1900, the British Commander-in-Chief, Field Marshall Lord Roberts, had promoted Col Robert Baden-Powell (of Mafeking fame) to Major General and appointed him to set up a new militarised police force, to be called the South African Constabulary (SAC), with the objective of enforcing peace and policing measures over the Transvaal and Orange River Colony. Members of the SAC were recruited from British subjects in SA and elsewhere (especially Canada). From an initial strength of about 800 (including 88 officers) in January 1901, the SAC eventually grew to a force of over 10 000 (with 272 officers) two years later. Initially, the continuation of the war as a guerrilla conflict meant that the SAC fought alongside other British units; indeed, it did so with great honour and suffered the second highest casualty rate (the Royal Artillery had the highest) of any British or colonial regiment in the war.[31]

But after May 1902 the SAC began to fulfil its original intent: A network of police posts was established across the rural areas and manned such that each farm could be visited by a patrol at least once a week. This commitment and presence did much to build relations between the returning Boer farmers and the British authorities. At its peak, in 1903, the SAC manned 210 police posts or stations in 28 districts across Transvaal, Orange River Colony and Swaziland. But gradually, as peace returned to the region, the size of the force was progressively reduced, until by the time of its dissolution in June 1908 – to be replaced (in the Transvaal) by the Transvaal Police force – the SAC numbered only 1742 men, two-thirds of them in the Transvaal.

During 1903 a Compensation Claims Commission was established for the

Waterberg District to consider claims from Boer farmers. Major Robert French of the 2nd Gloucestershire Regiment was seconded to the commission as its military member; and in January 1904, he submitted a summary of his experiences to British Military Intelligence. The following are extracts from his report:

> The roads throughout the District, except those which have been taken in hand by the Public Works, are simply mere tracks. The road, however, north from Warmbaths to Nylstroom and Potgieters has been laid, and is metalled [gravelled] in most places. The road west from Nylstroom through the Zand River poort is also being laid and should by now be nearly complete. It extends for about 35 miles [i.e. almost as far as where Vaalwater would shortly be established].
>
> I was fortunate enough to be able to mess with the South African Constabulary [headquartered in Nylstroom] instead of having to live at the wretched hotel while I was in Nylstroom, and I had a good opportunity of seeing their work generally throughout the District while trekking. They are scattered about the country in small posts of three or four men, except at headquarters or sub-districts, of which there are two in the Waterberg, one at Potgieters and another about sixty miles west of Nylstroom [Rhenosterpoort, later Rankin's Pass], where there are a few more men and an officer. They patrol the whole of their districts regularly, know all the farms and the people living on them, what stock, fowls &c., they possess and their circumstances generally, also whether they are industrious or otherwise. They are a great assistance to the Burghers generally in letting them know the various government ordinances, what farms or roads are closed on account of quarantine for lung sickness or Redwater, and they are very obliging in helping them in any way they can, taking messages or parcels for them. I have always heard them very well spoken of by the Dutch people I have met in the District. People in large towns may grumble at their cost, but they are doing a great deal of good work ... that no amount of money can repay.[32]

Although there were SAC posts at Nylstroom and Rankin's Pass, it seems that a police post was only established at Zandrivierspoort with the formation of the Transvaal Police in 1908 – by which time a small settlement was already in place on the farm Vaalwater, beyond the poort. In November 1909 H.C. Bredell (former Presidential Secretary to President Kruger)[33] prepared a report on the status of the various police posts in the Waterberg District for the Commissioner of the Transvaal Police. His report covered a total of nine police posts in the District, of which four (Brakfontein, now Dorset, Geelhoutkop, Zandrivierspoort and Rhenosterpoort, now Rankin's Pass) were located in what we know today as the Waterberg plateau. Included in the report was the following description of the police post at Zandrivierspoort:

> MEN's QUARTERS – Wood and Iron building 11 x 20 [feet] with a very low roof, used as an office and sleeping quarters for the men, which is unsatisfactory. The building is so buggy and rat-infested that one of the men will not sleep therein, preferring a tent.
> MESS ROOM – Of reeds, made by private enterprise.
> STABLE – Of wood and iron with earth floor in a very bad condition. Water 500 yards away. Considering the climatic conditions existing here and besides being a Fever Area and also the advisability of shifting the Station to a more convenient spot nearby, I beg to recommend that buildings of stone be put up and provision made for the proper housing of the men. At the proposed new site, water will only be about 130 yards away from the station and plenty of stone for building purposes close at hand. All buildings including stables should be Mosquito Proof.

FIGURE 8.9: The Zandrivierspoort police post, 1909. Illustration drawn by C. Bleach from a photograph in report by H.C. Bredell, 1909.

> Police signboards and a flagpole should also be put up to give the Station an official character and enable a stranger to find the Police Office without much difficulty.

In an addendum Mr Bredell recommended that the police post at Zandrivierspoort (which was manned by "two Europeans and one Native constable") should be abandoned and "concentrated" to Brakfontein (Dorset) and Rhenosterpoort (Rankin's Pass) posts, because

> there are about 100 Whites and 300 Native adults in this Police Area, averaging something like three Police cases per month, consisting mostly of Petty Crimes committed by Natives, and I do not consider that circumstances warrant the continuation of this Police Post. The buildings here are in a bad state of repair and unsuitable for permanent habitation. [34]

This was a surprising conclusion, because the map accompanying the report shows that the Zandrivierspoort Police Area included both the Vaalwater and Sanddrift settlements as well as a large area to the south of the poort along the open valley of the Sand and Klein Sand Rivers, including the settled properties on the farms Rietfontein and Heuningfontein. As it turned out, the police post was not closed then, but remained in operation at least until 1935, according to records of the general store at Zandrivierspoort.

Zandrivierspoort: The Chaney era

On the first day of August 1908, a 27-year-old immigrant from Donington in Lincolnshire joined the newly formed Transvaal Police as a constable (No. 1500), having previously served with the SAC since March 1904 and before that during the South African War with the Imperial Yeomanry.[35] His name was John William Chaney. He was the son of Annie and the late William Chaney, a farmer at Northorpe near Donington, who had died in 1885 when John was only three. The following year, Annie married John Sills, a widower in the town.

After joining the Transvaal Police, John was initially based at the Nylstroom station where in September 1909 he was appointed as Acting Public Prosecutor for the Court of Sub-Native Commissioner at Nylstroom.[36]

Four months later he was posted as officer in charge of the police post at Rhenosterpoort (Rankin's Pass). The resident magistrate in Nylstroom, in advising the Law Department in Pretoria of the transfer, requested that Const. Chaney be appointed clerk of the court at Nylstroom and also public prosecutor for that court with immediate effect. Despite a clerical query in the margin of the application letter as to the applicant's legal qualifications and knowledge of Dutch (both of which were wanting) the response was favourable:

> I, Jan Christiaan Smuts, Acting Attorney General for the Transvaal, do hereby nominate and appoint John William Chaney to appear before the Court of the Resident Justice of the Peace to be holden at Rhenosterpoort in the District of Waterberg, and for me, and in my name, to prosecute all such Criminal cases as shall be therein pending, from the 15th day of January, 1910. This delegation of authority is issued to John William Chaney in his capacity as Acting Public Prosecutor at Rhenosterpoort for the time being.
>
> Given under my hand at Pretoria, this 15th day of January, 1910.
> Signed: JC Smuts, Acting Attorney General.[37]

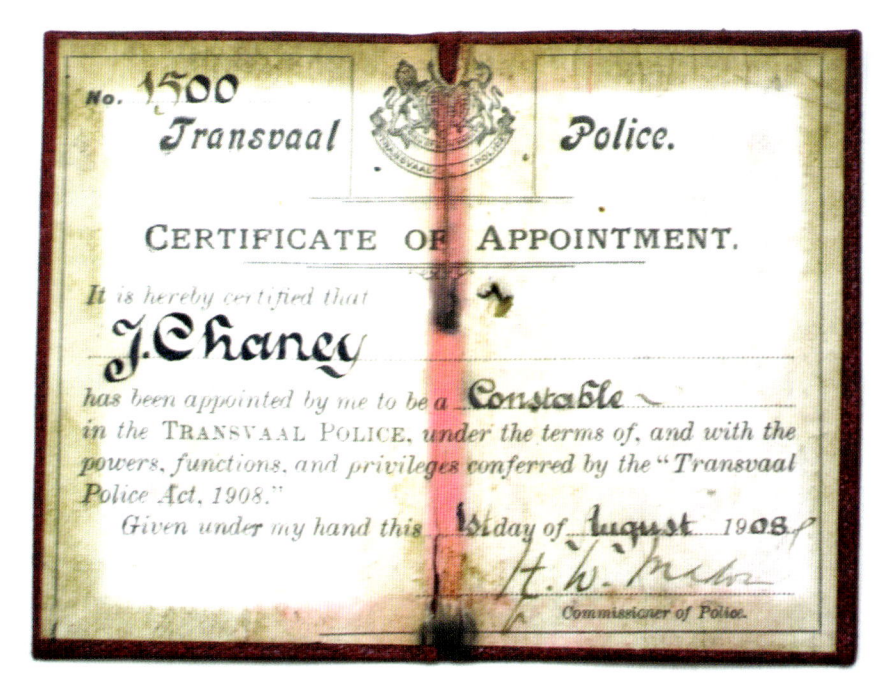

FIGURE 8.10: Certificate of appointment of John Chaney as constable with the Transvaal Police. SANA TAB TP 9 1500.

In 1908, according to memorabilia in the possession of Chaney's granddaughter, Anne Howe, a Vaalwater resident, Chaney became a member of the Freemasons, at the Pyramids Lodge in Nylstroom, where the Head (Worshipful Master) at the time was the prominent Nylstroom storekeeper, Emil Tamsen, who had also founded the Lodge (see Chapter 5).

At the end of December 1911, Chaney, now a Lance Corporal, signed an agreement to extend his service for a further two years from the completion of his current contract on 28 March 1912. However, this was not to be and Chaney left the Police after all when his first contract ended. What had happened that would change his mind evidently within days or weeks of having committed himself to an extension? The following record might provide an explanation.

In its issue of 28 January, 1913, the English regional newspaper *Lincolnshire Free Press* published a photograph and letter from J.W. Chaney:

> ZANDRIVERSPOORT, VIA AYLSTROOM [SIC], TRANSVAAL, DATED DEC. 27TH 1912:– As a regular reader of your paper … I thought the following would probably interest you. I am a native of Donington and have the Free Press sent to me every week. … Serving in the Transvaal Police, which I left a year ago, I was in charge of an area the size of Lincolnshire and was Public Prosecutor for that area and Postmaster for a considerable portion of it. I now have a general store and hotel in my old area, where I am well known. As you will notice in the

FIGURE 8.11: Chaney's hotel and store at Zandrivierspoort (Nylstroom to the right), with outspanned wagons. Photograph obtained from the archives of the Lincolnshire Free Press.

CHAPTER 8: THE GAP THROUGH THE HILLS   *273*

photo, I am residing at the foot of a mountain range, the pass being three miles long before one descends into the immeasurable bush veldt, the home of every sort of game that South Africa produces, from a hare to a lion. You will also notice the wagons which are "outspanned" outside the store. The photo was taken a long way from the house and some way up the mountainside, which has undoubtedly made it look small.[38]

Perhaps an opportunity suddenly arose that allowed Chaney to purchase Carl Forssman's hotel and store at Zandrivierspoort, persuading him to withdraw from his contract with the police.

There are several other images of Chaney's hotel and store dating from those early days, as shown below:

FIGURE 8.12 (TOP LEFT): A photograph of the back of the store pasted into one of John Chaney's diaries. The hill in the background is the one from which the photograph in Figure 8.11 was taken. Photo courtesy of Anne Howe. (TOP RIGHT): The front of the store, probably taken in the 1920s. Photo courtesy of Pierre de Wet. The Zandrivierspoort Hotel, then (BOTTOM LEFT) and now (BOTTOM RIGHT). Photographs courtesy of Pierre de Wet.

For many years, John Chaney's general store and hotel seem to have flourished, as can be seen from the ledgers recording his customers' accounts and the hotel register, still in the possession of his granddaughter, Anne Howe.

In 1919 the recently founded North Waterberg Farmers' Association chose to hold one of its meetings at the hotel. This led to a letter being written to the Department of Agriculture for the appointment of a locally resident vet, an extension officer and for wire for fencing off poisonous plants (*gifblare*).[39]

In 1928 Chaney took transfer of Portion C of the farm Zandrivierspoort (two morgen) from the State, which had previously reserved this piece of land for "public purposes", on which the hotel and store were built.[40]

In the period from 1925 until 1938, when the store relocated to Vaalwater (where Chaney had purchased the old Kirkman & Armstrong business), regular customers of the Zandrivierspoort Store included members of the nearby police post: Ted Davidson, John Mortimer and others from 24 Rivers; employees of the New Belgium Estate; Willem van Heerden, A. Knoble and John Rankin of Rhenosterpoort (Rankin's Pass); the Kirkmans of Vaalwater; and the Websters and Middletons of Sondagsloop.

Generally, the store seemed to have been able to secure payments from most of its account holders – although the ledgers show a few whose debts were eventually written off. The store sold a wide variety of goods, from grains, coal, oil and fuel, to liquor, cigarettes and blankets; truly a general store. The postal agency received mail deliveries from the South African Railways, which conveyed it three times a week by road from the nearest siding at Tweestroom.[41]

The hotel register (for the period 1925 to 1929) reflects the visits of numerous local residents as well as travellers and sales representatives to or through the region – and their often amusing or illuminating comments. Many refer to the poor state of the road. One complained about the impertinence of the desk clerk. Another, E.C. Holloway, described his address as "from Back of Beyond". The Griegs (who had bought Hermanusdoorns near Bulgerivier in 1913) were frequent overnight visitors, sometimes with their friends the Theesens and Ralstons from Johannesburg.

The Davidsons of 24 Rivers, too, often stayed or stopped at the hotel en route to or from Pretoria during this period. One of Ted and Molly Davidson's four daughters, Elizabeth Clarke (1912–1954), later wrote in her memoirs:

> At the foot of the Nek, and with a lovely view of the Seven Sisters, was a country store and inn, and many a story could that rambling,

white-washed building tell, for after the perils of the Nek, every trekker would relax and restore his nerves with liquid refreshment. Many a cold morning have we gone into the little sitting room and drunk boiling hot coffee. Meantime my father would disappear with his coffee through a door opposite to that to the store, and which we were forbidden ever to open. There would follow the clinking of glass, low husky murmur of men's voices, followed by gusty guffawing, and my father would return a little pinker in the cheeks and was always in good humour for the rest of that trek.[42]

There are several register entries for personnel involved with the construction of the railway line from Nylstroom to Vaalwater (the line was opened on 1 October 1925) and a surprising number of military visitors, often of senior rank. One was Lt Col A.H. Geyser, head of the Potgietersrust Command, who accompanied a Lt Col Ledger from Nairobi on a visit in July 1925. Back in November 1914, a Union Forces commando under then Commandant Geyser had suffered severe casualties in a skirmish with rebel forces opposed to South Africa's entry into World War I, on Zandfontein, south of Rankin's Pass, some 45 km southwest of the hotel (see Chapter 15).

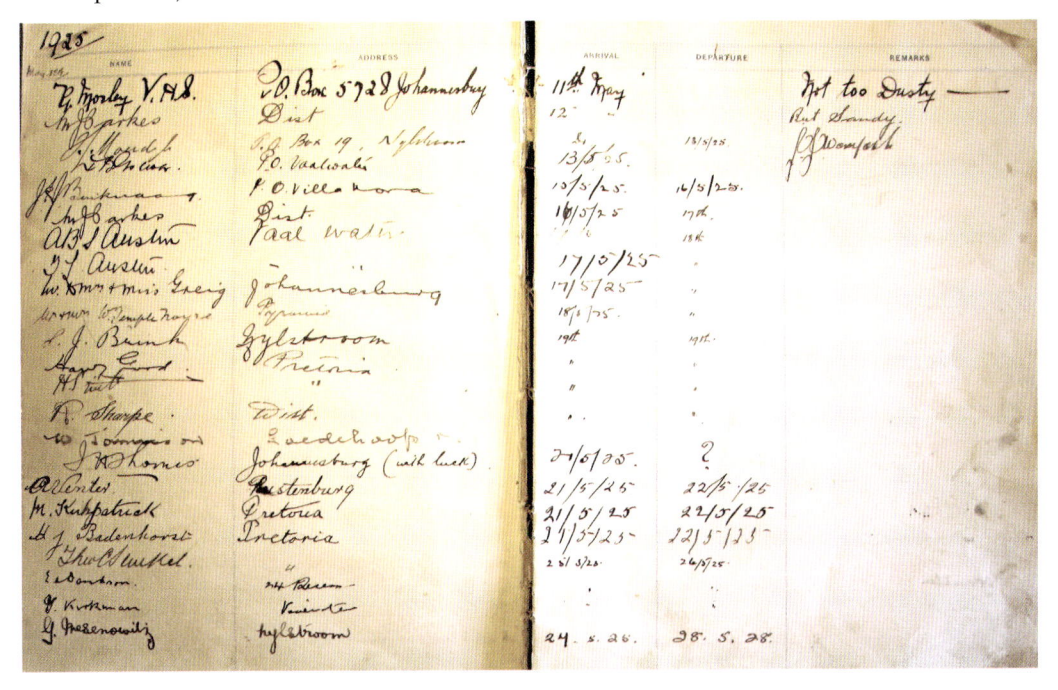

FIGURE 8.13: Zandrivierspoort Hotel Register page, May 1925. (Photo courtesy of Anne Howe.)

Louis Nel, a resident of Melkrivier and whose father was the head of the school there, records that in the early 1930s cars were few and far between and were not built to withstand the rough roads of the Waterberg. A journey from Vaalwater to Nylstroom, even with one of the best cars of the time, would easily take four hours.[43] Today, the 60-kilometre journey is much faster: Rumour has it that a descendant of one of the pioneer families has covered the distance in only 20 minutes …

In 1937, Mr Chaney, who some years before had married Miss Winifred Frances Armstrong, moved from Zandrivierspoort to Vaalwater where he bought the mill and store from the Kirkmans. In December 1939, Chaney purchased Portion B of the farm Vaalwater (500 morgen) from Messrs J.B. and G.R. Kirkman for £500; and five years later, sold it again, to Volkskoper Mark Bpk for £8100 – a shrewd investment.[44] The close relationship that existed between the Chaneys, Kirkmans and Armstrongs is illustrated by the fact that the Chaneys' joint will of 1940 was witnessed by J.B. Kirkman and R.M. Armstrong. According to notes made by their daughter Mary and held by their granddaughter Anne Howe, John and Winifred lived in the grand old house near the mill, which had possibly been built by Gilbert Kirkman in 1930–1935. The move from Zandrivierspoort to Vaalwater is reflected in the store letterheads.[45]

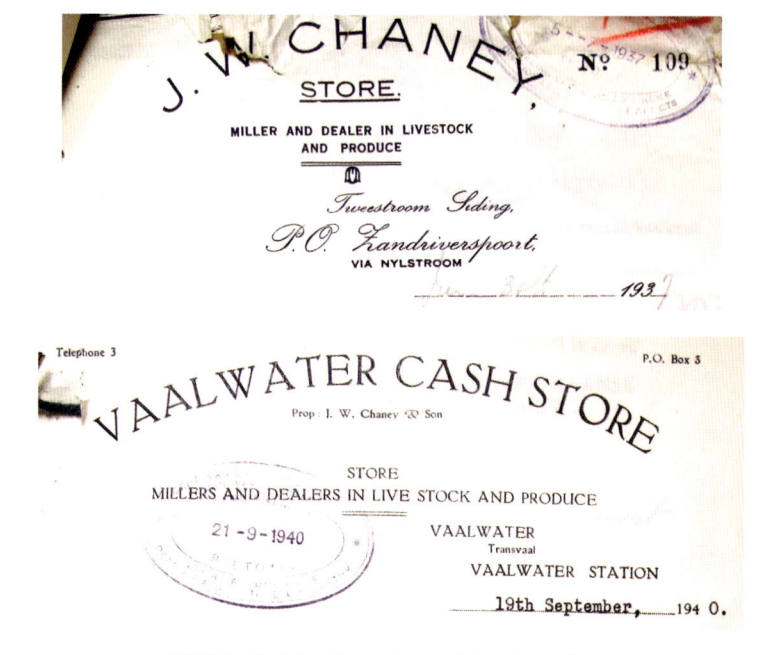

FIGURE 8.14: John Chaney's store letterheads from Zandrivierspoort (1937) and Vaalwater (1940).

John Chaney died, aged 70, on 19 August 1952 on the family farm Goedehoop, 10 km out of Vaalwater on the road to Bulgerivier. He left his widow Winifred and three adult children: John Everard, Frances Mary, and Margaret (who became a Perridge).[46] Winifred died in Nylstroom, where she had lived with her unmarried daughter, Mary, in February 1976, aged 93.[47] Mary died in 2005.

After the Chaneys had moved their business from Zandrivierspoort to Vaalwater, the hotel continued operating until the outbreak of World War II, when it too, closed, never to reopen. The postal agency founded by Forssman in 1891 and which since 1920 had been part of the general dealership licence, continued to offer a public service until May 1965 when it finally closed following the death of the last agent, J.J. Joubert.[48]

The hotel buildings still stand among the blue gums between the Klein Sand River and the base of one of the buttresses of the poort entrance, and are still occupied and lovingly maintained by the current owners. The old police post was only demolished at the beginning of the present century. It had been disused for many years, being occupied occasionally by school groups from the Mashudu resort as an overnight camping stop – when children were reportedly thrilled (and a little scared) by the manacle rings still fixed to the walls of the holding cell.[49]

More recently, the main tar road (the R33) was realigned to cross a new bridge over the Klein Sand River about 400 metres downstream and to the west of the original drift – in so doing, creating a hazardous bend that has been the scene of numerous accidents. The outspanned wagons seen in the *Free Press* photograph of 1913 (Figure 8.11) would have occupied the area between the new road and the hotel, an area which until recently was filled with blue gum trees. North of the bridge, the road through the poort has become a four-lane highway, with the result that traffic speeding through these hills offers little opportunity for passers-by to appreciate its strategic historical importance or the events that unfolded there over the last 150 years.

CHAPTER NINE
The New Belgium story

Along the northern edge of the Waterberg plateau, in what is arguably the wildest, least developed part of the UNESCO-accredited Waterberg Biosphere Reserve, and included within the rugged wilderness areas of the Lapalala and Kwalata Nature Reserves, lies the farm New Belgium.[1] Although the farm, which seems to be divided into at least two separated parts, is very sizeable (approximately 20 000 ha), the observant map-reader will notice that no homestead corresponding to its name is to be found on any map.

The property of today with the name New Belgium is but a small remnant of what was, until 75 years ago, a huge piece of Waterberg wilderness encompassing 116 870 ha. A typical example of the sixty farms that were to comprise the New Belgium Block was Uitkomst, which lay in the north of the block, straddling the northern Waterberg escarpment immediately east of where the Lephalala River cuts through the hills into the Bushveld. Uitkomst was originally inspected by J.A. Smit in April 1869 when it was given its name in the vast Zwagers Hoek Ward of the Waterberg District.

It may have been occupied informally after this, but it was not until 7 February 1888 that Uitkomst, like all the other inspected farms in the area, was awarded (*toegekend*) by deed of grant (*grondbrief*) to a named individual, by order of the State President, S.P.J. Kruger, for the payment of £1.10. The first lucky owner in this case was one Jan Marnitz. But not for long: The very next day, the title was transferred from Marnitz to the estate of Goosen Johannes Verdoorn (erstwhile postmaster at Nylstroom) for the payment of £7.10 and a day later, from this estate to Henry Pasteur, on behalf of Oceana Transvaal Land Company Limited of London. Henry Pasteur was at that stage chairman of Oceana, which was one of several land speculating outfits that had sprung up in the aftermath of the discovery of gold on the Witwatersrand in 1886. The purchase price is shown as being £100. Not bad appreciation over three days!

THE NEW BELGIUM COMPANY
Complex deals

But better (or more questionable) transactions were to come. Less than two years later, ownership of Uitkomst was again transferred, from Oceana to Edward Hawley, who was to hold the property (and another 59) in trust for the New Belgium Transvaal Land & Development Company Limited. Each of the 60 farms in question had followed a very similar route apart from the name of the very first, brief owner similar entries (including mention of a payment of £40 000 for the group of farms) occur against each property in the Deeds Register. This transfer (from Oceana to Hawley) was registered on 9 December 1889. Six months later, on 30 June 1890, all 60 properties were transferred from Hawley into the name of the New Belgium Company itself – for no additional fee – and on 26 August of the same year, the New Belgium (Transvaal) Land & Development Company Limited (NBTL), which had just been constituted in London, became the registered owner of 60 contiguous properties, together covering almost 117 000 ha of pristine Waterberg wilderness.[2]

The total consideration paid by the newly formed NBTL to Oceana was £182 000, of which £97 000 was in cash, £75 000 in fully paid-up ordinary NBTL shares and the balance of £10 000 in fully paid-up deferred NBTL shares. Given that the total issued share capital of the company was 260 000 shares, valued at £1 each, this meant that Oceana, in addition to receiving a hefty cash payment, retained a 33% interest in the property.

The London journal *The Statist* reported in late 1891 that Oceana had derived an enormous benefit from its trade in land in the Transvaal: In 1888/89 it had purchased a total of around 1,9 million acres of land (over three-quarters of a million hectares) for an average price of 1s/3d per acre. A mere two years later, thanks to the Witwatersrand gold boom, the average value of its holdings had increased more than tenfold, to 11s/4d per acre. In addition, Oceana owned some 185 000 shares in NBTL.[3] The second annual report of the NBTL (for the year ending 30 September 1891) contains a balance sheet of the company's business since its inception.[4] It states, among other matters:

> The Company's property, consisting of 60 farms, or about 380,000 English acres, at Purchase Price of £182,000.0.0. Payments to government, charges in respect of the formation of the Company and Transfer of Properties: £2,115.0.0.

Clearly then, despite the entries in the Deeds Register, an average of a little over £3000 was paid per farm – still a very large sum of money for soil of untested agricultural and unknown mineral potential.

In this same report, Henry Pasteur, Oceana's chairman, is shown as being one of the five directors of NBTL (which is not surprising, given Oceana's shareholding). This brings to mind another issue that was a frequent characteristic of mining, exploration and land companies in South Africa from the discovery of diamonds in the 1860s right up until the implementation of more stringent companies' legislation and governance practices (the King codes) well over a century later: complex cross-holdings of one company in another, and directors who sat on the boards of several companies ostensibly in competition with one another and therefore in potentially conflictual positions.

Indeed a celebrated legal case illustrating the conflicts that can arise in such situations actually involved NBTL – and another Waterberg landowner, the Transvaal Lands Company (TLC). In 1912 two directors of TLC, Samuel and Harvey, convinced their fellow directors of the merits of acquiring shares in Lydenburg Gold Exploration, a company owned by NBTL, which by then was little more than a shell. But in so doing, they failed to disclose that they also had interests in NBTL (Samuel was a director and Harvey administered a family trust which held NBTL shares). TLC successfully challenged this acquisition in court on the basis of the conflict of interest involved – a case that is still cited in modern case law, both in South Africa and internationally.[5]

The size of the estate

The *true* areal extent of the properties making up the New Belgium Block was not known at the time of these early acquisitions – the first annual report refers to a total of 380 997 acres (equivalent to 154 184 ha).[6] Because of the cursory, even questionable *Inspectie* work undertaken by J.A. Smit and his peers in 1869/70, the ZAR government decided in 1887 in terms of "Law 3" of that year, to create *Speciale Commissies* to revisit the properties in an attempt to locate the original beacons and to create new inspection plans.[7]

All the farms comprising the New Belgium Block, as it had come to be known, were inspected by a Speciale Commissie chaired by F.A. Grobler, with C.S. Potgieter and F.S. Watermeyer as fellow inspectors. (Grobler was to become well-known in the Waterberg as the hero of some battles against black chiefs in the early 1890s, and later as the ill-fated Commandant General of the Waterberg Commando at the beginning of the second South African War.) In its report of August 1888, the Commissie concluded that Uitkomst, for example, was 2300 morgen (1970 ha) in extent, rather than the 3000 morgen arbitrarily assigned to it originally. Six years later, a more rigorous survey under the management of E.N. Ferguson recalculated the area of Uitkomst to be 2327 morgen, 87 square roods (equivalent to roughly 1993 ha).[8] This is the generation of surveyed areas that still applies to all the properties concerned. The result is that the total area of the 60 farms was actually more like 289 000 English acres (about 117 000 ha), or only 76% of the extent stated in that annual report of 1891 (see details in Figure 9.2).

The New Belgium Land & Development Company Limited came into being during 1890, and within the year had acquired a large spread of essentially uninhabited land in the Waterberg – although there were certainly a few scattered villages in the region.

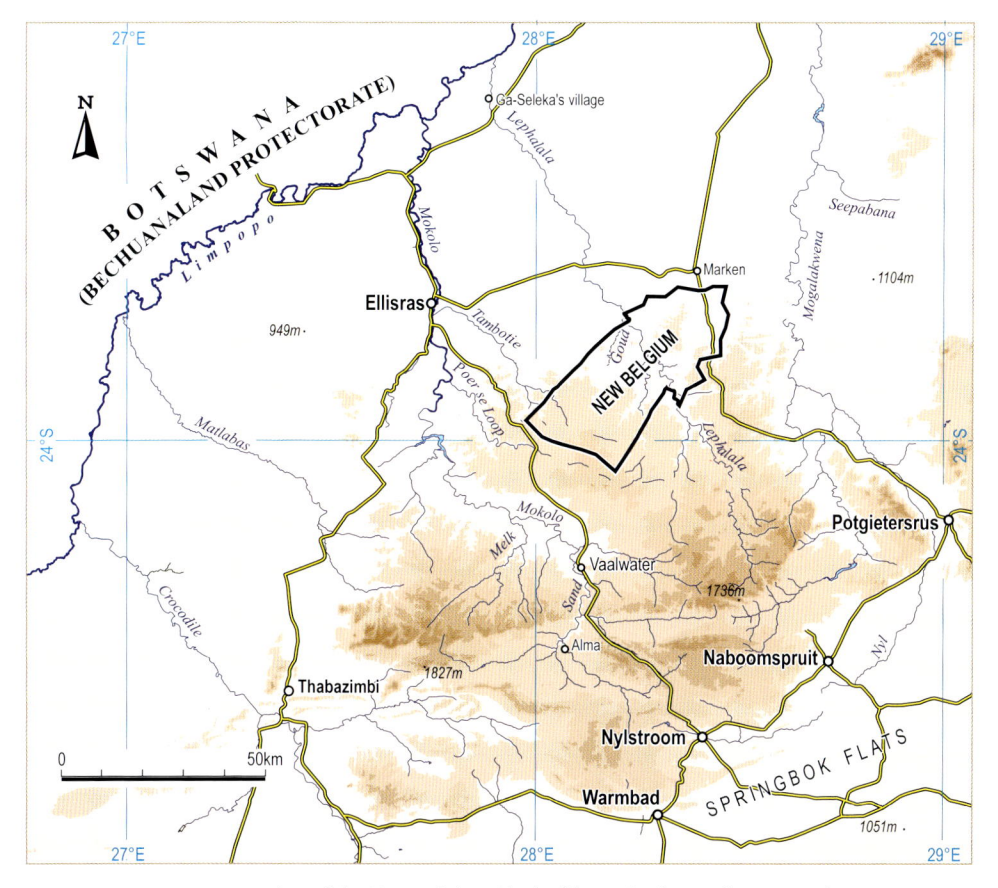

FIGURE 9.1: Location of the New Belgium Block of farms in the northern Waterberg.

CHAPTER 9: THE NEW BELGIUM STORY

THE NEW BELGIUM BLOCK								
Original farms and Evolution of Ownership								
Original farm Name	Original No.	Second No.	Current No. (if issued)	Approximate Area (Ha)	Incorporated in Kwalata	Incorporated in Lapalala	Incorporated in Touchstone	Incorporated in Keta
De Pont	2216	1175		1589	x			
Bouwlust	2169	1143		1859	x			
Schoongelegen	2170	1144		1524	x			
Doornfontein	2215	1174		1619	x			
Blinkwater	2195	1157	604 LR	1751	x			
Witwater	2197	1159		2141	x			
Noord Brabant	2218	1177		2168	x			
Werkendam	2199	1161	628 LR	2090	x			
Bouwland	2166	1140	603 LR	1317	x			
Akkervlyt	2174	1147	632 LR	1546	x			
Papkuilsvley	2212	1169		1457	x			
Wynkeldershoek	2219	1173		1112	x			
Zorgvliet	2175	1148		2105	x			
Goergap	2168	1142		2052	x			
Weltevreden	2167	1141		2395		x		
Nooitgedacht	2183	1150		2255		x		
In-den-Berg	2192	1154		2390		x		
By Uitzoek	2172	1146	600 LR	1694		x		
Ongegund (Boschplaats)	2210	1168	598 LR	1625		x		
Dubbelwater	2191	1153		1469		x		
Landman's Lust	2194	1156	595 LR	1766		x		
Wildenboschdrift	2213	1172	599 LR	1479		x		
Kliphoek	2220	1179	636 LR	1514		x		
Moerdyk	2221	916	593 LR	2630		x		
Haasjesveld	2207	913		2695		x		
Alem	2208	914	544 LR	2698		x		
Lith (split)	2185	907	541 LR	2494		x	x	
Gorcum (split)	2188	910	577 LR	2339		x	x	
Dordrecht (split)	2179	902	578 LR	2977		x	x	
Mooimeisjesfontein (split)	2187	909	536 LR	2380			x	x
Biesjeskraal	2178	901	540 LR	2272			x	x
Kirstenbosch	2186	908	497 LR	1902				x
Vleedermuisfontein	2182	905	497 LR	2769				x
Koekemoerskraal	2181	904	497 LR	1819				x
Koenap	2205	912	497 LR	2149				x
Keta	2262	911	497 LR	1957				x
Ganzekraal	2176	1149	* 608 LR	2084				
Spreeuwal	2204	1165	* 608 LR	1781				
Biccardsdrift	2203	1164	* 608 LR	1798				
Poenskopdrift	2214	1173	* 608 LR	1959				
Darling	2206	1166	* 608 LR	1554				
De Brak	2193	1155	* 608 LR	1764				
Derdekraal	2196	1158	* 608 LR	1348				
Bellevue	2201	1163	* 608 LR	1545				
Oldensfontein	2200	1162	* 608 LR	1916				
Berglust	2217	1176	* 608 LR	1982				
Wellust	2171	1145	* 608 LR	2304				
Tambootirivier	2190	1152	619 LR	1574				
Burgerspoort	2189	1151	620 LR	1736				
Giesendam	2198	1160	12 KR	1604				
Diggersfontein	2209	1167	15 KR	2198				
Sliedrecht	2222	1180	638 LR	1904				
Graaflust	2223	1181	637 LR	1744				
Burgersvlei	2180	903	496 LR	2066				
Lilyfontein	2170	899	506 LR	1855				
Klein Denteren	2221	915	495 LR	1816				
Weltevreden	2164	891	508 LR	1873				
Lhea	2184	906	534 LR	1951				
Groot Denteren	2173	898	533 LR	2523				
Uitkomst	2165	896	507 LR	1993				
Total Areal Extent of original New Belgium Estates (60 farms):				116870	Hectares			
* Remaining New Belgium Block (608 LR): 20 035 Ha								

FIGURE 9.2: Table of the 60 farms comprising the New Belgium Block compiled from the Deeds Register and a list contained in governmental correspondence of 1941. SANA SAB LDE 1774 32357: Letter dated 10 May 1941 from Secretary of Lands, F.E. Hayward, to Secretary of Mines, in connection with a request for a geological and water resource evaluation of the New Belgium Block.

The origin of the name "New Belgium"

Where did the name "New Belgium" come from? We don't know – but there's no doubt that the company NBTL was named for the block of land rather than the other way around. We know this because the map of the Transvaal by Friedrich Jeppe drawn in 1879, eleven years *before* the formation of NBTL, already shows a trapezoidal block of land named "New Belgium", straddling the Lephalala River and lying to the south of the line marking the southern limit of tsetse fly, but partly north of the Waterberg plateau – pretty much where the block actually lies (Figure 9.3).[9] However, the block is not shown on the Jeppe & Merensky Transvaal map of 1868,[10] nor, more importantly, on that of Merensky's 1875 edition.[11]

Therefore, it seems likely that the New Belgium Block was created and demarcated sometime before Jeppe's map of 1879, maybe even before the annexation

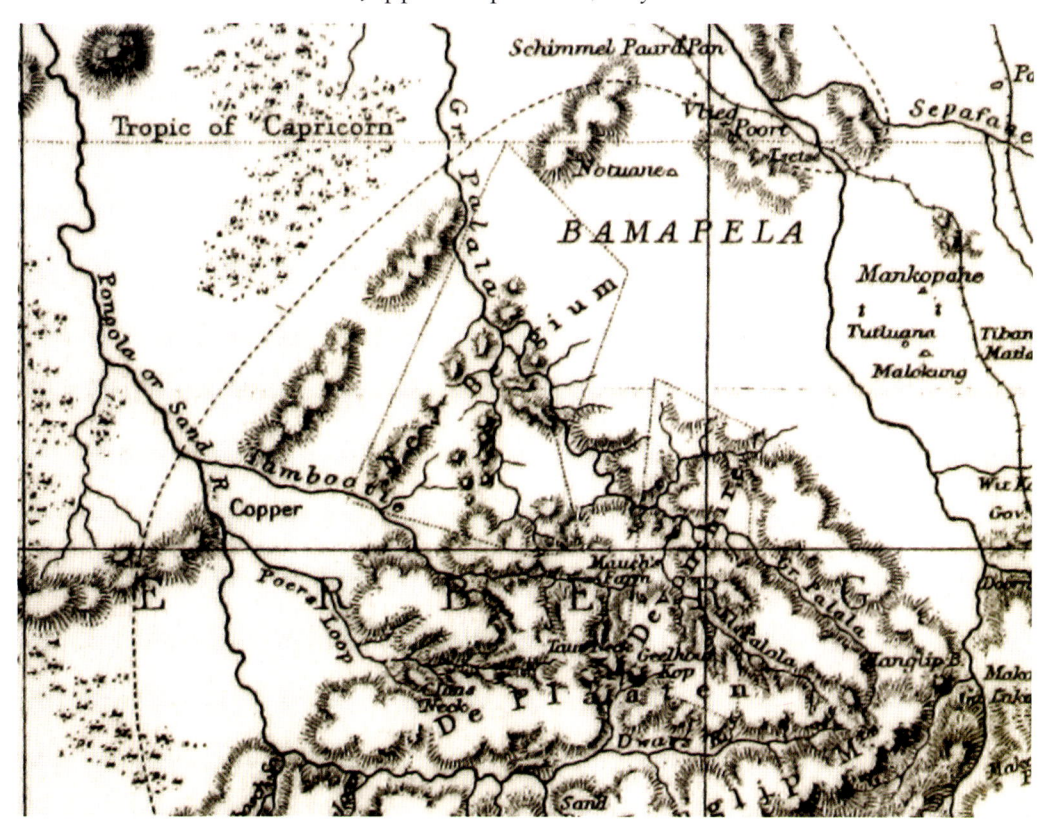

FIGURE 9.3: A portion of Jeppe's map of the Transvaal of 1879, showing the approximate demarcation of the New Belgium Block in the northern Waterberg. The dashed line is the southern limit of tsetse fly (see Chapter 12). Jeppe, F. (1879): *Map of the Transvaal and the Surrounding Territories*. Published by S.W. Silver & Co., London.

of the Transvaal by Britain in 1877, when Jeppe was still Postmaster General of the Transvaal – but after 1875.

In mid-1875, the then President of the ZAR, Thomas François Burgers, left on a fourteen-month visit to Europe where he hoped to improve perceptions about the stability of his Transvaal Republic and thereby to raise the loan finance he needed to build up its small economy and fund a rail line to Delagoa Bay. (He also needed to escape from a furore he had created at home by deciding to alter the Republic's coat-of-arms and change its flag.)[12] Among the countries his group visited was Belgium, where in Brussels a treaty was concluded between the Kingdom of Belgium and the ZAR in February 1876. The Treaty was ratified in August that year but, due to the intervention of British annexation followed by the First Anglo-Boer War (1881), was only finalised, with amendments, in April 1888.

The Treaty, written in Dutch and French, formalised the establishment of cordial relations between the two countries. It focused on the rights of citizens of each country while resident in, or visiting the other, made provision for favourable treatment of business ventures in terms of taxation and provided for the right of legal protection, the handling of estates, etc.[13] It said nothing about the exchange of land, let alone the granting of a block of land by the ZAR to Belgium or the Belgian sovereign.

However, given the timing of the Treaty relative to the first appearance of the New Belgium Block on maps of the Transvaal, it seems probable that the two are related. Possibly President Burgers pledged the Belgian state or its king (Leopold II) a piece of land in the northern part of the republic, either for the hunting enjoyment of the king (who was notorious for his feudal lifestyle), or for the settlement of Belgian settlers.

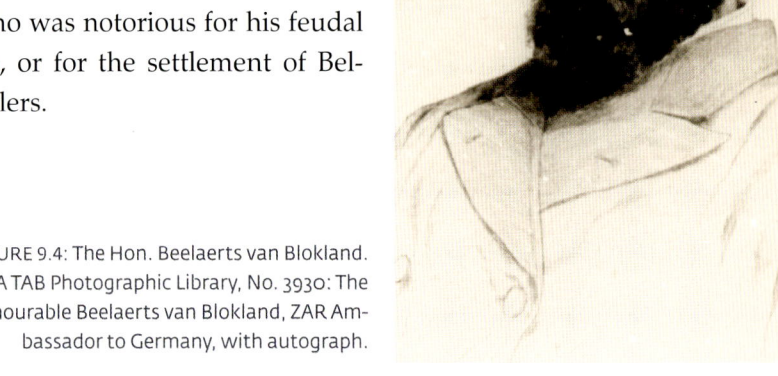

FIGURE 9.4: The Hon. Beelaerts van Blokland. SANA TAB Photographic Library, No. 3930: The Honourable Beelaerts van Blokland, ZAR Ambassador to Germany, with autograph.

The latter possibility is supported by a letter to the State Secretary of the ZAR from the Belgian Consul in Cape Town in October 1884, enquiring about the receipt of land "with an eye to the establishment of Belgian emigrants in this country".[14]

Another piece of information that might lend support to the idea that the name New Belgium derived from President Burgher's Treaty of 1876 is that when the treaty was eventually finalised in April 1888, the signatory on behalf of the ZAR was the Honourable Beelaerts van Blokland, then the country's ambassador to Germany. One of the principal rivers running through the New Belgium Block is called the Bloklandspruit.

From the outset, NBTL struggled to earn a satisfactory return on its investment. In the report to shareholders at the second annual meeting of the company in December 1891, the secretary noted that in the first year prospecting had been conducted on the Block, but that whilst results had been encouraging, "no satisfactory discoveries of minerals have yet been made." He also mentioned that efforts were made to develop land on the Block for stock breeding and agriculture and that it was hoped to attract settlers to the Block, especially as "a new road from Pretoria to Mashonaland ran right through" it.[15] (This was somewhat of an exaggeration.)

The same report (covering the year to 30 September 1891) named Mr Harry Wilkinson Peacock as the General Manager – presumably resident on the Block. It stated that up to that date, a total of £7723 had been spent on prospecting activities, and £4986 on salaries, wages, travel and office maintenance. Assets were listed as £3098 in livestock; £1080 in transport stock; and £714 in wagons.

By the following year, cost controls had been implemented, salaries had been reduced, the manager replaced – and directors had halved their usual fees. By December 1893, an even more sombre note was sounded: Two successive poor rainy seasons had reduced the feed available in pastures, merchandise sales from the company store had fallen because customers had less money, and there had been no discovery of payable minerals on the company's land.[16]

The comprehensive *Goldmann's South African Mining and Finance* (1895–96) painted a depressing picture in its description of the fortunes of the NBTL in its early days: It reported that the net losses incurred annually by the company since its inception had been £21 178 (1891); £5990 (1892); £3189 (1893); and £4126 (1894), and that by the end of this period, all prospecting operations had been abandoned and agricultural operations suspended. The property held about 800 head of cattle and ran a small, modestly profitable store.[17]

In April 1896 the dreaded rinderpest crossed the Limpopo in its rapid migration down the African continent. It can only be assumed that the cattle herds of the big land companies in the Waterberg, like New Belgium Land & Development Company Limited and its neighbour, the Transvaal Land & Exploration Company Limited, did not escape unscathed, although no specific mention of the disease is made in the summary reports.[18]

Communications between the farm and the Company's representative in Pretoria would not have been easy or quick at this time: A postal agency at Zandriviersport was only established (by Carl Forssman) in 1891, following a petition

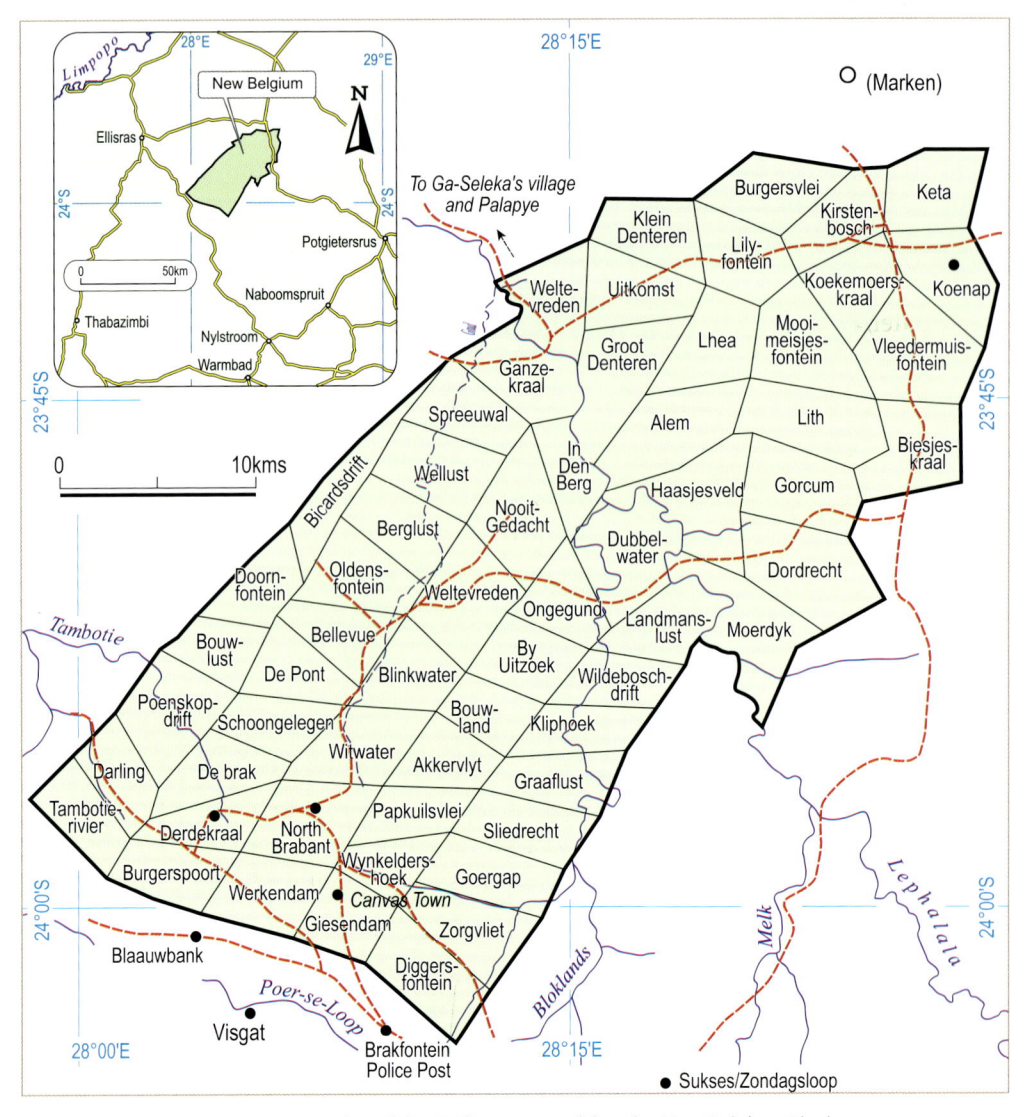

FIGURE 9.5: Location of the 60 farms comprising the New Belgium Block.

supported by over sixty residents of the country beyond the poort (see Chapter 8). Until then, the nearest post office was at Nylstroom. The signatories to the petition had included both A.F. (Arthur) and H.W. (Harry) Peacock, as well as L.V. Desborough, whose signature was embellished with two stamps – one saying "New Belgium (Transvaal) Land & Development Co. Ltd", and the other, below his name saying "Manager".[19] From the available correspondence, it seems that Desborough had been the first resident manager of the New Belgium property and that Peacock, mentioned above, replaced him.

In September 1890, Desborough, in his capacity as manager of the NBTL's "Palala" farm, had applied to the State Secretary for a permit to import 12, "No. 12 central-fire, choke-bore, double-barrelled breech-loading shotguns by James & Strand, London into 'the province' through Natal."[20] Presumably, this request was granted, because in January 1891, Desborough wrote to the Commandant General in Pretoria to ask if he could present one of these guns as a gift to Chief Selika [sic], whose village was about 50 miles from the company's property and from whom Desborough obtained his labour force. This application was denied.[21] A day later, Desborough wrote again to the State Secretary, this time in relation to an application for a road to be built from Warmbaths (then known as Hartingsburg), through part of the Block, to the confluence of the Lephalala and Limpopo rivers. He asserted that his company had already built a part of this road at a cost of £1000, in the vicinity of what became the settlement of Beauty (later Ga-Seleka).[22] This must have been the "road to Mashonaland" referred to in the Company's report of 1891.

Lawrence Desborough

A fair amount can be gleaned about Lawrence Vivian (or Laurence Vivion, depending on the record) Desborough from various physical and online sources. He was born on 4 December 1844, the youngest of ten children, to Henry and Mary Desborough, originally of Beckenham, Kent, and later of Pilton, near Barnstaple, North Devon. A remarkable photograph included on a website about the history of Pilton shows the whole Desborough family in 1858 – which makes this photograph one of the earliest portraitures taken in Britain. Young Laurie, then aged 14, is shown standing on the left at the back of the group.[23]

His father, Henry Desborough (1785–1862) was an insurance broker and secretary of the Atlas Assurance Company of Cheapside, London. Lawrence appa-

FIGURE 9.6: The Desborough family at Barnstable, Devon, in 1858. Lawrence is on the far left. This image was kindly provided by John Burnell and Xandra Houston, custodians of the Desborough family archives, and was further enhanced by Patrick Bonior.

rently trained as an accountant and may have worked under the supervision of his father, given his later roles in the insurance sector.

A fascinating aspect of the personality of the young man was revealed by a study about the history of association football: In the autumn of 1863, when Lawrence would have been 19, a group of 43 leading football players of the time held a series of six meetings to discuss the formalisation of the young sport. Lawrence was a member of the delegation from Crystal Palace, one of the newly formed London clubs. The meetings culminated in the acceptance by the Football Association of "The Laws of the Game", as debated and finalised by the club representatives on 8 December 1863.[24] These rules still govern the world's most popular sport.

The next record we have of Lawrence Desborough is from New Zealand where by December 1877 he had established a business in Southbridge, a small town about 45 km south of Christchurch in the Canterbury region, South Island. Here he started an agency for the Union Insurance Company and also sold equipment and later property. He quickly made a name for himself as a businessman of some stature. Within a year, he was elected to the board of the Ellesmere Domain, and soon afterwards to the Ellesmere Sports Committee and to the Southbridge Road Board. The Canterbury daily paper *The Press* contained numerous references to his appointments and advertisements about properties and equipment that he offered for sale.[25]

On 31 January 1878 Desborough married Mary Ann Little (1853–1927), a recent immigrant to New Zealand from England. They had two children: Henry (born 1879) and Jessie Mary Vivian (born 1887 in Christchurch).[26] By 1885 Desborough had become a farmer in addition to his other activities and even won a prize for the best breeding sow at an agricultural show. He was a member of the Southbridge school committee and also the Southbridge town board.

But trouble was brewing: As early as August 1883 an order had been issued against him by the Supreme Court in respect of some monies owing. In 1884 there began a prolonged dispute over a consignment of damaged wheat that Desborough and his partner had allegedly been involved in passing off as good and this eventually ended up in court too. Sadly, this expensive case ultimately resulted in Desborough filing for bankruptcy in early 1886, bringing to an ignominious end his rapid climb up the business and social ladder of Canterbury society.

All we know currently is that at some point after the birth of his daughter in July 1887, Desborough left New Zealand for good and that by February 1890, he was living in the Transvaal, maybe already as the manager on New Belgium. Perhaps he responded to an advertisement for the position in a local paper and travelled directly to South Africa; or did he first make the long, expensive voyage back to England? It does seem that he left his family in New Zealand, for both his wife and his son died there (in 1927 and 1955 respectively).[27]

Desborough must have left the employ of NBTL not long after the correspondence referred to earlier (that is, in early 1891) because, as mentioned, the Company's Annual Report for the year to September 1891 already named Harry Peacock as its estate manager. The Royal Colonial Institute, however, included L.V. Desborough, "of New Belgium Company, Nylstroom, Transvaal", as one of its Non-Resident Fellows in its Proceedings of July 1891 – but the same journal a year later reported him as being resident in Bombay, India. The following year he was not listed at all.[28] And indeed, an entry in a New Zealand newspaper in January 1893 reported that Desborough had died from a fever on 11 October 1892 at the British Hospital in Port Said, Egypt, on his way home to England aboard the P&O vessel SS *Orient*, from Bombay, where he'd been working for the Equitable Life Assurance Company.[29] He was 48.

Harry Peacock

Less is known about Harry Wilkinson Peacock, who seems to have succeeded Desborough as manager of New Belgium. He was born in Caistor, Lincolnshire, in April 1850, the third of nine children of Wilkinson Affleck Peacock (Rector of Croxby, Lincolnshire) and Isabella Hannah Smith.

The UK Census of April 1871 records Harry still living at his parents' home, aged 21, employed as a "rectory farming pupil".[30] An older brother, John, went out to South Africa as a Trooper with the 13th (Prince Albert's Light Infantry) Regiment and died on 3 July 1879 during the Battle of Ulundi at the end of the Zulu War. A younger brother, Arthur, who was to become the manager of the Transvaal Consolidated Land (TCL) properties in the Waterberg – and whose niece Molly would cofound the Baber dynasty in the area – was born in December 1851.

It seems that after his service with New Belgium – the duration of which was no more than a year or two – Harry was bitten by the prospecting bug, for the

FIGURE 9.7 (ABOVE): The Peacock family in the 1870s. John and Harry are on the left in the back row; Arthur is in the middle with his future wife Mary Katherine Fawssett on his right. Image provided courtesy of Elizabeth Hunter. FIGURE 9.8 (RIGHT): Harry Peacock's gold nugget. Image from the Barberton Museum website.

only official records about him that have come to light deal with his applications for the renewal of gold claims licences in the Kaapsche Hoop area (Barberton) in 1908–1910.[31] One interesting document from this collection is the application in November 1909 by Peacock to send a nugget of alluvial gold weighing 19 dwt (just under an ounce, or 29.5 g, worth $20 in 1909, and about $1350 in 2018) to his 17-year-old niece "Elsie" (Catherine Isabel) Ellis back in Spilsby, Lincolnshire.[32] The application was granted.

Three years later, on 3 July 1912, Peacock discovered a very large gold nugget on the farm Coetzeestroom between Kaapsche Hoop and Ngodwana in Mpumalanga. The nugget weighed 179.8 ounces (5,1 kg). He sold it to the National Bank, which in turn sold it to a speculative consortium for £800 (equivalent to about £86 000 in 2018). Apparently, the nugget lay for some time in the Johannesburg home of one of the speculators – where it was wrapped in green baize and disguised as a door stop – before being taken to England and sold to the Bank of England. A model of the nugget is displayed in the British Museum. Of Harry nothing further has been heard. It seems that he did the smart thing after the discovery: At the age of 62, apparently a bachelor, he disappeared from view in order to enjoy his long-awaited wealth in private.[33]

The record of William Caine

During the 1890s many of the numerous speculative land and exploration companies that had sprouted in the aftermath of the discovery of Witwatersrand gold began to founder, as the mineral potential of the properties they had acquired was found to be unexciting, and as drought, and especially the rinderpest epidemic of 1895–1896, took its toll on any fledgling farming activities. New Belgium, though, was one of the survivors.

Sometime after the departure of Harry Peacock, a new manager was appointed to look after the interests of NBTL. He was William Henry Attwood Caine, another enigmatic individual, although his ancestral background is well documented. Caine was born in Hong Kong on 19 December 1859, the second son of Emily and George Wittingham Caine.[34] The latter was a British consular official and a son of the illustrious Lt Col William Hull Caine, an Irishman, career officer with the 26th (Cameronian) Regiment of Foot in India, founder of the Hong Kong police force, the colony's first magistrate and, at the time of William's birth, its Lieutenant Governor. William's father, George, was later transferred to the Chinese city of Hankow as acting Consul. There he succumbed to the temptation to make illicit private use of Her Majesty's funds, was found out, arrested (in Sydney, Australia), tried and convicted for fraud and larceny and jailed in Shanghai where he died in 1874 aged 42.[35] He left behind a widow and eight children, of whom five-year-old William was the second-born (there were five sons and three daughters). Having grown up and being schooled in Belgium and England, young William joined the Royal Merchant Navy for at least the period 1881–1887, during which he served as ship's mate aboard several vessels of the P&O line on the Australian route. He is recorded as arriving in Cape Town from Southampton on the SS *Spartan* (Union Line) on 20 March 1891.[36]

Caine's name first appears in the Annual Report of NBTL for the year ended September 1894, so it is presumed that he was appointed to replace Peacock shortly beforehand. He lived at what appears to have been the principal, if not only permanent structure on the property at that time, on the farm Noord Brabant, which lay in the southwestern corner of the block. He may even have built the residence.

Caine was a bachelor, living alone in this remote part of the country, managing a herd of approximately 1000 cattle on behalf of the company. He seems to have been the ideal manager: honest, literate, self-contained, stable and diligent. He was meticulous in his record-keeping – and it is this trait we must thank for the detailed, 57-page diary he kept of events that took place during the course of the South African War from October 1899 to December 1901.

Whether or not Caine maintained a journal before this time is unknown – but it would have been consistent with his character had he done so. The diary seems to have been written for his mother in England. It is possible that when he could, he sent her instalments of it. Caine turned 40 in December 1899. He records his birthday in his diary:

> 19th. My birthday, rode after the boys early in the morning, caught them up and showed them where to make my camp and the cattle kraals. Rode back and got home just at dark to find the post just arrived with letters from home, Hurrah! What cared I, the heavy storm breaking just five minutes before I reached the house: I had letters to read from dear ones, a good birthday present.[37]

Pause for a moment to consider just how extraordinary are the implications of this casual entry: Caine is living alone on an isolated property more than 60 km

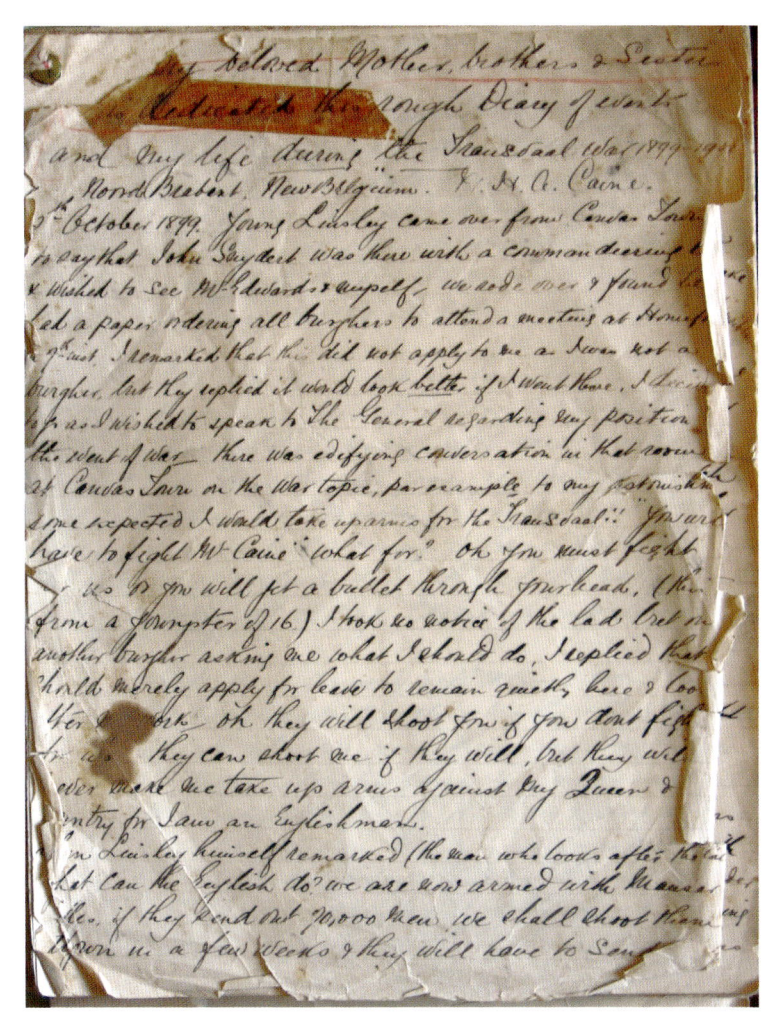

FIGURE 9.9: Front page of William Caine's diary.
Caine (1902). Courtesy of Rose Cochrane (see note 37).

from the nearest post office at the time (Zandrivierspoort), with no proper roads, only a rutted and frequently impassable track linking him with civilisation, two months into a war between the government responsible for his postal service and the country from which the letters he has just received had been sent (and to which he has vowed allegiance). And he clearly doesn't think that it is remarkable to receive such mail!

Caine's nearest neighbours were Arthur Peacock (the younger brother of Harry, the former manager of New Belgium), Peacock's wife (and cousin) Katherine, née Fawssett, her maiden sister Edith and their young niece Mary (Molly), who was

18 in 1899. The story of the Fawssett/Peacock family and their descendants in the Waterberg has been told in the unfinished memoirs of Elizabeth Clarke, Molly's daughter, extracts of which were published in a beautiful annotated collection of rare photographs of the period by Elizabeth Hunter, Elizabeth Clarke's niece, in 2010.[38]

Arthur Peacock had come out to South Africa in 1886 to manage cattle ranches in the Waterberg region belonging to the forerunner of the Transvaal Consolidated Land & Exploration Co. Ltd (TCL), one of the several land and prospecting companies that had sprung up since the discovery of gold on the Witwatersrand in that year. Liz Hunter's book refers to the rinderpest epizootic which killed off so many of the cattle of Peacock's employer that the company decided to shut down its operations and terminate his employment. In compensation, it gave Peacock a lease on one of its properties, the farm Blaauwbank, and the family moved there in 1892. Blaauwbank was about 10 km to the southwest of Caine's place on Noord Brabant. The two properties were separated by a ridge (1460 metres amsl) that forms the watershed between the catchment areas of the Goud River (which flows into the Lephalala) to the north and Poer-se-Loop (flowing into the Mokolo) to the south. The Noord Brabant homestead was situated close to the headwaters of the Goud River, while the homestead on Blaauwbank lay on the north bank of Poer-se-Loop.[39]

According to one of Jeppe's maps (published posthumously in 1899) the nearest road into the area from the south at that time is shown as running parallel to the south or left bank of Poer-se-Loop, crossing into Blaauwbank in its western corner.[40] This road would have come down the spruit from the ZAR police post – soon to be a Boer laager – on the farm Brakfontein where the Dorset police station stands today. The only other road into the New Belgium Block shown on the Jeppe map ran northwards from Sandriviersport across the Dwars River and up towards where Sukses would later be established. It crosses the Lephalala River on the farm Muisvogelkraal and continues northwards along the eastern boundary of the block – essentially the route followed today by the old road past Melkrivier school. This was the new road to Mashonaland mentioned in the NBTL report of 1891. Both routes are referred to in Caine's diary.

Caine referred frequently to a settlement he called "Canvas Town" which was a short (less than an hour) ride from Noord Brabant. This has so far not been possible to locate, but it might have been on the farm Wynkeldershoek, an NBTL property adjacent to Noord Brabant in the east. At one point, Caine mentions that someone went from Canvas Town to the laager (on Brakfontein) a journey of about "10 miles"; and elsewhere, the diary makes clear that Canvas Town lay within the Block.

The Peacock family were the only other English-speaking people with whom Caine regularly came into contact, apart from his assistant, Smith and his wife. This, together with their relative proximity – and, no doubt, the presence of two pretty unattached women (although Caine would have hastened to deny this!) – meant that Caine and the Peacocks frequently exchanged visits and gifts of vegetables, fruit etc. The diary entries make it clear, however, that Caine did not hold Arthur Peacock in particularly high regard. Peacock considered himself a burgher and as soon as hostilities broke out, he had gone off to register as one; whereas Caine, notwithstanding his father's experiences, was unequivocally loyal to the British Crown, although he had no wish or intention to be drawn into the conflict and sought only to be allowed to continue performing his duties and obligations to his employers.

Within a few weeks of war breaking out, Peacock had secured his *burgherskap* (ZAR citizenship) and had been posted to a commando on the Crocodile (Limpopo) River. However, within a fortnight, he'd been sent home (because, Caine implied, he was both a rotten shot and a poor horseman!).[41]

Caine, meanwhile, had to travel to Nylstroom and on to Pretoria to apply for and obtain a letter from the Commandant of the Waterberg, allowing him to remain a neutral person on New Belgium for the duration of the conflict. This he received on 30 October 1899.[42]

It was not long, however, before Caine was ordered to move to within range of protection of the Boer laager on Brakfontein and so in the week before Christmas, he moved with his herds to a temporary camp south of the laager on the farm Zondagsloop, close to where the (inappropriately named) settlement of Sukses would later be established.[43]

Ever philosophical about his position, Caine remarked that at least,

> the post card from Nylstroom to the laager passes by my tent, so I get the mail twice a week. By permission of Mr Forssman [Carl, now in his role as the Commissariat Officer at the Head Laager], they bring my mail; the only paper I can get is the "Standard & Diggers News", and a most scurrilous edition it is. The lies invented against the English is [sic] appalling, so that I only read the telegrams." [44]
> He continued, "I have pitched my tent away some 400 yards to one side of the road on purpose not to be in the way of the Boers riding

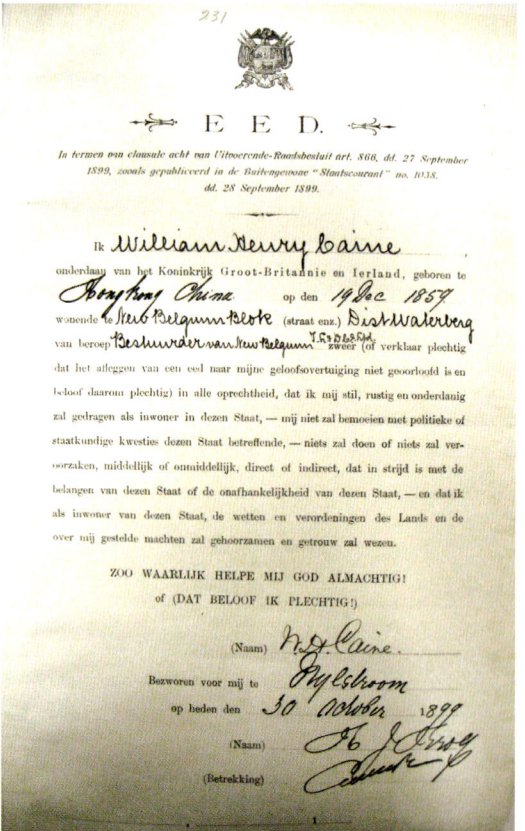

FIGURE 9.10 (TOP): Katherine and Arthur Peacock (left) and Edith Fawssett (right) outside their first home on Blaauwbank. Photograph courtesy of Elizabeth Hunter. FIGURE 9.11: Letter dated 16 October 1899 from Commandant H.S. Lombard, recognising Arthur Peacock as a burgher (LEFT); authority from the ZAR for Caine to remain on the New Belgium property in his capacity as its manager (RIGHT). SANA (left) TAB CJC 7 CJC49: Claims for compensation, British Subjects Transvaal, Waterberg, Arthur Frederick Peacock, 1902/03; and (right) TAB SS 8268: Landdrost Waterberg. WH Caine Vraagt permit om hier te Blyven 30.10.1899.

by, by which of course I mean that I did not want them to make my camp a resting place, but it was no use: they found it was a short cut past my tent and Aha! – they could always get a cup of coffee with milk and sugar in it.

In March 1900 Caine was permitted to return to Noord Brabant with his stock, but two months later he was again ordered to move (by Commandant Lombard, who'd granted Peacock his burgherskap). This time, he took the opportunity to move his herds down to winter grazing on a property (Hartbeestlaagte) owned and loaned to him by Carl Forssman, about 10 km to the southwest of the present-day town of Mookgophong (Naboomspruit). This journey of some 150 km via Nylstroom was accomplished in just under two weeks without loss (see Chapter 9).[45] By now, Caine was starting to be approached by representatives of the Boer forces with instructions to commandeer some of his animals, horses first and then cattle. True to his diligent nature, he insisted on an official requisition in each case and a signed receipt from the person taking the animals; all these documents, carefully kept, would be presented to the British authorities at the end of the war in substantiation of a claim for compensation for losses incurred by NBTL during the conflict.

FIGURE 9.12: William Caine at tea on Blaauwbank with Edith Fawssett and Katherine Peacock. Photograph courtesy of Elizabeth Hunter.

It was also becoming apparent that with the formal ZAR government preoccupied in fighting a war against the British invasion, normal administrative and policing enforcement was becoming lax. This vacuum was being exploited by some of the "subjugated" black chiefdoms, in particular that of Chief Hans Masebe (or Masibi) to the northeast of the Waterberg, whose followers were repeatedly alleged to be committing atrocities against occupants of villages not supportive of his rule or affiliated to his clan.

This feud derived from a succession battle to replace Chief Masebe III of the Langa, who'd committed suicide (maybe accidentally) in May 1890 (see Chapter 3). Two brothers, Hans and Bakenberg Masebe, each claimed the right to succeed their father, who by favouring each of them in turn had been responsible for building confusing expectations among his sons. The unhappy result was the partitioning of the formerly stable, strong Langa Ndbele clan into two chiefdoms, each under one of the two brothers, while each continued to assert his right of succession.[46] After partition, the brothers continued to quarrel, but until the outbreak of the war, the ZAR administration had managed to contain the dispute and prevent serious armed conflict.

From September 1899 all this changed. The first armed clash between the rival clans (or, more correctly, subclans) appears to have occurred on or close to New Belgium in November or December 1899 – which accords well with the timing of the instruction to Caine to move within the protection of the ZAR police laager. The two chiefs were summoned to Nylstroom to be reprimanded, but by March 1900 they were once again raiding each other's territories. Shortly afterwards, the capture of Pretoria by the British removed any form of administrative control until Pietersburg fell to the British in April 1901 – a year during which the warring factions had free rein to attack each other and Boer settlements alike.[47]

On 14 July 1900, Caine heard a report from one of his wagon operators, who had returned to Noord Brabant to collect salt, of how Hans Masebe's men had murdered several men and male children at kraals on the company's property and stolen their livestock.[48]

In September when the first British troops were taking control of Warmbaths and making forays into Nylstroom, Caine began to move his herds back to their spring grazing grounds on New Belgium, being harassed along the way by men of Commandant Barend Vorster's commando, who, being reluctant to see their food supply slip out of their control, made repeated requisitions of cattle. This was followed by an order from Gen. F.A. Grobler to supply 50–60 head of breeding cattle.

Upon his return to Noord Brabant, Caine set off to tour the New Belgium property. This five-day journey on horseback took him to the northeastern corner of the block, where the Keta Game Ranch is today, and then back across the Lephalala River along the western boundary of the property. He found the company house on Kirstenbosch and the store on Koenap deserted; at a kraal, the local *kaptein* told him all the men and boys had either fled into the hills, or had been taken by Hans Masebe's men to fight against his brother.[49] In October Caine learned that a pack of up to 40 wild dogs had been active on the property and had killed several donkeys before moving on.

News of the war was intermittent, conflicting and partisan: In October 1900 the Peacocks told Caine they'd heard that President Kruger and General Botha had gone to Europe to make peace [actually, to seek mediation and obtain more funding]; that General Grobler in the Waterberg had been replaced by General Beyers [true]; that England was withdrawing all her troops from South Africa because of a war with China; and that at Machadodorp the English had been driven back to Pretoria, leaving 18 000 dead.[50]

At the beginning of November, Caine received an order from General Beyers to release all his remaining cattle to the Boers "just as I was sitting down to a nice breakfast of liver and bacon (having killed a pig last night). [The news] completely spoiled my breakfast; and I thought that if this is true, I shall ask to go out myself, not seeing of what further use I could be here to the Company".[51] Fortunately, he was able to remove and hide a quarter of the herd before the General's agent came to look at the animals. As it happened, the agent only took about half of the remainder, leaving Caine with some 500 animals. However, before the end of the month, all the cattle not hidden had been taken, apart from a few milk cows.

In addition to cattle, Caine grew maize on New Belgium (he had to hide the harvested mealies to prevent it from being commandeered too) and tobacco. His diary recorded that in the season to end-April 1901, he'd cut and hung over a thousand sheaves of tobacco in his shed. He was a remarkably self-contained bachelor, making his own bread, jam and brandy – the latter from *stamvrug* (stem fruit, *Englerophytum magalismontanum*), which, however, he found too dry to produce anything other than an "indifferent liquor". He was also a keen vegetable gardener and wrote frequently of progress with his seedlings.

Caine spent a lot of time out on the veld on horseback while travelling on New Belgium or between his home and neighbours or Canvas Town, nearly always with a rifle or shotgun and frequently with one or more of his dogs. There are many diary entries describing attempts to shoot something for the pot, but apart from the very occasional duiker, steenbuck or pig (warthog), his "bag" was limited to guinea fowl, "pheasant" (francolin), the odd duck and even a korhaan. There seemed to be no larger antelope left in the area. This accords with anecdotal evidence that by the mid-1890s there were very few wild animals left in the Waterberg. This conclusion is supported by an analysis of debates of the period in the Volksraad, which repeatedly considered petitions and proposals to introduce legislation to protect species from extinction.[52] The Volksraad member for Waterberg, J. du P. de Beer, was especially vociferous in opposing moves to restrict the rights of burghers to hunt any game on their own properties (see Chapter 16). He claimed that in his district there were many destitute burghers who depended on hunting for their livelihood and asserted that those principally responsible for the rapid decline in game numbers were blacks.[53]

However, there was considerable evidence that, while the black population certainly did hunt game for meat, irrespective of the fact that they were forbidden to do so at any time, the principal culprits were white hunters who hunted for commercial gain from both meat and hides, as well as those burghers who hunted on their winter farms for sport. The rinderpest epizootic of 1896 had done much to decimate game as well as livestock (especially ungulates) across most of the Transvaal. Although cattle had been particularly vulnerable to the disease, the conclusion had quickly been reached that wild animals were carriers – with the result that all legislation intended to protect game was relaxed, while these perceived carriers were exterminated. Moreover, rural residents, black and white alike, having lost most or all of their livestock in the wake of the rinderpest, had no option but to hunt game to obtain meat. Therefore, by the time of the outbreak of the South African War, the Waterberg, in common with most other privately owned parts of the Transvaal, was severely depleted of game (kudu for example) – and the war relegated enforcement of protectionist legislation to a minor role. Consequently, it was not surprising that Caine, despite numerous local hunting outings, had to make do with game birds and a rare small antelope.

Christmas Day 1900 was spent with the Peacocks:

> We were all up by sunrise, enjoying the cool air of the morning. Breakfasted at 7am. Then went for a short walk with two of the ladies until 10am. They held divine service at 11am, after which Miss Molly Fawssett sang two songs very nicely: "Home Sweet Home" and "The Last Rose of Summer". I had killed an ox and sent them some beef, so we had roast beef for dinner with potatoes and cab-

bage, cabinet pudding, gooseberry fool [sic] [Cape gooseberries] and for dessert, plantains and grapes.

Later that afternoon, the Peacocks accompanied Caine in riding across to Canvas Town, "where I had to exhibit the Magic Lantern – which amused all, although it greatly disturbed the sleep of one of the children." [54]

January 1901 brought with it many conflicting stories about the war. Although it was apparent that British units were making more frequent forays into the Waterberg from their bases in Warmbaths and sometimes Nylstroom (distant cannon fire was heard more than once), equally, there were reports that they had been repulsed with heavy casualties. For Caine and the Peacock women, British subjects in "enemy territory", hopes were raised and dashed repeatedly, though fortunately without much impact on their daily lives, surrounded as they were by essentially benign and equally distressed Boer neighbours who themselves were becoming increasingly fatigued by the war, who also lacked food, clothing and other items and who were just as nervous about losing their stock or being forced to take up arms again. In mid-January, another two hundred head of cattle were commandeered, leaving Caine with little more than a hundred from his original herd of over a thousand. He also heard a rumour that he could himself be raided, and so he hid his diary in a despatch box and buried it.

Late in February 1901 Caine retrieved his diary to update it. He reported hearing – and disbelieving – a rumour that Queen Victoria had died (this was true; she had died on 22 January). During the month, he sent his assistant, Smith – an Englishman but, like Peacock, a burgher – down the Lephalala River to the village of Chief Selika (later Beauty, now Ga-Seleka) near the confluence with the Limpopo, to sell some of the tobacco crop. While there, Smith was invited across the river by the British magistrate at Palapye, who had urged him to take advantage of the protection that could be offered by the Protectorate of Bechuanaland. He sent a message inviting Caine to do the same. Caine decided that if he were to leave New Belgium, his black workers would all desert, as they feared the wrath of Hans Masebe and that there would be nothing to prevent the Boers from ransacking the place. However, he insisted that Smith should go with his wife (a Boer) and family. He sent them off together with a company wagon and 30 cattle, telling everyone else, including Smith's brother-in-law, that they had merely gone to sell more tobacco along the river.[55] Owing to the high level of the Limpopo River, it would be another month before Caine would learn that Smith and his family arrived safely in Palapye.

In March Caine was visited by his old friend Carl Forssman, who had been sent from the Laager Commissariat to commandeer his mealies – and to inspect his books. Forssman also said that someone would be along to take Caine's donkeys. Caine remarked in his diary on the irony that Forssman had been an Englishman in the war of 1881, though not under arms, and was now a burgher. He apparently could play the violin well and played Caine's violin when visiting, Caine having difficulty in doing so because of arthritis. On the second evening of this visit, Caine and Forssman were invited to the Peacocks' for dinner and to stay the night:

> So at 4pm, we harnessed the mules and drove over; I drove the mules and Forssman used the whip. We got there just at 6 pm and found the Peacocks had almost given us up. The ladies were dressed very becomingly and so stylishly we could have imagined we were in town instead of hundreds of miles in the bush. The dinner passed off very pleasantly. Forssman being a good talker and Miss [Edith] Fawssett can talk enough for four people, which is often a relief to a man like me. Miss Molly Fawssett, the niece, was very quiet – evidently shy. (There follows an affectionate, if unflattering description of Molly Fawssett, which is reproduced in Liz Hunter's book about the family.) After dinner, we sat on the verandah smoking whilst Mrs P and Miss Molly sang a few songs to the harmonium and even Forssman was persuaded to play ... I had taken a little brandy of my own making so we gentlemen sat up until 10.30pm, the ladies having retired at 9.30. [56]

By April Caine was hearing more news of the gradual northward advance of the British, securing the railway line to Pietersburg as they went, as well as the towns along the line (including Nylstroom). The first of several clashes occurred in the vicinity of Zandrivierspoort. An exodus of Boer families towards the Bechuanaland border was noted in the form of caravans of wagons and livestock moving slowly through Canvas Town and on the two roads to the Limpopo. In anticipation of a peace agreement, many Boers in the region abandoned the fight, came back to their families and either tried to stay on their farms without attracting attention, or began trekking towards the border. The Commandant of the Waterberg, General Beyers, who was determined to continue to resist the British occupation, became ever more demanding and threatening in his attempts to stem the exodus and to force his burghers to continue the war. Caine records many conversations with burghers torn between whether to stay with the commando or to make their escape.

On 10 May 1901, this anxiety was fuelled further by the news that three Boers had been killed by Masebe's men somewhere across the Palala.[57] This grizzly murder, which was committed for the purpose of procuring traditional medicine, involved the removal of parts of the men's bodies (genitals, facial skin, arms) while they were still alive. The full story was revealed in 1906 when a detective investigating the murders posed as a herbalist and persuaded a sangoma called Kgano, the chief's personal rainmaker, to explain what had happened and to show him the evidence, which was stored in clay pots in a hidden cave. Nine men were eventually arrested, tried, found guilty and sentenced to death, yet in the end, none served more than six years in jail (see Chapter 3).[58]

The next three months were a difficult time for Caine and his neighbours, the Peacocks. Mail from Pretoria could no longer be delivered because of the removal of the ZAR government and the temporary absence of a replacement administration. Staples like wheat, sugar, oil, matches, ammunition (or even powder), candles and caustic soda (with which to make soap) were in extremely short supply and clothes and shoes were falling into holes and becoming too worn to darn or repair. The news of the war, such as it was, alternated between real or rumoured victories or defeats of one side or the other, accompanied occasionally by the sound of distant artillery fire.

Fortunately, by August 1901, Caine's assistant Smith, who had crossed into Bechuanaland in February, was able to set up an informal, clandestine courier service which enabled Caine to receive occasional newspapers, items of food and clothing and to renew correspondence with his distant family. He was particularly frustrated by a lack of communication from the NBTL company's local agent, Stephanus Meintjes, who ran the Transvaal Mortgage, Loan & Finance Company in Pretoria. Eventually, in late July he received – to his dismay, for he had been contemplating escaping to Palapye – instructions to remain on the property.[59] This was followed a couple of weeks later by another letter from Meintjes enclosing one from the NBTL head office in London, in which it is remarked how strange it was that they never hear from Caine, but only from Smith! Caine was suitably incensed …[60]

A few days later, a passing Boer, Carl Boshoff (a former Landdrost and member of the Volksraad), announced that Russia had invaded India and that as a result, Britain was withdrawing her forces from South Africa to go to India's defence. This plunged the Peacocks and Caine into despair, even as it lifted the spirits of the local Boer community. In a moment of patriotic fervour, Caine diarised that he had written to his Uncle Willy – a retired Lieutenant General in the Royal (late Madras) Artillery[61] – to ask how best to volunteer for service in defence of his country in its "grapple with the treacherous bear".[62]

By this time British Mounted Infantry units, accompanied by light artillery, were increasingly harrying Beyers's forces in the Waterberg from their bases in Warmbaths and Nylstroom. Numerous skirmishes occurred all along the Sand-riviersberg, from Hanglip in the northeast across towards Rankin's Pass in the west. The intention was to persuade the Boers to surrender with minimal casualties and many were those who did so, or who were taken prisoner together with their families and stock.

Edward Gordon, a young officer with the 12th Mounted Infantry battalion, maintained a daily journal of his experiences during this period and described numerous incidents in which he was involved.[63] For two weeks, the column patrolled in the Alma valley from Koppie Alleen in the east (near present-day Alma) through the Swaershoekberge to the Sand River in the west (southeast of the modern town of Thabazimbi), pursuing bands of Boers up into the hills above Trichardt's Pass, often with the assistance of pom-pom artillery. These exchanges could be heard on the high ground of Blaauwbank and Noord Brabant over 50 km to the north:

> [Caine's diary] SATURDAY 10TH AUGUST: Peacock said he heard 7 or 10 cannon shots yesterday morning in the direction of Zwagershoek. Now what does this mean? In the meantime, what about Russia? Russia may have invaded India all the same! Alas! We cannot tell, the truth does not reach us here and the suspense is almost unbearable.[64]

In late August Caine travelled down the Goud River to its confluence with the Lephalala River and then on again to Kirstenbosch, the farm at the northeast corner of the Block. Here he found the farmhouse badly damaged by vandalism and theft of windows, doors, shelving, etc. On his return to Noord Brabant a few days later, he was delighted to find a note from across the border reassuring him that the Russian invasion of India was another false rumour (what we might term "fake news" today!). The messenger also brought a sheaf of proclamations with the request that Caine distribute them among the Boers. "The proclamation stated that all Boer officers still in the field will be permanently banished from South

Africa if they do not surrender before the 15th September next". Caine supported the proclamation, believing that the only way to bring an end to the war was to strike at the officers.[65]

Lung sickness (contagious bovine pleuropneumonia) struck his depleted herd of cattle in September, forcing him to shoot some and then inoculate others with the virus, which he took from infected animals; several died, but gradually, the inoculations began to take effect.[66] His woes were aggravated when some Boers brought an order from Beyers to commandeer his last horse, leaving him only with a mare in foal. And then he succumbed to erysipelas (a streptococcus inflammation) which put him to bed for a few days. On the day he got up, another Boer delegation arrived with a letter from Beyers commandeering ten slaughter cattle – but, fortunately, the animals were too sick to take. The mare foaled, and Caine immediately sent mother and foal off to safety with Smith at Palapye across the border. Despite these setbacks, he still managed to plant out over 4000 tobacco seedlings.

By early October the increased activity by British patrols had caused a concentration of Boer families to laager across Poer-se-Loop south of the Peacocks' farm. On the 21st, Caine heard two large bangs in the distance; and the next day, Peacock rode over to tell him that the British had attacked and captured the laager and that the Commanding Officer, Lt Col Dawkins, on hearing about Caine, had said that Caine had no right to be living in the area without a permit.[67]

The capture of the laager was corroborated by Lt Gordon's diary: He had been left in charge of the British camp at Wolvenfontein at the northern end of Zandriviersspoort (where the turn-off to Vrymansrus is today), while the 12th and 20th Mounted Infantry under Lt Dawkins set off at dusk on 20 October to surprise the laager some 30–35 miles to the north. Next morning, as Gordon was outspanned with wagons, pom-pom, artillery and infantry while en route to the laager, he learned that it had been captured:

> The laager was rushed at about 5am. No fighting as the Boers had no suspicion of our coming. No. 4 Co. got Van Staden & 6 men in separate laager. 3 or 4 families in laager. 12th & 20th M.I. probably covered about 53 miles during the 24 hours. Bag: 56 prisoners, including Comdt. Van Staden & 2 Field Cornets & Schutte, ex-landdrost of Pretoria described as the "Steyn of the Northern Transvaal" & a great capture. He had really been a man of considerable influence & a bit of a firebrand. 12 horses (indicating

great shortage of horses!); 79 rifles; 14 waggons; 3 carts; 293 cattle. All very fine cattle including a lot of milk cows. 144 trek oxen.[68]

One has to wonder how many of these cattle had belonged to New Belgium!

The explosions Caine had heard were from the demolition of the ammunition dump at the laager. He prudently immediately sent one of his men off to Makapanspoort (Potgietersrust, now Mokopane) requesting a residence permit. He received it from the CO there, Major Hunter Blair, a few days later (with a note that the application from Peacock, a burgher, was still being considered). Shortly afterwards, a note arrived ordering Peacock to go into Warmbaths immediately. Peacock and Caine responded requesting protection for the ladies in the group. [69]

A fortnight later, Caine was sent a horse, together with a note from Lt Col Johan Colenbrander, who had been up on the Limpopo near Seleka's village with his column. He was now returning along the "road" south through New Belgium and wished to meet Caine at the Company's old store on Koenap the following day. Caine set off immediately and after a night spent in the open in the rain, he arrived at dawn at the store to find Colenbrander's camp fast asleep, so much so that his arrival was not even challenged!

FIGURE 9.13: Lt Col Johan Colenbrander. SANA TAB Photographic Library, No. 35440.

> Saw the colonel and was much taken with him: he is a man rather below medium height and stout, fairish beard and hair streaked with grey. He has a pleasant manner, gave him all the news in my power, then had a wash, which refreshed me considerably, and a cup of coffee. I enjoyed a really good breakfast with the colonel and his staff at 7.30 am and to my surprise a gramophone was started shortly afterwards and many of the latest pieces well played. Then the colonel told me he would send 100 mounted men under Capt. Castera with me at 4 pm to go round by Noord Brabant and Mr Peacock's as he must take us all into Nylstroom with all our livestock. I had lunch with Wilmot of the Intelligence Department. A very good lunch too, for they had a French cook. I was offered to drink either whisky, beer, wine or champagne, verily this is campaigning in luxury. I had beer, however, and after lunch on returning to the staff tent, I was pressed to take a glass of port!![70]

The troop under Capt. Castera then left and two days later, on Friday 15 November, rejoined Colenbrander's column at Zandrivierspoort in the company of Caine and the Peacock family with as many of their stock and possessions as could be transported. An engagement with a large Boer force in the vicinity was anticipated, but did not materialise. The next day, guarded by a unit of Gordon Highlanders on foot, the convoy made its way to Warmbaths, which it reached on 19 November, bringing to an end Caine's Waterberg sojourn, and his diary.

From Warmbaths, Caine proceeded to Pretoria, where, presumably with the cooperation of the agent Meintjes, he compiled an application for compensation from the military authorities for the stock losses suffered by NBTL during the preceding two years. This application, submitted to the Military Compensation Board on 30 January 1902, was supported by those made earlier by Meintjes (in May and July 1901) using information relayed to him from Caine via Smith in Palapye. The original claim was for 838 head of cattle, three salted horses and 68 donkeys, all of which had been commandeered by the Boers against Caine's wishes.[71] It was followed on 3 February 1902 by a supplementary claim, which related to the deaths from rinderpest of 36 head of cattle that had been taken by the British to Warmbaths when Caine and the Peacocks had been forced to move, and where they had been exposed to the disease. The remaining 29 head of cattle then had to be sold at a reduced price. Moreover, Colenbrander's Provost Marshall had commandeered the remaining 36 donkeys and 13 bags of mealies.

Warm Baths.
Feb. 17. 1902.

The Commandant,
  Warm Baths.

Sir,
  I have the honour to herewith hand you a Claim for Compensation for Cattle, hereafter enumerated, lost by death from Rinderpest under the following circumstances viz:—
  On the evening of the 13th November 1901 Capt: Castera of Lt. Col. Colenbrander's Column came to Blaauwbank (No. 25) District Waterberg, my husband's farm, and said that we were ordered to leave our home and go with him.
  On the morning of November 15th I handed my cattle over to him in a perfectly healthy condition on the understanding that the Government had taken them over and would pay me the full value for them.
  On my arrival at Warm Baths on November 19th I was told that I must take back the cattle. A few days after receiving them, they commenced dying of the Rinderpest owing to their having been put amongst infected cattle on the trek.
  I contend that these cattle would not have died if they had been left at Blaauwbank as that District was perfectly healthy.

I am,
Your obedient servant,
M. R. Peacock.

This effectively left Caine with no assets to manage, and soon afterwards he left for England, only returning to the Transvaal at the end of November 1902. Shortly thereafter, he had to resubmit the compensation claim (adding the fourth horse) to the Civil Claims Commission.[72] It is not clear from the records how much, if any, compensation was paid to the Company. In February 1902 Caine had also submitted a personal claim for £15 in respect of a Martini-Henry rifle that had been confiscated from him by the British and not returned. This claim was eventually disallowed.[73]

Caine was not alone in submitting claims for compensation: On 7 February 1902, Arthur Peacock, by then living on the farm 24 Rivers, which was purchased for the family by his sister-in-law Edith Fawssett from Peacock's former employer, TCL, submitted a claim for £207 against grain commandeered by the British, as well as for the loss of crops on his farm caused by his forced evacuation. Remarkably, Peacock swore under oath that he'd been *forced* to become a burgher at the outbreak of the war, whereas Caine's diary had said that he was keen to become one! Mrs Peacock and her sister Edith also submitted claims in respect of cattle lost due to rinderpest after they had been herded by the British together with diseased cattle. None of the claims were successful. There's a note from the Provost Marshall remarking that Peacock was "a very doubtful character" who had willingly become a burgher according to a Boer commandant.[74] Caine's diary was correct.

William Caine had lived a remarkable but solitary life on Noord Brabant. A confirmed bachelor (by his own admission), he comes across in his writing as having been a dour man, but capable of fending for himself and generally comfortable with his own company. He was extremely – often outrageously – critical of virtually everyone else who was not, like him, a proud advocate of British Imperialism. He was invariably scathing in his (private) description of the Boers he met; belittling and disdainful of his black employees; and dismissive of Arthur Peacock for whom he never had a complimentary word. Yet the Peacock women clearly enjoyed his company and he seemed to have been treated with respect by most of the burghers and their wives. One suspects that he was able to present an amicable, cooperative façade, while keeping his darker thoughts to himself – and his diary.

FIGURE 9.14: Claim for compensation submitted to the British authorities by Mrs Katherine Peacock in 1902. SANA TAB CJC 7 CJC48: Claims for Compensation, British Subjects, Transvaal, Waterberg; Mrs Mary Katherine Peacock.

Caine returned to the service of NBTL after his "home" leave in 1902 and is still named as General Manager of the New Belgium property in the Annual Report for the year ending September 1904. But by 1905, after more than ten years in the service of NBTL, he had joined three of his brothers farming at Limuru on the "White Highlands" of Kenya, about 30 km to the north of Nairobi. His eldest brother, G.W.L. Caine, had settled in Kenya in 1903, and was responsible for the establishment of the tea planting industry there, having imported seedlings from India. For several years, the estate of The Caine Brothers was the leading tea producer in the colony. Another brother, Lionel, became Kenya's Postmaster General. During World War I, William, now 55 but ever loyal to his kingdom, served first as a private, then as sergeant in the East Africa Ordnance Department, receiving the East Africa Star (1914–1915), the British War Medal and Victory Medal for his participation. He remained in Kenya until his death there in 1929, aged 70.

The first decade of the twentieth century

By the time hostilities finally ceased in May 1902, the economy of the Waterberg region had been devastated. Many former landowners or tenants had been captured and sent to distant prisoner-of-war camps, their families interned in concentration camps at Nylstroom, Pietersburg, Irene and elsewhere (where thousands were to die of illnesses and disease), their homesteads burned to the ground, their stock commandeered, consumed, or killed due to the resurgence of the rinderpest. Their black labour force had dispersed as a result of intimidation from local warlords (like the Masebe brothers) or the general lawlessness that filled the vacuum left by the absence of an administration.

A Compensation Claims Commission established by the new British administration toured the Waterberg during 1903. The military member of this commission, Major Robert French (2nd Gloucester Regiment), compiled a report of his experiences, in which he noted that "the country beyond the 35-mile radius of Nylstroom is mostly held by large land companies, whose headquarters are in Johannesburg. These companies possess blocks of farms covering large tracts of country. They do not work them, but only hold them for what mineral wealth they may contain."[75]

Late in 1900, when the British evidently believed the end of the South African War to be imminent, a commission was established to evaluate the potential for settlement of discharged British soldiers upon land in the Transvaal. Among the numerous submissions to this commission was a report by Lt Col Parrott of the New South Wales Corps of Engineers, who had spent some time in the region. In his description of the Waterberg, Parrott had the following to say about the New Belgium area:

> [T]his section lies between the Zoutpansberg border on the east and the Matlabas River on the west. (...) The mean altitude is about 4,000 feet. In all divisions of this section there are extensive areas of rich alluvial soil suited for the cultivation of sugar, coffee, tea, cotton and other tropical products as well as all kinds of fruits. The section north of this runs parallel with the Limpopo River and has a mean elevation of 2,000 to 2,500 feet. The soil on the plains is chiefly sandy, but in the valleys there are rich alluvial areas.[76]

The report attached a list of 805 farms in the Waterberg which were registered in the name of the Government, including the farm Vaalwater, which would shortly afterwards be awarded to settlers under the provisions of the Settlers' Ordinance of 1902 (Chapter 8). The report also tabled a list of 18 companies owning 1541 farms in the Transvaal. The largest of these were: TCL (656 farms); Oceana (238 in total); Transvaal Estates & Development (212); Transvaal Land Co (87); Balkis Land Co (66); Northern Transvaal Lands Co (65); NBTL (60); and Anglo-French (43). For some reason, NBTL was the only landowner among this group that failed to respond to a request from the Commission for expressions of interest in selling. Several land companies, including TCL and Oceana, offered some of their Waterberg holdings as being available (and well-suited) for ex-serviceman resettlement, on a tenancy basis.[77]

Certainly, by the end of the war, as much as two-thirds of the Waterberg was owned by four land and exploration companies: TCL, NBTL, Oceana (which had assisted the ZAR government in financing the railway line to Pietersburg immediately before the war) and Anglo-French. (See figure 13.2 in Chapter 13.) TCL (Arthur Peacock's former employer), which was to become the cornerstone of the Rand Mines group, recognised that having a significant proportion of its wealth locked into idle land was not delivering value for its shareholders. In its Annual Report for 1907, Chairman H.C. Boyd, for example, explained to shareholders how the company was attempting to attract settlement on the 340 farms it then owned in the Waterberg and Potgietersrust districts by establishing stock-raising operations on some of the properties:

That considerable progress is being made in this regard is shown by the fact that 130 of your farms are now occupied by Europeans – an increase of 85 in the last year. We are content to accept small rentals at first, which increase yearly as the tenants become better able to pay. Mere squatting on the farms is not permitted, and at the expiration of the leases, which are generally for a term of seven years, we may expect to find the properties materially improved in value.[78]

Mike O'Brien

The records have yielded little about developments on the New Belgium Block over this period: Summaries from NBTL Annual Reports reported in Walter R. Skinner's *Mining Manual* in the first decade of the twentieth century indicate that first exploration and then agricultural operations were suspended. However, there does seem to have been a resident manager on the property for a while after William Caine's departure. A letter from the Company's Pretoria agent, Meintjes, addressed to the Secretary for Agriculture and dated 5 January 1906 is of interest here for several reasons:

> Dear Sir,
> Re: Tobacco Growing
>
> With reference to ... your kind offer to permit Mr Altenroxel to inspect two out of sixty farms ... belonging to this Company, I beg to inform you that Mr W F O'Brien ... will meet Mr Altenroxel in Nylstroom. The only conveyance Mr O'Brien has will be a wagon and oxen. The journey from Nylstroom to the Block and back will occupy at least 14 days ... allowing ten days for travelling and four days for inspection.
>
> In travelling, Mr Altenroxel will pass through a good many of the farms and will be able to have a general idea of the value of our Block of farms for tobacco and cotton industries. Mr O'Brien is thoroughly conversant with all the farms and...Mr Altenroxel will be able to obtain all the information he may require from him. Mr O'Brien's address is Noord Brabant, New Belgium Block, PO Sand Rivers Poort, via Nylstroom.[79]

Heinrich Altenroxel, scion of a prosperous and capable German farming family, had come out to South Africa in 1889 with a friend, Konrad Plange. After working for a period for Messrs Tamsen & Natorp at their Tweefontein store near Nylstroom, the partners moved to the Woodbush area where they bought several farms. These were consolidated into a property they called Krabbefontein, where, with skill and energy, they grew tobacco, cotton, maize and even coffee. In 1903 the Milner administration purchased the property and established it as a government experimental farm, known as the Tzaneen Estate. Altenroxel was retained as its manager until 1906 when he returned permanently to Germany. Plange, his partner, used the proceeds from the sale of Krabbefontein to purchase another group of farms which he called Westfalia. This also developed into a highly successful, productive farming operation, until it was bought by the Randlord Lionel Philips as a wedding present for his son. The farm then declined rapidly, only being rescued and revived when acquired in the 1920s by Hans Merensky.[80] Altenroxel was undoubtedly competent to advise the owners of NBTL about the agricultural potential of their estate; unfortunately, no record of his visit or report has come to light.

The unpublished manuscript of Elisabeth Clark also refers to NBTL manager W.F. O'Brien, although she describes him (incorrectly) as Caine's assistant:

> Working for him [Caine] was an Irishman, Mike O'Brien, married to a very large Afrikaner woman known as the Sea Cow [*seekoei*, Afrikaans for hippopotamus]. In his endeavours to adopt his wife's language, O'Brien used to get very mixed up, especially when building, which was his chief occupation: "Bring up those there speakers [*spykers*, Afrikaans for nails] and bung up them there *vensters* [windows]" he would shout.
>
> He owned a cab and horse, of which he was very proud, but the Sea Cow took up the whole of the inside of the cab, and O'Brien had to perch ignominiously outside. His eyes were weak and on one occasion, he nearly shot his own son, having mistaken him for a warthog.[81]

By 1908 NBTL's representative office had moved from Pretoria to Pietersburg (the agent there being Geo. F. Hughes & Co.); no manager was listed in the annual report and any interested party was invited to conduct prospecting or agricultural operations on the property.[82] It must be presumed that cattle-ranching operations were not revived on the block following the loss of almost all the Company's stock by the end of the war. In contrast, TCL, in its Annual Report for 1911 noted a report from its land manager, Captain C.A. Madge[83], that in the Waterberg, "the Company's herds of cattle on the farms Vierentwintigrivieren [the farm where the Peacocks moved after the war], Gemsbokfontein and the Blaauwbank Block [the Peacocks' former farm] on the banks of Poer-Zyn-Loop, have shown satisfactory increases. The total herd of cattle by 31 December 1911 was 668, all in excellent condition".[84] (See Chapter 8 for more on Captain Madge.)

A year later, TCL had added two more cattle farms in the area to those mentioned above and by the end of 1912, the total head of cattle had risen to 837, despite a prolonged, crippling drought. In the accompanying report of the chairman (Boyd), the company noted that during the year, it had contributed to the erection of two churches, one about 60 miles from the railway line (St John's at 24 Rivers where TCL owned half the property); and the other some 45 miles away (possibly Settlers), both in "northern localities practically uninhabited by Europeans until recent years".[85]

Both the French and TCL reports referred to above, mentioned the potential for tobacco and cotton to be grown profitably in the Waterberg, as intimated in Meintjes's letter to the Secretary for Agriculture.

In June 1913 the Union government implemented the first and one of the most far-reaching pieces of legislation that would come to define the apartheid era: the Natives Land Act (No. 27 of 1913). This Act specified where natives (blacks) could and could not own land or lease it or in any way be associated with its ownership or benefit (for example, through sharecropping, then a widespread practice). Henceforth, blacks would be permitted to purchase, hire, or in any other way acquire land only within specified scheduled areas (and only blacks could do so in these areas). More specifically, these scheduled areas were all communal, under the authority of a chief or headman, which meant that blacks were denied the possibility of personal ownership or of using land as collateral for other investments. The scheduled areas listed in the Act amounted to less than 10% of South Africa's land area (later increased by amendments to about 13%).

In the Waterberg District there were seven listed scheduled areas: those of Zebediela (33 114 ha); Bakeberg Masibi [sic] (18 915 ha); Valtyn Makapan (14 541 ha); Marcus Masibi (Hans Masebe's brother and successor) (12 403 ha); Selika (5155 ha); Soloman Maraba (3152 ha); and Pie van Teneriffe (921 ha).[86] In total, these seven areas amounted to less than three-quarters of the size of the New Belgium Block.

Some commentators have argued that the passing of the Land Act precipitated decisions by land companies to sell off their holdings. However, this is not substantiated by the reports of the time. Instead, it is more likely that the companies, having failed (in the main) to discover significant mineral deposits, having struggled to manage their vast land holdings (TCL for example held over 1.3 Mha of land in 1913)[87] and having failed to secure tenants to the extent expected, especially through several consecutive drought years, decided instead to try and sell off their properties. TCL was to continue this process over many years, until by 1965 it was able to report that its land holdings stood at just 108 043 ha, less than a tenth of its holdings fifty years earlier.[88]

NBTL, no doubt under pressure from its disappointed shareholders, was able to pull off a single deal. The Deeds Registers record that on 29 December 1913 all 60 of its farms – together with their mineral rights – were transferred for an amount of £40 000 (the same amount it paid for the properties back in 1890) to two brothers: William and Edmund Hoyle Vestey.[89]

The property comprising the New Belgium Block was as far as can be ascertained the only significant asset of the New Belgium (Transvaal) Land & Development Company Limited. It is hardly surprising therefore to find an entry in the British Government's official publication, the *London Gazette*, in its issue of 20 January 1914, announcing a resolution taken by the Company's Board of Directors at a meeting held on 14 January, to wind up its affairs.[90]

So ended a period of 25 years during which a large tract of Waterberg land was held by a speculative London company named after it, and created specifically to exploit the riches it was perceived to contain. Would the new owners fare any better?

The Vestey era

Who were these Vestey brothers, and what was their interest in a block of demonstrably disappointing farming land (with no identified mineral potential) in a remote part of South Africa?

William Vestey, who was born in 1859, was the eldest son of a Yorkshire couple, Samuel and Hannah. Samuel ran a business in Liverpool, importing provisions mainly from the USA. The Vesteys had 12 children; Edmund, born in 1866, was their fifth. At the age of 17, William was sent to America by his father to source more goods for import. He realised that meat, especially beef, was readily available in America and could meet a shortage at home. Before long, he'd established a factory in Chicago to produce corned beef and in 1883 his younger brother Edmund was sent across to manage it. William expanded his horizons and, with the development of the ammonia-freezing process, established a freezing plant in Argentina from where he began shipping frozen meat to the family's cold store in Liverpool. In 1890 this business became the Union Cold Storage Company, for a while the largest cold storage company in the world. In 1906 the Vesteys began to import eggs and chickens from China, using a fleet of tramp steamers they had converted to refrigerated vessels. By 1911 this fleet had grown large enough that it was formed into a new company: the Blue Star Line. The company added more freezing works around the world, especially in Australia and New Zealand. It moved into retail butchery distribution at home, setting up the J.H. Dewhurst chain of butcheries which, by 1925, was by far the largest in the UK with over 2300 outlets. Blue Star, similarly, was the world's largest refrigeration shipping line.[91]

The Vesteys also expanded into cattle breeding worldwide, including the acquisition of large tracts of grazing property in Argentina, Australia, New Zealand, Russia, Madagascar – and in 1913 at New Belgium, South Africa. It seems that from the beginning they favoured the use of Aberdeen-Angus cattle from Scotland and no doubt contributed to the widespread distribution and success of this breed globally.

During World War I – when imported meat, courtesy of the Vestey Group, kept British troops and families fed – changes in English tax law resulted in UK-based importers becoming significantly disadvantaged relative to

Figure 9.15: William (left) and Edmund (right) Vestey in 1925. Photograph taken from the book *The Vestey Affair*, by Phillip Knightley (1981), published by Macdonald Futura, opp. p. 64.

foreign-based importers. After unsuccessfully lobbying the government to amend the legislation in their favour, the Vesteys decided to move their headquarters to Buenos Aires. They then set up an elaborate, although perfectly legal, trust arrangement based in Paris, which allowed them to return to the UK at the end of the war without incurring a tax liability. They became masters in the art of tax avoidance (rather than illegal tax evasion). By now, the family fortune had grown to be the largest in Britain after the Windsors and the exposure of its scheme to avoid taxation drew an outcry from tax practitioners and some media. This was insufficient, however, to prevent both brothers from being awarded baronetcies in 1921, or for William to purchase a peerage from Lloyd George's government (for £20 000) the following year, despite severe criticism from King George V.[92] For 60 years the Vestey Group was to enjoy protection via the ingenious tax avoidance system it had established. At its peak, the family business was estimated to be worth some £1,4 billion; it had over 23 000 employees and ranched with over a quarter of a million cattle worldwide.

William Vestey (1st Baron of Kingswood) died in December 1940, his position in the company being taken over by his eldest son Samuel (1882–1954), the 2nd Baron Vestey. Co-founder Edmund Hoyle Vestey, 1st Baronet, died in 1953 at the age of 87. He was replaced by his only son, Ronald (1898–1987), who took full control of the company upon his cousin Samuel's death in 1954, the year after his father had died. Samuel's only son had been killed in action in Italy in 1944, so it was his grandson, the 3rd Baron Vestey, born in 1941 (and also named Samuel), who inherited both the title and a role in leading the company. Ronald's son, named Edmund Hoyle Vestey after his grandfather (1932–2007), became chairman of the Blue Star Line for 25 years from 1971.[93]

However, markets were changing, the dynasty was losing its cohesion and the economic recession of 1987 precipitated a severe reversal of the Group's fortunes. Despite management assistance from outside, some iconic businesses could not be saved: Union Cold Storage, the Blue Star Line and the Dewhurst butchery business were all sold off, amid dissension in the family and severe criticism from the markets and the media about both the corporate and private aspects of the family. A

FIGURE 9.16: Edmund Hoyle Vestey II.

critical biographer wrote that "they did not live on the income, they did not live on the interest from the investment; they lived on the interest on the interest".[94]

Edmund Hoyle Vestey II died in November 2007. In the obituaries to him he was described as a deeply religious Presbyterian, shy, with a strong work ethic and a disdain for the trappings of wealth: "There's jolly little security in money", he is recorded as saying, "Security is in a home. All that money means is that you can have some nice pictures on the wall".[95] His cousin Samuel (3rd Baron Vestey), in contrast, maintained a much higher public and socialite profile, being a member of the partying, polo-playing clique surrounding the Prince of Wales. Despite the losses of the 1980s and 1990s, the family remains among the wealthiest in the UK and its members are still frequently mentioned in social pages.

From the little information available it seems that the Vesteys must have acquired the New Belgium Block in December 1913 with the intention of using it as one of their several far-flung cattle ranching enterprises. Yet there is little evidence on public record of their ranching activities on the property during their tenure. Chaney's store at Sandrivierspoort held accounts on behalf of several people (J.F. van Staden, C. van Heerden and R. Sharpe) in 1925–1936, whose addresses were given as being at New Belgium Estates, PO Vischgat, but the hotel register there does not reflect any New Belgium visitors.

John Todd at New Belgium

An early manager for the Vestey family was John Todd. He had been born in Edinburgh, Scotland, in 1889. He learned about cattle and their management at an early age: His father had been a cattle agent in Scotland where one of his grandfathers had owned a cattle-farming estate. Quite how he came to be the manager of the Vestey estate in the Waterberg is uncertain, but from the various shreds of evidence that have been uncovered, it is conjectured that he might have joined the family business as early as 1912 and have gone to Argentina, Madagascar and even Australia before coming to New Belgium. What is known for certain is that through the years he formed a strong friendship with the Vesteys, one that was to endure long after he had left their business to become a substantial farmer in his own right. As recently as 1951, John's youngest daughter, Mary, was invited to stay at Stowell Park, the Vesteys' country estate in Gloucestershire. In 2015 Mary – by then a sprightly 86 – vividly recalled having been met at the station in a Rolls-Royce. During her stay on the estate (it having been established that she knew how to drive), this "young farm girl from Roedtan" was invited to take the Rolls for a spin around the property! Later she spent almost a year as the guest of the Vesteys at their mews flat in central London.

John Todd married a pretty young Canadian girl, Alma Evadne Humphris, in April 1917, apparently having met her on board a ship between Cape Town and Durban. The antenuptial contract gives John's address at that time as being New Belgium Estate.[96] However, it seems that within a year or so of their marriage, John and Alma were able to establish themselves on their own property east of Naboomspruit, so it is presumed that John was only on New Belgium for a few years, perhaps from 1914–1918.

It was probably Todd who was responsible for the construction of a new headquarters for the property on the farm Derdekraal (now part of New Belgium,

FIGURE 9.17: John Todd (centre) with his wife Alma (left). Photograph kindly provided by John Todd of Nottingham Road, KZN, a grandson.

owned by Kwalata), immediately adjacent to and west of Noord Brabant. Here, later reports show that a substantial home and accommodation/administrative complex was built: a comfortable wood-and-iron bungalow of seven rooms with an enclosed verandah on three sides; attached offices and a building intended for a trading store; two living rondavels, kitchen, mess room and rear servants quarters; a detached five-roomed foreman's bungalow; brick-and-iron quarters for stockmen; a large surrounding garden and orchard; and a strong borehole with windmill and storage tanks. The whole compound was plumbed, and wired for electricity.[97] The same report described windmills and tanks having been erected on at least ten of the camps into which the Block had been divided by over 600 km of fencing, and served by 130 km of motor roads.

John Todd is remembered with great affection by his eldest grandchild John, the son of David Todd, who in turn was the eldest of John and Alma's four children. There is no doubt that John (senior) was very much a man's man: stocky, but extremely strong, a heavy smoker and whisky drinker, adored by his grandchildren and respected by the community in which he lived.

He had a deep affection for cattle and was one of the early importers of the Aberdeen-Angus breed from Scotland. After leaving the service of the Vesteys, he purchased some land around Immerpan, on the road/rail link between Roedtan and Zebediela, about 50 km east of Naboomspruit (Mookgophong).

He named the place Ballindalloch Farms, after the estate in Scotland which at the time was a leading breeder of Aberdeen-Angus. John (junior) remembers driving around Ballindalloch on his grandfather's lap, behind the wheel of a 1938 Ford truck, with a faithful dachshund-cross-Scottish

FIGURE 9.18: John Todd of Ballindalloch Farms, early 1950s. Photograph kindly provided by John Todd.

terrier, Augusta, always present. John also recalls how on one occasion his grandfather, growing tired of watching his labourers trying to manage an animal preparatory to some operation, stepped up to it, grabbed its horns and after twisting it briefly from side to side, suddenly threw the creature to the ground in order to proceed. He also remembers how his school friends (from the newly opened Capricorn High in Pietersburg) would be astonished to come across grandpa with his whole arm up a beast's rear, attending to one or other malady![98]

In her affectionate, unpublished memoirs of the Waterberg and her family's history there, the late Elizabeth Clarke recounted one incident at New Belgium involving her father, the redoubtable E.A. (Ted) Davidson of 24 Rivers:

> When my father began exploring that part of the world the Block had just been bought by Vestey brothers. John Todd, who feared not any man, and his young Canadian wife, were then in residence. John was one of the first pioneers to own a motorcar; when he got a puncture he would stuff the tyre with grass, and in the end all four tyres were running on grass. He used to go once a week to the bushveld post office to collect his post. Early in the morning John Todd liked to set out, and there came a morning when he was so early that the postmaster was still sound asleep. "So I just shot open the box with my gun and helped myself" said John, describing the incident.
>
> A Miss Robinson, relation of the Vestey brothers, came up to the estate and fell in love with one of the farm managers. She was extremely elegant and beautiful, and wildly extravagant. Tiring of the quiet monotony of the veld, she decided to give an enormous and unforgettable dinner. A whole band was booked and brought down from Johannesburg, cases and cases of every kind of rare delicacy, whisky, champagne, pickled hams and expensive tinned foods, were trekked along the thorn-lined roads. A parson was invited out to say the grace. The whole bushveld simmered and bubbled with excitement, such as never known before. Every type of bushveld dweller, invited or not invited, made their way to that dinner. "Never", quoted my father, "have I seen a stranger assortment of people quaffing champagne and eating pate de foie gras, for the first time in their lives.
>
> As a friend of Miss Robinson my father was a legitimate guest,

> and thoroughly enjoyed the macabre scene. But there was a wild haunted look in Miss Robinson's eye that night, and just as the guests were getting uproarious and unrestrained, wild words were heard from her as she sat talking to a certain farm manager. No-one could help noticing when she suddenly rose and rushed from the room. She returned, and in a hush of horror it was realised that she had a revolver in her hand. Quickly big John was up, and the tragedy was averted, but the evening ended in chaos, and to this day the bush still flutters with stories of that momentous night.
>
> For some time before this my father had been employed by Vestey brothers to collect rents on the block, and he had become very friendly with the lively Miss Robinson. When she left abruptly after the party, he was promptly dismissed by the Vestey brothers, who were quite certain that he had been one of her lovers.
>
> Meanwhile one of the younger managers was steadily stealing stock from Miss Robinson, and then re-selling them to her. My father discovered this quite by accident. When the unscrupulous young man knew he had been found out he invited my father to his camp. As his visitor approached he swiftly lifted a revolver, pointed it at him and fired. Once again tragedy was averted, for the young man's nerves were bad, and eyes faulty, so the bullet whizzed past my father's ear and landed in a haakdoorn tree.
>
> After Miss Robinson left the block she chartered a yacht and was planning a magnificent expedition, when the Vestey brothers tired of her extravagance and she was certified as insane.[99]

Unfortunately, it has not been possible to trace the lady in question or to verify this story, but then the Vesteys were – and remain – renowned for their ability to keep their private affairs out of the public eye should they choose to do so. Apart from the occasional newspaper report, celebrity event and several high-profile court cases in which the family defended its elaborate tax avoidance schemes, little was known about the Vesteys until 1981 when the respected investigative journalist Phillip Knightley published *The Vestey Affair*, which described in detail the financial and other affairs of the family.[100]

It is not known how successful the New Belgium Estates property was in relation to the Vesteys' other ranching operations around the world. Although no further outbreaks of rinderpest were noted in the Waterberg after the end of the South

African War, "Rhodesian redwater" (later called east coast fever) seriously affected cattle in the Waterberg and elsewhere in 1902–1904 and other tick-borne diseases remain endemic to the region. The sourveld that characterises most of the Block does not support a high stock density, especially in the absence of supplementary feed in winter. The area is also known for two types of plant that are especially toxic to cattle: gifblaar (*Dichepetalum cymosum*); and gousiektebossie (*Pachystigma pymaeum*). Gousiekte can also be caused by other plant species: wild date (*Fadogia homblei*) and tonnabossie (*Pavetta sp.*), both of which occur in the Waterberg.[101]

Years later, when the Block was being prepared for the settlement of returning servicemen, the Department of Lands received a letter from an officer who had been familiar with the estate in John Todd's time. The Vesteys, he alleged, had sold out all their stock following enormous losses from gousiekte during a period of drought (possibly that of 1933–1935). He expressed concern that the department had not assessed the agricultural potential of the property adequately before deciding to proceed with a settlement scheme.[102] His concerns would be vindicated in the years ahead.

During the Great Depression, many businesses around the world were forced to close or else to greatly reduce their scale of operations. The Vestey group of enterprises did not escape unaffected. In addition, its operations in China and Russia suffered serious setbacks related to sociopolitical unrest in those countries. At home, the family faced the continuous attention of the UK Department of Inland Revenue. Whatever the reasons, by 1937 the New Belgium Estate was on the market. An internal Department of Lands memorandum in March of that year noted that the property was for sale and recommended that it be evaluated for its potential as a government settlement project in terms of Section 10 of the 1912 Land Settlement Act.[103] Negotiations commenced between the government (Lands Department) and the Vesteys' local agent, a Mr Gishford. They culminated in an agreement in December 1938 for government to purchase the entire property – but not the mineral rights – for £50 000 (a mere £10 000 more than the Vesteys had paid 25 years previously). Although the mineral rights were excluded from the sale (the Vesteys evidently still believed there might be some undiscovered mineral potential on the property), the government was asked to agree to take transfer of these rights at any time in the future, at the sellers' option, for no additional fee. This rather bizarre condition, which was accepted, was the result of a fear (proved groundless) that the government was about to impose a tax on mineral rights, which might render them a liability instead of an asset!

THE GOVERNMENT AND THE NEW BELGIUM SETTLEMENT SCHEME
William James Averre

In January 1939, when the government took transfer of the New Belgium Estate, the Department of Lands appointed the last Vestey manager, Major William Averre, as resident caretaker. Averre, who lived with his second wife Eleanor (known as Annie) on Derdekraal, was already a man of 76.

He had been appointed as manager by the Vesteys in 1936, following the unexpected death (from a heart condition) of his predecessor Robert Sharpe, who was a Scot like John Todd, and who had been managing the property at least since 1927. Sharpe, who was 57 when he died, had only the previous year become married for the first time to the widow Ellen Nichols (née Wright).[104] They had no children and Ellen, now widowed again, left the area for Kimberley where she married a Ferreira, and died in 1954. No further information about Sharpe has come to light.

William James Averre, like several of his predecessors on New Belgium, was rather an enigmatic character. Born in Barking, Middlesex, in 1863, Averre married Kate Elizabeth Maria Overton late in 1882 and lived in West Ham, London, where various UK Census reports describe his occupation as "meat salesman" or butcher. He and Kate seem to have had two children, both boys: William Frederick, born in 1885 and Sidney H., born in 1890. But tragedy struck, possibly during or soon after the birth of the second son, because the Census of 1891 records W.J. Averre (28) as a widower, with sons aged four and one.[105] What is really puzzling, however, is that the next Census report in 1901 describes the household of W.J. Averre (widower, meat salesman) as comprising Virginia M. Averre (daughter, 14) and Sidney H. Averre (son, 11) – William seemingly having morphed into Virginia! There is a record of William James Averre marrying Mary Augusta Halls (25, of West Ham) in October of 1892, but no further reference to this possible second wife has been found. What is known is that William Averre, aged 38 (i.e. born in 1863), departed from Southampton for the Cape aboard the *Braemar Castle* on 3 August 1901, his occupation described as "cattle dealer".[106] Perhaps he had already become involved in the Vestey's business. No family members accompanied him then, although South African records show that his eldest son, William Frederick, died in Johannesburg, aged about 80, on 18 June 1965.[107]

In March 1918, aged 55, Averre had been appointed a Temporary Major in the South African Army Service Corps (backdated to 16 April 1916),[108] a position he held until his demobilisation on 21 February 1919.[109] In the massive post-War Birthday Honours List of King George V, issued between June and August 1919, Averre was amongst over 5000 British and Empire recipients of the OBE for services rendered during the Great War just ended. (Another Waterberg personality on the same Honours List, was Richard Granville Nicholson – see Chapter 16.) To his credit, Averre did not append this award to his name in any of the numerous papers reviewed during the course of research into his story. Prior to his appointment on New Belgium, Averre and his wife Eleanor had been in dire straits financially. He is recorded as having been employed in 1935 as a farm assistant (bywoner) on a property in the Krugersdorp district under the provisions of an agricultural rehabilitation scheme run by the Department of Labour.[110] In his application, Averre had understated his age by four years (as 68 instead of 72), no doubt to improve his chances …

Averre shared a number of characteristics with his distant predecessor and namesake William Caine: he held strong views about what he considered right and wrong; he spoke no Afrikaans (or Dutch) and was suspicious of and derogatory about Afrikaners; he did not consider that black employees or residents had any rights (a view shared strongly by his Afrikaner neighbours); and he was quick to put his eloquent pen to paper! Departmental files for the next few years are filled with his long, distinctive, handwritten missives to the Secretary for Lands: lobbying for game protection; complaining about conspiracies between local landowners or tenants and the Brakfontein/Dorset police; informing the Secretary of illegal hiring and movement of black labourers, of illegal tenant-operated stills and of political intrigue led by a certain J.P. Jacobs. The latter was a teacher and farmer at Vischgat. He was also chairman of the Sukses Boerevereniging and a staunch supporter of the young right-wing National Party Member of Parliament for the Waterberg, the Nylstroom advocate J.G. Strijdom (the "Lion of the North" and a future Prime Minister). In no time at all, Averre had made himself several enemies in the area. He gradually lost favour too with his employers, who, while valuing his factual communications (like his regular rainfall figures and his knowledge of the area), found themselves having to defend the department against hostile letters from the local magistrate as well as Mr Strijdom, Mr Jacobs and the regional police inspector.

Indeed, it was not long before the department realised that it had erred: As early as September 1939, the Secretary for Lands requested the Land Inspector at Potgietersrust to appoint someone else (T.J. Henning) to manage the northern part of the property from a base on Kirstenbosch. But Averre was to remain in a form of caretaker role for the southern part of the estate until mid-1942 when the Minister himself intervened, saying:

> We should explain to Major Averre that, owing to [his] advancing years, [he was then 79] he is unable to undertake the onerous duties of a caretaker to the satisfaction of the Minister, but the Minister does not want, on that account, to remove Major Averre altogether from the settlement. Major Averre will be allowed to reside on the farm Derdekraal and to enjoy all the facilities which he has hitherto been allowed, at a nominal rental of 10/- per month. He will receive no salary and all that will be required of him will be to prevent native squatters and unauthorised persons from settling on Derdekraal.[111]

In the meanwhile, Averre was harassed by local farmers (who, for example, placed tree trunks on the roads to hinder his passage) and received at least one serious threat which he attributed to three particularly lawless occupants of the farm Wynkeldershoek (where Canvas Town had been located 40 years before):

> Please accept this as a first & last warning ... that your presence in this area is most undesirable. You have proved it through your most severe lies and thieving. We Africaners [sic] Gentiles of this district are aware of the fact that your presence amongst us is only to belie and betheise [sic] and to persecute us as it was your principles from the beginning like that of Great Britton [sic] your Mother Country. We take it you was [sic] an accomplice to the murder of 26,000 women and children in the Anglo Boer war and that your attitude today proves that you are still the same leopard with the same spots ... If you don't leave this district within 30 days ... you can be assure [sic] as you are alive today that your death will be an unnatural one, take this from hundreds of Gentile Africaners who wishes to live in peace in their own mother country. Signed, The Black Hand.[112]

Needless to say, given their affiliations, the local police under Sergeant Rochat of Dorset were unable to identify the perpetrators of this threat. Averre had few allies in the area.

FIGURE 9.19: The beautiful but rugged valley of the Lephalala River.

The Thompson plan

The Lands Department now proceeded to design a programme for the development of the New Belgium Estate as a settlement. It brought in the expertise of Dr W.R. Thompson, then superintendent of the Vaal-Harts settlement scheme in the Northern Cape. Thompson was asked to evaluate the settlement and agricultural potential of the whole property and make recommendations as to its development. In a comprehensive report, dated July 1941, Thompson effectively divided the Block into three zones: a southern, sandy zone of about 27 000 morgen (23 100 ha) that included about 11 of the original farms, with good spring and summer grazing and a carrying capacity of about 12 morgen (10 ha) per beast; a central hilly section of 70 000 morgen (60 000 ha) and about 30 farms, with difficult, inaccessible terrain and poor grazing (20 morgen [17 ha] per beast); and a northern, lowveld zone of 40 000 morgen (34 000 ha) (17 farms), comprising sweetveld and a good carrying capacity of 8–10 morgen (7 ha) per beast. He recommended using only the northern and southern zones for development, redividing the farms where necessary into economic-sized allotments each with water and access.[113]

The upshot of Thompson's report was a decision to divide the Block of 60 farms into 54 allotments, each of around 2000 morgen (1713 ha), and each having some form of access to water. This was achieved by consolidating several farms in the northeast of the Block into the existing farm Kirstenbosch (now renamed Kirstenbos), followed by the division of this property into several portions. A similar process was followed in the southwest, where another group of farms was consolidated to form a new entity, called New Belgium, which too was then divided into several portions. As a result, some 14 allotments were defined in the southern part of the Block and another 23 in the north, with the remainder being situated in Thompson's "central hilly section". Another government memorandum of the time pointed out that whereas the intention had been to advertise and allot pieces of land in holdings of suitable size to potential settlers, the outbreak of hostilities (World War II) had forced the suspension of this process. Instead, tenants were being sought to lease the properties in the short term. It continued:

> In order to meet the anticipated demand for land at the conclusion of the war, it has been decided to deal with these farms under the Probationary Lessee Scheme (Act 38 of 1924); and the construction of Buildings, Dipping Tanks, Roads etc at this stage becomes an essential requirement for the contemplated Settlement.[114]

Thereafter, a flurry of memoranda and letters during 1941 resulted in an amount of £18 000 being made available to identify suitable sites for the drilling of water boreholes, to build dwelling houses, dip tanks and to build a road linking the northern and southern sections of the Block. A geological survey of the area north of the Lephalala River was undertaken and a report issued in May 1942.[115] A letter from the Secretary for Lands, F.E. Hayward, to his counterpart in Mines provided a list of the original 60 farms comprising the estates, pointing out in conclusion that the mineral rights to the land in question were still held by the Vestey brothers.[116]

William Vestey, the first Baron of Kingswood, had died on 10 December 1940; his son Samuel was now the representative of the family's interests in conjunction with his uncle Edmund. Interestingly, although Baron Vestey had died in England, a death notice was also issued in South Africa, presumably to finalise his estate here. The archival dossier includes an inventory of assets of the late Baron. There is but one: a half share of the mineral rights in the property known as the New Belgium Estates. Value: £500.[117]

But not for long: A resolution of the Board of Directors of the New Union

Goldfields Limited, dated 25 September 1942, agreed that "the Company purchases from Sir Edmund Hoyle Vestey (Bt), the Rt. Hon. Samuel, Second Baron Vestey, the Hon. George Ellis Vestey and Edward Brown … their interest in certain rights to minerals and precious stones on certain farms. Signed, Norbert Stephen Erleigh, Chairman".[118] The price is not stated. It would be interesting to know whether the company had been persuaded to pay more than £500 for these rights!

And so the involvement of the Vestey family with New Belgium finally ended and with it came the end of the story of the New Belgium Block as a discrete entity.

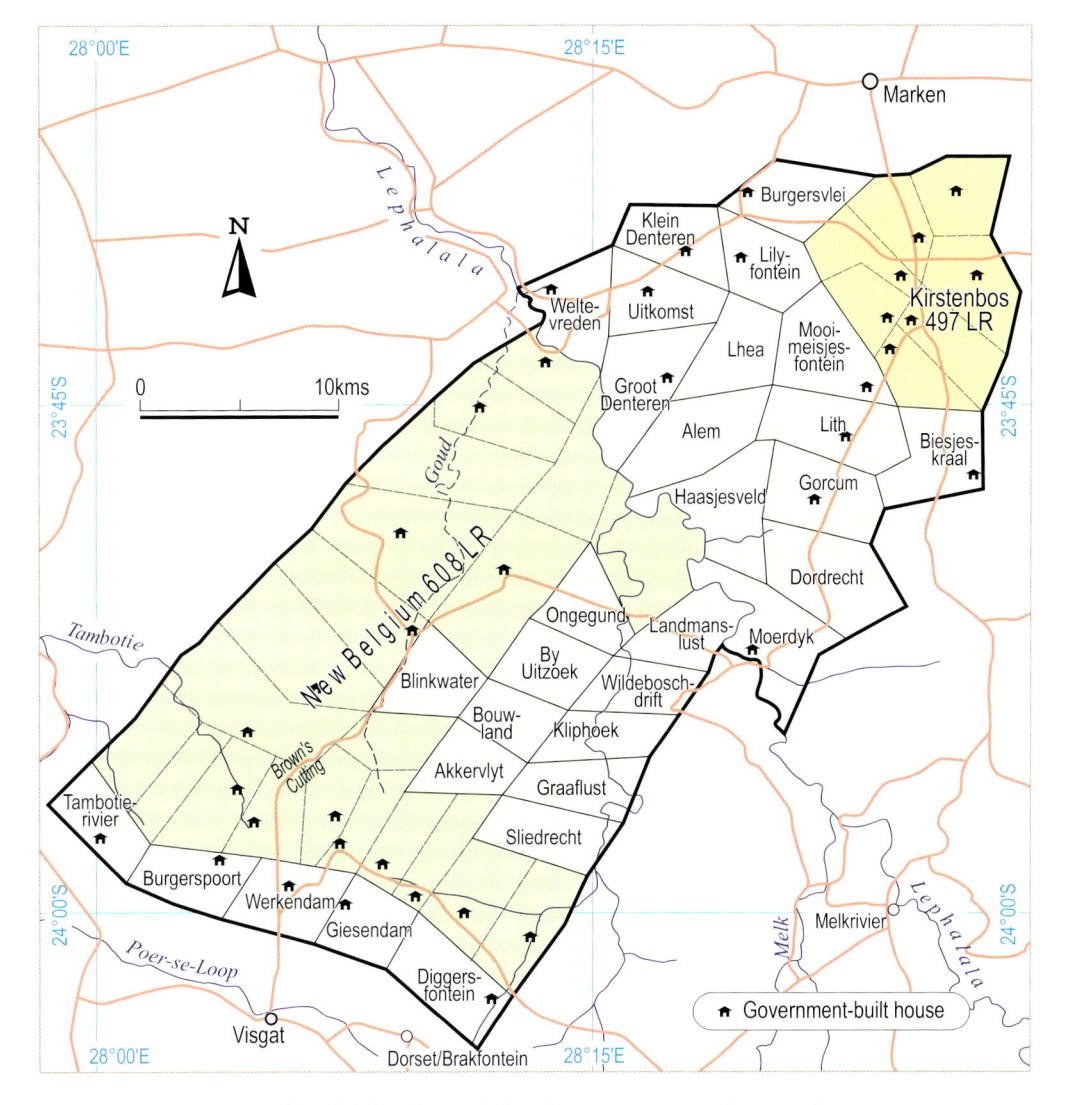

FIGURE 9.20: New Belgium Estate during the government settlement scheme era.

The Department of Lands decided to build simple houses of concrete blocks with iron roofs on 32 of the allotments. Work commenced in late 1941, under the overall supervision of Dr Thompson. In December Thompson reported that the foundations of 16 houses had been laid, with a total of 28 Europeans and 105 natives being employed on the project.[119] By March 1942, building of all 32 new houses was complete: £14 000 of the £18 000 allocated for the project had been spent, 150 miles (240 km) of drivable roads built or repaired, a viable low-level bridge (the one on Moerdyk near the original Wilderness School site) built across the Palala River, new boreholes drilled and several dip tanks either repaired or built. Thompson was requesting the appointment of a caretaker to manage the settlement, including the houses, as well as guidelines as to the number of blacks and stock to be permitted on each allotment.[120] Indeed, a remarkable amount of work had been undertaken in a short period of time within the budget allocated for it. For a project organised by a government department, this may be a unique record!

A recurring issue at New Belgium during these years was unskilled (black) labour: Firstly, the shortage of black labour in the area in general relative to the demand for unskilled labourers on farms to the south of the Block; and secondly, the policy regarding the number of black families to be allowed, with their stock, to reside on each farm or allotment within the Block in return for the work they would perform. This was the era of the *"nege maande"* arrangement in terms of which a black man was allowed to reside on a white-owned farm and grow crops

FIGURE 9.21: The first new house on the New Belgium development, 1942.

FIGURE 9.22: Original block house on Kirstenbos with lean-to additions, 2017.

and raise stock for his own use on that property, in return for which he would work for the landowner without wages, for three months of each year. Several farmers to the south of the Block, including the ever-critical J.P. Jacobs of Vischgat, complained that while they experienced labour shortages on their own properties, blacks were squatting idly on unoccupied farms within the Block. Some farmers even leased farms in the Block merely to gain access to the labourers permitted to reside on them.

With the development of the settlement project, the Department of Lands was concerned to ensure that for each property on which a house had been built, there should be at least one black resident appointed with the responsibility to look after that asset. But it was also concerned (and frequently reminded by Major Averre, whose retirement did little to dampen his enthusiasm for writing critical letters to the Secretary) that for the properties to remain attractive to prospective tenant/purchaser settlers and not become overgrazed, it was important to limit the number of blacks and their stock on each property. In May 1942 Thompson and the Minister announced that they had decided that there should not be more than three "native" families and a total of 60 head of cattle on each allotment. These families would be required to render "tenant labour" in the form of preparing firebreaks, maintaining roads etc., and to ensure that their private stock herds were dipped regularly.[121]

Following the enforced retirement of Major Averre, as well as the termination of the caretaking agreement with Mr Henning on Kirstenbos, the Department appointed a general caretaker for the whole settlement project, and provided him with a vehicle to enable him to get around the whole area. The appointee was C.W.P. Vermeulen, a young carpenter from Vaalwater, who had previously been one of Averre's stockmen. Vermeulen was instructed to provide a monthly report to the Secretary for Lands summarising conditions on the settlement: vegetation, soil and road conditions; numbers of black inhabitants and their stock; numbers and status of wild game; rainfall figures; and any other developments. This proved to be difficult to achieve because although Vermeulen seemed willing, he was barely literate and was really not capable of even obtaining all the information required from him, let alone committing it to a written report. Several of his "reports" – usually no more than a couple of lined sheets filled with almost unintelligible handwriting in blue crayon – survive in the files, as do a number of terse letters from the Department reminding him of his reporting obligations. His rainfall records, such as they are, reflect a fraction of the rainfall recorded (contemporaneously) by Major Averre on his farm in the south of the Block. For example, in his Annual Report of April 1947 Vermeulen reported that in the previous year, there had been 8" of rain (without any further detail); whereas at the end of 1947 Averre submitted a month-by-month record for 1946 and 1947 reporting totals of 21.44" and 20.38" respectively.

One has to wonder why those receiving the reports they continuously demanded from Vermeulen were not more critical of their content. Vermeulen also appears to have been somewhat in the thrall of the police station commander at Brakfontein (Dorset), Sgt Rochat. Unsurprisingly, Vermeulen was often a target of Averre's criticism – to which he responded very mildly and respectfully. The conclusion gained from an examination of the documentation on file is inescapable: Vermeulen, while no doubt doing his best, was completely out of his depth and easily intimidated by people he considered to be his superiors. The fault lay not with him, but with the departmental bureaucrats in Pretoria, who, after Thompson's departure or promotion, seem neither to have visited the settlement nor offered tangible support.

In July 1943 Vermeulen noted in one of his reports that the number of tenants on New Belgium had risen to eleven, including Averre on Derdekraal and Henning on Kirstenbos and that the total number of cattle (tenants and their black labourers) was 924. He also reported that the *"naturelle"* (natives) on the allotments no longer wished to abide by the *"nege maande"* arrangement. Instead, they wanted to be given some food (three sacks of maize per month) in return for their labour, given that they had to sell some of their own produce to raise money to pay for their cattle to be dipped, as well as pay taxes. The Department seems to have acceded to this demand without a quibble and Vermeulen was instructed to give each labourer 3 lb of mealiemeal per day as well as salt, during the three months of tenant labour worked. A year later, in July 1944, Vermeulen reported that there were ten tenants, who together with their labourers held 1088 cattle on the Block. By March 1945 the number of tenants and cattle had dropped to three and 160 respectively, owing to very poor rains and drought.[122]

Post-World War II development

Following the end of World War II and the influx of returning ex-servicemen, the idea of using New Belgium (as well as the Harmony block near Loskop Dam) as a settlement location for ex-servicemen was resuscitated. The total cost of developing the settlement as at March 1945 had been about £75 000, including the cost of acquisition and maintenance during the war. Two inspectors (De Jager & Geldenhuys) were despatched to the New Belgium settlement to evaluate all individual properties. This led to the compilation of a list of all 54 allotments, citing their characteristics, attributes and improvements with a monetary value being assigned to each. In the meanwhile, existing short-term tenants (by 1946 there were only two left apart from Averre) were given notice to vacate so that the properties could be taken over by the successful applicants.

The first 30 allotments were then advertised in a *Government Gazette Extraordinary* on 20 September 1946, inviting prospective tenant/purchasers to bid for the right to take over the allotments. Eleven of the allotments were readvertised in a later *Gazette* (of 20 December 1946), there having been no successful applicants in the first round. Further advertisements followed.

Purchase prices ranged from £989 for Gorkum (2338 ha in extent) to £2299 for Weltevreden (1872 ha, straddling the Lephalala River) according to the assessed quality of the land and its improvements. Successful applicants could lease a property for five years with the option to purchase it at any time during that period. For the first two years, the lease payment was zero, in year three it was to be 2% of the purchase price and in each of years four and five, 3¾% of the purchase price. If the lessee chose to purchase the property, the purchase price would be paid in 65 equal annual instalments, inclusive of interest.

By March 1947 a total of 17 allotments had been awarded and a further eleven were under consideration. By mid-1949, of the 30 allotments advertised, 24 had been awarded (including Derdekraal, rented by Averre, and Klein Denteren/Uitkomst, taken by C.W.P. Vermeulen). Ironically, not one of the neighbouring farmers (like Jacobs) who had been so vociferous in their criticism of the development of the settlement was among the names of the new tenant/owners.

During 1948 Vermeulen's wife and their four children were taken seriously ill as a result of suspected arsenical poisoning in their drinking water. Two of the children (twins) subsequently died, although the cause of the poisoning was not made clear in the files.[123] No suspicion of foul play was mentioned. Arsenic was widely used to combat locust infestations and poisoning from unsecured insecticides and herbicides in rural communities was not an uncommon occurrence at this time.

A year later, 22 of the original 54 plots remained vacant (unawarded).[124] Despite the very favourable terms offered by the government, prospective farmers recognised the marginal agricultural potential (and formidable terrain) of the remaining allotments and there was noticeably less haste to bid for them than there had been by the Department of Lands to acquire the Block before the war. The disappointing truth about New Belgium as agricultural property was at last beginning to sink in, after more than 60 years of ill-informed speculation had obscured common sense. Indeed, the area would remain largely blighted and subeconomic for another 30 years.

In December 1949 the Secretary for Lands instructed his inspector in Nylstroom to allow unalloted farms in the Block to be let on a temporary basis at the rate of 10/- per month.[125] Several local (nonresident) farmers applied in response to this initiative to obtain additional grazing for their stock. The consequences of this wholly ill-considered process would be borne for decades afterwards in the form of land severely overgrazed by tenants lacking any vested interest in it. During 1950 and 1951 discussions centred around an inspection report that described the unalloted properties as being essentially subeconomic on account of their poor sourveld grazing; the widespread occurrence of *gifblaar*; the shortage or absence of perennial water; the absence of all-weather roads; the distance of the properties from schools, stores, hospitals and infrastructure; and the general remoteness of the region.[126]

This debate was no doubt encouraged by the prevailing political ideology within government (the National Party had won the general election in 1948 and Advocate J.G. Strijdom, one of the most hardline members of the party, had requested and been given the portfolio of Minister of Lands, which he was to hold until he succeeded D.F. Malan as Prime Minister in 1954). It culminated in the decision in April 1952 to offer current tenants of New Belgium allotments the opportunity to purchase the properties at low prices on very favourable terms.

Feisty Major William Averre died on Derdekraal on 7 September 1951, aged 88, leaving his widow Eleanor and two adult sons from his first marriage.[127] An appraisal of his assets was undertaken by E.A. Davidson of 24 Rivers. The Averres had lived on New Belgium since 1936 – a total of 15 years. A stickler for protocol and procedure and a staunch opponent of corruption, although bound by the prevailing racial attitudes of his times, Averre had been a perpetual thorn in the sides of successive Secretaries for Land. But, unlike many of them, especially after 1948, he had always had the best interests of the Block at heart. His widow, Annie Eleanor Averre (née White), died on 14 August 1966 at the age of 82; she is buried at the St John's church graveyard at 24 Rivers.

NEW BELGIUM AFTER THE FAILURE OF THE DEVELOPMENT SCHEME
The Double R Ranch

Even in the aftermath of the failed settlement scheme, there were visionaries who saw the potential of this remote, rugged area for conservation: A man called Pearson had purchased two of the unoccupied properties, Ongegund and By Uitzoek, in 1950 and turned them into a nature reserve on which he kept locally indigenous roan antelope amongst other game. In 1967 another conservationist, a former big-game hunter by the name of Eric Rundgren, invested with a friend (Roberts) in two farms along the forbidding southern bank of the Lephalala River – Dubbelwater and In den Berg (properties which, like the two mentioned above, had been deemed correctly by the Thompson study to be unsuitable for agricultural development). They had purchased their 5000 ha spread for R9 per morgen (about R10/ha) and turned this too into a private nature reserve, the Double R Ranch (for Roberts and Rundgren). Around them were struggling conventional crop and stock farmers, many absentee owners and many involved in illegal hunting – an activity that had almost become a national sport. That unauthorised hunting could take place at all was due to the government's relaxed attitude towards the activity, especially in remote parts of the country. Around Marken, for example, in the Koedoesrand area just north of New Belgium, hunters did not even bother to conceal their trophies, carrying them away by the truckload. A little further to the northeast, the lower Mogalakwena River valley was still a favourite winter hunting area, as it had been since the 1930s.

Lapalala Wilderness

By the mid 1980s, parts of the former New Belgium Block were gradually being reconsolidated. Impoverished farmers, some of them descendants of the original investors of 1946–1950, others still in the process of paying off their purchase price, readily sold their overgrazed, poorly managed properties to wealthy conservation-oriented entrepreneurs of whom the late Dale Parker was a prime example. Within a few years between 1981 and 1997, Parker managed to put together portions of 15 former New Belgium farms, amounting to over 36 000 ha. He called this vast new estate, which included the most rugged properties straddling the Lephalala River (as well as the Pearson and Double R properties), the Lapalala Wilderness.

The Lapalala Wilderness Private Nature Reserve, now at almost 45 000 ha, is the largest privately owned conservation area on the Waterberg plateau and includes almost 40 km of the course of the Lephalala River within its boundaries. It was the inspiration of the talented artist, author and dynamic conservationist Clive Walker, who in 1981 was invited to visit the Double R Ranch: The owners, Rundgren and Roberts, wanted to sell and Walker was looking for a location for a wilderness school he wished to establish.[128] In the 14 years since they'd acquired the property, Rundgren and Roberts had allowed the land to revert to its natural state and by 1981 there was little evidence of any former farming activity. Walker was fortunate soon afterwards to meet Dale Parker, a Cape businessman and nature-lover and in no time persuaded Parker to purchase the 5000 ha farm on the Lephalala River, for around R100/ha.[129]

Immediately to the south of the Double R Ranch was the Pearson Nature Reserve, on which a rare herd of roan antelope had roamed, until they were caught and relocated to the Percy Fyfe and Nylsvley reserves by the Department of Nature Conservation, which had been concerned about their ability to survive. Parker purchased these properties in 1982, together with several neighbouring farms.

On one of these, Landmanslust, there stood the rusting hulk of a magnificent steam engine on the alluvial plain between the Lephalala River and the important archaeological site of Melora Hill. This machine made by Ransome, Simms & Jeffries of Ipswich, England, in 1923, had been discovered in a small Free State town by a former owner of the farm and brought here by rail and truck in 1961.[130]

FIGURE 9.23 (LEFT): The Ransome's steam engine standing amidst the dense bush on the banks of the Lephalala River. (RIGHT): Front hatch and nameplate of the engine.

At this time, the Lephalala River formed the northeastern boundary of Lapalala. Across the river, another conservationist, Hennie Dercksen, owned several former New Belgium farms. A few years later, these were sold to an American nature lover, David Methuen, to form part of the Touchstone Reserve. In 1997 most of this property too was purchased by Parker, bringing both banks of the Lephalala River into the reserve. By the time of Parker's untimely death in 2001, Lapalala Wilderness had grown to encompass 36 000 ha under the management of Clive Walker and his wife Conita. The land was a mixture of overgrazed, degraded farmland and rugged wilderness – but these visionaries saw its potential as both a unique nature reserve and as an ideal venue for the creation of a wilderness training school. With Dale's support, Clive established the Lapalala Wilderness School in 1985. Its capacity has grown steadily over the years, and its mission – to introduce young South Africans to the bush and to the sustainable value of environmental conservation – has never wavered. By 2018 over 90 000 youngsters and educators, the majority drawn from local rural schools, had enjoyed the magnificent wilderness scenery and participated in one of the several courses offered by the school – mostly subsidised through the sustained generosity of the Parker family and other benefactors.

In 1990 Lapalala became the first private reserve to reintroduce black rhino in the area and today it is one of the largest sanctuaries of both black and white rhino in southern Africa. It also focused on the environmental rehabilitation of the many degraded properties that had been acquired. For a few years, a leading professional ecologist and experienced former Natal Parks manager, Roger Collinson, was employed at Lapalala. He was responsible for the introduction of sound, appropriate land management programmes that are followed still. This is an ongoing and long-term project, but in the last 30 years there has been remarkable restoration of overused farmland to near-natural condition.

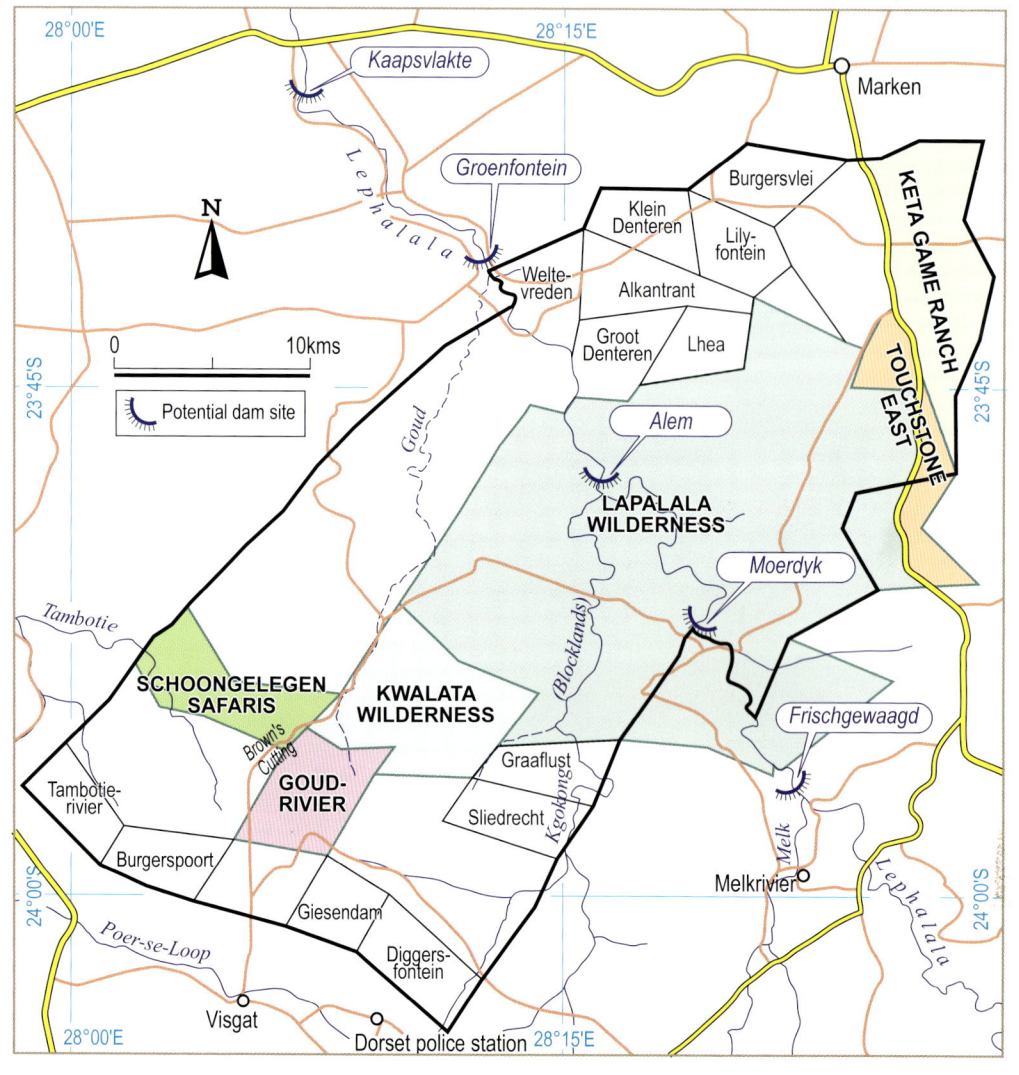

FIGURE 9.24: New Belgium in the early twenty-first century.

In Dale Parker's time, several rustic self-catering lodges were established on the reserve and many South Africans treasured the times they were able to spend enjoying the magnificent Waterberg wilderness at very affordable rates. Alas, this could never have become a sustainable business model; it was dependent on Dale Parker's generosity and passion for sharing his wilderness with fellow nature lovers.

Duncan Parker succeeded his father in taking responsibility for this unique family asset. He forged an association with a neighbouring property owner, Gianni Ravazotti, to expand Lapalala to about 45 000 ha and to create a development intended to sustain the property into the future. This involves the introduction of a formalised breeding project for roan antelope and Cape buffalo as well as the two rhino species, the reintroduction of elephant and lion to the reserve, the construction of a luxury commercial safari lodge and, most dramatically, the division of the whole reserve into 1400–1500 ha freehold "custodianship" sites, on each of which an investor would be allowed to build a private lodge. Investors would have traversing rights over the whole property.[131] The Wilderness School will continue, albeit in a new, rather more constricted location on the reserve's margin.

In the late 1980s Lapalala was joined by the creation of Kwalata Wilderness adjacent to its southern boundary, another consolidated block of almost 7000 ha made up of 14 of the original farms, including Caine's Noord Brabant and Averre's Derdekraal, under the stewardship of Dieter Heuser and his family. In the north, a third reserve, Keta, was established, based on the enlarged Kirstenbos allotments, comprising seven former farms and amounting to 15 000 ha.

Over two-thirds of the original New Belgium Block was eventually accounted for by these four properties, all of which focused on the rehabilitation of the overgrazed, overcultivated land; on the removal of exotic and invasive vegetation, rotting camp fences, dilapidated and ruined buildings; on the reintroduction and management of originally indigenous game species; and on sound conservation practices in general. How fortunate it has been that after so many years of poor management and neglect, the region should have benefitted more recently from the altruism of four far-sighted individuals and their families. It was this investment that prompted the concept of creating a biosphere reserve in the Waterberg according to the precepts defined by UNESCO, and in 2002 the Waterberg Biosphere Reserve was formally created and accredited.

Environmental and other threats

Of course, no land, however environmentally sensitive, is guaranteed immunity from the aspirations of developers, the schemes of infrastructure planners or the machinations of politicians. In 1991 the storage potential of the Lephalala River catchment was the subject of an investigation on behalf of the Lebowa Permanent Water Commission. Amongst other tasks, the study examined five possible alternative reservoir sites along the course of the Lephalala. Named from south (upstream end) to north (downstream), dam sites for these reservoirs were identified on Frischgewaagd, Moerdyk, Alem, Groenfontein and Kaapsvlakte (see Figure 9.24).[132] The first three sites lay within the New Belgium Block and most critically, each would impact enormously on the Lapalala and Touchstone private reserves, as well as transforming the rugged, pristine Lephalala River gorge. The consultants responsible for the project were aware of the environmental consequences and were at pains to point out to their client the opposition that would be encountered should any one of the three upstream sites be contemplated for development. By 1992 they were recommending that either of the two downstream sites could be developed without severe environmental impact, provided appropriate mitigatory measures were applied.[133] Fortunately perhaps, nothing came of the plans. Within a year Lebowa was no more and the country was distracted by preparations for the election of a new democratic government based for the first time on universal suffrage.

In 2002 Eskom needed to identify a route by which to deliver additional electricity from its Matimba/Medupi power station complex outside Lephalale to a substation at Witkop, located east of the N1 highway between Mokopane and Polokwane. An existing 400 kV transmission line already linked these two points, almost skirting the northern limit of the New Belgium Block, but traversing some northern properties including Uitkomst and Kirstenbos before running immediately to the south of the Masebe Provincial Nature Reserve. In the end, despite spirited opposition from affected parties, Eskom concluded that its environmental studies showed that the second 400 kV line should closely follow the track of the first, aggravating localised environmental degradation, but limiting its more regional effect.[134] The line was subsequently constructed.

During 2005–2006 at least six large land claims, amounting to a total of over 170 000 ha, were lodged across the northern Waterberg, ostensibly on behalf of communities allegedly dispossessed of their land since 1913, without compensation. The claims were lodged in terms of the Restitution of Land Rights Act

(No. 22 of 1994), one of the first statutes promulgated by the government elected earlier that year. Some of the claims were gazetted, even though due process had often not been followed as prescribed by the legislation. Virtually the whole of the former New Belgium Block was subject to at least one of these claims (they also overlapped in some areas). Titleholders formed groups to challenge the claims in each case, although some landowners chose rather to accede to the claims without contesting them and to sell their properties to the state. Only one claim had been tested in court at the time of writing: Judgment in the Land Claims Court in 2007 on the so-called Motse claim, which accounted for over 90 000 ha in the north of the area, found in favour of the landowners; it was apparent that the community named as the claimant didn't exist and was seemingly created specifically by the office of the Land Claims Commission for the sole purpose of lodging a claim. The judge took the unusual step of awarding costs against the Commission.[135] By 2018 challenges to several of the other land claims, considered equally fatuous by landowner groups, still awaited their day in court, well over a decade after proceedings commenced.

CHAPTER TEN
Sondagsloop and Tarentaalstraat: a Sunday stroll among the guinea fowl?

In the centre of the Waterberg plateau lies a basin, etched by the catchment of the perennial Melk River that flows northward to join the Lephalala. This iconic river in turn cuts through the layers of Waterberg sandstone, forming rugged gorges through the northern rim of the plateau on its journey down to the Lowveld and a confluence with the Limpopo. The Melk basin ranges in altitude from around 1200 metres amsl in the bed of the river to as high as 1600 metres on the surrounding lip – even to 1739 metres if one includes Geelhoutkop, on whose western slope the Melk has its source. The peak is named for the localised stand of lichen-bearded yellowwood trees (*Podocarpus latifolius*) that are to be found among the grotesquely eroded rocks on its crown. Together with the *krantz* aloe (*Aloe arborescens*) and silver sugarbush (*Protea roupelliae*) they reflect a montane microenvironment otherwise only found along the Sandriverberg range that marks the southern escarpment of the plateau.

The basin is bounded in the east by Tafelkop, so named for the flat-topped hill (called Sekungwe in Sepedi), famous for its reputation as a home for leopards and for the bushman paintings hidden in clefts at the feet of its craggy cliffs. To the west, the earth rises to a plateau upon which is located the Dorset police station, built on the uppermost layers of the Waterberg rock sequence. The southern lip is a lower ridge that in summer diverts or impedes the movement of rain-laden clouds driving in from the southwest.

The basin is known as the Sondagsloop – probably originally for the farm of that name, Zondagsloop, on its western flank, but also for two rivers bearing the same name: one flowing south from the Dorset area to join the Dwars just north of Vaalwater; the other running west from the Tafelkop catchment into the Melk.

The main tar road from Vaalwater north to Marken and Mokopane (the latter via the spectacular Kloof Pass – see Chapter 19) traverses the basin from south to north. In the middle, it passes the settlement of Sukses. Here a crossroad leads west to Dorset, Visgat and the southern part of the old New Belgium Block, and east along the Sondagsloop road to a T-junction with Tarentaalstraat. The latter, once a wagon track, later a well-maintained gravel road and today reverting in parts to its former status, links the hamlet of Melkrivier, 20 km to the north, with Modimolle, 75 km to the south. Its route cuts through the Sandrivierberg to the Alma valley down a magnificent kloof lined with water-berry trees (*Syzygium cordatum*). See Figure 10.5.

Tarentaalstraat

The name Tarentaalstraat formally applies only to this kloof, which, by ingeniously following several linked stream courses, enables a laden vehicle (formerly an ox-wagon) to negotiate the ascent of the plateau with relative ease. Previously, a much steeper route nearby had to be followed, necessitating the use of several spans of oxen. In 2005 remains of the old *wapad* could still be seen, with carefully packed stone retaining walls on the steepest sections and a well-defined outspan area in a saddle at the top.[1] This adjacent steep, well-forested kloof supports a remarkable variety of indigenous vegetation, including *Prunus africana* (red stinkwood), *Ilex mitis* (Cape holly), *Curtisia dentata* (assegai tree) and *Podocarpus latifolius* (common

FIGURE 10.1: Aloes, proteas and yellowwoods among the rocks on Geelhoutkop. Photographs courtesy of Lyn Wadley.

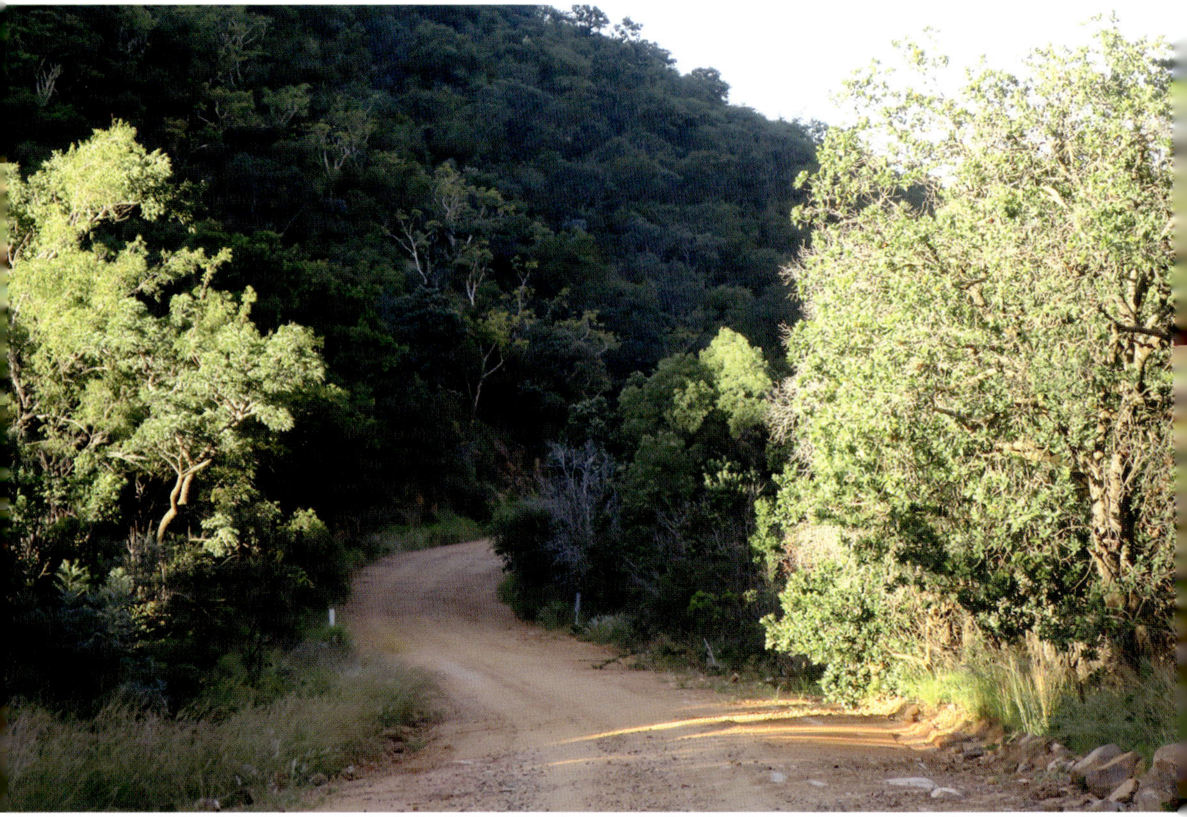

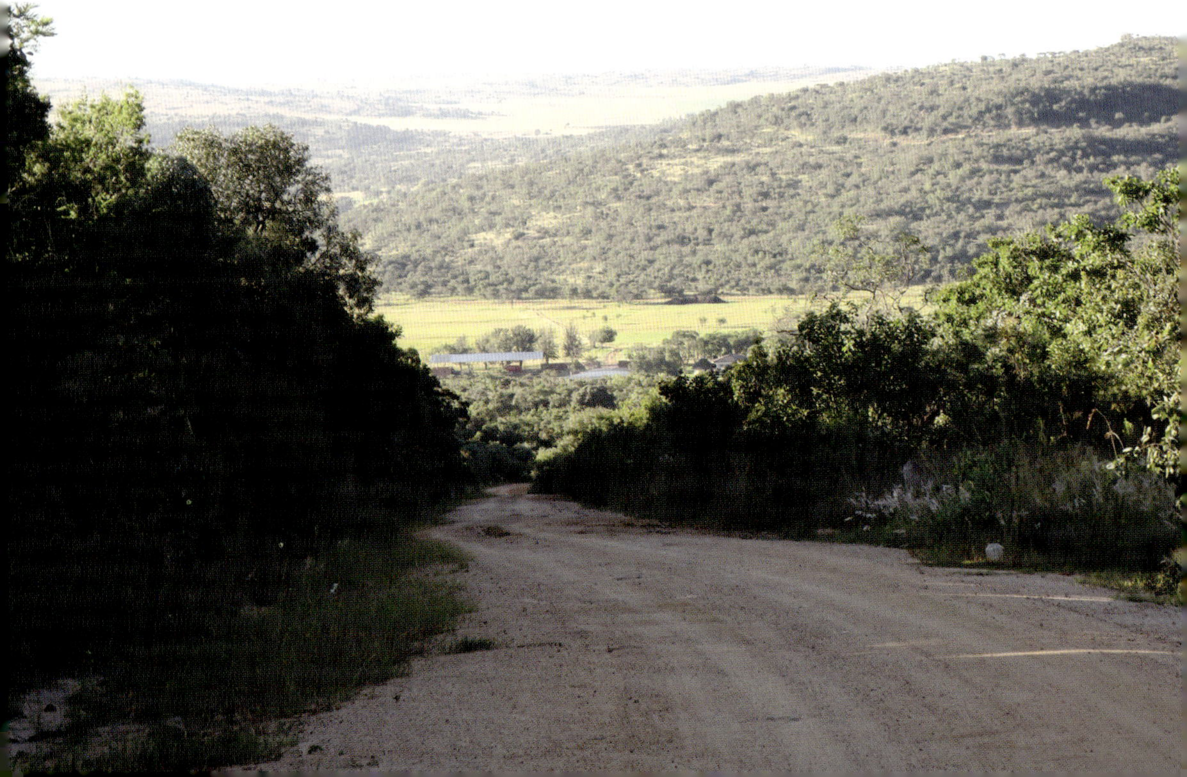

yellowwood). Discovery of the new pass is attributed to a reclusive inhabitant of the hills at the time of the South African War: "Rooi" Daniel Erasmus, a local transport rider.[2] Its name is derived, not from the helmeted guinea fowl that are frequently seen along the way, but, according to legend, from the acne-scarred complexion of the discoverer![3] Towards the southern end of the kloof stands a prominent rock adjacent to the road. This was known to former residents as Leopard Rock because on more than one occasion, a leopard had been seen relaxing on its surface, surveying the surrounding bush.[4]

The modern traveller along Tarentaalstraat kloof will be beguiled by its scenery. Not so Lieutenant Edward Gordon of the Royal Scots Fusiliers, who was seconded to the 12th Mounted Infantry battalion during the last guerrilla phase of the South African War. On 27 August 1901, he was instructed to take his company from Heuningfontein across the valley of the Klein Sand River and up the kloof to make the way safe for a column that was following. This they proceeded to do, with some trepidation:

> We were ordered to go straight through the poort to the other end, which we did. It was 8 miles long! [an exaggeration: from plain to crest, the kloof is 6 km (3,6 miles) long at most] & quite the worst place & most chilling place I have ever ridden thro', for there was cover enough to hide 100 men every 50 yds., without any chance of their being seen & if there had been any Boers in it we should have been fairly caught in a trap. To make it worse, we were sniped [at] from a long distance about [half] way.[5]

A few kilometres north of the kloof, the road climbs a second open ridge. This ascent is known as Papstraat, in acknowledgement of the food (*mielie* porridge) given to the teams of (white) labourers who excavated the pass during the Great Depression.[6] Thereafter, the road continues to climb northwards until it passes just below the crest of Geelhoutkop, where, close to the ruins of the school and police post that used to be here, there is a magnificent view in every direction. Just north of the crest, on the farm Kameelfontein, may be found the remains of the rough fortifications or *schanze* made by General Beyers and his men when they camped here in September 1901. From this point Tarentaalstraat winds northwards and

FIGURE 10.2: The scenic road on Tarentaalstraat (Tweebosch).
FIGURE 10.3: Papstraat Pass on Kaalvallei.

downhill, approaching the foot of Tafelkop via a narrow defile called Skrikkloof (fright gorge) on account of the jumpiness of horses and oxen smelling leopard as they travelled along it.[7]

Until the recent prosperous development nearby of Mpatamatcha, a game capture and auction facility with attached restaurant and zoo, Sukses was the epitome of a misnomer. Once the location of the Palmietfontein primary school, a store and a post office, it passed into obscurity as modern communications overtook it. Many local residents know Sukses as Tretsom, and the tiny store there as Tracson (a corruption of the former). The story goes that when local telephone exchanges were being rolled out in the 1950s, the postmaster at Vaalwater was at a loss as to what name to give the exchange built at this spot (presumably even then "sukses" was seen to be inappropriate). Eventually, he resorted to the use of his own name, Mostert, in reverse.[8] The only other known case of such a name inversion is Reivilo in North West Province. Today, the phone exchange has gone, but the multiple aliases of the settlement remain.

The tar road from Vaalwater to Melkrivier, where it is joined by Tarentaalstraat almost at the bridge over the Lephalala River, was opened in 1981. In fact, according to local residents Peter and Patricia Beith, it was opened precisely on the morning of 8 December that year, their wedding day.[9] A fine wedding present it has proved to be, halving the time taken to travel the 32 km from Sukses to Vaalwater, not to mention making the journey possible in all weather conditions.

The Dorset area

Two roads lead west from Sukses. The more southerly of the two climbs up the old pass known since Carl Mauch's visit in 1869 as Trouw Zyn Nek to a somewhat marshy plateau and the farm Brakfontein. Here, a police post had been established before the South African War. It has remained there ever since, making it the oldest police station on the Waterberg plateau. With the formation of the South African Constabulary by the British colonial administration early in the twentieth century, five police posts were created in the Waterberg: Brakfontein, Geelhoutkop, Sandrivierspoort, Rhenosterpoort and Clermont.[10] Geelhoutkop was the first to be closed, even before the outbreak of World War I. Sandrivierspoort followed, but the remaining three are still in operation.

The Brakfontein police post was manned by a Sgt Norman from February 1904 when a rental of £10 annually was paid to the owners of the property, Transvaal

Consolidated Land & Exploration Co. Ltd (TCL)[11] – which owned most of the other farms in the area south of New Belgium too, having purchased them in the 1890s from Alfred Beit, a property speculator who later became one of the country's leading businessmen and philanthropists. The government was habitually tardy about paying its annual rental and for almost every year there's a letter in the files from TCL's local representative, A.J. Mathewson (JoP) of Vrymansrust, or C.A. Madge of 24 Rivers, requesting payment before the financial year-end. In 1939 the government finally purchased the farm Brakfontein from TCL with the intention of adding it to the properties designated for settlement (like New Belgium immediately to the north).[12] In the process, an area of about eight hectares, which included a house some distance away occupied by Constable Van der Merwe and his family, was permanently reserved for a police station. The rest of the farm was then allocated to two tenants, F.J.L. van Loggerenberg and N.J. Roentgen, who were temporarily allowed to make seasonal use of the borehole on police property.[13]

In October 1932 the Brakfontein police post was officially renamed Dorset police station.[14] The Rhenosterpoort station became Rankin's Pass and the Clermont station today is called Villa Nora, far down the Lephalala River in the Lowveld.

From Brakfontein police station, there used to be a road heading off to the northwest along the watershed between the Tambotie River to the north and Poer-se-Loop to the south. Available maps of the early twentieth century note the existence along this road of Tomlinson's Hotel on the farm Blaauwbank (the TCL-owned property on which Arthur Peacock had been a tenant until he and his family were forced to leave by the British in 1901). The story of this hostelry is lost, unfortunately, apart from the knowledge that its founder and proprietor was Herbert Ernest (Bert) Tomlinson (1879–1954).[15] Tomlinson was a New Zealander born in Christchurch. He came to SA shortly after the end of the South African War and by 1904 had settled on two farms south of Settlers on the Springbok Flats, Bangor and Avondale, which he farmed with a partner called Booth.[16] In 1908 he was one of numerous signatories to a petition in Nylstroom (amongst many identical petitions from around the Transvaal) appealing to the Transvaal legislature to enforce laws restricting the movement of Asiatics in the colony.[17] Later, he moved to Blaauwbank as a manager on behalf of TCL. While there, in 1919, he corresponded with the Entomology Department about an outbreak of blow fly among his cattle, a condition attributed to the bont tick.[18] He also opened an establishment substantial enough to be called a hotel and be depicted on an official map of the area.

In 1924 Tomlinson was transferred to a large TCL cattle property called Toulon, in the Lowveld, adjoining the Sabie Game Reserve where the warden was Col James Stevenson-Hamilton. Two years later, on 31 May 1926, the Sabie and Shingwedzi reserves were merged into the Kruger National Park with Stevenson-Hamilton at the helm – and the following year Bert Tomlinson became the new park's first ranger. He immediately contacted an old friend from his Waterberg days, Walter ("Harry") Kirkman of Vaalwater, to tell him there was a vacancy for a manager on Toulon. Harry accepted the post and in due course followed Bert into a career as one of the more famous rangers of Kruger Park.[19] Tomlinson married Martha Amm in 1940 and they had two children before Bert's retirement from the park service in 1949. The couple retired to Somerset East where Bert died in 1954.

FIGURE 10.4 (RIGHT): H.E. (Bert) Tomlinson. Photograph taken, with acknowledgement, from Pienaar (2012), p. 509. Original photograph credited to the late D. Wolff of Pretoria. FIGURE 10.5 (BELOW): Spring on Steenbokfontein, with historical excavation for stock use.

The area around and to the south of Dorset is underlain by the uppermost unit of Waterberg sediments: a light beige-grey coloured, fine-grained siltstone that is markedly different from all other rock types on the plateau. This rock unit supports good supplies of shallow underground water, mostly alkaline (whereas groundwater elsewhere on the plateau is mildly to strongly acidic). As a result, springs are widespread, as is evidenced by the local farm names: Brakfontein, Dwarsfontein, Vischgat, Rietgat, Zandfontein, Bultfontein, Gunfontein and Steenbokfontein. These perennial springs had been used intermittently by humans for thousands of years; one on Steenbokfontein was excavated archaeologically in 2015 (see Chapter 2).

Most of the properties had been owned by TCL, which began to dispose of them after the end of World War I. Steenbokfontein, for example, was purchased from TCL by a deceased estate in 1948 and sold immediately to two brothers, Lodewyk Willem (Lood) and Erasmus Albertus (Bert) Swanevelder, whose sister had married J.P. Jacobs, the owner of the neighbouring property Vischgat. Jacobs

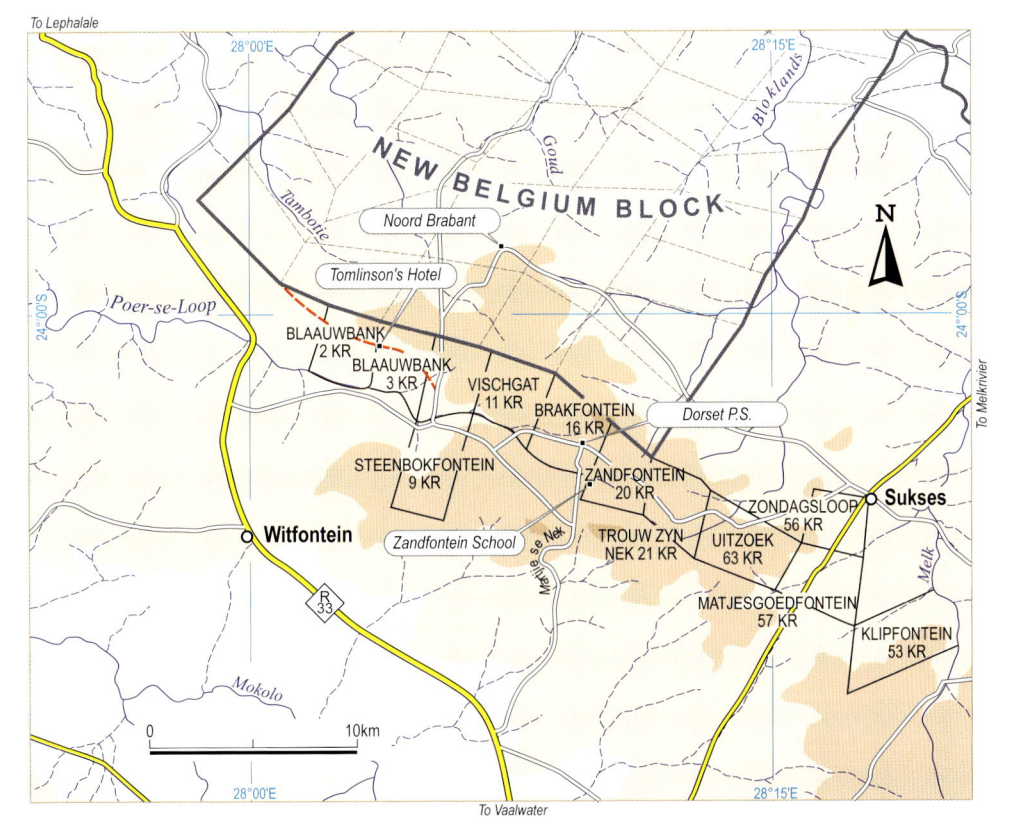

FIGURE 10.6: The Dorset area.

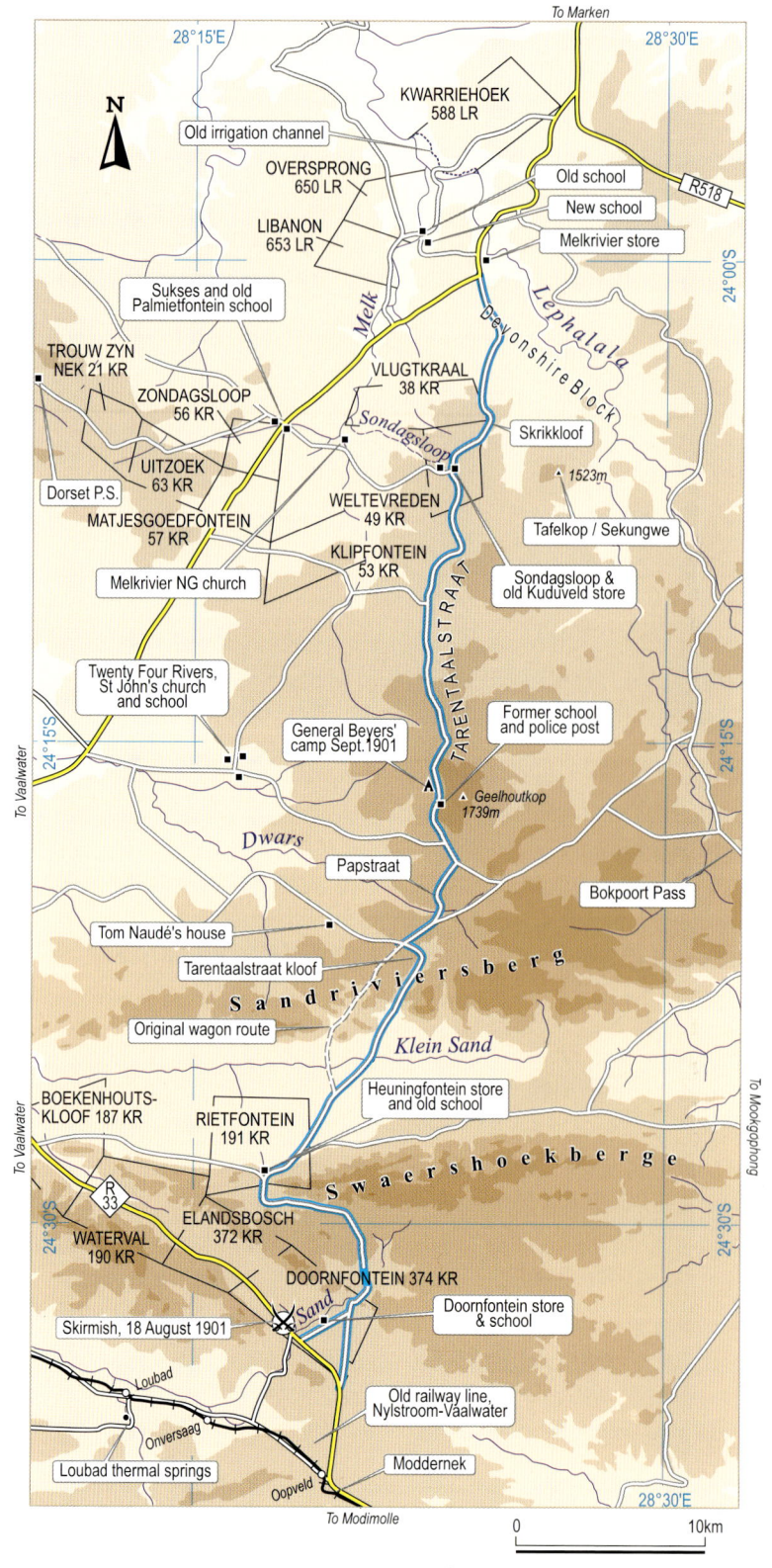

FIGURE 10.7: Tarentaalstraat.

was also the headmaster of the school on Zandfontein – and the nemesis of Major Averre of New Belgium. The Swanevelders' parents, Zacharias and Helena, had been bywoners on Jacobs's farm since 1937, after Zacharias's career as a diamond prospector had taken a turn for the worse.[20] He and Helena had at least eight children, six of them sons. Most resided in the area, marrying into the Jacobs and Venter families – with the result that in the 1950s apparently two-thirds of the approximately one hundred pupils at Zandfontein school were related through marriage![21]

A second road leading west from Sukses provides access to the southernmost properties of what once comprised the enormous New Belgium Block: Zorgvliet, Werkendam, Wynkeldershoek and North Brabant. On the latter, a T-junction allows the traveller either to turn right and follow the scenic Brown's Cutting next to the Goud River through Kwalata Game Reserve down to the Lowveld, or to turn left (south) towards Visgat (formerly the site of a post office and another school, (Zandfontein) and on to Dorset or the new tar road to Lephalale (the R33).

Sukses and surrounds

Farms in the Sondagsloop basin were demarcated in 1868–1870, although few were occupied even seasonally before the late 1880s. The land did not lend itself to farming or permanent habitation: the soil was poor sourveld. There was perennial water, it is true, but this brought its own problems – bilharzia and, in the lower reaches of the Melk/Lephalala systems, malaria. Seasonal flooding made the already difficult access to the nearest town, Nylstroom, even more arduous. Early settlers found virtually no availability of farm labour: The region was essentially unpopulated and those settlements that did exist towards the northwest and northeast at Ga-Seleka and Mogalakwena respectively, were reluctant to provide indentured workers. The settlers themselves were largely unschooled, unskilled and almost without resources other than their own physical and spiritual strength. Although they had been allocated the land virtually free, they were required to pay an annual stamp duty to the Transvaal government. Many could not afford this with the result that properties were either sold on to one of the many speculative land companies active in the region between the discovery of gold on the Witwatersrand (1886) and the outbreak of the South African War (1899); or returned to government ownership. Some settlers remained as tenants on land they had formerly owned.

Few people settled in the area around Sondagsloop. Most concentrated instead in the Melk River valley above or around its confluence with the Lephalala. Louis Nel, who was born at Melkrivier where his father was headmaster of the school from 1944 to 1970, thinks there may have been a dozen families eking out a living on the narrow floodplain on the farms Libanon and Kwarriehoek. (The farm immediately between them, Oversprong, was first issued by President Pretorius to the German traveller Carl Mauch in December 1868. Unlike the properties on either side, its frontage onto the Melk is a steep cliff.) A weir across the Lephalala upstream of its confluence with the Melk fed water into an irrigation channel, the remains of which can still be seen running along the north bank of the Lephalala. The channel carried water down to a series of narrow strips on Kwarriehoek, each with a small river frontage, on which irrigated crops were grown.

Another small concentration of settlers developed in the 1920s on the farm Klipfontein, also deriving their livelihood from the alluvial soils adjacent to a tributary of the Melk. Farmers in the area grew mealies, but also wheat, ground-nuts, tobacco and even cotton. Yields were normally not very good and most farmers struggled to make a living – except in one year, 1981, when there was a huge surplus of mealies. The silos at Vaalwater did not yet exist and heaps of bags of maize were stored on the ground under canvas in the town.[22]

Carel Dirker, who was born in 1935 on the farm Uitzoek, across the tar road from Klipfontein, recalled how his father, Samuel, who owned a wagon-mounted, steam-powered threshing machine and farmed on Matjesgoedfontein (later part of the Nyathi estate) would tow the machine around from farm to farm during the season, doing contract threshing for local farmers. The machine required to be kept constantly supplied with water and wood. One of the Swanevelder brothers (of Vischgat) used to work for Samuel Dirker during the threshing season. The machine was a Ransome with spoked iron wheels; it was pulled by a span of 16 oxen and was accompanied by a blue McCormick-Deering tractor which ran on paraffin (or "Voco", from Vacuum Oil).[23] Carel remembered that in addition to the four small schools in the area – Zandfontein, Naauwpoort, Zanddrift and Palmietfontein (he went to the latter) – there was a maternity home at Zanddrift in the 1950s where Sarah Furstenburg, the wife of one of the tenant-owners on New Belgium, was the midwife.

Carel recounted a tragic story about a pretty young girl – perhaps a teacher at Zandfontein school near Dorset – who disappeared without trace one day in the late 1940s. Her name was Hannetjie Joubert. Carel remembers that she used to carry a handbag with a long shoulder strap. Most people thought she had been murdered and many rumours circulated as to who might have been responsible, but since no corpse was ever found, her fate remains a mystery.

Carel's mother, Anna, ran the Sukses postal agency office from her home on Uitzoek. She would travel into Vaalwater to meet the train and bring the mail back to the farm for sorting. Residents of the area would then come to the Dirker home to collect their mail.

The Sondagsloop area

The road leading east from the crossroads at Sukses is called the Sondagsloop road and its junction 11 km further on with Tarentaalstraat is marked on some maps as being Sondagsloop, though there's no historical reason for this. On the way, the road crosses the Melk over a bridge notorious for being washed away in heavy rains. Just downstream is the original NG Kerk in the region, built in 1932 to serve the needs of the Waterberg-Noord community. It did so until 1960 when a new, centralised church was built in Vaalwater. Today, the old church has been restored and is used as a retreat and function venue for an East Rand Pentecostal community (see Chapter 14).

FIGURE 10.8: The old store at Melkrivier that is still in operation.

The Greers

In 1949 the church witnessed the marriage of a local couple: Patrick Greer (22), who'd arrived in 1947 to work on his parents' new farm at Weltevreden at the junction of Sondagsloop and Tarentaalstraat, and Ilva Webster (21). Ilva was born on Vlugtkraal to the north of Weltevreden, which her parents had bought several years earlier. Apparently, her mother, Martha Sophia (née Michau), a school teacher, had seen an article in a Johannesburg newspaper about a new school that had opened at Libanon near Melkrivier in the Waterberg and which was looking for a teacher. Ilva's father, Wallace, was working on the mines at the time and needed a change. He decided to travel by bicycle up to the Waterberg (a distance of about 300 km) to have a look around, and also at a farm nearby that was advertised for sale. The farm, Vlugtkraal, had been owned by TCL since 1892 (and before that by Herman Eckstein, the founder of the Cornerhouse Group which became Rand Mines and which in turn was the holding company for TCL).[24] TCL was selling off many of the hundreds of farms it owned around the Transvaal and the Websters succeeded in purchasing it in 1921 – it was registered in Martha Webster's name and the price paid for the 1700 ha property was almost £1500. Ilva recalled:

> [W]hen my father returned to Johannesburg, my parents sold everything but my mother's piano. They travelled to their newly-acquired farm by donkey cart. After days and days of being on the road, they arrived on their farm. They chose a huge syringa tree [*Burkea africana*] with sprawling branches to put up house, using tree stumps as chairs. The school [Libanon] where my mother was teaching was 12km from the farm. She walked to school every Monday morning, carrying her clothes and food for the week. On Friday afternoons, she walked home again. During the week, she boarded with a family who stayed closer to the school. As soon as he could afford it, my dad bought [her] a horse. In 1925, my sister was born [and] my father made a horse cart so that my mother could take the baby with her to school.[25]

H.A. van den Berg is about the same age as Ilva and lived on the neighbouring farm Klipheuwel (now incorporated into the Lindani Game Reserve). He described the Websters fondly as "the only English-speaking Afrikaners in the district!"[26] Mr Van den Berg also recounted to Sam van Coller of Lindani another story from his youth: Apparently, his grandfather, at the end of the South African War, was visited by British troops (actually in all probability the SA Constabulary, who were tasked with the chore) confiscating all weapons in the hands of the Boers. "Oupa" van den Berg sat on his stoep waiting for them with a Lee-Enfield (the famous .303) rifle on his lap. He instructed the visitors: "Tell Jannie Smuts, if he wants my weapon he must come and fetch it himself!" Whereupon the officers withdrew and the next day Van den Berg allegedly went off and hid the weapon in some crevice along the rocky escarpment behind his house. (The house no longer stands, but the main Lindani homestead, at the base of the striking line of cliffs on the farm, is very close to where it used to be.) The Van den Bergs later moved on and in the 1950s a family called Briel owned the property. One day, while exploring the cliffs behind the farmhouse, which incidentally also host interesting archaeological sites, Mr Briel discovered an old .303 rifle, carefully wrapped in greased cloth – but the stock had been damaged, possibly eaten away by termites. He had a new stock made and for many years the rifle was a talking point within the family, who had no idea of its background. When new firearms legislation came into force at the beginning of this century, the Briels decided to hand the unlicensed weapon in to the police to be destroyed – to the subsequent dismay of Mr Van den Berg (Jr), a frequent visitor to Lindani (the old family farm) when he came to hear of the story from his host.[27]

Louis Nel of Melkrivier also remembered the Websters of Vlugtkraal with great respect. He recalled how his father and Wallace were close friends, despite the fact that his father was a staunch Nationalist while Webster was a *"bloed-Sap"* (a dyed-in-the-wool supporter of the SA Party). Mrs Webster was soon transferred from Libanon to Palmietfontein, a school closer to home, where she taught for many years. Ilva remembered that eventually her father was able to purchase a car – a Model T Ford – of which he was very proud. Apparently, he went to Nylstroom to collect it, was shown how it worked and simply jumped in to drive it home, having never driven before! Later, many Palmietfontein pupils boarded with the Websters and would get a lift to school hanging onto the sides and roof of the car. Ilva also commented that in those days (before and through the Depression), the local farming community was very poor and suffered seriously from ill health, especially from pneumonia and malaria. Her mother used to say that "if a baby lives until the stamvrugte are ripe, it has a good chance of surviving".[28] Ilva grew up in the area and moved onto Weltevreden next door after her marriage to Pat Greer. Wallace Webster and his father and a brother-in-law, Gerald Michau, are buried on Vlugtkraal.

Tafelkop

Immediately east of Vlugtkraal lie the farms Macouwkuil and Tafelkop, both of which, like Vlugtkraal, were first registered in January 1870. Tafelkop was one of a group of 24 farms which together constituted the so-called "Devonshire Block". They included the three properties Koperfontein, Klipheuwel and Buffelshoek which, a hundred years later, would become the nucleus of the well-known Lindani private game reserve established by Sam and Peggy van Coller at the northern end of Tarentaalstraat. Other farms to the south of Tafelkop that were listed as being part of the Devonshire Block included Schurfpoort and Schikfontein. The first reference to this block was in correspondence (now missing) dating from 1877. In 1883 (that is, after the interregnum caused by the First Anglo-Boer War), the Surveyor General, Johann Rissik, submitted details of all the beacons defining the farms to the State Secretary, F.J. Eduard Bok, for approval by the President.[29]

For some reason, the farms did not make up a contiguous block, as had been the case with the New Belgium Block further west, but were scattered along the eastern escarpment of the Waterberg plateau. The block was shown (inaccurately) on Jeppe's map of 1879 together with the New Belgium Block (see Figure 9.3), but by the next edition ten years later, it had been removed. Deeds records show that the Devonshire farms, like those of New Belgium, were mainly held in the name of Oceana, a big land and exploration company, or one of its proxies. Unlike New Belgium, nothing further came of the Devonshire Block and by early in the next century, the farms were being sold off to various new owners.

Tafelkop was purchased from Oceana in 1920 by Christiaan Hendrik Boshoff. Upon his death in 1931, the farm (of about 1300 ha) was divided up between six children and three others, including Gert Pieter Brits. Macouwkuil also ended up being owned by members of the Brits and Boshoff families by 1930. The portions owned by the Boshoff family were informally consolidated into a single property, Sunnyside, and one of the descendants, Jacobus (Koos) Boshoff was born there in 1928. He was sent to school at Palmietfontein, where Mrs Webster was his teacher. At that time the farm, then owned by his father Japie, was renowned for its fine herd of Africander cattle, though later it diversified into growing high-quality air-cured tobacco.[30] The Brits family struggled to make a living from their portion of the combined property and one of their sons, Frikkie, went to work for the neighbours, Basil and Baby Barnes, on Weltevreden.

The Barnes family

At the junction of the Sondagsloop road with Tarentaalstraat, there used to be a general store (it was demolished early in the present century), run for many years by members of the Barnes family. The store was known as Barnes se Winkel, although by the time of its demolition a faded sign on its frontage named it as the Kuduveld Trading Store. (Ilva Greer gave it this name when she owned the store in the 1960s). George and Mary Barnes lived in a house, now also long demolished, around the corner from the store. Jacaranda trees still mark the site of the former home.

George Usher Barnes, born in Grahamstown in 1850, was the son of 1820 settlers. He farmed for a while in the Eastern Cape, married Mary Dorril Robinson (from another settler family) in the late 1870s, and moved to Klerksdorp some years later to start a dairy farm. Their herd was destroyed by the rinderpest of 1896, forcing the Barnes family (there were now five surviving children) to move to Maraisburg outside Johannesburg where the oldest son Lennox, by then 16, was employed in carting bags of coal to a nearby gold mine. After the South African War, which the family spent on the Barnes family farm near Grahamstown, George and Mary and their children returned to Maraisburg for a few years before migrating to the Waterberg. Here they settled as tenants on the farm Weltevreden, growing tobacco with Lennox and running their trading store. A second son, Basil, went off to the mines where he worked for several years and evidently contracted silicosis or tuberculosis.

FIGURE 10.9: The old Kuduveld Trading Store (formerly Barnes se Winkel) (now demolished).

FIGURE 10.10: George and Mary Barnes, c.1931. Photograph courtesy of Ilva Greer.

Weltevreden (1378 ha) was originally awarded to Hendrik Johannes Smith in 1870 by President Pretorius, having been "inspected" the previous year by the ubiquitous Johan Antonie Smit. It was unusual in that it was never owned by a land company.[31] In 1915 it was purchased by E.A. (Ted) Davidson of 24 Rivers fame, apparently for use as one of his cattle-fattening farms.[32] In 1920 ownership reverted to the state, probably due to nonpayment of taxes, and at about this time the Barnes family came to settle.

The farm was traversed by one of the old wagon trails that led eventually down along Tarentaalstraat to Nylstroom. This was the route used by early farmers when they went, three or four times a year, for nagmaal in Nylstroom – at least until the NG Kerk built its church on the Melk River in 1932.

Opposite the Barnes's store, on what is today part of the Waterberg Game Park timeshare resort, there used to be an outspan place on the banks of the Sondagsloopspruit; the late Kassie Steyn, a lifelong resident of the farm Libanon on the lower Melk, was born there in 1927 while his parents were on their way home from nagmaal (see Chapter 14).

George Barnes died on Weltevreden in 1932, his wife Mary following him almost ten years later in 1941; their graves lie beneath the shade of a huge fig tree on the property. In 1945 the two brothers, Lennox and Basil, jointly purchased the farm

FIGURE 10.11: The four graves under a fig tree on Moletadikgwa (Weltevreden). At the back (left to right): Magdalena Petronella Oosthuizen (née Joubert/Blignaut) (1844–1927); Mary Dorril Barnes (1855–1941) and George Usher Barnes (1850–1932). Front: Basil Barnes (1888–1951).

via a government grant. Two years later, a third of the property was sold to the Greer family and another third was sold to a neighbouring farm. Basil retained his portion until his death in 1951 at the age of 63. He is buried next to his parents.

At some point, Basil married a local girl, Magdalena Petronella Joubert, who had been born in 1907. Almost 20 years younger than her husband, she became known as "Baby" Barnes – although she was anything but small and slight. Basil and Baby continued to run his parents' trading store and to farm tobacco. Having no children of their own, they managed with the assistance of a niece, Dorrel Healey, and Frikkie Brits, the energetic young son of a neighbour, who needed work.

After Basil's death in 1953, Baby married Stephen Middleton, who was a member of the Jehovah's Witness sect, son of the former headmaster of the Palmietfontein school, and 17 years her junior. Several sources have stated that there was a flourishing Jehovah's Witness community around Sondagsloop at that time. Apart from the Middletons, both the Barnes and Boshoff families were said to have been members, together with others at Geelhoutkop. The Middletons sold their share of Weltevreden ten years later and moved to Middelburg, Transvaal, where Baby died in 1967. The Greers, Pat and Ilva, only moved out of the region in the 1980s.

FIGURE 10.12: Magdalena Petronella ("Baby") and Basil Barnes, with Frikkie Brits in front, c.1945. Photograph courtesy of Magda Streicher (née Brits) and Frikkie Brits, daughter and son of Frikkie Brits.

The Joubert enigma

Little has been found about Baby Barnes's ancestry, other than that she was one of twin daughters (the twin was Wilhemina), with at least seven other siblings. One of her sisters, Aletta, married Ben van Staden, who lived at Geelhoutkop. It was rumoured that she was a niece of the great General Piet Joubert – but this has never been substantiated. However, there is an intriguing clue to her history which has yet to be fully resolved.

A fourth grave lies under the fig tree on Weltevreden: that of Magdalena Petronella Oosthuizen, born Blignaut, who'd died in 1927 at the age of 83. (Baby would have been just 20 at the time, probably already married to Basil Barnes and living on the farm.) Research revealed that in 1911 Mrs Oosthuizen had married Gerhardus Oosthuizen, a widower from the farm Rietfontein along Tarentaalstraat where the Indian-run Heuningfontein store was located. She was then 67, certainly beyond child-bearing age. Yet the headstone on her grave refers to "our beloved mother and grandmother".

Clearly, Ouma Oosthuizen had been married before. Further investigation revealed that her first husband had been Jan Hendrik Joubert, who was born at Graaff-Reinet in 1819 and who she'd married at Uitenhage in 1860 when she was not quite 16 (she had also been born in the region and was baptised in Colesberg). Records thereafter become sketchy: The couple had at least two sons, one of whom, named after his father and, like him, born in Graaff-Reinet (in 1864), died in a Pretoria hospital on 15 October 1902 of wounds sustained during the guerrilla phase of the South African War. His death notice records that his mother, already widowed, was then residing at Geelhoutkloof in the Waterberg – not far

FIGURE 10.13: Headstone of Magdalena Petronella Oosthuizen.

to the east of Heuningfontein store. The record also stated that Jan Hendrik Joubert Jr left a widow and a single child, born in 1894, also called Jan Hendrik.

This widow, Christina Johanna Joubert (also, intriguingly, having Blignaut as her maiden name), had been interned in the Nylstroom and Irene concentration camps with her young son. Later, she returned to a devastated Geelhoutkloof where she lived close to destitution and submitted an impassioned letter to the British authorities appealing for financial support. Her application was refused. Whether and how she and her son survived is not known.

When Magdalena Joubert, the elderly widow, married the widower Gerhardus Oosthuizen in 1911, they drew up a combined Will, which survived and in which they specifically left assets to their respective families. Mrs Oosthuizen appointed a son, Pieter Gabriel Joubert, as her executor. Unfortunately, no further record of this son has been discovered. However, it is suspected that Pieter – or maybe another unknown brother – was the father of Baby Barnes, née Joubert (born in 1907) and who therefore shared the same names (Magdalena Petronella) as her paternal grandmother. (Jan Hendrik the third, the concentration camp survivor, aged 13 at the time of Baby's birth, would then have been her cousin.) This could explain the existence of the beautiful old grave of Mrs Oosthuizen in the Barnes family graveyard on Weltevreden.

Geelhoutkop

Twenty kilometres to the south of Weltevreden, the picturesque knoll of Geelhoutkop, the highest point along Tarentaalstraat, is actually located on the farm Driefontein. The Van Stadens (Ben being Baby Barnes's uncle) lived on the adjacent property Kameelfontein, a portion of which was sold to the Pretorius family who also owned Driefontein. According to the late "Boelie" Pretorius, who lived on the property until 2002, his family had acquired the farm in 1880. In the mid-1890s his grandfather (B.P. Pretorius) built a thick-walled mudbrick house near the road, but it was burned down by the British in 1901. Boelie used to go to the school at Geelhoutkop; he remembered that in the severe drought of 1962–1964, he had to trek with 370 cattle from the Koedoesrand near Marken all the way along Tarentaalstraat to the farm, a journey that took four days.[33]

South of Geelhoutkop, between Papstraat and the entrance to Tarentaalstraatkloof, the road crosses the upper reaches of the Dwars River, first described by Carl Mauch in 1869. The Dwars flows westwards, joining the Mokolo just north of Vaalwater. Nearby is the turn-off to Vrymansrus and along this road is a wood-and-iron home originally built in the 1960s as a hunting lodge for Tom Naudé, who was then President of the Senate.

Jozua Francois (Tom) Naudé was born in Middelburg, Cape, in 1889, the son of a local storekeeper. He matriculated there and in 1907 became articled to a firm of attorneys in Pretoria. He opposed SA's entry into World War I and was interned for three months as a result. He joined a Pietersburg legal practice, served as a town councillor and was elected to Parliament as the United Party MP for the new constituency of Pietersburg in 1920, a seat he held for 40 years until his retirement in 1960. Upon the outbreak of World War II, he joined Hertzog in leaving the United Party. He served for many years as the National Party's chief whip. In 1948 he was elected Speaker of the House of Assembly and became recognised as an expert on parliamentary procedure. He had a beautiful farm called Geluk near Polokwane and was an enthusiastic collector of succulents. He became a freemason in 1913 but resigned in 1921 after entering politics.

From 1950 to 1954, Naudé was Minister of Posts & Telegraphs; from 1954–1956, Minister of Health; from 1956–58, Minister of Finance; and from 1958–1961, Minister of Internal Affairs. He entered the Senate in 1961 and was immediately elected its President (1961–1969). In this capacity, according to the Constitution, he was called upon to act as State President for ten months from 1 June 1967 to April 1968 after Dr Eben Dönges, who had been appointed to succeed the late Dr C.R. Swart, fell into a coma and died before his inauguration. On 10 April 1968, Dr Jim Fouché was formally elected State President and Naudé retired; he died in May of the following year.[34]

A local story in Pietersburg went that on the day of Tom Naudé's funeral in that town, it happened that an armoured Standard Bank security van arriving from Ellisras loaded with cash somehow got itself caught up in the funeral procession. A Jewish businessman (there were still a few Jews in Pietersburg in 1969) turned to his friend, saying "You see, you *can* take it with you!"[35] (Tom Naudé should not be confused with his namesake, also J.F. Naudé (1873–1948), who was also born in Middelburg – they may have been cousins – and who became a prominent NG Kerk theologian and Afrikaner republican, cofounding the Afrikaner Broederbond in 1918. His son, the well-known T.F. Beyers Naudé (1915–2004) followed closely in his father's footsteps until 1961 when he formed the Christian Institute to oppose the racial conservatism of his church and party. Forced to resign from the AB and from his position as Moderator in the Church, Beyers was eventually banned for eight years from 1977. In 1985 he succeeded Archbishop Desmond Tutu as General Secretary of the SA Council of Churches, creating a platform for his unrelenting criticism of the National Party government in its waning years.

Southern Tarentaalstraat

The lower entrance to Tarentaalstraatkloof opens onto the wide valley of the Klein Sand River, which flows lazily westwards until its confluence with the Sand at Sandrivierspoort. From the kloof, the road heads south across a broad, sandy plain to Heuningfontein. This was the name given to the southernmost portion of the farm Rietfontein on the southern edge of the Klein Sand valley and at a point where the only access through to the valley from the south emerges from a winding kloof through the Swaershoekberge.

By early in the twentieth century, Rietfontein had been divided into numerous small portions to take advantage of the river and relatively good soil and supported quite a large farming community. There was a school on the property too (officially

FIGURE 10.14: View north from Geelhoutkop.

CHAPTER 10: SONDAGSLOOP AND TARENTAALSTRAAT

called Rietfontein school, but known to all as Heuningfontein) until 1960, after which its remaining pupils were bussed south along Tarentaalstraat to the school at Doornfontein. From Heuningfontein roads went west to Alma and east towards Naboomspruit. In February 1911 the resident magistrate at Nylstroom received a petition from farmers in the area appealing for the reopening of Tarentaalstraat, its closure that summer having been a preventative measure against the spread of east coast fever. Both the magistrate and the Secretary for Agriculture turned down the appeal because of the prevalence of the disease in the district.[36]

The Doornfontein settlers

The farm Doornfontein at the southern end of Tarentaalstraat occupies a special place in the history of the Waterberg. It is one of a small group of farms in the immediate vicinity that were first settled in 1861 by Boers who had been children during the Great Trek and who had travelled with their parents to various destinations: Phillipolis and Thaba Nchu in the Free State, Utrecht in Natal, or even, with Potgieter to Ohrigstad in the northern Transvaal. Twenty years after the settlers at Eersbewoond on the farm Tweefontein (between the modern towns of Modimolle and Bela-Bela), the newcomers were seemingly attracted by the newly surveyed and registered properties being laid out amongst the foothills of the Waterberg. They included Carel Frederik Boshoff (1827–1878), who settled on the 3000 ha farm Doornfontein and Petrus van der Merwe (1819–1895), who settled on Elandsbosch. Van der Merwe had fought at the Battle of Blood River in 1838 and trekked with Potgieter to Ohrigstad before coming to the Waterberg. The house he built on Elandsbosch in 1869 still stands, as does his grave nearby. Other early settlers here included Johannes Jurie ("Lang Hans") van Staden (1853–1936) – Ben van Staden's father – on Boekenhoutskloof; and Petrus van der Merwe's two sons, Lucas Cornelius (Kerneels) (1857–1932) on Waterval, and Louwrens Petrus (Louw) (1878–1942) on Elandsbosch.[37] Deeds records show that only Boshoff was an original owner of his farm, the others only acquiring their rights later.

In fact the founder of Nylstroom, Ernest Olferman Collins, was effectively the first owner of Waterval, which he held from January 1862 until October

FIGURE 10.15 (TOP): The old Indian store at Heuningfontein – detail of painting by Lyn Wadley. FIGURE 10.16 (MIDDLE): The road to Alma from Heuningfontein with Sandriviersberg in the north. FIGURE 10.17 (BOTTOM): Field of cosmos in the valley of the upper Sand River on Doornkom.

1866. Interestingly, Collins is also listed in archival deeds records as being the first owner of Elandsbosch, which was granted to him by President Pretorius in February 1870 – even though he'd died in October 1868! Several months later, the joint estate of Ernest Collins and his wife Emma sold Elandsbosch to Thomas du

FIGURE 10.18: Sketch of Petrus van der Merwe's original home on Elandsbosch, redrawn by C. Bleach from a photograph by R. Wadley, 2017. FIGURE 10.19: Sketch of the grave of Petrus van der Merwe (1819–1895) (middle).

Toit. It was not until 1875 that the property passed into the hands of the Van der Merwe family, whose descendants still own it.[38] These four farms – Boekenhoutskloof, Waterval, Elandsbosch, and Doornfontein – form a line with a northwest-southeast orientation along a valley flanked by two hill ranges (Figure 10.7). The main road (R33) between Modimolle and Vaalwater runs along this valley, which is known to the local community as "Die Moot" on account of the way in which the hills funnel the wind along its length. For years the centre of a viticulture industry (both table and wine varieties) which underpinned Nylstroom's annual Grape Festival, the valley is now almost entirely given over to peach orchards, whose orderly plantations are bedecked with pink blossoms in early spring, overlooked by the grey, leafless hills on either side.

Although belonging to different Dutch Reformed sister churches – and each was a prominent member of his particular church – the early Boer settlers of Die Moot had in common their large families, their determination to succeed on their new properties and their participation in the various wars against black tribes in the 1880s. (There was even a large laager built on Doornfontein to safeguard women and children from attack by marauding blacks while the men were away on commando.) They and their families participated in the South African War: Several had family members who were interned at Nylstroom, some dying there. Some of the men were interned overseas; many returned to find their treasured farmsteads burned down and their stock and crops killed or destroyed.

Frederik (Groot Freek) Boshoff, one of the original settler's sons, lived north of Doornfontein on Magalakynsoog. He had been a member of the Volksraad for Waterberg in the years prior to the outbreak of war in 1899, had fought through the conflict as a bittereinder with Beyers and, like his leader, managed to evade capture (though note the comment about Boshoff by William Caine of New Belgium in Chapter 9). In 1902 he decided that there was an urgent need to provide grieving, devastated members of his community with an opportunity to come to terms with their losses (of family and assets), to commiserate with and help each other rebuild their lives and to renew their faith in their religion. Accordingly, he and his nephew, Willie Boshoff, organised the first post-war Geloftefees (Festival of the Covenant) to commemorate the Battle of Blood River on 16 December that year. The event was held on Doornfontein; it was endorsed by all three sister churches, was attended by virtually every settler in the area (those who by then had returned to South Africa) and helped the families re-establish their roots.

The festival has been held every year since. As pointed out by the Waterberg writer C.R. Prance, the symbolism of the Geloftefees changed after the South

African War: having begun as a commemoration of the punishment exacted by the Boers upon the Zulus in retribution for the 1838 murder of the trekker Piet Retief and his family by Dingaan (hence the popular name Dingaan's Day), it became a day on which Afrikaners would recall all the real and perceived injustices heaped on them, from the Slagtersnek incident of 1815 to the present. Increasingly, the ceremonies would reflect anti-British, and by extension, anti-English-speaking sentiments in step with the rise of the National Party and Afrikaner republicanism.[39]

Unsurprisingly, the settlers – and their numerous sons – became committed to this growing movement. They were the nucleus of the political tide that would sweep the newly formed National Party into control of the Waterberg in 1915 – one of only four Transvaal seats won by the party in that election. Their descendants would remain party stalwarts until – and beyond – the dawn of the universal democratic franchise in 1994. Carel Boshoff, for example, the great grandson of the original Doornfontein settler, married Anna Verwoerd, a daughter of the assassinated Prime Minister, became a professor of theology at the University of Pretoria, chaired the Afrikaner Broederbond in the 1980s, cofounded the right-wing Freedom Front Party and eventually helped establish Orania, the white Afrikaner *volksgebied* (homeland) along the Orange River.

But for all their conservatism and preoccupation with the rights and culture of the Afrikaner, these men, their sons and later arrivals like Pieter Prinsloo, Frederik van Deventer, Francois du Toit and Willem Stander, were also passionate about education. In 1893 they established with their own meagre funds a school at Doornfontein on a morg of land donated by the Boshoffs for this purpose. The story of the Doornfontein school is described elsewhere in this book (Chapter 17). Suffice to say here that it was not only the first school to be established in the Waterberg, but that it is one of very few still to be in operation 125 years later – albeit with a very different ethos and demography from that intended by its founders!

South African War skirmish on Doornfontein

During the South African War, Doornfontein was the scene of one of numerous skirmishes between British and Boer forces during the bitter guerrilla phase of that conflict. It typified the differences that still characterised the Imperial soldiers, often brave, but generally unfamiliar with the terrain and constrained by the strictures of a formal army, and their Boer opponents who were completely at home on horseback in the veld, individualistic, highly mobile, having a keen eye for opportunity but lacking military discipline.

A British supply column on its way back to Nylstroom from the Rankin's Pass area under the escort of a detachment of Kitchener's Fighting Scouts had been attacked by a strong Boer force in the open valley of the Sand River some 15 km to the southwest of Doornfontein. There had been no losses, but the expectation was that the replenished convoy would be attacked when it returned. At 3:30 am on the morning of 18 August 1901, therefore, the four companies of the 12th Mounted Infantry (MI) battalion were despatched – from their temporary base well to the west of where Alma lies today – to intercept and escort the expected convoy. In the meanwhile, the main column with supply wagons and artillery under the command of Lt Col Grenfell would lumber slowly towards Nylstroom along the Sand valley, following more or less the same route to be taken by the railway line in later years. The battalion, who had left their baggage with the column in order to be more mobile, were directed onto the road leading from Zandrivierspoort to Nylstroom. Reaching it at Boekenhoutskloof (the western end of Die Moot) at sunrise, they rested for a few hours (at Van Staden's "white house") before deploying into four separate units and advancing rapidly down the road past Waterval and Elandsbosch towards Doornfontein (see Figure 10.20).[40]

Before long, one of the units was fired upon by a group of four Boers; the troops managed to capture one, together with three horses. One Boer (alleged by the prisoner to have been General Beyers himself) escaped on horseback and the other two did so on foot. As the four MI companies advanced, more cautiously now, eastwards down the long open kloof towards where the Sand River crosses the road near the Loubad turn-off on Doornfontein, they came under fire from a line of Boers up along the ridge to their left (north). At the same time, a group of Boers on horseback emerged from behind Doornfontein homestead on the far side of the Sand River and galloped across the road ahead to take up positions on a low knoll on the right (south) of the road. The company under command of Lt Gordon, which had been deployed on the left flank of the advance, briefly took cover in a *donga* running parallel to the road. Finding, however, that this severely limited their ability to see the terrain, they left the donga to join their colleagues, who were now scattered across the right (southern) slope of the kloof where they sought cover from the increasing fire coming at them from across the valley.

As the afternoon advanced, horses and guns (the MI battalion carried light 37 mm Vickers-Maxim artillery known as pom-poms) were gradually moved southwards over the ridge of the slope and out of sight or range of the Boers, but the troops were effectively pinned down along the slope. Their situation was aggravated by the arrival from the west (i.e. from the direction of Boekenhoutskloof) of another group of Boers on horseback; these ran straight into the donga from which Gordon had earlier extricated his men – and soon found, as had Gordon, that this prevented them from taking any further action. On the contrary, the Boers in the donga were now unable to leave it without attracting fire from the British troops on the hill slope to the south of them!

This stalemate continued until dusk, neither side being able to move without risking being fired upon by the other. At length, as night fell, the British slipped south over the back of the ridge, attempted to regroup in the pitch darkness and made their rather disorderly escape into the valley of the Sand River towards Alma. They were unnerved by the many fires lit by Boer groups encamped in the surrounding hills. Just before dawn, they succeeded in rendezvousing with the column near Tambotierand, a frequent camping place. On the way, such was the difficulty of maintaining contact on the rocky earth in the middle of the night, one of the MI units had briefly been fired on by another, fortunately without consequence. Their objective of reaching and defending the anticipated convoy had been thwarted – and abandoned. (As Gordon makes no mention of the convoy in subsequent entries, it presumably was unharmed.) Two British men had been wounded; the Boers had one fatality (J.J. van Wyngaard) and one wounded. In his diary Lt Gordon estimated the number of Boers to have been about 400, though Boer estimates put their own number at no more than 80, while claiming that the MI battalion would have been about 400! (In fact, the MI battalions, while nominally comprising four companies each of 100 men, seldom came close to being at full strength.)

Loubad

Loubad is a natural thermal spring (*lou* means tepid) – actually a group of springs – located about a kilometre south of the former railway siding of that name; water from the springs flows north under the railway line to join the Sand River. The springs are situated on the farm Middelfontein, a property that was divided into numerous portions many years ago. The springs straddle several of these portions, but the principal one has been owned since 1852 by the Lewis family and their Viljoen ancestors.

According to a study undertaken in the 1940s, the Loubad springs could have a combined yield of over two million litres per day if the various sources were to be opened up. The average temperature is about 33 °C, the water is mildly alkaline and rich in carbon dioxide, but contains no other unusually high mineral concentrations.[41] Unlike its better-known commercial counterparts at Bela-Bela and around Mookgophong, the resort at Loubad has always been under discrete private ownership, retaining much of its original character.

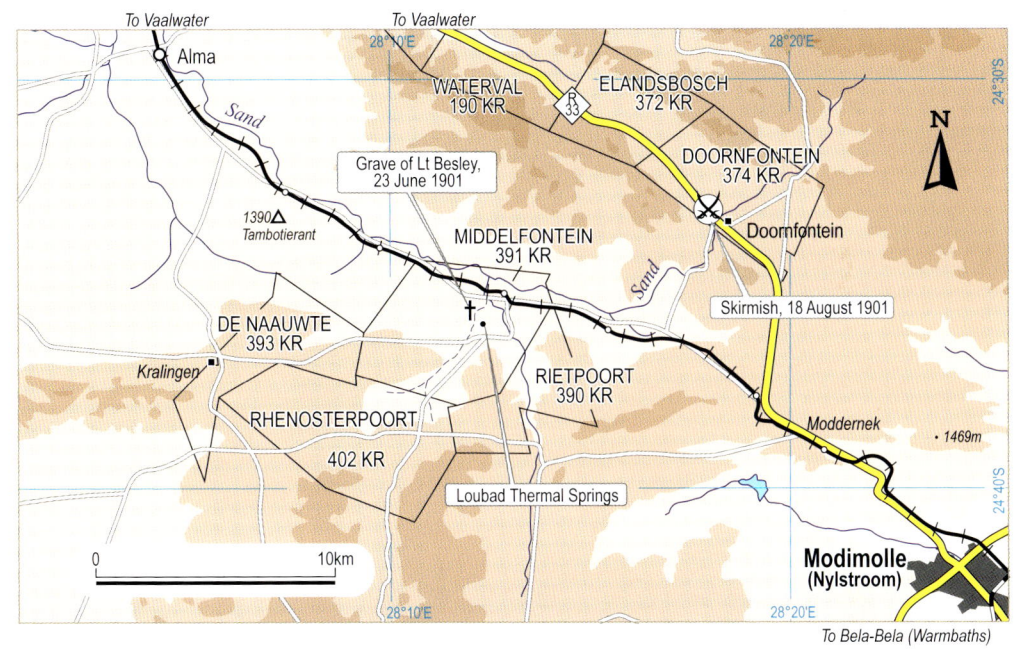

FIGURE 10.20: Loubad and surrounds.

Major P.J. Pretorius

Loubad's most famous son must be Major P.J. Pretorius, the hunter extraordinaire best known by the name given to the title of his popular autobiography *Jungle Man*.[42] Born in 1877 on the farm Rhenosterpoort in the valley of the Sand River just south of the Loubad springs, Phillipus Jacobus Pretorius was the eldest of eight children born to Gerhardus Petrus and Gertbreggie Andriesina Pretorius (née Van der Walt), possibly related to the trekker Andries Pretorius. (Other members of the family lived on neighbouring farms.) At the age of 12 he was sent by his father to work on transport columns taking food to Southern Rhodesia.[43] This led to his joining the newly formed British South Africa Company (BSA Co) in

Matabeleland as a transport rider. Later he ignored appeals from his family to return home and continued to find odd work around Mashonaland.

In 1899 when he'd saved enough to equip himself with a rifle, several hundred rounds of ammunition and trade goods, Pretorius set off north towards the Zambesi valley. There he remained for several years, living on both sides of that mighty river as well as along the Kafue, selling ivory and skins to the Portuguese at Zumbo and replenishing his stock of ammunition from remote Rhodesian police stations. He claimed that on one such visit he was informed that in his absence the South African War had begun and ended in his homeland.[44]

After several years involving many scarcely credible adventures along the Zambesi valley in "Portuguese Nyasaland", the Congo, Ruanda and Tanganyika, Pretorius began farming close to the coast south of Dar es Salaam. During the next few years he came to know the coastal waterways intimately, bought other farms with the proceeds of his hunting and was imprisoned for a time by the German colonial administration for poaching and allegedly killing several tribesmen. These experiences left Pretorius with a loathing for the German authorities. He eventually found his way home via northern Mozambique to the family farm outside Nylstroom shortly after the outbreak of World War I.

He'd no sooner arrived home – to discover that all his family members were supporting or participating in the rebellion against the Union government[45] – than he was approached by the British Government with a request to assist the Royal Navy in locating a German cruiser, SMS *Königsberg*. This vessel had been active in disrupting allied shipping in the Indian Ocean and was believed to have gone into hiding in the Rufiji delta, south of Dar es Salaam, while undergoing repairs to its engines. The delta, a maze of channels too shallow to allow the approach of blockading British warships, was well-known to Pretorius, whose confiscated farm had been nearby. In due course, he was able to pinpoint the position of the cruiser, which had moved far upstream to be out of range

FIGURE 10.21: Major P.J. Pretorius. *Pretorius* (1948): Frontispiece, photograph of the author. Reproduced with the kind consent of his daughter-in-law, Mrs Corrie Pretorius.

of enemy fire. In July 1915 under cover of an offshore bombardment from British warships, two shallow draught vessels armed with incendiary cannons managed to navigate a delta channel to within firing range of the *Königsberg* and set it alight. The vessel was then scuttled by its captain after removal of its ten guns. The naval guns continued to see service for the remainder of the war in East Africa under the command of the legendary Col Von Lettow-Vorbeck. One of three surviving guns, a 10,5 cm (4.1") calibre naval cannon, was captured by SA forces in March 1916 and may be seen today mounted near the western entrance to the Union Buildings in Pretoria.

Pretorius remained in East Africa for the duration of the allied campaign, serving under Generals Smuts and Van Deventer as a highly effective scout behind German lines. In November 1917 he was instrumental in, if not responsible, for the isolation and capture of a large German force near the Rovuma River which forms the border between Tanzania and Mozambique.[46] He ended the war with the rank of Major and was awarded a CMG (Order of St Michael & St George) as well as a Distinguished Service Order (DSO) with Bar for his highly meritorious services.

After the war, Pretorius returned to the family farm at Rhenosterpoort near Loubad, but remained restless. In 1919 he was commissioned to reduce the elephant population in the Addo forest near Port Elizabeth, a task declined by other

FIGURE 10.22: One of the guns from the SMS *Königsberg* mounted at the western entrance to the Union Buildings, Pretoria.

hunters, including Selous, on account of the impenetrable, thorn-filled bush. The elephants were considered to have become "rogue" in that they were continually harassing surrounding farmers. In the following eleven months, Pretorius killed 120 of the animals, before joining concerned members of the public in appealing to the authorities to spare the surviving 16 specimens so that they could perpetuate the now decimated herd.[47] This brought to 557 the number of elephants Pretorius claimed to have shot during his hunting career.[48]

A few years later, Pretorius married Susara Hendrika Nel from Nylstroom (his third wife), to whom he gave the farm De Naauwte, adjacent to Rhenosterpoort.[49] They had eight children (six girls and two boys), the last being born in November 1943. Towards the end of his life, Pretorius engaged in some stunts that involved shooting leopard and lion at very close range in front of a movie camera. He was then employed to control problem animals on the Nuanetsi Ranch in Southern Rhodesia and ran a Johannesburg cartage company before eventually retiring to his farm, where he died from stomach cancer in 1945 aged 68, when his youngest son was only two years old.[50] He is buried in Modimolle in the cemetery adjacent to the concentration camp cemetery there.

FIGURE 10.23: The grave of Maj. Pretorius in the old cemetery at Modimolle. FIGURE 10.24 (ABOVE AND OPPOSITE): The grave of Lt Arthur Besley.

Grave of Lieutenant Besley

To the northwest of the Loubad thermal springs resort, but still on the farm Middelfontein, immediately adjacent to the Pretorius farm, lies a lonely grave alongside the road to Alma, close to the crossing of the Loubad River.[51] It is that of 2nd Lieutenant Arthur Charles Gordon Besley (20) of the 2nd Battalion, Royal Fusiliers (City of London Regiment). Besley was born in April 1881, the second son of Emily Gordon-Moore (granddaughter of the 9th Marquess of Huntly/5th Earl of Aboyn, an ancient Scottish noble family)[52] and Charles Robert Besley, who until his death in 1896 was a Governor of St Thomas's Hospital and Commissioner of Lieutenancy of the City of London.[53]

Arthur Besley was educated at Wellington College, an eminent private school, before enrolling at the Royal Military College, Sandhurst. He graduated on 11 August 1900 and was posted to his regiment. On 16 March 1901, he and several other officers were "seconded for service with the Mounted Infantry in South Africa".[54] They embarked the following month. On arrival, he was posted to the 4th Mounted Infantry Battalion on account of his being a "first-rate rider and a good shot".[55] On 23 June 1901, he was leading a patrol through a narrow defile in the Loubad area when he was shot from some distance – allegedly 1300 yards – by a sharpshooter (said to be Carel Boshoff) assisted by General Beyers himself with binoculars.[56] The Boers then withdrew and there were no other casualties in this brief action. A monument to the 81 officers and men of the Royal Fusiliers who fell during this war was erected in London's Guildhall in 1907.[57] A memorial to Lt Besley was placed by his fellow officers in the church of St John the Baptist, Palmeira Square, in Hove, East Sussex, the home of his widowed mother.

CHAPTER ELEVEN
Making tracks to the plateau

First-time visitors to Vaalwater since 2008 could be forgiven for not realising that until the turn of the millennium, the village used to be linked to the rest of the country by a railway line. When the first stretch of the road (R33) from Modimolle (Nylstroom) was refurbished early that year, the tracks across the road were lifted and their attendant "Railway Crossing" signs removed. And with them went the khaki-clad traffic cops, who used to earn a lucrative income from hiding behind the nearby marula trees and catching local motorists who ignored the "Stop" signs, knowing there hadn't been a train on the line for years. Now, although nearly all of the line has been removed (illicitly), there are still a few places along the road where glimpses of track or embankment can be discerned amongst the long grass.

A railway line to Vaalwater? Why? The answer is a long and interesting one, encompassing as it does, many aspects of the political and economic development plans for the country almost a hundred years ago. On 31 May 1910, the four British colonies in the southernmost part of the African continent – Cape Colony, Natal, Transvaal and Orange Free State – combined to create a single new state, the Union of South Africa. (A fifth, Southern Rhodesia, opted out.) This was accompanied by the unification of the various railway systems in the country and the formation of the South African Railways and Harbours Administration (SAR&H).

The advent of Union came on the heels of the devastating consequences for the rural agrarian economies of the Transvaal and Orange Free State of the rinderpest epidemic (1896–1898), the South African War (1899–1902) and the desiccating drought that followed immediately afterwards. In particular, the "scorched earth" policy implemented in 1900 under British military commander Lord Kitchener in his attempt to end the guerrilla war after the capture of Pretoria, resulted in widespread devastation of farmland and infrastructure, in the shattering of settled rural (white) farming communities and, indirectly, in the deaths of over 27 000 women and children in the concentration camps.

Unification focused the minds of politicians and planners on ways in which to revive the agricultural sector to stabilise populations in rural areas and to strengthen the country's ability to feed itself.[1] One way in which to achieve these objectives whilst also providing much-needed employment for displaced (especially white) peasant farmers was to construct railway branch lines into the remote rural districts from existing trunk routes. Indeed, this process had already commenced under the British colonial administration immediately following the South African War: Lines from Pienaars River to Settlers and from Pretoria to Rustenburg, for example, were opened in 1906.[2]

In 1911 a group of farmers from the Waterberg District (Palala) sent an elegantly worded petition to the Minister for Railways and Harbours urging the government to consider the construction of a new railway line from Nylstroom northwards via Zandrivierspoort across the Waterberg plateau and down into the Palala valley.[3] Although the Minister assured the farmers that attention would be given to their proposal, no evidence of this can be found in the records. Three years later, the realisation of any plans that might have been contemplated was interrupted by the outbreak of World War I.

Immediately after the Armistice in 1918, the national railway building programme was resumed with vigour. By November 1919 the office of the General Manager of the SAR&H was able to submit for the attention of the Minister and the Railway Commissioners a report from the Engineer-in-Chief, Mr Whitehouse.[4] He had been instructed to explore alternative rail routes to the north and northwest from Nylstroom (or other points along the line to Pietersburg). In his study, Whitehouse had investigated alternative routes using the only two passes through the mountains: Zandrivierspoort and Tarentaalstraat. Having concluded that the Tarentaalstraat route, although passing through intensive prime agricultural land in the vicinity of Rietfontein (Heuningsfontein), would prove too difficult and expensive to negotiate, Whitehouse settled on the alternative through the Zandrivierspoort via "Zwagers Hoek":

> This route leaves the Pretoria-Pietersburg line at the bridge over the Little Nile River near Nylstroom and proceeds along the valley of that river, thence along the main road to Zwagers Hoek, through Modder Nek, up the left bank of the Sand River and through the Pongola River Poort to a terminus on the farm Vaalwater. The length of the line will be 47 miles and [I] estimate the cost of construction with second-hand 45lb rails and steel sleepers at £189,395.

FIGURE 11.1: Alternative routes inspected by Whitehouse in 1919. SANA SAB/MVE/416/16/48: GM Memorandum.

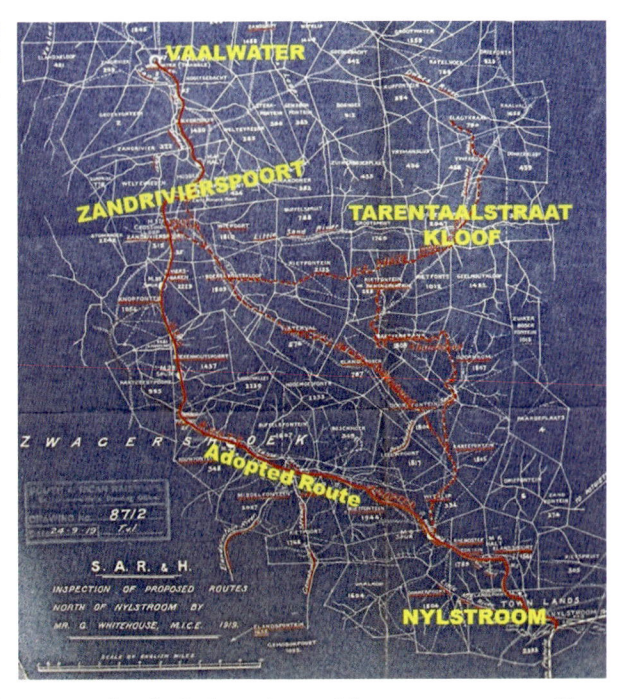

The rationale for building this line was clearly in keeping with government policy:

> The farming community is at a great disadvantage in transporting their produce because of the heavy sand roads and the distance from the railway. The line laid down will satisfy the immediate needs of the Northern Waterberg. Throughout its length, it is located in the vicinity of perennial streams and through a populated district where considerable farming is being undertaken. It will be of considerable help to the farmers at Zwagers Hoek; and at Vaalwater, the grain growers on the Pongola River, Dwars River and the Twenty Four Rivers and upper Palala districts will have rail communication.[5]

Whitehouse reported that the farmers of the area were "generally not very interested in statistics" and that he was unable to obtain much information about crop yields, acreage under cultivation etc. (In retrospect, this finding should have rung some alarm bells.) He forecast that the traffic would consist mainly of grain, although there were valuable citrus and other fruit farms along the route and these might become important. He also expected a "large and increasing trade in cattle and timber from the terminus".

Whitehouse's report concluded with the interesting statement that:

> Although a terminus on the farm Zand Drift [where today, the tar road to Melkrivier crosses the Dwars River] would be more central and would satisfy practically all the farmers, [I] decided to fix the terminus on Vaalwater as this would be a more suitable point for an extension northwards should either the Pongola River or Palala River route be chosen.

Note that at this time the river we now know as the Mokolo or Mogol was still referred to as the Pongola. So, even as early as 1919, consideration was being given to extending the branch line northwards towards the Limpopo.

News of these internal reports in the Railways Administration didn't take long to filter out into the agricultural community. On 26 March 1920, E.A. Davidson, then secretary of the North Waterberg Farmers' Society, wrote to the secretary of the Railway Board in Cape Town:

> Sir, Being informed that the Ministry is now preparing a programme of Railway construction, we must earnestly beg that our appeal for the construction of a line from Nylstroom to Zand River Poort will be given due consideration. Again during the last few months, great strides have been made north of the Zand River Mountains and there will be many thousands of bags of grain to be transported over 50 to 60 miles of bad road to the railway this season.[6]

Evidently, Davidson did not receive a reply to this plea, for in September, he again wrote to the Railway Board:

> Dear Sir,
> I wrote some little while ago asking if your Board would be so good and grant an interview to a Deputation of our Society re: the construction of the surveyed line from Nylstroom to Zandrivierpoort. I explained we are in a very difficult position having a great quantity of grain to produce and no means of transporting same.
> PS I think my previous letter must have gone astray.[7]

(He was too kind – it hadn't. The letter is in the Board's files).

Davidson generously offered to form a joint delegation with the Zwagers Hoek Farmers' Society, whose secretary, C.E. Read, had written indignantly to the Board when he learned that a proposed deputation from his group had been rebuffed.[8] A meeting of the farmers and members of the Railway Board eventually took place in May 1922,[9] but in the meanwhile, the farmers were not going to allow this possible opportunity to slip away: In May 1921 the new (temporary) secretary of the North Waterberg Farmers' Society, I.B. Grewar, wrote to the Minister of Railways applying more pressure:

> The members of the NWFS beg to remind the Minister that we were encouraged to hope for an early extension of the railway line from Nylstroom to Vaalwater. Acting on this, they largely increased their productions [sic] and have an accumulation of produce on their hands which it is impossible for them, as things stand at present, to put on the market.
>
> While fully recognising the financial difficulties of present position, they earnestly beg that the plan of construction may not be abandoned. As they derive little benefit from the taxes they have to pay, having neither good roads, telephone nor railway, they trust you will consider that they have a claim to early consideration in the matter of transport. They also beg to draw the Minister's attention to the large numbers of unemployed in the country to whom relief could be given by employing them on the construction of this line.[10]

This appeal was endorsed by a letter from the Village Council of Nylstroom, addressed to Railways Minister Jagger by its chairman, G.V. Willson and fellow councillors, which added that:

> When you reach Vaalwater you get to the fringe of a great, good, fertile and generous Hinterland well worth supporting. Our little village will grow into an important depot when you decide upon building the line which will tap the resources of the Waterberg Hinterland. Might we possibly suggest the extension of that line to a point say forty miles beyond Vaalwater?[11]

Meanwhile, in April 1921, Minister Jagger and his Railways Council had received a further report from the General Manager in which the potential traffic and economic benefit resulting from the construction of the line was described in detail even more effusive than that contained in the letters quoted above.[12] For example, the report referred to the country to the west of the proposed line (i.e. towards Rankin's Pass) as

> [P]roducing mealies, wheat and oat forage. It also carries a large number of cattle.
>
> When the line is built, it is the intention of the Pretoria Iron & Steel Company to work an iron deposit on the farm Beechwood, which they own. [Beechwood is a small property 3 km south of Kralingen and 14 km south-west of Loubad. The small vein deposit of iron mineralisation there is of no economic significance.]
>
> On the farm Rhenosterpoort 78 ... there is a mill which grinds 5,000 bags of mealies and 1,000 bags of wheat per annum.
>
> Between mileage 41 and Vaalwater, there is a large number of trees of a suitable size for mine props and it is anticipated that 1,000 tons of props will be despatched during the first twelve months after the line is opened.
>
> Six ranches have been established beyond Vaalwater and two more are in the course of development. The ranches are extensive, the area of one [New Belgium] being 180,000 morgen. I understand that ultimately there will be 36,000 animals on the ranches [although] it is not anticipated that slaughter stock will be despatched in large numbers until about 4 or 5 years.
>
> It is understood that several Rustenburg farmers have taken up farms in the Dwars River district for the purpose of growing tobacco.
>
> As an indication of the progress that has been made in the Waterberg District, about twelve years ago the settlers were so scarce that 80% of the farms were unoccupied, whereas at present [it is understood that] there are only two farms which are unoccupied.

The report estimated that in its first year the rail line would carry some 6000 single-fare passengers, at least 11 000 tons of general goods in each direction and about 6000 head of cattle. In the face of such an optimistic report, and enthusiastic lobbying from the community, what politician could resist!

And so it was that on 26 July 1922 the Railways Construction Act (No. 30 of 1922) was approved by Parliament and signed into law by the Governor General.[13] The

Act authorised the construction, subject to various conditions, of several branch lines across the country, including: "Nylstroom to Vaalwater (agricultural railway), approximate distance 43¾ miles, Estimated Cost £187,468".

This legislation was enacted during a turbulent period in the history of the new Union. The Armistice of November 1918 led to South Africa's appointment by the newly created League of Nations as custodian of the former German territory of South West Africa. In 1919 the death in office of the first Union Prime Minister, General Louis Botha, brought General Jan Smuts to the helm and with him the merging of two political parties (the Union Party and the South African Party) in an attempt to stem growing support for the National Party led by Boer General J.B.M. Hertzog.

The gold mining industry was in turmoil: A collapse in the gold price caused companies to retrench many of the unskilled white workers who had flooded to the Rand from rural areas and to replace them with lower-paid black labour. The white workers, supported by remnant elements of old Boer commando units and infiltrated by communist agitators, began a series of increasingly violent demonstrations that culminated, in February and March of 1922, in an attempt to take control of Johannesburg. The revolt, known as the Rand Rebellion, was eventually quashed with the assistance of the army and air force which bombed and strafed the rioters, causing considerable loss of life and property. Great ill-feeling was aroused against Smuts and his government for its handling of the situation and the die of legislated, colour-based job reservation had been cast. The events of these months were shortly to form the basis of a profound change in the political direction of government in South Africa.

The government was also accused of having used an unduly heavy hand to deal with two other outbreaks of unrest. The first was in the Eastern Cape in 1921 where an evangelical religious sect who called themselves the Israelites and who prophesied the end of the world in that year, had attracted widespread support. A huge gathering, on a farm called Bulhoek, ran out of control and the police were called in, leading to the deaths of over 200 followers.[14] The second event occurred in southern South West Africa (in May 1922), where a Khoi-speaking Nama kapteincy, the Bondelswarts, had objected to the imposition of a dog tax and refused to cooperate with the administration. A group of about 500 offered violent resistance to the police contingent sent to pacify them. The Administrator immediately called in the air force, which bombed the group, resulting in its surrender and over a hundred deaths. A survey found that 90% of the white population in South Africa was unhappy with the way the uprising had been handled.[15]

In the light of all these distractions, it is perhaps not surprising that almost two years were to pass before the following recommendation was made, on the Prime Minister's letterhead to the Governor General on 16 May 1924:[16]

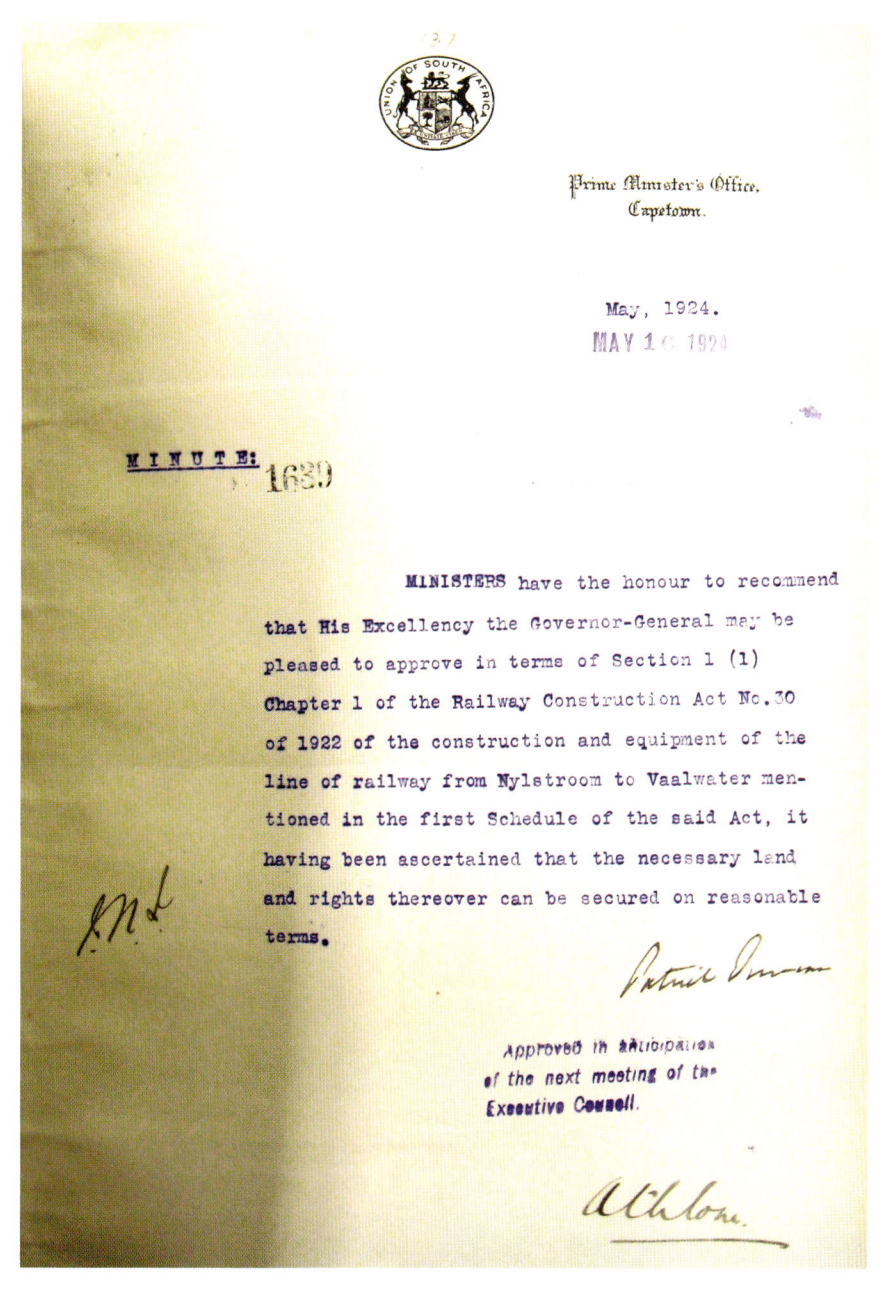

> Prime Minister's Office.
> Capetown.
>
> May, 1924.
> MAY 16 1924
>
> MINUTE: 1639
>
> MINISTERS have the honour to recommend that His Excellency the Governor-General may be pleased to approve in terms of Section 1 (1) Chapter 1 of the Railway Construction Act No.30 of 1922 of the construction and equipment of the line of railway from Nylstroom to Vaalwater mentioned in the first Schedule of the said Act, it having been ascertained that the necessary land and rights thereover can be secured on reasonable terms.
>
> Approved in anticipation of the next meeting of the Executive Council.

FIGURE 11.2: The Minute approving the construction of the line. SANA SAB/MVE/416/16/48: Minute 1639 from Prime Minister's Office to Governor General, 16 May 1924.

At this time, Duncan (later knighted by King George VI and appointed the 6th Governor General of SA on the recommendation of Prime Minister Hertzog) was the Minister of Education, of the Interior, and of Health in the South African Party government under Jan Smuts. He had signed the bill on behalf of the Prime Minister. "Athlone" was Major General Alexander Cambridge, the 1st Earl of Athlone, who only three months previously had arrived in South Africa to take up his appointment as the country's fourth Governor General. The signing of this Minute was timely, because within the next month, Athlone was to witness the defeat of Smuts's SA Party in a general election and the accession to power for the first time of the National Party under the leadership of General Hertzog.

Almost immediately after approval to proceed with the construction of the line had been granted, an urgent application was made for a deviation in its proposed route. Taking advantage of the imminent general election, the chairman of the SA Party at Zwagers Hoek, Gert van Heerden (of Rankin's Pass – see Chapter 4), wrote a personal letter to Prime Minister Smuts on 28 May 1924 urging his approval of a deviation of the line:

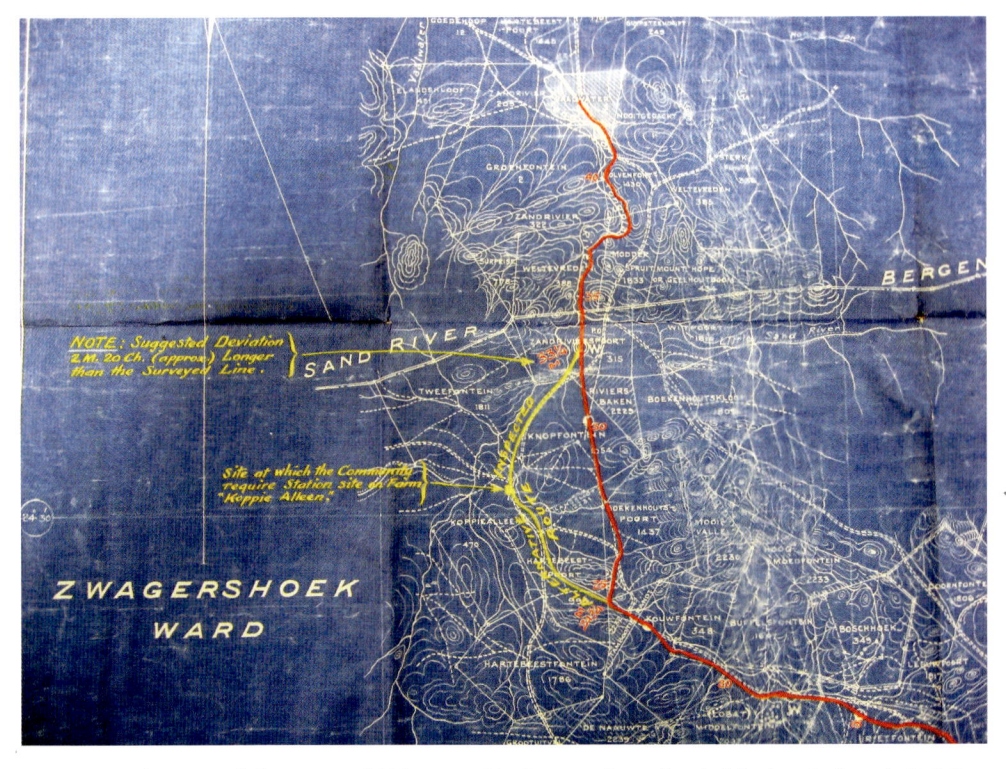

FIGURE 11.3: The route of the proposed Nylstroom–Vaalwater railway line (original route in red, deviation in yellow). SANA SAB/MVE/416/16/48: Map of proposed Deviation of Railway from Nylstroom to Vaalwater.

> What we require is the Railway to come on the South-West side of Zandspruit till a point near the farm "Homestead" of Koppie Alleen and a station there and [thereafter] the line can go direct through to Zandrivierspoort as at present surveyed.[17]

The writer pointed out that whereas the existing route went on the eastern side of the Sand River to a station at the foot of Zandrivierspoort – which would be of benefit to only three farmers and a store – a deviation as requested would save over 100 farmers the cost of having to haul their produce an extra nine miles to the nearest station and incur additional railage fees. Mr Van Heerden concluded his letter with the warning that

> I am sure you will use your influence to meet the [needs of the] majority of farmers here. Indeed, they have expressed themselves so strongly that they will abstain from registering their votes and many will vote for the opposite party [unless you do so]. I will be pleased to have your definite assurance that the matter will have your immediate attention so that I can inform our members.

Needless to say, the deviation was hastily approved – but, as it turned out, to no avail for the SA Party: The (presumed) endorsement of the mighty Zwagers Hoek community proved insufficient to offset its losses elsewhere in the country. In the General Election of June 1924, a coalition of the National and Labour Parties, under the leadership of J.B.M. (Barry) Hertzog, won with a comfortable majority and Smuts lost office.

The deviation demanded by the Zwagers Hoek farmers, while logical in that it would serve a larger community than the original route, also meant that the Sand River would have to be crossed twice. However, the overall effect was the addition of a mere 2¼ miles to the overall length of the line, taking it to exactly 46 miles (74 km). In the end, perhaps to offset the additional cost of the deviation, only an unmanned siding – rather than a station – was built at Koppie Alleen (despite further lobbying). For obscure reasons (see Chapter 5), the name was soon changed to "Alma".

Construction work on the line must have commenced promptly, because little more than a year later, traffic began to move along it. A line that had taken over five years of investigating, planning, lobbying and approving took only a year to build!

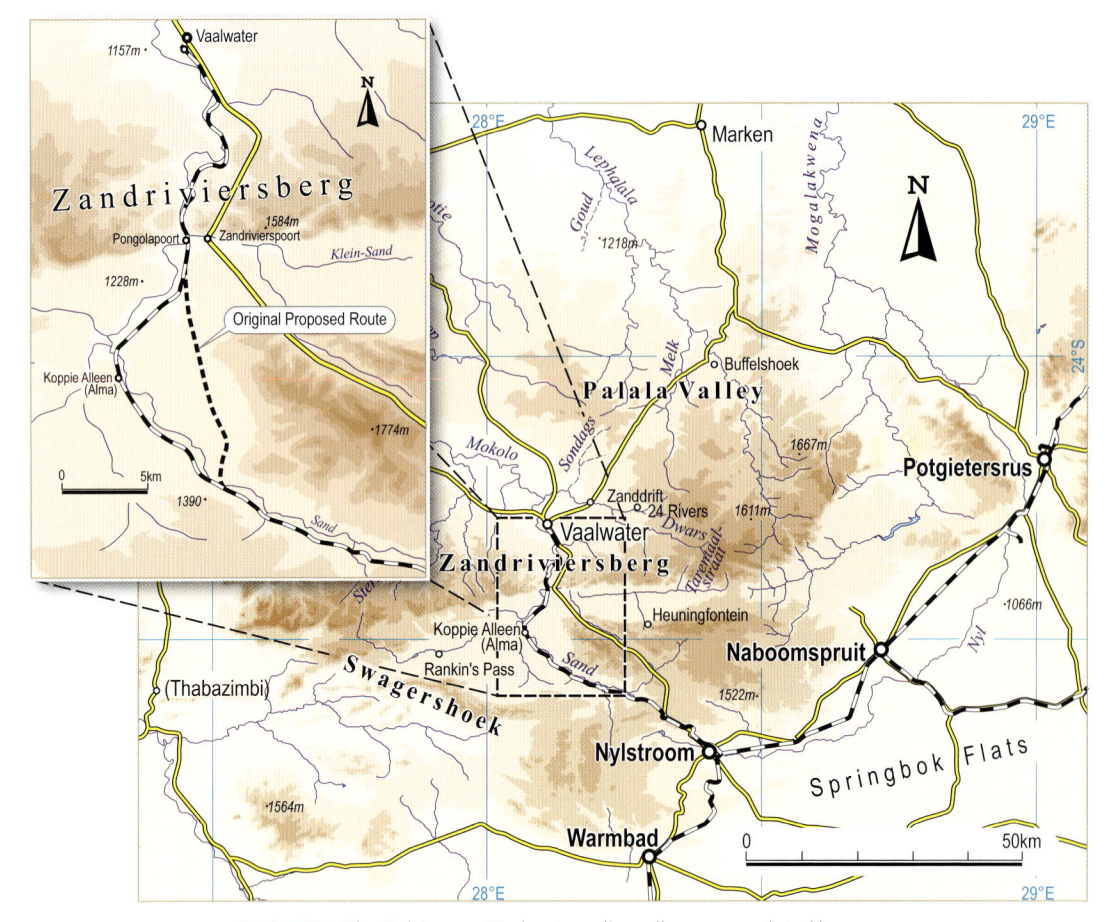

FIGURE 11.4: The Nylstroom–Vaalwater railway line as completed in 1925.

This was achieved despite unseasonable rains, which delayed construction and increased costs by £20 000.[18] The November 1925 issue of the *SA Railways & Harbours Magazine*[19] reported that the line was originally intended to be built using black labour; but ultimately, only white workers (over 1000 of them) were involved – an early consequence, one suspects, of the recent change in governing party.

The Nylstroom–Vaalwater line was officially opened on 1 October 1925 with a large ceremony at the Vaalwater terminus.[20] The officiating officer was C.T.M. Wilcox, a member of the Railways Board (and later Administrator of the Orange Free State), who attended on behalf of Minister Malan. Two packed trains conveyed over 1500 passengers from Nylstroom to Vaalwater for the event, whilst another picked up farmers and their families along the route in open wagons and delivered them to the gaily decorated terminus.[21] One later account reported that

FIGURE 11.5: The official opening of the Nylstroom–Vaalwater line on 1 October 1925.
SAR&H Magazine, November 1925.

the day was marked by the extremely unusual occurrence of three inches of snow; and although the accompanying photograph of the opening appears to support this contention, the Railways magazine doesn't mention it.[22] The Weather Bureau also has no record of the weather in the area being unduly cold that day and no local resident remembers hearing about what would have been as much of a once-in-a-lifetime event as the arrival of the first train. Perhaps this was just another of those "rural legends" for which Vaalwater would become famous!

Mr Wilcox urged the members of the community to make full use of the line and justify the investment that had been made in it. He said that it was purely an agricultural line in that no revenue was expected to be derived from minerals or industry. He described the country beyond the railhead as "The Land of Canaan – a land of considerable promise". According to the *SAR&H Magazine*:

> Looking on the throng of sturdy men present, one felt that these hardy pioneers, imbued by the spirit of the old Voortrekkers, would make the line pay, and the sincerity with which Mr Wilcox expressed his pleasure at meeting farmers of such a type evoked hearty acclamation. "You deserve your railway", added the speaker,

"and it gives me great pleasure to open it for your needs". The cheers that greeted these words burst forth anew when handsome tribute was paid by Mr Wilcox to the skill and devotion of Mr Jaffray, the construction engineer, and the work performed by the men under him.[23]

The new MP for the Waterberg constituency, P. le Roux van Niekerk, took the opportunity to point out to Mr Wilcox that at Nylstroom the station stood a mile outside the town – merely because the wife of the Landdrost at the time of its construction had objected to the prospect of noise from trains – and that this had

FIGURE 11.6: C.T.M. Wilcox, member of the Railways Board (right), shown tightening the last fishplate on the new line. Caricature of C.T.M. Wilcocks by Daniel Cornelis Boonzaier (1930), by which time Wilcocks had been appointed Administrator of the Orange Free State. The caricature, which is signed "Nemo", a pseudonym used by Boonzaier for his contributions to the magazine *The Sjambok* during 1929–1931, had the caption that Wilcocks was "fortunate in having some Kosher Afrikaans Christian names [Carl Theodorus Muller] as an antidote to the English title Wilcocks". Daniel Cornelis Boonzaier (1865–1950) was a prolific professional cartoonist who specialised in caricature. He drew for *Die Burger* from 1915–1940, but his work also appeared in many other publications. This caricature was reproduced as No. B941 in the collection *Catalogue of Pictures in the Africana Museum* (Volume 1, p. 210), compiled by R.F. Kennedy and published by the museum in 1966.

proved to be most inconvenient for the townsfolk. He appealed to Mr Wilcox to build a new station in the centre of Nylstroom. The magazine continued:

> Mr Hewson, Chairman of the North Waterberg Farmers' Association, ... speaking from wide experience, assured his hearers that the country now tapped by the new line, all down the Palala to the Limpopo, had a value which none could believe who had not seen it. The Association's Secretary, Mr Davidson, asked the assembly not to forget the late Mr Johann Rissik, who with Mr JW Jagger, Colonel Mentz and others had granted the construction of the line.
>
> Catering arrangements were carried out with commendable thoroughness. To feed 2,000 people on the veld, miles away from a village, is no easy task, and if the engine enjoyed the champagne it spilt in smashing the time-honoured bottle slung across the decorated arch [over the line], the onlookers must have wished it joy on the exuberance of their own ample supply of good cheer.

Mr Wilcox, it transpired, was not going to have an easy day. Immediately following the opening ceremony, he was obliged to attend at least two meetings of lobbyists. The first was from a group pushing for the extension of the line northwards to the Palala. Messrs Van Niekerk, Loubser, Botha, Mortimer and Davidson made a strong case that this option was far preferable to the alternative route under consideration, which was along the Pongola (now being referred to as the Magaul).

> Mr Botha said they had 130 schools [on the Palala route? – an unlikely number] and in the other direction there were only two. The Magaul area had poor sand, whereas along the Palala direction, the soil was red and black. Where the Palala came out of the mountains, a great irrigation scheme could be entered into. The Magalaquin area was a great area. He said the Palala was a river which ran like the Vaal River today. If they could get a line to the bank of the Palala, it would be 40 miles [from Vaalwater] to the Farm called Buffelshoek [on Lindani, near where the tar road to Marken crosses the Palala today].[24]

In his reply, Mr Wilcox said he could not hold out any hope that an extension to the line would be possible within the next five years, given the large capital outlay just incurred. Next came a deputation on behalf of the Magaul [Mokolo] area who expressed the view that *their* route was the best and cheapest option for an extension:

> It was one level flat from Vaalwater to the Limpopo [they said]. In the Zoutpan ward [where Lephalale/Ellisras is today] there was coal. It was about 20 miles from the Limpopo. As regards the settlement in the Zoutpan District, there were hundreds of farms which could be given out. Then there was the Tuli block, with more cattle than the whole Waterberg.[25]

This deputation argued for a line all the way from Vaalwater to Mahalapye in Bechuanaland, a distance of over 300 km, and proposed that the state itself invest in coal mines in the region. Indeed, the state, in the form of Iscor, would eventually do exactly that when it opened the Grootegeluk coal mine in the 1980s and linked it by rail to Thabazimbi (see Chapter 19).

In 1926 the Swaershoek (now modernised from Zwagers Hoek) Farmers' Association, through its secretary, R.S. Read of Rankin's Pass, appealed – again unsuccessfully – for the siding at Alma to be upgraded to a station.[26]

And so the lobbying continued. In the early 1930s, as the country sank into the Great Depression, traffic on the line declined to such an extent that the government contemplated closing it. In a desperate effort to keep it alive, Ted Davidson, according to his grandson Charles Baber, went so far as to pawn his gold watch to pay for a train ticket all the way to Cape Town to lobby in Parliament on behalf of the local farming community. He not only succeeded in this mission, but

FIGURE 11.7: Tweestroom siding in 2015. Illustration by C. Bleach from original photograph by R. Wadley.

his proposal to relook at an earlier idea of a Road Motor Service from the railhead at Vaalwater to service the Bushveld (Stockpoort) was accepted and led to the revival of the line.[27]

Even as late as 1938, in a letter from its secretary – still the indefatigable E.A. Davidson – the North Waterberg Farmers' Association passed a resolution:

> That whereas this meeting has been informed that a Government Commission has been appointed to investigate the extension of railway lines for defensive purposes, we ask that the said Commission investigate the extension of the Nylstroom–Vaalwater line, as this meeting is convinced that it will be highly important for defensive purposes.[28]

As we now know, none of the desired extensions to the Nylstroom–Vaalwater line was ever built. Instead, a line from Rustenburg was constructed northwards to Northam and then in 1934 on to Kwaggashoek (see Chapters 18 & 19). There, the town of Thabazimbi (Hill of Iron) had been established in 1930 to enable iron ore from a new mine to be delivered to the purpose-built steelworks in Pretoria West, on behalf of the state-owned SA Iron & Steel Corporation (Iscor). In the late 1970s, this line was further extended to Ellisras (Lephalale) to serve Iscor's coking coal mine at Grootegeluk and the associated Eskom power station at Matimba.[29] From this time onwards, agricultural and mineral traffic from the Lowveld, previously road-hauled to the railhead at Vaalwater, could be despatched directly by rail from Ellisras.

Did the Vaalwater–Nylstroom line ever live up to the potential claimed for it by its early proponents? Probably not, although there was a period when it did carry a remarkable amount of traffic. According to one source, in the beginning only three trains a week ran between the two towns. By 1966 there were 13 trains per

FIGURE 11.8: Goods train crossing Loubadspruit bridge at Loubad. Photograph courtesy of Nathan Berelowitz, *Friends of the Rail*.

week, increasing to as many as 16 during the height of the watermelon season. For the most part, the trains carried produce and cattle rather than passengers.[30] As the Gedenkalbum of the NG Gemeente Vaalwater (1930–2005) remarked: "Die spoorlyn tussen Vaalwater en Nylstroom is wel in 1925 geopen maar was op daardie stadium redelik onbekostigbaar vir die arme Waterbergers".[31] Most Waterbergers could not afford the high rate they needed to pay in order to attend nagmaal – held four times a year – at the nearest church, which was in Nylstroom. It is said that this led to the decision in 1930 to build a new NG church for the North Waterberg Gemeente on the Melkrivier near Sondagsloop.

In Vaalwater a very large stockyard – reputed to be the second largest in the country after Vryburg in the Northern Cape – was built to handle the movement of cattle to the abattoirs in Pretoria. The line also used to bring coal from Witbank to Vaalwater where tobacco farmers would collect it to fuel their drying flues until the new coal mine at Grootegeluk near Ellisras opened in the late 1970s. From the railhead at Vaalwater, SAR&H ran a Road Motor Service to Ellisras and the Bechuanaland border post at Stockpoort. The huge RMS workshops near the station are still in use, albeit for other purposes.

CHAPTER 11: MAKING TRACKS TO THE PLATEAU   393

The light gauge line could not carry large locomotives or haul long trains: Only locomotives of Classes 7, 19D and 24 were used on the line, the latter running until the first diesel units (Class 35) were introduced in 1985, following a modest upgrade of the line.[32]

In its last years, especially after the re-emergence of South Africa into the global market, the line carried ever-decreasing traffic in the form of livestock and produce. The tobacco industry collapsed (see Chapter 13) and cattle could be transported to the abattoir in Pretoria more efficiently by road. A festival train ran for

FIGURE 11.9 (OPPOSITE): Two Class 24 locomotives about to leave Vaalwater station with a long freight train. Photograph courtesy of Dick Manton, *Friends of the Rail*. FIGURE 11.10 (TOP): Class 24 with freight at dusk on the line to Nylstroom. Photograph courtesy of Dennis Mitchell (Australia), *Friends of the Rail*, taken on 2 April 1985. FIGURE 11.11 (BOTTOM): Vaalwater station in 2010.

several years on the occasion of the annual Nylstroom Grape Fest: Guests would be taken up the line to Vaalwater for a picnic lunch before returning to Nylstroom in the afternoon. The last train to carry maize from the silos at Vaalwater and Alma travelled down the line in 2000. In 2003 it was finally closed to traffic.

Early in 2010, Exxaro, the owner of the Grootegeluk mine near Lephalale that supplies not only Eskom's Matimba and Medupi power stations but also a variety of other products for the local and export markets, announced that it was investigating a range of expansion options.[33] This was prompted by the award to Grootegeluk of a contract to provide feedstock for Medupi. This contract would yield several million tons per annum of high-quality by-product coal for export. Exxaro was examining alternative ways of transporting this additional tonnage to the coast at Richards Bay.

One option was to build a new line from a junction near Matlabas on the existing Lephalale–Thabazimbi line to Vaalwater, and then to refurbish and upgrade the old line from Vaalwater to Modimolle. If this option had been pursued, it would at last have brought to fruition those earnest appeals from the local farming community of 85 years ago – and once again, residents of the town might have listened at night to the mournful, but reassuring sound of passing trains. But it was not to be. The idea was soon abandoned in favour of the more obvious alternative: upgrading and electrifying the existing line through Thabazimbi.

CHAPTER TWELVE
Pests and pestilence

In common with most parts of the world, the Waterberg region has long been exposed to various pests and the diseases they spread. But because the carriers of some of these diseases require or include residence in the human as part of their life cycle, their presence was minor until human populations began to grow and became more settled. Human and animal communities with a long history of living with some diseases also seem to develop a degree of immunity to their consequences, whereas recent settlers, or stock introduced from elsewhere, are far more susceptible. The first European travellers (and their animals) passing into or through the region early in the nineteenth century were especially vulnerable and soon learned which areas or routes were safe and which to avoid. They came from a Europe where the Industrial Revolution had caused the migration of millions of rural folk into congested, poorly ventilated, unsanitary urban environments. Very soon, around a quarter of all deaths could be attributed to an ancient, slow, wasting lung disease they called consumption, but which in agrarian times had been far less important as a cause of mortality. Although the scientific name of this disease, tuberculosis (TB), had been coined as early as 1839, only in the 1880s was it realised that the disease was contagious and caused by a bacillus; an antibiotic to combat TB was not discovered until 1946.

In subtropical Africa – and in other parts of the world where Europeans ventured – new pests and diseases awaited travellers and settlers, and initially, almost nothing was known about their origins, let alone how to treat them. Preoccupied as we are these days with the alarming incidence of HIV/Aids on the African continent and the consequences of this condition for our population – including its implications for the development of active (as opposed to latent) TB – we are inclined to forget that, historically, there were many other pests and diseases that determined, or substantially influenced where people could live. A discussion of the historical impact and present status of diseases – on humans, domestic stock and wild animals – is vital to an understanding of where communities settled and the livelihoods they pursued on and around the Waterberg plateau.

Tsetse fly

Two of these diseases, human trypanosomiasis or sleeping sickness, and African animal trypanosomiasis or *nagana*, are borne by the tsetse fly. Human sleeping sickness, which is concentrated in but not confined to West Africa, is estimated to have killed at least 10 million people over the last century, and is still responsible for the deaths of up to 50 000 people every year.[1] Thanks to concerted efforts by programmes run through the World Health Organisation (WHO), newly reported cases fell below 10 000 for the first time in 2009, although many cases continue to go unreported.[2] There is no vaccine for sleeping sickness. Early treatment can lead to full recovery, but untreated cases have a 100% mortality rate.

Nagana, derived from the Zulu word *unakane* meaning "to be depressed", is much more widespread, reaching right across sub-Saharan Africa into Zimbabwe. It not only results in the loss of about three million head of domestic livestock (mainly cattle) every year, but in the process deprives rural pastoralists of the means to plough their fields and transport goods, and precludes the expansion of herders into many areas that would otherwise be well suited to grazing.[3]

Human sleeping sickness was never a serious condition south of the Limpopo River, although it was at one time widespread along the Zambezi valley in Zimbabwe and still occurs there; but nagana, in one form or another, was a major impediment to the settlement of cattle farmers in the Limpopo valley up until the end of the nineteenth century. It was also a serious disease in northern KwaZulu-Natal (KZN) until the 1950s and even to this day occasionally causes stock losses in that region.

The virtual extermination of the tsetse fly from South Africa by the second half of the twentieth century was the result of intensive, concerted efforts by veterinary and agricultural authorities – often using well-intentioned programmes later found to be counterproductive or inappropriate. These were undertaken in largely fortuitous conjunction with other natural or inadvertent human activities that together had the effect of firstly halting and then reversing the advance of the fly. There is concern that recent changes in land use patterns in areas where the tsetse fly was formerly endemic, aggravated by the creation of transfrontier game parks and the decline in technical capacity and interest on the part of responsible government authorities, could lead to conditions amenable to reinfestation of the tsetse fly and the diseases for which it is a vector. Because tsetse fly have, for the time being at least, been almost eliminated from South Africa, we are reliant on the records of early travellers to describe the historical distribution of the fly in and around the Waterberg region and the constraints its presence imposed on the movement and settlement of people here.

The tsetse fly and its diseases

The tsetse fly of the genus *Glossina* (derived from the Greek for tongue, in reference to the insect's prominent proboscis) occurs as 22 species, all of them confined to sub-Saharan Africa. The current distribution of *Glossina* is shown in Figure 12.1. The limits are determined primarily by the climate and secondly by

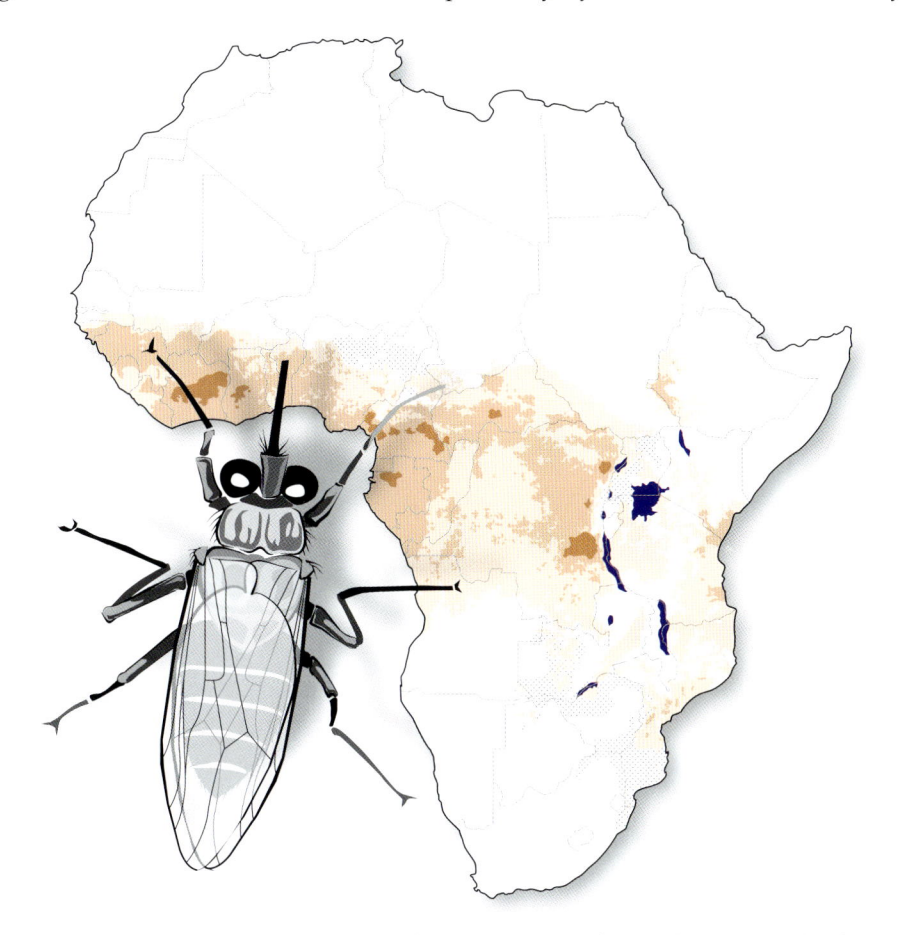

FIGURE 12.1: Tsetse fly at rest (note prominent proboscis and folded wings) and distribution of *Glossina* spp. across sub-Saharan Africa (historical distribution shown in dotted pattern). Illustrations by C. Bleach, modified from photograph of tsetse fly by Nash (1969) and distribution maps provided by Bengis (2011) and Environmental Research Group, Oxford (2011): http://ergodd.zoo.ox.ac.uk/livatl2/tsetse.htm

the vegetation, which can often mitigate the severity of climate. In general, tsetse are rarely found in areas having an annual rainfall of less than 750 mm – except where denser vegetation along watercourses has facilitated their penetration into drier areas.[4] They thrive at temperatures between 15 °C and 35 °C; they may be found at altitudes up to 1800 metres amsl at the equator, but rarely above 1200 metres in southern Africa.[5]

Only four of the 22 species of *Glossina* have been recorded from South Africa and of these, two are important here: *G. morsitans,* the species prevalent historically in the Limpopo valley; and *G. pallidipes,* which was the dominant species occurring in eastern Kruger Park and southwards into northern KZN until its eradication in the 1950s. Two other species have been found to occur sporadically in northern KZN and were responsible for an outbreak of nagana there in 1990.[6]

All species of *Glossina* are essentially similar in appearance, varying only in size and in abdominal markings. Their overall colour is dark- or yellowish-brown, often with smoky wings. They are easily distinguished from other medium-sized flies by two characteristics: a) When alive and at rest, "the abdomen is completely hidden by the wings, which fold over one another like a closed pair of scissors and project well beyond the tip of the abdomen, giving the fly a streamlined appearance";[7] and b) The proboscis which sticks out in front of the head, as shown in Figure 12.1.

In the course of feeding on the blood of an animal, tsetse flies may become infected with small, single-celled parasites called trypanosomes, whose life cycles require time spent within both the flies and their vertebrate hosts. Once infected, a fly is able to transmit these parasites to uninfected animals, including humans; some trypanosome species cause human trypanosomiasis (sleeping sickness), an infectious disease, whereas other species cause nagana. The trypanosomes are injected into vertebrate muscle tissue of the bitten animal, but make their way first into the lymphatic system, then into the bloodstream, and ultimately into the brain. The disease causes the swelling of the lymph glands, emaciation of the body, and eventually leads to death.

Historical distribution in South Africa

The name "tsetse", used to describe the species of fly whose bite causes the death of cattle, appears to be of Setswana origin. The first record of the name can be found in the book by the hunter Gordon Cumming, *A Hunter's Life in South Africa,* published in 1850.[8] During 1846–1849 Cumming made several expeditions along the upper reaches of the Limpopo River. On one of these trips, he visited the kraal of Chief Selika (now Ga-Seleka, on the farm Beauty, on the banks of the Lephalala River where it is crossed by the R572 between Lephalale and Swartwater). Here, on 21 August 1847, Cumming's local black guides "told me that I should lose my cattle by the fly called 'Tsetse'; and there was also reason to believe that the country in advance was not very healthy for man".[9] Wisely, Cumming heeded his guides' advice and turned for home – but too late, as described below.

Another traveller to the region, a few years earlier, had been Johan August Wahlberg, the Swedish naturalist and hunter. Wahlberg also referred in his diary to "the deadly fly", which he called *Tsatasa* (= *Zeze*) and which he encountered along the Aari River (an early name for the Ouri, Limpopo or Crocodile), in the vicinity of present-day Thabazimbi.[10]

The earliest references

The treacherous "fly" and its consequences for travel with horses and oxen had been noted by several even earlier travellers. The entomologist Claude Fuller documented in great detail the records of several early references to the fly.[11] He considered the earliest written reference to the fly (as yet unnamed) to have been that of the explorer David Hume, who travelled along part of the Limpopo River in 1835–1836 and who, on his return to Grahamstown, gave an account of the problems it caused him. Later that year, another traveller, Captain William Cornwallis Harris, an engineer on furlough from the East India Company, wandered through the area lying between the rivers now called the Crocodile River and the Marico, just south of their confluence to form the Limpopo. Having trekked downstream along the Crocodile from the gorge that now contains the Hartebeespoort Dam, it was not long before Harris encountered the fly, probably close to where Brits lies today:

> During the rainy season especially, [the area is] infested by a large species of gadfly, nearly as large as a honey bee, the bite of which, like a similar fly in Abyssinia, proves fatal to cattle. A desire to escape the officious visits of these destructive insects, whose persecutions relieved us of two of our oxen, soon obliged us to abandon the willow-fringed river.[12]

The Mural Mountains

Interestingly, Harris went on to talk about the "great range of mountains known as the Mural"[13] which he could see to the east of the Crocodile. Fuller presumed this to be a reference to the sheer escarpment of the southwestern Waterberg range, now within the Marakele National Park, but pointed out that for years after Harris's description, even after the Waterberg range had been identified, cartographers faithfully included the fictitious Mural Mountains on their maps of the region! Even the great cartographer Carl Jeppe showed the Murals lying between the Crocodile and the Marico Rivers on his Transvaal map of 1868,[14] although in his updated map of 1879[15] the range had been relocated to between the headwaters of the Matlabas and Pongola (Mokolo) rivers – much closer to the actual position of the western Waterberg range. But this was not the last move for the "migrant" Mural: A map produced by the London cartographers Samson, Low and Marston in 1903 to illustrate the *Times History of the War in South Africa* showed the Mural range lying close to and parallel with the Limpopo River between its confluences with the Matalabas and the Mokolo.[16]

The Vlieëpoortberge

The next significant references to the presence of tsetse fly in the western Limpopo valley were those made by two hunter-naturalist-adventurers who separately travelled into the upper Limpopo lowveld in the 1840s: Adulphe Delegorgue and Johan August Wahlberg. Wahlberg's account has already been mentioned, but that of Delegorgue[17] is particularly interesting for two reasons. Although Delegorgue travelled down the Crocodile (which he called the Ouri) from the gorge through the Magaliesberg, and at a similar time of year to Harris, he did not encounter tsetse fly until much further downstream, when he negotiated the rapids and gorges of the river as it cut through the two ranges of hills immediately west of present-day Thabazimbi. It appeared that the fly no longer occurred as far upstream as it had done in force only ten years previously. However, in his several trips through the gorges west of Thabazimbi, Delegorgue repeatedly suffered there from the depredations of the fly, which seemed to occur in localised concentrations in thick bush along the river. It was for good reason that these hills became known as the Vlieëpoortberge – a name by which they are still known today.

The importance of not getting wet

Wahlberg and Delegorgue – the latter known more for his wanton slaughter of any wild animal crossing his path than for his contribution to natural sciences – were closely followed by Cumming and two other hunter explorers, Frank Vardon and Cotton Oswell, all in the period 1846–1848. An account given by Cumming provides the first evidence linking fatalities in cattle to their exposure to moisture after having been bitten by the fly:

> The rains still poured down, rendering the country impossible to travel, and my oxen died daily of the tsetse bite. At length I came fairly to a stand, not having sufficient oxen left to draw one wagon … and in a few days more, all my cattle had died. I did not then know to what cause to attribute this sad, and to me important, change in their condition. Alas, it was now too evident that nearly all of them were dying, having been bitten by the fly "tsetse" at the mountain. The rains of the last three days have made this melancholy truth more strongly manifest.[18]

Many subsequent writers have referred to the fact that, having been bitten by the fly, cattle and horses continued to behave normally for several days until they became wet, either from rain or mist, or through fording a river, whereupon they died within hours.

Andrew Smith's expedition

Further evidence of the role played by moisture in the manifestation of nagana in animals bitten by tsetse can be found in the account by the medical doctor Andrew Smith[19] of his expedition into the Limpopo valley in 1835 – the year before Harris. Although Smith made brief mention of having to detour away from the upper Crocodile for a period (in the Brits area perhaps) because of a troublesome fly, his account contains no other reference to the insect. However, he spent some days travelling close to the west (Botswanan) bank of the Limpopo River between its confluence with the Notwane and the Tropic of Capricorn (roughly where the Stockpoort border post is today). Shortly after crossing the Tropic on 2 September 1835, Smith noted that his cattle were losing condition – which he attributed to the poor grazing in the area – and decided to turn around and head back to his base at Moffat's mission at Kuruman. He then experienced the first rain showers of the season – and within days his trek oxen began to sicken and die. Although Smith does not mention seeing tsetse fly on his animals, it is most likely that the fly was the cause of their deaths.

The location

Vardon's and Oswell's trip along the Limpopo in 1846 was especially important for one reason: In the vicinity of the "Siloquana Ridge", a minor topographic feature some 20 km beyond where Smith had turned around, lying in the angle formed by the confluence of the Limpopo and the Pongola (Mokolo) Rivers (probably the low ridge today named Witkop on the farm Bellevue), Oswell and Vardon collected the actual specimens of the fly ultimately used for its formal definition. As Oswell (quoted in Fuller) said: "On the low Siloquana Hills … we made our acquaintance with the tsetse fly, which we were the first to bring to notice, Vardon taking or sending to England some he caught on his favourite horse. The fly infests particular spots and never shifts."[20]

The specimens were sent to Prof. J.O. Westwood, head of the Zoology Department at Oxford University, who named the species *Glossina mortisans*. In a letter accompanying the specimens, Vardon wrote: "I have ridden up a hill and found 'setse' increasing at every step, till at least forty or fifty would be on my horse at once. The specimens you saw cost me one of the best of my stud. He was stung by some ten or dozen of them, and died in twenty days."[21]

The legendary commando path

Another story of interest at about this time concerns the legendary "Commando Path", a cleared corridor that was said to have been cut by a fleeing column of early Voortrekkers who attempted to cut off the retreating Matebele leader Mzilikazi from his refuge in the Matobo hills of southwestern Zimbabwe. According to Fuller, the hunter Frederick Courtney Selous, when giving evidence to the British South Africa Company's Sleeping Sickness Commission in 1913, referred to this corridor as follows:

CHAPTER 12: PESTS AND PESTILENCE

When the Boers followed Umzilikazi [sic] from the Transvaal in 1840 and crossed the Limpopo, they chopped their army road. It was a quarter of a mile broad, and they trekked with their oxen through that cleared space and did not lose their cattle by the fly. I only heard about it from the Boers, and that was a long time ago, so I cannot say exactly the width.[22]

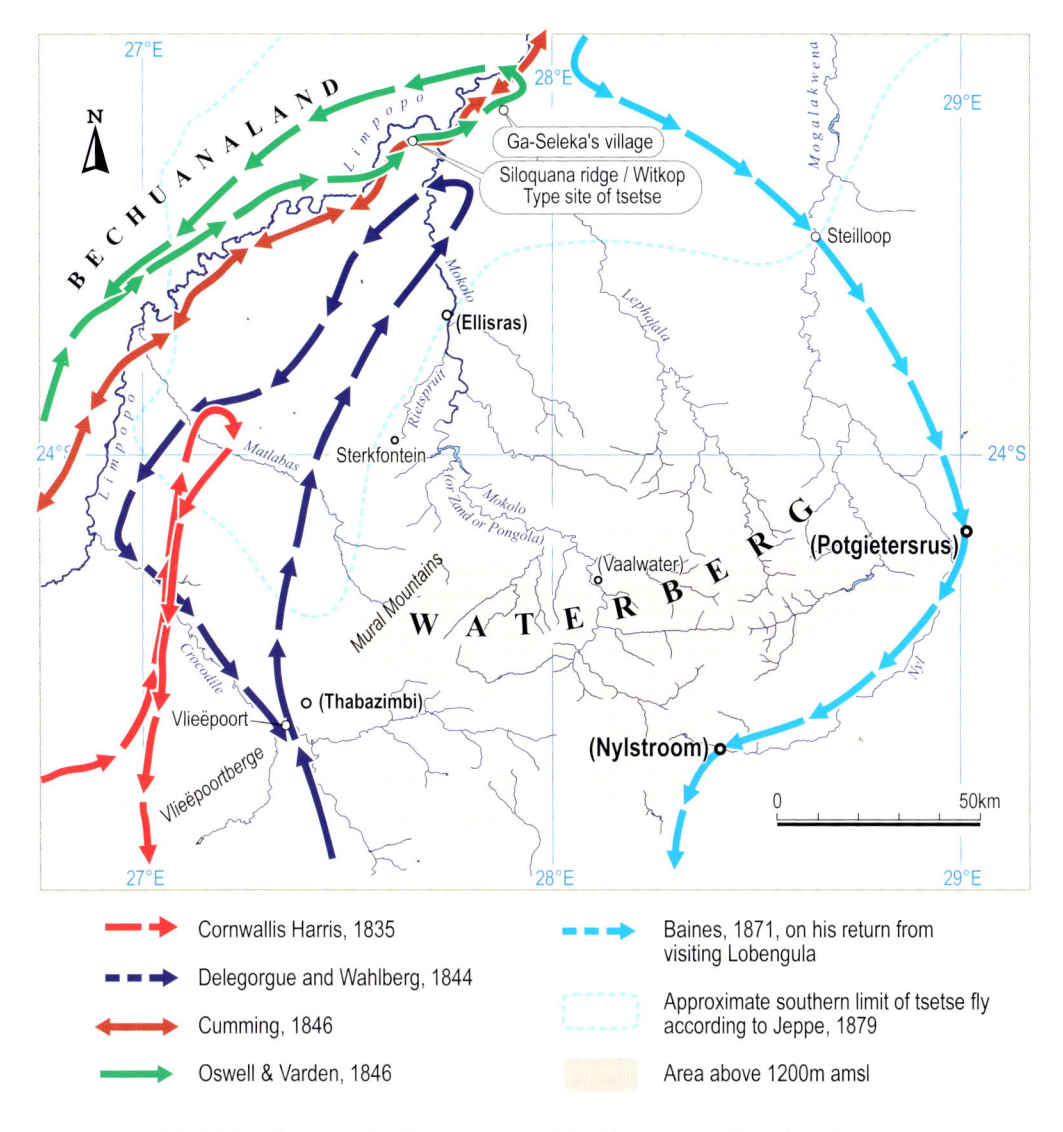

FIGURE 12.2: Early travellers' journeys around the Waterberg and Jeppe's tsetse zone.

Fuller, having found no substantive evidence for this cleared path, noted that "it is greatly to be feared that so remarkable a feat existed only in the imaginations of some too communicative descendants of the pioneers". However, the path is also mentioned by Thomas Baines:

> We were in hopes of finding the "commando path" of the Emigrant Boers, where in old times they cleared away the bush, 100 yards on each side of the road, but we found reason to believe that this was still east of us, and we knew of no safe way of reaching it.[23]

The experiences of Thomas Baines

In October 1871, when returning from one of his several visits to Matebeleland and the Tati goldfields, Baines and his party chose to return to the Transvaal via a new route to the north and east of the Waterberg massif rather than the usual, tsetse-free route down eastern Bechuanaland. Baines had decided to attempt the new route at the request of the Matebele chief Lobengula (Mzilikazi's son), with whom he'd established a good relationship:

> He told me that a more direct route existed to the southward, through the Tsetse Fly, but that some of his father's [Mzilikazi's] people knew of passages by which the danger might be avoided, and to his great gratification I consented to open the road by going that way.[24]

Baines crossed the Limpopo into the Transvaal on the farm Clanwilliam northwest of the present settlement of Swartwater, travelled upstream along the eastern bank for about 24 km and then turned south, away from the river. They encountered patches of fly all the way, and so travelled only at night.

When they reached Madlala Pan (on the farm Kameelfontein along the present N11, near Tom Burke), the group turned towards the southeast. From here until the crossing of the Mogalakwena River (at Steilloopbrug), a distance of some 75 km, Baines's party travelled through or close to numerous patches of fly and again were forced to trek mainly at night:

> [Local people] told us the pan in front was called Madlala, and a day's journey beyond it was "Schimmel Paard Pan" [on the N11], there was fly between them, but no water, and also fly beyond, but we must leave [the pan] before dark so as to be able to get into a definite track, and yet not so soon as to rush into the fly until it has retired for the night.[25]

Baines noted that Schimmelpaard Pan was so named after an incident involving an earlier traveller, Joseph Macabe, "who, refusing to believe that a ridiculous little fly could kill so large an animal, had outspanned by the pan and deliberately rode a valuable horse into the infested country. In a few hours after his return, the steed was dead".[26] (The word *schimmel* referred to the dappled colouring of the horse.) Having crossed the Mogalakwena River, and turned south towards Nylstroom, the Baines party left the fly-infested region behind them. Fuller noted several points from Baines's account: that the area west of the Limpopo was free

FIGURE 12.3: Wagons crossing the Limpopo. SANA KAB Photographic Library, No. J1714: Wagons crossing the Limpopo Drift, 1894.

of tsetse; that the Limpopo River itself was virtually free of fly and that it was only when moving into the area of low rocky ridges, densely covered in bush ("bosrande" as he called them) that the party encountered the fly.

The relationship between tsetse and game

Fuller also mentioned various reports that associated the occurrence of the fly with buffalo. This association was later developed in detail by Selous,[27] who made a powerful case, based on his long and extensive experience of the country south of the Zambezi River, that wherever buffalo had been hunted out, the tsetse fly soon disappeared too; that it was very rare to find tsetse in the absence of buffalo, but that the latter could tolerate a wider habitat range than the fly, which was restricted by altitude, temperature and vegetation types. Therefore, it was possible to find buffalo without tsetse, but not the other way around. More recently, it has been shown that all spiral-horned antelope (bovids that are related to buffalo and cattle), wild suids (pigs) and elephants are among the fly's preferred hosts.[28]

The distribution of tsetse fly in northwest Limpopo

From the discussion above, it is clear that from around 1830 to 1875 tsetse fly occurred widely in the region between the Limpopo River and the northern and western limits of the Waterberg massif and that the distribution of the fly periodically extended southwards up the Crocodile (or Limpopo or Ouri) River possibly as far as the point where the river cuts through the Magaliesberg at Hartebeespoort. Comparison between descriptions of different travellers who traversed the same routes over this period, and in different seasons, strongly suggests that the distribution of the fly was not constant but extended and receded both seasonally and from year to year. These variations may have been the result of both climatic changes and game movements. In the Kruger National Park, for instance, a study of numerous Iron Age sites (fifteenth to eighteenth centuries) concluded that many Iron Age communities kept cattle in areas that later became endemic for tsetse and therefore unsuitable for cattle.[29] Fuller argued that tsetse preferred certain vegetation habitats – the *bosrande* – and Selous and others have linked the fly to particular game species, especially buffalo.

FIGURE 12.4: Historical and current distribution of tsetse fly in South Africa. Map compiled from those included in Fuller (1923), Du Toit (1954), Kappmeier et al (1998) and Wint (2008).

From around the 1850s the prevalence of guns in the Limpopo region increased rapidly as a result of increasing incursions by hunters, travellers, traders and settlers, and because of a growing possession of firearms by local black communities. Fuller and Selous argued that this inevitably caused an accelerated rate of elimination of game in the region, which in turn resulted in the retreat of the tsetse fly to areas further north where their preferred hosts still abounded.

Jeppe's map of the Transvaal dated 1879 shows a line indicating the limit of tsetse occurrence in the region (Figure 12.2). His later map of 1899 shows the same line, although by then numerous accounts testify to the fact that the fly no longer existed in much of the area indicated. As an example, Harry Zeederberg's *Veld Express*, which describes the story of the Zeederberg coach service,[30] makes no mention of tsetse being encountered south of the Limpopo River (and west of Pilgrims Rest) during the 1890s. The well-known Nylstroom resident and trader Emil Tamsen wrote to Fuller in 1923 that

> in 1880, there appears to have been fly in the mopani (bushveld) beyond Zandrivierpoort down to the Limpopo. I do not remember hearing of much loss of cattle. The farmers, knowing where the fly was, never trekked until they shot all the game away. When, in 1882, I trekked into Rhodesia, I struck the fly only near Zimbabwe ruins and had to turn back: thus the Waterberg and Zoutpansberg must have been clear in those days.[31]

The rinderpest

Some writers have suggested that the disappearance of the fly was related to the arrival of the rinderpest in the mid-1890s. Fuller's view was that in the sector of the Limpopo valley west of Beit Bridge, tsetse fly had largely been eliminated before 1895 as a result of the drastic diminution, through hunting, of host game species.

Rinderpest was a viral disease (related to measles) of cloven-hoofed animals, characterised by fever, gastroenteritis and high mortality.[32] In epidemic form, it was the most lethal plague known in cattle. Susceptibility was particularly high in African buffalo, giraffes, wild pigs, large antelope and certain breeds of cattle; moderate in wildebeest and East African zebus; and mild in gazelles and small domestic ruminants. Evidently, rinderpest had not been known in South Africa prior to 1896, although it had periodically invaded Egypt, possibly from Arabia or India.[33] It is speculated that following one of these invasions the disease was carried southwards along the Nile by the Italian army in its conquest of Abyssinia in 1887 and that it gradually spread further south through Uganda and into Tanganyika, reaching Lake Nyasa in 1892. By early 1896 large numbers of cattle and game were reported to be dying from an unknown disease on both sides of the Zambezi River. The disease reached Bulawayo where it was positively identified as rinderpest in March 1896, whence it spread rapidly south via transport oxen. It crossed the Limpopo into Transvaal in April, reaching Mafeking (now Mahikeng) at the same time, decimating cattle and game as it went. It also advanced rapidly into Bechuanaland and German South West Africa (Namibia today).

In an effort to prevent the migration of the disease further south, the Cape government erected fence lines and established a cordon of mounted police to stop infected cattle passing south. Later a 1600-km-long barbed wire fence pa-

FIGURE 12.5: The rinderpest epizootic in South Africa, 1897. Photograph attributed to G.R. Thomson, used inter alia by *New York Times*, in its article on 27 June 2011.

trolled by police was erected south of the Orange River between Basutoland and Bechuanaland. However, the disease eventually breached even this defence in early 1897. By this time, the immunologist Harold Koch, working in Kimberley, had confirmed the efficacy of an inoculation process involving the use of the bile of infected cattle. By the end of 1899 over two million cattle had been immunised and the disease was (temporarily) brought under control.

During this virulent epizootic, rinderpest caused ruin and devastation over extensive stretches of country, destroying not only the majority of domestic bovines along its route but also considerable numbers of indigenous antelope. It was estimated that more than two-and-a-half million head of cattle succumbed to it in South Africa alone.

In May 1901 during the South African War, rinderpest made a second appearance, probably the result of infected cattle imported into Basutoland from German South West Africa where the disease had never been fully eradicated. This was the last time the disease was encountered in South Africa, although it persisted in pockets in East Africa for many more years.

In June 2011 the United Nations announced that rinderpest had finally been eradicated from the world – only the second time in history that this has occurred (the first was smallpox).[34] This achievement was the result of intensive campaigns using modern vaccines – but still based on Koch's bile vaccine – in the last areas of northeastern Africa where it had managed to survive.

Eradication of tsetse fly in South Africa

Much has been made of the role played in the eradication of tsetse fly by the rinderpest epizootic. Both Fuller and Selous believed the influence of rinderpest to have been overstated, at least in the central and western Limpopo valley. Fuller acknowledged that in the eastern part of the Transvaal the rinderpest was responsible for the final eradication of tsetse, but was convinced that in most of the western part of the province the fly had already gone by 1896. E.J. Carruthers,[35] in looking at the history of game protection in the Transvaal, also concluded that the fly was already receding from the region by the time the rinderpest arrived. Carruthers also made the important point that up until around 1860, the prevailing view among hunters and settlers was that there was an inexhaustible supply of game and that there was no necessity for game conservation. Written accounts of these hunters and adventurers – many of whom were obviously poor shots – clearly reflects that there was little or no concern about the game being hunted, with innumerable wounded specimens simply being abandoned to their fate. Trade with the indigenous black population also included the (illegal) sale of firearms, with the result that black communities who had long been "market hunters" as well as subsistence hunters, also increased the rate at which they killed off game close to where they lived.

Many conservationists have remarked that it is thanks to the prolonged prevalence of tsetse fly, horse sickness and malaria in the northeastern Lowveld of the Transvaal and in far northern KwaZulu-Natal (KZN), that these regions were largely spared the successive predations of black communities, itinerant hunter/adventurers and later white settler farmers, until they were eventually formally incorporated into national parks like Kruger, Ndumo, Mkusi, Umfolozi and Hluhluwe. In the latter areas of northern KZN, tsetse – and therefore nagana – persisted into the 1950s before a combination of chemical and physical methods (essentially the obliteration of hundreds of thousands of head of "host" game species) eliminated the fly from that region too.

The future

The records of the last 180 years show very clearly that the distribution of tsetse fly over the southern part of the African continent has been far from constant. Even before the decimation of game by adventurer-hunters began in earnest, the boundaries of fly infestation were fluid, varying with seasons and possibly other factors, including the migratory behaviour of preferred host game, especially buffalo. There is also plenty of evidence to demonstrate that if management or control programmes are suspended, the fly readily returns to areas where it had previously been endemic, but from which it had been evicted.

In the western part of the Limpopo valley, changing economics have resulted in large tracts of land settled, cleared and developed for conventional crop agriculture in the late nineteenth and early twentieth centuries reverting to bush for the purpose of ecotourism and/or hunting operations. This process is being accelerated by the tendency of landowners to limit the extent and frequency of veld fires which is the principal natural method of bush clearance. Game densities on these properties are also increasing again, especially of high-value species like buffalo – the tsetse's most preferred host. Indeed, it might be argued that game densities today are higher than at any time in the past. Therefore, many of the conditions that formerly rendered the Limpopo valley and Tuli Block of Botswana appropriate for tsetse infestation are being regenerated. In addition, it is almost certainly true that the relevant government departments in the countries concerned no longer have the same institutional capacity or budgets to mount control campaigns as effectively or promptly as in the past.

The risk of reinfestation of tsetse fly in suitable areas of the Lowveld – including Kruger National Park (KNP) – is aggravated by the otherwise visionary plans to create multinational or transfrontier game parks. For example the Great Limpopo Transfrontier Park (which incorporates KNP, Gone Re Zhou and the Mozambique Limpopo Park) and the Limpopo-Shashe Transfrontier Conservation Area (which includes Mapungubwe National Park in South Africa, the Northern Tuli Game Reserve in Botswana and the Tuli Circle and other properties in southern Zimbabwe) are two very large areas, the management of which falls under several different national jurisdictions of varying competence and capacity. These parks carry substantial risk that they might unwittingly act as "corridors" for the migration of infected tsetse-bearing game into areas from which the diseases they carry have long been removed.

The scourge of malaria

> The fever that has been so prevalent this year in the Waterberg district is that known as Intermittent Fever or Ague, which is due to Malaria – a term applied to emanations or invisible effluvia from the surface of the earth, chiefly found in marshy lands, and supposed to be produced by decaying animal or vegetable matter, particularly both … [T]he district of Waterberg is a typical fever country … [T]he site of the town of Nylstroom is particularly unhealthy … out of five families resident here, which is the entire population, not one escaped the fever.[36]

This report about malaria in Nylstroom during a particularly bad outbreak in 1879–1880 described a disease that was endemic in the Nyl valley in those days. Many local residents were "saturated" with the disease according to doctors who attended to them during the South African War. The relationship between malaria and a parasite carried by mosquitoes was not discovered until the end of the century and the fever was still thought to be caused by noxious gases emitted by swamps – hence the name, from the Italian for "bad air". So bad was the outbreak that had affected the village of Nylstroom between 1878 and 1880 that serious thought was given to relocating it to a healthier site. The epidemic had brought to an end the practice of holding a once-monthly sermon-reading in the church hall for the benefit of residents. The *dominee*, Ds. Bosman, also had to announce that the twice-yearly nagmaal services would have to be suspended. The local school, built in 1879, was also badly affected by the fever and there were numerous fatalities among the population.[37]

Malaria had of course long been associated with the coastal regions of Natal and Mozambique, as well as with the Lowveld of the eastern Transvaal and with the Limpopo valley. The death of almost half the members of the Louis Tregardt (Trichardt) trek, including its leader in 1837–1838, is perhaps the first well-documented account of its depredations. Many of the mid-nineteenth-century travellers into the Limpopo valley fell victim to malaria, although most survived. The problem, as many commentators noted, was that because the disease was so widespread in tropical and subtropical environments around the world with its cause attributed to bad air emanating from swampy terrain, about which little could be done, attempts to prevent or treat it were few.

C. Louis Leipoldt, the talented author, poet, natural historian and highly qualified medical doctor, served as the Transvaal's first Medical Inspector of Schools from 1914 until 1923. He referred to malaria as "the curse of the Bushveld". Because it was more often debilitating than fatal, communities and health authorities tolerated the condition almost as though it was an occupational hazard, failing to appreciate the long-term negative effects it had on the wellbeing of those infected, especially children.[38] Leipoldt considered the combination of malaria and bilharzia to have had a devastating deleterious effect on generations of Lowveld children, stunting their growth, sapping their energy and numbing their intellects.

The recognition that malaria was an infectious disease transmitted by a particular genus of mosquito, *Anopheles* sp., when infected with a particular parasite of the genus *Plasmodium*, developed in stages during the last two decades of the nineteenth century. Two doctors, Charles Laveran (a French doctor working in Algeria) and Sir Ronald Ross (in Calcutta), received Nobel Prizes for their work that demonstrated that four or five species of *Plasmodium*, especially *P. falciparum* and *P. vivax*, required human and mosquito hosts to complete their life cycle, and that the disease was only transmitted by (predominantly female) *Anopheles* mosquitoes. A bite from an infected *Anopheles* injects the parasites into the blood stream and they travel to the liver where they grow and reproduce, causing the development of a fever in the host. Rather like the tsetse fly, another *Anopheles* mosquito may become infected by drawing blood from an infected human host, thereby repeating the life cycle and spreading the disease.

In the seventeenth century Jesuits working in the Peruvian Andes collected seeds of a local tree *Cinchona pubescens*, the bark of which was used by indigenous people to treat fever.[39] The Jesuits took the seeds back to Europe – where malaria was endemic across much of the Mediterranean region – and successfully cultivated the tree. A preparation of the bark became widely recognised as an effective treatment for malaria fever, although its active ingredient, quinine, was not isolated until 1820.

Early treatment for malaria

The first documented use of quinine for the treatment of malaria specifically was reported to the Imperial Malaria Conference held in Simla, India, in 1909 following successful trials on inmates of the Delhi jail in 1887.[40] In the interim, the medicine had been used informally, especially by the British Army. Quinine remained the treatment of choice well into the next century before it was replaced first by chloroquine and, more recently, by drugs of the artemisinin family. Artemisinin is also an ancient drug: derived from the herb *Artemisia annua*, it was known to Chinese herbalists over two thousand years ago. Dr Tu Youyou, the Chinese doctor who demonstrated the efficacy of artemisinin, was the third malaria researcher to win a Nobel Prize in 2015. However, the use of quinine was not without problems: A minority of patients suffering from *P. falciparum* malaria on treatment with quinine or other similar drugs were found to develop a serious reaction known as blackwater fever. It seems that under certain circumstances, there is an interaction between the malarial parasite and ingested quinine. This causes acute renal failure and the bursting of red blood corpuscles, leading to the release of haemoglobin into the urine, staining it dark red or black. The mechanism is still not well-defined, but since the evolution of malaria treatment from quinine-based drugs to those containing artemisinin or its derivatives, the incidence of blackwater fever has fallen dramatically.[41]

By the time of the South African War in which many participants suffered from the disease, the (informal) use of quinine had become widespread, although it was expensive and often in short supply. Malaria was a common condition in some of the British concentration camps, aggravated by the fact that many of the inmates who arrived at camps were already infected with the disease.[42]

In the "Middleveld" (i.e. 800–1100 metres amsl) that characterises the Springbok Flats region, Nylstroom and the areas to the west and north of the Waterberg plateau, malaria was soon recognised as being a disease of an epidemic (periodic) nature, as opposed to the eastern Lowveld and northern coastal regions of Natal where it was endemic (regular, or always present). Inhabitants of endemic malarial areas seemed, over time to develop some level of resistance to the disease, (so-called "acquired immunity") although this was accompanied by a degree of torpidity – as evidenced by Leipoldt's observations of drowsy, unresponsive Lowveld schoolchildren. Its periodic outbreaks were triggered by a combination of climatic factors – principally high rainfall and warm temperatures – with altitude also being a qualitative determinant. As a rule of thumb, those areas of the northern Transvaal at an elevation above 1100–1200 metres amsl seem unable to support sustainable breeding colonies of *Anopheles* mosquito, the essential vector for the transmission of the *Plasmodium* parasite. (Altitude also affects the species of parasite that predominates, with *P. vivax* found at higher altitudes than *P. falciparum*, which is more prevalent in the Lowveld.[43])

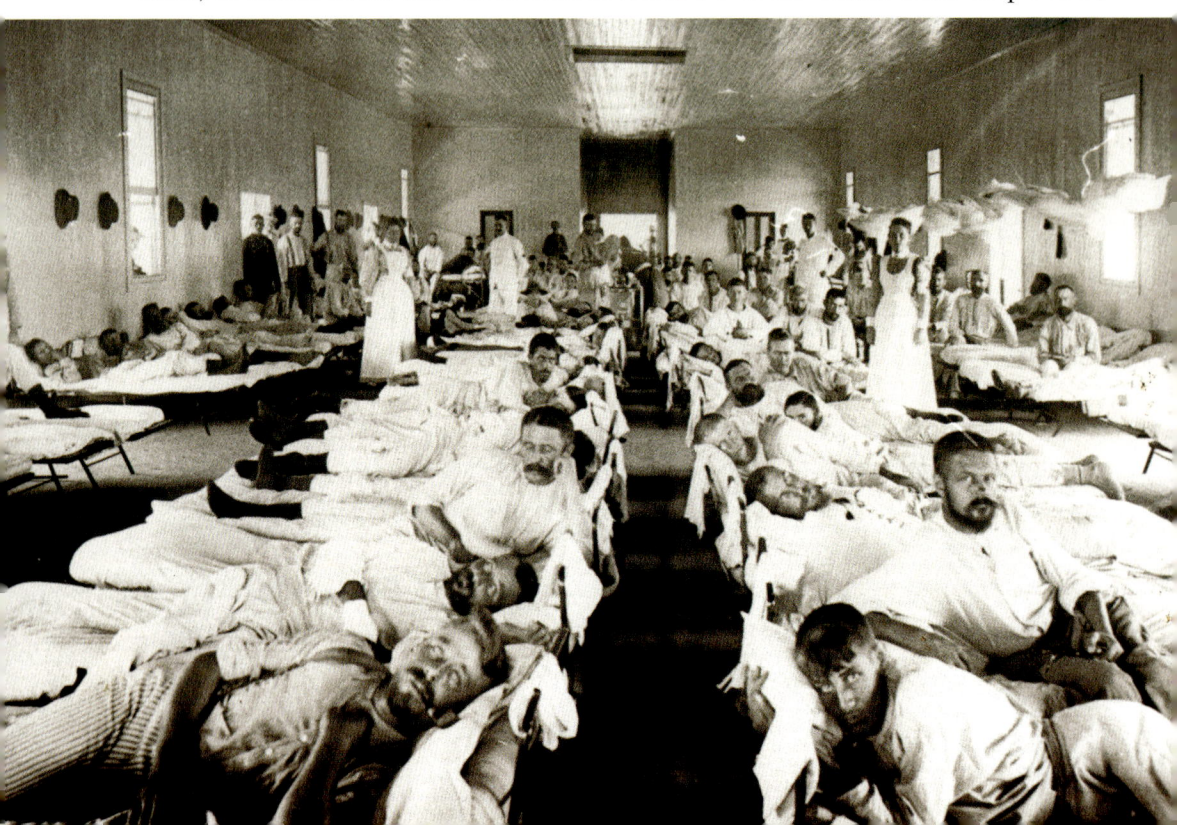

In 1909 there was an outbreak of malaria in Pretoria amounting to over 200 cases, of which about half were considered to be "local" rather than "imported", prompting severe criticism of the government's uncaring attitude towards the disease.[44] This also led to the realisation that infected people and infected mosquitoes might be "imported" into nonmalarial areas, with trains being identified as an important agent. The result was the implementation of a rigorous programme to spray and disinfect trains travelling to the Highveld from Mozambique and Rhodesia. Generations of subsequent passengers (including the author) on trains from these destinations would come to associate train travel with the pervasive odour of Jeyes Fluid!

Malarial research bears fruit

In April and May 1920, Dr K.J. Dekema, the respected District Surgeon in Nylstroom, reported a serious outbreak of malaria on the Springbok Flats. After visiting about 300 white families and over 2000 "native" households and distributing over 150 000 quinine tablets, he'd not found a single family that was free of the disease.[45] In August 1921, Col Deneys Reitz, Minister of Lands in Smuts's cabinet, paid a visit to the northern Transvaal to assess the incidence of malaria in his Department's land settlement schemes.[46]

In May 1923 Dr Dekema reported to his superiors in Pretoria that in the last few weeks 80 per cent of the white population of the region and at least a third of the black population in just one location had suffered from the disease. He listed the names of 45 whites who had died as a result, despite his liberal distribution of quinine, and predicted that more would succumb before the end of the season.[47] Three years later, Dekema reported that "malaria is rampant along the Zand [Mokolo], Crocodile and Matlabas Rivers as well as in the pan veld between them. The rivers, with the exception of the Zand, are dried up, leaving huge ponds [which are] swarming with mosquitoes by the million." He had recommended the use of mosquito netting for the protection of homes, but most families were too poor to afford it. He also advised people to clear vegetation from around their dwellings. His attempts to inoculate people with quinine had been resisted; they maintained that it had no effect, but the real problem, he said, was that they were not compliant.[48]

FIGURE 12.6: Boer prisoners of war being treated for malaria in a Lourenço Marques hospital. SANA TAB Photographic Library, No. 12936.

Botha de Meillon and Siegfried Annecke

In response to these reports, the SA Institute for Medical Research (SAIMR) despatched a young entomology research officer, Botha de Meillon, to the Waterberg in August 1926 to undertake a survey of the prevalence of malaria in the region. This led to the publication the following year of a seminal report by De Meillon and Dr Alex Ingram (his superior) on the incidence of malaria in South Africa, and to the launch of De Meillon's long and illustrious career in combatting malaria in the country.[49] The report described sites within the Waterberg where species of *Anopheles* mosquitoes were collected, including the Pongola (Mokolo) River at Vaalwater and the Sterkstroom a little further downstream. The largest population by far was found on the farm Sterkfontein below the present-day Mokolo Dam where the Rietspruit River emerges from marshy ground to flow into the Mokolo.[50]

In 1931 De Meillon succeeded Ingram as head of the Department of Entomology at the SAIMR – a position he would hold for the next 29 years. He was the first to identify the powerful insecticidal properties of pyrethrum. Having realised that female vector mosquitoes tended to rest indoors for several days after having had a "blood meal" before they left to lay their eggs, he advocated indoor spraying of walls with pyrethroid pesticide ("indoor residual spraying" or IRS) as a means of limiting the transmission of malaria. This had immediate positive results on the incidence of malaria among communities where the spraying programme was implemented. With the development of dichlorodiphenyltrichloroethane (DDT), a residual pesticide, towards the end of World War II and its increasing availability thereafter, government-funded programmes of residual spraying of homes in malarial areas with DDT became widely adopted, in addition to programmes aimed at spraying oils on larval (mosquito breeding) sites along streams.[51]

In 1932 a national malaria control centre was established at Tzaneen with De Meillon responsible for research and his colleague Siegfried Annecke in charge of control programmes. The efficacy of these programmes continued to depend on the cooperation of farming and other rural communities. In 1935 E.A. Davidson of the North Waterberg Farmers' Association proposed to the Minister of the Interior, "that the northern portion of the [Waterberg] district – or the whole district – be included as a malarial district and so receive the [same] benefits as regards free malarial attendance as that granted to Potgietersrus".[52] After a devastating malaria epidemic in 1943, which resulted in so many casualties that there was insufficient labour to harvest the crops, in turn taking farmers to the brink of bankruptcy, a meeting was called by the Farmers' Union on the Springbok Flats to discuss ways in which to prevent a recurrence. Under Annecke's management, a massive programme of knockdown indoor spraying with pyrethroids and oil-based larvicides outdoors was implemented in 1944.[53]

Meanwhile De Meillon published a detailed description of all the species of anopheline mosquitoes present in southern Africa. This differentiation enabled subsequent researchers to plot the distribution of the various species of malarial vector mosquitoes across the continent. By the turn of the twenty-first century of almost 500 *Anopheles* species that been identified worldwide, around 70 were considered to be competent vectors for human malaria and 13 were categorised as dominant vector species. Of these, the most important in the South African context are the *A. gambiae* complex (comprising five subspecies) and *A. funestus*.[54]

Schalk Jacobus Botha de Meillon was born in Prieska, Northern Cape, in 1902 and died in the USA in December 2000 at the age of 98. The career of this remarkable South African scientist spanned 70 years from his graduation with a BSc (Hons) degree from the University of the Witwatersrand in 1925 until 1996. Unable to find a teaching position, he joined the SAIMR in 1925, obtained his MSc the following year and in 1934 his DSc (both from Wits) "for his outstanding work on the taxonomy of the anopheline mosquitoes of South Africa". He headed the Department of Entomology at the SAIMR from 1930 until his retirement in 1962, after which he worked for the World Health Organisation (WHO) in Brazzaville (Congo), Burma and latterly in the US where he headed several key government and privately funded mosquito and malaria research projects.[55]

In 1927 De Meillon and his then boss, Dr Alex Ingram, discovered that a preparation of crushed pyrethrum flowers in kerosene made a toxic poison for mosquitoes and flies. This development, added to his discovery that female mosquitoes rested indoors after feeding but

FIGURE 12.7: Botha de Meillon (80) in 1982. Photograph kindly supplied by Prof. Maureen Coetzee; as published in Coetzee et al (2013), p. 771.

before laying eggs, meant that the life cycle of the malaria vector mosquito could be interrupted through indoor spraying. This simple, practical breakthrough led to IRS of insecticide becoming a universal malaria control method, first in South Africa and later throughout the world, with the lives of millions saved as a result. It is rather surprising that De Meillon was not nominated for the Nobel Prize for this work – although he received numerous local and international awards in recognition of his contribution to medicine and entomology. He also worked extensively on describing the taxonomy of southern African anopheline mosquitoes as well as that of several other insects of medical importance.

Botha de Meillon was an outstanding scientist who was instrumental in saving the lives of many, many of his compatriots, who brought great credit to his country, yet who was hardly known by people outside his discipline.

In 1938 the Department of Public Health published a map showing the different categories of malaria risk in the country.[56] This indicated that the Highveld as far south as Pretoria (as well as the Natal midlands as far south as Durban) carried only a minor risk of experiencing occasional epidemics of malaria. The Middle-

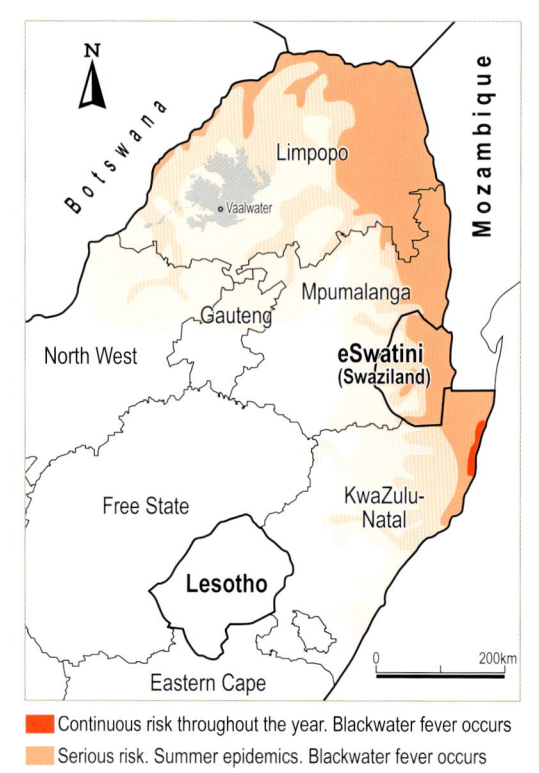

■ Continuous risk throughout the year. Blackwater fever occurs
■ Serious risk. Summer epidemics. Blackwater fever occurs
■ Moderate risk. Milder summer epidemics. Blackwater fever NIL
■ Slight risk. Only occasional epidemics

FIGURE 12.8: Distribution of malaria risk in northern South Africa, 1938 (Source: Department of Public Health).

veld areas surrounding the Waterberg plateau and along the eastern margin of the Highveld were of moderate risk, with the possibility of mild summer epidemics. The Limpopo valley as well as the whole of the eastern Transvaal below the Drakensberg escarpment, eastern Swaziland and northern Natal, were at serious risk of experiencing summer epidemics, including blackwater fever. The only area of South Africa where malaria and blackwater fever could be expected to occur throughout the year was in a small region along the far northern KZN coast. Of interest in the context of the Waterberg is the finger of high malarial risk that is depicted extending south from the Limpopo River along the Mogalakwena River valley as far as the northern edge of the Waterberg plateau (Figure 12.8).

Even as recently as 1948, the Senior Malaria Control Officer in the Department of Health, based in Tzaneen, had this to say about the incidence of malaria in the Waterberg:

> Malaria does occur in the Nylstroom, Vaalwater and Warmbaths districts. The incidence increases westwards towards the Bushveld and eventually to the Limpopo river, where [it] is very high. For the past three years, malaria was controlled in these parts. Very few cases occurred amongst residents in the bushveld, where the Limpopo and its numerous tributaries, [of which] the Palala, Mogol and Matlabas are the most important, form a permanent danger. Extensive control measures are adopted … and anybody prepared to do his share of domestic prophylaxis can live malaria-free in these parts. Knockdown insecticide, spray pumps, quinine and malaria oil are available at all malaria depots, of which a large number exists in the areas under reference.[57]

Malaria distribution since World War II

The concerted efforts made by the authorities to control and even eliminate malarial risk in South Africa were hugely successful over the period following the end of World War II and the areas remaining at risk contracted significantly. In the Waterberg periodic systematic larval and adult spraying campaigns reduced the incidence of malaria significantly – although "imported" cases, now referred to as "taxi" or "Odyssean" malaria continued to occur as movements of people into and from malarial areas grew. This was aggravated by the migration of labour to the Witwatersrand gold mines, by the influx of cross-border refugees and by the decline in management of malaria in neighbouring countries like Mozambique (due to civil war) and Zimbabwe.

In 1989 a few malaria cases were reported from the Vaalwater area, probably from along the Mokolo River, but there was no recurrence.[58] During the 1980s synthetic pyrethroids were developed to counter the growing resistance of malarial vectors to pyrethrin and as an alternative to DDT. Following its widespread use as an agricultural pesticide, DDT was increasingly being discouraged owing to its long-term environmental consequences and its tendency to persist in the food chain. In 1995, in response to pressure from the World Health Organisation (WHO) and environmental lobby groups, the SA government abandoned DDT in favour of pyrethroids.[59] However, in the 1999–2000 season, there was an unexpected, severe outbreak of malaria in KwaZulu-Natal and eastern Mpumalanga, with over 64 000 cases documented. It was discovered that *A. funestus*, which had been eliminated in the 1970s, had developed resistance to pyrethroids and was staging a comeback in the region. The SA government sought and obtained approval from the WHO to resume spraying with DDT in this area. Within a couple of seasons *A. funestus* had been eliminated once again.[60]

Occasional cases of malaria are still reported on the Waterberg plateau. By far the majority are of the imported type, but there have been a few isolated cases documented of children living along the Mokolo River west of Vaalwater (which lies below 1100 metres amsl) contracting malaria, presumably from local sources. Such cases must by law be reported by medical practitioners to the malarial control office in Tzaneen, which then despatches teams to spray the location where the cases have occurred. However, these cases are rare; it is considered that the region no longer supports the sustainable breeding of vector mosquitoes that could pick up and transmit any parasites that might be imported. This allows the Waterberg plateau to declare legitimately that it is "malaria-free".

However, the same does not apply to lower lying parts of the region: in February 2017, for example, and again in March 2018, there was an outbreak of malaria along the lower Lephalala River downstream from Villa Nora. Other cases were reported from the Thabazimbi area. These are potential risk areas for the disease. Anti-malarial spraying teams stationed in the region seemed to have been caught off-guard, allowing the outbreaks to occur.

Globally, malaria is still one of the most serious diseases afflicting humankind. In 2015 over 200 million new cases were reported and over 40% of the world's population is at risk of contracting the disease. By far the majority (88%) of cases occur

Bilharzia or schistosomiasis

Schistosomiasis (or bilharzia, to use its lay term) was first discovered in Egypt in 1851 by the German physician Dr Theodor Bilharz. The first documented cases in South Africa by Dr J. Harley in 1863 were those found among boys who had been swimming in local rivers. In the 1920s Dr Louis Leipoldt (the first Transvaal Medical Inspector of Schools as mentioned earlier), was dismayed by the prevalence of the condition – which was known as "redwater" due to the discoloration of urine by blood – among children at the schools he visited in the northern bushveld. Most were boys, apparently because they were more likely to swim in rivers than girls.

> Most of the little patients had a characteristic appearance – a peculiarly anaemic, pallid complexion, with sunken eyes and widely dilated pupils. One could diagnose the condition from the smooth velvety skin, with a tendency to weal formation on the slightest pressure, and the intense fatigue it caused … So great was its incidence among the boys of the schools that it was almost impossible for the teacher to get any effective response from his pupils. They were far too fatigued to do school work. Frequent headaches, sickness, lassitude, cramps in the limbs, disturbances of vision and intense weakness were the main symptoms. [Because] everyone knew these were the concomitants of redwater, and that redwater was incurable, no great attention was paid to them.[61]

Waterberg school journals also reported cases of bilharzia through the years. For example, after a visit in August 1953 to the Laerskool Vaalwater by the school nursing sister, who reported numerous children affected with the disease, parents were sent the following notice (translated):

> To Parents … regarding the Bilharzia (Redwater) Danger: You are surely all aware of the fact that our rivers are infected with Bilharzia. I would like to bring the following to your attention:
>   That Redwater is a dangerous disease which in time, if not

treated promptly, can become just as serious as cancer, with just as much risk of causing death. Thus if you have any suspicions that you might be sickening with this disease, it is advisable that you contact your nearest doctor as soon as possible. You can contract the disease from swimming or walking in our rivers, even from washing your hands in, or drinking river water, or from fishing.

The water in our neighbourhood contains few of the minerals that are very important for your health. We recommend that iodised salt be used at our tables and in our food in place of ordinary salt. You can buy it at any chemist or from Mr Farrant.

With kind regards, your friend, S Hoffman, Headmaster, Vaalwater School.[62]

Bilharzia is caused by a group of blood flukes (or trematodes) called Schistosomatidae. The predominant species present in South Africa are *Schistosoma haematobium*, which infects the kidneys, bladder, urinary and female genital tracts, and *S. mansoni*, which is less prevalent and infects the intestines and rectal area. Their life cycle is dependent on two hosts: a principal human host and an intermediate,

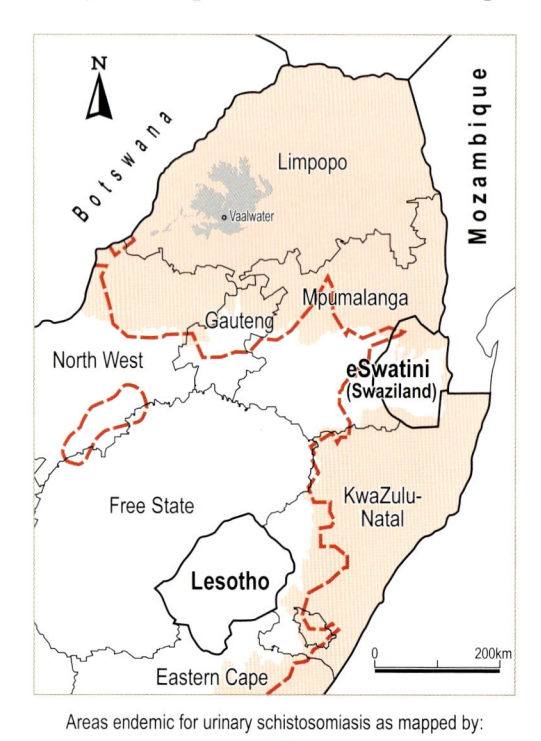

FIGURE 12.9: Distribution of bilharzia risk in South Africa. Map compiled from information included in papers by Gear et al (1980), Moodley et al (2003), Appleton & Miranda (2015) and Magaisa et al (2015).

shallow freshwater snail host. They occur in still or slow-moving water which is contaminated with human urinary or faecal waste and which is used by humans for washing, swimming or drinking. In Limpopo, three species of snail act as intermediate hosts for the parasites.[63]

Several studies over the years have used medical reports of bilharzia incidence, physical identification of vector snail species and the temperature tolerance of both snails and parasites to produce maps showing the distribution of the disease in South Africa. In 1938, 1980 and most recently in 2003, almost all of Limpopo Province (including the Waterberg), the northern part of North West Province as well as the Lowveld of Mpumalanga and KwaZulu-Natal have been shown to be regions where bilharzia is endemic.

The importance of bilharzia

In terms of its impact on global public health, bilharzia is the second most important tropical disease after malaria. Some 207 million people are infected with the disease globally, 85% of them in sub-Saharan Africa. An estimated 280 000 people die of the disease each year. In South Africa, as many as five million people either have the disease or are at risk of contracting it, with the highest prevalence (and the most intense infection) being amongst children of schoolgoing age.[64] Yet, despite its severity, bilharzia seems not to be regarded by the South African medical authorities as a serious public health issue.

The most attention given to it was in the 1990s when the new national government endorsed a World Health Organisation programme to address the disease. In accordance with WHO guidelines, a manual was drawn up to instruct local and primary health care authorities on how to identify, treat and monitor bilharzia patients, but the programme was never fully supported in practice or rolled out to relevant clinics.

Tick-borne diseases

> Ticks rank [below] mosquitoes as global vectors of human diseases, but are the most relevant vectors of disease-carrying pathogens in domestic and wild animals.[65]

It has been argued that the indigenous cattle which populated South Africa until the nineteenth century had long adapted to ticks and tick-borne diseases together with most of the other indigenous diseases and pests that characterised the subcontinent. The practice of dipping had not yet developed, yet most cattle survived. This changed dramatically when the rinderpest epizootic of 1896–1897 destroyed over three-quarters of the national herd. Breeds imported from Europe to rebuild the herds lacked the immunity of the indigenous species and were immediately susceptible to disease.[66] Although ticks had long been recognised as being deleterious to their hosts, their removal was haphazard and primitive: hand-picking, the application of insecticidal plant material or turpentine, burning with hot irons, etc. The first attempt to control ticks through dipping began in 1893 and accelerated after the end of the South African War and in the light of the discovery in the US in 1889 that redwater disease in cattle was transmitted via ticks.

There are almost 900 tick species recognised worldwide and about 10% of them occur in southern Africa. In the Waterberg alone, over 30 species of tick have been identified. With respect to domestic livestock, however, a few species are responsible for the majority of disease-carrying vectors. Rickettsial parasites cause gall sickness in cattle and are also the principal cause of tick-bite fever in humans; the bont tick is the predominant carrier (vector) for this parasite in the Waterberg region.[67] Heartwater, a disease caused by the rickettsial parasite *Ehrlichia*, also using the bont tick as its principal vector, is a serious, usually fatal, endemic southern African condition. It is estimated that 35% of the total cattle population (of 15 million), 54% of the six million goats and 12% of the country's 25 million sheep are at risk.[68] In addition to domestic stock, the adult tick makes use of giraffe, buffalo, eland, warthog and rhino as hosts. It is a three-host species and in its immature (larval and nymphal) stages, it can be hosted by scrub hares, tortoises and helmeted guinea fowl. The first historical reference to heartwater was probably made in 1838 by Louis Tregardt, who lost some of his sheep to a fatal tick-borne disease when trekking through what is now Limpopo Province. It is likely, though not certain, that heartwater is indigenous to Africa. The disease can be treated with tetracyclines; animals that survive it develop immunity. An effective vaccine against the disease was developed at Onderstepoort in the 1950s.[69] Blue ticks are the principal vectors for *Babesia*, the group of parasites responsible for redwater, an endemic cattle disease that places over a third of the national herd at risk in the northern part of South Africa. Redwater is more widespread than heartwater, because its vector ticks are more prevalent.

East coast fever (ECF) is another serious, often fatal disease in cattle. Although long recognised as a disease prevalent along the East African coast, it had not been recorded in South Africa prior to 1902 when it is thought to have been introduced from the Tanganyika Highveld via Mozambique by cattle imported from East Africa and staged at the coast before being brought inland.[70] Many early deaths in Rhodesia were attributed to a virulent strain of redwater disease until the vector tick was identified as the brown ear-tick, and the parasite was defined as *Theileria parva* in honour of Sir Arnold Theiler, first director of Onderstepoort and the person who recognised ECF as being an entirely new disease. By 1910 the movement and importation of cattle through and from coastal regions resulted in the rapid spread of the disease. It is estimated that between 1910 and 1914 almost a million head of cattle died from ECF in South Africa alone.[71] In the absence of any prophylactic or treatment medication at the time, the government's initial reaction was to impose stringent restrictions on cattle movements, the compulsory reporting of all deaths, the wholesale slaughter of infected herds and attempts to eliminate infected ticks on pasture land. In the Waterberg the severe restrictions on stock movement made it difficult for farmers to practise their traditional seasonal migration from summer to winter pastures. On 20 July 1910, for example, the new Union Prime Minister, General Louis Botha, wrote to the Governor General requesting that the wards of Zwagershoek and Palala in the Waterberg District be declared restricted areas because of the prevalence of ECF there.[72] To enforce the restrictions, special police constables were established or existing posts reinforced at access points including Tarentaalstraat, Geelhoutkop and Zandrivierspoort.

Buffalo living in Lowveld areas were soon recognised as important hosts of a related *T. parva* vector tick, the Lowveld brown ear-tick. An extreme form of ECF associated with these animals in northern Natal was named corridor disease, because it was first identified in buffalo traversing the corridor that used to exist between the Hluhluwe and Umfolozi Game Reserves. It is for this reason (together with foot-and-mouth disease) that the translocation of buffalo into other parts of the country, for example the Waterberg, is strictly controlled and that buffalo have to be certified as "disease-free" prior to being relocated.[73] The practice of strict quarantine and dipping was found to be effective against ECF, but only if undertaken with high frequency – at one point every three days, later, at seven- to ten-day intervals. The first dips in the period from the turn of the twentieth century up to World War II used arsenical compounds to kill ticks on stock, but towards the end of this period, resistance to arsenic was becoming apparent. Traces of arsenic were also being discovered in foods. At the end of the war, DDT and BHC (benzene hexachloride) were introduced with great effect, but within a decade they too had prompted resistance in the tick population. In addition, compliance amongst farmers – adherence to the specific dipping frequency advocated by manufacturers – was variable, leading to an increased likelihood of resistance to the pesticides developing. Moreover, the indiscriminate use of toxins, including home-brewed pesticides, to treat stock against ticks was having a serious adverse impact on the population of birds whose principal diet had been ticks on domestic and wild animals. This eventually led, for example, to the local extinction in the Waterberg of the red-billed oxpecker. (Fortunately, a change in formulation of cattle dips has enabled the reintroduction of the birds in the last decade.)

There is still no treatment for ECF, although the use of tetracycline may reduce the severity of the condition. Mortality among affected animals is above 90%. In South Africa the combination of intensive tick control, quarantine and slaughter of infected animals appears to have eradicated the disease.

African horse sickness

African horse sickness (AHS) is a viral condition caused by a member of the genus *Orbivirus* of which nine variants or serotypes are recognised, four of them present in South Africa. The disease was first recorded in the subcontinent in the seventeenth century. It is endemic to low-lying areas of the summer rainfall region of sub-Saharan Africa, is highly infectious but noncontagious, and its hosts are limited to the equids: horses, mules, donkeys and zebras. A major outbreak occurred in 1855 when an estimated 70 000 horses died – more than 40% of the total population of the Cape Colony.

AHS was well known to early settlers on the Waterberg plateau. In the Swagershoek area, farmers found that by keeping their horses on the top of a prominent nearby hill (appropriately named Perdekop) overnight, they could reduce the incidence of the disease – perhaps because this was cooler and windier than the surrounding terrain.

AHS is most prevalent in seasons when heavy rains are followed by warm, dry spells in summer rainfall areas, disappearing with the first frost, or a little later in frost-free areas. There have been many epidemics of AHS recorded in South African history, for example in the years 1780, 1801, 1839, 1855, 1862, 1891, 1914, 1918, 1923, 1940, 1946, 1953 and 1959. The latter epidemic resulted in the deaths of an estimated 300 000 equids across the continent. There has been some research to suggest that epidemics coincide with the development of the warm phase of the El Niño phenomenon.[74]

The main biological vector of AHS is a small midge, *Culicoides imicola*;[75] the disease is transmitted by bites from the midge. There are two types of AHS: a pulmonary type (known as "*dunkop*") and a cardiac type ("*dikkop*"). Horses are especially susceptible to the effects of the virus with mortality exceeding 90% in the case of dunkop and around 50% with dikkop.

The virus first appears in early February each year, peaking in March/April and disappearing in May/June. The *Culicoides* midge breeds and nests in damp soil rich in organic material or dung; it is most active between sunset and sunrise. There is no specific treatment for an animal that has contracted the virus other than rest, but a polyvalent vaccine (which caters for all nine serotypes) has been available since the 1930s. It is not always effective, but regular annual vaccinations appear to greatly reduce the risk of an animal succumbing to the disease.

Summary

Nagana and malaria, two major historical causes of sickness and death in stock and humans respectively, have now been almost entirely eliminated from the Waterberg region, although there are very occasional outbreaks (epidemics) of malaria in lower-lying regions. Bilharzia remains endemic and although modern drugs can cure it quickly, their cost means that many poor members of communities continue to live with its debilitating effects, mainly because the SA government, unlike its neighbours, has not yet adopted readily available cheap generics. Tick-borne diseases are still a major concern for domestic cattle and their owners, although regular spraying keeps these diseases under control. AHS is also prevalent, with limited vaccination efficacy.

The reforestation of the Waterberg – a consequence of fire control – and the enormous increase in the population of wild game, including important disease vectors, is recreating conditions appropriate for the resurgence of both the tsetse fly and the *Anopheles* mosquito. Increased vigilance and prophylactic action on the part of health authorities will be essential if the region is to remain free of these pests.

CHAPTER THIRTEEN
Fitting the land to its purpose

The Waterberg plateau today is most closely associated with nature reserves, game farms, hunting outfits, breeders of rare and exotic animal species and ecotourism. But this is a recent phenomenon. Up until the 1980s, the region was characterised almost exclusively by commercial agriculture, a combination of beef and stud cattle breeding and the cultivation of a wide variety of crops, ranging from groundnuts through maize and tobacco to grapes, deciduous and citrus fruit, and tree nuts. In the last 30 years, though, conventional farming of domestic livestock or crops, while still actively practised, has become a subordinate land use in all but a few areas. This transformation has had profound demographic, cultural, economic and environmental consequences for the region.

The soils of the Waterberg plateau, being derived mainly from the erosion and weathering of a thick sequence of interlayered conglomerate, sandstone and locally minor siltstone, are generally coarse, impoverished in clay content, acidic and leached – so-called *dystrophic* soils.[1] Grasslands are dominated by species that are relatively less palatable, especially in winter, resulting in most of the plateau being characterised as sourveld.[2] Only in the limited areas where the sandstone sequence has been intruded by younger igneous rocks like diabase/dolerite sills and dykes (where older underlying volcanic rocks are exposed or where the uppermost, silty strata are preserved) has a more clay-rich, alkaline and nutrient-enriched soil profile been developed, with associated sweet grasses. Most of the plateau is at an altitude of greater than 1100 metres amsl (its highest point, Aasvoëlkop, on Kransberg is at 2088 metres) and receives an average of around 600 mm of rainfall per annum. The southern escarpment can receive in excess of 900 mm, but the lower lying areas towards the north receive an average of below 500 mm. In common with other summer rainfall areas in South Africa, the Waterberg is subject to periodic crippling droughts. In the last century, the worst of these occurred in 1904–1906, 1932–1934, 1962–1964 and 1991–1993.[3] The drought of 2015–2017, which continued into 2018, ranked as severe as that of the 1930s. Agricultural records are replete with references to these drought periods and their consequences for a farming community that has struggled to prosper in the best of times.

The overall vegetation across the plateau is classified predominantly as Waterberg Mountain Bushveld. This is defined by mountainous or rocky terrain, generally thin, gravelly soils, populated by *Faurea saligna – Protea caffra – Burkea africana – Terminalia sericea – Diplorhynchus condylocarpon* tree vegetation with moderate to well-developed grass cover. Some parts, including the Bulgerivier–Vaalwater region and the Alma valley from Rankin's Pass all the way along the Sterk River to Doorndraai Dam, is classified as Central Sandy Bushveld. Altitude here is generally below 1100 metres, soils are deep, sandy and neutral to slightly alkaline; and vegetation is typically dominated by *Burkea africana – Terminalia sericea – Combretum* spp. – *Acacia* spp. – *Ziziphus* spp., and a grass-dominated herbaceous layer.[4] This vegetation unit with its associated better soils is inherently more amenable to sustainable crop-based agriculture than the Waterberg Mountain Bushveld, and consequently, has not been as affected by the transformation to game-based land use that is evident on the plateau.

The point to be drawn from the descriptions above is that the Waterberg plateau as a whole has always been, and remains, of marginal potential for conventional agriculture. The soils are coarse and poor in nutrient and moisture content, limiting stock-carrying capacity and requiring the addition of relatively large quantities of fertiliser for crop farming to succeed. While numerous streams draining the hills often yield adequate perennial supplies of (rather acidic) surface water for domestic use and small-scale irrigation operations, the underlying rock formations typically host small, "fracture-fill" aquifers and offer very limited opportunities for underground water storage. Periodic severe droughts have proved to be the final straw for many farmers. The region is notorious for its density and diversity of ticks (see Chapter 12) and susceptibility, when overgrazed, to invasion by plant species alien to the local ecology (such as the sinisterly named "bankrupt bush", *Seriphium plumosum* [= *Stoebe vulgaris*]).

Early settlement

The pre-white settler inhabitation of the Waterberg plateau appears from the archaeological record to have been sparse and intermittent as discussed in Chapter 2. Crop farming by pastoralists in the fifteenth to nineteenth centuries, such as it was, seems to have been concentrated on vleis or alluvial terraces in proximity to perennial water. Stock farming was limited to areas of sweet veld – with consequent denudation due to severe overgrazing, followed by the invasion of species like sickle bush (*Dichrostachys cinerea*).

Early white settler agricultural practices, not surprisingly, were little different. The first trekker-settlers who arrived in the area in the early 1840s either as part of, or in the wake of the principal Voortrekker groups, were (unlike many Voortrekker leaders) ill-educated, invariably lacking in resources and dependent on themselves and their families for labour. Essentially peasant farmers, their focus was on subsistence: They were too remote from a commercial market to have an outlet for crop or stock production and their only sources of cash were in the form of ivory, hides and possibly, dried meat (*biltong*), derived from their hunting activities; and, in the west, from the collection of salt.

This was a hard, unforgiving, largely ungoverned and sparsely populated region – but it was well suited to the tough, simple and fiercely independent people who first settled here, and for whom solitude, their religion and freedom from authority were paramount. They were described as difficult, untrustworthy and indolent by many who encountered them. According to the acclaimed historian Hermann Giliomee, in 1858 the Rev. Van der Hoff, who founded the Hervormde Kerk, described the Transvaal burghers as "not very energetic, and prefer everything to be done by the Kaffers". Where black labour was available, the trekkers were inclined to turn responsibility for farming over to their black employees "and adapted to their [black] customary agricultural methods rather than imposing their own".[5]

Until the mid-1850s the land occupied by white settlers, by their Ndebele predecessors (whose invasions of the region had occurred in several waves over the previous century) and by pre-existing Tswana/Sotho communities, was unmapped, with only vague and ephemeral territorial demarcations. The arriving Voortrekkers had deemed that every burgher should be entitled to ownership of land. Initially, this was to take the form of up to two freehold farms and one town plot; but as land was occupied, the allocation of farmland was reduced to one per burgher.

The process by which this would happen was that a burgher, having decided on a piece of land unoccupied by someone else, would submit a description of the location to the local Landdrost. In due course, a senior official, usually a field cornet, would be assigned the job of inspecting a group of such land claims to demarcate boundaries and resolve any disputes.[6] Once this was done, a farm could be granted as freehold to the burgher concerned, upon payment of the costs involved in the inspection and drawing up of the title deeds. The usual way in which such a property was defined was by the "*een uur gaans overkruis*" or "one hour duly crossed" principle – the distance covered by a man on horseback in half an hour from the centre point of a claimed piece of land to each of the four principal points of the compass – in other words, forming a square of land of about an hour's ride on each side. This was considered to yield a block of land of approximately 3000 morgen (subsequently revised to 3600 morgen).[7]

The original Deeds Register entries show that many Waterberg farms were first inspected and described mainly by or under the supervision of Jan Antonie Smit, a field cornet in the ZAR administration. The process was often not only poor in quality and implementation, but lent itself to abuse. L. Braun, who made a study of early survey practices in the Zuid-Afrikaansche Republiek (ZAR) commented that:

> Prime examples of truly shoddy inspections and the ones frequently invoked by those calling for the reform of inspection and survey regulations were those carried out between 1869 and 1873 in the district of Waterberg [ZAR]. In one particularly egregious example, Veldcornet JA Smit tendered reports for 577 plaatsen ostensibly inspected between 4 October and 4 November 1869, a pace of over 18 farm inspections per day, seven days a week.[8]

Clearly, this was impossible to have achieved, given that each inspection should have taken several hours to complete. But the fee of £1.10/-. payable per inspection would have been a powerful inducement to overstate the number of inspections completed. Braun moreover made the point that the resulting cartographic output of these *inspecties* may have been of value from an administrative and taxation perspective, but was totally worthless as a source of mapping information. Smit, who had been born in 1830, came from the newly formed village of Nylstroom where he died in June 1910. He took advantage of his familiarity with the inspection process to purchase, or otherwise acquire, several properties for himself – a common practice, it seems, among surveyors of the period. The Surveyor General of the ZAR from 1866, Magnus Forssman, was often paid by his employer, the ZAR government, in land, because the government was perpetually short of cash. His brother, O.W.A. Forssman (a Potchefstroom-based businessman), by his own admission at one time owned 118 properties in the Transvaal Republic, including 58 in the Waterberg[9] (see Chapter 5). A later holder of the position of

Surveyor General, Johann Rissik, touted several farms recommended by his surveyors to the Transvaal Consolidated Land & Exploration Company Limited (TCL), of which he was a director (and after 1902, chairman).

In 1858 the Volksraad decreed that all land not already allocated as freehold would become state land. The following year, it advised that each burgher, whether or not already the owner of a freehold property, could lay claim to one or more *leningsplaatsen* or quitrent farms from this state ground upon the payment of a tax or quitrent fee of 30 shillings annually.[10]

Farms formally demarcated

The first Waterberg farms were "inspected" and title deeds issued from the late 1850s, although most were only demarcated in the following decade as quitrent farms. By 1870 most of the land across the Waterberg plateau had been divided up into inspected farms – but in several cases the properties had already reverted to

FIGURE 13.1: The original ledgers recording Waterberg title deeds from 1865. SANA TAB RAK 3058-3060: The first three ledgers form Series 1, including all farms registered up until about 1897, when the books were full. A second series commenced (RAK 3061-3073), in which the records in these first ledgers were re-entered and augmented by subsequent deeds issues up until about 1949.

the state due to nonpayment of fees. A notable exception was the area now occupied by the Welgevonden Private Nature Reserve and the Marakele National Park. Gustav Troye's map of the Transvaal, published in 1892, shows the western limit of demarcated and named farms on the Waterberg plateau as being the Pongola or Zand (now Mokolo) River; west of Sandrivierspoort, no farms had been laid out to the north of the Zandriviersberg range. The Welgevonden/Marakele region was depicted simply as "hilly country, very rugged" with no detail apart from the Malmanie, Sterkstroom and Taaibosch rivers.[11] However, this soon changed: Jeppe's last Transvaal map, produced posthumously in1899, shows all the farms of today complete with their numbers, although some are still outlined rather tentatively.[12] As has been described, the initial inspections were rudimentary and of little or no use on the ground. From 1888, the ZAR government, having long recognised this shortcoming, was able with the benefit of income from the recently discovered Witwatersrand goldfields to implement legislation that would result in more stringent, quantitative surveys of all the farms in the Transvaal.

Surveyors or inspectors employed by the ZAR from the 1860s to lay out farms throughout the northern part of the republic seem to have had the freedom to append names of their choice to the properties they defined. Many were mundane, unimaginative and repetitive – there are, for example 147 Rietfonteins listed in the old farms list of the former Transvaal; another 270 start with Klip (-bank, -drift, -fontein, -pan, etc.); 213 begin with Doorn (-bult, -fontein, -hoek, -kop, -plaat, -spruit etc); 125 farms are prefixed with Leeuw-; and 95 with Buffels-. There are also any number of Nooitgedachts, Roodepoorts, Sterkfonteins, Uitkyks, Vlakfonteins, Welgevondens and Weltevredens. Other surveyors were more imaginative and assigned names that evoked distant homelands, for example (London, New York, Hanover, Berlin, Amsterdam, Melbourne), or romantic memories (Kiss-me-quick-and-go-my-honey, Liefde-en-Vrede, Landmans Lust, Mooimeisjesfontein), or less happy emotions (Last Hope, Eenzaamheid, Deadbeat, Ongelukskraal, Verloren).[13] Contrary to popular belief, all these farms had been named at the time of their registration in the 1860s and 70s and owed their origins to events or fantasies far older than either of the two Boer freedom wars.

Burghers lose their land

For many poor burghers in remote and marginal areas like the Waterberg, the ownership of freehold land did not enable them to carry on a sustainable farming activity, especially if their principal source of income had been hunting, which required greater mobility than that offered by a fixed piece of land. In such cases, freehold titles were readily and extensively traded for cash, with land companies or some wealthy individuals, often government officials, snapping up farms for very low prices and accumulating vast tracts of land or, like the Forssmans, numerous farms across the Transvaal. Similarly, many burghers who took advantage of the later quitrent programme also found themselves unable to pay the annual fee or tax on these properties, which consequently also re-entered the market. Moreover, by the 1880s most of the game species preferred for hunting were becoming greatly reduced in numbers, or even extinct on the Waterberg plateau, as a result of extensive and unregulated hunting, forcing those whose livelihood depended on hunting to move further and further north across the Limpopo.

One of the consequences of this trend was a growing population of "landless" burghers who ended up working as employees of landowners, or as increasingly impoverished tenant farmers (bywoners) on land owned by others – or as job-seeking migrants to the developing Witwatersrand gold mining industry. By the end of the century, half the burgher population of the Transvaal Republic had become landless.[14] The rinderpest epidemic of 1896–1897, coupled with a drought in the latter year, resulted in the deaths of more than half of all the cattle in the Transvaal (and in the process, a severe disruption in oxen-drawn wagon transport to rural areas like the Waterberg), leaving many burghers destitute. In the aftermath of the devastation caused by the scorched-earth policy of the British towards the end of the South African War, followed by the privations of World War I and the Great Depression, these displaced people and their descendants contributed to the "poor white" population which grew to alarming proportions in the first part of the last century.

Among the buyers of freehold and quitrent farms from financially distressed burghers, particularly in the decade after the discovery of gold in 1886, were numerous, mainly London-registered speculative land (and exploration) companies, most of them established with this specific objective in mind. In 1897 the Transvaal Registrar of Deeds reported that of 11 045 surveyed and inspected farms in the ZAR, just half were owned by the original burghers, 22% by the government and 1087 (10%) by foreign companies whose head offices were in England, Germany or France.[15] By 1899 over 60% of the farms in the Waterberg were owned by a handful of such companies, including New Belgium Estates, Transvaal Consolidated Lands (TCL), Oceana and Anglo-French.[16] In some instances, the former owners were hired as managers or tenants on the corporate farms, but most were administered by managers imported from England, like Peacock and Madge for TCL (see Chapter 7) and Caine on New Belgium (see Chapter 9).

Few, if any, of these properties were successful agricultural developments; even fewer were found to have any significant economic mineral potential – the prospect of which had been the principal reason for their acquisition. After the end of the South African War, most of the farms were fed gradually back into the property market for sale, a process that was to continue into the 1960s.

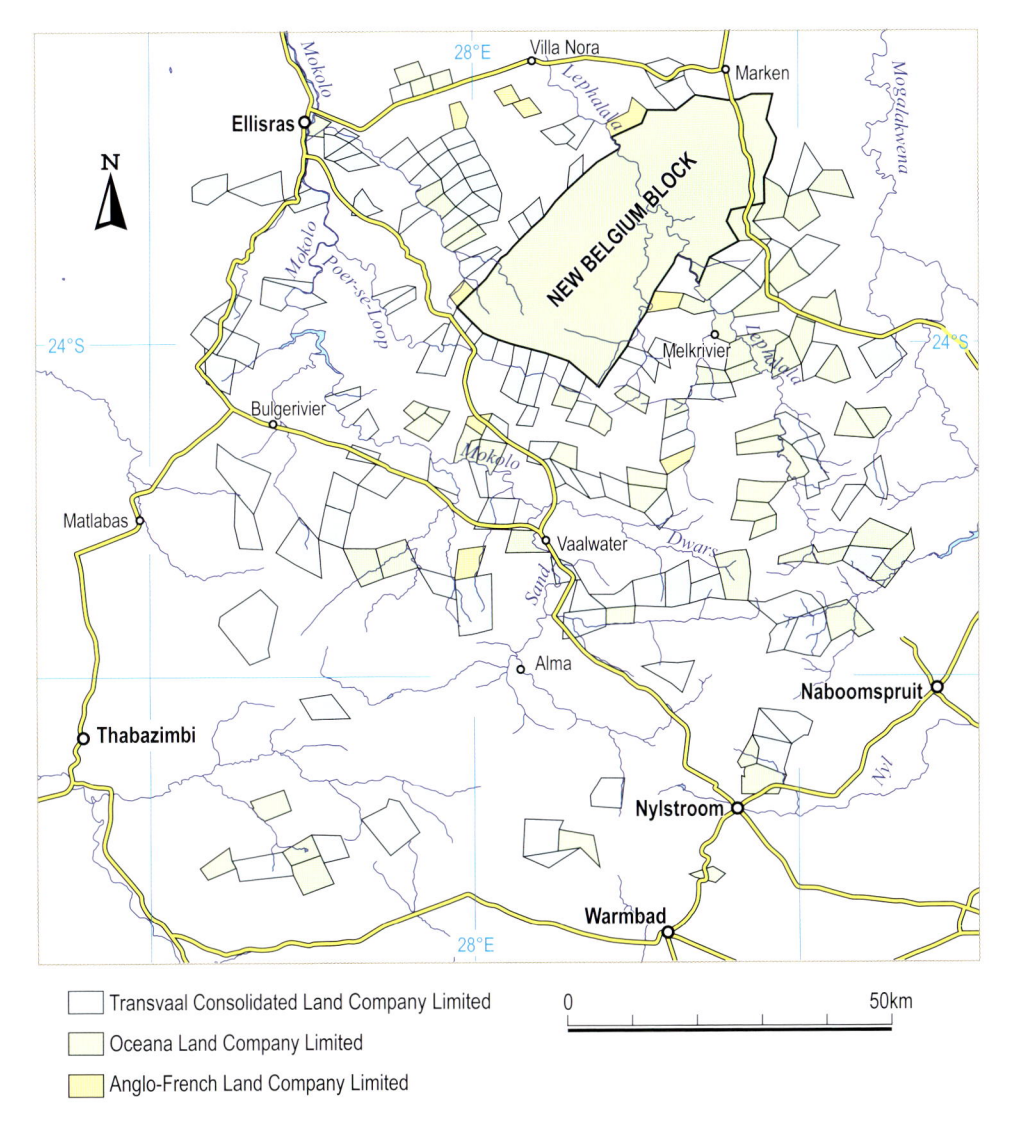

FIGURE 13.2: Company landownership across the Waterberg, 1899. Compiled from an examination of Jeppe's 1899 map of the Transvaal.

British post-war settlement schemes

In the aftermath of the South African War, the Waterberg and surrounding areas was an agricultural wasteland. The minority of land that was still privately owned had lain untended for two or three years, buildings had been burned down or vandalised, equipment stolen, stock commandeered or slaughtered and the owners imprisoned, injured or killed. In the Transvaal as a whole, 80% of cattle, 75% of horses and 73% of the sheep were destroyed.[17] On the other hand, the railway line to Pretoria, completed just as war broke out and still intact at its end, created market opportunities for crops and stock where there had been none before.

The British administration, for all its shortcomings, viewed the stabilisation and rehabilitation of the *platteland* as a priority. Reparations in the form of monetary compensation, although grossly inadequate, were made to many burghers who had lost assets in the conflict. Communication infrastructure was revived and improved, law and order was imposed and formal schooling was introduced, in some cases for the first time. In the belief that this would contribute to stability and unification among whites, the British High Commissioner, Lord Milner, sought to "Anglicise" the rural areas through schooling, policing and administration (using British military personnel) and through the settlement of British and colonial ex-servicemen as farmers – initially as tenants, but with the right to purchase their properties.

Examples of these early English farmer families include the Farrants, Furmans, Armstrongs, Davidsons, Mortimers and Grewars of Vaalwater, Chaneys of Sandrivierspoort and the Doyles, Weirs, Mansons, Grahams and many others on the Springbok Flats and around Naboomspruit. While they may have lacked the long veld experience and ingrained hardiness of their Boer neighbours, they were generally much better educated and they came unencumbered by large families, crippling debt or grudges against the new administration. They also brought with them an awareness of modern farming methods, market sophistication and a willingness to tackle the hardships of the "backveld" innovatively.

In 1903 a Compensation Claims Committee was formed to investigate claims

FIGURE 13.3 (TOP): Trekking in the Waterberg Bushveld. Museum Africa Photographic Library, No. PH2002-934. Photograph, numbered 120, by an unnamed photographer. FIGURE 13.4 (BOTTOM): A Boer family at their winter camp in the Bushveld, 1888. National Museum of Cultural History: Photographic Library No. HKF8303, from the collection of H.F. Gros.

CHAPTER 13: FITTING THE LAND TO ITS PURPOSE

for reparations in the Waterberg District. The committee comprised three members: the resident magistrate in Nylstroom (A.R. Orsmond), a prominent burgher of the former ZAR (Commandant Marthinus van Staden), and a British military representative, Major Robert French of the 2nd Gloucester Regiment. Major French submitted a report on his experiences to the British Colonial Office in January 1904. It contains a number of pertinent comments. He reported that the region under investigation (Pienaars River to the Zoutpansberg District and Springbok Flats to the Bechuanaland border) was inhabited by about 1500 white families (almost all within a radius of about 35 miles around Nylstroom). Nearly all of them were Dutch and their principal occupation was the breeding of trek oxen, sheep only rarely being viable. He considered the Waterberg to be

> one of the poorest and most backward districts of the Transvaal ... Where water can be got, grain grows very well, but the burgher as a rule only grows enough for his personal use. Fruit growing, especially oranges, do very well. Tobacco is also cultivated to a limited extent and grows well. The great drawbacks to the farmer are horse-sickness among horses and mules, and lung sickness and Redwater among oxen ... The country beyond the 35 mile radius of Nylstroom is mostly held by large Land Companies whose headquarters are in Johannesburg. These companies possess blocks of farms covering large tracts of country. They do not work them, but only hold them for what mineral wealth they may contain. They are visited occasionally by their agents or managers. There are a few farms in this part of the district owned some by Englishmen and some by Dutch, which are being worked.
>
> From about 130 to 150 miles west of Nylstroom the country falls about 1000 feet below the normal level, and from here to the Crocodile River is known as the low country, or the Bush Veld. In the cold weather, when the grass becomes scarce on the higher ground, a number of Boers trek down into this country with all their cattle and stock, and live a gipsy life at suitable places where they can get water. This country does not seem to belong to any individual, the Boers seem to camp wherever they please. The roads down through the passes from the higher ground are very rough.[18]

The advent of inorganic fertilisers

At this time the manufacture of inorganic chemical fertilisers still lay in the future; the only way in which the poor soils of the Waterberg plateau could be sustained was through the addition of kraal manure, guano (where available) and bone meal. Dips too were still to be invented and stock were at the mercy of ticks and flies and the diseases they brought. The importance of potassium, nitrogen and phosphorus in the promotion of crop and feed growth had been recognised for most of the last century, but it was not until the development of the Haber process for the manufacture of ammonia in about 1900 that commercial technologies to make chemical fertiliser began to appear. In 1902 the Ostwald process to make nitric acid from ammonia was invented and this, together with urea (itself produced from ammonia), created the feedstock for all nitrogenous fertilisers since then. Superphosphates had been manufactured in England (by Fisons) since the 1850s, and although these became widely exported by the 1870s, they were well beyond the means of poor farmers in the South African hinterland.

The first domestic manufacture began in 1903 when phosphate fertiliser was produced from bone meal by the SA Fertiliser Company (SAFCO) at its plant in Durban. However, the breakthrough for fertilisers in South Africa came as a result of the demand from the burgeoning gold mining industry for explosives. The manufacture of explosives produced sulphuric acid as a by-product and sulphuric acid could be used to produce phosphate fertilisers from imported rock phosphates. In 1919 and 1921, two plants were built, one outside Durban, the other in Somerset West (primarily an explosives factory) to produce superphosphates. Under the ownership of African Explosives & Chemical Industries Ltd (AECI), they became leading sources of fertiliser in the country.[19]

In 1951 the IDC developed Foskor at Phalaborwa to extract phosphate rock for the production of phosphoric acid and phosphate fertilisers. In the 1960s and 1970s, numerous plants manufacturing fertilisers were built in SA and, with the benefit of price control and other supportive programmes for agriculture, the fertiliser industry prospered throughout the post-war period into the 1980s. This subsidisation of both agriculture and one of its principal input materials enabled commercial agricultural crop cultivation to advance into areas – like the Waterberg plateau – that were inherently marginal for such activity and much better suited to cattle farming.[20] Clearly, this trend was not sustainable in the long term as subsequent events proved.

Price controls on fertilisers were lifted in 1984. This, combined with a severe drought and a deep economic recession, had a major adverse impact on many marginal crop producing areas. Most Waterberg crop farming operations, especially in groundnuts, and with the exception of those in proximity to rivers like the Mokolo, Sterkstroom and Sand, came to an end at around this time. The domestic fertiliser industry also suffered and experienced a phase of closures and rationalisation, with the number of players being substantially reduced. Sasol emerged as the dominant manufacturer. The lifting of price controls on products (administered by innumerable government boards) also rendered many crops unprofitable in the Waterberg.

Organised farmer representation

One of the first moves made by the farming community post-1902 was to form associations to represent the interests of farmers in addressing market issues and in lobbying government. In fact, this type of organisation had already been established in Natal just before the war and also in the Transvaal where the Transvaal Agricultural Union (TAU/TLU) had been formed in 1897. In December 1903 members of the community argued strongly for the creation of a single national body that could represent the interests of farmers. In July 1904 the Intercolonial Agricultural Union of South Africa, soon renamed the South African Agricultural Union (SAAU, or SALU in Afrikaans), was formally constituted with subunion members in each of the four provinces as well as Rhodesia.[21]

Local unions, affiliated to SAAU either directly or via district unions, soon followed. The North Waterberg Farmers' Society (later changed to Association, so NWFA), established in 1917 and based at Vaalwater, was one of the first such local unions in the northern Transvaal. Prominent among its founding members was the 24 Rivers farmer E.A. "Ted" Davidson, who acted as its secretary for over 30 years. The first chairman was Rufus Kirkman of Vaalwater. Within a short period the association had almost 40 members.

The oldest surviving correspondence of the NWFA relates to a meeting – probably the first – of 16 January 1917.[22] The secretary was A.J. (Arthur) Mathewson, also of 24 Rivers, where he was a Justice of the Peace and land manager for TCL. After the meeting, Mathewson wrote several letters pursuant to resolutions passed at the meeting: to the Department of Agriculture (about branding of cattle, pounds, stock census procedure and the need for qualified stock inspectors); to the Waterberg Road Board (suggesting an alternative route to the north and requesting that the top of The Nek in Zandrivierspoort be lowered); to the Minister for Railways (lobbying, not for the first time, for a line to be laid from Nylstroom to Vaalwater); to the Minister for Lands (arguing for better management and control of the use of unoccupied government land); and to the Minister for Posts & Telegraphs (asking that all police posts and postal agencies be linked by telephone). Needless to say, he received unhelpful, though prompt and courteous replies to all these requests. The Waterberg was a "backveld" region and destined to remain a low governmental priority for decades to come. Evidence of this is provided by a similar set of correspondence following a NWFA meeting over 40 years later in 1958 by which time R.A. Grewar was secretary. [23]

Jack Farrant and T.B. Grewar both served the association for many years before being followed by their sons Rupert and Alan respectively, and in Farrant's case by his grandson Dr Peter Farrant. Davidson was also succeeded by his grandsons,

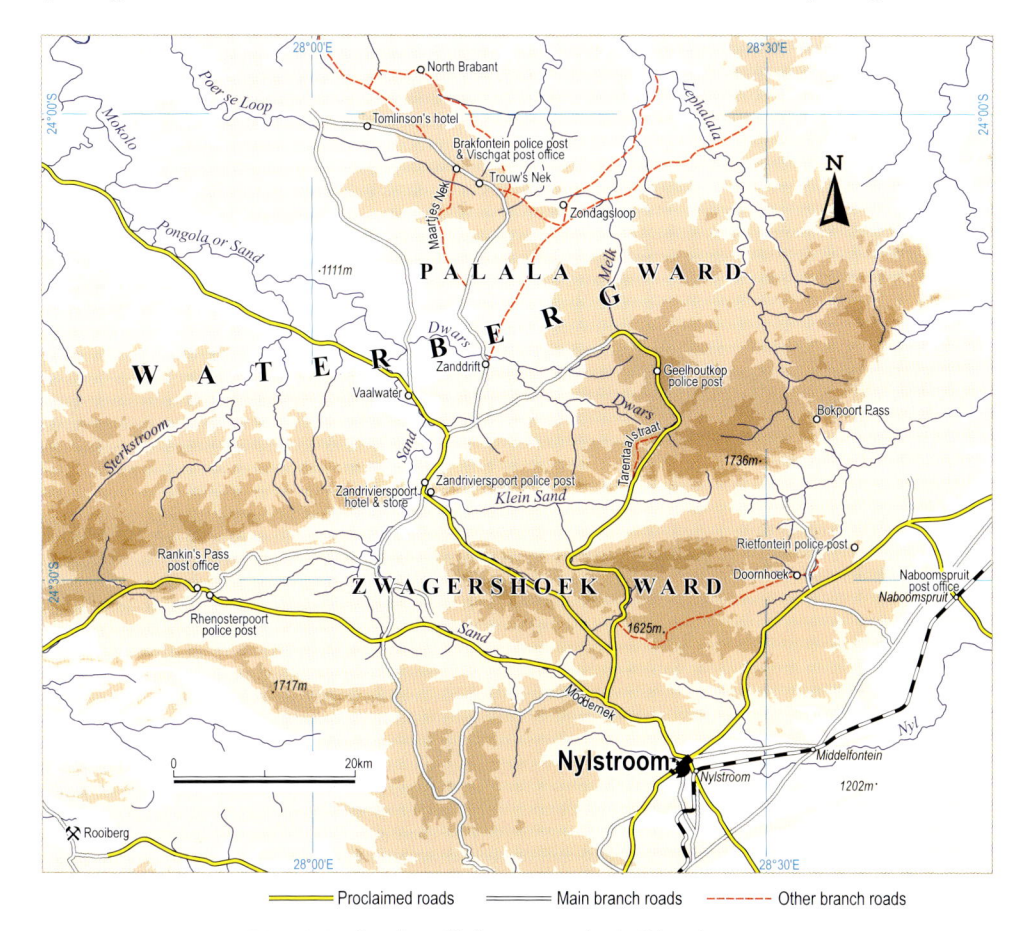

FIGURE 13.5: Roads and infrastructure in the Waterberg, 1910.

Colin and Charles Baber, both of whom were longstanding dynamic members of the local agricultural community, together with men like Herman Willemse, Ben Vorster, Newlyn Clover, Alex Furman, Jan Lourens and more recently, Louis Trichardt and Ben Mostert.[24] Rupert Farrant was a member of the NWFA from 1946 when he returned to the family farm from highly decorated service with the South African Air Force in North Africa until his death in 2002, and was its chairman for 19 years (1965–1984).[25] Colin Baber served as the NWFA's secretary for many years. Charles Baber also served both the NWFA and the local farmers' union for several decades, was named the Transvaal Farmer of the Year by the agricultural reporters' association in 1993 and Personality of the Year in 1995 by the Transvaal Agricultural Union, the same year in which he became chairman of the NWFA for the first time. He was tireless in urging local farmers to support their associations. In 1997 he pointed out that the NWFA had been responsible,

FIGURE 13.6: Attendees of possibly the first meeting of the North Waterberg Farmers' Society (later Association) on 16 January 1917. Many records have been examined in an attempt to identify the participants. Some possible members, for example Charles Madge of 24 Rivers, had been killed in France in 1916 and others, like John "Vaal" Kirkman and his cousin Harry, were still in service, so were unlikely to have been present in January 1917. BACK ROW, L TO R: unknown; unknown; Bert Tomlinson; unknown; unknown; unknown; unknown; Gilbert Kirkman; John Chaney. MIDDLE ROW, L TO R: Arthur Mathewson; unknown; EA (Ted) Davidson; Jack Farrant; unknown. FRONT ROW, L TO R: unknown; Rufus Kirkman (Chairman); unknown; John Mortimer. Others who may be present in the photograph include T.B. Grewar, Herman Willemse, Frans Lensly, William Greig, the Van Emmenis brothers, Eric Willet, Horace and/or "Pard" Wale. All were resident in the area at the time. Photograph from the collection of Elizabeth Maud (previously Hunter).

to a greater or lesser extent, for the introduction of all the infrastructure now taken for granted by the community: road and rail links, telephone and postal services, electricity, radio and TV, schools, medical and veterinary facilities, etc. It had always been and remained the strongest of all the associations affiliated to the district union.[26]

The NWFA was one of more than a dozen local farmers' associations that in turn belonged to the larger Waterberg Central Agricultural Union (later renamed the Vaalwater District Agricultural/Landbou Union, VDLU). This had been formed in 1943 when the local unions north of the Zandriviersberg decided to leave the Waterberg Agricultural Union, based in Nylstroom, and to establish a new district union based in Vaalwater. In addition to North Waterberg, members included New Belgium, Melkrivier, Saamtrek, Bulgerivier, Diamant, Kransberg, Gedenk, Haakdoringboom, Sukses, Sondagsloop, Geelhoutkop, Libertas, Ellisras, Steenbokpan, Limpopodraai and Klipspruit. Eventually, unions in the Lowveld split off to form another district union centred in Ellisras. The VDLU remains active to this day and is still involved in lobbying for benefits for the greater Vaalwater community.

The local and district unions proved to be extremely valuable community forums, not only in respect of agricultural matters, but also in the absence of elected local government officials, as representatives of community needs. They lobbied for the railway, bus services, tar roads, for telephones, electricity supply, for schools, clinics, district surgeons, vets, policing, and for the creation of formal black townships. They harangued the NTK co-op or its predecessors for better service, a wider range of materials, better storage facilities and better prices. At the same time, they focused on agricultural issues. Study groups were created for almost every activity: cattle, pigs, citrus, cotton & castor oil, oil seed, tobacco, maize, soil conservation, with each local association nominating a representative to attend these forums at district level. Open days were held on various farms, with specialists from government research departments, co-ops and elsewhere attending to advise farmers on trends in farming practice and markets. The proceedings were faithfully recorded, eloquently and at great length by secretaries (always male farmers) fluent in both English and Afrikaans. All these activities were carried out, year after year, with great commitment and diligence, all voluntarily and in addition to the day-to-day responsibilities of trying to make a living out of mixed farming on poor agricultural land.

But it was not all work. Early in 1952 the Entertainment Committee of the NWFA (Charles Baber was the chairman) set about raising funds with which to build a community hall on the site (opposite the Total garage outside Vaalwater) set aside for this purpose and made available by the Farrants. They decided to hold a *boeredag* in May, "by which time the crops would have been reaped and sold". Events were to include horse racing, .22 shooting, a pellet-gun range, an Aunt Sally, beer and Coke fishing, tug-of-war competition, bicycle race, darts, chicken chase, greasy pig chase and, of course, many stalls. In the evening there would be a braai (requests were made for donations of meat), followed by a dance on tarpaulins borrowed from the co-op, with music to be supplied by the local *boereorkes*. United Tobacco would loan a public address system, but batteries to power it had to be borrowed (there was no electricity then). Some talented local wives would design and draw up posters to be placed at strategic points around the district (as far as Ellisras and Naboomspruit); Mrs Horn of the Nylstroom Hotel would be asked to run a bar and *Die Morester* to place adverts. Coopers & Nephew were to be arm-twisted into sending a lecturer with a projector and its new film about soil conservation.

Evidently, the day was a great success, because when the committee (now the Hall Committee) met in November of that year at the Farrants' home with R.A. Grewar as chairperson and Alex Furman as secretary, the total cost of the proposed hall was estimated to be £2054.7.2d, of which £1135.7.2d had already been raised or pledged, leaving only £879 to be found! It was decided that there would be monthly dances at various members' homes to raise the shortfall: Mrs Clover volunteered to hold the first one at their home in January. Herman Willemse had loaned his "Little Wonder" machine to make concrete bricks, but it had been found that the rate of manufacture (600 per day) by the worker gang had been much slower than anticipated, with the result that the 23 000 bricks made so far had cost an exorbitant £5 per thousand. Eustace Kirkman said he had a stock of 20 000 burned bricks and offered to donate half of these and to charge £2 per thousand for the balance.

Members were asked to assist in making doorframes from SA pine and to donate scrap iron for use as reinforcing material. Construction started on the hall in March 1953, a second boeredag was held in July to raise some more funds for furnishings and in May 1954, a huge dance with catering for 300 was held to celebrate the opening of the hall. In 1955 Alex Furman drew up the final expenditure account for the building and equipping of the hall: £2506.16.5d, plus another £128.2.9d for turntable, records, speakers and amplifier. In the years that followed, many were the memorable dances, meetings, weddings and other events that were held in this venerable hall, which over sixty years later still stands awaiting guests.[27]

FIGURE 13.7: A 1954 advertisement for Little Wonder Brick Machines. Alexander, Norman (Compiler) (1954): *The Farmer's Handy-Book/Die Boek vir die Boer, 1954 Edition/Uitgawe*. Pan African Publications, Johannesburg, p. 380.

Although it was apparent that even before the turn of the twentieth century, virtually all large wild game had been eliminated (hunted out or chased away) from the Waterberg, occasionally a report would appear to remind the population that this region remained a remote and wild place. In the *Rand Daily Mail* of 13 December 1948, for example, was a short news item headed "Lions Raid Cattle on Waterberg Farm". The article went on to say that:

> Lions are increasing in number in the Waterberg bushveld and have accounted for many head of cattle. Cattle grazing on one farm were chased for six miles yesterday by a lion which made a kill and got away to the adjoining hills. Recently, a lion was shot on the farm Melkboschkraal [about 15 km northeast of where Lephalale stands today] by Mr M .van Staden, after it had accounted for 13 head of cattle.[28]

Such incidents were rare though, and largely confined to the Lowveld areas.

The rise of the co-operations

Even as the farmers' associations addressed issues of concern to their members, the relationship between agricultural producers and their markets was found to be skewed in favour of the markets, at the expense of the producers – who already had to cater for the vagaries of weather and pests. To strengthen their position, farmers lobbied government for the creation of a statutory framework for co-operative bodies that could intervene to establish stable prices that would allow farmers to plan their future crop and stock production with less risk and greater confidence. Such co-operative structures had been found to work well in England and Europe to the benefit of both producers and consumers. Eventually, the Co-operative Agricultural Societies Act (No. 17 of 1908) was promulgated in the Transvaal Legislature with J.M.B. Stilling-Anderson as its first Registrar. This led to a meeting of farmers in Nylstroom on 15 May 1909 and then to the formation under this new legislation of the Waterberg Landbouwers Koöperatieve Vereniging (WLKV) on 29 May 1909.

There were 103 foundation members of the co-op, including W.R. Kirkman and F.D. & N.E. Lensly from Vaalwater, A.H. & G.S. van Heerden and J.J. Enslin from Swaershoek, C.F. Seegers and several Heysteks.[29] The first chairman of the WLKV was Mr Pieter ("Oom Piet") Wynand le Roux van Niekerk, who had been born in Wellington in the Cape in 1878 but farmed locally. He occupied the chair until his election as the National Party MP for Waterberg in the general election of 1915. His deputy was Stephanus J.M. Swanepoel, a Waterberg *seun* born in 1873 on the farm Knopfontein (on which the settlement and rail siding of Alma would later be sited). Swanepoel remained an active member of the co-op for the next 30 years. His father, Christoffel Bernardus Swanepoel (1846–1918) was also a founder director, as well as a leading member of the Gereformeerde Kerk. Other early directors of the WLKV included John Gilfillan, Frederik Boshoff (former *Tweede Volksraad* member for Waterberg), John Rankin, J.J. Enslin and Jooste Heystek (Sr), who replaced Van Niekerk as chairman in 1915.[30]

The co-op had as its principal functions the creation and maintenance of orderly, viable commercial markets for its members' produce, including storage facilities, and a collective procurement agency for farmers' needs: seed, implements, bags, etc. From the outset, the WLKV focused on crops like corn, sorghum and maize (later diversifying into others). It played no role in the marketing of tobacco and specifically did not get involved in domestic stock.

In 1911 an outbreak of east coast fever in the district resulted in the temporary prohibition of the use of oxen to haul wagons of maize and other produce from outlying farms into Nylstroom. This was almost sufficient to cause the collapse of the fledgling co-op, but the crisis was averted through the use of donkeys; even a steam tractor was considered as a possible option. A shortcoming of the co-operatives' legislation was that each member of a co-operative association was individually responsible for a share of his organisation's debt; when agricultural or market conditions deteriorated, as they did in 1917 for example, members resigned en masse, leaving those remaining to bear the liability. In 1923 the legislation was amended to resolve this problem, but in the meanwhile, the Land Bank (established in 1912) was empowered to lend to the co-operatives, albeit at punitive interest rates.

In 1919 Theodor Würth was appointed as secretary (effectively GM) of the WLKV and served productively for eleven years in this role as a loyal and hard-working employee. In order to make recruiting house visits he needed a car, which was purchased on his behalf by the co-op in 1924 for £125. Würth personally paid £25 up front and the remainder was paid for at the rate of £25 per annum, of which Würth had to pay half. At the end of his period, he became the owner of the car. He also had to pay for its maintenance and upkeep, apart from tyres and an annual service. For trips out of town, he was reimbursed at the rate of 9d per mile.[31]

Groundnuts

Groundnuts were first introduced to farmers in the Waterberg District in 1923 and a separate organisation, the Northern Transvaal Nut Growers Association, was formed to manage their interests. Initially, though, the crop enjoyed limited success. It was only when Roelof van Ryneveld joined the WLKV in 1930 as Würth's successor that the groundnut business began to prosper. Van Ryneveld, a Free Stater (and member of the family that included Sir Pierre van Ryneveld, the famous pilot and founder of the SA Air Force) had joined a bank after matriculation and was sent to its Nyasaland (Malawi) branch. Soon he left to join two of his brothers who were farming in Kenya. There he learned about the business of groundnut farming. Upon his return to South Africa and a job with the WLKV in 1928, Van Ryneveld was determined to use his Kenyan experience to promote the groundnut business among the co-op's members. He personally ran several experiments using machinery that he designed and modified himself to pluck and dehusk the crop. By 1937 he had developed a process to wash and bleach unshelled nuts. This was followed by a machine that washed, shelled and graded nuts – the breakthrough needed to produce nuts suited to the export edible nut market. This equipment was brought into use in 1941 with the assistance of Marcel Brucoporko, an Italian prisoner of war.[32]

The Waterberg groundnut industry flourished. By 1943/1944 over 90% of the country's groundnut production was being marketed via the WLKV. By 1945 it had become one of the WLKV's principal income generators. Arthur Pitout, a nut farmer on the Springbok Flats, succeeded in organising the amalgamation into the WLKV of various small nut co-ops that had developed. The railway to Marble Hall had just been built and passed through his peanut plantation, so a siding built on his farm was called Nutfield. Another pioneer groundnut farmer – from the Cullinan area – was Wyatt Vause Raw, who joined the WLKV board in 1949, but left soon afterwards for Natal where he would become a prominent politician for the South African/United Party.

In 1946 a group of representatives, including Van Ryneveld, visited the USA, Canada and Britain to promote exports. This was not very successful, but it enabled Van Ryneveld to inspect and order some equipment for oil pressing (from groundnuts, cottonseed and castor oil seeds) and peanut butter manufacture. The construction of 13 pneumatic suction pumps commenced, designed to grade the nuts. These machines were placed at strategic points where groundnuts were received: they included Pietersburg, Potgietersrus, Naboomspruit, Nylstroom and Vaalwater. An order was also placed for 150 dehusking machines for distribution to members.[33]

The WLKV commenced construction of a facility in Nylstroom to process groundnuts and early in 1949 the Nylstroom factory began to press nuts to produce peanut butter as well as salted, roasted and bottled roasted nuts. In its first full year of operation, the peanut butter factory converted 14 tons of nuts to butter and pressed another 1300 tons of peanuts and 947 tons of sunflower seed to produce oil. The peanut butter brands "Apie" and "Nyl" would soon become household names around the country.[34] By 1950 the plant was producing 150 000 13-oz bottles of peanut butter a month, as well as many tons of butter in drums for resale. Most of the production was for export.[35]

The building was located in Station Street. It turned out that the noise made by the pressing machine was a major problem for the co-op's head office personnel above. To overcome this problem, a one-way loudspeaker intercom system was installed to link the General Manager's office on the top floor with that of other officials. On one occasion, the head of the grain section was so incensed by being criticised over the system by both the GM and the auditor – without the ability to respond – that he ripped the speaker from his wall and stormed upstairs to the GM's office where he threw the equipment on the floor in front of them saying: "If you want to tell me something, then summon me!"[36]

In 1945 the WLKV had begun a transformation of its business by creating a new organisation, the Waterberg Ko-operatiewe Landboumaatskappy Beperk (WKLM) which, over the next few years, would absorb and ultimately replace its predecessor. A decade later, in 1957, the WKLM morphed into yet another company, the Noord-Transvaalse Koöperasie Beperk (NTK), by which name it is still known today.

In 1952 the co-op's business was still dominated by groundnuts. Commission from nut sales amounted to 30% of gross income, with a similar proportion being derived from sales of processed groundnuts and sunflower seeds. At its

FIGURE 13.8: The old headquarters of the NTK near the railway station in Modimolle with the peanut butter factory beneath it.

FIGURE 13.9: Picture of *Apie* peanut butter redrawn by C. Bleach from photograph in Snyman (ed.) (2008), p. 8.

peak, WLKV/WKLM handled around half the national groundnut harvest. But trouble was on the horizon. By 1956 new producers of groundnuts elsewhere in the world (especially Israel) – some having better, more consistent quality than the SA product – meant that SA ranked only ninth among exporters of nuts in their shells. Worse was to follow: In July 1959 the governments of the West Indies islands announced that they would ban the import of all SA products due to its government's apartheid policy. These countries had represented two-thirds of the market for SA peanut butter. In 1960 the boycott was joined by countries in the Middle East.[37] At about the same time, the regional crop began to develop a growing concentration of aflatoxin[38] which had an adverse effect on nut quality and was found to be carcinogenic. By the early 1970s there was insufficient production in the region to operate the butter plant.[39]

The NTK by now had established branches at several places around the Waterberg, and concluded an exclusive fuel supply agreement with Shell for some of these, including Vaalwater where there was still only a depot. This arrangement has endured for over 50 years.

Groundnuts, of course, was not the only crop managed by the co-op for its members, although it dominated from 1942 to 1959. Maize remained a principal crop for most members (of whom there were over 14 000 by 1969). The handling, management, storage, transport and marketing of maize were tasks best performed on behalf of a group of producers, who could share the costs. Initially, mealies were stored in bags under sails, then under *afdakke*, later in storage sheds and, eventually, in silos. The first silos were built by SAR&H at Pienaarsrivier in 1924 and at Settlers in 1927. The Settlers silo, a strange square concrete structure that held 1700 tons, was built over a rail siding like those constructed later. Both the silo and the railway line beneath it, now long disused, can still be seen in the village. For many years, Vaalwater had only an afdak (lean-to) shelter under which to store mealies. Mills were built to process the maize, one of the earliest being that purchased by the WLKV in 1910: the Mielie Dorschmachine.

In the 1970s SABS-approved steel silos were built at several sites, including at Alma, but the big concrete silos at Vaalwater (30 000 tons) and Warmbaths (40 000 tons) were only built in 1984, a year that saw 14 new silos completed, with a combined capacity of almost half a million tons. But within a year, renewed drought saw production plummeting from 300 000 tons (at about a ton/hectare) per annum in 1985 to a third of that the following season. In the 1990s South Africa returned to the international market, including a free market in which to sell maize through the mechanism of Safex, a grain-oriented trading exchange. This in turn resulted in the role of the NTK diminishing, as producers could sell and transport directly to the market, usually by road. The silo business declined as a result and many, including those at Vaalwater, have fallen into disuse.[40]

Other crops promoted by the NTK through the years included corn, grain sorghum, cotton, castor oil and grapes. Cotton was first grown in the area in 1917 when maize and corn prices were very low. In the next few years, it traded well and was grown widely across the Waterberg – including at Matlabas north of Thabazimbi. However, costs rose gradually, while international competition kept prices steady, and eventually the NTK withdrew from this sector. Castor oil plants had been grown from early in the century, the oil being used for medicinal purposes as well as in the production of high-performance lubricants (Castrol). In conjunction with Nola Oil Industries, the WKLM built a pressing plant in Potgietersrus in 1952 and by the end of the decade, the co-op had an effective monopoly on oil production in the Transvaal. But, as with groundnuts, cheaper, better varieties of the castor oil plant elsewhere ultimately eroded the co-op's market and production ceased.[41]

In 1934, largely as a result of the urging of Jan Smuts (who the previous year returned to government in a coalition with General Hertzog), two agricultural research stations were established in the northern Transvaal: the Towoomba Pasture Research Station, on a 590 ha portion of the farm Roodekuil, just south of Warmbaths; and the Mara Research Station on the farm of that name lying at the southern foot of the Soutpansberg, about 55 km west of Louis Trichardt (Makhado). Both were to play important roles in guiding and advising Waterberg farmers in the years ahead and often involved themselves in experimental crops or farming methods intended to improve the economic productivity of the farming community. Sadly, this is no longer the case. Both of these venerable institutions have now lost virtually all their professionally trained technical personnel and are today little more than low-level training colleges.

Tobacco

Tobacco has long been an important – probably the most important – crop in the Waterberg area, not least because the plant tolerates poor soil. The principal growing areas have been around Vaalwater, the Sterkrivier settlement area north of Mookgophong and around Lephalale. Although first mentioned in reports at the turn of the twentieth century, it seems that tobacco only became significant locally in the 1930s when virtually all tobacco grown in the region was of the air-cured, burley variety. Jack Farrant, on Hartebeestpoort outside Vaalwater, was an early grower.[42]

From the beginning, tobacco production in South Africa was sold through co-ops on a contractual basis; individual growers were allocated quotas that determined how much they could produce. All the growers on the Waterberg plateau sold their produce via the Magaliesbergse Koöperatiewe Tabakplantersvereniging (MKTV), a co-op that had been formed in 1909 and which was based in Rustenburg.[43] The crop would be taken by train from Vaalwater all the way to Rustenburg via Pretoria. There were at one time five other tobacco co-ops in the country, including one at Potgietersrus (Mokopane) (the PTK), but MKTV was the oldest and by far the largest.

In the 1960s and thereafter, the Waterberg area was settled by several tobacco growers who moved down from Northern and Southern Rhodesia (Zambia and Zimbabwe respectively). Burley tobacco was replaced by flue-cured Virginia (FCV) leaf. Again, Hartebeestpoort was a pioneer in this respect when Jack Farrant's son Rupert introduced FCV following a visit to the Centenary area of Rhodesia. Harry Parham, a tobacco farmer from both Rhodesia and Zambia – and whose father had been a prominent tobacco grower in North Carolina – came in to manage the Farrant operation. A whole generation of tobacco growers would learn their skills from Harry over the next 25 years. T.F. (Tillie) van der Merwe, who also arrived from Zambia, became an important tobacco grower too and with Charles Baber (who also visited Rhodesia to learn about farming the crop), was a director of the MKTV in the 1970s. Tillie died in a light aircraft accident outside Vaalwater in 1979, but one of his sons, Francois, became CEO and subsequently chairman of the Tobacco Institute of Southern Africa (TISA), an unusual forum of growers, marketers and manufacturers which represents the noncommercial interests of the industry players.[44] Another son, Janneman, established the large feedstock manufacturer, Driehoek Voere – Vaalwater's largest industrial enterprise by far – on Goedehoop, adjacent to Hartebeestpoort.

By 2004 the Waterberg area produced almost 60% of the total national tobacco crop of about 31 000 tons per annum with the Vaalwater area alone responsible for almost 8000 tons per annum, or a quarter of national production.[45] About 85% of national production is flue-cured. Vaalwater growers, of whom there were about a hundred in 2003, produced exclusively the FCV tobacco used in cigarette manufacture, whereas those in the Lowveld around Lephalale delivered air-cured burley leaf used mainly for pipe tobacco, snuff etc.

FIGURE 13.10: Tobacco barns at 24 Rivers. Pen-and-ink sketch by Karel Bakker, 1988. Reproduced by permission of Prof. Terri Bakker and the archives of UPSpace and the Department of Architecture at the University of Pretoria, where Bakker (1956–2014) had been an esteemed professor and departmental head.

Tobacco is a resource- and labour-intensive crop. At its peak in 2002–2003 when national production reached 37 000 tons per annum, the industry used a mere 15 000 ha of land, involved only about 600 growers, but employed a labour force of almost 50 000. Another 7500 workers were employed by the co-operatives and manufacturers.

Throughout the twentieth century, all SA tobacco production was consumed locally: neither exports nor imports were permitted by the government. This artificial, closely regulated environment certainly served to protect the local industry from competition from lower-cost producers elsewhere (like Rhodesia for example), but it also bred a culture of inefficiency and complacency, especially in the co-op organisations. Following the election of the country's first fully democratic government in 1994 and its re-entry into global markets, the South African tobacco industry all but collapsed. Competition and health and regulatory issues contributed to this disaster, but much of the blame could also be laid at the feet of management of the moribund co-ops, which were simply not competent to deal with the aggressive new market conditions, nor able to adapt. There were also allegations of widespread corruption in the sector. Early in the present century major restructuring of the industry took place to ensure its survival. The

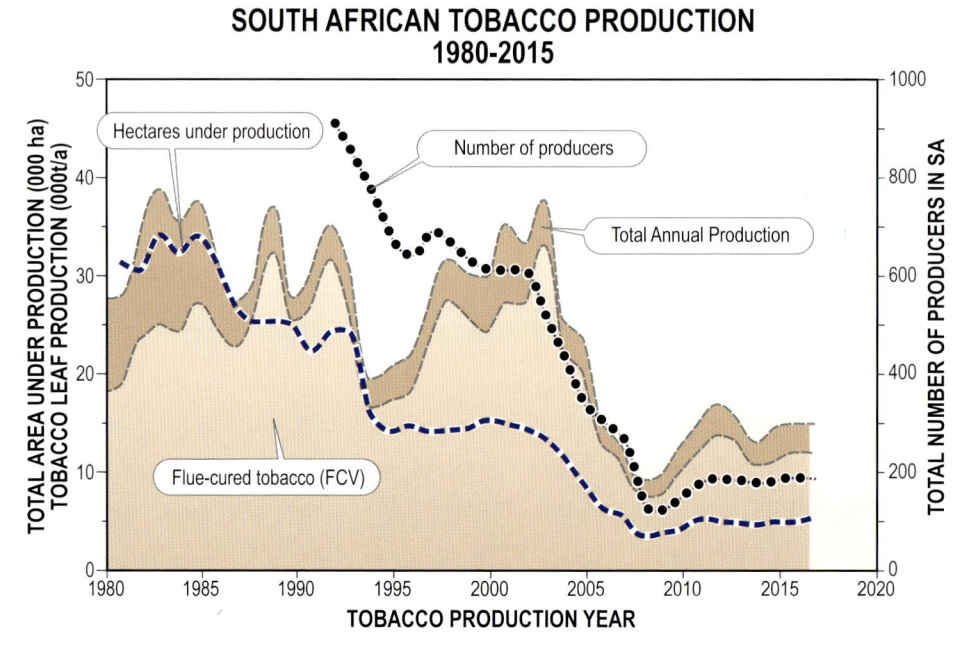

FIGURE 13.11: South African tobacco production, 1980–2015. Data compiled from Anonymous (2011), DAFF (2016) and Du Preez (2017). Figure drawn by C. Bleach.

number of processors was reduced to a single plant for flue-cured Virginia (FCV) tobacco, located in Rustenburg (the site of the former MKTV), and the old facility in Mokopane was closed.[46] Today, the Rustenburg operation processes all FCV leaf in the country; it is administered by Limpopo Tobacco Processors, recently reconstituted as the black-empowered company Tobacco Producer Development Company (TPDC).[47]

By 2008 national production had plummeted by two-thirds to below 10 000 tons per annum, grown on a mere 3700 ha owned by only 150 growers. The decline in production occurred disproportionately amongst producers of air-cured and other tobacco types. Whereas in the early 1980s a little over two-thirds of total production was FCV, by 2010 this had increased to almost 80%. Burley production ceased in the early 1990s.[48,49] Today, about half of total FCV and two-thirds of air-cured tobacco production is consumed domestically.[50] This led to a price increase for leaf and a slight recovery in production. Between 2011 and 2015, the Waterberg area (dominated by Vaalwater) produced an average of over 4500 tons per annum and enjoyed some of the best prices in the domestic industry (R37/kg in 2015). By 2017 production in the Vaalwater area had risen to almost 7000 tons (i.e. almost half of total national production); the crop was still being graded via TPDC and sold to the largest of the tobacco companies, British American Tobacco (BAT).[51]

Cattle

Cattle ranching is also a longstanding and important agricultural activity in the Waterberg and one that always stood outside the ambit of the NTK and its antecedents. In fact, the predominantly sourveld grasses of the plateau are anything but ideal for stock farming: they are not nutritious and are low in phosphorus content. Before the advent of supplementary licks, many farmers on the plateau used to run their cattle herds on a second farm down in the Lowveld, on sweetveld, and only move the herds up onto the plateau in times of drought. This necessitated that the cattle were able to walk long distances and tolerate harsh conditions. The Africander, a breed suited to this regimen, was the dominant breed in the area.

Research undertaken during and after World War II at the Mara Agricultural Research station west of Louis Trichardt by Prof. Jan Bonsma of Pretoria University eventually led to the derivation of a breed even better able to handle the environmental conditions of the region (hot, dry, tick-infested) than the Africander. Called the Bonsmara, this breed was first registered in the 1960s. Inspired by the hugely successful Santa Gertrudis breed of South Texas (3/8 Brahman, 5/8 Shorthorn), the Bonsmara comprises 5/8 Africander and 3/8 Hereford/Shorthorn. Its characteristics include a heavy, loose, but sleek chocolate to brown coat (no white), a low shoulder hump, square head and narrow hips. The breed is noted for its good meat and milk yield, docile nature and tolerance of heat and ticks. Although it is today the dominant breed on the plateau, others, including Brahman and Huguenot are also found, together with the indigenous Nguni, the latter being particularly well-suited to coexistence with game.

Cattle ranchers, both stud and beef, use local and regional auction yards for the sale and purchase of their animals. In Vaalwater, the Transvaal Lewendehawe Koöperasie (TLK) was for many years the organisation responsible for these auctions. It was later acquired by Vleissentraal, which is still the dominant local operator. In May 1972 the Vaalwater District Agricultural Union was plunged into crisis when the owner of the land on which the union's sale yard had been situated since 1958 announced that the property had been sold and that the yard had to be removed at short notice. A committee under the chairmanship of Ben Vorster, with the able assistance of Rupert Farrant, Ola Smith, Alex Furman and the Baber brothers immediately set to work to resolve the problem. Within two months,

FIGURE 13.12: Champion Bonsmara bull BEI 04 155A, bred by Peter Beith of Sondagsloop. Image courtesy of Peter Beith.

they had found a new piece of land (the present sales yard on the Melkrivier road), negotiated its lease, had the site approved by the conservation authorities, arranged for it to be surveyed, negotiated new terms under which TLK would administer the auctions and recover the loan it had made to the association, acquired some second-hand yard fencing and equipment and had this installed on site. It was a remarkable effort and a tribute to the energy and acumen of those involved.[52] Today, auctions are held at this yard almost weekly throughout the year.

Until the end of the 1980s, all cattle bound for the state abattoir in Pretoria would leave Vaalwater by rail. It is said that at one time, the sales yards at Vaalwater station were second in size only to those at Vryburg in the Northern Cape. As with virtually every other aspect of agriculture in the isolationist days of the former Nationalist government, cattle farmers were subject to a quota system, administered by a cumbersome government board, in this case the Meat Board.

As the prevalence of game farms adjacent to cattle farms increased, the incidence of cattle becoming infected with *snotsiekte* (bovine malignant catarrh) caused by a virus carried by wildebeest[53] also began to rise. In June 1983 Peter Beith, a national award-winning Bonsmara stud breeder in the Sondagsloop area, reported an outbreak of the disease among his herd, causing the death of at least four animals. The matter was raised at a provincial level and eventually resulted in the disease being proclaimed, with the further requirement that landowners wishing to import blue wildebeest onto their properties would have to apply for permits.[54]

Changes at co-ops and farmers' associations

In 1984 the NTK decided to build a grand new head office in Nylstroom, financed by the Land Bank. But its fortunes were already waning. In 2004 straitened circumstances forced the NTK to sell its head office building (which is now occupied by several departments of the Limpopo provincial government). More recently, it moved into the former offices of its financier, the Land Bank! During the first decade of the current century, the NTK, like so many other co-ops around the country, seemed to have become moribund. In Vaalwater, for example, the initials were said to mean "*Niks te Koop*" (nothing to buy), because the branch appeared to be undecided about whether it was a farmers' supply outlet, an outfitter or a supermarket – and it often lacked relevant stock in all these areas.

The NTK's salvation has been that it was acquired in 2010 by a Free State company based in Reitz, the Vrystaat Koöperasie Beperk (VKB). VKB, while coming from a similar background to NTK and being of a similar age, has brought a younger, more dynamic character to its Limpopo affiliate, reviving its image and reputation in the eyes of its members and the public.

From the above, it is apparent that for much of the previous century, organised agriculture in the Waterberg region (and for that matter, throughout the country) was tightly regulated by a wide variety of statutes, regulations, control boards and co-operatives. The overall effect of this rigorous, thoroughly socialistic control was to ensure the sustainability of the white farmer – whose vote kept the ruling party in power – through the maintenance of guaranteed quotas, artificial markets and prices unrelated to surrounding realities. Inevitably, this environment created an expensive, moribund governmental (and, as we've seen, co-operative) bureaucracy, tolerated mediocrity and inefficiency among farmers, limited the potential of those who could excel and encouraged agricultural development in areas inherently unsuited to this activity.

The dramatic transformation of government that took place in April 1994, while it has subsequently led to widespread deterioration in the quality and capacity of state-sponsored services (not to mention rural local municipalities), nevertheless had the healthy long-term consequence of exposing agriculture to the realities of the global market. Price controls had already started easing in the late 1980s, and gradually the various control boards fell away too. In the Waterberg, never a prosperous farming environment, the impact on commercial agriculture was serious and, in some respects, irreversible. The NTK silos at Vaalwater, completed only in 1984 (more than a decade after they had been mooted), would never be filled – yet only two years earlier, a bumper maize crop had led to a national surplus and at Vaalwater to mountains of stacked bags of mealies on the ground under tarpaulins (and an explosion in the local rat population).[55] Except for a few properties under irrigation along the Mokolo River, re-entry to the global market spelled the end of the maize sector in the area. The tobacco growing sector declined sharply and it would be almost a decade before it began to recover. Cattle ranching moved increasingly from beef production to smaller, higher value stud herds, intensively managed on scientific principles by smart farmers.

The advent of the new universal democratic government opened long-sealed wounds within the farming community, splitting it into one faction deeply conservative in nature and which was instinctively opposed to and resisted the changes accompanying black emancipation; and another faction that, albeit with some trepidation, was prepared to embrace the new dispensation and to try to work with the new government to secure the interests of farmers. These differences inevitably did not take long to manifest themselves in the structures of organised agriculture. In the Waterberg, evidence of growing tension over political differences was apparent as early as 1975 when Tillie van der Merwe told members of the NWFA that if farmers hoped to retain their workers in the face of competition from the surrounding mining industry, they would simply have to improve their working and accommodation conditions. They couldn't rely on legislation to protect them.[56] In 1981 the NWFA urged the District Union to insist that the conservative NTK relaxed its rules regarding access by black employees to sales counters. In 1990 the District Union was criticised for selecting its representatives to the newly formed Bosveld Regional Services Council according to their political views, and Z.F. van der Merwe expressed his concern about the growing politicisation of organised agriculture by both the District and Provincial Unions.

By 1995 matters were coming to a head. A new constitution for the national union (SAAU/SALU) was being drafted that would incorporate the National African Farmers' Union. The chairman of the Vaalwater District Union (Louis Trichardt, a member of the Conservative Party) called on all farmers to support the union in representing the interests of organised agriculture to a government no longer close to agriculture. In 1998 several members of the NWFA, including Dr Peter Farrant and Janneman van der Merwe, argued that the association could no longer benefit from its affiliation to the Transvaal Agricultural Union (TAU/TLU), which now carried too much political baggage to have any credibility with the authorities. It would be preferable for the NWFA to affiliate directly with the national parent body, the SAAU. The TAU stated that it was not prepared to allow this to happen. NWFA members were divided about how to proceed – but having been members of the same agricultural fraternity for so long, they managed to retain personal friendships despite the growing schism in their political outlook. The end of the NWFA came quietly during the year 2000 when the majority of members resigned and joined the new incarnation of the SAAU, to be called Agri SA. The remainder stayed with the District Union, which in turn strengthened its affiliation with the TAU/TLU. As Ben Mostert, the last chairman, and Harald Schaaf, the vice-chair of the NWFA wrote in a joint letter to members:

> Some members of NWFA preferred to resign and join Agri SA, which is their personal right to do. This caused a temporary lull

> in the functioning of our farmers association. There are however, members of NWFA who prefer to remain affiliated to TAU ... The NWFA has served the farming community for 70 years or more and is the forerunner of our present VDAU ... During the years many obstacles on a variety of issues were overcome – more so than those we are facing today. Through the years many individuals of NWFA have made their mark, not only in our community but over the whole of RSA. We cannot allow this proud heritage to be lost!

Members were invited to reaffirm their membership. The files include letters from the descendants of two founder members of the NWFA back in 1914:

> I must inform you that I have affiliated with Agri SA. I very much regret this split in the farming representation as I feel it can only be to our detriment. The matter is however ... beyond my control. I wish you well in your representations. (Dr Peter Farrant, 30 October 2001);

> Thank you for the notice of NWFA meeting to be held on 25 March. I regret that I will not be able to attend. I would have preferred to attend and submit my resignation in person. I have virtually retired from farming and even so, I find my allegiance more for Agri SA. I resign with many years of happy memories of NWFA and our activities. (Colin Baber, 19 March 2002).

Transformation to conservation-oriented activities

The combination of social, market and political changes that burdened organised agriculture in the Waterberg from the mid-1980s into the first few years of the present century inevitably created great distress within the local farming community, characterised as it still was by many farmers with limited education, few alternative skills, long-term debt and not much cash. Their future appeared bleak. Fortunately for them, though, another kind of transformation was taking place concurrently: From the early 1980s city people and foreigners with substantial disposable income were becoming increasingly interested in the environment, in the bush and in conservation. They were looking for locations not too far away from their places of employment and primary residence where they could invest in a lifestyle that realised these generally altruistic interests (few at that time were looking at commercial breeding of game). The Waterberg was unknown to most city dwellers, a fact which gave it an added aura of wilderness and mystery. This, together with its proximity to Gauteng, its dramatic scenery and relatively low land prices, soon began to define it as a primary investment region for conservationists.

The promulgation of the Game Theft Act (No. 105 of 1991) provided an incentive to landowners to exercise private ownership of game species on their properties and to install perimeter fencing designed to prevent animals from relocating. To export or import animals or to shoot them outside of the designated hunting season, landowners had to be in possession of a permit, issued by the provincial nature conservation authorities upon their being satisfied as to the condition of the property and its fencing. Today, about three-quarters of game farms around the country are believed to be in possession of such permits, and together account for almost 15% of the total land area.

FIGURE 13.13: "Rain storm near Hanglip, Waterberg" by Wenselois Coetzer (1993); oil on board. Reproduced by kind permission of the artist.

With few exceptions, activities involving a large, unskilled labour force gave way to those requiring a much smaller number of higher-skilled employees. There commenced a migration of displaced former employees (as well as refugees from other countries) to informal settlements surrounding the few urban centres, bringing increased unemployment, crime and health issues and imposing demands for basic services that local municipalities were (and are) ill-equipped to deliver. Many properties owned for generations by the same landowning families were sold to new individual or corporate owners from the city or overseas, who often do not reside on the properties they purchase and have little knowledge of, or interest in the community they have joined. Land values have risen substantially: A typical Waterberg property valued at R150/ha in 1980 could readily find a buyer in 2016 at R10 000/ha, whereas its inflation-adjusted 1980-value would have depreciated to about R3100/ha[57]. This three-fold real increase in property values enabled many distressed conventional farmers to sell up and move into other occupations or retirement. Subsequent introduction of legislation intended to limit foreign ownership of land in South Africa, together with growing domestic political uncertainty and a collapse in the market for colour variants (also called freak species), led to a shrinkage of the market for wildlife properties and a softening in prices.

More recently, subdivisions of farms into smaller ownerships, further subdivisions into stringently fenced breeding camps, concerted fire control, illegal dam construction and the dramatically increased concentration of endemic and exotic game species have all exerted pressure on the environment. Consequences include extensive afforestation of formerly open grassland (due to fire control); reduced run-off (because of dams and the replacement of grassland by forest); predator extermination or control, leading to a burgeoning baboon population; and the threatened extinction of some small mammal and reptile species (python, pangolin, tortoise, hedgehog) due to electrified fences (ground wires in particular). These trends continue despite growing environmental awareness and conservation-oriented landownership – although it has to be stated that in the past, commercial farmers were at least as concerned about soil and veld conservation.

Species that were locally extinct in the 1960s – red hartebeest, waterbuck, roan, sable, blue wildebeest, zebra – now populate the plateau in unprecedented concentrations, together with large numbers of kudu, impala and smaller buck that had survived generations of uncontrolled hunting. Predators such as leopard, brown hyena, jackal and caracal are gradually increasing in numbers once more, although all are preyed upon by hunters (often illegally) and investors who wish to protect their allegedly high-value genetic variants enclosed in electrified predator-proof camps. On some large properties, species such as white and black rhino, lion, buffalo and elephant – even spotted hyena – have been introduced or reintroduced to attract tourists whose interest in African wildlife is often limited to the so-called Big Five, irrespective of context. The rise in demand for rhino horn and pangolin scales as medicaments in Far Eastern communities (their value is mythical, they have no medicinal benefits) has led to an explosion in rhino poaching and brought pangolins to the brink of extinction locally. Control of poaching by government is ineffectual, leading to allegations that senior officials and politicians might be complicit in the activity.

At the same time, governmental enforcement of land-use and nature conservation policies all but collapsed, owing to the departure from the relevant departments of experienced, dedicated professional practitioners and their replacement by often inappropriately qualified, disinterested and underresourced cadre deployees. This absence of competent management of the environment by government created an opportunity for speculators and short-term investors, often behind a facade of conservation, to introduce species that were never endemic to the Waterberg plateau – blesbok, springbok, gemsbok, eland, even bontebok – and worse, to participate in genetic engineering programmes designed to produce numbers of "colour variants" (golden wildebeest, black impala, coffee-coloured springbok, etc.). These recessive-genetic freaks were not "consumed" either by trophy hunters or the meat market; instead, they were supposedly sold, at vastly inflated prices, to an opaque market that might include other breeders, or gullible "wannabe" breeders lured, as with any good pyramid scheme, by the unbelievable prices widely reported in the media. Inevitably, this fad could not be sustained and by 2017 market prices had fallen steeply, leaving early participants wealthy and recent entrants ruing their gullibility. Inevitably, the crash in the market for colour variants spilled over into that for regular game and even rare species like buffalo and sable. Land values too, already weakened by the state of the economy and political threats to private (white) ownership, were further deflated.

The transformation of land use from commercial agriculture to less conventional activities has also had many positive consequences. Many farmers through history were ignorant, impoverished or desperate and in attempting to make a living out of land essentially unsuited to conventional agriculture, they damaged it in several ways. Overgrazing was a common fault, which left a legacy of poor soil highly susceptible to encroachment by invasives such as *Dichrostachys cinerea* (sekelbos), *Seriphium plumosum* (=*Stoebe vulgaris*, bankrotbos) and *Lopholeana conifolia* (pluisbos). The planting of exotic flora including *Eucalyptus spp.* (gum trees), *Populis spp.* (poplars), *Melia azerdarach* (syringa), *Lantana camara* (common lantana) and *Opuntia spp.* (prickly pears) for various purposes – drying out the soil, fodder, firewood, ornamentation – crowded out endemic vegetation, choked water courses and degraded the environment.

Many of the new conservation-oriented landowners, often more affluent than their predecessors, are actively engaged in rehabilitating old lands, clearing water-courses and removing alien or invasive vegetation in the hope of restoring the land to a more natural condition. The government agency Working for Water has been quite effective in cooperating with landowners to remove infestation of alien vegetation along watercourses. Groundwater quality has improved and the valuable catchment area represented by the Waterberg plateau – at least four important rivers have their source in these mountains – is being better protected than before. Several programmes operate on the plateau to provide environmental education to local residents, school children and others. Over a million hectares are now incorporated in a UNESCO-accredited biosphere reserve (see p. 466).

Lapalala

The creation of the Lapalala Wilderness Private Nature Reserve within the New Belgium Block in 1981 has been described elsewhere in this book (Chapter 9). It began a trend in that area that has resulted in over 70 000 ha being transformed into carefully managed, ecologically-sustainable conservation land.

Welgevonden

Across the Waterberg plateau, in the southwest, a similar pattern was developing. In 1987 Pienkes du Plessis, then the owner of the farm Welgevonden (some 15 km west of Vaalwater) conceived the vision of acquiring neighbouring farmland, establishing a nature conservancy and returning the area to its former wilderness state, in the same way that Messrs Parker and Walker were doing further north. He eventually (in 1993) secured the support of the Rand Merchant Bank for his concept and with the guidance and professional advice of several players, especially Nick Diemont and Trevor Jordan, the new Welgevonden Game Reserve began to take shape. Eventually, it acquired some 30 000 ha, divided into about 60 sites each carrying 500 ha of freehold title with traversing rights to the whole property. A ten-bed lodge could be built on each site, most of the sites carefully located so as to be concealed from roads and from each other. A landowners association was formed and a land management company created to supervise the maintenance of the property. Later, more land was included in the reserve, which today covers some 38 000 ha with over 55 lodges built on it. Many of these are corporate-owned and some operate on a commercial basis. The reserve has long accommodated the "Big Five" and is popular with high-profile international guests seeking a discrete and private experience in the African bush. Welgevonden is also actively involved in a number of faunal and floral research projects of benefit to the wider community. [58,59]

As with Lapalala, Welgevonden soon became the nucleus around which several other large nature and game reserves clustered: To the east is the 10 000 ha Shambala Game Reserve owned by the insurance tycoon Douw Steyn; to the west and southwest it is bounded by the Marakele National Park, an area of 42 000 ha straddling the dry low-lying sweetveld north of Thabazimbi and a moist mountainous region of the western Waterberg escarpment. This had first been proposed by the Department of Nature Conservation in 1984 when it was to be called the Kransberg Nature Reserve, an informal conservation area.[60] It was eventually declared as a national park in 1994, but lack of funds crippled its ability to develop. To the north of the Park, a remarkable Dutch businessman, passionate conservationist and philanthropist, Paul van Vlissingen, began assembling a private nature reserve from properties straddling the Matlabas River catchment area. In a bold move, following a meeting with then-President Nelson Mandela, Van Vlissingen proposed a scheme in terms of which his properties would be amalgamated with the new national park and managed operationally under a co-management arrangement between SanParks and a company established by him – a so-called public-private-partnership. Sadly, Van Vlissingen died before his vision could be fully realised, but with the involvement and support from his family, the concept has continued to progress. Today, the overall Marakele National Park, including privately owned land, covers an area of 68 000 ha with no internal fences and contains both private and public tourism facilities. Together with Welgevonden and Shambala, there is now a contiguous block of about 116 000 ha of the southwestern Waterberg plateau that is a formally protected nature conservation area.

Paul Fentener van Vlissingen

Born in the Netherlands in 1941, he was the descendant of a prominent and wealthy Dutch family that owned a conglomerate called SHV Holdings (with interests in coal, gas distribution, oil exploration, scrap metal, renewable energy and retailing). As a child he escaped Nazi occupation with his parents and grew up in Scotland, which became his second home and where he would become a prominent landowner and conservationist. From the mid-1960s, after obtaining an Economics degree, he involved himself in the family business, eventually becoming CEO and Chairman of SHV in 1984. He was described as having an unconventional leadership style, famed for his ability to find innovative solutions to business problems and for his thoughtful, empathetic respect for his employees, to whom he readily delegated authority.

He was a philosopher and a poet with a ready sense of humour and preferred to seek solutions through dialogue rather than through conflict. In 1998 he retired to nonexecutive chairmanship of the company and commenced investment in African game conservation, firstly through Marakele and later via the African Parks Foundation, which he helped establish. He invested tens of millions of Euro in Marakele and elsewhere in sub-Saharan Africa with the objective of establishing commercially sustainable game reserves. He died in 2006.

FIGURE 13.14: Paul van Vlissingen. Image kindly provided by his family.

Waterberg Biosphere Reserve

These two very sizable blocks of conservation-focused land, the Lapalala/Kwalata Block in the north and the southern Welgevonden/Marakele Block, became the pillars upon which the concept of the Waterberg Biosphere Reserve (WBR) was founded in 2000. They would be the "core" areas of the Biosphere, to be surrounded by less-rigorously conserved buffer zones that separated and shielded them from "transition areas" in which lay towns, settlements, communications routes and fully agricultural land. The brainchild of Dr Annemie de Klerk, then a conservation officer at the Limpopo provincial department of Environment and Tourism (LEDET), and Clive Walker, the WBR attained coveted accreditation with UNESCO in 2001.

The vision of the WBR is stated as being to maximise the potential of the Waterberg area for conservation, sustainable development, social upliftment, research and education. Since its founding, the WBR has been involved in tourism projects, rhino protection and a wide range of skills training programmes, including security guarding, hospitality and guiding, under its chairmen Dr Rupert Baber, and more recently, Lesiba Masibe. As with many such organisations with a broad constituency base, it struggles to satisfy every aspiration and to secure sustainable funding. Initially, the WBR covered an area of about 654 000 ha, made up of a highly protected core zone (16%), a buffer zone (28%) and an outer transition zone. By 2018 plans were afoot to increase the size of the biosphere to include the whole of the Waterberg municipal district.[61]

Waterberg Nature Conservancy

From as early as the mid-1980s, Clive Walker, the indefatigable conservationist, promoted the idea of a conservancy in the Waterberg to bring together landowners sharing similar views and concerns about the environment and land management. The Waterberg Nature Conservancy (WNC) was eventually launched in June 1990 with Lapalala and Touchstone as its anchor members. In the succeeding 28 years, this voluntary organisation has gone from strength to strength under the leadership of several chairmen, the longest-serving of whom have been Sam van Coller and, especially, John Miller (seven years). Today, membership is approaching a hundred, representing ownership of almost 200 000 ha (though not contiguous) across the plateau. The WNC has a very active programme of meetings (at each of which there are lectures covering a wide range of topics), debates, newsletters and other activities, including community-funding initiatives. It represents the community as an interested party in development and mining projects as well as in local and regional government forums on environmental issues.[62]

CHAPTER FOURTEEN
Matters of faith

When white settlers first moved into the remote region that became known as the Waterberg, they were forced to rely largely on their own physical and spiritual resources as they struggled to eke a living out of the poor local soils. Schools, clinics, courts and other forms of community support were scarce or absent. For most of the original settlers, the only reading material in their possession was the family Bible – and for many, that was all they knew how to read (often by rote). It is hardly surprising to find, therefore, that Waterbergers, in common with most rural communities, have always been and remain generally pious folk and that churches have always played an important role in community life on and around the plateau. Despite the sparse population of the region, numerous religious institutions have been or are represented. Therefore, it is important to understand how churches evolved to meet the needs of the communities they served – especially the evolution of the Dutch Reformed group of churches, which was closely linked to the settlement and growth of the white population of the Waterberg.

The oldest and probably the best-known church on the Waterberg plateau is the quaint Herbert Baker-inspired St John's Church at 24 Rivers, which celebrated its centenary in 2014. However, the earliest formal services in the Waterberg preceded St John's by over 50 years. To understand the context in which development of these early churches took place, we need to review the circumstances that led to their formation, even before they made their first appearances in the Waterberg.

The formation of the three "sister" branches of the Dutch Reformed Church

The Dutch Reformed Church, or *Nederduitse Gereformeerde Kerk* (referred to hereafter as the NGK), came to the Cape from the Netherlands in the seventeenth century and by the time of the Batavian Republic (1803–1806) was well established there. Although Governor De Mist in his Church Ordinance of 1804 ensured that all religious bodies which acknowledged a supreme being were granted equal legal protection, he also made them subservient to the State. The NGK was the dominant church of the colony; the government provided and paid for ministers who served in it.[1]

This relationship between State and Church was maintained after the British took responsibility for the Cape in 1806, except that the British administrators were keen to appoint English-speaking Calvinist ministers to fill any vacancies that might occur within the NGK, as part of a campaign to promote English language and customs in the community. This process was assisted after 1820 by a shortage of Dutch-speaking candidates, necessitating the appointment of several Scottish Presbyterian ministers (for example, the Rev. Andrew Murray Sr); rather surprisingly, they were successfully assimilated into the NGK.

However, as might be expected, this attempted "Anglicisation" of the NGK was not well received by the Cape Dutch community, particularly in the outlying regions of the colony. Together with the other policies and legislation adopted by the British administration after it officially took control of the Cape in 1814 – in particular, the abolition of slavery in 1834 and the adoption of English as the official language – the Anglicisation of the church was among the principal factors that led to the Great Trek which began in 1835.

The Cape Synod of the NGK officially deplored the decision by disaffected Dutch settlers to leave the Colony and start migrating into the hinterland and it refused to allow its ministers to accompany the trekkers. The reasons for this included the fear that in leaving the developed religious environment of the Cape for the interior, the trekkers would "lose their civilization" – and also that the ministers were effectively employees of the Cape government which paid their salaries. Moreover, the NGK's jurisdiction did not extend legitimately beyond the borders of the Cape Colony. Consequently, the trek groups, for whom religion was a critical cultural element, lost their source of formal religious instruction and mentoring. It was not long before the services were solicited of a variety of other ministers, missionaries and even clergymen of the London Missionary Society to minister to their spiritual needs.[2] Chief among them was an American missionary with the Zulu, the Rev. Daniel Lindley. Once the Voortrekkers had broken the power of the Zulu people in 1840, they appointed Lindley to lead their community from his base in Pietermaritzburg. By 1842 he had established parishes in Winburg in the Orange Free State and at Potchefstroom in the Zuid-Afrikaanse Republiek (ZAR, or Transvaal); but after the British colonisation of Natal and annexation of the Free State, only the Potchefstroom parish remained. This independent church, which became the foundation of the *Nederduitsche Hervormde Kerk* (NHK), gradually expanded and by 1852 had parishes in Rustenburg, Lydenburg and Soutpansberg as well as at Potchefstroom.[3]

In 1852 *dominee* (Ds.) J.H. Neethling and Ds. Andrew Murray Jr of the Cape Synod of the NGK travelled through the Transvaal and Orange Free State with the offer that the NGK would supply ministers to the communities, provided the congregations agreed to be incorporated into the Synod of the mother church (Cape governmental control of the NGK having ended in 1841). In the Orange Free State, this offer was accepted and the NGK became the official State church from 1854. However, the Transvaalers were to prove less amenable. There was still concern that somehow an association with the Cape Synod would facilitate an extension of the hated British influence. In August 1853, Ds. Dirk van der Hoff, who had been the first-appointed permanent NGK minister for the Transvaal, attended a meeting at Rustenburg at which it was resolved that the NHK, rather than the NGK, would be the *only* church recognised in the Transvaal.[4] This resolution was subsequently formally adopted by the Volksraad of the ZAR at Rustenburg on 13 February 1858 under the leadership of President Pretorius. The Lydenburg parish objected to this decision and opted to remain associated with the Cape synod of the NGK, thereby highlighting one of the many internecine feuds that were to characterise the early years of the ZAR and bedevil its administration and its leadership.

Although there were no formal settlements in the Waterberg region of the Transvaal at this time, the NGK minister from Bloemfontein, Ds. Murray, on one of his travels to Lydenburg, conducted a service in 1850 for the local Boer community in the area at some hot springs. It was attended by 12 wagons from the local area and another 15 that came from the Soutpansberg. He is said to have baptised 36 children on this occasion.[5]

During the next few years there were several similar excursions into this little-settled part of the republic, and on each occasion, services were held and baptisms performed by a visiting minister. The baptismal register recorded that a mere 143 children from about 50 families were baptised along the road through the Waterberg to the Soutpansberg in the period 1853–1864.[6]

While the decision to formally exclude the NGK from the ZAR in favour of the NHK was initiated by church elders whose minds were no doubt greatly occupied by the matter, rank-and-file congregants were largely unaffected by it. This is reflected in the fact that the first NHK service to be held in what would become the

FIGURE 14.1 (LEFT): Ds. Dirk van der Hoff. SANA TAB Photographic Library, No. 32893: Ds. Dirk van der Hoff of the NHK. FIGURE 14.2 (RIGHT): Ds. Dirk Postma. SANA TAB Photographic Library, No. 2121: Ds Dirk Postma, Gereformeerde Kerk.

Nylstroom area on 14 June 1860 was conducted by Ds. van Heyningen, the NGK minister from Lydenburg. The largely trivial nature of the differences between the two churches, such as they were, would again become apparent 25 years hence when common sense would lead them to merge their activities in Nylstroom, if only for a few years.

The inclusion of hymn singing in the NGK and NHK (as opposed to the singing of Psalms) stemmed from early in the nineteenth century when the practice was imported from the Netherlands. By the time of the Great Trek, it was well established, although there were still several parishes in the Eastern Cape (e.g. Graaff-Reinet) where this was not the case.

In 1859 a new minister sent from the Netherlands to serve the NHK in the Transvaal arrived in Potchefstroom. He was Ds. Dirk Postma, born in the Netherlands in 1818. He was deeply offended by what he considered to be the blasphemous practice of singing hymns (rather than Psalms) in the church – because the words to the hymns, unlike those of the Psalms, were not taken from the Bible. His objections led to a special meeting of the church council in Pretoria at which it was concluded that hymn-singing should remain part of the normal programme, despite Postma's objections.

So incensed by this decision was Ds. Postma that he resigned and, with likeminded members of the congregation, established a more orthodox church, the *Gereformeerde Kerk in Suid Afrika* (GK), (also known as the Dopperkerk), based on the branch of the church in the Netherlands whence Postma had been drawn. The group was supported by the influential Commandant (later President) Paul

Kruger, himself a *Dopper*. Postma went on to establish a theological training college in Burgersdorp, Cape, and later a similar college in Potchefstroom. Indeed, he and his family came to dominate the GK for the rest of the century. Dirk Postma was married five times: He outlived his first four wives, but from among them came 20 children. Five of his sons became ministers in the GK, as well as three sons-in-law; two of his daughters grew up to become authors of biblical and childrens' stories. All were graduates of his Burgersdorp college.[7]

Ds. Dirk Postma died in Burgersdorp in December 1890. One of his sons-in-law was Dr Jacob Daniel du Toit, better known as Totius. Totius was not only a respected and prolific Afrikaans poet, he and his father were also largely responsible for the translation of the Bible into Afrikaans, the first edition of which was published in 1933.[8] (Ironically, considering the orthodox nature of his church, Totius is said to have written "poetical versions" of the Psalms, also in Afrikaans.) The GK in Nylstroom would later be proud to record that the marriage of Totius to Postma's youngest daughter Maria (from his fifth wife) took place in the local church in 1903 when the officiating minister was Ds. J.A. van Rooy – who was married to another Postma daughter!

FIGURE 14.3: The original Gereformeerde Kerk in Nylstroom (Modimolle).

The churches take root in the Waterberg

By the time of the foundation of the first formal settlement in the Waterberg region (Nylstroom, in 1866), the ZAR hosted three Dutch Reformed churches, of which the NHK was the official and largest body, followed by the GK. (The NGK had managed to re-establish itself in the republic and slowly from its Lydenburg base was beginning to regain ground lost in 1852.) When the English-speaking, Lancashire-born farmer and merchant Ernest Olferman Collins offered to establish a township on part of his property on the banks of the Nyl River near Kranskop in 1865 (see Chapter 6), it was logical that he should donate to the community he hoped to establish not only a church square, but also some property on which the official church, the NHK, could construct a church. A local parish of the NHK

FIGURE 14.4: Detail and cornerstone of the first church building.

FIGURE 14.5: Nagmaal on the square in front of the old Gereformeerde Kerk. Origin of photograph unknown.

was already in existence (having been established on 20 November 1865),[9] and although there was no resident minister, visiting *konsulente* (relieving ministers) would conduct services on nearby farms. One such minister, Ds. G.W. Smits from Rustenburg, seized the initiative on behalf of the local church board and led a campaign to raise funds and to commence building. The cornerstone of the church was laid on 14 April 1866. Mr Collins himself headed the building committee, but unfortunately he died on 16 October 1868 before building had been completed. As a result, completion of the church was delayed and the building was finally dedicated only on 6 December 1873 – appropriately by Ds. G.W. Smits.[10]

The GK also had a presence in the Waterberg from the time of Nylstroom's formation, but unlike the NHK, it did not have the means to build a church of its own in those early years. Instead, services were held occasionally on one of several neighbouring farms with relieving ministers from Pretoria, including Ds. Dirk Postma himself officiating. It was not until 12 April 1889 that the foundation stone of the new GK church building was laid. Building sped ahead and the church was opened seven months later on 23 November 1889 with State President Paul Kruger (a Dopper) as the leading guest and Ds. J.A. de Ridder of Rustenburg conducting the dedication. De Ridder, who had been the principal relieving minister since 1879, would continue in this role until 1895, a total of 16 years' service. This original GK church building still stands in the grounds of its successor and has been beautifully restored for service as a church museum. It also houses a very fine collection of Bibles dating back to the eighteenth century. It is the oldest surviving church building in Modimolle.[11]

The NGK had no significant presence in the Waterberg area in the 1850s and 1860s, its influence in the Transvaal being concentrated in the parish of Lydenburg. However, with the easing of relations between it and its sister church, the NHK, services began to be held on nearby farms. The population of Nylstroom and surroundings was still very small. The first survey conducted in 1869 indicated that there were 312 white people in the area, of whom 192 were practising members of the NHK. These numbers grew to 503 and 207 respectively eight years later. Ds. Smits's service to the NHK community ended in 1877. He was replaced as relieving minister by Ds. H.S. Bosman from Pretoria.

Ds. Bosman was to continue in this role for the next 13 years, becoming greatly revered by his parishioners. Through his patient moderation, an agreement was reached in 1885 in terms of which the NHK and NGK would merge to form the *Nederduitse Hervormde of Gereformeerde Kerk (NH of GK)*, or, as it became called, the *Verenigde Kerk* (United Church) – an outcome widely welcomed in the community,

FIGURE 14.6 (LEFT): Ds. H.S. Bosman. SANA TAB Photographic Library, No. 2004: Ds. H.S. Bosman of the NGK. FIGURE 14.7 (RIGHT): Ds. M.J. Goddefroy. SANA TAB Photographic Library, No. 12093: Ds. M.J. Goddefroy of the NHK.

although there was (inevitably!) a small group of *Hervormers* who did not support the merger. For a few years, this arrangement endured: the Verenigde Kerk used the NHK church hall for its services, while the "rebel" Hervormers met elsewhere.

In 1888 a group of Hervormers under the influence of Ds. M.J. Goddefroy, revived the NHK faction. They appointed as their leader one F. Boshoff who until the previous year, had served as a deacon with the Verenigde Kerk and was familiar with its activities. This revival created a new wave of bitterness and friction in the small community with some meetings even ending in blows. However, in due course, the Hervormers obtained a court order entitling them to the shared use of their former church. Clearly, having both the NHK and the Verenigde Kerk sharing the building was impractical. A commission established to look at the matter valued the church premises at £2000. It was agreed that if the rebel group paid the Verenigde Kerk half of this amount, they could resume sole control of the building. And so, in 1896, the revived NHK under Goddefroy re-occupied its old church, and the Verenigde Kerk, which comprised the NGK and the rump of the old NHK, was left without a formal place of worship.[12]

The ousted church immediately took steps to rectify this situation. Since 1890, it had enjoyed the benefit of a permanent minister, Ds. D.J. Viljoen, who lived in a pastorie built for him and who was paid the princely amount of £300 per annum. This was raised by levying £1 per annum upon each parishioner. Several could not or did not pay of course, and in one year the shortfall was covered by the generosity of Emil Tamsen, the owner of the first store in Nylstroom. By appealing to its members, the town and the district, the church was eventually able to raise enough money to commence construction of its own new church building. The foundation stone was laid by the venerable Commandant General Piet Joubert in April 1898. The dedication of the fine new building took place exactly a year later when a new resident minister, Ds. J.L. du Preez, was also installed. This church is the beautiful building, known to all as *die Wit Kerk* (with its black steeple) that still dominates the town of Modimolle, located close to the junction of the road to Lephalale with the old main road between Bela-Bela and Mookgophong. It has since the turn of the twentieth century been the principal home of the NGK in the town.

By the time of the outbreak of the South African War in September 1899, there were three variants of the Dutch Reformed Church in Nylstroom – the three sister churches each had its own church premises. The NGK and the GK churches had resident ministers (Ds. Du Preez and Ds. J.A. van Rooy respectively), while the NHK was served by Ds. M.J. Goddefroy, who had been visiting the town since 1887 when he'd led the split from the United Church.

FIGURE 14.8 (LEFT): The Wit Kerk, home of the NGK in Nylstroom since the early 1900s. Museum Africa Photographic Collection, No. PH2006-10102A. FIGURE 14.9 (RIGHT): The Wit Kerk today.

The role of the Berlin missionaries

Throughout the preceding 30 years, during which the Dutch Reformed Church had been subjected to the repeated *broedertwiste* (fraternal quarrels) described above – and which seemed to characterise much of Boer society at that time – another organisation had been quietly establishing itself and focusing on spreading the gospel in the northern part of the Transvaal amongst the black community.

This was the Berlin Missionary Society (BMS), which, organised and funded from Bremen in Germany, set out to spread Christianity and at the same time provide black communities with a basic education while taking care to avoid becoming involved in politics – a tightrope if there was one. In this respect, it differed sharply from the London Missionary Society that operated in the Cape and was highly politicised in its endorsement of British imperialism. Consequently, the BMS generally enjoyed the support of the ZAR government up until the outbreak of the South African War. The BMS was extremely successful too: From a single mission (Gerlachshoop, near Groblersdal) in 1860, with two missionaries and one baptised member, it grew to 30 mission stations, 105 outposts, 36 full-time missionaries and a total of over 20 000 baptised black converts by the end of the century.[13]

The BMS concentrated its activities in the far northern and eastern parts of the Transvaal where large black communities lived. However, in the very sparsely populated Waterberg, it had only a single mission station: Modimulle (as then spelt) on the farm Middelfontein near the present rail siding of that name about 15 km east-northeast of modern Modimolle. In August 1867 the missionary Heinrich Koboldt established the Modimulle mission station on a piece of land near the Kranskop hill donated for this purpose by a local farmer, Gert Lottering, who was well disposed towards missionary work. Lottering and his brothers, Frans and Cornelis, had themselves already undertaken missionary work among their black employees. Koboldt soon realised that the property donated by Lottering would be too small to support a mission station. Accordingly, on his own initiative, he purchased the nearby property of Middelfontein on the other side of the Nyl River – at some personal risk, because he was now obliged to ensure that he could generate sufficient income from the property to pay the BMS for it, or lose his job. But Koboldt worked hard and in time the Modimulle station on Middelfontein came to be regarded by the BMS as a valuable strategic asset.[14] The BMS needed a place of safety midway between Pretoria and its activities in the northern Transvaal and Modimulle fulfilled this role ideally. The station was attacked several times by black militants, but Koboldt achieved his first baptismal in 1870. It was with missionary Koboldt that the German traveller Carl Mauch stayed in November 1869 before making his well-diarised visit across the Waterberg in the company of a ZAR survey team (see Chapter 5).

Unfortunately, Koboldt contracted an illness and died in 1874 while still busy building his church. He was immediately succeeded by Emil Beyer, whose first task was to complete the construction. The results of work at the station were sufficiently encouraging to warrant the posting of a second missionary, Otto Posselt, to the station. Beyer stayed at Modimulle until his resignation in 1880 over a disagreement with the BMS administration. Such was the capacity of the BMS that Beyer was replaced without delay by Oswald Krause, who remained at the station until 1892. Krause's first wife died there from malaria in 1883, but the missionary stayed on to develop a school and subsistence farming operation in addition to his mission work. He was by all accounts popular and achieved much at the station, including the establishment of three outposts in the district. In 1886 he was honoured by an overnight visit from President Paul Kruger, who was greatly impressed by the neatness and productivity of the cultivated lands.

FIGURE 14.10: *Mission Station, Waterberg, Transvaal, 19 March 1885*. A pencil sketch by Dr Hermann Wangemann, Director of the Berlin Mission Society, while on a visit to South Africa. The sketch is among 106 of Wangemann's works in the collection of the Ditsong National Cultural History Museum, Pretoria (Acquisition No. HG 6089) which authorised this reproduction. Imagery kindly provided by Jan Middeljans.

FIGURE 14.11 (TOP): The Berlin Mission at Modimulle at the time of Rev. Krause. SANA TAB Photographic Library, No 16308: Modimulle Mission, photograph by H.F. Gros and included as No. 504 in his Pictorial Description of the Transvaal. Photograph taken c.1891.
FIGURE 14.12 (BOTTOM): Rev. Oswald Krause (left) and Rev. R. Jensen with his wife (right). SANA KAB Cape archives photographic library: AG8884 – Rev. Oswald Krause; AG8896 – Rev. R. Jensen and his wife.

When Krause was appointed as superintendent of the Northern Transvaal Synod of the BMS in September 1892, his position at Modimulle was taken by Rasmus Jensen. Jensen urged the people of the mission to work hard in the name of *Christendom*. This commitment enabled the station to survive the periods of drought and locust plagues that occurred in the 1890s, as well as a whooping cough epidemic in 1898 which caused the number of people at the station to fall sharply. Nevertheless, sufficient workers remained on hand to commence with the building of a new church which was commissioned in May 1899. The ceremony was attended by Commandant General Piet Joubert, an admirer of the BMS in general and the work of Jensen in particular. The latter was unashamedly pro-Boer, with the result that when the British occupied the area in April 1901 towards the end of the South African War, Jensen was forced to leave the station and move to Pretoria.[15] He only returned to Modimulle in July 1902. The community had grown to over a thousand in his absence and the station was flourishing.

In addition to the farms Middelfontein and its southern neighbour Zandfontein (the original Lottering property), the BMS acquired ground to the north and east of the mission so as to enable the growing resident community to generate an income from farming, both for themselves and the mission.

Unfortunately, this progress could not be sustained. During World War I and II, many German missionaries were interned, and the mission stations they'd managed suffered accordingly. Worse for the BMS, its Berlin headquarters fell into the Russian sector of Berlin at the end of World War II, cutting off both missionary recruits and funding. Many mission stations had to close.[16] Between the wars, the BMS had attempted to sell Zandfontein to members of the resident (black) community, but this was in conflict with the 1913 Land Act and

FIGURE 14.13: The ruined tower of the Modimulle mission church, 2008. Photograph provided by Jan de Beer, Kokanje, Modimolle.

was disallowed. Instead, it began to sell off portions of its properties to adjacent white farmers to generate sufficient income to maintain its operations at Modimulle. Many members of the community therefore found themselves no longer living on mission property, but as unwelcome tenants on privately owned white land. After the promulgation of the Group Areas legislation in the 1950s, their tenure on these properties became vulnerable; most were relocated to new, often unserviced stands in nearby fragments of land destined to become part of the Bophuthatswana homeland. Eventually, the mission station on Modimulle fell into disuse and was demolished. In 2004 descendants of some of the early residents on former BMS land succeeded in a land claim enabling them to return to Middelfontein and surrounding properties where they reside today.[17] The ruin of the old mission tower was still standing in 2008, but it and the remaining buildings were subsequently demolished.

The sister churches and the South African War

The war effectively forced the suspension of activities at the three Dutch Reformed sister churches in Nylstroom: Both Ds. van Rooy (GK) and Ds. Goddefroy (NHK) joined the Boers in the field, Van Rooy being attached to the command of General Beyers. Goddefroy was captured by the British while sick with fever and sent off to India as a prisoner of war. At the end of the war he was repatriated to the Netherlands, only returning to the Transvaal in 1903. Ds. Du Preez (NGK) fell ill and went to live with his family in Pretoria for the duration, as there were no funds to pay him a salary. Several members of the boards of all three churches were killed or taken prisoner during the war, and those who were found to have aided the British enemy were later expelled.

The British occupation of Nylstroom in 1901/02 further aggravated the situation: The new NGK church was commandeered as the local military headquarters and the NHK church across the road was converted for use as stables. The original GK hall was turned into a hospital (and its vestry into a mortuary) for both military and civilian wounded and sick, including the many patients (almost entirely women and children) who came from the concentration camp established nearby.

While it is inevitable (and necessary) in wartime for harsh military decisions to be made that are intended to subdue or punish the enemy and those who would provide him succour, the totality of measures adopted by the British against the Boers in this conflict (the burning of homes, confiscation of stock and property, transportation and internment of families under unhygienic, poorly administered conditions, exiling of men to distant internment camps, occupation and defilement of sacred buildings, grossly inadequate compensation for losses incurred) is difficult to justify or condone.

There is no doubt, with the benefit of more than a century of hindsight, that the Boers who chose to continue the war after the fall of Pretoria in June 1900 (the so-called *bittereinders*) had little, if any prospect of victory; on the contrary, they had a great deal to lose. Equally, however, the measures adopted by Kitchener, Hamilton and Milner in their efforts to hasten the end of the guerrilla phase of the war (1900–1902) were unjustifiably draconian, as the outspoken Emily Hobhouse repeatedly pointed out to her country-folk. Yet many English-speakers in South Africa remain astonishingly oblivious of the dire consequences of these actions for the Boer community. Hence, they fail to understand or appreciate the residual enmity that lingers in the Waterberg and elsewhere to this day – as evidenced, for example, by the content of a local (Modimolle) publication written to celebrate the 150th anniversary of the conservative GK in 2009.[18]

Rebuilding the communities after the war

The Treaty of Vereeniging signed on 31 May 1902 may have brought the tragic conflict that had been the South African War to a conclusion, but for the three Dutch Reformed sister churches the process of re-establishing themselves in the Waterberg would be long and hard. Many of their congregations had died during the conflict in battle, of sickness and injuries, and from factors related to their incarceration. Others had moved away to different parts of the country, and many men were still imprisoned in camps overseas in India, Ceylon, Bermuda and elsewhere. The return of these prisoners of war was a slow process, requiring first an oath of allegiance to the new government that many found hard to take. In December 1903 – 18 months after peace had been agreed – some 500 Boer prisoners in an Indian jail were finally persuaded, thanks to a personal visit by General De la Rey, to put the past behind them and sign the oath.[19] (Others, including, for example, the soldier, statesman and author Deneys Reitz, could not bring themselves to do so and remained in exile. Reitz only returned to South Africa from Madagascar in 1905, at the urging of his friend and mentor General Smuts.)[20]

The three churches dealt harshly with those Boers accused of having surrendered to the British; and especially with those who had joined the National Scouts, a corps of paid Boer spies and scouts established by Kitchener in December 1901 to assist him in hastening the end of the guerrilla war. By May 1902 there were over a thousand of these scouts active in the Transvaal, and half that number in the Free State. As it turned out, the Scouts contributed little to the war effort, but added substantially to the enmity of the Afrikaner population towards Britain. They were hated by the rest of the Boer community (those captured were sometimes summarily shot or mutilated) and they and their families had to be protected.[21,22] These so-called *hensoppers en joiners* were especially vilified in the northern Transvaal and Free State where much of the last action of the war had been fought. All three churches, the GK especially, were determined to punish their countrymen for their desertion from the Boer cause. The men (and a few women) concerned were effectively excommunicated from the churches until such time as they had formally apologised for their "sins against God and his community". Even then, there was lingering animosity towards them, and some even established a separate church for a period.[23]

The GK held its first post-war service in Nylstroom in November 1902, followed a month later by that of the NGK and in July 1903 by the first service of the NHK – which received its first resident priest, Ds. Brandt, in April 1904.[24] When Brandt left in 1908, he was replaced by Ds. Goddefroy who, since his return to the country, had been minister at Pietersburg and Potgietersrus. Goddefroy eventually retired in 1913 after 26 years' service in the NHK. At the GK, Ds. J.A. van Rooy returned from war to continue his ministry until July 1905 when he was replaced as *konsulent* by none other than his brother-in-law, the revered Totius himself, who, after serving with the Boer forces in the Cape during the war, had married Dirk Postma's daughter Marie in the Nylstroom church in 1903. Keeping things in the family, so to speak, Du Toit was succeeded in December 1906 by Ds. M. Postma, another of his many brothers-in-law. Ds. J.L. du Preez of the NGK returned to Nylstroom after the war, but his health remained poor; he died in 1906, shortly after moving to Pietersburg. For several years the church had to make do with visiting ministers.

The decade following unification in 1910 was remarkably difficult for the three sister churches and the country as a whole: The outbreak of war between Britain and Germany in August 1914 triggered a resurgence of anti-British animosity among large sections of the Afrikaner community, especially in conservative areas of the Free State, western and northern Transvaal. Rebellion broke out in September of that year with several former Boer leaders, including De Wet, De la Rey, Beyers, Kemp and Maritz resigning their commissions in the Union Defence Force to lead the rebels. This led in turn to a split within the community between those who supported the rebels and those who retained allegiance to the Union government under Prime Minister Botha. Even after the rebellion had been quelled by January 1915, resentment lingered. It was no doubt aggravated by the straitened national economic circumstances brought about by South Africa's participation in the war – in German South West Africa (1915), in East Africa (1915–1918) and in France. The battles of Delville Wood (part of the Somme Offensive) in July 1916, Ypres in September 1917 and Marrieres Wood and Messines Ridge in early 1918 accounted for most of the 12 000 South Africans who died in World War I. In late 1918, an epidemic of the so-called Spanish influenza struck: The total death toll in the country from the pandemic has been estimated at between 150 000 and 200 000, including some 11 750 whites.[25] In addition, there were numerous strikes between 1913 and 1920, the rise of increasingly militant African and Afrikaner nationalism, and the death on 27 August 1919 of the embattled, depressed and stressed Prime Minister, Louis Botha, aged just 57, of a heart attack.

FIGURE 14.14: The Hervormde Kerk in Modimolle (LEFT) with the commemorative plaque (RIGHT).

The cooperation of the sister churches

The NGK *raad* in Nylstroom was appalled to discover that the flu epidemic had left 174 children orphaned or homeless within its community alone. Despite having no funds available, the raad somehow arranged to accommodate and cater for the children, with an initiative that would eventually (in 1958) result in the building of a local branch of the Abraham Kriel Children's Home, which still operates in the town.[26] A few years later, the urgent need arose for the accommodation of children attending school in Nylstroom from outlying areas. No money was available from government to cater for this need, so the three sister churches cooperated to raise the necessary funds. Two hostels were built for a cost of £14 000 on properties near the school, donated by the municipality. An interfaith committee was established to manage them. In 1934 the Education Department assumed responsibility for the hostels; they are still in use today, albeit no longer as hostels (see section on Nylstroom in Chapter 6).[27]

New church buildings

In 1920 the NHK community was at last able to raise funds (from the sale of a farm) enabling it to commence building a new church to replace the original structure built in 1868 on the land donated by Olferman Collins – and control of which had been wrested from the Verenigde Kerk in 1896. The cornerstone was laid on 26 March 1921 and the building was consecrated at a service on 8 April 1922. This is the same building currently used by the NHK in Modimolle, although a tower was added later. The years following the consecration were spent in paying off the debt incurred in building the new church and a home for the resident minister. This process was prolonged due to the global Depression of the 1930s, followed by World War II. It was only in February 1956 that the community was able to celebrate the completion of the clock tower and steeple that graces the building today.[28]

The GK had also felt the need to replace its original church building which dated from 1889. In 1930 its present church with its attractive octagonal structure and tower was completed. It was designed by Gerard Moerdyk, the architect of the Voortrekker Monument, who had a home between Nylstroom and Warmbaths. The church was built adjacent to the original building, which fortunately was not demolished, becoming instead the charming museum that can be visited to this day. It is notable that throughout this period, indeed for 37 years from 1923 to 1960, the GK community in Nylstroom enjoyed the almost uninterrupted service of Ds. H.J. Venter, at a time when its two sister churches struggled to retain resident ministers. Also worth noting is that from the inception of the GK in Nylstroom in 1859 until 1908 – a period of 49 years – the community had been served by one or other member of the founding Postma family.

Nagmaal

Throughout the first 90 years of the existence of the three sister churches in the Waterberg region, all congregants who lived outside of Nylstroom had to travel into the town if they wished to participate in formal services – although many ministers also travelled through the district holding services on farms. It became traditional that on (at least) four occasions a year, families would trek – usually in

FIGURE 14.15: The Gereformeerde Kerk in Nylstroom, designed by Gerard Moerdyk in the 1930s. Museum Africa Photographic Collection, No. PH2006-10098.
FIGURE 14.16: The Gereformeerde Kerk in Modimolle today.

ox-wagons, because roads were poor or nonexistent and cars rare even until the 1930s – to their nearest town church where they would camp for a few days while they attended the church to celebrate Holy Communion, or nagmaal. These were major occasions in the lives of rural citizens: Not only were they able to renew, formally, their vows to their faith as well as baptise and confirm their children – and maybe even celebrate a wedding – but at the same time they could purchase much-needed supplies, learn about the news of the country and enjoy a wide range of social events, not least of which was the opportunity to court a future spouse. For some distant parishioners, nagmaal could occupy a full week: three nights camped on the *kerkplein* and two nights on the trail before and afterwards. For many farming families, these events represented the only times when they would leave their homes together.

Typically, families would attempt to arrive in town on the Thursday evening in time to set up camp on the kerkplein. Friday would be spent shopping and socialising, with a service, sometimes informal, held that evening. Proper ser-

FIGURE 14.17: Nagmaal weekend at the "Wit Kerk" in Nylstroom in 1941. *Nagmaal by die Witkerk, Nylstroom*, 1941. Watercolour by Erich Mayer. This work is in the collection of the Ditsong National Cultural History Museum, Pretoria (Acquisition No. HG 7014 404) and is reproduced with permission. The image was kindly provided by Jan Middeljans.

vices would commence on Saturday morning with confessional and marriage ceremonies, followed by community meetings on various subjects, elections of kerkraad members and resolution of financial matters. The afternoon was spent by adults in preparation for the next day, while youngsters would talk, play or practise choir singing. Sunday morning was the time of the formal communion service, with white table cloths, bread and wine, choir singing, etc. Dress was also formal and womenfolk invariably wore a *kappie*, or headdress. The communion service would be followed by a children's service and at midday by a thanksgiving service. Sunday evening would be spent in socialising and in taking leave of one's friends. Oxen were rounded up and made fast close to the wagons and the process of packing up the camp would begin. By late evening, everything would be ready for departure and a few hours' sleep would be sought before leaving. But early in the morning, especially under a full moon, the wagons would set out on their long trek home – some only reaching their destinations two or even three days later.[29] The late Oom Kassie Steyn of Melkrivier was born during such a trip on 11 October 1925, while the family was outspanned at what today is the Waterberg Game Park on Tarentaalstraat, on the way home from Nylstroom. The old wagon still stands outside the Steyn homestead on Libanon.[30]

Ds. Alheit was the NGK minister in Nylstroom from 1925 to 1931. In 1998, at the centenary of that church, his son recounted how during a nagmaal in 1925, he (the son) had been one of 52 babies baptised in the "Wit Kerk" that Sunday. As his father was putting things away in the vestry after the service, a young

FIGURE 14.18: The wagon on which Kassie Steyn was born in 1925.

man rushed in and asked him if he would mind carrying out one more baptism. When the minister asked why the baby hadn't been included in the service just concluded, the man responded that the baby had only just been born in a tent on the kerkplein – and that if he wasn't baptised now, they'd have to wait another three months![31]

Establishment of Dutch Reform churches in outlying areas

Matlabas

Enjoyable and memorable though these quarterly trips into town surely were, they were insufficient to provide religious support to the communities the rest of the time. Moreover, the district was becoming more populated and the need grew to develop more churches. Primitive structures were common, at least in the early years. In June 1918 the GK established a community at Matlabas, below the western escarpment of the Waterberg. The first service was held under a *tambotie* tree, but eventually on 9 July 1932 the cornerstone was laid of what became, the following year, a fine church for the Matlabas Gemeente (see Figures 19.5 and 19.6 in Chapter 19).[32] After the Magol [sic] (later Ellisras/Lephalale) community of the GK was established in 1942, nagmaal was held beneath a tent tabernacle erected on each occasion until a church was built there too in 1958.[33]

Warmbaths

In 1909 the NGK was awarded a plot of land in Warmbaths for the purpose of building a church. Immediately, three sisters (Catherina Potgieter, Aletta Grobler and Johanna McCord – they were all Potgieters) sought approval to build a small hall on the property in which to hold services. This was granted and building commenced without delay. The hall, which became known as the *Sustersaal* (sisters' hall) was commissioned the following year and remained in use for over 40 years, until the present church (*die Moederkerk*) was commissioned in 1951.[34] In due course, this congregation grew so large that a second congregation, Warmbad-Wes, was established in 1969, eventually opening its own church in 1975.

In the area of Het Bad, (Warmbaths), early GK members would travel to Rustenburg (the seat of the GK) or to Pretoria for services. Ds. J.A. de Ridder, the *dominee* from Rustenburg, was *konsulent* for the area from 1879 to 1895. Thereafter, it fell under the jurisdiction of the Waterberg parish in Nylstroom, until a Warmbad Gemeente was established in 1954. This change was preceded by the decision in 1943 to build a facility for services in Warmbaths. Building began immediately after the end of the war; the cornerstone was laid by Ds. H.J. Venter of Nylstroom and the hall was taken into service on 28 July 1945.[35]

Naboomspruit

The town of Naboomspruit (now Mookgophong), lying on the railway line north from Nylstroom to Potgietersrus (Mokopane) and Pietersburg (Polokwane), developed from the settlement that formed around the rail siding established on the farm of that name in 1898. It was preceded by a small settlement (and Vroom's store) on the adjacent farm Vlakfontein where the first NHK service, led by Ds. Goddefroy, had been held in 1888. In the years following the South African War, Ds. L.E. Brandt visited the area to hold services, the first being in a tent on Kotie Engelbrecht's portion of Vlakfontein. After the establishment of the town in 1910, the population grew sufficiently for the NHK to set up a Sunday school in 1911, attended by children from all three sister churches. This continued until 1927. Services were held in private homes with visiting *dominees* from Waterberg and Potgietersrus. During World War II, an abandoned garage was purchased and this became the temporary home of the church. After the war, the size of the congregation grew; by 1962 it was over 200 and a new church was built, opening in 1964, although the original hall was restored and is still used today.[36] The new

FIGURE 14.19: The Hervormde Kerk in Mookgophong (Naboomspruit).

church, unusually modernistic in design, is set in a garden renowned for its fine selection of indigenous trees, amongst which are memorials to the South African War and the Great Trek. Today, the congregation totals almost 700.

Unlike the other two churches, the NGK in Naboomspruit had in 1939, upon the establishment of a community in the town, acquired a hall of its own in which to hold services – a building it had purchased from the SA Party, which no longer needed it. This hall, called Anderson Hall after the man who had built it in 1917, was the only one in the town, and for a while the congregation allowed it to be used for civic functions unrelated to the church. Dr Moorrees was appointed as the first resident NGK minister, and a pastorie was built to accommodate him. In 1920 the construction of a branch rail line to Zebediela – and the placing of a contingent of railway maintenance personnel in the town – promoted the growth of Naboomspruit. In 1939 NGK Naboomspruit became a congregation in its own right and by 1951 it could boast a membership of 850.[37] A proper church building – the one used today – was taken into service in 1954.

The GK Gemeente of Naboomspruit was established in November 1957 through the efforts of Ds. G.B.S. Pasch from Potgietersrus with services initially being held in the NGK building. The first resident priest was J.T.C. Snyman, a young and dynamic leader. A church building was commissioned in November 1958, and by 1959 there were 215 members of the church. On 6 December 1958, the GK established a community in Vaalwater where seven practising members of the church had settled. This number quickly grew to 82 and the following January, an agreement was negotiated with the Vaalwater community in terms of which it would take responsibility for a portion of the salary and transport costs of the dominee based in Naboomspruit, enabling him to attend to its needs. Ds. Snyman became the first GK dominee of Vaalwater in addition to his role in Naboomspruit. The first elders of the Vaalwater community were Messrs J.P. Bekker, T.J. Kloppers, P.S. Vermaak and J.C. Coetzee.[38]

In January 1960 the Vaalwater GK community decided that it needed to build a church for itself. No time was lost: In March, Ds. Snyman brought a building commission to visit the proposed site (made available by the private township developers) and placed the first peg. Under the leadership of Bekker, members of the community commenced construction. This was completed in a mere five months and on 9 July 1960, the GK Vaalwater with a capacity of 160 members was opened. It was the first church to be built in the village. It had a detached tower on the top of which was placed not the usual weathercock, but a trowel, as a reminder of the further work to be done.[39]

Ds. Snyman was replaced in 1962 by Ds. L.A.S. van Wyk and in 1966 by Ds. G.J. van Staden-Roos until 1975. From then until 1985, Naboomspruit and Vaalwater were led by Ds. P.G.L. van der Walt. A hall was built at Naboomspruit in 1980. The Vaalwater community continued to share dominees with other small Waterberg communities up until 2003 when it was able to appoint its first resident, full-time priest, Ds. Albert Berg. By 2017 when Ds. Berg left the ministry to go farming, the GK community in Vaalwater had grown to about 140.

Lowveld/Ellisras

The NGK also began to develop communities to the north of Nylstroom. The first of these followed a meeting at the school on Zanddrift (10 km north of Vaalwater) on 6 May 1930. It was decided there to form a new NGK community to be called Waterberg-Noord with Ds. Alheit from Nylstroom doing additional duty as its first konsulent. This was an enormous area to cover and Ds. Alheit soon had to split it into two: the *Bo-berg* or Vaalwater area on the Waterberg plateau; and the Bushveld area to the north. Each would meet separately at an agreed convenient location.

FIGURE 14.20: The Gereformeerde Kerk in Vaalwater and its cornerstone (INSERT).

The northern or Bushveld area had always been so far away from Nylstroom that it had already formed several informal communities that would meet at homesteads. From 1915 these small communities began to receive the occasional services of a visiting konsulent from Nylstroom. One such congregation would meet regularly on the farm Oranjefontein where a wagon shed belonging to the owner, Koos Geldenhuys, was used for services. Soon, however, a more appropriate building was needed and a site was selected on the farm Hoornbosch. This church, the so-called Mispa Kerk (*Mizpah* from the biblical symbolism of a bond, witnessed by God, between people widely separated) – still within the parish of Waterberg-Noord – was built in 1940. Six years later, it was granted its own status as the Albertyn community in acknowledgement of the generous services rendered to the parishioners by Ds. J.R. Albertyn, the secretary of the *Armsorgraad* (Poor Relief Board). However, by 1950 the original structure was becoming dilapidated and funds were raised to build a new church about 10 km further south on the farm Waterkloof where there was already the embryo of the settlement that would soon become Ellisras (Lephalale). The NGK Albertyn community of Lephalale has been housed in this new building since 1957.[40] In 1945, Steenbokpan (50 km to the west) was granted its own church board and soon built a church of its own there too. Similarly, the NGK built a church at Marken in 1953.

Melkrivier

Meanwhile, back on the Waterberg plateau, members of the Bo-berg group voted on three alternative locations for their church in May 1931. The result was: Vaalwater (15 votes); Zanddrift (20); and 85 for Sondagsloop along the Melkrivier on Nooitgedacht.[41] This result reflected the concentration of farmers in the Melkrivier area at the time of the poll. The local farmers' association and one of the NGK's Bo-berg parishioners, Bezuidenhout, donated 100 morgen (86 ha) from Nooitgedacht on which to build a church. The cornerstone was laid on 7 May 1932 by Ds. J.A. Koch (who had succeeded Ds. Alheit at Nylstroom) and building proceeded apace, being completed late in the year. The building was rectangular, oriented east-west, with a *konsistorie* (vestry) at the eastern end, entrances on the western and southern sides, and large wooden windows along each side. By 1937 a modest pastorie had been added; its first occupant was the newly ordained Ds. J.C. Botha, who for the previous two years had been boarding with members of the congregation.[42]

Many are the memories and stories told about life around the church on the Melkrivier. Louis Nel recalls the intensity of *katkisasieklasse* (confirmation classes), held for days before nagmaal in tents set up on the kerkplein where the ministers would do their best to inculcate the essentials of religious doctrine ahead of the confirmation ceremony. He also remembers their value as a means of meeting potential marriage partners. The annual *Geloftefees* held on Dingaan's Day (16 December), was also an occasion filled with social and sporting events as well as services of thanksgiving. At Sondagsloop a huge roofed, flagstone-floored, but open-walled structure had been built to host the various functions. On one occasion, the resident dominee and his *ouderlinge* (elders) were returning from having conducted nagmaal in the Lowveld when a cold front blew in over the plateau. Those sitting in the back of the dominee's open *bakkie* soon became very cold, being unsuitably dressed for such weather. Eventually, their situation became so dire that they were forced to resort to the only solution at hand: to drink the left-over communion wine! Another time, a habitually irascible dominee complained bitterly about the quality of his drinking water. At first, the elders put this down to his usual bad temper but, eventually, one of them went to the well from which the water was pumped and discovered that lying in the bottom of the well were the remains of two massive black mambas! Some said that the mamba-water was the cause of the minister's temperament ...[43]

Throughout the history of the NGK's Waterberg-Noord church on the Melk, from 1933 to 1962, the job of treasurer had been in the hands of one man, Samuel Dirker, who lived on a farm between Sukses and Dorset. Upon his retirement, the head of the Waterberg community in Nylstroom, Ds. Horak, said of Oom Samuel

FIGURE 14.21: The NG "Melkrivier" Church at Sondagsloop on Nooitgedacht in 2000 before renovation. Photograph courtesy of Magda Streicher and Frikkie Brits.

that he had never come across a *"kassier"* with such bad writing, or one who so flaunted the rules of bookkeeping; but who (unlike many other treasurers), was absolutely trustworthy, whose figures could always be relied upon to the last cent, and whose books were always flawlessly accurate. Oom Samuel and Tante Anna Dirker were devoted and much-loved members of the congregation.[44]

By the mid-1950s, the growth of the village of Vaalwater encouraged proposals to move the home of the NGK Waterberg-Noord there from Sondagsloop. First, a new and more comfortable pastorie was built in Vaalwater and services began to be held at the home of one of the elders, Mike Maas, who then owned the mill in the town and lived next door to it. This shift was not welcomed by residents of the Sukses and Melkrivier areas, who now had much further to travel to church. The communities north of the Lephalala River on the New Belgium properties of Gorcum and Koenap, transferred their support to a new church in Marken (see Chapter 19). On the other hand, the communities around Bulgerivier and Hermanusdoorns, which had been linked to the Albertyn ward at Ellisras, now joined Waterberg-Noord at Vaalwater. In 1962 work commenced on a church building in Vaalwater under the authority of a new, energetic resident minister, Ds. J.A. Dreyer. The new building was taken into use on 26 October 1963. Electricity had not yet reached outlying towns like Vaalwater and the church was powered by a 5 kw Lister diesel generator, while hot water was supplied from a wood-fired donkey geyser. In the year 2000, extensions to the church were commissioned.

FIGURE 14.22: The NGK church at Sondagsloop after renovation in 2005 (LEFT) and the cornerstone (RIGHT).

FIGURE 14.23: The NG Kerk in Vaalwater.

FIGURE 14.24: The Nederduitsche Hervormde Kerk in Vaalwater.

After the move of the NGK community to Vaalwater, the little church on the Melk fell into disuse. It was not until preparations were being made to celebrate the 75th anniversary of the founding of the Waterberg-Noord community that the then owner of the farm on which the church was located, Morkel Munnik, set about its refurbishment – with the intention of creating a wedding venue on the property. The renovated church hosted part of the anniversary celebrations in 2005; a second cornerstone was laid in commemoration on 8 May 2005 by Ds. J.J. Swart – the *"kwaai dominee"*, who'd been at Sondagsloop from 1957 to 1959. The hall served subsequently as a venue for country weddings. In 2013 it was purchased as a rural retreat for the 12 000-strong congregation of the *Woord en Lewe* charismatic church at Sunward Park on the East Rand. Under its new ownership, the old hall has been lovingly restored and cared for and is used almost weekly for church services, weddings or retreats. The old pastorie has been incorporated into the accommodation for 60 guests that has been developed around the church.

The Hervormde Kerk (NHK) was the last of the three sister churches to build a formal place of worship in Vaalwater. The decision to create a new community was taken on 26 August 1967 and plans were made to raise funds to build a church. Before then, the dominee from Nylstroom would occasionally visit the area to hold services. Frans Faber remembers services being held at the school on Witfontein by Ds. Conradie. Another resident of the Witfontein area, Piet Vermaak, was still trekking in the traditional way to Nylstroom several times a year for nagmaal in the 1950s, on a wagon pulled by a span of mules.[45]

Construction commenced in 1971, with voluntary labour and contributions from the community, including Rudolph Faber (Frans's father) from Witfontein and Herman Willemse from Zanddrift, under the leadership of ouderling F. van der Merwe. The new church was taken into service on 26 February 1972 and the pastorie three years later.[46] The first resident *predikant* was Ds. J.P. Verster. Initially, all *wyke* (wards) north of the Sandriviersberg were included in the Vaalwater community; later, those of Alma and Rankin's Pass were also incorporated.

The NHK Waterberg Gemeente celebrated its 150th anniversary in 2015.[47]

In the years following the end of World War II, all three Dutch Reformed sister churches in Nylstroom undertook renovations to their church buildings and invested in community projects such as old-age homes. The two largest churches, the NGK and NHK, also expanded into missionary work among the black communities, establishing facilities in local townships and elsewhere. More recently, each has grown such that there are now at least two communities for each of the three sister churches in the town of Modimolle.

A fourth sister church is born

In 1982 the NGK, which, in common with its sister churches maintained that the policy of apartheid was justified by biblical references, was expelled from the World Alliance of Reformed Churches which declared apartheid to be a sin and its theological justification a heresy.[48] The resulting isolation of the NGK from the global community of Dutch Reformed Churches eventually led to the NGK partially retracting its commitment to apartheid with the publication of a document entitled *Kerk en Samelewing* (Church and the Community) at its General Synod meeting in October 1986. At the same time, some members of the Dutch Reformed Churches adopted what was called the Belhar Confession, which was a rather poorly patched-together list of tenets by which these churches ought to be governed. Essentially, this document sought to remove the segregationist practices that had long been part of church policy.

This was not well-received by the more conservative members of the NGK clergy, and led speedily to the founding of a new church, the Afrikaanse Protestantse Kerk (APK) at the Skilpadsaal in Pretoria on 27 June 1987.[49] Almost immediately, other congregations of the APK were established, especially in traditionally conservative communities of the Transvaal and Orange Free State. In the Waterberg region, the first APK communities to be formed were in Naboomspruit (on 5 July 1987) and Tuinplaats (on the Springbok Flats), followed quickly by those in Warmbad, Nylstroom and Thabazimbi. The first minister admitted to the APK in Naboomspruit was Ds. Willie Jooste in September 1987; six months later, Ds. Roelof Marè was appointed to serve three communities: Waterberg (Nylstroom), Warmbad and Tuinplaats. The Vaalwater community of the APK originated as a ward of the Waterberg community and members had to travel to Nylstroom for services.

Ds. Marè gave the first "visiting" APK service for the Vaalwater community at the Farmers' Hall on 5 June 1988 when there were about 25 registered members.[50] By 1989 membership of the APK in the Vaalwater area had grown to the point where a local home for the church was needed. A leading elder, Buks van Tonder, purchased and then donated two plots in the town to the community in 1990. Immediately, plans were made to build a hall and church using members' own skills and resources. The building was quickly completed and on 12 May 1991, the Vaalwater community of the APK was formally established with Ds. Dirk Malan as the visiting minister.

Today, the APK can boast over 220 congregations across South Africa – although largely concentrated in the former Transvaal and in the Free State – with around 35 000 adherents. According to its website, the Waterberg communities of the APK in 2013 amounted to 373 in Mookgophong, 273 in Thabazimbi, 248 in Modimolle (the Waterberg Gemeente), 208 in Lephalale, 165 in Bela-Bela, and 51 in Vaalwater. (It is perhaps a reflection of the overall conservatism of the church that on its modern and topical website it still lists all the above towns using their former names, under the name of the former province of Transvaal.)[51]

The NG Kerk in Afrika

From the middle of the nineteenth century, the NGK in the Cape had been proselytising to members of the Cape coloured community, but it was only towards the end of the century that these efforts were extended to black communities in the Free State and later, the Transvaal. In the Waterberg region, the first NGK missionaries to black communities were Eerw. Stephanus J.G. Hofmeyr in 1881 and John N. Murray (a son of Andrew Murray) in 1892; the latter ventured as far as the community of Chief Seleka on the lower Lephalala River. After the disruptions caused by the South African War, Eerw. G.H.J. van Rensburg was appointed to the district. At Eersbewoond, a facility of the Manne Sending Bond (MSB) or Men's Missionary Society operated for several years, led from 1949 by Eerw. P.J. Joubert, who was transferred from Ga-Seleka.[52] The NG Kerk in Afrika (NGKA) or Dutch Reformed Church in Africa (DRCA) was formed in 1963 to consolidate the separate NG mission churches in each of the four provinces. The NGKA built a new church in Naboomspruit in 1966.[53] Since 1994, the NGKA and its successors have struggled to retain cohesion amidst the changing politics of the country as well as the rise in popularity of independent African churches.

African Independent/Indigenous/Initiated Churches (AIC)

It is unlikely that there is anyone resident in Limpopo Province who has not heard the name Zion City Moria. This is the headquarters of both factions, or components, of the Zion Christian Church (ZCC), which is by far the largest religious institution in the subcontinent with a combined membership estimated to be somewhere between 10 million and 20 million. (The churches do not divulge accurate membership figures; the last census to have listed religious affiliation, in 2001, gave total ZCC membership as 4,97 million; in 2014, membership was reported to be 5,5 million.)[54]

How did this young institution (it celebrated its centenary in 2010) come about, survive and grow to its present gargantuan size, especially given the turbulent national historical context in which it has done so?

Zionist churches in southern Africa owe their origin to the Christian Catholic Apostolic Church (CCAC), founded at Zion City in Chicago, USA, by John Alexander Dowie in 1896. "The CCAC was characterised by the use of faith healing, adult baptism by threefold immersion in water, the belief that the Son of God's second coming is imminent, as well as prohibitions, amongst other things, on the use of medical doctors, drugs, pork, alcohol and tobacco."[55] Immediately after the end of the South African War, the teachings of the CCAC, together with those of another American church, the American Pentecostal Church (or Apostolic Faith Mission, AFM) were introduced to South Africa and attracted numerous converts, both black and white. The activities of charismatic faith healing evangelists led to rapid growth in both the number of conversions and in the number of splinter churches across the region. Whilst details of their credos varied, all had in common the basic features enunciated by the CCAC above.[56] Some, especially those aimed predominantly at rural black converts, incorporated aspects of traditional African customs, including recognition of the role of ancestors and acceptance of polygamy. Since then, a large number of these indigenous (as opposed to imported) Christian-based movements have been formed. Most of these African Initiated Churches (AICs) have had short lives, based upon the personal charisma of a particular leader; others are parochial in nature, with a localised, language- or community-specific congregation. But some have endured and developed a large, nationwide following. None is more important than the ZCC: By far the largest of the AICs (and also one of the oldest), it was founded outside Polokwane in Limpopo Province by a local Mopedi man close to its present headquarters at Moria.

Ignatius Barnabas (Engenas) Lekganyane was born in the Mmamabolo Pedi chieftaincy at Thabakgone – some 45 km east of Pietersburg (Polokwane) on the R71 road to Tzaneen – in about 1885. In the early 1900s he was baptised in Johannesburg into the Scottish Presbyterian church and sent back to his village as an evangelist. Legend has it that in 1910, whilst standing atop the prominent hill (Sethethong sa Modiki) to the south of his village (now called GaMahlanhle), Engenas experienced a revelation from God to the effect that he would come to be followed by a multitude. This moment has since been taken by the ZCC as the time of its founding,[57] although the first congregation under the name of ZCC was not held until 1924. Engenas was prompted by this vision to return to Johannesburg to join the newly formed AFM there. He was baptised into this faith and worked for the movement, was baptised for a third time (in 1915) and ordained as a minister. At his third baptismal, a childhood eye disorder was suddenly resolved, fulfilling a prophesy he'd received some years before. However, he became unhappy with the local AFM and went to Basutoland (Lesotho) where he joined the Zion African Church (ZAC) in 1920, and was ordained as its Bishop responsible for the northern part of South Africa.

Returning to his home village, Engenas began his missionary work and soon became known for his charismatic evangelism, which emphasised the use of faith healing and oral testimony. When he took a second wife, he courted the disapproval of the ZAC, which was opposed to polygamy. This (and other differences) prompted him in 1924 to set up his own church, which he called the Zion Christian Church (ZCC), based at Thabakgone. The following year, he applied for the registration of the church (with its tax-exempt benefits), but this was turned down by government, which was of the view the organisation did not yet fulfil its perception of a church.[58]

FIGURE 14.25: The only known photograph said to be of the founder of the ZCC, Ignatius Lekganyane (source unknown).

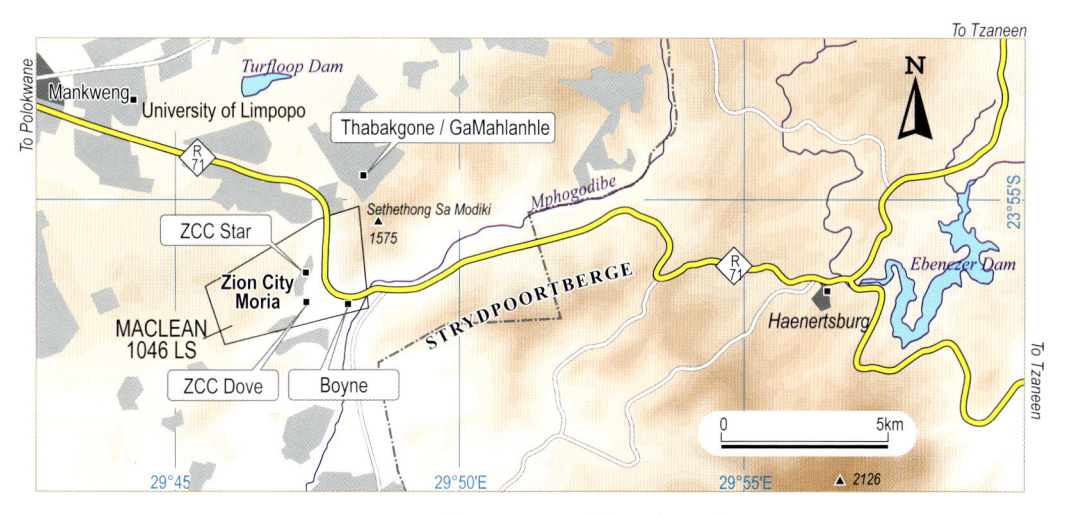

FIGURE 14.26: The area around Zion City Moria.

In 1928 Engenas introduced a simple badge, a dark green vertically-oriented piece of cloth, by which members could identify one another. By now, his popularity and that of his church was becoming substantial, to the extent that relations between him and his chief (Mmamabolo) became strained. Before long, it became necessary for the church to move. In 1938 Engenas was able to purchase a farm called Warmburg against the southern slopes of the Strydpoortberge, overlooking the villages that would later come to be the Lebowa capital of Lebowakgomo. For a while, this was the home of the church. In 1942 the church purchased two farms at Boyne immediately south of the hill on which Engenas had experienced his revelation in 1910, and on one of these, Maclean, Engenas established the new permanent headquarters of the ZCC which he named Zion City Moria. However, he retained the original site at Warmburg and it would later remain an important subsidiary nucleus for the church. The origin of the name Moria is unclear. It is quite likely to have been derived from the important village of Morija (also spelled Moriah) south of the Lesotho capital of Maseru, for it was here that missionaries of the Paris Evangelical Mission Church (PEMS) established the first Christian church in that kingdom in 1833, at the express request of King Moshesh.[59]

In the same year (1942) application was again made for the registration of the ZCC as a church. The report of the Native Commissioner in Pietersburg to the Secretary for Native Affairs in Pretoria was unfavourable:

> It is very doubtful whether this organisation, which I am not prepared to call a religious sect, is above board. I have heard so many rumours of Lekganyane, that I feel sceptical as to the bona fides of the whole concern ...

The Commissioner considered Engenas to be a fanatic and warned that his organisation was likely to become troublesome.[60] In support of the ZCC's application, however, its attorneys countered that membership of the church now exceeded 27 000 and that there were 55 congregations around the country, including one in each of Basutoland, Rhodesia and Bechuanaland. Most were in the then Northern Transvaal, including Warmbaths.[61] It seems that registration of the ZCC as a church was then granted. Within a year, membership exceeded 50 000.

Engenas adhered to the founding principles of his movement in practising faith-based healing, prophesising and oral testimony, all centred on his personal charisma and his exceptional skill as a healer. With his new headquarters at Zion City Moria, his position became comparable to that of a traditional chief with the difference that his followers were defined by their religious beliefs rather than by their ethnicity. Some researchers have described Engenas as a true *Kgoši ya Masione* (King of the Zionists) and the ZCC as a "supra-ethnical tribe" or "church-tribe".[62]

It was hardly surprising therefore that when he died on 1 June 1948, the succession would be based on traditional Bapedi custom: The eldest son of his senior wife would inherit his position. However, this candidate – Barnabas, who would have been a very popular successor – died a few months after his father and before being confirmed in his role and so the position of head of the church then fell to one of the two remaining sons: Edward and Joseph. Edward (named in memory of Engenas's mentor Edward Motaung) was the older of the two and therefore the logical candidate, but he had been living in Durban and Johannesburg, whereas Joseph had remained at Zion City Moria where he had worked for his father. Joseph took advantage of his presence on site to assume control of the existing headquarters and to become de facto head of the church. This did not please many elders among the membership of the ZCC, especially those from urban areas (Joseph's support being more rurally based), who appealed to Edward to return and establish his own branch of the church with his own headquarters on the same farm. Therefore, in 1949 a second Zion City Moria was born about 1,6 km to the northeast of the original complex; henceforth there would be two sections or factions of the ZCC operating from essentially the same location and adhering to very similar practices and beliefs.

The ZCC "Star"

FIGURE 14.27: The badge of ZCC Star.

Bishop Edward Lekganyane, born in 1926, was well educated, a flamboyant character with a tertiary qualification in divinity. He continued with his father's work, although he gradually guided the church into more biblically-based practices and in so doing attracted an increasingly urban support base. However, he remained an advocate of polygamy (but not promiscuity!) and was said to have had a total of 23 wives around the country.

He identified his section of the church by introducing a new badge: the well-known silver five-pointed Star of David with the letters ZCC engraved upon it, the star being pinned on a circular piece of black cloth which in turn was pinned onto the pre-existing rectangle of dark green (a piece of blue cloth is hidden between the two, symbolising the cleansing water of the church).[63] The church became known colloquially as ZCC Star.

Edward also introduced the practice of *mokhukhu:* groups of (mainly younger) men in khaki uniforms who would lead singing, perform dances and be responsible for motivating the many far-flung congregations. The origin of this term is conjectural: It can be interpreted to mean "person of the shacks", a reference to the informal townships that were springing up around the metropolitan areas, as opposed to the traditional rural villages, whose residents supported brother Joseph's branch of the church.

The organisation of the church remains a male-run function with a hierarchy comprising family members and close associates all under the ultimate control of the Bishop. The principal weekly Sunday service is for men, although women can attend without participating – their own service is on a Wednesday. Uniforms are designed for both men and women: the men's *mokhukhu* dress of khaki trousers and jacket, white boots and black military-style hat with the star, and their formal dress of dark green suit with braided jacket, white shirt and black shoes.[64] Women too have two dress styles: a formal dress of calf-length bottle-green skirt with long-sleeved yellow blouse, green head scarf and dark shoes; and a choir uniform of a bright blue long-sleeved dress with brown tie, dark green beret and dark shoes. Only baptised adults may wear the formal uniforms. Men are not permitted to grow beards.

FIGURE 14.28: The badge of ZCC Dove.

Bishop Joseph Lekganyane managed his section of the ZCC in a similar way to that of Edward. To differentiate his church from that of his brother, he renamed his branch of the church St Engenas ZCC in memory of his father. In 1965 he introduced a badge with a silver dove surrounded by a laurel wreath, and the words "St Engenas" at the top and ZCC in a blue bar beneath. The church became known as *ZCC na leebana* (ZCC of the Dove). Its members have adopted uniforms fairly similar to those of ZCC Star; but the men, while also dressed in khaki (except on formal occasions) march rather than dance.

Edward Lekganyane died in 1967 at the age of 41, having suffered a heart attack. Just before his death, he'd completed a three-year evangelical course offered by a nearby DRC college. Wisely, he'd already nominated his eldest son Barnabas (to whom he gave the praise name Ramarumo (father of the spears) as his successor before his death.[65] But Ramarumo was only 13 at the time, so until he came of age, the traditional Bapedi custom was followed of appointing a "regent" – in this case called a superintendent. The first superintendent had to be dismissed by the elders for attempting to take control of the church, but the second cared for it well until Ramarumo (Barnabas) succeeded to his father's position in 1975. He still occupied this position in 2017, aged 63.

After Joseph Lekganyane died in 1972, he was succeeded as head of the ZCC Dove section in 1975 by his second son, also called (Engenas) Joseph.[66] He has only taken a single wife.

The ZCC Star and ZCC Dove churches continue to be run firmly by their two leaders, cousins and descendants of a shared family dynasty. This has led to accusations of "Messianism", but one researcher, Retief Müller, has argued:

> Rather than seeing the Lekganyane dynasty as a reincarnation of the historical Jesus, it seems to me that the Lekganyanes are venerated by ZCC members as real and powerful local ancestors within a constructed indigenous tradition, which also happens to stand within a global tradition of historical Christianity. [67]

The relationship between the two churches is unclear. Their common origin, the schism caused by the two sons of the founder and the leadership currently in the hands of two cousins, co-existing on the same property and essentially competing for members in the same "market" must create an interesting dynamic. Membership numbers are equally opaque, but it is apparent that ZCC Star is about twice the size of ZCC Dove with perhaps five to ten million versus three to five million members respectively. A total of up to a million members from the two churches attend the annual Easter event (although media reports commonly assert an order of magnitude more). The extent to which there is cooperation or conflict between the two organisations is not public knowledge.

Both sections of the ZCC are committed to peace; this is emphasised in particular by the symbol of the dove chosen by the St Engenas section. A typical exchange of greetings between members might be "*Khotso!*" (Peace!), to which the response could be "*A e ate!*" (Let it spread!).[68] Both organisations still advocate no

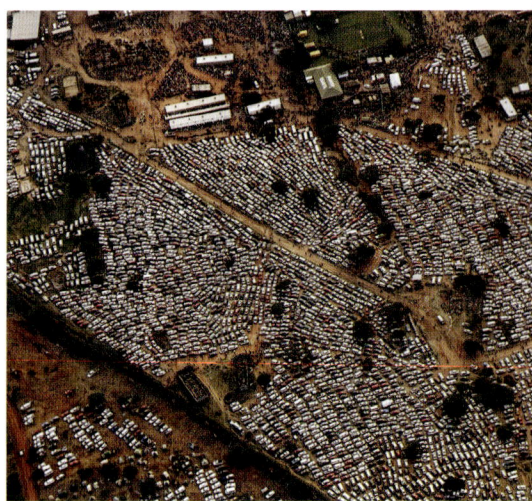

FIGURE 14.29: Zion City Moria at Easter. Aerial imagery kindly provided by the professional photographer Ivan Muller (Ivanmullerphotography@gmail.com).

FIGURE 14.30: Bishop Barnabus Ramarumo Lekganyane with State President P.W. Botha at Moria, Easter 1985. SANA Pretoria Photographic Library, No. SAB 17349.

FIGURE 14.31: Bishop Joseph Engenas Lekganyane of ZCC Dove.

smoking, drinking of alcohol, eating of pork or promiscuity – although polygamy is still permitted. Members are drawn from across southern Africa. However, the origin of the church, its location and the ethnicity of the majority of its members results in the principal language of communication at Moria being Sepedi (North Sotho).

The ZCC is avowedly apolitical and takes no official political stand. Nonetheless, it clearly has an interest in politics and respects political leadership. In 1985 then Prime Minister P.W. Botha attended the 75th anniversary celebrations of the church at Moria.

In 1992 Messrs Nelson Mandela, F.W. de Klerk and Mangosuthu Buthelezi addressed the Easter ceremonies. More recently, EFF leader Julius Malema and then President Jacob Zuma attended the Easter 2012 event.[69] President Cyril Ramaphosa (whose ancestral background is Venda) was a guest at the services of St Engenas ZCC in both 2016 and 2017. During Easter 2017, Bishop Ramarumo Lekganyane of ZCC Star said that the church would not participate in the national groundswell calling for the resignation of the former President, Jacob Zuma. It is said that the leaders of the two churches do not participate in elections and that their priests (*baruti*) are also not permitted to vote.

English-speaking Christian churches

Much has been written in the chapter about the evolution of the Dutch Reformed Churches in the Transvaal, but what of the English-speaking and other communities? The fact is that prior to the South African War, there were very few residents in the Waterberg who would have considered themselves Muslims, Anglicans, Presbyterians, Methodists, Catholics or Jews; ministering to their religious needs would have been practically impossible.

Roman Catholicism

Until 1870 Roman Catholics, for example, were forbidden to practise their faith in the Transvaal[70] and it was only after the short-lived annexation of the Transvaal by the British in 1877 that the first Catholic mass was celebrated in Pretoria. The following year, a convent (the Loreto Convent) was opened in the town, staffed by some nuns from Brittany. During the First Anglo-Boer War (1880–1881), Pretoria was besieged by Boer forces and the convent was occupied by defending pro-

British troops, who caused considerable damage to the building in converting it to a military redoubt. To the relief of the nuns, the return to Boer rule in April 1881 did not result in their eviction. On the contrary, the convent grew from strength to strength, and by the end of the Second Anglo-Boer War, there were almost 400 children attending its school, including some Dutch-speaking pupils.

The Roman Catholic presence in the Northern Transvaal focused on Pietersburg (now Polokwane). The first resident priest there (from 1904) was Father Stephen Hammer, a man fluent in Dutch, who built a simple wood-and-iron structure to cater for his small congregation. His death two years later left a gap in the presence of the church in Pietersburg until 1910 when a Benedictine Prefecture was created. The first apostolic (or missionary) was Dom Ildefons Lanslots, who, despite having served many years in India, found this new environment very difficult. The Benedictines had little money, agricultural conditions were different from their experience and they encountered considerable anti-Catholic prejudice – the so-called "*Roomse Gevaar*" (Roman peril) – from the Afrikaner community in the town.[71]

In the Waterberg, the first Catholic presence was established by Father Laurence Schüling, a German Benedictine monk. A priest serving the Northern Transvaal diocese, this remarkable man spent 19 years in the region, travelling by bicycle – and later by motor bike – from one village or farm (or kraal) to the next, carrying all his camping equipment and supplies with him. He would baptise and minister to his congregants at whatever venue was available for the occasion. Several of his early parishioners were Portuguese-speaking residents from Mozambique.

In November 1929 the church purchased five hectares and a house on the farm Het Bad for £600. This property, which would now be on the corner of Moffat and Meininger streets (just above St Vincent's) in Bela-Bela was then in the bush between the new town of Warmbaths and the abandoned original settlement of Hartingsburg. A Belgian priest, Fr Humbertus Smetrijns, "architect, builder, parish priest, cook, all combined",[72] commenced the construction of a mission station, which was consecrated on Ascension Day, 29 May 1930. The design of the church is interesting: The nave comprises two wings which share an altar located at their junction. Each wing has its own entrance. This design was to accommodate the Church's principle that the building should be open to all races, whilst still complying with the prevailing segregationist laws. It enabled black and white parishioners to attend the same service, albeit seated separately, in areas accessed by separate entrances.[73] Four young Sisters were sent from Belgium in April 1930 to staff the new church and to start a school. According to the church records,

> [T]heir first days were spent chopping pathways through the bush from the house to the church [which was under construction]. As they had no money or income, they had brought tinned spinach, dried beans and peas from Belgium and set about making a garden. Their Sundays started with Mass at 7 am; [after which] they packed a basket of food and walked kilometres in all directions with no regard to fences or ownership of ground, searching for black children for the school they intended starting.[74]

The school, which was established in October 1930 in the Belabela black township, was operated by the Sisters of Charity until 1964.

Fr Humbert was succeeded by Fr Clemens van Hoeck, who served the parish for 24 years before his election as abbot-bishop to Pietersburg in 1954 (where he was to serve until his retirement in 1974).[75] In 1936 Fr Clemens founded the St Vincent's Private Hospital in Warmbaths (Bela-Bela), under the supervision of the Sisters of Charity.[76] The hospital had accommodation for black and white patients (very unusual in those days), as well as for enrolled nurses and midwives. Another four sisters were sent from Belgium to staff the hospital. After receiving training for four years in Durban and Pietermaritzburg, they arrived in Warmbaths where

FIGURE 14.32: Father Humbert's Catholic Church in Warmbaths/Bela-Bela.

the original hospital comprised two wards and six beds. Sister Henry was the first Matron and the church records provide a vivid description of conditions she and her team encountered:

> [T]he first dispensary was a lean-to against the fowl-house. The equipment was very scant: two broom sticks and a sheet served as the first screen, and a paraffin lamp gave light while an empty soap box was the chair. The water tap was 200 feet away.[77]

Many older white residents of the Waterberg were introduced to the world via the maternity ward of St Vincent's, which for years was regarded as the best such facility north of Pretoria.

From 1958 until his death in 1986, a total of 28 years, the parish priest at Warmbaths was Fr Emmanuel Goossens. A Belgian Jesuit who joined the Benedictine Order of monks, Fr Emmanuel had been born in 1909. His application to come to South Africa was thwarted by the outbreak of World War II when he was drafted into the Belgian army, but captured and taken prisoner. After the war, he succeeded in coming to South Africa where he was first posted to missions near Pietersburg before eventually arriving in Warmbaths. He was a natural linguist and became fluent in Dutch, English, German, Northern Sotho and Afrikaans.

Fr Clemens – who was also the architect of the cathedral in Polokwane – returned to Warmbaths as Bishop Clemens in 1964 to bless the new church of Our Lady Mediatrix, which was built immediately adjacent to the structure of Fr Humbert (which in 2015 was still in good condition and was being used as a hall). In the 1990s the new Emmanuel church, which can seat over a thousand worshippers, was opened in Belabela (black) township and blessed by Bishop Fulgence le Roy.[78]

The Catholic churches in Bela-Bela remain the only representatives of that faith in the Waterberg region.

The Anglican Church

The first formal Anglican presence in the Transvaal arrived in the form of Henry Brougham Bousfield in January 1879 during the brief British administration of the region, with Bousfield having been consecrated as the first Bishop of Pretoria at St Paul's Cathedral the previous year. According to historian Vivien Allen, Bousfield "was a handsome, strongly built man with a fine speaking voice and an impressive presence, standing 6 feet 2 inches in his socks. He was a man of firm opinions who did not suffer fools gladly".[79] (One may ask what business Ms Allen had to see the bishop in his socks).

Bousfield faced a formidable task: His new diocese covered the whole of the Transvaal Republic, with scattered, sparsely populated settlements separated by large distances and poor roads.

In the first volume of his autobiography he listed these villages as including Heidelberg, Potchefstroom, Zeerust, Rustenburg, Klerksdorp, Standerton, Lydenburg, Middelburg, Wakkerstroom, Utrecht, Spelonk, Woodbush (near Tzaneen) and Pilgrims Rest.[80] In his first six years, he managed to tour his diocese twice, while at the same time establishing a church and two schools in Pretoria. One of these, the Diocesan School for Girls, is still in existence.

Records collated from early church papers suggest that in the years between the two Boer Wars, occasional Anglican services were held in the home of the Tamsen family in Nylstroom by clergy travelling to and from the north. It is unclear whether Bishop Bousfield himself was ever present at one of these services.[81] The Tamsen home in Nylstroom became well known for its hospitality and gracious accommodation (see Chapter 5).[82]

In 1892 the Fawssett sisters, Katherine and Edith, whose brother was a Lincolnshire vicar, came out to South Africa to join Katherine's fiancé, Arthur Peacock (the couple was married in Bloemfontein). At the time, Peacock was the manager of a ranching property in the remote northwestern part of the Transvaal, some 150 km northwest of Nylstroom. On arrival, the sisters began to hold church services regularly every Sunday in their home – they and their niece Molly would have been almost the only members of the congregation – and would travel to Pretoria once a year to celebrate Easter communion.[83]

After the occupation of Pretoria by British troops in June 1900, the bitter guerrilla phase of

FIGURE 14.33: Bishop Bousfield, first Anglican bishop of Pretoria. SANA TAB Photographic Library, No.2270: Bishop Bousfield.

the South African War commenced, and was to prolong the conflict for another two years until peace was finally achieved through the Treaty of Vereeniging on 31 May 1902. During this period, increasing numbers of British troops were stationed in and around Nylstroom and the need arose for Anglican services to be held. The end of the war was followed by Milner's settlement programme, which brought numerous English-speakers to settle in the district. Bishop Bousfield had died (in Cape Town) in April 1901 and was succeeded by Bishop William Carter, who appointed the Rev. Godfrey to conduct services in Nylstroom. These were held initially in the Tamsen home (the Tamsens having returned from wartime exile in Pretoria), the first being on 18 September 1904 when the Rev. C.S. Carey officiated.[84]

St Michael's in Nylstroom

In the same year, the government granted a stand on which a church could be built and fund-raising for this project commenced. In the meanwhile, services were held in the Court House and even the Masonic Hall, as well as at the Tamsens. The first marriage to be held was that of Mary (Molly) Fawssett and Edward (Ted) Davidson, on 30 November 1905. By May 1907 plans for a church had been drawn up and the foundation stone was laid on 11 August of that year. The name chosen for the church was St Michael and All Angels; building was assisted by members of the South African Constabulary (a remarkable irregularity), and the church was dedicated by Bishop Carter on 12 November 1907. Over a hundred people attended the ceremony – a number not matched at the church since. The church had cost £820 to build – it took the small congregation 20 years to repay the loans incurred.[85]

Despite the building of St Michael's Church, the small resident Anglican population of Nylstroom and the rudimentary roads linking outlying communities meant that it was seldom worthwhile to hold services at the church. In addition, although a succession of priests was appointed to officiate in the parish, there was no permanent rector. Instead, the priests would tour the district, conducting services in various farmhouses. A resident priest had to find his own accommodation until 1923, when the first rectory (or vicarage) was built. A small congregation, two world wars and the Depression of the early 1930s all contributed to the tribulations faced by the church of St Michael and All Angels in Nylstroom.

In several years, there was no resident priest, and many, for one or other reason,

CHAPTER 14: MATTERS OF FAITH   *513*

FIGURE 14.34 (TOP): The Church of St Michael & All Angels, Nylstroom, 1907. Museum Africa Photographic Library, No. PH2006-10101. FIGURE 14.35 (BOTTOM): St Michael's United (formerly Anglican) Church in Modimolle, 2017.

were resident for only short periods. In the first century of its existence, for example, at least 30 different priests served the congregation, several of them on a visiting basis. The longest-serving resident priest was Rev. Hubert Lewis (1953–1961), followed by Rev. Horace Gaylard, who served the congregation throughout World War II and who in 1942 married Jane, the third daughter of Ted and Molly Davidson of 24 Rivers. The second Davidson daughter, Elizabeth, also married an Anglican priest, the Rev. Frank Clarke, who was based at Pietersburg but occasionally held services at 24 Rivers. Rev. Gaylard started a parish magazine, the first edition of which was both written and entirely funded by himself. It included the following comment: "The Parish car is the first in this district to run on charcoal. I hope this is going to solve the petrol question and it will be possible to do more visiting in the Parish."[86]

An interesting resident priest was the Rev. Ernest Prance, who served the parish during 1920–1922 and again in 1925. Prance was a bachelor and lived with his sister Edith on a farm outside the village. He had served in France with the Royal Army Chaplains' Department during the Great War before coming out to South Africa to join his brother, Cyril Rooke Prance, a former member of the SA Constabulary and now farming in the west of the Waterberg (see Chapter 4). Church records note that on 16 September 1923, a wooden font made by Cyril Prance and the gift of Mr Elliott of the New Belgium Block, was consecrated by the Bishop Talbot.

An old parishioner of St Michael's, Freddie Gibb, who'd arrived in the area from India after the war, remembered the following:

> Vicar ER Prance was a popular padre ... no cars ... he rode horseback ... came to the different farms for the day, often losing his way. The congregation was always waiting for him – or he for the congregation. Later, he got a car with a native boy to accompany him. When he christened Eva Allen [Gibb's neighbour's daughter], he arrived without a prayer book or surplice, but thought it would be OK if he

FIGURE 14.36: Rev. Ernest Prance on active service in France, c.1916. Photograph courtesy of Ian Valentine, who in 2016 wrote a biography of Edith Prance and her association with the Boy Scout movement.

wore a night dress belonging to Mrs Allen. The only other person at the service was Mrs Blandy – who was Jewish. Prance often used to play cards in the afternoons at the various houses.[87]

For seven years, between 1947 and 1952, the Parish enjoyed the resident services of Rev. Stanley Andrews, who did his best to serve the large district for which he was responsible. In his annual report to Bishop Parker for 1948, he noted that "services were held at Nylstroom, Naboomspruit, Rietbokspruit, Crecy, Warmbaths, Settlers, 24 Rivers, Vaalwater, Alma, Pogietersrus, Zebediela, Radium, Hammanskraal, Pienaars River, Rooiberg, Tuinplaats and Nutfield". Not surprisingly, he appealed for the Waterberg parish to be given boundaries, and for Potgietersrus and Zebediela to be incorporated into the Pietersburg parish.[88]

For a time, Anglican parishioners had to be content with services provided by the Railway Mission; the church itself began to fall into disrepair, owing to lack of use and shortage of funds. In terms of Anglican convention, a church building may only be consecrated (as opposed to dedicated), once it is free of debt. Yet such was the weak support for the Nylstroom parish (which was called The Chapelry of Nylstroom in the Waterberg District)[89] that it was not until 19 September 1966 – almost 60 years after its completion – that the church of St Michael and All Angels was finally consecrated by Bishop Edward Fisher. The building had been restored and refurnished, and a new rectory built for the occasion, which coincided with Nylstroom's centenary celebrations.[90]

St John's at 24 Rivers

At 24 Rivers, on the Waterberg plateau not far from the small settlement of Vaalwater, Mrs Katherine Peacock, her sister, Edith Fawssett, and their niece, Mary (Molly) Davidson, were the nucleus of a small but growing English-speaking, Anglican community which did not take long to conclude that it needed a church of its own. Ms Fawssett donated some land, and Herbert Baker, an architect who had just completed the design for the Union Buildings in Pretoria, was persuaded to design the church. (Baker went on to design many prominent buildings in South Africa, including several churches, leading private schools and Groote Schuur, the Prime Minister's residence in Cape Town, before going to India where he designed the Secretariat Buildings and other structures in Delhi. He was knighted in 1926). The church at 24 Rivers was built with materials available locally and was dedicated as the Church of St John the Baptist by Bishop Michael Furse on 15 July 1914 – a mere two weeks before the outbreak of World War I. The Transvaal Consolidated Land Co. Ltd (TCL), a neighbouring landowner, contributed substantially to the cost of construction of this first church to be built on the Waterberg plateau. It has been a vigorous, well-supported church ever since and celebrated its centenary in July 2014, by which time it had evolved into an interdenominational "charismatic" congregation – although the land remains the property of the Anglican diocese. The story of St John's is described in *Pioneers of the Waterberg* by Elizabeth Hunter, a granddaughter of Molly Davidson.[91]

In Warmbaths, an Anglican Ladies' Guild operated successfully for many years under the leadership of Mrs J. Earle. Their tireless fund-raising efforts enabled the St Mary's Church to be built there and consecrated in October 1954 during the term at St Michael's (Nylstroom) of an elderly, frail bachelor, Rev. Hubert Lewis, whose ministry was assisted by the local Ladies' Guild. Evidently the parish car suffered many accidents during the tenure of Rev. Lewis, until eventually a chauffeur was hired to drive him around![92] St Mary's is still operational, its current priest also doing duty in Modimolle.

Rev. Lewis also oversaw the building and consecration of the All Saints Church in Naboomspruit in 1960. The church was built on land donated by the

FIGURE 14.37: St John's Church and cemetery at 24 Rivers.

municipality, the building being constructed by the Erasmus brothers. The cornerstone reads: "Consecrated by Rt Rev. Lord Bishop of Pretoria Edward Knapp-Fisher; incumbent Rev HC Collin Lewis, 18 September 1960". In the 1980s the congregation numbered about 30 and was still served by a visiting priest, at that time the Rev. Theo Schmidt of Nylstroom.[93]

In 1969 St Michael's in Nylstroom honoured Mary (Molly) Davidson of 24 Rivers with the Order of Simon of Cyrene in recognition of her more than 70 years of service and worship in the Waterberg. Molly, who'd been married in the church in 1905, was 88 at the time of the award. The family remained closely involved with both the St Michael's and St John's (24 Rivers) churches: Molly's daughter, Lois Baber, took over the sacristy work at St Michael's from her great aunt (Katherine Peacock) in 1935 and continued to fulfil this role until her death in 1988 – having passed on the role of treasurer of 24 Rivers to her niece, Ros King, only in 1986. Lois Baber's two sons, Charles and Colin, were Lay Readers in the church: In the 22 years until 1988, their names are listed in the Service Register for Vaalwater/24 Rivers/Ellisras almost 90 times.[94]

The late 1960s were good years for the parish: debt for the new rectory built in 1962 was cleared; the church managed for the first time to cover all its costs without recourse to the diocese; and in 1972 it was able to purchase from the state the land on which the church and rectory stand – for R7000.

The All Saints Anglican Church in Naboomspruit was part of the parish of Nylstroom. In the early days, the priest from Nylstroom would visit all the various chapelries by Cape Cart about once every three months, wherever members could get together. He usually made his trip from Nylstroom to 24 Rivers where a service was held at the St John's Church. After spending the night there, he would reach Naboomspruit at about 14:00 the next day, and (until All Saints was built) an afternoon service would be held at the home of Eric Jackson of Rietbokspruit. The next day he would return to Nylstroom. One of the first Anglican marriages to have taken place in Naboomspruit, in 1928, had been that of the pioneer botanist Ernest Galpin, who'd settled at Mosdene on the Springbok Flats in 1918. The Anglican community in the town was always small and struggled to remain viable. At the end of 2000, the Lutheran Church acquired partial ownership of the Anglican Church property and the latter ceased to hold regular services, leading several members to move across to the Methodist Church.

The first Anglican services in Ellisras (for whites) were held in 1974 at the home of an important and generous benefactor there, E. Gill, and continued for the next decade.

In 1987 a new Anglican Diocese, that of St Mark the Evangelist, was inaugurated, with its seat in Pietersburg (now Polokwane) and Philip le Feuvre as its first Bishop. Consequently, the Parish of Waterberg (which included St Michael's, St John's at 24 Rivers, All Saints at Naboomspruit, St Paul's at Settlers, the Rooiberg community and St Mary's at Warmbaths) became part of the new diocese, ending its 80-year-old link with Pretoria. The parish itself came to be called the Parish of St Andrew.

During the 1990s the prosperity of the Nylstroom parish declined as the number of congregants in the town diminished; so much so that when a new resident priest, Rev. Richard Menees was appointed in 1994, he chose to accept the offer of a rectory at 24 Rivers rather than the one next to St Michael's, on the grounds that there was a bigger congregation at St John's.[95] On 31 October 1999, following continued financial difficulties among both the Anglican and Methodist congregations of Nylstroom, a service was held at St Michael's to celebrate the unification of the two congregations, the church itself to be renamed the St Michael's United Church. More recently, St Michael's United Church embraced the local Presbyterian community too, each of the three participating churches being granted a rotational five-year tenure.[96]

Methodism

The Methodist congregation of the Waterberg predates the arrival of both the Catholic and Anglican churches in the area, having existed continuously since the establishment of a Methodist presence in the region in 1885 when the Rev. Owen Watkins from Pietermaritzburg met and travelled with a number of black evangelists operating in the region. Watkins was impressed with the area south of Hartingsburg (Warmbaths) as a location for the church's missionary work and sent Rev. Isaac Shimmin to "evangelise" the district on its behalf.[97] In 1889 a large property (20 000 ha) called Syferkuil was acquired about 10 km west of the present-day Springbok Flats settlement of Radium. It was named Olverton and Shimmin was able to start a training college there.[98] Later, a mission church was established on the farm Roodepoort, which today falls within the Bela-Bela townlands. A school for black children operated from this mission until sometime in the 1940s when a large gum tree fell onto the building during a storm. As another school had been built within Belabela township by then, it was decided sell the property rather than to repair the seriously damaged building.

CHAPTER 14: MATTERS OF FAITH

519

Meanwhile, in 1909, the Methodist congregation acquired a wood-and-iron building on Luna Road. There are two versions as to how this occurred: The first is that the Warmbaths Town Council had run into serious financial difficulties and struggled to pay its debts. One creditor was the prominent Pretoria merchant and philanthropist, T.W. Beckett. In settlement of what was owed to him, Mr Beckett accepted a recently built structure that had served as the town community hall. Being a Methodist himself (he had been very active in the development of Pretoria and its Methodist Church), Mr Beckett then generously donated the building to the Warmbaths Methodist community.[99] The alternative version is that several prominent members of the local Methodist community met in September 1909 to form a church hall building committee. Mr Beckett, already known for his generosity towards the church in Pretoria, was prevailed upon to assist. He agreed to provide all the materials needed for the construction on credit and to accept payment as and when the Warmbaths church could afford to do so. Either way, the church was acquired, and Mr Beckett was behind it. The first resident minister to use the new building was Rev. H. Milton Brown (1909–1910).

A further story regarding the original church is that when the town was being formally laid out and stands surveyed, it was discovered that the church building straddled two erven (at the time the church hall had been built, there was no town plan). The result was that the church acquired both erven. A further twist was that the structure was oriented at a strange angle to the road. The story goes that this was because the English surveyor had decided to lay out the streets according to the design of the Union Jack![100] Certainly, the modern town of Bela-Bela still

FIGURE 14.38: The original wood-and-iron Methodist Church in Warmbaths. Image reconstructed from an old wedding photograph by Harold Nicolson of Eersbewoond, a life-long parishioner of the Methodist Church in Warmbaths.

displays an unusual street plan in which Mentz and Kretzschmar Avenues could have represented the diagonals of the flag of St Andrew, with Chris Hani (formerly Voortrekker) Road as one of the flag borders. (While on the subject of Bela-Bela streets, there is also a Beckett Street in the town, suggesting that its namesake was responsible for more than the donation of the church hall.)

In 1956 the original church was replaced by a new, attractive brick structure with steeple, the present home of the Warmbaths/Waterberg circuit of the Methodist Church. In 1959 the silver jubilee of the construction of the original church was celebrated with the opening of the Jubilee Hall, built on the site of the old wood-and-iron structure. Plaques in the wall commemorate the event on 7 June 1959 with stones laid by the resident minister of the time, Rev. James Hardwick, and the Circuit Steward, Charles Mathews.

In 1915 it seems that the Mission function of the church in the Waterberg was separated from that of the church, which became the Warmbaths European Circuit. Unlike many of the other churches in the district, the Warmbaths Methodist Church managed to retain an almost continuous line of resident ministers since the arrival of Rev. Shimmin in 1887. Most ministers resided for only one

FIGURE 14.39: The present Methodist Church in Luna Road, Bela-Bela, built in 1956.

or two years, but the Rev. C. Denis Choate and Rev. James Barrow were each in office for eight years (1936–1943; and 1968–1975 respectively), and the Revs Kenneth Coggin, John Gardiner and Roland Watson each for six years (1944–1949; 1982–1987; and 1991–1995).[101]

Rev. Coggin's term ended unexpectedly. The notes of the Women's Auxiliary meetings of the time mention that Rev. Coggin and his wife – who had both been very active in promoting the church to its parishioners – had needed to take a fortnight's rest in Rustenburg and that they'd requested bedding from the community.[102] According to Harold Nicolson, one of the parishioners of the time, the reason for Rev. Coggin's sudden departure was that he'd been defending his decision to allow the visiting husband of one of his black staff to spend the night on the church grounds, contrary to the prevailing black curfew regulations, and had his jaw broken by a white police officer for his trouble.[103] It is well known that in the 1950s to 1970s, there was no love lost between the ruling National Party and its police force, and the various English-speaking churches, which were considered by government to be fomenting unrest and resistance among the black population.

Another long-standing member of the Warmbaths Methodist congregation from that era was the late Brian Stone. Born in 1920, Brian had grown up at Settlers, the post-Great War settlement established for returning servicemen on the Springbok Flats. Being only 20 km from Warmbaths, Settlers was served by the Warmbaths church. Brian not only corroborated Harold Nicolson's story about the hapless Rev. Coggin, but also provided several reminiscences of his own. He had a particularly high regard for the contribution made by the Rev. Reginald Spink, who served the parish from 1960 to 1963. Brian said that Rev. Spink was so charismatic that he was able to squeeze money out of a Stone (Brian)! More productively, he secured the support of Douggie Rens, a landowner in the Hammanskraal area and a parishioner of the church, and this enabled the building of a new manse adjacent to the new Jubilee Hall. Unfortunately, during the course of the move from the old manse to the new one, a large proportion of old church records, temporarily stored in a rondavel, were destroyed in a fire. Rev. Spink's daughter, Rev. Dorothy Spink, was Minister at Bela-Bela in 1988–1989 and later served at the Circuit Office in Polokwane.

Brian recalled that originally the Settlers community used to hold church services at Chester, the home of his parents, and later his own home, adjacent to the mill his father ran there. Once a month, the minister would come over from Warmbaths to conduct the service. Brian remembered the Rev. Walter Goodwin arriving in his Model A Ford in the 1930s. Later, a cabinet minister, Colonel Stallard, donated the old dining hall of the local Settlers Primary School as a chapel and even equipped it with pews and an organ. But by then, Warmbaths was easier to reach and attendance began to decline. Brian also recounted with great affection the tenure of Rev. John Gardiner (1982–1987). Having been quite a reprobate in his young days, John Gardiner turned over a new leaf when he joined the ministry and became an excellent, fondly regarded leader of the congregation, writing several new hymns for the Methodist hymn book in the process.[104]

The Methodist congregation of Naboomspruit is part of a circuit that includes Mokopane, Polokwane and Groblersdal. The first local service was held in June 1948 in the school hall, with a congregation of about 20 members led by Rev. Tom Parker of Potgietersrus, according to Willy Law and Donny Stratford, two founder members.[105] Later, other venues, including the town hall, the lounge of the hotel and the NGK and Anglican churches, were used; until in December 1972 when the Methodist community built its own hall on land donated by the municipality. It was opened by Rev. Lee, President of the Church, accompanied by Rev. Dudley Goodenough of Pietersburg and Rev. Brian Oosthuizen of Potgietersrus. Later, services were led by the well-loved Rev. John Gibson of Potgietersrus, who was totally blind. His guide dog would lead him through the church to the pulpit and would then take his place next to the pulpit during the service, facing the congregation throughout. Knowing the words of the Benediction to mark the end of the service, the dog would then stand and take his place next to his master, ready to lead him out of the building. The congregation in the 1980s comprised some 30 members. By 2010, following the termination of Anglican services in the town, the congregation had risen to over 70 members.

The only other formal Methodist presence in the immediate vicinity of the Waterberg was a short-lived attempt at the turn of the present century to establish a community in Ellisras (now Lephalale). However, this had to be served by a minister from Warmbaths, and a combination of poor attendance and difficult communication soon brought about its demise. In 1999, as already described, the Methodist and Anglican communities of Modimolle combined to form the United Church of St Michael's in that town. In Bela-Bela there is still a fairly small but enthusiastic and thriving Methodist community.

Judaism

From the time of the earliest travellers, Jews have been present in small numbers throughout what is now Limpopo Province. The largest Jewish community grew in Pietersburg where the first religious ceremony was conducted (in the home of Marcus Rosenberg) in 1893 and the first Jewish cemetery was established in 1896, on land granted by President Paul Kruger. Charlotte Wiener, a former resident of Polokwane and a prominent member of the town's Jewish community, now resident in Israel, recounted the following delightful story: When, during a visit to the town in 1896, Kruger was approached by members of the Jewish community to grant a plot of land for a cemetery, he responded that he would approve a piece of land half the size of the Christian cemetery, on the basis that the Jews only believed in half the Bible! He subsequently relented (almost certainly, given his renowned dry sense of humour, with a twinkle in his eye), saying that the Jews were after all, good and law-abiding citizens.[106] He could also have reflected on the Old Testament focus of his own church. Later, Jewish cemeteries were also established in Warmbaths, Potgietersrus, Louis Trichardt and Messina.[107]

In 1897 the Zoutpansberg Hebrew Congregation, which included both Pietersburg and Louis Trichardt, was established, but by 1912 the Jewish community in Pietersburg had grown large enough for the town to found its own congregation. It built its first communal hall (which also served as the synagogue) in 1921 and a new synagogue in 1953 (20 years after a decision to do so had been taken).[108] Jews of the region relied on the Pietersburg minister for the provision of Hebrew education for their children, for the supply of kosher meat, and for arranging cultural events. A burial society (Pietersburg Chevrah Kadisha) was established to perform burial ceremonies. Jewish ministers were employed in the town from 1914 until 1992, but funds were always a problem and many ministers were not prepared to work with very few facilities and members. The longest-serving minister in Pietersburg was Rev. J.I. Levine (1931–1963), reportedly a humble, caring man who was deeply involved with his congregants and whose ministry, which extended to many outlying communities, was used as an example for other rural areas.[109]

The closest formal Jewish community to the Waterberg was that of Nylstroom, where the first *minyan* (a quorum of ten men required to hold a service) was held in a private home on *Yom Kippur* in 1918. This led to the establishment of a congregation in the town in 1926 and the Rev. Nathan Dove acted as minister from 1925–1929. After his departure, Nylstroom Jews travelled to Warmbaths for services and later, in the 1960s, to Pietersburg.[110] An early prominent member of the community was Major Joseph Huneberg, a German-Jewish veteran of the Franco-Prussian War, who was the Waterberg Landdrost in 1880.[111]

Until the 1970s there was also a small Jewish community in Naboomspruit (Mookgophong), mainly of merchants. The earliest, a Kaufmann – after whom the settlement was originally named – established an inn, stables (which housed the horses of the Zeederberg coach service), and a trading store before 1910.[112]

The only other formal Jewish community in the Waterberg region was at Warmbaths, where several Jews had settled in the early years of the twentieth century as farmers, hoteliers, traders and miners (at Rooiberg tin mine west of the town). The first Jewish farmer recorded in the area was A.V. Sacke. In 1913 the Bellevue Kosher Hotel was established in the town by the Cohen family. A later hotelier was Mr Kerzner – his son was the better-known Sol Kerzner, founder of the Southern Sun Group – who owned the Carmel Hotel.[113] In 1928 the Warmbaths Hebrew congregation was established with Rev. Rootshain as its first minister. Until a communal hall/synagogue was built in 1945 (next to the Methodist Church on Luna Road), services were held at various venues, including the Methodist church, hotel dining rooms and the town hall.[114] In 1948 a house was built for the resident minister on land adjacent to the hall. Over 180 Jewish families were associated with the congregation during the course of the next 50 years; the

FIGURE 14.40: The former Hebrew Communal Hall in Luna Road, Bela-Bela.

community was most active in the 1950s when about 36 families attended and services were held nearly daily, led by Rabbi Chazdan, who served there almost continuously from 1948 until 1960. He also ministered to the smaller Jewish communities of surrounding settlements, under the authority of Rev. Levine of Pietersburg.

As roads in the province improved, so the need for small rural hotels and general stores (both often run by Jews) began to diminish, and there was a migration by Jewry to the towns, especially to Pietersburg, and later to Gauteng. The Limpopo region's Jewish population peaked during World War II and in 1943 there were almost 400 Jews in Pietersburg, with smaller communities in Louis Trichardt (where a synagogue was also built in 1938), Warmbaths and Potgietersrust. However, from the 1950s, the Jewish population of the region began to decline, until by 2003 there were only about 12 Jews left in Pietersburg, forcing the closure and sale of both the communal hall and synagogue there. (This decline had occurred even earlier in Louis Trichardt where the synagogue was sold in 1978, only 40 years after it was built.) In Warmbaths, the Jewish population had dropped to only 37 by 1977. The construction of the state-owned spa resort (later part of Aventura) at the hot springs in 1978 accelerated the closure of the town's mainly Jewish-owned hotels and the departure of the owners to other towns. The last formal functions in the communal hall were held in the late 1990s and by 2002 there was a single Jewish family left in the town.[115] The hall and minister's home were finally sold only in June 2015.

Throughout the first 30 years after the founding in 1893 of the first Jewish congregation in what is now Limpopo Province, Jews were readily accepted by other members of the communities in which they lived. Whilst they were generally careful to remain apolitical, given their minority status, most became fully bilingual (or trilingual) and many identified with the desire of the Afrikaner for self-determination. It is a little-known fact that well over 300 Jewish males served with the Boers during the South African War.[116] Only after World War I did anti-Semitism become a significant factor, principally within the Republican-oriented sector of the Afrikaans-speaking community, and almost certainly linked to the identification of some Afrikaners with Germany and the rise there of Hitler and National Socialism. Even then, the only serious incidents of anti-Semitism occurred in Nylstroom[117] – which of course was the home of ultra-conservative Afrikaner nationalists like J.G. Strijdom (overtly anti-Semitic as well as anti-Indian), his legal partner F.H. "Fox" Odendaal, and later men like Roelof Odendaal, the partisan Afrikaner author. The town also hosted an active branch of the Afrikaner

Broederbond (of which Fox Odendaal was the local head), and during World War II the Ossewabrandwag, a Nazi-affiliated organisation. No doubt this antipathy towards the dwindling numbers of Jews in the small country towns hastened their migration to the cities. Community records showed that there were still 15 Jews living in Nylstroom in 1939, but the number dropped to 12 by 1941; by the 1960s, there were none.[118]

Having said all the above, it is interesting to note the comment made by one historian of the 1950s, Leo Marquard, who, having grown up in a village in the Orange Free State, observed that

> Social discrimination against Jews is far more common, and more humiliating, among English-speaking than among Afrikaans-speaking South Africans. The Afrikaner has never been a thorough-going anti-Semite. There have been periods when 'the Jew' has been used as a bogey in Nationalist political propaganda; but the Afrikaner has always respected the Jews and their religion and has, indeed, had a kind of fellow-feeling for them.[119]

He noted that what really soured relationships between many Afrikaners and South African Jewry was the alignment of the National(ist) Party with the policies of Nazi Germany – and later, Israeli support of African and Asian states opposed to South Africa's apartheid ideology.

Today, there is no formal resident representation of Judaism in Limpopo Province, although spiritual services continue to be provided from Johannesburg by a minister dedicated to serving country communities. The longest-serving incumbent has been the ebullient, passionate and charismatic Rabbi Moshe Silberhaft ("the Travelling Rabbi"), who has ministered to rural Jewish communities in most of sub-Saharan Africa, Madagascar and Mauritius since 1993.[120] Sadly, his task has often had to include officiating over the closure and deconsecration of rural synagogues and communal halls; but he offers a vibrant, essential link between isolated Jews and their church.

Islam

Islam is the last of the Abrahamic[121] faiths included in the religious history of the Waterberg region. The advent of Muslim Indians, migrating inland from Natal and the Cape, was constrained by legislation in the Transvaal and Orange Free State republics that denied Indians citizenship and required them to live in their own locations or townships. (Note that until the partition and independence of India in 1948, Pakistan and Bangladesh [originally East Pakistan] did not exist: Indians of Hindu, Muslim and other faiths were all citizens of India. Most Muslim migrants to South Africa came from Gujarat State, which is still in western India, adjacent to the border with Pakistan.)

The date of arrival of the first Indians in Nylstroom is not known, but the writer Eugène Marais knew of three itinerant traders there from 1908. One of them, Ebrahim Ravat, used to supply Marais with the morphine he needed to feed his addiction to the drug (see Chapter 6).[122] By 1910 there were at least 30 Indians living in the town.[123] Among the earliest families were the Suleimans, Hassims, Mossads and the Saloojees. Originally, the Indians were concentrated at the western end of the town, where the main road from Warmbaths to Naboomspruit passed through (the road exiting to Zandrivierspoort followed a different route then). One of them, Ahmed Mahmoud, who was known as Taraj after the village near Surat that he came from, donated some land among the shops and houses on which to build a mosque.[124] The initial structure (erected in 1909) was very primitive, but in due course, a brick building was built on the site.[125]

FIGURE 14.41: The new mosque in Modimolle.

For many years, the Muslim community, including some Indian families settled near Kranskop, formed part of the greater urban community of the town, although in 1908 a petition circulated lobbying for increased regulation of the movement of Indians.[126] After the promulgation of the Group Areas Act in 1950 and its associated Separate Amenities Act in 1953, pressure mounted on the local authorities to insist on the removal of Indians to their own residential and business area. Eventually, in about 1956, this legislation was enforced, and the whole Indian community was forced to relocate to Asalaam, an area at the eastern edge of Nylstroom on the road to Marble Hall. Only the mosque was exempted from removal. Most Indian residents of Modimolle still live in this modest suburb, even though their mosque – a new structure having been built in 1987 on the site of the old one – is at the other end of the town.[127] Most Indian-owned businesses, including that of the descendants of Ebrahim Ravat, are also still to be found at the eastern end of Modimolle.

The first Indians arrived in Warmbaths in the 1930s. Prominent among them was the Bham family, after whom the road linking the "white" town and the Indian township of Jinnah Park, established in the 1960s, was named. Initially, the spiritual needs of ten Muslim families in the town were served by a house that was used as a prayer room, or *jamaat khana*. By 1972 the Muslim population of the town had grown sufficiently that a mosque, with minaret, could be built in the segregated township of Jinnah Park (also a product of the Group Areas Act) in a precinct that also accommodates the residence of the Imam and a jamaat khana for women.[128] For several years until 2006, Younus Lorgat, the brother of the current Imam of Bela-Bela, was a popular mayor of the town. This was prior to his deployment by the ANC as Speaker of the Waterberg District Municipality, a role he filled until his death in 2009.

CHAPTER FIFTEEN
*A futile broedertwis**

Just to the south of a range of hills known as the Hoekberge that lie below Rankin's Pass and form the southernmost rampart of the Waterberg, there's an old transport road (once called the Bulawayo Road) skirting the mountains. Exploiting a kloof cut through the hills by a tributary of the Sand River, which flows west to join the Crocodile River south of Thabazimbi, the road provides the shortest link between Bela-Bela (Warmbaths) and the western bushveld. In the last years of the nineteenth century and the first quarter of the twentieth century, this was a popular route followed by travellers, ranchers and hunters wanting to access the bushveld without having to go via Rustenburg. About 20 km east of the kloof and midway between Rankin's Pass and the old Rooiberg tin mine, the road is straddled by a farm called Zandfontein.[1]

On 8 November 1914, there was a skirmish in this area between a unit of the Union Defence Force (UDF) and a rebel group, one of several that were opposed to South Africa's entry into the war against Germany and supportive of other rebel initiatives around the country, particularly in South West Africa. This so-called Maritz Rebellion had been fomented by several former Boer leaders then serving in the Union Army, including Lt Col Manie Maritz and Major Jan Kemp, as well as by Captain Josef (Jopie) Fourie and retired Boer War hero General Christiaan de Wet.

The rebellion, which had originally been planned in August/September 1914 as a coup by disaffected Boers to regain their independence, commenced in October. Initially caught off-guard and preoccupied with mounting a force to dislodge the Germans from South West Africa (in terms of a commitment given to Britain by Botha and Smuts), the UDF scrambled to call for volunteers, especially Afrikaans-

---

* *Broedertwis*: a fraternal quarrel. This chapter is a shorter, edited version of a paper originally published in the December 2014 edition of the *Military History Journal* 16 (4) pp. 143–149, entitled "Forgotten Casualties of the 1914 Rebellion: Zandfontein 476KQ, 8 November, 1914". The publishers of the journal, Ditsong Museums of South Africa, are thanked for permission to reproduce that paper.

speaking men, to counter the rebellion. Several engagements took place across the Free State (where rebel units were led by De Wet), Transvaal and Northern Cape, but by mid-December 1914, all the rebel groups had been defeated, captured or dispersed – with the exception of one near Upington, which held out for another few weeks.[2,3]

Maritz, a serving UDF officer, had been involved (with great reluctance, having covertly reached an agreement with the Germans to support the planned Boer coup) in a battle on 26 September 1914 which took place in southern South West Africa at Sandfontein (also known as Zandfontein) near Warmbad, about 20 km north of the Orange River. There a column of heavily outnumbered UDF troops had surrendered after having their horses captured and shot, sustaining about 70 casualties.[4,5] Shortly afterwards, on 9 October, Maritz and a group of over a thousand men formally crossed the border to join the German forces, thereby triggering the rebellion. The severity of this unexpected military setback in South West Africa, as well as the misleading similarity of names, Zandfontein and Warmbad, to those of the Waterberg incident six weeks later may have contributed to the latter being overlooked in most literature.

To add to the confusion, on Sunday 8 November 1914, the very same morning on which the battle at Zandfontein in the southern Waterberg took place, a group of rebels under De Wet clashed with a UDF force on the Sand River between Winburg and Ventersburg in the Free State (almost at the site of the famous Sand River Convention of 1852).[6]

Zandfontein, Sunday 8 November 1914

The events of the Waterberg Zandfontein skirmish have escaped the notice of most historians; they were summarised in passing in a biography of General De Wet which, however, described the incident very much from the perspective of the government.[7] Evidently, a large group of rebels, variously reported to have been under the leadership of ex-Major Pienaar, ex-Captain Fourie or Commandant Jan Harm du Plessis and Field Cornet Koos Niemandt, gathered in the southern Waterberg in early November and *laagered* on Zandfontein.[8,9] As will be seen later, there is no evidence to suggest that either Pienaar or Fourie was present at the time of the actual battle. Perhaps they had already moved on; or, at least in the case of Fourie, maybe mythology had a role to play.

According to the official Defence Headquarters communiqué, released on Monday 9 November 1914 (the day following the skirmish) and published the next morning (Tuesday 10 November) in the *Pretoria News*:

> Lt. Col. Dirk van Deventer,[10] with his commando, the Waterberg Commando under Commandant Geyser, a Pretoria commando under Commandant Jones and a portion of the commando under Acting Commandant Theunis Botha, reports as follows: "Yesterday [Sunday, 8 November] we drove the enemy from Weltevreden over Zandfontein in the direction of Rietfontein and took a portion of their laager at Zandfontein, about thirty miles west of Warmbaths. The fight took place at about 3 pm and about 120 of the enemy were killed or wounded, and 25 were captured. The loss on our side was 12 killed and eleven wounded, three slightly. The officers and the men conducted themselves very bravely. The heaviest fighting took place under Commandant Geyser of Waterberg and a portion of the commando under Acting Commandant Theunis Botha, under Veld Cornet Hoffman and Captain van Coller. The left wing was held by Commandant Jones and the right by Commandant Geyser and Acting Commandant Botha. All the men were smart and brave. Veld Cornet Hoffman, although wounded, led his men throughout the engagement. I wish to make special mention of the officers and men killed. Of our officers and men who were killed all, with a few exceptions, were shot with dum-dum bullets and suffered frightful wounds. Details of our casualties are being furnished to the Press today. This rebel commando is supposed to be under 'Jack' Pienaar, an ex-Major of the Permanent Force Staff."[11]

The following day, Wednesday 11 November, the paper published the officially released list of UDF casualties, as follows:

> WEST OF WARMBATHS, ON SUNDAY NOVEMBER 8:
> KILLED
> Captain E.J. Geldenhuys, Lieut. A.W. Wrighton, F.J. Botes, Piet Cronning, F.J. du Preez, H. Hitchcock, W.S. Johnson, J.D. Nel, and A.C. de Wilde (all of Commandant Geyser's Commando), Jan Nagel (Commandant Jones' Pretoria Commando), J.C. van der Merwe, jun. and D.J.E. Opperman (Commandant Theunis Botha's Commando).

WOUNDED

Sergt. M. Morris (Commandant Geyser's Commando), A. Butcher, D. Linsley, and G.H. Roberts (Commandant Geyser's Commando), J. Chain and H. Lorentz (Commandant Jones' Pretoria Commando), P.G. Prinsloo, J. van Schalkwijk, J.C. van der Walt, and D. Woolonsky [Wolaitsky?] (all of Commandant Theunis Botha's Commando), and Veld Cornet Hoffman (Commandant Van Deventer's Commando).[12]

A genealogical researcher, Annamarie Folkus, whose husband is a descendant of Jeremiah Daniel Nel, one of the burghers buried at Zandfontein,[13] located the battle site in July 2006 and photographed ten graves there which she posted on the website of the Genealogical Society of South Africa.[14] A more recent visit to the grave site in March 2014 revealed that the cemetery actually has two parts, one containing eight graves; and the other, some 330 metres to the west, a single

FIGURE 15.1: Commandants Dirk van Deventer, Jan Smuts and Manie Maritz, comrades-in-arms in 1901. SANA TAB Photographic Library, No. 5734.

grave with the headstones of two soldiers. The main cemetery site comprises two groups of four graves, neatly laid out with headstones, each group surrounded by a stout low palisade fence. They contain the remains of the following men: Captain E.R.J. Geldenhuys and burghers F.J. Botes, B.P. Cronning, H.J. Devanner, A.C. de Wilde, F.J.J. du Preez, J.D. Nel and J.C. van der Merwe.

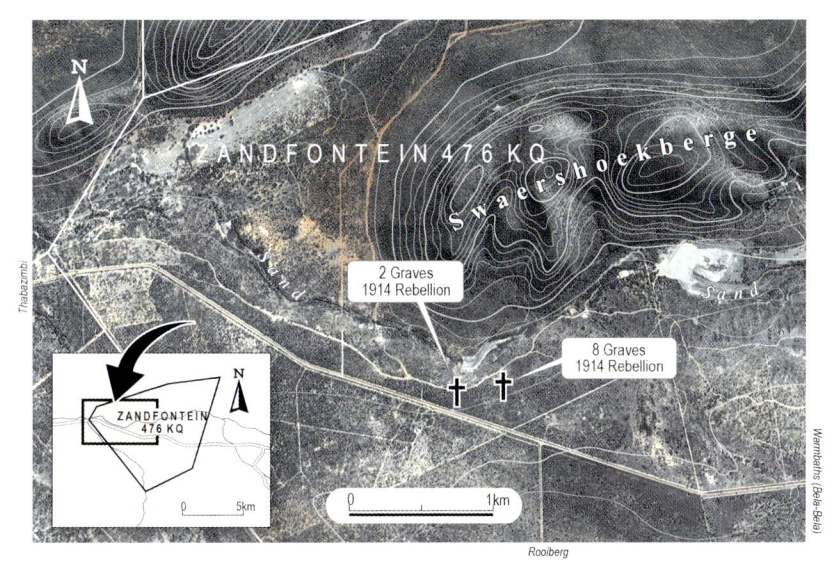

FIGURE 15.2: The 1914 Rebellion graves of Union Forces burghers on Zandfontein. (Image modified from Google Earth, 3 June 2016)

FIGURE 15.3: The main cemetery.

To the west of the cemetery lies a single grave containing the headstones of two men buried together: Lieutenant A. Wrighton and burgher J. Nagel.

Note that whereas the official report stated that there had been 12 UDF fatalities on 8 November, there are only ten graves at the site – and one of these is of a burgher (Devanner) who'd died two days before the skirmish. The graves of the three other casualties from 8 November (burghers A.S.A. Opperman, Hitchcock and Johnson) are to be found elsewhere, as described below.

The Zandfontein battle site and cemetery is now located within a large private game reserve (the Meletse Game Reserve) on a separate concession within the reserve. Some information about each of the ten men buried on Zandfontein has been collated from various sources and is detailed in the end notes.[15,16] In addition to these men, nine of whom died in the battle on 8 November, there were three other fatalities that day, buried elsewhere: Lieutenant D.J.E. Opperman and burghers H.H.H. Hitchcock and W.S. Jackson.[17]

FIGURE 15.4: The joint grave of Lieutenant Wrighton and burgher Nagel.

FIGURE 15.5: The grave of burgher Albert Opperman next to the road on Rhenosterpoort.

Two other fatalities in the Warmbaths area at about the time of the Zandfontein battle were also originally buried at Warmbaths, but later re-interred, together with Johnson in Polokwane. These were burghers G.W. Mapham and H.J. Dewrance.[18]

In addition, there were at least two other UDF casualties in this part of the Waterberg region in the course of the rebellion: Burgher A.S.A. Opperman, who died on 25 November and Sergeant F.J.J. Scheepers, who died on 20 November.[19]

When it is considered that there were 12 Union Defence Force casualties (including two officers and seven men of Geyser's Waterberg Commando) in the Zandfontein incident of 8 November 1914, as well as a large, though unknown number of rebel casualties, it is quite remarkable how little information about or reference to the battle can be found in the literature. For example, the historian Scholz in his definitive, if partisan, book on the rebellion, does not even mention the Zandfontein skirmish.[20] Not a single report about the engagement could be found in the National Military History Museum. Unconfirmed anecdotal information is that many records relating to the 1914 Rebellion were subsequently destroyed, because they reflected several instances of families being divided by the conflict, and that this was not considered conducive to nation-building.

Eyewitness accounts of the skirmish

The rebel perspective

Only two eyewitness accounts of the incident have come to light. In an article published in the journal *Historia* in 1960, J.H. ("Oom Jan") van Wyk, a survivor of the battle from the rebel side, remembered aspects of that day when in 1955 he was brought to the farm Zandfontein, which happened at that time to be owned by his son.[21] The account (translated) has been reproduced in full here because the original journal is difficult to locate:

> On a still Sunday morning in November 1914, the rebel commando under Commandant Jan Harm du Plessis (Oom Rooi Jan) with Oom Koos Niemandt as field cornet, was outspanned in the bed of the Sand River on the farm of the same name. They were busy burying a certain Roos, who was killed in action the previous day near Warmbaths. His body was later exhumed and re-buried by his family on the farm Rooikoppies in the Brits district.
>
> The ground was very hard and they couldn't dig the grave deep

enough. Field Cornet Niemandt instructed that enough stones be stacked on the grave "want môre is miskien my beurt" (because perhaps tomorrow is my turn), he said.

Suddenly, they received news that the Government forces were approaching. They immediately set off along the road but nevertheless, the Government forces overtook them; and, at the upper end of a cultivated field, the battle broke out. Christiaan Liebenberg, Frans Roos, Willem Cronje, Manie du Plessis and a certain Grobler, were wounded during this battle. The latter died the same night and it is assumed that it was he who was buried on the neighbouring farm Weltevreden.

The late Liebenberg, who was also wounded here, was a brother-in-law of Oom Jan van Wyk, the narrator of the story, who came out of the battle without a scratch. Several Government troops were killed in action during this battle and many were wounded. Mr van Wyk was of the opinion that many more of the 'other side' [the rebels] died in the battle, but because it was the last battle of the rebellion in this area, many family members came to the site after the fight to collect their relatives for burial elsewhere.

FIGURE 15.6: J.H. (Oom Jan) van Wyk in 1955. Van den Berg (1960), p 53. Photograph reproduced by courtesy of the Historical Association of South Africa (HASA).

During the battle, the utterance of Field Cornet Niemandt during the morning's funeral came true – and he died five paces away from Oom Jan van Wyk.

A large portion of the rebel force had already been taken prisoner. The rest of them fled into the mountain and fired back at the attackers from the crest of the ridge. In order to establish a position on the mountain, they had to run across a bare cultivated land and Oom Jan told how "die koeëls soos warm vet om hom gegons het" (the bullets spattered around him like hot fat). It was nothing less than a miracle, he believed, that he reached the mountain without injury. The [surviving] rebels headed back along the mountainside towards their laager.

The rebel laager had consisted of approximately eight hundred able-bodied men, with wagons, oxen and horses. None of the forces had artillery at their disposal; and all the casualties were the result of small-arms fire.

None of the rebel graves have been located.

The government forces' perspective

The second eyewitness account is from the opposing side of the battle. In its edition of Friday 11 December 1914, the *Pretoria News* published a letter it had received, dated exactly a month earlier:

Rooiberg, 11/11/14.

> My dear Father, –
> Just a line to let you know I'm still kicking. Since we left Krugersdorp the column has been moving about practically all the time and this is the first opportunity for mailing that has come my way.
>
> From Krugersdorp we went to Brits station where we tried to get in touch with the rebel commando under Rooi Jan du Plessis. We shelled a couple of his men out of the kloof at Wagen Drift (on road to Kameeldrift), but besides knocking up our horses and ourselves we did nothing. Most of the men living along the river in the vicinity of Brits are rebels. Rooi Daniel (Opperman) [Lt Opperman of Botha's Commando] and I fed our horses on Rooi Jan du Plessis' forage and annexed a light buggy of his, but Col V.D. [Col Van Deventer] would not let us take it. From there we went to Warmbaths and arrived a couple of hours after the rebels had passed through the place. We followed them up and came up with them one evening. We

> drove them from a very good position, but as darkness fell, we were unable to follow up our advantage.
>
> We took spoor again and followed through very dense and treacherous country. We came up with the rebel commando on Sunday at about 2 pm. Their position seemed impregnable. However, Col. Van Deventer decided to charge and I went back and gave orders to charge. We charged and broke them and they cleared after about three hours. Our charge only drove them from their good position. As you will have seen by the papers, our loss was 13 killed and 12 wounded. They shot us with soft-nosed bullets like one would shoot vermin. Rooi Daniel was killed.
>
> Their loss we estimate at 120 killed and wounded. We took 26 prisoners. Had we seen what was waiting for us when we returned [across] the ground over which we charged, there would have been no taking of prisoners. As it was, we wanted to shoot them in cold blood when we saw our dead. However, we are all determined to settle the account when we catch up again ...
>
> Love to yourself, Concie, Frank and Aunt Mabel.
> Edward.[22]

The author of this letter is unknown, although he is likely to have been a fellow officer with Lt Opperman in Theunis Botha's Commando. Many of the details correspond with those of the official report, although he numbers one additional fatality – which could mean that one of the wounded men subsequently died (the official records show 12 dead). The alleged use of soft-nosed expanding, or "dum-dum" bullets, if true, was entirely contrary to all agreed military engagement conventions – and was unlikely to have been authorised by the rebel leaders, who themselves had a military background.[23]

An editorial in the *Rand Daily Mail* of Wednesday 11 November 1914, had the following comment to make about the rebellion:

> The fuller reports now being supplied by the authorities indicate very clearly the hopeless nature of the revolt. As far as the Transvaal is concerned, all that remains to be done is the dispersal of the bands of malcontents still wandering aimlessly about the more remote areas. That this task is being effectively undertaken by loyal

CHAPTER 15: A FUTILE BROEDERTWIS

commandoes is shown by the details of the gallant little fight at Zandfontein in the Warmbaths district. A burgher force attacked the rebels in a most spirited manner, and though unfortunately they sustained twenty-three casualties – twelve men being killed, a high percentage indicative of the fierce nature of the battle – they routed the enemy with far heavier loss and captured a portion of their laager. The doings of the past week have proved conclusively that the mass of the Dutch people themselves are resolved to stand by the Government, and maintain law and order in this country. They appreciate the fact that they have enjoyed the widest possible

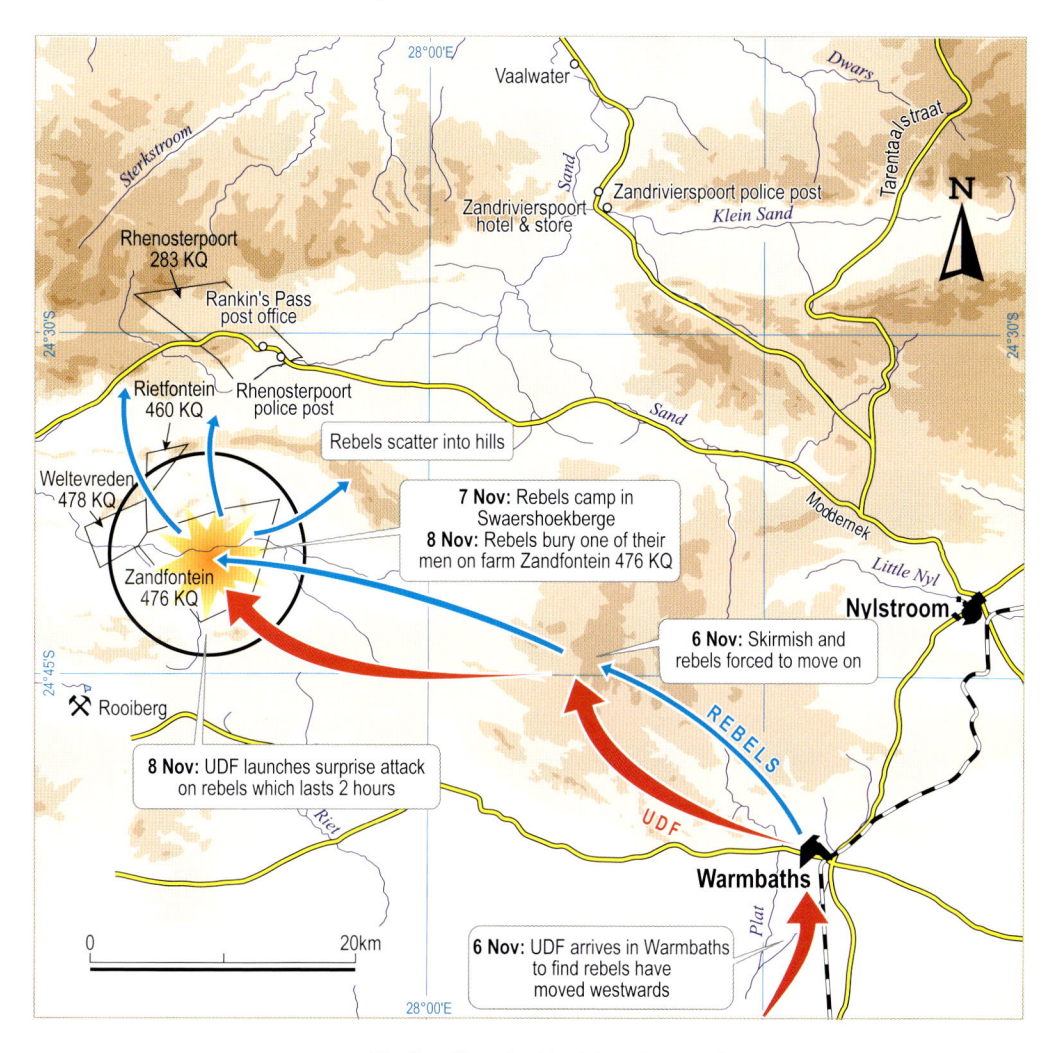

FIGURE 15.7: The Zandfontein skirmish on 8 November 1914.

form of self-government. The Union Government was elected by them and they are determined that its authority shall be upheld. Rebellion under the patronage of Germany does not appeal to them. Painful as the task must be to them, they are ready to do their duty and suppress a movement which threatens their country with anarchy.[24]

The significance of the rebellion

Although small and of limited duration (in total, some 11 400 citizens rebelled and 32 000 troops were used to suppress it; 190 rebels and 132 Union men were killed over a three-month period),[25] the rebellion and incidents associated with it caused serious and lasting damage to efforts by former Boer leaders like Botha and Smuts to build a united nation around the concept of the Union. It stoked the fires of Afrikaner Nationalism in Transvaal and the Free State, adding to the still powerful legacy of bitterness over the consequences of Kitchener's draconian policies (concentration camps, burning of homesteads, etc.) in the closing phase of the South African War little over a decade before. These grievances provided ideal material for writers of emotive, pro-Nationalist propaganda (Gustav Preller, for example), using the newly born written medium of Afrikaans (as opposed to Dutch or English).

The hugely popular General De la Rey, sympathetic to, but not actively part of the rebellion, was accidentally killed on 15 September 1914 when the car in which he and General Beyers were travelling failed to heed a roadblock set up by police to intercept a notorious gang of thieves (the Foster Gang) on the run outside Johannesburg. A shot fired by a policeman apparently ricocheted off a wheel and entered De la Rey's heart, killing him instantly. He was 67. Many rebels believed the (false) rumour that De la Rey had been killed deliberately by orders from Botha or Smuts – a myth perpetuated by Afrikaner republican writers.

It is worth noting that the car carrying Generals Beyers and De la Rey had already ignored three earlier police roadblocks that evening, in the process knocking one policeman off his feet, because Beyers instructed the driver not to stop.[26] Beyers later asserted that he was unaware of the recent activities of the Foster Gang (who had shot and killed three policemen earlier that day). De la Rey was not the only casualty that evening. A little earlier, a car driven by a Dr Grace (of the cricketing family of that name) had inadvertently failed to stop when challenged near Germiston and the doctor was killed when the police opened fire.

De la Rey was renowned for his superb military skills, gracious character, sense of chivalry and natural leadership, notwithstanding his inclination to be guided by an eccentric soothsayer, one "Siener" van Rensburg. On 7 March 1902, in one of the last military actions of the South African War, a motley relief column under Brigadier General Methuen had been surprised by De la Rey near Tweebosch in the western Transvaal. This engagement inflicted on the British the worst defeat of the guerrilla phase of the war: 200 men killed or wounded, and 600 captured. In the course of the battle, Methuen was severely wounded in the side and thigh, an injury compounded when his horse fell onto his leg, crushing it. When De la Rey found Methuen, he ordered that he be taken to the nearest British field hospital. Under a flag of truce, he despatched a message requesting that a telegram be

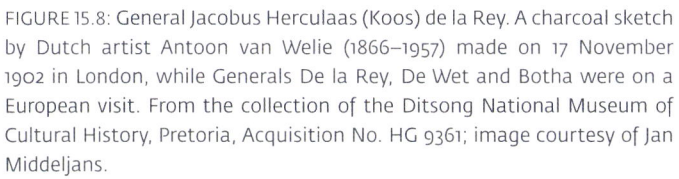

FIGURE 15.8: General Jacobus Herculaas (Koos) de la Rey. A charcoal sketch by Dutch artist Antoon van Welie (1866–1957) made on 17 November 1902 in London, while Generals De la Rey, De Wet and Botha were on a European visit. From the collection of the Ditsong National Museum of Cultural History, Pretoria, Acquisition No. HG 9361; image courtesy of Jan Middeljans.

FIGURE 15.9: General Beyers. SANA TAB Photographic Library, No. 31916: General Beyers.

sent to Methuen's wife expressing his concern about the gravity of her husband's wounds.[27] De la Rey and Methuen had crossed swords before: At the battle of Magersfontein (11 December 1899), it was De la Rey's innovative, pioneering use of trenches that had halted Methuen's attempt to relieve Kimberley. Methuen lost almost a thousand men and was forced to beat a temporary retreat.[28]

Beyers was, like De la Rey, sympathetic to those of his countrymen who opposed entering the war on the side of the British against the Germans in South West Africa. However, his loyalties were split between his affection for his fellow Boer rebels and his devotion to the new country being forged under the leadership of former comrades Botha and Smuts. He had, after all, been appointed the first Speaker of the national Parliament and then as Commandant General of the new Union Defence Force in 1910 (he resigned from this position on the fateful day he and De la Rey drove to Potchefstroom). The attorney H.S. Webb, who defended General De Wet in his trial after the rebellion, recounts a fascinating story that epitomises the internal conflict endured by Beyers at this time:

> 9 November 1914 [the day after the battles at Zandfontein and Sandrivier]: Beyers met General Hertzog at Onze Rust [the residence of the ailing former Free State President Steyn]. During the visit to Onze Rust, General Beyers had a conversation with General Hertzog in the grounds near the house, in the course of which Gen. Hertzog pointed out the dangers of a fratricidal civil war. Gen. Beyers pointed to his bandolier, which held its full complement of cartridges, saying, "I have not fired a shot yet, and by God's help, I never shall". His body, when found [following his drowning in the Vaal River on 10 December 1914, while attempting to evade a government patrol], still had his bandolier with the full number of cartridges. He had never fired a shot.[29]

The story is corroborated by a similar account of the meeting, in Scholtz's biography of this fine South African patriot and leader.[30] However, the English media of the time did not share these sentiments: The *Pretoria News*, in a special edition on 11 December 1914, merely announced "the death, by drowning in the Vaal, of the rebel Beyers". This prompted an immediate letter of admonishment from Tielman Roos, a leading Transvaal Nationalist, pointing out Beyers's honourable behaviour – which in turn led to a flurry of hostile correspondence from both the editor and other Union loyalists who accused Beyers of treachery and worse.[31]

Maritz, the instigator of the rebellion, fled to Angola for the duration of the war after Botha's successful operation against the Germans in South West Africa.[32] Generals Kemp (who helped Maritz to capture Upington briefly on 25 January 1915) and De Wet were captured, tried and jailed, the latter after an exciting car chase across the southern Kalahari.

Captain Josef (Jopie) Fourie, who, with Major J.J. Pienaar, may not even have been in the Waterberg in early November, was captured with 43 other rebels on 16 December in the Magaliesberg at Nooitgedacht (ironically, the scene of the famous battle in December 1900, when combined Boer forces under De la Rey and Beyers had inflicted a crushing defeat on a British column led by General Clements). On 18/19 December, Jopie and his brother Johannes were court-martialled on a charge of treason because they alone amongst UDF officers had failed to resign their commissions before joining the rebels, and had foolishly continued to wear their UDF uniforms. Jopie was found guilty of treason and was shot by firing squad on 20 December. (His brother's sentence was commuted to five years' imprisonment with hard labour.)[33] So did this otherwise unremarkable junior officer – although he had earned some fame after being seriously wounded in the knee and captured in November 1901 whilst trying to take horses from a British camp north of Pretoria – become a martyr and hero in the eyes of his sympathisers. While Fourie's execution was undoubtedly militarily justifiable under the prevailing martial law conditions, it turned out to be a serious political mistake by Smuts, who lost valuable support from the Afrikaner community, a loss which would ultimately contribute to costing him the election of 1924.

In January and February 1915, all 400 captured rebels were released from various prisons. Three-quarters came from the Free State, home of the feisty De Wet. There were only five from the Waterberg region, including Jo-

FIGURE 15.10: Captain Josef ("Jopie") Fourie (1878–1914); this is the best-known photograph of him, taken on his wedding day in 1903, when he married Susanna Engelmohr. SANA TAB Photographic Library, No. 1404: Capt. Jopie Fourie.

hannes Hendrick Grobler from Groothoek and Jan Johannes Liebenberg from Bulgerivier.[34] Kemp and De Wet were sentenced to seven and six years' imprisonment respectively, but were released within two years. Maritz eventually returned to SA in 1924. He was tried and sentenced to three years in jail, only to be released a few months later when the new Nationalist government under General J.B.M. Hertzog (in which General Kemp was a cabinet minister!) came to power.[35] De Wet retired to his Free State farm where he died of ill-health in 1922, aged 67.

And what of the UDF Commandant Geyser, whose commando suffered such heavy casualties at the battle of Zandfontein? He was Commandant Andries Hermanus Geyser who as a young field cornet had been among the signatories to an emotional statement presented to General Beyers by his Boer officers in the Waterberg on 9 June 1902, shortly after peace had been declared. This opened with the following words:

> Wij ondergeteekenden, Officieren van de districten Zoutpansberg en Waterberg wenschen van deze gelegenheid gebruik te maken UEd. Gestr. namens onszelven en onze burgers onzen hartelijken dank toe te brengen voor wat gij gedurende dezen zwaren en noodlottigen oorlog voor ons geweest zijt.[36]

In 1918 Major A.H. Geyser was one of several UDF officers to receive a Distinguished Service Order (DSO) from King George V "for distinguished service in the field and in connection with the campaign in German South West Africa, 1914–1915". In 1920 Geyser received a similar honour – the *Dekoratie voor Trouwe Dienst* (DTD) – for his service as field cornet for the Boer forces during the South African War.[37] This was the same A.H. Geyser, by now a Lieutenant Colonel in charge of the Potgietersrust Commando, who on 27 July 1925, stopped in to sign the register at the Sandriviersport Hotel between Nylstroom and Vaalwater (45 km northeast of the Zandfontein battle site).[38] Lt Col Geyser died in Potgietersrust (now Mokopane) in September 1956, aged 85.[39]

The skirmish on Zandfontein was a bitter, violent and ultimately (like so many such battles) quite pointless incident in which many men lost their lives and from which there were no positive consequences, but many, lasting negative impacts. Little did anyone know, at that time, of the horrendous events that were to unfold in Europe over the next four years: the most devastating war in the history of humankind, and about which much the same conclusions would later be drawn.

CHAPTER SIXTEEN

Domain of the Northern Lions

The Waterberg has long been rightly regarded as having been one of the most politically conservative – if not *the* most conservative – enclaves of white South Africa prior to the democratic elections of 1994.

Even since then, political conservatism has manifested itself from time to time among disaffected elements of the white community (for example, the *Boeremag* of the 1990s). This is not really surprising: The hilly region that lay behind the settlements of Warmbaths, Nylstroom, Naboomspruit, and later Thabazimbi, was rugged, difficult to traverse, sparsely populated, poorly mapped, barely administered and characterised by poor soils.

It attracted, as such areas always do, only those for whom a solitary, hardy and austere lifestyle was either an eccentric choice, or a necessity. The legend – proudly repeated by old Waterbergers – that Paul Kruger used to offer troublemakers a piece of land in the Waterberg to get them out of his hair might not be based on fact, but it retains a ring of authenticity nevertheless. The early settlers of the Waterberg were loners, backwoodsmen, close to the ground on which they eked out a living, convinced of their right to do with their own land what they wished and resentful of attempts by the outside world to bring order, formal property boundaries, taxes, and regulations to their mountain fastness. This attitude lives on amongst a number of residents.

The first (white) settlers along the southern foothills of the Waterberg were the Van Rensburg and Van der Merwe families, who in 1837 established themselves on what would later become the farm Tweefontein. Their settlement was called Eersbewoond (first occupied); today it is a rail siding along the old coach road that runs north from Bela-Bela towards Modimolle. Within a few years, another small settlement had sprung up around the marshy hot springs, known as Het Bad, a few kilometres to the south of Eersbewoond.

The first white political governance of the Waterberg

Until the creation of the Transvaal Republic in December 1856 and the election of its first President, Marthinus Pretorius, a month later, governance of the region north of the Vaal River was patchy and divided into four small republics: Potchefstroom, Rustenburg, Zoutpansberg and Lydenburg. The appointment of the Voortrekker hero Andries Pretorius as Commandant General of the two southern republics in about 1850 began a tortuous process that culminated in the creation of the Transvaal Republic six years later under the leadership of Andries's son, Marthinus. Resistance to the amalgamation of the four republics was led by the leaders of the two northern states, Stephanus Schoeman of Zoutpansberg and Piet Potgieter of Lydenburg. In the meanwhile, administration of the Waterberg region was controlled from Rustenburg or Potchefstroom (Pretoria only being founded in 1855), and in 1852 H.J. van Staden was appointed as the first field cornet for the area.[1]

However, no sooner had Marthinus Pretorius assumed the presidency of the fractious Transvaal than he began to make overtures to members of the Volksraad of the Orange Free State to incorporate that republic into his Transvaal (as had been his father's vision). Eventually, following the resignation of the Free State's President (Boshof) in 1859, Pretorius was indeed invited to take over the position, notwithstanding that he was currently also President of the Transvaal. This bizarre situation persisted for eighteen months until the Transvaal Volksraad insisted on his resignation and appointed C.J. van Rensburg as Acting President in his stead. This appointment was contested by Schoeman from Zoutpansberg, who enlisted the support of Pretorius to oust Van Rensburg, a move strongly opposed by Paul Kruger, the young Transvaal Commandant. Several skirmishes ensued, until eventually some sense prevailed: A convention was held, followed by elections (albeit three rounds!) which resulted in the formal election of Pretorius as Transvaal President for the second time on 10 May 1864. (He had resigned from the Free State post the previous year.)[2]

J.H. Viljoen had been appointed as the first Commandant of the Waterberg in 1858.[3] Only after a local farmer, Olferman Collins, offered some of his land along the Nylspruit for the development of a town was the first formal settlement, Nylstroom, established in 1866. This triggered the appointment of the first magistrate for the Waterberg region, J.J. Prinsloo, and the formal proclamation of the Waterberg as a legal district. Field Cornet Van Staden was elected to the Volksraad as the District's first representative.⁴ He was followed as field cornet by Willem Stellen in 1870,⁵ and by G.J. Verdoorn (later the first postmaster at Nylstroom) in 1875.⁶

Annexation and resumption of independence

In November 1871 Pretorius was forced to resign the Transvaal presidency (for the last time), following a series of judgemental and governance errors that cost him the support of his burghers. He was replaced (despite opposition from Kruger) by a mild-mannered, liberal-thinking NGK minister from Hanover in the Cape, Thomas François Burgers. Unfortunately for Burgers, amongst the Transvaalers were ambitious men who were never likely to tolerate his ideas for long and did everything possible, including abandoning their president in the battlefield against Chief Sekhukhune, to discredit him. His weakened administration provided the ideal pretext for the British at the instigation of Sir Theophilus Shepstone to annex the republic on 11 April 1877. Burgers left for the Cape, where, penniless, he died on a farm near Richmond four years later after his horse tripped and fell on him.⁷

The annexation of the Transvaal by Britain was short-lived. Within a month, the burghers had sent a deputation, led by Paul Kruger, to London to lobby their case for independence. This was followed by several other attempts to negotiate the return of independence peacefully, but all were rebuffed. Eventually, in desperation, the Boers prepared to seize their independence by force. The opening salvo of this war was fired in Potchefstroom on 16 December 1880. It was followed by battles at Bronkhorstspruit, Laing's Nek and, finally, at Majuba Hill on 27 February 1881, where a large British force under General George Colley was overwhelmed by a much smaller group of Boers led by General Piet Joubert. This crushing defeat led speedily to a peace agreement a month later, to the restoration of Transvaal's independence, and to lasting esteem for Joubert.

The following year, C.S. Potgieter was elected to the Volksraad as the Waterberg's representative.⁸ It seems that he remained in this role for two terms, because the next reference is the election to the position of Jan du Plessis de Beer in 1892.⁹ De Beer was the son of another early Boer settler family in the district and farmed at Tweefontein between Nylstroom and Warmbaths.¹⁰

The composition of the Volksraad

During the period before the South African War of 1899–1902, political representation in the Volksraad of the Transvaal Republic (ZAR) had been based more on individual personalities than on any formal political groupings. However, in the mid-1880s, a group calling itself the "Progressives" was formed and in due course it came to have 12 or more members in the Volksraad (which at that time had a total of 26 members). Among its members were the future Union Prime Minister Louis Botha, Koos de la Rey, the future Boer General, R.K. Loveday (the member for Barberton and the founder of the party), Lukas Meyer, the Lodewyk brothers and Carl Jeppe, the latter being a former public prosecutor in Nylstroom and now the member for the newly formed town of Johannesburg. They were inclined to support General Joubert as a presidential candidate rather than the incumbent Kruger.[11]

In 1889, in belated response to the growing animosity among the foreigner or *uitlander* community on the Witwatersrand (and Barberton area) because of their exclusion from the franchise (despite their obligation to come to the defence of the republic), President Kruger proposed the creation of a second (or Lower) house in the legislature to be called the Second Volksraad. This would have the same number of seats as the original chamber (now to be called the First Volksraad), but eligibility to vote for it would be less stringent, allowing uitlanders (at least, white males of 16 years or older with at least two years' residence) to participate. The powers of the Second Volksraad were mainly limited to matters pertaining to mining and the Witwatersrand; in any case, no resolutions passed by it would become law unless endorsed by the First Volksraad and the President. In 1890 Kruger's proposal was put to the Volksraad, which reluctantly approved it, and the bicameral legislature was introduced the following year.[12] Voters and members could not participate in both chambers – they had to choose which one to support. The bicameral arrangement would continue until the termination of the republic in 1902. However, given its lack of real authority, the Second Volksraad was not well received by the uitlander community for whom it was intended and their support for it was poor.

The year 1893 was a Presidential election year. Paul Kruger was standing for a third five-year term, his opponent then and in each previous case having been General Piet Joubert. In 1883 Kruger had won with 3431 votes to 1171 for Pretorius, and in 1888 by an almost doubled majority.[13] By 1892, though, Kruger's popularity had begun to wane, mainly because of his attitude towards the uit-

FIGURE 16.1: President Paul Kruger. SANA TAB Photographic Library, No. 21490: President Paul Kruger.

landers and because of his penchant for hiring Hollanders (in particular, Dr W.J. Leyds, who'd been appointed as State Secretary in 1888) to fill senior positions within his government. Joubert, though ageing (at 61, he was six years younger than Kruger) and uninspiring, began to regain popularity.

This trend was vociferously and provocatively encouraged by Eugène Marais, the young journalist who in 1889 had taken over the editorship of *Land en Volk*, a Pretoria newspaper. In part to revive the flagging fortunes of the paper but also to communicate his own antipathy towards Kruger, Marais began to write highly critical, colourful and occasionally scurrilous articles about Kruger and his administration, while simultaneously promoting the "progressive" philosophy of the Joubert camp. To safeguard the paper's income from government sources and to avoid litigation, Marais developed the device of "creating" correspondents, whose vituperative letters his paper published (and which he wrote under pseudonyms). One of his best-known and most radical anonymous correspondents was "Africanus Junior". When pressed for the identity of this writer, Marais led the enquirer to believe that it was one J. de V. Smit from that backwater, the Waterberg![14] *Land en Volk*, and the views it projected in endorsing Pretorius and his Progressives undoubtedly assisted him in building support for his election, despite his own rather lacklustre performance, absence of policy and poor public persona.

The 1893 election drew keen public interest and the result was a close call: Kruger won, polling 7911 votes (52% of the total), but Pretorius drew 7246 votes and a third candidate, Chief Justice Kotze, received 85 votes.[15] Pretorius won in half of the wards – Waterberg was among them – but Kruger's narrow victory was helped in particular by powerful support for him in the large ward of Rustenburg, his home constituency.[16] Marais and his newspaper criticised the result, crying that there had been foul play and that objections had been treated with disdain – but the result was approved by the incoming Volksraad and the country moved on. Indeed, such was the growing concern among citizens of the republic (most uitlanders were effectively unable to become citizens) about the aspirations of Rhodes and the English in the Cape that Kruger was able to move quickly and effectively towards closing the rift that had developed between his supporters and those of Joubert – a sense of unity that was finally cemented by the ill-conceived invasion of Dr Leander Starr Jameson and his ragtag force in December 1895.

Jan de Beer

Throughout this period, the Waterberg District continued to be represented in the Volksraad (now the First Volksraad) of the Transvaal Republic by Jan du Plessis de Beer. De Beer had made his mark in the chamber as early as 1884 when (with another Waterberg representative, C.C. van Heerden, from Zwagershoek, where Rankin's Pass is today), he had contested the Government's assertion that existing game protection legislation was adequate, "contending that the game law was quite ineffective, he himself being aware of the illegal slaughter of animals by hide hunters in his district".[17] This view – that hunting legislation needed to be revised – was supported by all the landdrosts of the Transvaal with the surprising exception of the Landdrost for Waterberg, who considered existing legislation to be satisfactory! De Beer was vehemently opposed to the proposal that there should be a closed season for hunting, especially for owners on their own farms. It seems as though the Landdrost might have been more in tune with the burghers than their Volksraad representative: that "they were content with the game law as it stood, precisely because it imposed no restrictions on their freedom to hunt".[18]

But the nub of the issue was that the majority of the members of the Volksraad were in favour of introducing measures that would curb the widespread and indis-

FIGURE 16.2: Jan du Plessis de Beer, Volksraad member. Redrawn by C. Bleach from a poor photograph: SANA TAB Photographic Library, No. 4330.

criminate decimation of wild game in the Transvaal by hunters generally, whereas conservative representatives like De Beer wished to see such measures applied only to visiting trophy and hide hunters – and blacks. When, in 1889, President Kruger endorsed the proposal to create the Pongola Game Reserve in the area bounded by the Pongola River, the Swaziland border and the Lebombo range (just to the north of today's Itala Reserve), De Beer was one of only two members of the Volksraad to oppose it – ostensibly on the basis that it was unclear whether blacks would also be prohibited from hunting there.[19] Throughout the many debates about wildlife in the Volksraad in the years until the South African War, De Beer's attitude was consistent: He wanted to prevent blacks from having the right to hunt game at all, while protecting that of white landowners to do so at all times on their own properties and of bywoners to do so for subsistence purposes. His racist and self-interest ideology presaged that of many of his successors as parliamentary representatives. In his later years, De Beer would become a director of the local farmers' co-operative, WLKV.

The Waterberg representative in the Second Volksraad (1890–1900) was a Nylstroom resident, Frederik Boshoff. Boshoff was born in Colesberg, Cape, in 1846, and moved as a child with his family to the farm Doornfontein – a store of that name marks the farm's location along the tar road to Vaalwater today. Boshoff became a prominent farmer and politician in the area (see Chapter 10). He was one of the founder directors of the WLKV and a post-Boer War member of the Provincial Council. He died in 1918.[20]

The Het Volk party

Serious party-political representation in the Transvaal only began to take form during the British administration that followed the Treaty of Vereeniging in May 1902, which promised the two former republics some form of representative government. In May 1904 in the wake of acknowledgement by the Administrator, Milner, that there should be a move towards self-government, the party Het Volk (The People) was formed in the Transvaal. It was led by Boer Generals Botha, Smuts, De la Rey and Beyers. Its objective was to bring about conciliation between Boer and British communities. In the Free State, Generals J.B.M. Hertzog and Christiaan de Wet, together with Abraham Fischer, would establish the Oranjia Unie party two years later.[21]

The Transvaal elections of 1907

The move towards self-determination was accelerated by a change of government in the UK in January 1906 when the Liberal Party, which favoured self-government for the South African colonies, displaced Balfour and his Unionists. A new constitution was speedily drawn up and the first elections in the Transvaal were scheduled for 20 February 1907, for a Legislative Assembly of 69 seats.

In May 1903 Milner had used the Land Settlement Ordinance 45, passed the year before, to establish a settlement scheme on the Springbok Flats for British and colonial soldiers willing to remain in South Africa. Nine large farms were divided into 37 smaller properties each of about 400 ha; applicants were invited to hire the properties with an eventual right to purchase them (each had an average value of about £565).[22] A year later, most of the properties were occupied and Milner's objective of ensuring a British voice in the community was accomplished. This block of farms was centred on the hamlet of Settlers that fell within the Waterberg constituency of the Transvaal. A similar, though much smaller, settlement of Eng-

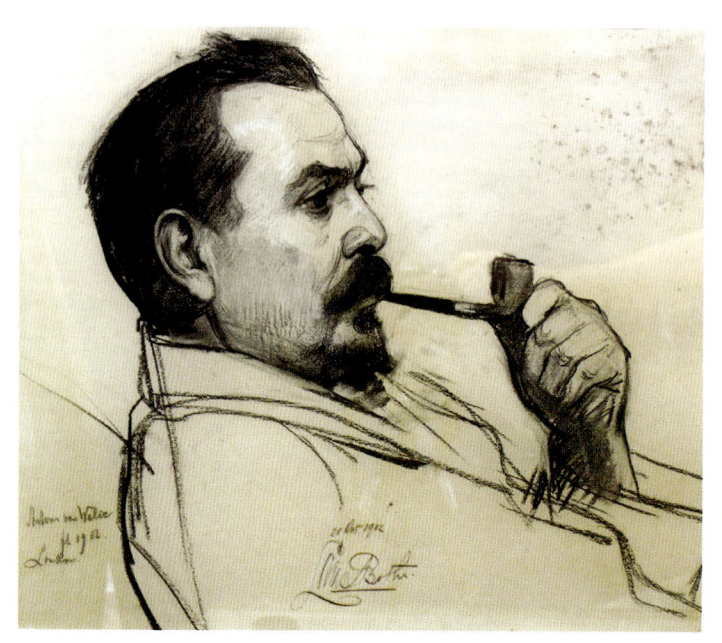

FIGURE 16.3 (LEFT): General Louis Botha, Prime Minister of Transvaal (1907) and first Prime Minister of the Union of SA (1910). A charcoal sketch by Dutch artist Antoon van Welie (1866–1957) made on 20 October 1902 in London, while Generals De la Rey, De Wet and Botha were on a European visit. From the collection of the Ditsong National Museum of Cultural History, Pretoria, Acquisition No. HG 9374; image courtesy of Jan Middeljans. FIGURE 16.4 (RIGHT): General C.F. Beyers. SANA TAB Photographic Library, No. 15245: General C.F. Beyers, Speaker of Transvaal Legislature, later first Commandant of Union Defence Force.

lish-speakers was established at this time in terms of the same Ordinance on and around the farm Vaalwater, also in the Waterberg. At the time, Nylstroom had a total of 361 adult white residents: 218 men and 143 women. According to the records, there were also 217 "nonwhite" adults in the town.

General Beyers, whose popularity with the Afrikaner community of the Waterberg had not dimmed since his leadership during the war, was chosen to represent Het Volk in that constituency and he campaigned vigorously in Nylstroom, Potgietersrust and Pietersburg in the run-up to the election. Other parties contesting the event were the Progressive Party (Percy Fitzpatrick, George Farrar and Abe Bailey); the Responsible (or National) Party led by H.C. Hull and Sir Richard Solomon, Attorney General of Transvaal; and the Labour Party. The result was a conclusive victory for Botha's Het Volk: It secured 37 seats in the new Transvaal Legislature to 21 for the Progressives, 6 for the Responsible/National Party and 3 for Labour. In the Waterberg constituency, Beyers received 779 votes, compared with 217 for his opponent, the Nylstroom stalwart Emil Tamsen of the Progressives.[23]

(In November 1907 the first election in the Orange River Colony (Free State) was held: Oranjia Unie won 31 of the 38 seats; Abraham Fischer became Prime Minister, with Hertzog as Minister of Education and Attorney General).[24]

During the campaign, Beyers had argued forcefully for the rights of the Afrikaner, especially regarding religion, culture and education. His views, though, were often seen as offensive by English-speaking supporters of the party. Because of his popularity in the Waterberg, Beyers and his supporters fully expected him to be appointed to Prime Minister Botha's first cabinet. But Botha and Smuts were determined to bring about reconciliation between the two former enemies. In a clever stratagem, they appointed Beyers as Speaker of the new Parliament, thereby effectively silencing their outspoken colleague. Though angry and disappointed, Beyers accepted the role with magnanimity and rose to the occasion: At the end of the Parliament three years later, upon the formation of Union, he was universally acclaimed for the professional, dispassionate and fair way in which he'd managed the affairs of the Assembly. Botha also appointed Solomon and Hull of the Responsible (National) Party to his cabinet.[25]

Richard Granville Nicholson

Beyers's appointment as Speaker necessitated a by-election in the Waterberg constituency. Het Volk nominated Richard Granville Nicholson from Pietersburg as its candidate; he was duly elected later that year.[26] Richard Granville Nicholson was a remarkable personality. He was born at Hondeklipbaai on the bleak Western Cape coast in 1859 to an Irish Catholic couple (his father was a trader and guano exporter). This remote outpost was described at the time in the following words:

> Of all the dismal places that I have seen, Hondeklip Bay as it was in 1855 is the worst, consisting as it did of black wooden buildings standing irregularly in a desert of white sand, the blinding glare from which was insufferable. There was no vegetation; there was no water save such as was brought over sea from Cape Town, and as vessels arrived irregularly, the supply often ran short.[27]

Nicholson was sent to school in Rondebosch and at the age of 17 went to Kimberley to find employment. In 1876 he joined a volunteer militia called the Kimberley Light Horse and saw action in the eastern Cape and Griqualand, as well as in Zululand, before becoming a burgher of the ZAR and transferring his allegiance to the Lydenburg commando. With the latter, he participated on the side of the Boers at the Battle of Majuba Hill in February 1881. He then went hunting in the northern Transvaal where he met a hunter called Soloman Marais from whom he bought a farm, Majebas Kraal (also spelled Matibaskraal) about 20 km east of Pietersburg (Polokwane). In 1885 he married one of Marais's daughters, Judith Johanna (Sannie), a "strong, self-sufficient, hardy and resourceful" woman.[28] However, because he had married a Protestant, Nicholson was effectively excommunicated by the Catholic Church, leading him to swear that he'd never set foot in a church again, although he encouraged his children as they grew up to choose their own religion. This led to the unusual result that two of his children became Catholics (they were brought up by an aunt), one an Anglican, one a Presbyterian and the remaining three, members of the Dutch Reformed Church.[29]

In 1890 Nicholson was invited by Cecil Rhodes to join the officer corps of the Rhodesian Pioneer Column because of his considerable knowledge of Mashonaland, his fluency in local languages and his skills as a horseman and tracker. He was promised a farm in Mashonaland in return for his services. Ever the adventurer, Nicholson signed up. As the column approached its intended destination, predetermined by its scout Frederick Courtenay Selous, Nicholson was tasked by Selous to lead the way and to establish a camp at the chosen spot. As Nicholson recounted the story, a miscommunication between the two led to Nicholson setting up camp about 20 km to the south of the hill (Mt Hampden) intended by

Selous, at the site (by a smaller hill) now occupied by the Zimbabwean capital of Harare (formerly Salisbury).

Foregoing the opportunity of settling in the new country, Nicholson returned to his farm where he and Sannie produced a total of nine children, seven of whom survived to adulthood. The fifth-born, who arrived in 1892, was named Richard Granville after his father. During the 1890s Nicholson participated in many of the battles against militant black polities, probably as a member of the Zoutpansberg Commando. This alignment of Nicholson's loyalties with those of his fellow burghers made it inevitable that upon the outbreak of the South African War in 1899, he would join them in their opposition to the British. He initially joined the Northern commando under the leadership of Commandant "Groot Freek" Grobler, but becoming quickly disillusioned with the latter's leadership, applied for and was granted leave to join the Boer forces in the Tugela valley of Natal.

FIGURE 16.5 (TOP LEFT): Richard Granville Nicholson, MP for Waterberg (1910–1915). SANA TAB Photographic Library, No. 2222: R.G. Nicholson, MP for Waterberg. FIGURE 16.6 (TOP): Officers of the Rhodesian Pioneer Column, 1890. In the front row, second from the left is R.G. Nicholson and F.C. Selous is in the middle row on the right. *Officers of the Rhodesian Pioneer Column, 1890*. Photograph kindly provided by Nicholson's granddaughter, Shelagh Nation. FIGURE 16.7 (BOTTOM LEFT): Judith Susanna (Sannie) Nicholson (née Marais). Lewis, Thomas H. (ed.) (1913): *Women of South Africa – a Historical, Educational & Industrial Encyclopaedia & Social Directory of the Women of the Sub-Continent*. Le Quesne & Hooton-Smith, Cape Town, p. 198.

Later in the war, after the fall of Pretoria in May 1900, Nicholson rejoined the Transvaal Boers under General Beyers who were harrying the British advance north towards Pietersburg.[30] In the meanwhile, his wife Sannie, who had been a nurse with Commandant Grobler's abortive attack on British forces at Rhodes Drift, tending to the wounded on both sides, returned to manage Majebas Kraal in Nicholson's absence.

In August 1901 Nicholson was apparently persuaded by Beyers to surrender in order to protect his family. He was charged with treason on account of his historical cooperation with the British in Mashonaland, but after pleas for clemency had been made by family members in the Cape, he was sent to Ladysmith to take an oath of neutrality and thence to the concentration camp at Howick where Sannie and family joined him in September. They spent the rest of the war there and at a camp at Wentworth.

After peace had been declared, the family returned to their farm, which had been totally destroyed by the British. Nicholson then involved himself in politics, becoming a founder member of Het Volk. He was a logical successor to his former leader, General Beyers, as MP for Waterberg. After his election, Nicholson and family took up residence in Nylstroom, this being the heart of his parliamentary seat.

The advent of Union

After the elections for self-government in the Transvaal and Orange Free State, several parties advocated for a closer relationship between them and the other two colonies of Natal and the Cape. This culminated in a National Convention in 1908/09, which drafted a proposed constitution to govern a union of the four colonies. The resulting South Africa Act, having been approved by the parliaments of the Cape, Orange River Colony and Transvaal, by a referendum in Natal and by both Houses of the British Parliament, was signed by King Edward VII on 20 September 1909 and came into effect on 31 May 1910.[31] The legislature comprised of two houses: an Upper, Senate and a House of Assembly.

The first Governor General, Herbert Gladstone (a son of the former British Prime Minister), was tasked with appointing the first Prime Minister of the new Union. He selected General Louis Botha, who then appointed a cabinet and set a date for the first national elections: 15 September 1910. A reconfiguration of political parties took place to contest these elections. The governing parties in the

Cape, Free State and Transvaal – the South African Party, the Oranjia Unie and the South African National Party (a Het Volk/Responsible Party merger) respectively – formed a loose alliance named for its Transvaal member: the National Party. The Unionist Party was a merger of the Transvaal Progressives, the Cape Unionists and the Constitutional Party of the Orange Free State. And the Labour Party, under the leadership of F.H.P. Creswell, was the third contestant. The result of the election was an overwhelming victory for the governing National Party alliance: it won 67 seats, versus 39 for the Unionists, 4 for Labour and 11 for Independent candidates.

In the Waterberg, R.G. Nicholson, the member of the governing party member in the previous Transvaal legislature, was elected to the national parliament.[32] Johann Rissik, the Dutch-born former Surveyor General of the Transvaal Republic and chairman of the mining and land speculation company Transvaal Consolidated Lands, TCL (which still owned numerous properties in the Waterberg), was appointed the first Administrator of the Transvaal. (Rissik retired from this position in 1917 and died in 1925).[33]

The rise of the Afrikaner nationalists

Dissent within the ruling Nationalist alliance (soon to be renamed the South African Party), was quick to surface, given the ideological differences between Botha (the conciliator) and Hertzog (advocate of Afrikaner nationalism). Hertzog's insistence on developing a separate, but equal educational, linguistic and cultural policy for the Afrikaner, which the Constitution allowed him to do in his province the Free State, brought him into conflict with Botha and his desire to forge greater cooperation with the largely English-speaking Unionists. At a meeting in Bloemfontein in January 1914, Hertzog formed the National Party, the membership of which was boosted by the unpopular decision of the Union government to support Britain upon the outbreak of World War I.

In the general election held in November 1915 (by which time the total number of seats had been increased to 130), Hertzog's NP won 27 seats, compared with 54 seats for the governing South African Party, 39 for the Unionists, 4 for Labour and 6 Independents. The SAP no longer had an absolute majority, although its opponents were ideologically on either side of it.

Among the governing party members to lose their seats was R.G. Nicholson of the Waterberg, who was defeated by P.W. le Roux van Niekerk of the National Party – by a margin of just three votes![34,35]

At the outbreak of World War I, Nicholson, saddened by the involvement of good friends like De la Rey and Beyers (and Jopie Fourie) in the rebellion, joined General Botha's campaign to remove the Germans from South West Africa. He was awarded the rank of Major with command of the 3rd Mounted Brigade and his efforts ultimately led to his being awarded a military Order of the British Empire (OBE) "for distinguished services during the GWSA campaign" in the Birthday Honours List of King George V, issued between June and August 1919. At the end of that successful campaign, aged 56 and no longer MP for Waterberg, he accepted an appointment as Crown Lands Inspector for a vast region that included the Kalahari, northern Natal and Pietersburg. He fulfilled this role for several years before final retirement to his farm Majebas Kraal.[36]

Nicholson died in 1931, and Sannie in 1943. Like a number of bureaucrats and community leaders of that time, Nicholson was completely fluent in both English and Dutch (and the new language of Afrikaans), as reflected in his correspondence. His son of the same name went on to become Chief Inspector for Provincial Audits; he died in Johannesburg in 1948, having also given one of his sons the family name.[37] A great-grandson, also named Richard Granville Nicholson, became a businessman and farmer in Modimolle.

"Oom" Piet le Roux van Niekerk had been born in Wellington in the Cape in 1878. He studied agriculture at Stellenbosch before moving to the Springbok Flats where he later purchased the farm Leeuwarden in 1941. Van Niekerk was the first chairman of the Waterberg Landbouers Koöperatieve Vereeniging (WLKV), which he had helped to found in 1909 and which would later become the NTK.[38] In 1914 he joined up with rebels opposing the Union government, was arrested and jailed.[39] Van Niekerk's election in 1915 to represent Waterberg in the National Assembly was the beginning of a 72-year domination of this constituency by the National Party. Van Niekerk himself retained his seat until 1929 when he was elevated to the Senate and replaced by Advocate J.G. Strijdom, who had a legal practice in Nylstroom and owned a portion of the farm Elsjeskraal a few kilometres northeast of Settlers.

Advocate J.G. Strijdom

Johannes Gerhardus (Hans) Strijdom was born on the family farm near Willowmore, Cape, on 14 July 1893. After graduating in the study of languages at Victoria College, later to be Stellenbosch University, he began farming with ostriches, a career terminated by the collapse in the feather market. In 1915 he joined up to serve in the Medical Corps of the Union Defence Force sent under General Louis Botha (who was also the Prime Minister at the time) to capture South West Africa. He was later a corporal in Helgaardt's Scouts. At the end of the campaign he moved to Pretoria University to complete his law degree. He was fully expected to settle in the city and to practise as an advocate there.[40]

In 1918, shortly after being admitted to the Bar, Strijdom was caught up in a whirlwind romance with a pretty young actress with the stage name of Marda Vanne. Her real name was Margaretha ("Scrappy") van Hulsteyn; her father, Sir Willem van Hulsteyn, had come to South Africa from the Netherlands as a boy; he became a lawyer, served as an adviser to Lord Milner during and after the South African War, and joined the prominent legal practice in Johannesburg that still bears his name. Her mother Margaretha was the daughter of Dr Izaak Pronk of the Orange Free State, a family with other Waterberg connections (see the section on Ellisras in Chapter 19).[41] Hans and Margaretha (Jr) married, but within months the marriage came to an end. Vanne left for England, where, after a wildly promiscuous life she formed a life-long relationship with the famous

FIGURE 16.8: P.W. le Roux van Niekerk, MP for Waterberg (1915-1929). Nel, J.J. et al (eds): *Nylstroom 1866-1966*. Published by Nylstroom Town Council, p. 101.

FIGURE 16.9: Advocate J.G .Strijdom, MP for Waterberg (1929-1958), Prime Minister of South Africa (1954-1958). SANA SAB Photographic Library, No. 19555: J.G. Strijdom, Prime Minister, 1954-1958.

English actress Gwen Ffrangcon-Davies, member of a circle that included John Gielgud, Ivor Novello, Noel Coward and others.[42] Strijdom moved to Nylstroom to lick his wounds. The Afrikaans media then and later were at pains to play down Strijdom's first marriage, let alone to speculate on the reason for the divorce, presumably in order not to prejudice his political career.[43]

Strijdom eventually remarried in 1931; his second wife, Susan de Klerk (1910–1999), 17 years his junior, was a young music teacher in Nylstroom where her older brother Jan taught history. When Strijdom became Prime Minister in 1954, he controversially appointed Jan, by then a headmaster in Springs, both to the Senate and as Minister for Public Works in his cabinet. Jan de Klerk (1903–1979) would later become President of the Senate. One of Jan's sons, Frederik Willem (F.W.) – Strijdom's nephew – became a lawyer and eventually State President and joint winner of the Nobel Peace Prize with Nelson Mandela. Another son, Willem (Wimpie), was a journalist, political analyst and remarkably, considering the family's conservative background, a cofounder of the Democratic Party. The De Klerks and Strijdoms were staunch members of the GK, or Dopperkerk.[44]

Over the next 25 years following his election as the Member for Waterberg in 1929, the principled, but uncompromising, anti-Semitic and outwardly humourless Strijdom earned the epithet "Lion of the North" for the forceful advocacy of his views, in particular with regard to Afrikaner republicanism, Christian National Education (CNE, or CNO) and the separate development of the different race groups. For example, it is recorded that in 1948, "he unflinchingly demanded what he called *eendersdenkenheid* (uniform thinking). He declared that opposition to apartheid was as treasonable as refusing to defend one's country if it was at war".[45] (This comment should be seen in the light of Strijdom's voluntary participation in the South West Africa campaign in 1915 despite being politically opposed to the liberal Smuts/Botha group.) Strijdom's strength as a leader lay more in his principled commitment to an Afrikaner republic than in any conventional political skills, where he was uninspiring. Even a sympathetic biographer would later describe him as "the only dead normal person among all the brilliant people" who had become prime ministers of South Africa.[46] In the late 1920s he built a fine house (later to be converted to "neo-Cape Dutch" style by the local architect Gerard Moerdyk) in Church Street, Nylstroom. This would be his home until his appointment as Minister of Lands and Irrigation in 1948. Today the house is a national monument and a museum of Strijdom memorabilia – although it is seldom found to be open.

Formation of the Afrikaner Broederbond

In 1918 an organisation called the Afrikaner Broederbond (brotherhood) (AB) had been formed to safeguard and promote Afrikaans culture and the Afrikaans language at a time when growing numbers of Afrikaner nationalists were concerned about the seeming affinity of the Smuts administration for Britain. Within three years, it had become a secret society and increasingly political in nature, determined to ensure the dominance of Afrikaans and the ideology of CNE in schools and the placement of Afrikaans-speakers at all levels of government. Advocate J.G. Strijdom was among its high-ranking members and the partner in his Nylstroom practice, F.H. "Fox" Odendaal, later headed the Waterberg branch or cell. By 1944 the AB had 2 500 members, of whom a third were teachers, with other members being drawn from the church, civil service and farming organisations. By the 1980s, when the AB was at its peak, there were over 12 500 members, of whom at least a quarter were educators.[47] Cells were established in several Waterberg towns: Koedoesrand (Marken), Mogol (Vaalwater), Nylstroom, Naboomspruit, Warmbaths and Ellisras.[48] Despite its strong ideological, nationalistic, ethnic character, the AB maintained that it had no involvement in party politics – although probably all its members would have been National Party supporters, if not actually members. It claimed its objective was "to ensure that our public life and our upbringing and education are Afrikaans in a Christian-national sense, while allowing for the inner development of all sections of the volk, insofar as this does not imperil the state."[49]

In 1933 Generals Hertzog (National Party) and Smuts (South African Party) agreed to form a coalition government (the so-called Fusion Government). Some Nationalists, under the leadership of D.F. Malan – and including Strijdom – objected to the coalition and broke away to form the Gesuiwerde (Purified) National Party. In 1934 the Fusion coalition contested and won the general election as the United Party. That year and again in 1938, Strijdom, re-elected as Member for Waterberg, became the sole Transvaal representative of the (Purified) National Party in Parliament.

In 1936 the Nylstroom pharmacist Gerrit Bakker was elected as the Waterberg representative on the Provincial Council. During his 13 years in office Bakker would, among other achievements, arrange for the construction of the pass named after him down the southwestern edge of the Waterberg escarpment towards Thabazimbi. In 1938/39, the road from Pretoria north to Pietersburg via Warmbaths, Nylstroom and Naboomspruit was tarred for the first time.[50] Gerrit

Bakker was also elected mayor of Nylstroom for a record 14 years from 1924–1930 (when he was replaced by Ludvig Field – see Chapter 5); and again from 1935 until 1943. (His son Andries was also Nylstroom's mayor for several years: in 1960 and again in 1966/67.)[51]

Anti-Semitism

During the late 1930s Afrikaner nationalists became increasingly anti-Semitic (and anti-Indian) in their attitudes. In part this trend was fuelled by the arrival of refugees escaping from Hitler's campaigns to evict Jews from Germany; in part it reflected a legacy of bitterness from the days when rural farmers became heavily indebted to the Jewish traders who had provided them with supplies on credit.[52] Strijdom, his fellow ardent republican H.F. Verwoerd, and a retired diplomat, Eric Louw, led the drive to victimise and denigrate the Jewish community. In 1936 their efforts were assisted by the passing of legislation introduced by Hertzog (astonishingly, with the support of Smuts) that curtailed the immigration of more Jews from Europe.[53] In 1940 Strijdom was even able to introduce a clause in the constitution of the National Party barring Jews from party membership.[54] The ascendancy of Strijdom, Verwoerd and other Afrikaners of similar views in the years ahead would lead inevitably to the migration of Jews away from the hostile rural towns of the Transvaal and Free State into the cities (see Chapter 14).

In November 1940 a congress was held in Bloemfontein to establish the Herenigde (Reunited) National Party, bringing together the Hertzogites from the United Party and Malan's Nationalists, in opposition to the ruling United Party led by Smuts. However, Hertzog himself could not be forgiven for having formed a coalition with Smuts and was soon forced to withdraw permanently from politics. (As the historian L.M. Grobler noted in his account of the Springbok Flats, a popular saying at the time was that when two Englishmen get together, they form a club; when two Jews do so, they form a company; but when two Afrikaners get together, they form two political parties!)[55]

In 1938 an alliance between the AB and Malan's National Party was formed to use the forthcoming centenary of the Battle of Blood River (16 December 1838) as a platform upon which to build support for Afrikaner Nationalism and unity among Afrikaans-speakers. There would be a symbolic re-enactment of the Great Trek with two custom-made ox-wagons making the journey northwards from Cape Town, one destined for the site of the Blood River battle, there to lay a foundation stone for a commemorative monument, and the other for Pretoria, which had been selected as the site for a national monument to the Voortrekkers. In a highly successful, carefully choreographed sequence of ceremonies, the cornerstones of the Voortrekker Monument and the Blood River Monument were laid simultaneously on 16 December 1938, accompanied by the arrival of the ox-wagons, before rapturous crowds numbering tens of thousands. Fiery speeches were made affirming the sacred role of the Afrikaner trekker forebears, the threats posed by black economic advancement and the importance of defending Afrikaner nationalistic values.[56]

The Voortrekker Monument that was built on the hills overlooking Pretoria from the south (unilaterally renamed Voortrekkerhoogte by General Kemp, Minister of Defence, from the previous name of Roberts Heights), was designed by the architect Gerard Moerdyk, who owned a farm near Nylstroom, where his father, Jan Moerdijk (Dutch spelling), a former schools inspector, had settled. Indeed, Waterberg legend has it that his idea for the design of the monument came from the silhouette of the lone sandstone hill Kranskop (or Modimulle) outside Nylstroom – and certainly, from the west, there is a close superficial resemblance. Others point out that the Voortrekker edifice strongly resembles a monument in Leipzig, Germany.[57] There is little doubt, however, that Moerdyk, whose parents had emigrated from the Netherlands shortly before his birth in 1890, endorsed the ideology represented by the monument: He was a member of the AB.

The Ossewabrandwag

This resurgence in Afrikaner nationalism soon manifested itself in the formation of an organisation designed to capture the enthusiasm generated by the centenary and to use this to further promote the interests of Afrikaner nationalism. It was called the *Ossewabrandwag* (Oxwagon Sentinel) (OB), and from the outset, early in 1939, it had a militaristic edge to its structure, being linked to citizen force (*burgermag*) structures in the Orange Free State under the control of a Colonel J.C.C. Laas of Bloemfontein. (So concerned was the Smuts government about the rapid growth in support for the OB that within months of its formation, army employees were banned from joining it.) Laas, together with the prominent Bloemfontein NGK minister C.R. Kotzé, and N.G.S. van der Walt, a journalist from *Die Volksblad,* were seen as the founding members of the OB. The Broederbond was always at pains to distance itself from the more militant OB – and in time to come, Malan would force National Party members to resign from the OB – but there is some evidence that behind the scenes the OB enjoyed the Broederbond's tacit support.[58]

On the platteland as well as in urban areas, the attraction of the OB had much to do with what it seemed able to offer Afrikaners who had been economically marginalised by poor education, debt, competition from work seekers of other races and the declining fortunes of agriculture. Only a few years earlier, in 1932, the respected Carnegie Commission had reported on its investigation into the plight of a growing Afrikaner underclass – the so-called "poor whites". It had estimated that 17,5% of all [white] families with children at school could be classified as "very poor", i.e. largely supported by charity in cities, or who subsisted in dire poverty on farms. High birth rates and child mortality rates were characteristic, as were poor hygiene and diet. Some whites in the Rustenburg/Waterberg area worked for prosperous blacks in the locations. Dr Ernest Malherbe of the Commission identified 27 000 children whom he categorised as "retarded", being more than two years behind the norm for their age group. Over half did not complete primary education and in going to the cities, they "had no prospect of earning a living".[59] (Malherbe, a dedicated civil servant, outstanding educator and Smuts loyalist, would go on to become a popular Vice-Chancellor of the University of Natal (Durban) for 20 years, ending in 1965. He died in 1982, aged 87.)

The Depression of the early 1930s did little to alleviate either the poor white situation or that of struggling farmers in rural Transvaal. The OB was able to capitalise on this, bearing in mind that since 1934, Strijdom had been the only representative of conservative Afrikanerdom to represent the region in Parliament. Moreover, once World War II began in September 1939, and the Germans seemed to be quickly gaining the upper hand, republican Afrikaner nationalists opposed to anything related to their former oppressor concluded that an Afrikaner republic outside the British Empire was a growing possibility and that the OB represented a vehicle epitomising that vision.

By 1940 the OB could claim a membership of 130 000 in the Transvaal alone; of these, 32 000 were members of the Northern Transvaal branch, which included Pretoria and the Waterberg. Within six months of its founding, the organisation was thought to be represented in every *dorp* in the province, with especially strong support from DRC ministers and cultural organisations. It had become a mass movement with whole families, communities and organisations (like parishes and farmers' associations) joining it en masse, attracted by its Afrikaner-supportive policies and efficient paramilitary structure.[60] Countrywide, the OB may have had over 200 000 members at its peak during 1941, although estimates vary widely, from 170 000 to 400 000.

The continued growth in support for the OB led the Smuts administration to use wartime ordinances to ban membership of the organisation by various categories of civil servant: the police in November 1940 and teachers and other civil servants in March 1941. While some of these members simply went "under cover", many others heeded the bans and gradually the movement lost its momentum.

Ultra-right-wing militancy

In the meanwhile, early in 1940 an elite military subgroup of the OB had been quietly formed with the specific intention of undertaking subversive violent activities against sensitive government targets. This group, known as the *Stormjaers* was organised in small military-styled units with limited if any oversight from the parent OB. At one time there were said to be as many as 8000 men in the Transvaal, divided into 15 "battalions". Covert communication was maintained with German military agents in the belief that Allied defeat was imminent. Stormjaers cut phone wires, blew up pylons, power lines and rail tracks, as well as post offices, shops and banks. Many police were implicated in this anti-war activity and in 1942 alone, 314 policemen (and 59 railway police) were relieved of their duties and arrested.[61] There can be little doubt, given the political attitude prevalent in Nylstroom, that a Stormjaer cell was based in the town during the early 1940s, although no references have been discovered of any incidents of sabotage being carried out in the area. There was an unconfirmed report of a farmhouse near Immerpan on the road to Roedtan owned by an English-speaker, John Todd (a former New Belgium manager, see Chapter 9) being burned down, but this could not be corroborated.

Robey Leibbrandt

The most famous German agent was Robey Leibbrandt, a popular South African light-heavyweight boxer and former policeman from Potchefstroom, who travelled to Berlin to represent his country in the 1936 Olympics. A long-time admirer of Hitler and Nazism, Leibbrandt actually met the Führer during his German visit and was soon recruited as an agent on behalf of the regime. In April 1941 Leibbrandt returned clandestinely to South Africa (by a yacht which dropped him off along the west coast) with the objective of assassinating Prime Minister Jan Smuts as a prelude to a coup to be undertaken by militant members of the OB. However, Leibbrandt's ego and bravado quickly got the better of him and the rump of the OB soon dissociated itself from him and his activities. He was embraced by the Stormjaers, though, and with their assistance travelled widely around the northern Transvaal, including the Waterberg, fomenting anarchy among them before being lured into a police trap on Christmas Day in 1941. In March 1943 Leibbrandt was found guilty of high treason and sentenced to death by hanging. But at the end of that year, Smuts commuted the sentence to life imprisonment (in part because Smuts had fought with Leibbrandt's father in the South African War and in part because he had no wish to create a second martyr as he had done with another colourful extremist, Jopie Fourie, after the 1914 Rebellion).[62] The publicity surrounding Leibbrandt's arrest, and the trove of contacts and names that were discovered in its wake combined to sow the seeds of the demise of the Stormjaers and created a growing distance between its parent, the Ossewabrandwag, and the National Party.[63]

FIGURE 16.10: Robey Leibbrandt after his arrest with a Hitlerian moustache, a conceit common amongst Republican Afrikaners at that time and later.

The National Party rises to power

Despite the resurgence of Afrikaner Nationalism in the late 1930s, Smuts's United Party easily won the national elections of 1938 (111 seats out of a total of 153) and again in 1943 (89 seats, plus 16 from coalition partners, against 48 for Malan's Nationalists and Independents). However, Strijdom's majority in the Waterberg increased from 242 in 1938 to 984 in 1943 when his UP opponent was a Bezuidenhout – and to 2448 in the elections of 1948 when Malan and the Nationalists swept Smuts out of power for the last time, winning a total of 79 seats against 74 for the UP and its allies. The UP secured over 50% of the popular vote in the 1948 election compared with less than 40% for the Nationalists and Afrikaner Party, but the weighting of country seats had secured victory for the latter.[64] Malan became Prime Minister and he appointed Strijdom to his cabinet as Minister for Lands and Irrigation – a minor portfolio that allowed Strijdom time to build the National Party's strength in the Transvaal.[65]

Strijdom's partner in his Nylstroom legal practice, Fox Odendaal (1898–1966), was elected as the Waterberg representative to the Provincial Council. He fulfilled this role until 1954 when he was appointed as Administrator of the Transvaal, as his legal partner ascended to the Premiership. The hospital in Nylstroom was named in honour of Odendaal, who, amongst numerous achievements during his political career was responsible for arranging the prefabricated classrooms at the Melkrivier school (see Chapter 17). Odendaal died of a heart attack in 1966, while still in office. For 16 years, he chaired the National Parks Board and also led a commission into the investigation of the development of South West Africa (Namibia).[66]

In the elections of 1953, the Nationalists increased their majority to 29 seats and Strijdom his majority in Waterberg to 3417. Strijdom represented Waterberg until his death in 1958 (including the last four years as Prime Minister, following the resignation of Malan in 1954, aged 80); he was in turn replaced by Jooste Heystek, until 1970.[67]

FIGURE 16.11: F.H. "Fox" Odendaal, Member of Provincial Council for Waterberg (1949–1958), Administrator of the Transvaal (1958–1966). Photograph from Nel, J.J. et al (1966): p. 103.

Key apartheid legislation passed

The 1950s saw the passing of key legislation that served to define and enforce the ideology of apartheid: the Prohibition of Mixed Marriages Act in 1949; the Immorality Amendment Act, the Population Registration Act and the Group Areas Act, all in 1950; and the Separate Amenities Act of 1953. More stringent amendments and attendant regulations to these laws followed in the next few years. It seems that the government selected rural towns in two particularly safe Nationalist constituencies, Ventersdorp and Nylstroom, as pilot sites for the implementation of Group Areas legislation. So it was that, beginning in 1956, the long-standing, civil and respected Indian community of Nylstroom was forced to abandon its established homes and premises in the western part of Nylstroom and move to a newly laid-out suburb, Asalaam, on the town's eastern boundary. Later, the black population, already "safely separated" from the town by the railway line, was relocated to Phagameng. Similar moves followed in Warmbaths (see Chapter 6).

The implementation of the Group Areas Act, together with the Bantu Authorities Act of 1951, the Promotion of Bantu Self-Government Act of 1959 and related legislation saw the creation of "homelands" and "territorial authorities" for the various black population groups identified or fabricated by government. In respect of the Waterberg region, the relevant territorial authority became that of Lebowa (for the so-called North Sotho people) via Proclamations 115 and 202 of 1969. By 1969 Lebowa, consisting of three portions, encompassed a total area of about 1,7 million hectares.[68] (See also Chapter 3.)

Jooste Heystek, who was born in the Waterberg in 1897 (and whose grandfather had been the auctioneer for the first stands sold in Johannesburg in 1886, before buying a farm near Settlers) trained to become a teacher. He taught for 40 years, becoming head of Hoërskool Nylstroom and superintendent of its girls' hostel, Ons Hoop. Louis Trichardt recalled being taught by this fine, but strict teacher. He remembered an occasion when Heystek announced in class one morning that on a recent visit to Pretoria he'd been shocked to see banana skins being flung from the window of a car with a TAH (Nylstroom) registration plate as it travelled down Church Street. What was worse, added Heystek, the young girl responsible was sitting in this very class!

FIGURE 16.12: Jooste Heystek, MP for Waterberg (1958–1970). Nel J.J. et al (1966): p. 104.

During these years, Heystek served on the management board of the provincial teachers' union as well as on the Nylstroom Town Council. He was also a member of the Gereformeerde Kerksraad in the town for 40 years. As MP for Waterberg from 1958 to 1970, he was instrumental in building two schools in Nylstroom and in the commissioning of both the Glen Alpine Dam on the Mokgalakwena River and the Hans Strijdom (now Mokolo) Dam on the Mogol River. He died in 1985.[69]

Decline of the United Party

The South African or United Party remained active in the region for many years after losing the Waterberg seat in 1915 and maintained an office in Nylstroom until the 1960s. In the early years, its national or provincial candidates included B.D. Budler of Nylstroom and J.J. McCord of Warmbaths (1919); E.P. van Deventer and W. Creswell (Naboomsprruit) in 1924; and in 1943, L.M. Baxter (Alma), Bezuidenhout (who stood against Strijdom) and Mrs A.H. Curlewis (Naboomspruit). The Minister of Justice in Smuts's wartime cabinet, Dr Colin Steyn, spoke on their behalf at a public meeting in Nylstroom on 2 October 1942, but to no avail.[70]

After the war, when the United Party was defeated by the National Party in the general elections of 1948, its local support waned further. The triumphal progress of the Nationalists from this point onwards required that their opponents, especially if they came from the Afrikaans-speaking community, had to be resilient and thick-skinned. The UP candidate against Strijdom in 1948 had been D.R.J. van Rensburg, a farmer from Klippoort near Sandriviersspoort. He was followed, equally unsuccessfully, by H.S.E. Jordaan, a farmer from Haaskloof near Naboomspruit.

The UP's support lay predominantly among farmers along the Nyl valley and at one point in the 1940s, party members even built a meeting hall (on the farm Zandfontein), the ruins of which can still be seen next to the road south of the river leading to Nylsvley. In 1949 a farmer from Cullinan who would become one of the party's leading office bearers played a brief role in the region: W. Vause Raw was elected to the Board of the WLKV. As one of the first groundnut farmers in the area, he was very supportive of efforts to promote the crop in the northern Transvaal. But in 1950 he moved to Natal where his political career as the MP for Durban Point, first for the United Party and later for the New Republic Party, would follow.[71] In the Waterberg during the heyday of Afrikaner Nationalist government, when successive ministers of land (or agriculture) ensured tangible support for rural constituents in the form of infrastructure, price controls, low-interest loans (the Land Bank, a Broederbond initiative, was created in 1934) and effective policing, liberal-minded opponents drew almost no support, even from English-speakers.

It is also important to realise, with the benefit of hindsight, that in most respects other than Afrikaner Nationalism, religious affiliation and to some extent, language, the policies of the UP differed little from those of the Nationalists, especially when it came to matters involving white supremacy, the franchise, perceived security threats and job reservation. Nowhere has the charade of the UP as an opposition party been better reflected than in the memoirs of Helen Suzman, Progressive Party member for Houghton from 1953 until 1989, and for 13 of those years (1961–1974) the sole representative of that party – and the only effective opposition voice – in the national parliament.[72] She and her meticulously factual revelations were disliked as much by UP members of Parliament as by their National Party colleagues.

In a by-election for the Waterberg constituency in June 1971, the National Party candidate was Dr Andries Treurnicht. His opponent was not from the official opposition, the United Party (which by now had effectively given up on the region), but from a new party that had recently broken away from the right wing of the ruling party: Jaap Marais, of the Herstigte (Reconstituted) National Party, or HNP. The HNP had been formed in 1969 in opposition to the (reluctant) decision of the Prime Minister, B.J. Vorster, to allow Maori (i.e. nonwhite) players and spectators to participate in the 1970 rugby tour of South Africa by the All Blacks. That this decision was so hard to take, and proved sufficiently offensive as to warrant the formation of a new party, reflects the level of racism that pervaded the governing regime of the day.

The new party was led by Albert Hertzog, the son of General J.B.M. Hertzog. Jaap Marais, who had represented the Innesdale constituency for the National Party in Parliament since 1958 was one of four Nationalist MPs to support Hertzog. Vorster called a snap election in April 1970, over a year ahead of schedule, to forestall further support for this new threat on his right. His tactic was successful: The Nationalists won 118 of the 166 seats available – a drop of 8 from the previous election in 1966; the United Party 47 (up from 39 previously) and the Progressives 1 (the indomitable Helen Suzman). All 78 HNP candidates, including Hertzog, were roundly defeated; only Marais retained his deposit.[73]

Jaap Marais had, unsuccessfully, challenged Hans Campher (the NP's replacement for Jooste Heystek, who'd retired) for the Waterberg seat in the 1970 election, and caused quite a stir in the process. Campher had been the constituency's representative on the Provincial Council during Jooste Heystek's time in Parliament, and was the logical candidate to replace Heystek upon the latter's retirement. As provincial councillor, he'd been responsible, among other things, for ensuring that the prefabricated classrooms used to build the first school in Ellisras were relocated to the Melkrivier school and Laerskool Vaalwater when a new school was built in Ellisras. Campher had no difficulty in defeating Marais in the 1970 election – but died of a heart attack during a parliamentary session several months later, thereby necessitating the by-election of June 1971.[74]

Dr Andries Treurnicht and the rise of the right wing

As the only HNP candidate to retain his deposit in the general election, Marais was nominated to challenge Treurnicht for the Waterberg seat in the by-election a year later. So concerned was the National Party about the threat posed by Marais that B.J. Vorster himself visited the constituency to give an impassioned speech in support of Treurnicht at Naboomspruit on 17 June.[75] But he needn't have worried; even in the conservative Waterberg, Marais had no chance of victory against such a formidable candidate, although he did surprisingly well. The result of the June 1971 by-election in the constituency was 5456 votes for Treurnicht and 2182 for Marais.[76]

This was Treurnicht's first election, although he was not new to National Party politics. Born in Piketberg, Western Cape in 1921, Andries Treurnicht was a brilliant student of theology, gaining an MA from Stellenbosch and his PhD from Cape Town (he was fluent in English and Afrikaans). He was short in stature, played scrumhalf for South Western Districts and rose quickly through the ranks of the

FIGURE 16.13: Dr Andries Treurnicht, MP for Waterberg (1971–1993). SANA SAB Photographic Library, No. 18000.

NGK after becoming a minister. In 1967 Treurnicht was appointed editor of the National Party mouthpiece *Die Hoofstad* and was a staunch party supporter and Broederbonder (he became chairman of the AB in 1972).

In 1976 Treurnicht was appointed Deputy Minister for Bantu Administration and Education. In this capacity, it was his instruction to implement policy requiring black pupils to be taught half in Afrikaans and half in English that triggered the student riots in Soweto of June that year, an event that would become seen as a watershed in South African politics. Undaunted (and seemingly unconcerned by what he had precipitated), Treurnicht went on to occupy the roles of Minister of Public Works, Tourism, State Administration and Statistics.

In 1982 he led a large group of Nationalist MPs who opposed Prime Minister P.W. Botha's initiative to create a tricameral Parliament in an attempt to accommodate the political aspirations of coloured and Indian citizens (the rights of black South Africans having been assigned to one of ten ethnic homelands, or Bantustans). The group, comprising 23 MPs, left the National Party to form the Conservative Party (CP) with Treurnicht as its head. In the elections of 1987 (by which time Botha had become the country's first Executive State President), the Conservatives romped home to become the official opposition with 22 seats and over half a million votes. Treurnicht easily defeated his opponent in the Waterberg constituency, Johan "Oom Poem" Lampbrecht of the National Party, bringing to an end that party's control of the district. Soon Treurnicht inherited the title "Leeu van die Noorde" (Lion of the North) from his predecessor, J.G. Strijdom.

In December 1990, Treurnicht, the once-prominent theologian, vigorously opposed the decision by his church (the NGK) to renounce apartheid, earning for himself the epithet "Dr No" on account of his unrelenting opposition to any political reform. In March 1992, true to form, he led the "No" campaign in the referendum called by President F.W. de Klerk (who succeeded P.W. Botha in 1989), proposing negotiations to bring an end to apartheid rule. Although the "No" campaign garnered 875 000 votes, it was swamped by an overwhelming 2:1 support for De Klerk's initiative.

Treurnicht died in April 1993, aged 72, while still the sitting representative for the Waterberg constituency. The respected Afrikaner historian, author and staunch opponent of apartheid, Prof. Hermann Giliomee, had this to say of Dr Treurnicht:

> His greatest service to South Africa was in restraining violence in right-wing ranks and committing his party to negotiations despite the fact that the new political order coming about in South Africa

represented the nightmare which had haunted him his entire life. A dapper, courteous man, he refrained from using racist rhetoric in public.[77]

By arrangement between the parties, Treurnicht was replaced without a by-election by fellow CP member and western Waterberg farmer Louis Trichardt (descendant of the Voortrekker) for the remainder of the parliamentary term.[78] Louis would be the last parliamentary representative for the Waterberg constituency. The new government elected in April 1994 by universal franchise for the first time, would be based on proportional, rather than constituency-based representation in the state and provincial legislatures. The CP chose not to contest the 1994 election and later merged with other right-wing groups to form the Freedom Front Plus, which in 2016 still had 4 seats in the 400-seat National Assembly, reflecting the 0,9% of the electorate who had voted for it at the last general election in 2014.

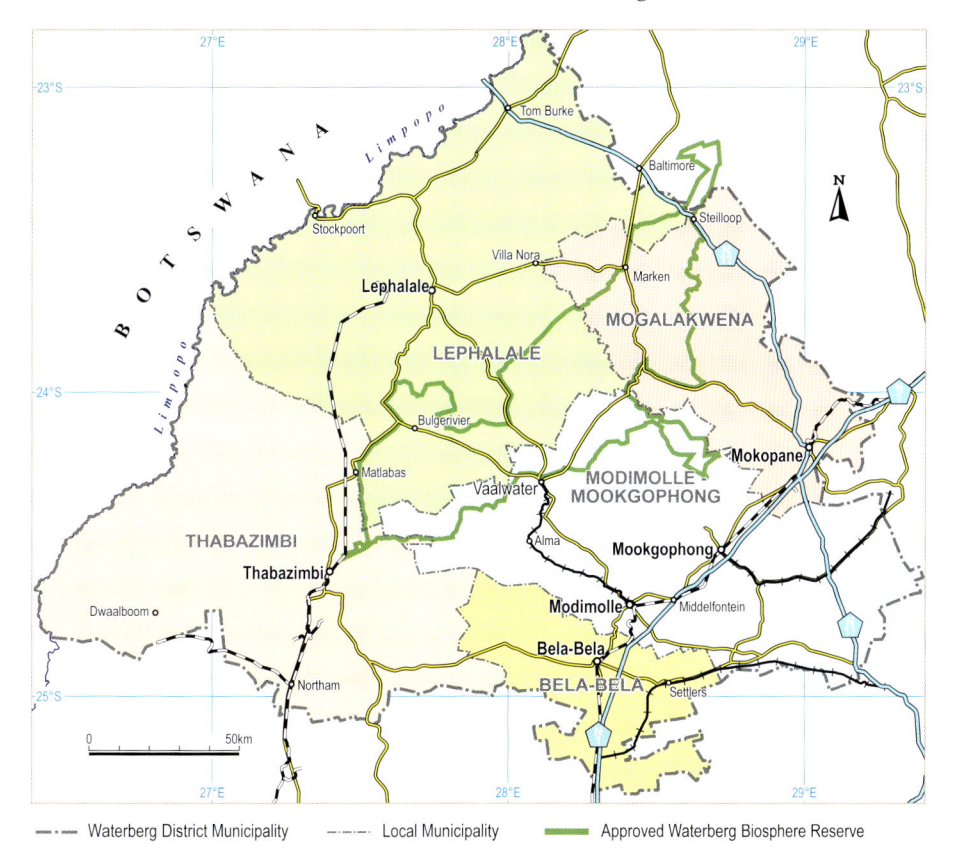

FIGURE 16.14: The Waterberg District Municipality with its five local municipalities (2017) and outline of the Waterberg Biosphere Reserve.

Politics of the Indian community

The evolution of black politics in the Waterberg region before the elections of 1994 is described in Chapter 3, but it is appropriate here to say something of the political life of the Indian community in the area up until this time. Most Transvaal-resident Indians were not related to the indentured labour that had been imported (mainly from south India) to work on the Natal sugar plantations; rather, they were mainly traders and business people who came to South Africa by choice and at their own expense, predominantly from the Gujarat area of western India – and were mostly of the Muslim faith. They were described at the time as the "Passenger" or "Arab" class of Indians to differentiate them from the indentured Hindu workers who formed the majority of the Natal population.[79] In 1904 the total Indian population of South Africa was estimated at 122 000, of whom about 10% resided in the Transvaal.[80]

In 1906, the implementation of the Asiatic Amendment Ordinance (No. 29 of 1906) subjected all Indians to compulsory registration, fingerprint identification and to the carrying of passes at all times, in addition to the existing regulations limiting where they could live or work. There was an immediate outcry and the initiation of protests of "passive resistance" or *satyagraha*, introduced by Mahatma Gandhi, a London-trained lawyer who'd been in the country since 1893 when he came out to Durban to represent a wealthy Indian client in court. An appeal by Gandhi and others to the Colonial Secretary in London, Lord Elgin, led to the withdrawal of the legislation – but the following year, when Transvaal obtained self-government, Jan Smuts reintroduced the Ordinance and refused to entertain any objections from the Indian community.

Further negotiations were held after unification in 1910, but once again, the government failed to honour concessions it had agreed to implement and resistance continued. Intercession from representatives of the Indian government helped conclude an agreement in 1914 between Gandhi (on behalf of the National Indian Congress) and Smuts that traditional marriages would be recognised and that a poll tax levied upon Indians would be abandoned. However, the government also insisted on the disenfranchisement of Indians in the Transvaal, that Indians could not own property in either Transvaal or Orange Free State – or even be allowed to live in the latter. Gandhi left South Africa the same year.

In 1925, D.F. Malan, the new Minister of the Interior, introduced the Areas Reservation, Immigration and Registration Bill, in terms of which all Indians in the country were to be defined as aliens and subject to repatriation (i.e. deportation).

He was persuaded to withdraw the Bill two years later. Indian activists continued to resist discriminatory legislation and many ultimately gravitated into the embrace of the African National Congress. Some rose to prominence in the early ANC-led governments after 1994. In Nylstroom for example, Zunaid Mosam was among the group of young Indians to pursue this route.[81]

The 1994 elections brought to an end the era of constituency-based politics at a national and provincial level in South Africa. Local municipal elections are still constituency- or ward-based, with elected individuals, rather than party-list appointees ("cadres") representing the electorate. In the Modimolle Local Municipality, which included Vaalwater, the overwhelming majority of councillors were members of the ruling ANC, although there was also a sprinkling of councillors from the Freedom Front Plus and the Democratic Alliance (DA). In 2015 the government decided to merge the local municipalities of Modimolle and Mookgophong into a single entity, ostensibly to optimise the limited available qualified human resources. Fierce opposition to this proposal, on the grounds that there was no evidence to suggest a merger would improve the quality of service delivery in either area, was ignored and in 2016 a new entity, the Modimolle-Mookgaphong Local Municipality, was created.

Local government elections in 2016

In the municipal elections of 3 August 2016 there was a shock result: Two local municipalities in Limpopo Province, Modimolle-Mookgophong and Thabazimbi, were won by an unlikely coalition of the DA and the Economic Freedom Fighters (EFF). The ANC was forced into the unfamiliar position of being in opposition. The reasons for the eviction of the ANC were simple: Failure to deliver basic services to the communities or to manage the municipal administration. There was a belief that corruption and patronage played a role in these failures, although lack of relevant technical competence, for example in administration, finance, engineering and town planning were principally responsible. At the time of writing, it remained to be demonstrated that the new administrations could do better, especially in the face of crippling accumulated municipal debts (hitherto concealed) and obstructive cadre-deployed municipal management.

CHAPTER SEVENTEEN
Gaining the right to read

As we have seen from previous chapters in this account, white settlement of the land north of the Gariep (Orange) River had taken place gradually since the late eighteenth century by farmers from the eastern boundary of the Cape Colony in search of additional grazing. Some of these migrants were landowners (under the loan farm system), some migrated only seasonally and others were nomadic tenant farmers (bywoners). Collectively, they became known as *trekboers* – and it is estimated that by 1834, there were a thousand white trekboer families living outside the borders of the Cape.[1] As a consequence of their nomadic lifestyle, trekboer children could not enjoy the benefit of the conventional (and often good) education available to the families of burghers in the Cape and were reliant on their parents or itinerant teachers for whatever schooling they obtained.

In the decade after 1835, the trekboer community was joined by the arrival of over 2500 Afrikaner families (possibly 17 000 individuals) who emigrated from the eastern part of the Cape Colony in the course of an exodus that came to be known as the Great Trek.[2] Some Voortrekker leaders (for example Tregardt, Potgieter and Maritz) were themselves educated and prosperous farmers at the time they decided to leave the Colony, but many of those who chose to join them were less so. And everyone, like the trekboers before them, left behind not only their churches, but their schools. Life on trek was austere, hard and dangerous, survival often being the principal concern. Education, other than in the teachings of the Old Testament, became a luxury few could afford and for which opportunities were almost nonexistent. Consequently, the culture of formal education was quickly lost among the trekker communities. By 1877 only 12% of white children of schoolgoing age in the Free State attended school. In the Transvaal, the figure was 8%.[3] If it is assumed that the majority of children living in towns like Potchefstroom, Pretoria and Rustenburg would have gone to school, the formal schooling of rural white children, for example in the Waterberg, must have been close to zero. Even in Nylstroom, founded in 1866, there was not a single school before 1879.

President Burgers promotes education

Thomas Burgers, who acceded to the presidency of the Zuid-Afrikaansche Republiek (ZAR, or Transvaal) in 1872, recognised this shortcoming in a speech to the Volksraad:

> The question of education is foremost. All our needs taken together do not outweigh the need for the education of our youth ... this Republic has no greater danger than ignorance ... Experience has taught that our present law is completely inadequate for the purpose of sound education.[4]

Though flawed in many other ways, President Burgers proceeded to push through reforms that included the establishment of an education department, a museum and a library. But his reforms didn't extend beyond the towns. During the period of British annexation of the Transvaal (1877–1881), several private schools were also established in Pretoria, but none elsewhere.

After regaining its independence, with Paul Kruger as its new President, the ZAR appointed Ds. S.J. du Toit as Superintendent of Education. He was a dedicated educator and intellectual (he also began the work of translating the Bible from Dutch into Afrikaans, a task ultimately concluded by a team that included his son Totius as described in Chapter 15. Du Toit implemented new education legislation based on Christian national principles, which found favour with the Boer communities and their churches. In January 1882 there were nine state-funded schools in the whole Transvaal Republic accommodating 206 children. By the end of the year, thanks to Du Toit's intervention, the numbers had risen to 43 and 872 respectively.[5] In the Waterberg the only school started was one in Nylstroom, a so-called *dorpskool* that offered primary education to those children within reach of it. Education policy also provided funding for *wijkscholen*, or "ward" schools, which were small estab-

FIGURE 17.1: President Thomas Burgers.

lishments on outlying farms, generally served by a single teacher – most of whom were unqualified and who were often drawn from the surrounding community. Many ward schools were founded during Du Toit's administration, including across the Waterberg, but most were short-lived and records of their existence were anecdotal. By 1891 there were two town schools and eight ward schools in the district, serving a total of 190 pupils.[6]

The first rural Waterberg schools

The Education Department's register of town and ward schools in the Waterberg in 1890–1899 records ward schools only from 1896, the majority of them in the Zwagershoek ward. Several schools moved location, usually with the teacher,

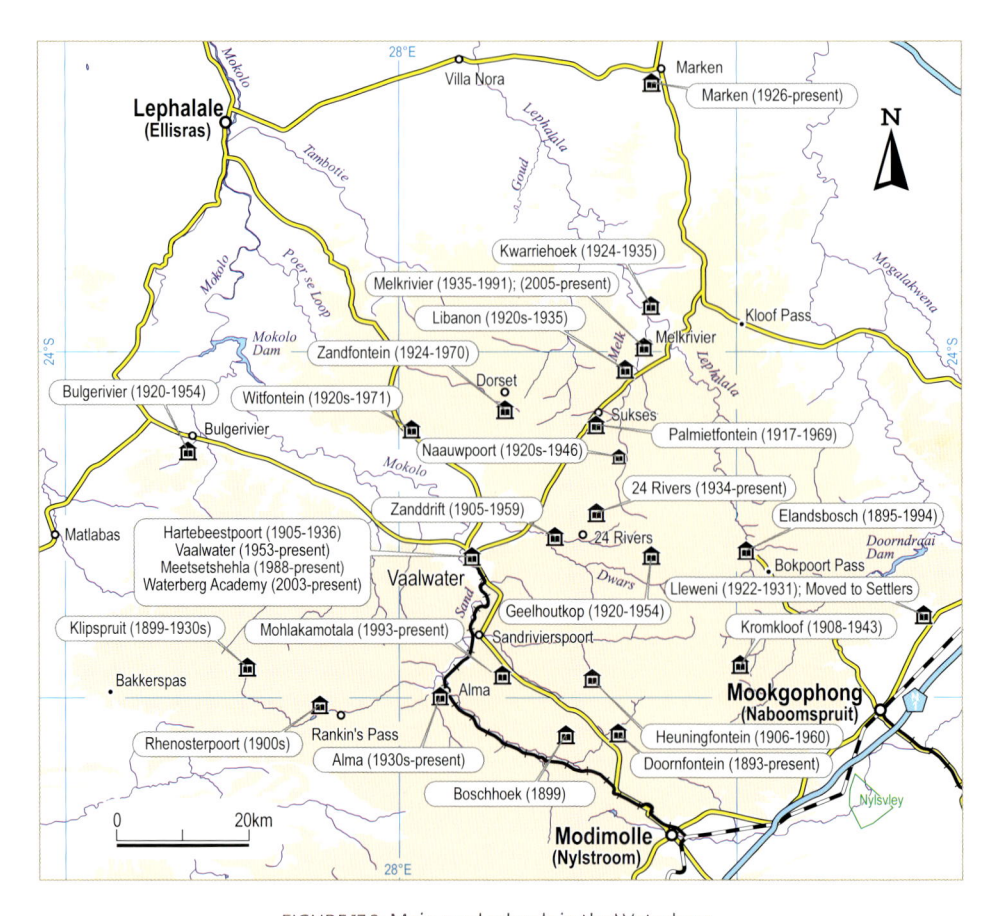

FIGURE 17.2: Main rural schools in the Waterberg.

who also often doubled as the secretary of the local school committee. One of these schools was one in which F.W. Stork was the teacher. Having opened at Buiskop near Hartingsburg (the forerunner of Warmbaths) in August 1896, it moved a year later to the farm Kariefontein and then in January 1899 to Boschhoek in the Zwagershoek ward, still with Stork in charge. At Rhenosterpoort in the Zwagershoek ward, on the farm adjacent to Rankin's Pass, a school opened in February 1897 with J.P. Jurrius as the head teacher and Mrs E.P. Eloff as his assistant. The school committee comprised S.J.N. van Heerden (chairman), John Rankin (treasurer) and P.A. Eloff as secretary. By 1898 there were 15 ward schools in the Waterberg District as well as the two town schools, with a total pupil enrolment of 388. Inspectors had been appointed to monitor and improve the standard of education at the various schools.[7]

Schooling after the South African War

Following the outbreak of the South African War in October 1899, rural schooling was suspended with the result that for almost three years, children received no education, in addition to the other deprivations they endured during that

FIGURE 17.3: Former teachers' house on Rhenosterpoort, near Rankin's Pass.

period. When peace returned in June 1902, the education of the populace (that is, white people) was a priority of the new British colonial Transvaal administration that was determined to introduce a more liberal curriculum, based on English culture and, provocatively, the English language. Understandably, this change was not well received amongst rural Afrikaans-speaking communities, such as those predominating in the Waterberg, who saw this as a deliberate attempt to undermine their *volksidentiteit* (national identity).

Inspectors J.A. Richey and S.O. Purves set about trying to rectify the deficiencies in the educational system. Afrikaner farmers reacted by creating private schools that subscribed to Christelik-Nasionale Onderwys (CNO) (Christian National Education) principles, but the general penury in the aftermath of the war meant that most of these institutions soon fell away. In December 1905 outlines for the administration of schools were published and in June 1906 a new Waterberg school board was formed. Two years later, its members were J.C. Krogh (chairman), N. Carwell (secretary), G.F. Grobler, C. Maggs, P. Pock, J. Rankin, P.J. Steynberg, E. Tamsen and D.P. van Rooyen. The department appointed two officers, P. Burger and A.P. Swanepoel to the area to enforce school attendance. Gradually, formalised education was restored, both in the towns and in the outlying districts. By 1910 there were 25 government schools (all for white pupils) in the Waterberg, staffed by 38 teachers and serving 872 pupils. Expansion of town schools, including provision of hostel facilities, resulted in almost a doubling of the total number of teachers and pupils to 61 and 1693 respectively by the end of World War I at 29 schools across the Waterberg District.[8]

About 17 km north of Nylstroom on the road to Zandrivierspoort, a ward school opened before the war next to the upper Sand River on the farm Doornfontein, then owned by W. Boshoff. Initially, classes were held under a tree, then in a barn. In 1905 the first classroom was built, followed in 1912 by the buildings that are still in use today. Several headmasters ran the school over the years, including Hetmeyer, Ackhurst, Boswell, Stork, Olivier, Rob Ferreira and Jooste Heystek. Ferreira and Heystek would rise to prominence as educators and politicians.

On the Waterberg plateau one of the first post-South African War rural schools was on the farm Kromkloof, about 20 km west of Naboomspruit – the farm on which the Union tin mine would later be established. The Kromkloof school operated from 1908 until 1943 when a departmental "centralisation" policy resulted in its closure. Among its first pupils were Naboomspruit luminaries Koot Engelbrecht, Bennie Marais and Petrus Geyser.[9]

Zanddrift school

Even older than Kromkloof was the school on Zanddrift on the Dwars River, less than 10 km northeast of where Vaalwater was being established. Zanddrift opened its doors in 1905 and remained in operation for 55 years. In the early years, it catered for a large population of pupils from the host farm which had been divided up into numerous smallholdings (see Chapter 7). So many children were born in this area that there was even a maternity home on the property for several years. The earliest preserved school journal is from 1930 when there were 72 pupils with W.J. Louw as the head, assisted by two other teachers. Throughout the 1930s and 1940s, pupil numbers ranged between 60 and 80.[10]

Louw was succeeded by G.J. van Wyk, followed by J.H. Strauss. In 1949 a bus service brought pupils in from Vogelstruisfontein in the west, lifting pupil numbers to over a hundred. The peak was reached in 1952 when there were 152 pupils registered at the school, with J.A. du Plessis as head and two longstanding teachers, Messrs A.S. van Staden and Bosman as his assistants. The following year, Laerskool Vaalwater opened and pupil numbers halved, causing one, then two teaching positions to be terminated. In the late 1950s, high staff turnover, which seemed to be a characteristic of departmental policy at this time, aggravated the school's performance and more children left for Vaalwater. H. Williamson was the last head of Zanddrift and in 1959 his report lamented the decline in the standing of the school. By the year's end, numbers had fallen to 34 and closure was inevitable. The last note in the journal, dated January 1960, acknowledges receipt of the school's documentation by S.J. Hoffman, the headmaster of Laerskool Vaalwater.

FIGURE 17.4: The old school on Zanddrift. Photograph taken on the farm by courtesy of Jannie Grové.

Elandsbosch school

There were several other farm schools in the eastern part of the Waterberg plateau through the years, for example the one on Elandsbosch (which dated from 1895) on the upper Palala River – which the children of the "Purekrans" van Rooyen family and others attended. The school was located on government land, but across the Palala from where many children lived, making access difficult. Several children had attended other small farm schools in the area, on Kafferskraal and on Vlakfontein. They all relocated to Elandsbosch after a footbridge over the river had been built by the local community in 1912 during the time when a Hollander, Van Brink, was principal. By 1914 the school was led by a British ex-serviceman, Cochrane; soon he had an assistant, Faul, because there were almost 80 pupils. In 1915 Dr Louis Leipoldt was recorded as visiting the school on a health inspection and in the same year, J.L. Moerdijk, the school inspector, also paid a visit. Both these men were exceptional personalities whose names crop up in many Waterberg school journals and who did much to enhance the educational environment of the rural Transvaal.

Louis Leipoldt

C. Louis Leipoldt was truly one of the most underappreciated Afrikaner intellectuals of his time. Born in Worcester, Cape, in December 1880, he grew up in Clanwilliam where his father was the local NGK pastor. His mother decided that the four children would be better educated at home than at school, with the result that Leipoldt never attended a school – though he became fluent in Dutch, English, German and French. He studied history and classics, including Latin and Greek, under the tutorship of his father. He developed an early and comprehensive interest in natural history. This led him, while still a youth, to start making collections of plants in the Cedarberg, which he sent to the botanist Dr

FIGURE 17.5: Dr C. Louis Leipoldt. National Cultural History Museum Photographic Collection, No. HKF4146: C. Louis Leipoldt.

Harry Bolus of Cape Town (some specimens are now housed in the famous Bolus Herbarium at the University of Cape Town). The two developed a close friendship and corresponded regularly until Bolus's death in 1911.

At the end of the South African War, Leipoldt left for England to pursue a career in journalism, which he had begun in Cape Town as a correspondent for local and international papers. While in London, Leipoldt was lent money by Bolus to study medicine at the prestigious Guys Hospital in the city. In due course, Leipoldt won the medical school's gold medals for both medicine and surgery – a rare achievement. He then filled a succession of posts with hospitals in London and became a county medical inspector of schools there. He continued to write as a journalist and novelist throughout this period. Upon his return to South Africa in 1914, Leipoldt was appointed to the new position of Medical Inspector of Schools in the Transvaal, a role he fulfilled for the next eight years, apart from service as a captain with the Union forces in German South West Africa during World War I (where he was General Botha's personal physician). He wrote a number of books and poems during this period, establishing himself as a champion of written Afrikaans. His work *Oom Gert vertel en ander gedigte* (1911), a collection of patriotic and nostalgic poems, was received with acclaim. Leipoldt's memoirs of his experiences in the rural Transvaal were recounted in his book *Bushveld Doctor*, first published (in English) in London in 1937.[11]

In the course of his numerous tours of rural Transvaal schools, Leipoldt discovered "an appalling state of disease among the children, especially from malaria, 'the curse of the bushveld', and bilharziasis" (see Chapter 12).[12] His memoirs discretely avoided mentioning any names of people, schools or places, but provide graphic and sobering accounts of the impoverished communities he encountered and stubborn parental resistance to palliative care procedures.

Jan Moerdijk

J.L. (Jan) Moerdijk had been a school inspector since 1897, and in the 1920s wrote text books on maths and language for primary schools. He was the father of Gerard Moerdyk, the architect. (The biographies of both have already been described, in Chapter 6.)

By the 1950s the depopulation of the region meant that pupil numbers at Elandsbosch had fallen to around 50. This led to a sensible decision in 1954 to close the Geelhoutkop school on the farm Driefontein about 15 km to the west, which had already been losing pupils, and to merge its pupils with Elandsbosch where A.G. van der Westhuizen was the head. The head of Geelhoutkop, P.R. Fouche, and the remaining 21 pupils were transferred, bringing the numbers at Elandsbosch back up to 75. At the same time, funds were allocated to the enlargement and refurbishing of the school. Sadly, Van der Westhuizen, a most popular principal in the district, was killed in a car accident on the road to Nylstroom in 1972. By 1985, with G.J. Swart as head and two assistant women teachers, the school had only 34 names on its list and it closed soon afterwards.

Geelhoutkop school

Before its closure in 1954, Geelhoutkop school had been in operation for many years, although no formal records earlier than 1939 have been discovered. Located on one of the highest points of the plateau (the crest behind the school is at 1739 metres amsl), the site of the school enjoys clean air, an outstanding view towards the south and remarkable vegetation, including aloes and the yellowwood trees (*Podocarpus latifolius*) that gave it its name. The farm Driefontein was owned by the Pretorius family and in the 1940s their children attended the school. A grandson, Barend Pretorius, who still owns the property, became a teacher and eventually head of the Meetsetshehla Secondary School in Vaalwater, before moving to head the science department at the neighbouring Waterberg Academy, an independent school in the town.

The Inspector's report of June 1939 recorded the head of Geelhoutkop as being P.J. Pretorius (no relation to Barend Pretorius's family) with 30 pupils, a library of 200 books, and two classrooms, a teacher's residence, koshuis and outbuildings, all in good condition. However, the inspector remarked that the local community was poor, that malnutrition was a concern and that the personal hygiene of several children "leaves much to be desired".[13] This concern was borne out by a medical inspection of the school during 1942 (at which time C.P. Seegers, a member of the Jehovah's Witness sect, was the acting head), when it was found that 25 of the 29 pupils attending had some medical condition; 19 had bad teeth, eight suffered from ENT-related problems and three were significantly malnourished. A feeding scheme was supposed to be in place, but instead of receiving cheese and raisins in summer and soup in winter, the children had only been given tea.

The Geelhoutkop ridge is notorious for lightning: In December 1946 two brothers, Jacobus (14) and Michiel (12) van Staden, who were pupils at the school and members of the large family that lived on the farm Kameelfontein nearby, were killed by a lightning strike.

Palmietfontein school

In the central part of the plateau, one of the earliest schools was on Palmietfontein at Sukses. It opened in January 1917 with one teacher, J. Barnard, and 45 registered pupils, according to the school's official journal, which is more complete than that of most of its peers.[14] Its story typifies the conditions and issues faced by most such schools on the Waterberg plateau.

The first school committee was chaired by J.L. Botha, a local parent. Inspector Moerdijk paid his initial visit to the school in February. On a visit a year later, he reported that the rudely built walls of the two-roomed building had been damaged by rain to the extent that they had "moved out of the perpendicular". He also noticed that there were no latrines. He instructed that all instruction, with the exception of English lessons, should be in Afrikaans. In August 1918 Moerdijk's report noted that a fine room had been added to the building. Attendance, however, was weak (only 33 pupils present):

> Many parents in this neighbourhood do not appear to have the slightest idea of their responsibility in this respect. Ten children left [in April], apparently in the hope that this would prompt another school being built closer to them. There is disagreement between the school committee and the Principal. In particular, the latter (Barnard) has sometimes stepped beyond his powers and taken action in corporal punishment in dealing with girls too, which must be strongly disapproved. He has been warned previously about this behaviour and his future as a teacher is in jeopardy. On the side of the parents, little co-operation is shown and by their arbitrary interference in the regular course of work, the work of teaching staff is seriously hampered.

Three months later, the school (in common with many others) had to be closed temporarily because of the risk posed by the Spanish flu epidemic that year.

A year later, the school journal records that pupil numbers had improved to 43 and that Inspector Moerdijk had appointed an assistant teacher (R.J. Adlum) to assist with the junior grades. The principal, Barnard, was finally replaced in November 1919 by A.H. Middleton, "who makes a favourable impression".

A record from March 1921 states that

> [T]he school building is steadily falling into ruins; the roof leaks and thus the walls are spoilt and one of the side walls is bending outwards. As the school is showing signs of vitality … a new building should be erected. There are still no latrines, which especially in a bare district like this is a serious drawback. Attendance is very poor – seemingly because there is whooping cough in the district and parents are forbidding their children to come to school if they've not had it.

Palmietfontein school developed a tradition that the opening assembly of each term would be accompanied by the singing of Psalm 84 (verses 1&3):

> How amiable are thy tabernacles, O Lord of hosts!
> Yea, the sparrow hath found an house, and the swallow a nest for herself, where she may lay her young …

In 1926, Martha Webster, a farmer's wife from Vlugtkraal (later part of the Born Wild property), joined the school to run needlework classes. She soon reported that one of the boys had hit a girl so hard that a doctor had to be called to attend to her (the journal does not record what discipline was administered to the boy). In November that year, the principal, Middleton, advised the parents of two pupils in Standard 5, Magdalena Adlum and William Wakeford, that if their children were to advance to secondary school, they would have to leave Palmietfontein at the end of the year. As neither family could afford to send their children away to boarding school, the parents opted to leave them at Palmietfontein for Standard 6 in 1927, so as to write the Junior Certificate examinations that would bring their formal education to a close. The 1926 Junior Certificate examinations for the district were held at Geelhoutkop; of five Palmietfontein pupils who wrote the exams, four – Johannes Botha, Martin & Anna Barnard and Michiel de Kock – passed with orders of merit.

Advancing to high school could be more than a matter of affordability: In 1953, for example, a Standard 5 pupil at Geelhoutkop, Aletta Victor, an exceptionally bright daughter of the chairman of the school's committee, applied to attend the secondary school in Nylstroom, but could not do so because the girls' hostel there (Ons Hoop) was already full. She had resigned herself to stay at Geelhoutkop in 1954 to write the Junior Certificate at the end of Standard 6 when fortune smiled and a vacancy in the hostel enabled her to proceed to high school after all.

In June 1927 a new school building (with toilets) was opened and the pupils and staff marched there from the old building singing Psalm 84. There were 45 pupils in total: 21 in the junior grades (up to Standard 2) under Mrs Webster, and 24 in Standards 3 to 6 under Mr Middleton. In November Mrs Webster had a mishap:

> Vandag is Mev Webster afwesig. Haar perd het op pad na skool geskrik met gevolg dat sy afgeval het en haar rug beseer het. Sy is verplig gewees om terug te kom na haar huis. (Today, Mrs Webster was absent. On the way to school, her horse got a fright and she fell off, hurting her back. She was assisted back to her home.) Fortunately, the injury was not too serious and she was back at work the following week.

It is of interest to note the demographics of this small school. At the end of 1927, for example, of the 33 pupils listed, half were girls; only seven, based on their names, may have been English-speaking. All four pupils in Standard 6 wrote and passed their Junior Certificate exams that year, two with merit. The school received a letter from the Nylstroom school board congratulating it on the good results.[15]

In 1929 the inspector noted that the river from which the school drew its water (the Sondagsloop) was full of bilharzia and that a borehole was needed. (It was eventually drilled six years later). He also reported that four pupils travelled 22 miles (35 km) daily by donkey to get to school and four others at least 13 miles (21 km). Some form of boarding facility was needed; in the meanwhile the school board would provide a subsidy of 1/6d per day for each of the first group and 1/-d per day for the second. It would be another 17 years before a hostel was built; until then, some pupils made arrangements to stay with families closer to the school. There are several references in the journal to attendance during the rainy season being adversely affected by children unable to cross swollen rivers.

In 1934 there were 53 children at the school, 31 of them in Standards 3 and above, with nine candidates for the Junior Certificate exam, of whom four passed. On several occasions now, pupils wrote examinations for departmental bursaries to enable them to go to secondary school. In 1937 the school held a concert to raise funds to purchase books for its library. A total of £6.16.6d was collected and enabled the school to order 194 books, most of which arrived during the next year.

In February 1937 a meeting was held at the request of the department to discuss the consolidation ("centralisation" was the term used) of rural schools in the district, following a drop in pupil numbers. A combination of boarding facilities and bus transport was discussed.

No journal is available for Palmietfontein for the period 1938–1944. The first entry in 1945 was for the opening of the second term (accompanied as usual by Psalm 84), when only 26 pupils were on the roll. The journal entry for 8 May 1945 noted that the radio news the previous night had announced the end of the war in Europe and that the next day had been declared a public holiday, followed by two school holidays. On 15 August came the news of Japan's surrender, marked by another two school days off.

At the end of 1945, and again in the middle of the following year, the school held successful fund-raising events ("*tiekieaand met vleisbraai en konsert*"), each attracting between 200 and 300 participants. The funds were to be used for the acquisition of a piano (which arrived in January 1947) and for the building of a koshuis. A Mr & Mrs Roos were signed on as *koshuisvader en -moeder*; the hostel, which was really just a house, was built nearby.

The "centralisation" issue was a recurring concern for the department and parents, especially droughts, economics and urbanisation caused depopulation of

FIGURE 17.6: The old school at Palmietfontein with the teachers' house at the back.

the area. Several proposals were debated around school merger configurations in the area north of Vaalwater with consolidation at Palmietfontein, Zanddrift and Melkrivier being most favoured.

In July 1949 Arthur Middleton formally retired from Palmietfontein school (and from teaching) because of poor health after 30 years' service there. He was succeeded by L.C. Kotze in January 1950; Mrs Louw was his assistant, with 54 pupils on the roll. However, parents immediately mounted an objection: Within the first week of term, 15 pupils were removed from Kotze's class, because the parents were opposed to Kotze's religious beliefs. Like Middleton before him (and Seegers at Geelhoutkop), Kotze was a Jehovah's Witness – although no complaint had been recorded against the former. Several families in the district – for example the Boshoffs and Barneses – were also Witnesses. Although parents were eventually persuaded to allow their children to return to classes (they had remained on school grounds in the interim to comply with attendance laws), the crisis continued until H. Brits, a teacher from Zandfontein, was seconded to Palmietfontein as acting head in the place of Kotze who had to leave. There followed a succession of unsettling staff changes over the next few years that added to parental dissatisfaction. By 1966 the principal was S.J. Prinsloo, with 38 pupils. The Inspector's report described the classrooms as deteriorating wood-and-zinc structures with old, unhygienic toilets (though by now there were some trees!). The decline continued: When school opened in January 1970 there were 20 pupils, but the number declined to 12 within a week. Parents met to discuss the situation and concluded that the school was no longer viable. It finally closed its doors on 21 January 1970 after 53 years, most of the children moving to Vaalwater and others to Melkrivier.

Melkrivier school

This school, some 20 km north of Palmietfontein, had been operating since October 1935 when it was built to accommodate the merger of two even earlier schools, Kwarriehoek and Libanon (it was built exactly halfway between them!). The first head at Melkrivier, H.J.T. Steyn, was previously head of Kwarriehoek. However, the master who became synonymous with Melkrivier school was Jan Christoffel Nel, who was posted there as a teacher in 1937 and became its head in 1944, a position he held until his retirement at the end of 1970.[16]

By the 1950s Melkrivier was educating about 60 pupils, its further growth constrained by the limited boarding and classroom accommodation. Eventually, in 1958, after intervention by the new Prime Minister (and Waterberg MP), Advocate J.G. Strijdom, and his legal partner F.H. Odendaal (then Waterberg's Provincial Council representative), a new hostel was constructed on a small (eight morgen) portion of the adjacent farm Muisvogelkraal. A bridge was built over the Lephalala River to assist pupils from the Kwarriehoek side to reach the school during the rainy season.

In 1964 a new school was built on the Muisvogelkraal property; it could cater for most of the increase in pupil numbers, although classrooms at the old, privately owned building (which later became Clive Walker's Rhino Museum until taken over by land claimants in about 2007) were still used for a while. Matrons' quarters and sportsfields were also built. The latter enabled Melkrivier to host several regional athletics and sports meetings; several of its pupils went on to excel at district and provincial level. Jooste Heystek, Strijdom's successor as Waterberg MP arranged for some redundant prefabricated buildings from the original school at Ellisras to be re-erected at Melkrivier, thereby providing the latter with a hall. The closure of Palmietfontein school in January 1970 resulted in an immediate increase in pupil numbers at Melkrivier: initially to 79 and by the end of the year to 91.

Through the years bilharzia was a perennial problem: In 1951 the school and hostel were placed under quarantine for two weeks because of suspected scarlet fever – later diagnosed as measles; in 1955 there was a polio epidemic; and in 1957 one of Asian flu. During the rainy season – except in drought years such as those of 1963 and 1964 – flooded rivers frequently disrupted attendance by pupils and teachers alike.

In July 1973 the school's population peaked at 105 pupils, 82 of whom were boarders. By then, though, the frequent changes in personnel (many of them temporary rather than permanent appointments) imposed on schools by increasingly dysfunctional management, were having a negative effect on teaching, on the spirit of the schools and on parental support. Several senior staff exacerbated the problem by deciding to leave the profession, perhaps from frustration over this instability. At Melkrivier, numbers declined to between 60 and 70 by 1983, with by far the majority of pupils being boarders. Although there was some recovery a few years later, the registration of only 48 pupils in January 1991 soon triggered a decision to close the school at the end of the second term. Most pupils were transferred to Vaalwater, making use of a comprehensive bus service that had been put in place. Others went north to Marken. So ended the proud academic and sporting record of what for many years was an extremely well-administered and highly regarded school.

About 15 years later, Melkrivier school reopened and was still in operation in 2018, although the boarding facilities had been abandoned in favour of bussing children in daily from the surrounding district.

Bulgerivier school

Another early government school on the Waterberg plateau was Bulgerivier on the farm Bulge River, 45 km west of Vaalwater. The school was originally established in July 1920 as a private school on the property of the owner, A.S. van Emmenis, who not only built the facility but furnished it and paid the salary of the first teacher, H.L.B. Helberg, for the first six months. In 1921 the school was taken over by the Department of Education, which retained the appointment of Helberg as a temporary head, despite his lack of certification. Bulgerivier school could accommodate up to 22 pupils and within its first year there were 20, in grades and the first five standards. It was visited by Inspector Moerdijk in 1921; he described it as one of the best-built schools in the Waterberg – lots of space, well-lit and properly ventilated. He commented that Helberg, although unqualified, had experience, a good temperament and was enthusiastic and that all the pupils succeeded in advancing to the next level.

Helberg was followed in 1922 by a fully qualified teacher. He was J.S. (Faan) van der Walt, whose descendants still own and live on the farm.[17] Some families had several children at the school concurrently. For example, in 1928, there were five Erasmus, three Nel and three Van Staden pupils among the total of 23. Originally, there had also been a small, privately run school at Schoongelegen a few kilometres to the south of Bulgerivier, but soon the pupils there were transferred across to provide economies of scale. Bulgerivier school operated at its original location until 1959, although only partial records remain.[18] Then a new two-teacher government-built school opened on the main road. The building still stands, although it is disused. In the early 1970s, when the Strijdom (now Mokolo) Dam was under construction, children from the construction camp were bussed in every day; the numbers necessitated some additional classrooms and a doubling of teachers to four for the duration of the construction period.[19]

Zandfontein school

Up on the high plateau, near Dorset police station, was the school at Zandfontein. Its most well-known principal was J.P. Jacobs, who, in the 1930s, purchased the nearby farm of Vischgat from the Transvaal Consolidated Land & Exploration Company (TCL). Jacobs was also chairman of the local farmers' association and was known in the 1940s and 50s as a staunch Afrikaner Nationalist, Broederbonder and supporter of Strijdom. He married a Swanevelder, whose brothers, Lodewyk and Erasmus, had also purchased a nearby farm (Steenbokfontein) from TCL. The Swanevelders, Jacobses and a third family, the Venters, formed an extended family with many children between them. According to descendant Pieter Swanevelder, there was a time when, of almost a hundred pupils at the school, 63 were related to the Swanevelder family![20] (See Chapter 10 for more details about this area.)

FIGURE 17.7: The Zandfontein-Vogel school (1934) (LEFT). Zandfontein school with headmaster J.P. Jacobs (RIGHT). Images provided by Magda Streicher and Frikkie Brits from their collection.

Witfontein school

Another school to the north of Vaalwater was Witfontein, on the present farm of that name on the R33 to Lephalale. Witfontein school first opened in about 1922, with J.R. "Meester" Oosthuizen as its only teacher. Oosthuizen, who had qualified in the Cape (both factors making him a rarity in the area) remained attached to this school throughout his career, most of the time as its head. He retired from this position in 1964, but remained on the staff until the school's eventual closure at the beginning of 1972. At the time of Oosthuizen's retirement, the school board was chaired by W.P. Louw, a local farmer, with his son-in-law Ben Vorster as secretary. There being no quarters for staff, Oosthuizen built a home for himself on an adjacent portion of Witfontein. Later, when the original mud-walled school proved inadequate, he and a neighbour, Rudolph Faber, donated a portion of Witfontein on which they built a new school. Finally, a third government-built structure was constructed nearby. By the late 1960s, when C.A. Holtzhausen was head, the school accommodated about 30 pupils, a prefabricated teacher's house and hall had been built and an athletics track laid out. This positive development of the school was not supported by pupil numbers however, and in January 1972 it failed to open its doors, the children being transferred to Vaalwater instead.[21]

Hartebeestpoort school

Just outside the village of Vaalwater is the farm Hartebeestpoort, which had been the home of Jack and Rhoda Farrant since their marriage in 1912. Rhoda was the eldest daughter of Vaalwater pioneer Rufus Kirkman and had been trained as a teacher – a most unusual qualification for the area in those days. Jack and Rhoda had three children: Marjorie (born in 1913), Kathleen (1914) and Rupert (1917). In the absence of a school in the area, Rhoda began to teach her children at home. This eventually led to the decision to build a classroom that would enable neighbouring children to be schooled too and inevitably, Rhoda soon had to start a boarding establishment to accommodate those who lived too far away to make the journey daily. In due course, the Transvaal Education Department rented the buildings, appointed a teacher and took over responsibility for the school's administration.[22]

The first available public record of Hartebeestpoort school is the school journal commencing in January 1923 when the institution reopened after the holidays under P.B. Bosch as head, with 27 pupils from Grade 1 to Standard 5 – including Marjorie and Kathleen Farrant in Standards 3 and 2 respectively. Unusually for schools in the Waterberg, English-speakers predominated, at least initially. Indeed, in his inspection report that September, the ubiquitous Jan Moerdijk noted that this was the only English-medium school in the district. The acting head at the time, Mrs M.M. Ross, was a qualified teacher and Moerdijk hoped she would be appointed permanently in order to keep her at the school. His report complimented Mrs Ross on her teaching skills, her administration and her ability somehow to handle seven classes on a dual-medium basis without assistance.[23]

Mrs Ross was again complimented in 1924, although Moerdijk suggested that the school should accommodate pupils up to Standard 4 only, because the task was too much for a single teacher to handle. That winter, the head lamented that her appeal to the school board in Nylstroom for the provision of heating lamps and cocoa to help stave off illnesses had been ignored. The following year, the youngest Farrant child, Rupert, was recorded as attending Standard 1, among a total of 26 pupils at the school.

Unfortunately, by June 1926, Mrs Ross was succeeded by a new principal, M.G. Green, who came under fire from the inspector for failing to maintain administrative records and for allowing teaching standards to fall. One of his first crises occurred when school opened after the winter holidays to the attendance of only eight pupils, the rest having been stricken by a measles epidemic. The demographics of the school were also changing: Of a total of 25 pupils on the roll, only nine were now English-speaking and the predominant medium of instruction had become Afrikaans.

Mr Green was evidently impervious to criticism because he was still at the school five years later when the new inspector, W.T. Hurter, having listed all the shortcomings of the school's administration, concluded that Green was clearly capable of good work, but had simply allowed himself to relax! The inspector also criticised the headmaster for not providing any leadership or mentoring to his inexperienced assistant.

By 1933 numbers had fallen to 20, only six being English-speaking; they continued to fall in the following years until, when school opened in January 1936, there were between 12 and 16 pupils present. (The journal notes that the school was closed on 22 January to mark the death of King George V of England). In April parents met to discuss the future of the school. Their worst fears were confirmed when notification was received from the Department in May that the school would close on 26 June 1936 – although it finally did so a month later when the keys were returned to the landowner, Jack Farrant. It is interesting to note that the school building survived and formed the temporary nucleus around which the pre-primary phase of the Waterberg Academy independent school was established in 2002, before its own buildings were completed.

The stories of Palmietfontein, Melkrivier, Geelhoutkop, Bulgerivier and Hartebeestpoort recounted above are similar to those of other government schools on the Waterberg plateau, although none has surviving journal records that are as comprehensive. It is not even clear how many rural schools there were: at least those referred to above, as well as schools at Alma, Rankin's Pass, Doornfontein, Wolwefontein and Heuningfontein – possibly 15 to 20 at the peak, serving a population of 500 to 800 children up to Standard 6. The only relatively nearby secondary schools were those in Nylstroom, Naboomspruit and Potgietersrust, all of which offered Afrikaans-medium instruction. The nearest English-medium secondary schools were Capricorn at Pietersburg and the Pretoria Boys' and Girls' High schools, where most English-speaking children were sent to board, by train.

By the late 1970s most rural government and private primary schools had closed; some, like those at 24 Rivers, Melkrivier and Alma, continue to operate today, catering for mainly black children who are unable to attend one of the larger schools in nearby towns. In addition, there were several privately operated "farm" schools for the children of black employees, which never became part of the Department of Education. One of these, on Vrischgewaagd, was run for many years by Ansa Nel, the wife of Louis Nel, whose father Jannie had been the headmaster of the nearby Melkrivier school.

Laerskool Vaalwater

Proposals for a school at Vaalwater dated back to 1951. Even before the township had been formally proclaimed, a school had been built, although not on the stand eventually negotiated for it. This was largely thanks to the energetic efforts of a local school committee led by J.P. Bekker, the owner of the hotel and other properties surrounding the developers' land. Laerskool Vaalwater opened its doors on 7 January 1953 to 60 pupils, initially under the supervision of a single teacher, F.J. van Dyk, the acting headmaster. The occasion was attended by Bekker, the Schools Inspector, W.H. Böhmer, dominees Viviers and Venter of the school board and Rupert Farrant among other dignitaries. Ds. Horak from Nylstroom officiated. The following week, an assistant teacher arrived: Mrs M.M. van Pittius. The classes were then divided between the two staff members with Van Dyk taking the senior classes up to Standard 6. A few months later, the school committee was elected, under the chairmanship of J.P. Bekker with R. Farrant as a member. At the end of July 1953 the newly appointed permanent head arrived. He was S.S.B. Hoffman, and he would occupy this position for the next 23 years. [24]

In January 1960 the school moved into new premises – those still occupied by it today – and pupil numbers had risen to 146 due to the closure of the nearby Zanddrift school. Work commenced on developing sports fields and by 1963, a pavilion had been financed and built by the parent body, followed by a house for the head. Hoffman was highly regarded as an efficient and dedicated teacher and administrator. In keeping with the times, though, he was an ardent supporter of the ruling National Party and a member of the Afrikaner Broederbond into which he was admitted in 1964.[25] The school journal of the period (which he maintained) is filled with newspaper cuttings of the activities of party political figures and events unrelated to education – including the deaths of Messrs Malan and Strijdom and the assassination of Dr Verwoerd.

Rupert Farrant had fought hard – successfully – to have Laerskool Vaalwater designated as a dual-medium school. But this conflicted with the ideology of the Broederbond, which subscribed to the philosophy of Christian National Education (CNE) that schools would accept the Holy Scriptures as their basis, use mother-tongue as the medium of instruction and espouse what was termed the "national principle", by which was meant a love for one's own language, culture and history.[26] Certainly, the school journal offers little evidence that the dual-medium policy was being adhered to under Hoffman's leadership. Having said that, however, it must be acknowledged that the English-speaking pupils at the school would always have been a small minority: In 1972, for example, on the first occasion when such numbers were documented in the school's journal, there were only five English-speakers out of a total of 154 pupils – yet every standard needed to provide for their presence, which was obviously impractical.[27]

It is also true that the English-speakers kept to themselves and were not prone to assimilation with the larger Afrikaans-speaking community. Colin Baber has been cited as a rare example of a locally resident English-speaker who interacted extensively with his Afrikaans-speaking neighbours and who encouraged his family to do likewise.[28]

FIGURE 17.8: S.S.B. Hoffman, head of Laerskool Vaalwater (1953–1977); photograph in foyer of school offices.

The English were inclined to send their children off to board at schools like Lleweni and later at Settlers, going on to high school in Pretoria or further afield. As a result, there were few opportunities for the young of the two language groups to establish lasting friendships; marriages between English- and Afrikaans-speaking residents locally were extremely unusual.

In January 1970, as recorded above, Palmietfontein school closed, followed two years later by Witfontein school on the Beauty Road; most of their pupils transferred to Vaalwater, lifting pupil numbers there to 154, with seven teaching staff. Tribute was paid to "Meester" Oosthuizen, who had been head of Witfontein for the best part of 50 years. Later that year (1972), the selective conservatism which then characterised South African education was evident in a speech given to a course for school principals by the Director of Education, Dr A.L. Kotzee in launching the department's Youth Preparedness Programme. Christianity, he said, was facing the greatest crisis in its history, due to five principal elements of permissiveness that had been allowed to creep into society: the drug menace; the collapse of sex morals; the breakdown of loyalties and of authority; and permissiveness in general. He lamented "the treason being conducted against the youth by some intellectuals, writers, thinkers and philosophers, who write and speak in such a way as to make, for instance, sex immorality and drug abuse seem normal … we are only 3 ½ million whites and for the time being we still have to render many services to other population groups in our country."[29]

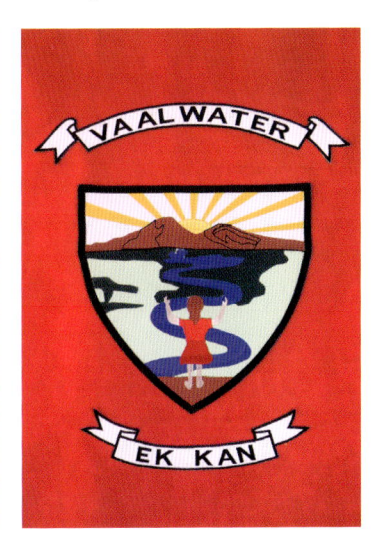

FIGURE 17.9: S.J.M. Swanepoel, head of Laerskool Vaalwater (1977–1992); photograph in foyer of school offices (LEFT). P.B.E. van der Westhuizen, head of Elandsbosch school (1993–1998) (MIDDLE). The Vaalwater school badge (RIGHT).

Hoffman retired as head of Vaalwater school in December 1976 and was replaced by S.J.M. Swanepoel (also a *Broeder*)[30], who fulfilled that role for the next 16 years. This tendency of headmasters to remain in position for over a decade had contributed greatly to the stability and reputation of the schools they had led. Unfortunately, as we've seen in the case of the district schools already mentioned, it seems that a combination of factors (decline in departmental support and oversight, attractive alternative job opportunities outside of education, difficulties in securing quality junior teaching personnel) began to result in higher turnover of principals, and in turn, of subordinate staff. By the end of Swanepoel's term of office, Vaalwater school had been under the leadership of just two heads over a period of 39 years. In the next 20 years, it would see at least seven permanent or acting principals come and go. While most of these heads no doubt did their best to continue to protect and build the school, journal entries reflect the increasingly difficult and stressful conditions under which they had to operate. Not least among these were the repeated policy changes accompanying the new democratic era after 1994. Worst was the abandonment (at the behest of teachers' union SADTU) of the long-established school inspectorate whose highly experienced members, ever since the days of Jan Moerdijk, used to provide valuable support and guidance (and discipline) to heads and staff. School journal entries across the Waterberg record the names and detailed reports of people like W.T. Hurter, W.R. Böhmer, N.J. Bornman, G.J. du Toit and J.W. Lotz. In the Vaalwater area, the last of these was the respected Sam Mafora, who later, as a Circuit Manager, helped guide successive heads through the difficulties of transformation prior to his retirement in 2012.

Lleweni Private School

The decline in the number of English-speaking families in the Waterberg region during the 1920s and 1930s was of great concern for parents seeking mother-tongue primary education for their children. For some, the solution came with the founding of Lleweni. In 1922 a Welsh immigrant, Henry (Harry) Thelwall Wingfield Salusbury, who had come to the Waterberg area as a Milner settler after serving with the British during the South African War, decided to build a private school on property he had purchased about 12 km north of the little village of Naboomspruit. The school was to educate his own three children (Eileen, Tom and Gerald) and those of other English-speaking residents of the district. Salusbury had been employed by the Mines Department after the end of the South African

War, completing his service while based on the farm Doorndraai, now beneath the reservoir of the same name on the Sterk River. His own farm was Driefontein just to the south of Doorndraai.

Salusbury named the school he built there Lleweni after the estate near Denbigh in north Wales that had been the seat of his illustrious forebears since the time of William the Conqueror.[31]

Lleweni school included a boarding house. Among its early pupils was Brian Stone from Settlers. Interviewed in 2015, aged 95, Brian readily recalled his time in the late 1920s with Salusbury and his wife Geraldine de Lyon Wentworth Salusbury (née Dillon) who the children knew as "Mop". Initially, the school had only 18 pupils (including the Salusbury children), who were taught in a thatched house with two classrooms. Within a year, that number had doubled and two dormitories, one each for boys and girls, had to be built to accommodate them. Salusbury turned out to be not only a talented carpenter and craftsman, but also an excellent teacher, painter and musician.

He used to make regular trips to Naboomspruit (12 km south) by bicycle or ox-wagon to fetch supplies and collect the post. Much was needed because nearly all the pupils were boarders from Sandrivierspoort, Vaalwater, Hanglip, Crecy, Singlewood and Settlers. They travelled to the school by cart, ox-wagon, on horseback, or even by train. They saw their parents at least once a quarter. Later, Salusbury got a second-hand Model T Ford in which to transport them all.

In the beginning, Salusbury was the only teacher and taught all the standards during the mornings. In the afternoons, pupils played games, did reading and sewing (overseen by Mrs Salusbury), walked and rode donkeys, learned woodwork, took piano lessons and played soccer, hockey and cricket. On Saturday

FIGURE 17.10: Harry and Geraldine Salusbury.

evenings, there was dancing to music from a gramophone. On Sundays, church services were conducted, usually by Salusbury himself, but occasionally by the visiting Bishop of Pretoria, Talbot.[32]

By 1926/27, numbers having increased to between 40 and 50 pupils, a new school building was constructed and a second teacher, Miss A. du Toit, came to the school. By 1930 the pupil numbers may have risen to 90, many now coming from Settlers. Salusbury would collect children from Settlers in a donkey cart – and later in a Ford van – for the long ride (over 80 km) across to the school via Nylstroom.

The move to Settlers

Before long, so many of Lleweni's pupils came from Settlers that the community there persuaded Salusbury to move the school to Settlers where his wife, Geraldine, purchased four morgen of the farm Knapp from A.J. Williams on 29 January 1932. Knapp was adjacent to the village. A.J. had asked a local builder/farmer, Edgar Gillbanks, to build a simple wood-and-iron school structure on stilts on the property. The school was built at the end of 1931 and opened on 23 January 1932. The remainder of the farm Knapp was eventually sold by A.J. to the government in 1964, for the erection of the Settlers Agricultural High School.[33]

According to Brian Stone, about 60 children attended the new Lleweni school, many of them as boarders, with Mrs Salusbury employed as the matron. Mr Salusbury taught Standards 3–6 and a Miss MacDonald the lower grades. Pupils came from all over the district, including the Waterberg. The Salusburys were very popular with their pupils, despite being strict. Mrs Salusbury was an accomplished pianist and had a beautiful voice. She taught music at the school, while Mr Salusbury, who was an excellent cabinet-maker, taught woodwork. He was also a talented water colourist and encouraged pupils to take up painting.[34] Towards the end of 1932, though, the Salusburys left the school and Harry was succeeded as principal by Mr Meyer.

Ownership of Lleweni school in Settlers was transferred in 1935 to the government, which had in the meanwhile acquired 80 morgen of adjacent land. The name of the school was changed first to Springbok Flats School and later to Lord Milner Farm School, in commemoration of the British administrator whose idea it had been to create a settlement of English-speaking farmers on the Springbok Flats after the end of the South African War. From 1935 until 1937, the principal was Jensen. He was replaced by a German-speaking teacher called Ross Bresler, who had been teaching in the Witbank area since 1922. Bresler did a great deal to expand the school's facilities, especially the grounds, with much of the manual work being undertaken by the pupils. By the time Ross Bresler left Lord Milner in 1943, there were over 150 boarders, most of them from Pretoria, Johannesburg and the Reef. It was the only English-speaking farm school among the nine that had been built in the Transvaal since the Depression.[35] Bresler was credited with much of the school's growth in capacity and reputation.

Brian Stone remembered an incident in which the son of the doctor in Warmbaths, Mickey Cohen, was beaten by Bresler, who was accused (probably unfairly) of anti-Semitism and applied for a transfer (to Benoni). Stone was instrumental in persuading the Department of Education to build the Agricultural High School at Settlers in 1969 to complement the Lord Milner Primary School.[36]

Harry and Geraldine returned to their farm Lleweni on Driefontein after leaving Settlers, but later sold the property to the Lacanti family.[37] Salusbury died in Port Shepstone in December 1957, aged 77.[38] He and Geraldine had two sons, Thomas and Gerald, and a daughter Aileen. Gerald died in service in Libya in 1941, aged 23. Thomas later moved to Rhodesia. He was a guest at Lord Milner School's 60th jubilee celebration in 1982. The site and buildings of the original Lleweni school on Driefontein would later become the locus of a resort development in the early twenty-first century.

The Waterberg Academy

The only other truly private (now referred to as "independent") school on the Waterberg plateau was, and remains, the Waterberg Academy, located just outside the town of Vaalwater. The Academy was founded only in 2002 by local residents concerned about the future of education in the region and anxious to ensure that their own children obtained a high quality of education. The initiative was led by Janneman and Cathy van der Merwe and the first junior grade classrooms opened in 2003 on a portion of the farm Hartbeestpoort adjacent to the Meetsetshehla Secondary School. A new class was added each year and by 2007 the decision was taken to proceed with expansion to a secondary school, accompanied by boarding facilities.

By 2018 the Waterberg Academy was educating around 140 pupils of all races, from pre-primary to matric and had enjoyed an unbroken 100% matric pass rate (under the auspices of the Independent Examining Board, or IEB) since inception.

Almost a third of the pupils were beneficiaries of financial assistance and several were boarders. Because of its small numbers, the school had chosen to focus on excelling in small-team or individual sporting activities with remarkable results. In 2015 and 2016, for example, over 10% of its learners obtained provincial colours in archery, mountain biking or horse riding.

BLACK RURAL EDUCATION

Mission school education

The discussion until now has dealt with the schooling of white children on the Waterberg plateau in the past. But what of the education of the black community prior to the relaxation of apartheid in the late 1980s? Information on the provision of schooling for blacks in Nylstroom, for example, in the first half of the twentieth century is scant – for the simple reason that there was virtually none. Until 1953 the education of black children was provided exclusively by missionary or church bodies (or, very occasionally, by private individuals). The only state involvement was the disbursement of relatively modest funds to the churches and missions as contributions to their costs.

This state of affairs dated back to the Cape administration of the early nineteenth century when education became increasingly segregated: Lord Charles Somerset, the British Governor of the Cape in the 1820s, introduced free, government-funded education for burghers of the colony, the term "burgher" referring to children of European descent only. Initially, there were a few coloured and slave children who attended these government schools, but by the time of the Great Trek, this was no longer the case. Instead, those slave and coloured children in the Cape who went to school at all were educated by missionary societies, whose schools also accommodated poorer white children. Indeed, by the close of the nineteenth century, a third of white children in the Cape attended mission schools.[39]

Neither the British administration in the Transvaal after the South African War nor the Union government initially focused much attention on the education of black people, especially in rural areas, given the enormous demands they faced in rehabilitating the education of whites – and their general disregard for the upliftment of black people. Such black education as existed was almost entirely due to the efforts of missionary groups and a few progressive private individuals. (The exception was in Natal where state-run schools for black pupils were established as early as 1918.) Certainly, until the establishment of a Union in 1910, there were scarcely any schooling opportunities for black people in the Transvaal.[40]

In the Waterberg region, mission or church schools were particularly scarce because of the small and scattered black population: They were limited to the school built by Rev. Oswald Krause in 1883 at the Berlin Missionary Society's Modimulle mission station at Middelfontein, about 15 km east-northeast of Nylstroom; the Methodist training institution started by Rev. Isaac Shimmin at Olverton south of Warmbaths in 1885; and much later, the school opened at the Catholic Church in Warmbaths in 1930.

Early Union government policy

Long before the advent of formal apartheid, marked by the National Party's election victory in 1948, political representatives of the white population, irrespective of party, had been of the view that it was unnecessary and pointless – even potentially dangerous – to encourage the education of black children. Instead, successive governments from 1903 provided limited financial assistance to various missions, assisting them to build, staff and operate schools (there were 4500 mission schools nationally by 1949).[41] However, this assistance was inadequate: In 1935 fewer than 30% of all black children in the Union received any education at all, very few of those who did, progressed beyond the first two standards and less than 2% of pupils were in high school. White primary education was compulsory and free, whereas black children had to pay for school fees and books (and their parents were subjected to a poll tax, some of the revenue from which went towards subsidising the missions). Government's gross contribution towards the education of a black child was one tenth of that for a white child (and on a per capita basis, one fortieth).[42]

In 1919, Charles Loram, a member of the Native Affairs Commission and a Chief Inspector of schools, had abolished the teaching of algebra and geometry in black schools in favour of hygiene, nature study and other subjects with "practical and demonstrable value", based on his views that black people's future lay in a rural environment and that they would mainly perform manual and agricultural work.[43] The curriculum introduced for black schooling in 1922 enforced compulsory instruction in the vernacular in all primary classes, with an emphasis on "practical skills" like gardening, handiwork, agriculture, housecraft and needlework.[44]

In 1936 the Union government received the report of the Welsh Commission, a body that had been tasked with an investigation into black education, which since 1910 had been relegated to a provincial responsibility.⁴⁵ It reaffirmed the fact that, whereas coloured, Indian and white children went to free government schools, the majority of black children (apart from those in Natal) who received any education at all attended schools controlled, with limited funding, by church and mission groups. The Welsh Commission urged that all education, including black education, should become a national government responsibility and reside under the administration of the Minister of Education. This was opposed by another commission (the Native Affairs Commission), which insisted that black education should be the responsibility of the national Native Affairs Department that administered other aspects of the black population. This view, which was adopted by government despite the Welsh Commission's protests, reflected the segregationist agenda that dominated white thinking at the time. However, party politics and the intrusion of World War II meant that little was done to implement the transfer of responsibility for black education to the central government. In the Transvaal at least, rural black schooling continued to be provided by churches, missions and occasional enlightened private landowners.⁴⁶

National Party black education policy

In 1949 the newly elected Nationalist Prime Minister, D.F. Malan, wasted no time in commissioning another enquiry (the Eiselen Commission) into the nature and status of black education, this time with his party's apartheid ideology as the guiding principle. The chairman, Dr W.W.M. Eiselen, son of Berlin missionaries, a German-trained anthropologist, member of the AB and staunch supporter of apartheid, has been described as "a leading fascist intellectual".⁴⁷

The commission's report tabled in 1951 concluded that mission education had "achieved nothing but the destruction of Bantu culture … nothing beyond succeeding in making the native an imitation Westerner".⁴⁸ This led to the promulgation of the Bantu Education Act (No. 47 of 1953), which, under the zealous administration of Dr Verwoerd, then Minister of Native Affairs, enforced the racial segregation of education. The Minister withdrew governmental support for mission schools and created a new department intended to provide education to black people "in conformity with their own traditions and needs" as opposed to that of missions, where "frustrated potential agitators were bred who found no popular outlet for their limited talents".⁴⁹ Verwoerd was quoted as asking "What is the use of teaching a Bantu child mathematics when it [sic] cannot use it in practice? That is absurd."[50]

One result was the closure, immediately, or shortly, of nearly all mission schools, because they lacked the means to operate without state assistance. By 1957 the vast majority of mission schools had either closed their doors through lack of money, or else their facilities had been handed over to the state. This vacuum was filled, in time, with the construction and staffing of schools under the administration of the Department of Native, later Bantu, Affairs, to which responsibility had now been transferred from the provinces. The curriculum was designed in accordance with National Party ideology and intended to equip blacks with the skills needed to serve and augment the white-dominated economy – but not more.

The National Party government extended its segregationist philosophy through the imposition of the principles of Christian National Education (*Christelik-Nasionale Onderwys*, or CNO) on all state schools. In respect of black pupils, this mandated the use of mother-tongue instruction throughout all phases of education; the preparation of black people for their (inferior) station in life, principally in a rural environment; preservation of the cultural identity of black people, who were considered to be in their cultural infancy and required mentoring by senior white trustees; and the teaching of religion, preferably Calvinism, at schools. Subjects including nature study, geography, history and civics, were to be taught in accordance with Calvinistic tenets.[51]

The transfer of responsibility for black education to the Department of Native/Bantu Affairs was ideologically motivated, but it did have one benefit: In the seven years from 1953 to 1960, the number of black learners nationally doubled to over 1,5 million. However, state expenditure per pupil fell from R17,08 to R12,46 per annum and the state of school buildings deteriorated rapidly, teacher-pupil ratios increased, teacher morale fell, drop-out rates rose and pass rates fell.[52] Parents, already struggling to make a living in the impoverished rural locations to which they had been banished (for example, Steilloopbrug in the case of Vaalwater residents), found it difficult to afford school fees too. In addition to the problems raised above, the government sought to control the curriculum according to which black pupils were taught in such a way as to deliver graduates suited to the roles required of them within the white economy, or in a black rural environment (Verwoerd's "hewers of wood and drawers of water").

Vaalwater's first black schools

Several state schools for black children were established in and around Vaalwater under the new legislation. One of the first was the Vaalwater Bantu public school, which was located almost next door to the original Laerskool Vaalwater on the main road into the village from Nylstroom. Its first headmaster was Godfrey Moima and an early pupil there (1954) was Sam Mafora, who went on to become a well-loved teacher in the area and later an inspector before retiring in 2012.[53] Amongst other "black" primary schools that were established in the area in the 1950s were those on Gemsbokfontein on the Vrymansrus Road, where the first principal was Mr Makwela, and at Diephiring on the farm Olievenhoutfontein about 20 km out of Vaalwater on the Beauty Road, with Mr Mmamabolo as its first head.[54] In the 1980s more primary schools were established by the state on private farms; the owner provided the building while the state provided the teacher(s).

Asian schooling in the Waterberg

Given the small Asiatic population of the Waterberg and the prevailing attitude by successive governments towards the education of children other than whites, provision for the schooling of Indian children was not made until the establishment of primary schools by the Nationalist government from the 1950s onwards. For example, the Nylstroom Primary School was built in the new Indian suburb of Asalaam to serve the children of that Group Areas creation.

Indian residents of the Waterberg had no option but to send their children to distant schools for secondary education, to the extent that these existed at all. In 1910 the Transvaal as a whole, for example, had an Indian population of about 11 000,[55] there were 12 schools in the province serving "coloureds" (which included Indians) in 1909, with a total enrolment of just 1664. The first government primary school specifically for Indians opened in Fordsburg, Johannesburg, in 1913, followed three years later by two in Pretoria. In the 1930s the latter were combined into the Pretoria Indian School and classes extended to Standard 8 (Grade 10).[56] The first Indian matriculants from these schools were in 1944 (in Johannesburg) and 1948 (in Pretoria). Later of course, following the creation of the Indian suburbs of Lenasia south of Johannesburg and Laudium west of Pretoria in terms of the Group Areas Act, new schools were built and the original ones closed.

Notwithstanding its generally negative attitude towards the basic education of black people, the National Party government did at least establish a tertiary education facility for blacks in the northern part of the country (although this was a corollary to its decision to terminate or severely restrict the attendance of blacks at "white" universities). On 1 August 1959, the University College of the North at Turfloop was founded on the farm of that name, some 30 km east of Pietersburg. Initially, Turfloop was a college of UNISA. In September 1970 it became the independent black University of the North. In 2005 it merged with the local campus of Medunsa (Medical University of SA) to become the University of Limpopo.[57]

The quality of black education

The disparity between the amounts spent on education per black child compared with a white child was huge: about R14 per black child per annum in 1968, for example, versus R330 per white child.[58] On the other hand, the state had greatly increased access to *some* schooling for black children: Whereas in 1954, a total of 0,94 million black children attended some 5700 schools nationwide staffed by 21 500 teachers, by 1967 these numbers had increased to over two million pupils, 9000 schools and 35 000 teachers.[59] The problems lay with the quality of the education offered, the ideological rationale for insistence on mother-tongue education throughout and the low numbers of children proceeding to high school. For example, of a cohort of black children entering junior primary school in 1955, less than 1% reached the matriculation class 12 years later.[60]

In the aftermath of the Soweto riots in June 1976, which were precipitated by Dr Treurnicht's insistence that half of instruction be undertaken in Afrikaans, a commission was established to investigate all aspects of black education. This De Lange Commission issued its report in 1981. It made numerous recommendations, the most significant of which were that there should be equal opportunities for education, including equal educational standards for every inhabitant of the country, irrespective of race, creed or gender; and a single Ministry of Education responsible for overall educational policy.[61]

Despite many continuing shortcomings in the government's attitudes toward black education, there were some remarkable advances. For example, the growth in pupil numbers had been enormous over the previous decade: a fivefold increase in the number of black pupils completing Standard 10 (Grade 12, Matriculation). More black pupils completed Grade 12 in the period 1980–1983 than the total number in the *whole previous history of black education* in South Africa. A major impediment to the quality of education was the poor training of teachers. Over half the total (black) teaching force of 120 000 in 1983 was found to be under the age of 30, some 17% were professionally unqualified and untrained (compared with 2% among white teachers) and less than a quarter possessed even a Senior Certificate themselves (vs 97% of white teachers). There were only 1651 black graduate teachers in 1983. These deficiencies meant that by 1984 there were nationally "over 5 million adults who either did not go to school at all or [who had] 'dropped out' before becoming functionally literate".[62]

The comments above refer to black education nationally. However, it must be noted that at this time, two thirds of black pupils came from rural environments such as the Waterberg, where the numbers (both results and teacher standards) would have been far worse than the national averages. In 1983 half a million black children were pupils at so-called "farm schools", which would have been of highly variable quality.

Despite the hesitant reforms introduced in black education by the National Party government, including a substantial increase in expenditure and an alignment of curricula and examination standards with white education policy, resistance among pupils continued to grow in both urban and rural areas. By 1986 over 70 schools had been closed or were dysfunctional because of absenteeism, much of it due to political agitation.[63]

E.A. Davidson Primary School, 24 Rivers

Occasionally, altruistic members of the white community recognised the inadequacy (or absence) of educational facilities for the children of their employees and took steps to address these shortcomings. One such individual was E.A. (Ted) Davidson of 24 Rivers, who established a "farm" school for children of his black employees in 1934.[64] Former inspector and teacher Sam Mafora taught at the 24 Rivers school from 1973 until 1988. The school, subsequently relocated, upgraded and eventually transferred to the control of the state (although it is still on private property and a descendant of the family is a member of the governing body), is still in existence and is regarded as one of the best feeder schools in the district for pupils advancing to secondary school in Vaalwater. As a tribute to its late founder, the school was formally renamed the E.A. Davidson Primary School.

Another altruistic local resident has been Dr Peter Farrant, a fourth-generation resident of Vaalwater where, at the time of writing, he practises as a paediatrician and also runs, with his son, the family farm Hartbeestpoort outside the town. In 1985 Farrant secured the assistance of the local farmers' association in lobbying the government to both upgrade the small Matlasedi state primary school for black children in Vaalwater and to build a secondary school. Apart from anything else, farmers realised that the provision of education for their employees' children would assist in stabilising the local labour market.[65]

Meetsetshehla Secondary School

A year later, Farrant reported to the association that although approval for a high school had been obtained from both the Peri-Urban Board and the Department of Education & Training, land in the town was a problem and so he had decided to build the school on his own property, close to the boundary it shared with Vaalwater. It seemed that the Department favoured some kind of technical school for Standards 8–10 (Grades 10–12), on the assumption that farm schools would provide facilities for the first two years of secondary education.[66] By January 1988 Matlasedi (secondary) school had been established on Hartbeestpoort and the first three years of secondary education were being offered there, despite many difficulties with staffing and state bureaucracy. The following year, Farrant was able to report that the second phase of building had commenced at the school, now to be called Meetsetshehla ("beige water") to differentiate the secondary phase of education from the primary phase, which would continue at Matlasedi.

Meetsetshehla almost immediately fell victim to the protests that had afflicted other "black" schools (although Meetsetshehla was nonracial) and the decision was taken to close the school – which was after all on private property and privately owned, albeit staffed by the Department. In January 1990, following appeals from parents, the school reopened with Deon Buys as its headmaster; several white children were among the 400 pupils. Over the next few years, the school grew and began to benefit from donations from private benefactors in addition to the Farrant family via the Northern Education Trust (NET) created specifically for this purpose. In 1994 political unrest at local government level forced the school to close briefly for a second time.[67]

Since that time, Meetsetshehla has continued to grow and prosper and has served as a fine example of a private-public partnership, in terms of which the State provides staff and some financial subvention, while the private sector, in the form of the NET, has provided the comprehensive suite of facilities, equipment and additional staff. By 2016 the school accommodated over 800 learners (all as day scholars) and enjoyed a good reputation for the quality of its academic and vocational education. Its matriculation pass rate was well above the provincial average and several of its graduates went on to complete university degrees in disciplines like medicine, actuarial science and engineering. This quality of education would have been impossible were it not for the combination of private donations and school fees. Some members of the local community, fuelled by political agents, were opposed to fees payment and their militancy briefly threatened the future of the school. Fortunately, wise heads among the school's governing body were able to persuade parents that the sustainability of the school's standards was dependent on the nominal school fees being paid.

Mohlakamotala Secondary School

Another bright example of private-state cooperation in the provision of secondary education in the Waterberg is the Mohlakamotala ("green vlei") High School, some 25 km from Vaalwater on the road to Modimolle. Mohlakamotala was the brainchild of a local landowner, Brian Merz, who decided in 1992 to build a school next to the R33 on his farm Boekenhoutskloof in the hope that it would attract business to his general dealership there. The school opened in January 1993 for learners in Grades 8 and 9, in eight classrooms with four teachers, the latter paid for by the state. The first headmaster was Harry Malan, who had previously taught at Settlers. The school's first matriculants graduated in 1996 when an 85% pass rate was achieved.

Merz sold his property two years later and to prevent the closure of the new school, a trust involving members of Malan's family purchased the farm and school and continued its operation. It soon became apparent that boarding facilities were essential for the school's survival and these were quickly built and managed by the trust. Today, virtually all of the school's approximately 300 learners stay in the boarding hostels (named Brian and Val Merz after the founding couple).

Harry Malan retired in 2011, but in 2017 he could still be found on the campus, a dedicated educator continuing to offer his teaching skills and experience without remuneration. His wife and some family members remain on the staff.[68] For several years, the school was able to boast a 100% matriculation pass rate. Mohlakamotala was designated by the Department as one of a few technical schools in the province (another is the highly regarded Hans Strijdom school ["Hansies"] in Mookgophong). Consequently, it offers a remarkably wide selection of subjects to matrics: Sepedi (home language), English and Afrikaans as additional languages, mathematics and maths literacy, physical and life sciences, geography, business studies, agricultural science, technical maths and science, hospitality and tourism. A total of 14 academic staff are employed.

State education in the Waterberg since 1994

The election of a new democratic government brought with it a wide range of policies, ideologies and aspirations regarding education, and in particular programmes designed to uplift the quality, conditions, availability and standards of education available to black children to those levels enjoyed hitherto only by white pupils.

In the Waterberg, the results of these commendable ideas have generally been disappointing in terms of pupil and teacher attendance, the quality of teacher skills, adherence to curricula, administration, the provision of facilities and infrastructure, discipline and results. Many smaller schools have been closed and pupils and staff consolidated at a few larger venues in an effort to optimise resources. This policy has largely proved beneficial, but at the cost of community involvement and the expense involved in transporting children to the new centres. Government spending on education is high by comparison with other countries, but has not been rewarded with improved results. Teacher absenteeism remains an issue and the forced abandonment of an inspectorate owing to pressure from the dominant teachers' union has had an adverse effect on teaching quality. According to statistics from the Department of Basic Education, only 45% of pupils (of all races) nationally who commenced school in 2003 completed Grade 12 in 2015 (12 years later) – and of those, only 71% achieved even a basic matriculation pass (at least a 40% mark in three subjects and 30% in another three). A mere 4% of the 2003 cohort obtained university entrance passes in 2015.[69] If high-performing urban schools (such as the former Model-C institutions) are excluded from these figures, the numbers for rural areas like the Waterberg will be even more dismal.

The inevitable result has been the creation of a large population of schoolleavers, with or without a School Leaving Certificate, whose poor educational records can never enable them to achieve their aspirations or expectations in the job market. For example, a survey conducted for the Waterberg Welfare Society in 2010 found that almost three quarters of Vaalwater's black school-leavers (including matriculants) in 2008 were still sitting jobless at home a year later; 15% were serving as unpaid volunteers gaining work experience in the private sector; only 9% had entered some form of post-school training; and a mere 2% were employed. It is probably true to state that many, if not most, rural black school-leavers are effectively as ill-equipped as were their parents and grandparents to secure sustainable employment with decent wages in an economy that is becoming increasingly sophisticated and skills-dependent.

CHAPTER EIGHTEEN
Rocks, dynamite and dreams

In the village of Vaalwater, that quintessential rural rumour mill at the heart of the Waterberg plateau, stories keep circulating about mineral discoveries and imminent mining operations in the Waterberg. One problem is that the word "Waterberg" means different things to different people. The Waterberg District Municipality, for example, includes in its area not only the whole of the plateau, but also the important mining towns of Lephalale (coal), Mokopane (platinum), Thabazimbi (iron ore), the old tin mines of Rooiberg (near Bela-Bela), Union (Mookgophong) and Zaaiplaats (Mokopane) and the defunct Buffalo Fluorspar mine outside Mookgophong. And since these places effectively surround the plateau, it is understandable that some people assume the plateau, too, must be richly endowed in valuable minerals just waiting to be discovered and extracted. This expectation is fuelled by the unfortunate names given to some mineral deposits: the Waterberg coalfield and the Waterberg platinum deposits, for example, when in fact the Waterberg mountains are barely visible on the horizon from such places.

Other people, having no idea about geology, little interest in minerals and no intention of actually digging a hole in the ground, merely exploit the gullibility of the public to invest in one of their myriad "get-rich-quick" scams. Then, having secured considerable funds for a proposed exploration programme, such operators disappear into cyberspace, their "office" having never amounted to more than a website, an e-mail address and a pay-as-you-go mobile phone number. During 2006–2012 in the midst of an international commodities boom, over 30 applications were made for prospecting licences in respect of one or other mineral across the Waterberg. Most of them had no techno-economic merit whatsoever.

It is hoped that the following simplified description of the geology of the region and the history of the mines which were developed around the plateau disabuses landowners, tourists, bureaucrats and potential investors alike of the notion that within the hills and valleys of the Waterberg plateau there lurks yet an undiscovered Eldorado.

The Waterberg sequence

The Waterberg plateau consists of a mass of sandstone rocks that define an area bounded more or less by the following points moving in a clockwise direction: Marken – Masebe – Kloof Pass – Hanglip – Bokpoort Pass – Modimolle – Rankin's Pass – Bakker's Pass – Matlabas – D'Nyala (on the new Lephalale road) – and back to Marken. This area includes the whole of the currently defined Waterberg Biosphere Reserve; all of Welgevonden, Lapalala and the Moepel Farms; most of Marakele National Park; the entire catchments of the Lephalala, Melk, Dwars, Matlabas, Blocklands and Mokolo rivers; Vaalwater, Alma, 24 Rivers, Bulge River, Mokolo Dam, Melkrivier, Tarentaalstraat, etc. In other words, the plateau can be said to incorporate all terrain above about 1100 metres amsl in the region.

The Waterberg sandstone rocks were laid down by a river system that drained from a mountainous region to the northeast, more or less where Tzaneen is today, 2000–1600 million years ago. In the course of their long journey, the sediments carried by these rivers became clean, well sorted and almost entirely winnowed or leached of any useful minerals that they might have contained when they started their journey, apart from minor amounts of iron and ilmenite (an iron-titanium oxide). The result is that a huge pile of sandy material – with a few pebble bands but very few clay horizons – accumulated over a long period of time in a slowly subsiding elongated, fault-bounded basin that extended southwestwards towards the present-day Kalahari.[1,2] Only at the base of this sequence and again possibly at the very top are there occasional sediments that might be attributable to a shallow marine or lacustrine origin. Everything in between was carried here by rivers.

The directions in which the streams flowed carrying the vast volume of sediments that accumulated over many millions of years to form the massif of today can readily be identified from traces left in the rocks of small- and large-scale wave structures: from ripple marks to trough cross-bedding. Several thousand readings of the orientation of these traces were taken across the Waterberg plateau. These "palaeocurrent" traces indicated that throughout the long depositional history of the sandstone sequence, the rivers transporting the sands flowed predominantly towards the south and west, from a periodically rising highland far to the northeast.[3] This highland may have been located close to where the northern part of the Kruger Park is today and probably also gave rise to the slightly older, but very similar sandstone formations of the Soutpansberg.

CHAPTER 18: ROCKS, DYNAMITE AND DREAMS

615

FIGURE 18.1: Trough cross-beds (TOP) and ripples (BOTTOM) in the Mogalakwena Formation sandstones.

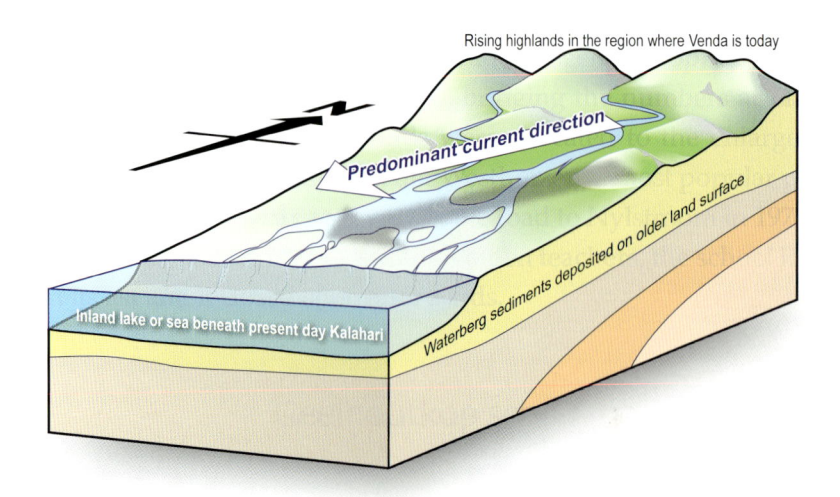

FIGURE 18.2: Schematic depositional model of Waterberg sediments.

The oldest "red beds"

Throughout the thick sequence of fluvial (river-derived) sediments that make up the Waterberg plateau, but especially towards its base, the sandstones are characterised by a reddish-purple coloration. This is the most significant single feature of the sequence. The colour is the product of oxidation (rusting) of iron grains contained within the sandstones, rusting having occurred before the grains were transformed into rock. For iron to rust, two conditions must be satisfied: moisture and free oxygen must be present. The Waterberg sandstones, in common with their very few peers (equivalents in age) elsewhere in the world represent the oldest rocks on our planet that provide evidence of rusting, and therefore, of the presence of free oxygen in the atmosphere. Until there was atmospheric oxygen, no oxygen-dependent life would have been possible.

The Waterberg sediments were formed at a time when the only life on earth consisted of single-celled, carbon-monoxide/dioxide respiring organisms: There were no plants, and no animals – and so there were no organisms either to form soil or to suppress the erosion of weathered rock by rivers. There can also not be any fossils. It can be expected that the rate at which material was carried away by rivers would have been considerably greater than is typically the case today. Modern parallels are therefore difficult to find, but one could be the southward seasonal flow of snowmelt from the Moroccan Atlas Mountains onto the arid, almost vegetation-free plains of the western Sahara desert.

FIGURE 18.3: Red coloration in cross-bedded sandstones near Marken.

Isolated mineral occurrences

Waterberg sediments have not been found to contain any concentrations of minerals approaching economic levels. Nevertheless, there are a few occurrences of minerals on the Waterberg plateau which have been explored or even mined briefly. Along the southern escarpment, between Trichardt's Pass and Bakker's Pass, several ancient pits and trenches have been described, probably associated mainly with the extraction of iron from the Waterberg sediments for small-scale Iron Age smelting operations (of which there are many scattered across the plateau). After the discovery of gold on the Witwatersrand in 1886, prospectors scoured the country for conglomeratic sandstones similar to those hosting the Witwatersrand mineralisation. The conglomerates near the base of the Waterberg sequence seemed to fit the description and so were subjected to intense investigation.

In his book *Lost Trails of the Transvaal*, T.V. Bulpin recounted a story he'd heard about two prospectors in the 1880s, Cohen and Widder, who, having been shown a lump of gold-bearing rock said to be from the Rankin's Pass area, set out to find the source. Allegedly, they discovered the lode but were then killed by the local farmer, who visited their camp with a celebratory, but poisoned bottle of brandy. Unfortunately for the farmer, he was never able to locate the discovery site – nor were the numerous subsequent prospectors who arrived to pursue the rumoured strike. Bulpin added that in 1931 a prospector called David McKerral is said to have found an uneconomic occurrence of gold in the area.[4] Official geological reports of the region, however, make no mention of any of these personalities or their discoveries, which are therefore presumed to be among the myriad of typical prospectors' tales.[5]

Nooitgedacht lead mine

The only mineral deposit of any consequence discovered on the Waterberg plateau was the Nooitgedacht lead mine in the extreme northwest corner of the farm Nooitgedacht, about 16 km north of Vaalwater.[6] In 1928 two 15–20-metres-deep shafts were sunk into the rock hosting the mineralisation, but exploitation of the lead/zinc/copper-containing ore only commenced when a Johannesburg company, Consolidated Lead & Copper Company Ltd, took over the property in 1948. The following year it was acquired by Monarch Cinnabar (Pty) Ltd, which developed the ore body and operated a mine there for two years. No production figures could be ascertained. Later, the rights were acquired by the Afrikander Proprietary Gold Mines and, in 1966, by the Transvaal Consolidated Land & Exploration Company Ltd (TCL) a long-standing owner of properties in the Waterberg. Eventually, the rights passed to the international mining company BHP Billiton, which, having assessed that the deposit had no further potential, was saddled with the responsibility for closing the shafts and rehabilitating the surface in about 2010. To its credit, virtually no trace of the old mine can be seen today.

The mineralisation at Nooitgedacht was contained in a quartz vein, which in turn is associated with one of the many diabase dykes and sills that have intruded into the Waterberg sandstones during the 1,5 billion (1500 million) years since they were deposited.[7] The dyke associated with the Nooitgedacht deposit can be seen very clearly on Google Earth imagery: Although only about 20 metres wide, it can be traced for almost 2 km in an east-southeasterly direction. Occurrences of localised copper mineralisation related to diabase intrusives can also be found for example at Melora Hill on Lapalala and near Dorset, none of them having any economic potential. Some of these dykes and sills also have had the effect of remobilising and concentrating iron contained in the sandstones they intruded, leading to localised accumulations of small pebbles of magnetite. These were collected and smelted at numerous sites on the plateau by Iron Age communities. Other minerals known to occur on the plateau, albeit in very small quantities, include surface, or "placer" accumulations of heavy minerals like ilmenite and zircon, which have been eroded out of the sandstones (for example on Boschdraai); and alluvial tin and thorium, which occur towards Alma and are remnants of source rocks that used to exist further south.

Diamonds

Another mineral that *could* occur on the plateau is diamond – although none has been found yet, as far as is known. Diamonds occur in kimberlites, relatively young rocks which emanate from deep within the Earth's mantle and which, very occasionally carrying diamonds, intruded volcanically through many different older formations. The location of individual kimberlite pipes or fissures was influenced by a combination of deep-seated structural features of the earth's crust. Among these features are the pronounced regional fractures that crisscross the Waterberg plateau, easily seen on aerial photographs and satellite imagery and which are also important controls for underground water. Examples of economically mineralised kimberlites in the region include the Venetia mine near Alldays; a cluster of weakly mineralised pipes at Marnitz, north of Marken; and the Klipspringer-Marsfontein field north of Zebediela about 35 km east of Mokopane. Marsfontein, in its short life (1998–2001), was the richest diamond pipe ever mined anywhere on Earth! Several companies have prospected for kimberlites on the Waterberg plateau over the last 30 years, but there is no evidence that any were found – and anyway, only about one in 200 kimberlites contains any diamonds at all.

Economic minerals surrounding the Waterberg plateau

It seems, then, that the Waterberg plateau itself doesn't have any mineral wealth worth exploiting. But if this is so, how can it be that the plateau is encircled by mines? The answer is that all these mines extract minerals occurring in rock formations that are either older or much younger than the Waterberg sandstones and which have very different characteristics. It is appropriate to take a brief look at these deposits and the history of their development, as this influenced – and continues to affect – the communities of the Waterberg plateau and its surroundings.

Iron ore at Thabazimbi

The banded ironstones and limestones of the Thabazimbi area were deposited in a shallow marine environment (a sea) about 2400 million years ago, in a series of linked basins that extended all the way from Sabie in the east to Prieska in the west. All the limestones which occur in the northern part of South Africa – at Sabie, Mokopane, in the Magaliesberg, around Kuruman and even at Potchefstroom – were laid down as chemical sediments in this huge marine basin, immediately beneath the iron (and manganese) formations which gave rise to the mineable deposits at Thabazimbi, Sishen and elsewhere. These were later overlain and buried by shales and sandstones that were deposited in the basin as it filled up.

The discovery of the iron ore deposits around Thabazimbi is attributed to a prospector, J.H. Williams, who apparently came across them in about 1903 while camping near the *kloof* carved through a line of hills by the Crocodile River, some 130 km north of Rustenburg. Williams pegged claims on the farm Kwaggashoek in 1916. The hill range known as the Witfonteinrant or Vlieëpoortberge, and the valley through it, the Vlieëpoort, were so named because of the prevalence of tsetse fly in that area in the 1840s when several travellers had followed the river down from the highlands to the south (the Witwatersrand) on their way to the Limpopo River (see Chapter 12). Considering how numerous the travellers' accounts are of their experiences along this route, and especially their encounters with tsetse fly in the kloof, it is surprising that none of them mentioned the presence of iron ore, or even the distinctly red coloration of the soils. Perhaps they were too preoccupied in chasing down game and swatting flies to notice the riches at their feet!

The deposits were evaluated by the famous geologist Percy Wagner during and after World War I. In 1921 he published his findings about the "Crocodile River" iron deposits.[8] The discovery immediately elicited the keen interest of Cornelius Frederik Delfos, a Dutch-born engineer and entrepreneur who had immigrated to the Transvaal in 1890 and who, in 1909, had launched the Pretoria Coal, Power and Iron Ore Syndicate. He mined ore quarried from the Magaliesberg hills west of Pretoria and generated power from Witbank coal, but the business was not successful, because the ores, although of high grade, proved difficult to process. Undeterred, and convinced of the merits of an indigenous steel producer, Delfos went on to establish the South African Iron & Steel Corporation in 1919. The discovery of the Crocodile River deposit provided exactly the feedstock he needed. He purchased the rights to the deposits from Williams on the farms Kwaggashoek, Wachteenbietjiesdraai and Buffelshoek, and set about raising the capital needed to establish the complex mining, infrastructure and steel-making facility he envisaged.[9]

It proved impossible to excite sufficient interest among foreign financiers to contribute the funds needed to establish Delfos's dream project. Even the South African government was a reluctant investor – at least, until the election of the National Party under Hertzog in 1924. Then, state participation in (white) job creation, support for rural farming communities and the rise of Afrikaner nationalism all contributed to a decision by the new government to proceed with the development of a strategic domestic steel industry. Eventually, on 30 March 1928, the Iron & Steel Industry Act was pushed through Parliament, despite vigorous objection from Smuts's South African Party, which had serious doubts about the viability of an export-oriented steel industry (it was proved correct) and was of the view that the state should not be involved in industry. These objections were overruled and the first board meeting of the new state-owned enterprise, the SA Iron & Steel Corporation (Iscor, or Yskor) was held in August 1928. The chairman was Dr Hendrik van der Bijl, a 40-year-old talented engineer, scientist and inventor, who, after graduating with distinction from Stellenbosch, had gone to work in Germany and the USA, before being persuaded by Smuts in 1923 to return to South Africa to take up the chairmanship of the Electricity Supply Commission.[10]

Construction of a steel plant at Pretoria West commenced in May 1930. The extension of the railway line north from Rustenburg reached Boshoek in 1925, Northam in 1929, and Thabazimbi in February 1934,[11] by which time mine infrastructure and the beginnings of a township there had been established, thanks in part to some enterprising wagon transport operators who had hauled materials the 45 km from Northam. The crossing of the Crocodile River was a particular challenge until the completion of the Helpmekaar Drift, so named on account of an agreement reached in terms of which the mine would pay for the building of the drift if trek oxen could be made available for mine use on completion of each journey.

Mine development had begun in March 1931 with the excavation of the Mollie Adit (named after the wife of the first manager). Production commenced as soon as the rail link had been completed and the first ore from the mine arrived at Pretoria West in December 1934. The first millionth ton of ore was mined in June 1937. The name chosen for the mine was a combination of the Sechwana word for hill (*thaba*) and a corruption of the Nguni word for iron (*ntsimbi*). But the name adopted, Thabazimbi, has a meaning closer to "the mountains that are evil" than to "hill of iron".[12]

During World War II demand for steel and therefore for Thabazimbi iron ore (some to be exported) grew dramatically. Production increased to half a million tons a year. The end of the war brought greatly increased demand for domestic steel production, because foreign steel plants had either been destroyed or were committed to providing steel to rebuild domestic industries. Iscor took the decision to build a new steel plant on the Vaal River to be called Vanderbijlpark, in memory of Iscor's founding chairman. Thabazimbi's output exceeded one million tons per year in 1947. In order to supply the new steel mill, opencast operations, revised

FIGURE 18.4: The "hill of iron" or Sovereign Hill at Thabazimbi in July 1932 with first excavations and Mollie Adit (top, right). Image kindly supplied from mine archives by Heilet Hattingh of Thabazimbi Iron Ore Mine Safety. This image originally appeared in Barnard (2007), p. 45.

underground mining methods and other improvements were adopted. By 1953 cumulative despatches of iron ore from the mine had exceeded 14 million tons, but even the enhanced production rate was insufficient to supply the demand of the new steel plant at Vanderbijlpark. A new mine at Sishen in the Northern Cape was developed by Iscor to meet this need.

In 1963 Iscor obtained the mineral rights to the farm Donkerpoort, lying to the west of the original workings and along strike from them. Opencast mining of the Donkerpoort reserve was approved in 1969. This added both life and production capacity to the mine at a time when the original underground workings were reaching the end of their life. By 1972 Thabazimbi had produced 58 million tons of iron ore for the steel works, and by 1994 annual production reached 2,4 million tons.

In July 1980 the railway line was extended 112 km north to Ellisras (now Lephalale) in order to enable the new coal mine there (Grootegeluk, then also owned by Iscor, now by Exxaro, the successor to Kumba Resources) to rail semi-soft coking coal to Iscor's steel plants.

In November 1989 Iscor became the first state-owned company to be privatised (as Smuts had argued 60 years earlier should have been the case from the outset) and it listed successfully on the Johannesburg Stock Exchange. Eleven years later, in 2001, Iscor's mining business (including Thabazimbi) would split off from the steel business to form a new company, Kumba Resources, creating a vast increase in the combined market value of the old Iscor assets. A condition of this split, imposed by government, was that Kumba was obliged to supply the Iscor steel business (which soon after the split was acquired by an Indian steelmaker, Arcelor Mittal) with a fixed annual tonnage of iron ore on a cost-plus basis. This constraint, while to Kumba's detriment, actually worked in favour of Thabazimbi, whose operational costs were such that it was probably never viable on a stand-alone basis. But because it could build its production costs into the price of the ore supplied to Arcelor Mittal, it was able to extend its existense for at least a decade. The mine finally announced its closure during 2015, by which time it had despatched about 125 million tons of iron ore to the steel plants (including a small tonnage for export during the war) in its 80-year life.

About 25 km to the east of Thabazimbi, a strike extension of the original iron ore body was discovered years ago on a prominent peak called Meletse Mountain. More recently, an Australian company, Aquila, proposed to extract ore from this deposit for processing at Thabazimbi's (now defunct) facilities. However, very strong resistance from local communities, heritage groups and the environmental community, accompanied by a sharp drop in iron-ore prices, forced a suspension of activities at Meletse for the time being.

It is not impossible that further, unknown deposits of iron ore occur beneath the southern Waterberg where they might be buried by the younger sediments of the Waterberg Group, but they would be too deep even to discover, let alone to mine commercially with current technology.

Platinum mineralisation

About 2050 million years ago, a most remarkable event occurred. An enormous volume of molten material, called magma, from deep within the earth's crust, intruded – probably in several discrete pulses – into the thick (in places only partially consolidated) sediments of the Transvaal Supergroup, as the package described above is known, and the overlying volcanic rocks of the Rooiberg Group. As it cooled, the magma separated into layers of differing composition to form the Rustenburg Layered Suite of the Bushveld Complex. A later, more granitic phase followed. This is the largest formation of its kind in the world and one of the planet's richest depositories of valuable metals. Most of the global resources of platinum group metals, chrome and vanadium, as well as significant amounts of nickel and magnetite occur here. The precise geometry of the Bushveld Complex is still to be understood, but it occurs in at least three distinct "lobes", or limbs: one in the west straddling Brits, Rustenburg and Northam; a second in the east from Dullstroom to Burgersfort; and a third, extending north and south from Mokopane. All three lobes are still being intensively mined or explored for the minerals mentioned.

The so-called northern limb of the Bushveld Complex emerges from beneath the cover of Karoo rocks near Roedtan in the northern Springbok Flats and strikes northwards through Mokopane for over 120 km before it again disappears, this time diving beneath sediments of the Waterberg Group that form the Blouberg massif to the northeast of the Waterberg plateau. As the original platinum deposits around Rustenburg have been depleted or become more expensive to exploit, so the Mogalakwena (previously Potgietersrust) Platinum Mine just north of Mokopane, commissioned as an open pit in 1993, has risen to prominence as the world's largest single producer of the metal. Near the old Grass Valley chrome mine next to the N1 freeway 20 km south of Mokopane, a new platinum mine has been proposed on the farm Volspruit. However, its location on the flood plain of the Nyl/Mogalakwena River poses serious environmental concerns for water consumers downstream: The town of Mokopane would be most vulnerable to any pollution of groundwater that could result from mining operations. At the time of writing, the developers had not yet received approval to proceed with their project.

North of the operations of the Mogalakwena Platinum mine, numerous parties are either actively exploring the rocks of the Rustenburg Layered Suite – for platinum, vanadium and titaniferous magnetite – or are talking of doing so. The only project to go ahead so far is almost 90 km north of Mokopane, where platinum-rich horizons were discovered a few years ago just as they become obscured by the overlying Waterberg Group sandstones. This mine, misleadingly and inappropriately named Waterberg Platinum – because it is not even in the Waterberg District, let alone exploiting Waterberg rocks – was under development when this was written and was expected to be commissioned shortly.

Of interest here is that the Bushveld Complex has a fourth lobe too. Poorly exposed and evidently not significantly mineralised, but definitely comprised of similar rocks, the so-called Villa Nora lobe of the Bushveld Complex is exposed immediately west of Marken on the road to Lephalale. Its presence certainly introduces the possibility that extensions of the platinum-bearing rocks could occur beneath the younger Waterberg plateau. Platinum and its related minerals, unlike iron ore, are extremely high-value commodities, so they can support expensive, deep underground mining operations. However, the deepest deposits currently being evaluated (not yet mined) for extraction anywhere are around 2000 metres below surface – and any that might lie beneath the Waterberg plateau are almost certainly at least that deep. This not only renders them uneconomic targets in the foreseeable future, but also extremely difficult and expensive to locate, even assuming that they exist.

But the first platinum discovered and worked in the Waterberg region preceded the discovery of the mineralisation in the Bushveld Complex by three years. Early in 1923 a well-known local tin prospector and one time friend of Eugène Marais, Adolf (Dolf) Erasmus (see Chapter 6), discovered platinum on the farm Welgevonden about 15 km northwest of Naboomspruit. He followed the "reef" eastwards onto Rietfontein (an old Marais haunt) and west to Doornhoek, where the Union tin mine operated. The discovery was reported on and described in detail by Dr Percy Wagner of the Geological Survey, and Tudor Trevor, the Inspector of Mines, in a paper published in December 1923. They described Dolf Erasmus as "a prospector of exceptional enthusiasms and great originality".[13]

The platinum mineralisation occurred in quartz breccia (the main lode) that occupied a fault and could be traced continuously for a distance of almost 4 km. On the boundary of Welgevonden and Rietfontein, a branch lode ran south onto the latter farm and was found to contain the richest mineralisation (370 grams/ton platinum and 171 grams/ton palladium).[14] A minor prospecting rush followed and at least six companies, including the Transvaal Platinum Ltd and Doornhoek Platinum Mines Ltd, were floated in short order.[15]

On Welgevonden a shaft reached a depth of 46 metres with an intended depth of 152 metres. Johannesburg newspapers were reporting intensive shafting, trenching and other prospecting activities:

> ... on Doornhoek the company [Doornhoek Platinum] has opened up one of the most impressive platiniferous ore bodies so far found on any of the much impressive platinum fields of the Transvaal. Doornhoek has for a long time past been regarded as one of the most highly mineralised farms in South Africa, and on it, cassiterite [tin] occurrences were opened up some years ago by the Doornhoek tin mines. That this farm is on the line of the platinum-bearing belt has been known for some time past, but it was only about a month

FIGURE 18.5: Adolph Erasmus, who discovered the platinum mineralisation on Rietfontein in February 1923, sits on the extremely rich "elephant" outcrop of the branch lode. Wagner & Tudor (1923): Plate II opposite p. 592. Image obtained by courtesy of the National Library of South Africa, Pretoria.

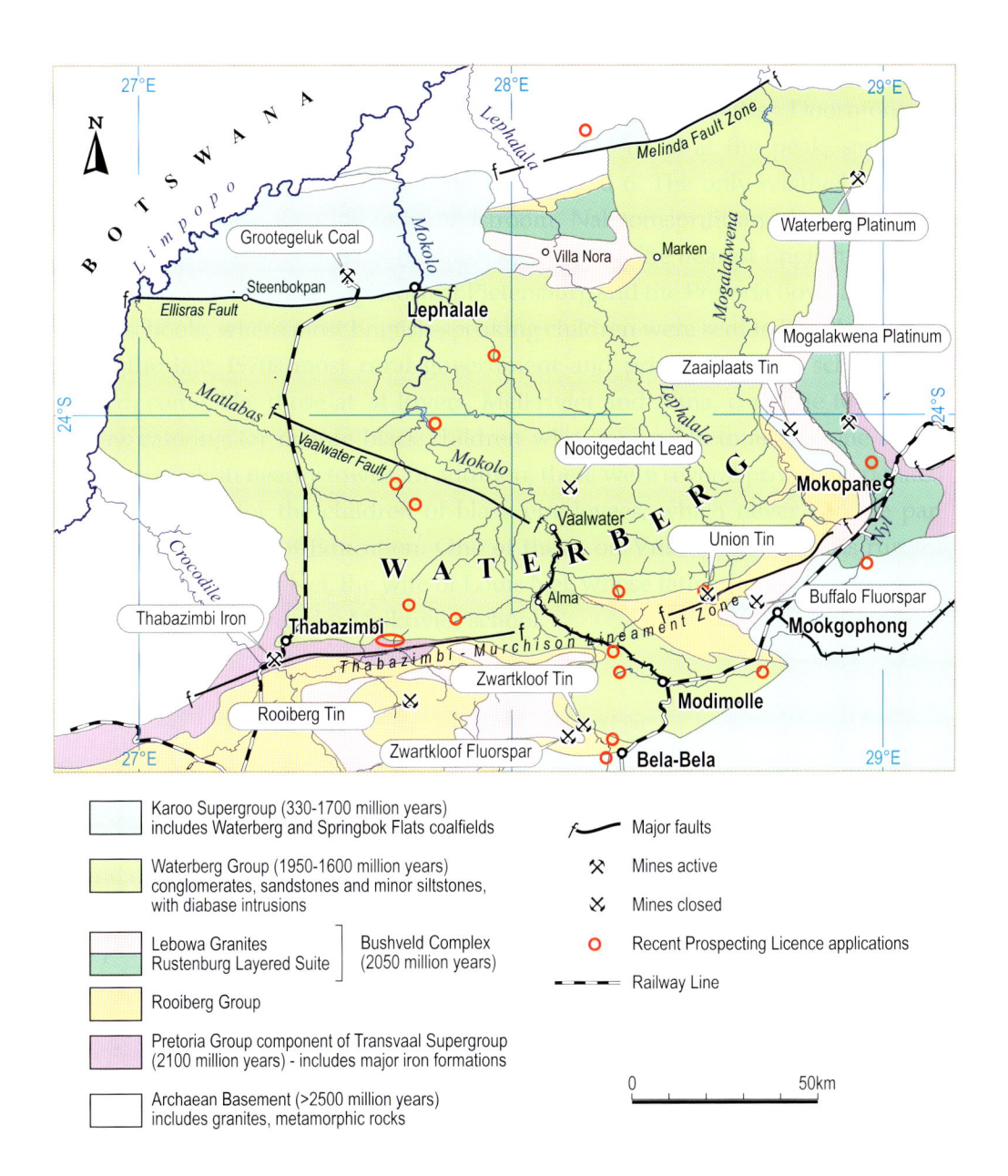

FIGURE 18.6: Geology and mining history of the Waterberg.

ago that a very large lode was located on the south side of the hill by Mr Dolf Erasmus, the well-known discoverer of platinum on the Transvaal Platinum Company's properties [on Welgevonden], which was the first platinum discovery of economic importance in South Africa.[16]

But by August 1926 all operations had been suspended after it was realised that the economic potential of the properties was far lower than the promoters had asserted, that the mineralisation was erratic, and that the process of extracting platinum from the ore was difficult and expensive. Moreover, in June 1926, discovery of the huge plantiniferous deposits of the Merensky Reef in the eastern Transvaal had been announced, thereby causing a slump in the price of the metal. The Naboomspruit deposits were investigated again sporadically in the 1930s and in the past few years there have been further attempts to interest investors in the exploration of the area.

The Bushveld tin and fluorspar deposits

The tin deposits mined historically at Rooiberg, Union and Zaaiplaats, as well as the numerous fluorspar occurrences around Mookgophong and west of Bela-Bela, whilst found in differing hosts, had a similar origin at a similar time. The heat caused by the intrusion of the granitic magma of the final phase of the Bushveld Complex caused hot solutions enriched in elements like tin, molybdenum and fluorine – even platinum, as on Welgevonden – to migrate into fractures in surrounding rocks, most of which were Transvaal Supergroup sediments or Rooiberg Group volcano-sedimentary rocks.

Rooiberg

During 1906 prospectors in the bushveld 60 km west of Warmbaths came across very old mine workings and dumps which they initially assumed to have been linked to ancient iron or copper mining. However, after traces of cassiterite (an oxide of tin and its principal ore mineral) were discovered in the dump material, a search was made which led to the discovery of extensive ancient underground excavations.[17] These were subsequently concluded to have been the work of people of the Late Iron Age (fifteenth to nineteenth century – see Chapter 2); they not only mined and beneficiated ("washed") ore selectively, but also extracted the tin from the host rock by smelting. The ore was accessed by means of narrow vertical and inclined shafts, from which veins were followed using low, narrow drives. Ore was broken out of the rock using iron or steel chisels, often in extremely cramped spaces – which makes it all the more remarkable how efficiently the operations seem to have been undertaken. Broken ore was moved to the surface in baskets where it was ground and beneficiated (possibly by panning before being smelted.[18] An estimated 180 000 tons of ore (corresponding to about 1000 tons of tin metal) were extracted by these early miners.

Remains of ancient furnaces, with slag, have been discovered at several places in the vicinity of Rooiberg and at least seven tin ingots of one form or another have been recovered from various places in the former northern Transvaal region. Analyses of these ingots revealed tin contents exceeding 90%, a three- to six-fold

FIGURE 18.7: Iron Age tin mining at Rooiberg: the pit (LEFT) and tools used (RIGHT). Mason Revil (1962): *Prehistory of the Transvaal*. Wits University Press, Johannesburg. Figures 236 (p. 417) and 239 (p. 420). The sketch of the mining pit was drawn by the late Judy Mason. The figures were reproduced by courtesy of Prof. Revil Mason.

enhancement of the content in the ores from which they were extracted. It cannot be said with certainty that all the ingots found came from Rooiberg, but no other local site of that age has been discovered and only at Rooiberg have moulds used for ingot manufacture been found. There are records of Portuguese traders in the early eighteenth century purchasing tin ingots in Delagoa Bay, and there is widespread evidence of the existence of well-defined trade routes in the region prior to the arrival of the first white travellers.[19] It has been argued that Iron Age tin mining operations continued until the communities and trade agents involved were dispersed as a result of the Matebele-led difaqane[20] in the early nineteenth century. The deposits were then abandoned until their rediscovery a century later.

The tin discoveries at Rooiberg in 1906 were made on the farms Hartbeestfontein and Olievenbosch. These were among 225 farms, 127 of them in the Waterberg, that had been acquired by the Oceana Transvaal Land Company Ltd since 1886 for a shilling an acre. Oceana and other companies saw the Waterberg region as offering cheap land that was little known or explored and which therefore offered the potential for a good return on the capital invested.[21] In order to create a separate vehicle to develop the tin deposits, Oceana formed the Rooiberg Minerals Development Company in 1908 with issued capital of £150 000, and William McCallum as its chairman. Two years later, administration of the company was taken over by the Anglo-French Exploration Company, which had purchased a controlling stake in Oceana. Anglo-French was registered in London but with French participation. William Dalrymple became chairman – a position he would hold for 30 years.

Tin production at Rooiberg commenced in 1907 and continued until 1993 when rising mine costs and a prolonged depression in international tin prices finally forced the mine's closure. Rooiberg was the last active tin producer in South Africa.[22] The mineralisation at Rooiberg was in the form of cassiterite, which had crystallised in sediments of the Pretoria Group of the Transvaal Supergroup, having been injected into fractures in the sediments by hot, ore-forming fluids associated with the emplacement of Bushveld Complex granites. The geometry of the mineralisation was complex, occurring in a variety of fracture types and mine planning was always difficult.

In 1960, after nearly 50 years of the "boom & bust" cycles so common in the mining industry and which repeatedly threaten the survival of small independent mines, Rooiberg was acquired by a major mining house, Gold Fields of South Africa, and was able to benefit from the comprehensive technological and management skills and access to capital brought by this new owner. By the time of Rooiberg's closure, it had developed three separate mines in the area, of which the last, on the farm Leeuwpoort about 14 km south of Rooiberg village, was the most productive. A private railway line brought ore from Leeuwpoort to the plant at Rooiberg. Eskom power arrived on the mine in 1955, helping to reduce operating costs. In 1979 Gold Fields invested in a smelter at Rooiberg to convert its tin concentrates as well as those from its sister mine, Union Tin near Naboomspruit, to metal. Until then, both mines had sent their concentrates for smelting to a third mine, Zaaiplaats, near Potgietersrus, which was owned by another company. By 1982, ten years before closure, the complex had produced a total of 57 000 tons of tin in 75 years of production, earning it an overall pre-tax profit of R107 million – not a lot, over such a long life.[23] By the time it closed in 1993, the Rooiberg complex had produced a total of 84 000 tons of tin metal.[24]

Rooiberg was interesting because from the time of its establishment, it had to be a self-supporting township with its own power and water supplies, effluent

FIGURE 18.8: Restored five-stamp mill at Rooiberg. Photographed by Alison Rose in 2015, on behalf of the SA Micromount Society (SAMS). Image supplied by Stephen Gomersall.

disposal system, hospital, housing estates, etc. Until the establishment of Thabazimbi, 45 km away, in about 1930, the nearest town was Warmbaths via a very poor road, which took more than a day to reach by ox-wagon or mule cart. Animals used to be replaced (staged) at a halfway house called Mabula – today an upmarket residential game estate. A police post and a branch of the National Bank opened in 1911. A telephone line was installed in 1913 when the first post office was also built, and a formal school building opened in 1915. Black unskilled mine labour was nearly always in short supply. At the start of the mine's life, some Chinese workers were sent briefly from the Witwatersrand. In 1908 Anglo French, the managers, arranged for the provision of workers from a group of Namibian Herero refugees, who had been allowed to settle on a company-owned farm in the northern part of the Waterberg (see Chapter 20). This experiment too, did not succeed.[25]

After Rooiberg mine closed in 1993, its owners sold it to another mining company, Metorex, which wanted the mineral rights but did not reopen the mine itself. However, so insular and dependent on the company had some of the village's residents become that when the mine finally closed, they had nowhere else to go and chose to stay on in their homes on the property. That this was possible was due to the unexpected arrival in 1992 of members of Koevoet, the former South African Police covert unit in Namibia, who needed accommodation following their disbandment and withdrawal from that country. The rental paid during their two-year stay on the mine property enabled the town to survive the mine closure and departure of Gold Fields. Metorex, the new owner, raised funds through the sale of mine houses to individual local families (some of whom were already living in the town) and other buyers. In 1996 Rooiberg became a town, nominally within the ambit of the Bela-Bela Local Municipality (it was transferred to Thabazimbi in 2002). The development of the Mabula Game Reserve in the area in 1979 had started a shift away from commercial farming to game reserves, and by the mid-1990s, Rooiberg was able to offer sporting facilities – like a golf course – that appealed to international visitors at the nearby lodges. One lodge was even established in the town itself using the 100-year-old building that had been the town's hotel. The local school, the survival of which had been achieved by a very determined headmistress, continued to attract new, mainly black residents from the surrounding farms, a process accelerated by the government's decision to build houses in the town in 2001.[26]

Union tin

Following the "rediscovery" of the tin deposits at Rooiberg, prospectors scoured the region in which sediments of the Waterberg and Karoo sequences were separated by exposures of older rocks like those at Rooiberg. A prospector called Van Tonder and his assistant, Isak, are attributed with the discovery in around 1908 of tin on the farm Doornhoek, about 20 km west of the Naboomspruit rail siding (the town of that name having not yet been founded). The news was passed on to Dolf Erasmus, a well-known prospector of the area – and who, almost 20 years later, would discover platinum on the same property. Erasmus teamed up with Eugène Marais, who had just arrived in the Waterberg (see Chapter 6). The opening lines of *The Road to Waterberg*, Marais's book about his eight-year sojourn in the foothills of the Waterberg, recount his arrival in the region: "An old acquaintance, Dolf Erasmus, met me in Nylstroom with a cart and four horses and for the first time in my life I entered Waterberg, the mystery region of my boyhood."[27]

The two formed the Marais Tin Syndicate to promote the development of the deposit. A German company became involved, systematic prospecting was undertaken, the property was purchased from its owner, Christiaan van Staden, and in April 1908 the South African Tin Mines Company Ltd was floated, with production commencing the following year. The initial operation was primitive and the first manager was C. Fred Thomas. By 1910 a road had been built to the mine from Naboomspruit station and stamp mills, workshops and staff houses erected. A severe shortage of black labour impeded development until rock drills were adopted under a second manager, an American called Ernest Ford. Philip Cohen of Naboomspruit started a private hotel and shop at the mine. An outbreak of east coast fever in 1910 caused the mine to be quarantined until it switched to the use of donkeys to haul products to and from the railway line.[28]

Owing to its poor economics, the mine closed down in February 1914. All equipment was removed and the shaft allowed to flood. Total production in its short life (1908–1914) was 388 tons of pure tin metal (at a price of £163/ton), derived from 900 tons of concentrate and 36 000 tons of ore mined. The total labour force comprised 21 whites and 150 black labourers.

During the 1920s, Dolf Erasmus renewed his prospecting activities in the area, now supported by Max Baumann, manager of Rooiberg at the time. More evidence of tin mineralisation was found, but despite the rise in the price of tin over the period, none of several investors was able to make the property work. The last of these, Anglo-French, made some effort (1933–1935) in developing the mine, but abandoned it. In 1950 the property was introduced to Anglo-Rand Mining Finance Corporation by Jack Thorland, a speculator. Anglo-Rand floated a new company, Union Tin Mines, which immediately benefited from the rise in tin price caused by the Korean War. The mine was dewatered, and fortunately, much of the existing infrastructure and equipment was found to be in reasonable condition. In 1954 the company was restructured and became a subsidiary of Goldfields. By 1979 some 432 tons of tin metal had been recovered from 578 tons of concentrate, extracted from 124 445 tons of mined ore. The price of tin then was R10 000/ton; by 1985, the price had doubled and the mine was still in operation.[29] However, changes in the tin market soon affected Union, as had been the case with Rooiberg, and by the early 1990s the mine had closed. In its long but chequered life, Union had produced a total of close to 10 000 tons of tin metal, about a tenth of the output of its stablemate Rooiberg.[30]

Unlike the Rooiberg deposit, cassiterite mineralisation at Union occurred in shale, a sedimentary unit sandwiched within volcanic rocks of the Rooiberg Group, the youngest phase of the Transvaal sequence, deposited immediately before the injection of the layered rocks of the Bushveld Complex. The mineralisation was

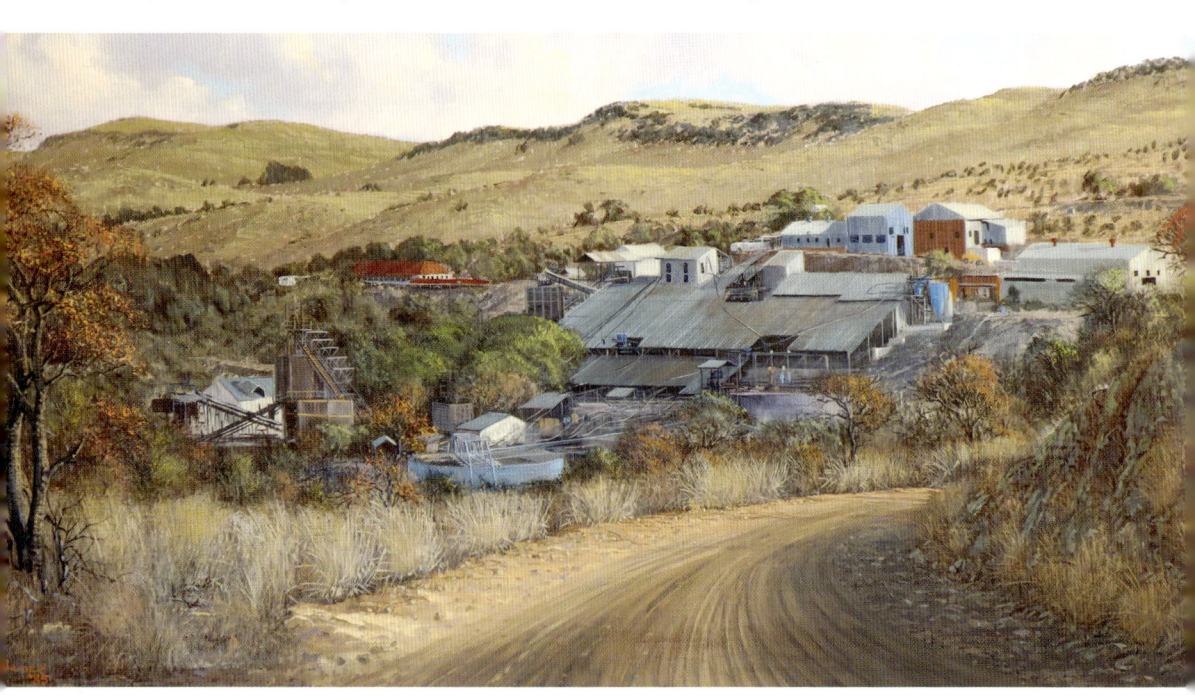

FIGURE 18.9: The reduction plant at Union Tin Mines. The original oil-on-board painting of the plant, 1985, by the late Roy Gardener, is owned by Ian Pringle of Bela-Bela, formerly chief geologist at Union Tin. It is reproduced by permission of both Ian Pringle and Dave Gardener, Roy's son.

confined to "a multiplicity of fractures", only a few of which were ore-bearing, deposited there by fluids mobilised by the related granitic intrusions. The exploitation of this complex mineralisation required considerable mining skill and experience.[31]

Zaaiplaats tin field

In 1906 a prospector called Moore discovered cassiterite on the farms Roodepoort and Groenfontein in the hills about 27 km northwest of Potgietersrust (Mokopane), leading to the establishment of the Groenfontein Tin Mine there. Shortly afterwards, two other prospectors, Munro and Maggs, found the mineral on the adjoining farm Zaaiplaats where a mine of that name was opened. Several other occurrences of tin were subsequently found on neighbouring properties in the district, forming the Zaaiplaats tin field.[32] The small settlement of Tinmyn with its police station, hotel and general store was established on the main road from Potgietersrust to Marken (the R518), although mine personnel lived in housing on mine property.

Many small-working operations in the first two decades after discovery produced ore for treatment at the two larger mines. The Groenfontein mine set a record production of over 1000 tons of tin concentrate in its second year of operation, but closed during the Depression of the early 1930s (when Union Tin was also dormant and Rooiberg went to the brink of closure). In later years, the Zaaiplaats Tin Mining Company produced ore from Groenfontein under tribute from its then owner, the ubiquitous Transvaal Consolidated Lands (TCL). Zaaiplaats also built a smelter to convert tin concentrate to metal and for many years also processed concentrates from Union and Rooiberg, until Goldfields built a smelter of its own at Rooiberg in 1979. Total tin metal production from the field by the time Zaaiplaats itself closed in the early 1990s amounted to about 40 000 tons, half the total from Rooiberg.

Tin mineralisation in the Zaaiplaats field occurs predominantly as lenses, pipes or disseminations at structurally controlled locations within the young granites of the Bushveld Complex. The lenses and pipes would contain high concentrations of the metal, but the dominant ore type at Zaaiplaats was the lower-grade disseminated ore.[33] Exploration for further tin deposits in the Zaaiplaats field continues today as newer models are developed to explain the style and controls of mineralisation.

Buffalo Fluorspar mine

Fluorspar is the name of the ore of the mineral fluorite, which is the principal source of fluorine. It is produced in several different qualities: Metallurgical-grade fluorspar is used as a flux in the iron and steel industry, acid-grade is the feedstock for hydrofluoric acid, an important chemical; ceramic-grade is used in the manufacture of enamels and opaque glasses; and optical-grade fluorspar is extremely clear and is used in specialist optical applications.[34] There are numerous occurrences of fluorspar mineralisation in association with granites of the Bushveld Complex in two areas close to the Waterberg: along an extensive zone to the west and north-west of Bela-Bela and in a small cluster north of Mookgophong. In the first area, two deposits, Zwartkloof and Rheno, were exploited briefly in the 1970s, and in the second the only operation has been that of the Buffalo mine.

In 1903 a British immigrant, John Hotine, settled on the farm Buffelsfontein, just 8 km north of the rail siding at Naboomspruit (the founding of the town lay seven years into the future). The following year his son Jack and his family came out from England to join him. Jack recognised some unusual rock on the surface of the farm, possibly realised it was fluorspar and contacted J.G. Gubbins of Zeerust, at the time the owner of the only significant producer of that mineral in the country. Under an agreement reached between them, they began to develop a mine on Buffelsfontein. However, Hotine struggled to raise sufficient finance to fund the operation and in 1917 he abandoned the mine and moved to Pretoria. In the 1920s the deposit was "rediscovered", but it seems that no mining took place.[35]

During World War II, the farm was owned by a man called Gladwin, who was also a petrol controller with an office in Potgietersrust. He used to visit his farm for hunting at weekends, travelling by train to Naboomspruit to be met there by Ben Gerber, a local garage owner, who would then drive him out to the farm and return him to the station on Sunday afternoons. It happened that on one occasion when Gerber was wandering around the farm, he came across the old fluorspar workings. He sampled the material, found it to be of good quality, applied for a permit with Jacob Smuts, a clerk at Skolsky's shop in Naboomspruit, and began to exploit the loose surface material in a pick-&-shovel operation. The production was sent to a Johannesburg company by rail at the rate of about a truck per week. The surface resource lasted about a year, but the pair lacked the funds to go into a proper mining operation and Gerber was in any case busy running his garage in the town. Whether Gerber, the miner, and Gladwin, the landowner, continued to speak to each other after this venture is not recorded.

A couple of years later, in about 1943, an entrepreneur called Percy du Toit appeared on the scene with the intention of opening up a proper mine. In 1948 the company Frank Martin & Co. (Pty) Ltd commenced production, with Du Toit as the first resident manager (he may also have been a part owner). He and his wife Lulu lived in a primitive zinc hut on the property, while his assistant, Botha, travelled out from Naboomspruit daily. Gradually, additional buildings were established and a small panel van was acquired to bring goods, including dynamite, from town. The vehicle was not licenced for this purpose and it became necessary on occasion for it to follow curiously circuitous routes to and from Naboomspruit to avoid roaming explosives inspectors!

During the 1950s production increased and further processing facilities were installed, making use of the new dam on Welgevonden (Frikkie Geyser Dam), which was commissioned in about 1957. In 1964 General Mining & Finance, a subsidiary of Gencor (later BHP Billiton), acquired control of the mine and conducted further exploration. At the time, production was about 800 tons per month, with a labour force of 10 whites and about 120 blacks. In 1971 a total of 35 new houses were built for employees in Naboomspruit. A concentrator plant was installed in 1973 with a capacity of 150 000 tons per year, making Buffalo Fluorspar the world's largest producer of acid-grade fluorspar. By the 1980s, under the management of Sydney Absolom, production had increased to 240 000 tons per year; almost 500 personnel were employed. The mine worked closely with the Naboomspruit Municipality to ensure that there was adequate water and electricity for both parties. Product from the mine was transported to the rail siding at the rate of 3000 tons per month initially, rising, by 1985, to 35 000 tons.

At Buffalo, purple fluorspar occurs as a stockwork of numerous veins present in large fragments of a metamorphic rock called leptite, which in turn occur as inclusions (xenoliths) in the Bushveld granite. It was concluded that the fluorite

FIGURE 18.10: A flooded pit at the old Buffalo Fluorspar mine.
Image kindly provided by Prof. Bruce Cairncross of University of Johannesburg.

was introduced into the fractures by hot gaseous fluids during the time of granitic intrusion.[36]

Buffalo Fluorspar closed in 1994 due to a depressed market for its product, although substantial reserves of fluorite remained. Today, the dumps at the mine are processed for the production of building and road aggregate.

Grootegeluk and the Waterberg coalfield

In the area between Lephalale and the Limpopo River lies South Africa's richest remaining coalfield from which Eskom hopes enough coal can be extracted to underpin the country's electricity demand for much of this century. Already home to one of the world's biggest collieries (Grootegeluk) and two of the largest power stations – Matimba and Medupi (still under construction in 2018) – the Waterberg coalfield has the potential, within the next decade, to support a *quadrupling* of its current electricity output, as well as possibly the country's first coal liquefaction plant. The westward extension of the field into Botswana has also been subjected to intensive evaluation for coal and methane gas mining.

Coal was first discovered in the Ellisras area in 1920 when the owner of the farm Grootegeluk was drilling for water. A few more holes were put down and the cores analysed. Then, probably because of the remoteness of the area, no further drilling was done for coal until 1941 when the Geological Survey, in conjunction with the Irrigation Department, began a comprehensive evaluation programme that lasted for over a decade and involved mapping the entire area and the drilling of 143 boreholes across the field.[37] Iscor took an interest in the Waterberg coalfield, as it became known, because of the potential for the extraction of a coking quality product (suitable for its blast furnaces) from the uniquely complex coal measures that had been identified by the drilling programme.

The coalfield occurs in Ecca shales of the Karoo Supergroup, laid down about 260 million years ago in a series of vast basins that covered most of subequatorial Africa, as well as parts of India, South America, Madagascar, Antarctica and Australia; all these pieces of the earth being, at that time, part of a single continent we call Gondwana. These youngish sediments (more than a billion years younger than the youngest Waterberg sandstones and almost two billion years younger than the rocks hosting the iron ore at Thabazimbi) contained a high concentration of plant material (which formed the coal); they would have been deposited on top of, or next to, the earlier Waterberg Sequence, much of which had eroded away during the intervening eons. It was only in the early 1970s that Iscor (and, elsewhere on the coalfield, Sasol) turned its mind seriously to the exploitation of the huge Waterberg coalfield lying to the west of Ellisras. This was because the complexity of the numerous, closely packed coal seams (resembling a vertical barcode) necessitated the implementation of a sophisticated beneficiation process to extract the coking coals. In addition, a commercial outlet was needed for the remainder of the coal, which was of lower quality and suited only for power generation. So was born the partnership between Iscor and Eskom in terms of which Iscor's mine – named Grootegeluk for the original discovery site and the farm on which operations commenced – would provide low-cost feedstock to a new, custom-built Eskom power station to be called Matimba (a Tsonga word for power) in close proximity to the mine. Iscor would then be able to rail its highly beneficiated coking product to its steel plants. Hence the extension, in 1980, of the railway line north from Thabazimbi, mentioned earlier.

This massive, multifaceted project, of a size that could only be contemplated by state-owned entities at the time, commenced in the mid-1970s. Grootegeluk produced its first coal in 1980, reaching full production two years later. Construction of Matimba commenced in 1981. The beneficiation plant was – and with recent additions remains – the biggest coal processing plant in the world. The open pit from which coal is removed has become one of the largest on the planet, even though waste material is returned to be stored in its mined-out areas. The low-ash coking product began to be shipped out to the steel plants as soon as the railway line was operational, while feedstock for Matimba was carefully stockpiled until the first generating set of the country's biggest, most advanced, partially air-cooled power station was commissioned in 1988. When fully operational in 1993, almost 12 years after construction commenced, the six generating units at Matimba had a rated capacity of almost 4000 megawatts and consumed some 14,5 million tons per year of coal from Grootegeluk.

Water for the project came from a new dam, the Hans Strijdom (now Mokolo) Dam on the Mokolo River between Lephalale and Vaalwater. A new tar road (the R33) was built from Vaalwater to carry all the requisite plant and equipment. New residential townships (Onverwacht and Marapong) were built, the former initially to accommodate white Eskom employees. Grootegeluk mine's production continued to expand as additional markets, including exports via Richard's Bay, were secured.

In 2007, following an electricity crisis caused by prolonged central governmental indecision and political interference, Eskom was belatedly authorised

Figure 18.11 (TOP): The original box cut into the multilayered coal formation at Grootegeluk, November 1975. (BOTTOM): The open pit (May 1989) with Eskom's Matimba power station partially in operation in the background. Both images of Grootegeluk kindly provided as colour slides by Claris Dreyer, longstanding chief geologist at Grootegeluk and leading authority on the Waterberg coalfield.

to commence construction of Medupi, a station similar in design to Matimba, located on an adjacent property, but with a higher projected output of nearly 4800 megawatts. Medupi refers to a Sepedi term to describe "rain that soaks parched lands".

Coal to supply this new station would come from a doubling in the size of Grootegeluk to a total annual coal output of over 33 million tons per year (29 million tons per year of which is committed to Eskom), which will make it the biggest coaling operation in the world. The mine expansion was completed in 2014, but protracted delays in the power station project, related to labour disputes, management issues and further political meddling (this project has a value well in excess of R100 billion), resulted in the first of its six generating sets being fully commissioned only in mid-2015, several years behind schedule. The project is now not expected to be fully operational before 2020.

In the meanwhile, upgrading of the railway line to the south will allow the mine, as well as several other aspirant miners in the Waterberg coalfield, which still contains enormous untapped resources, to sell product to other domestic,

FIGURE 18.12: The Grootegeluk complex in 2017 with Matimba power station on the right and Medupi at the bottom. (Modified from Google Earth imagery collected on 25 May 2017). Details on image confirmed by Claris Dreyer.

as well as international customers. Considerable potential remains for the development of other large coal projects in the area surrounding Lephalale, the principal constraint being the shortage of processing and cooling water. A 200 km pipeline is planned, for completion by 2020, to bring water from the Crocodile River (from a weir point downstream from Thabazimbi) to augment increased extraction from the Mokolo Dam. This water, in turn, will be derived from treated acid-mine drainage (AMD) water fed into the catchment area of the Crocodile River in the western part of the Witwatersrand.

These developments potentially have several environmental consequences in addition to actual mining operations, including the risk to the riverine ecology of the Crocodile River, pollution from power station smoke, the proliferation of transmission lines and water pipelines, and the expansion/extension of the existing railway line between Grootegeluk and Thabazimbi. Technology exists to manage all of these issues satisfactorily, but at considerable cost.

Economic mineral potential on the Waterberg plateau

Is there potential for coal deposits to occur on the Waterberg plateau? No. If ever there had been Karoo-age sediments on the plateau, lying on top of the Waterberg sandstones, they've long been removed by erosion, for there's no trace of them today. It's even possible that the uppermost layers of the plateau were never covered by Karoo sediments.

But then, why are coal beds preserved around Lephalale, which is over 500 metres *lower* than the plateau? Well, there's a simple explanation for this: Along the northern boundary of the Waterberg escarpment is a very large regional fault system called the Melinda fault zone, which with its extensions, including the Eensaamheid fault, can be traced from eastern Botswana in a northeasterly direction all the way to Crook's Corner at the top of the Kruger Park. The hot springs at Tshipise owe their existence to it. The effect of the fault zone has been to displace all the rocks to the north of it *downwards* relative to the rocks to the south of it – by at least several hundred metres. The result has been that young Karoo rocks, including the coal beds, are preserved to the north of the fault, but not to the south. The line of the fault crosses the main road from Vaalwater to Lephalale (the R33), just about where that road goes over the Mokolo River outside Lephalale.

A similar, parallel group of faults forms the southern boundary of the Waterberg escarpment too, this time with the downthrow to the south – which is why one can find Karoo-age sediments (containing both coal and uranium) from the outskirts of Modimolle, Mookgophong and Bela-Bela all the way south to Pienaarsrivier (near the Mantsole weighbridge on the N1 freeway), but none to the north. This major fault zone, called the Thabazimbi-Murchison Lineament (it runs all the way to Gravelotte in the northeast), is marked by a series of thermal springs which include those at Radikwekwe east-northeast of Thabazimbi, Bela-Bela (Warmbaths), Loubad, Die Oog, and the group of springs to the northwest and north of Mookgophong.

The net effect is that the Waterberg plateau stands like an island in a sea of younger rocks to the north and south and older rocks to its east and west. Its thick pile of barren sandstones *may* be underlain, at great depth (too great to detect currently, let alone mine) by formations that could host platinum, iron and tin mineralisation, but it is not covered with coal-bearing Karoo rocks – and no kimberlites have been reported from it. Most recent prospecting rights applications over properties on the Waterberg plateau for minerals as diverse as iron, manganese, vanadium, andalusite, platinum group metals and chrome owe more to ignorance, greed, deceit or fantasy than to geological facts. One applicant was advised that he stood a greater chance of discovering a whale in the Mokolo Dam than the minerals listed on his prospecting application!

Thermal springs

A discussion of the geology of the Waterberg would be incomplete without at least some reference to the thermal springs that are strung out along its southern margin, related, as mentioned above, to the set of fractures that comprise the Thabazimbi–Murchison Lineament. The best known of these springs is of course the resort at Warmbaths/Bela-Bela, but there are several others lying to the west of Mookgophong, as well as a small group at Loubad, north of Modimolle.[38] To a greater or lesser extent, it is claimed that the waters emanating from all these springs have some therapeutic and/or healing powers, by virtue of the minerals said to be dissolved in them. Consequently, many of the larger springs support resorts, offering both recreational and therapeutic benefits.

A thermal spring is loosely defined as producing water of a higher temperature than the average ambient temperature – in the case of South Africa >25 °C is generally used as the definition of a thermal spring.[39] Thermal springs may be either plutonic (related to recent volcanism) or meteoric (related to natural atmospheric rainfall) in origin. All South African springs fall into the latter category. The temperature of meteoric thermal spring water is proportional to 1) the depth to which meteoric water has percolated downwards into the earth via fractures, faults etc. (temperature increases at a rate of 2–3 °C per 100 metres of depth); and 2) the rate at which the heated water returns to surface. The flow rate of water at a spring is a function of several variables, the most important of which are the rate at which the system is recharged by meteoric water, and the aperture of the point at which heated water issues from the ground. Most springs are located in topographically low places with their recharge zones topographically above them. Most thermal springs are, in addition, intimately associated with local geological structures like faults or major folding of strata.

In the southern Waterberg, some eight to ten thermal springs have been identified, depending on whether outlets are separated or combined: Two, Loubad and Buffelshoek (a small spring southwest of Vaalwater), are defined as tepid (25–37 °C); four (most of the springs west of Mookgophong) as hot (38–50 °C); and two, (Warmbaths and Libertas in the Mookgophong cluster) as scalding (>50 °C). All except Loubad have fluoride levels that render the water unsuitable for drinking; and all are characterised by high sodium carbonate and chlorine contents as well as anomalous levels of titanium and strontium.[40] None of these characteristics seems likely to imbue the waters with any especial beneficial elixir any more than a tub of bath salts – both are just pleasant and relaxing to sit in (although of course the home bather has greater control over the content and temperature of the bath water …).

CHAPTER NINETEEN
On the periphery

Although the Waterberg plateau itself has not been found to contain any mineralisation of sustainable economic value, the same cannot be said for its surroundings. Two towns in particular, Thabazimbi and Lephalale (Ellisras), owe their existence to the mines established close by – indeed, by the same parastatal company, Iscor. Both much younger than Vaalwater, they quickly became much bigger, more prosperous and more influential than the village they soon eclipsed. And in so doing, they began to influence the development of the Waterberg plateau. Thabazimbi today, even after the closure of the mine there, dominates the economy of the southwestern part of the plateau and Lephalale, by virtue of its enormous growth and the traffic that flows between it and Gauteng, has fuelled the modest growth of Vaalwater and even Modimolle, through which virtually all that traffic runs.

Between these two mining towns lies an all but abandoned settlement at the foot of the plateau: Matlabas. And in the north, between Lephalale and Mokopane, sleeps a similar, though in fairness, slightly larger, village: Marken. Despite being (only just) within the Waterberg District Municipality, Mokopane has never been considered part of the Waterberg region, and so its own fascinating history is barely mentioned in this book. However, the stories of Thabazimbi, Matlabas, Lephalale and Marken – and especially the area to the south of that hamlet – are integral to, and inseparable from the history of the plateau.

THABAZIMBI
Altitude: 980 metres amsl
Population (2011): 32 500[1]

Thabazimbi lies below the southwestern limit of the Waterberg plateau, although still falling within the Waterberg District Municipality. The western parts of the plateau, as well as the Marakele National Park, are accessed through the town.

It is much younger than the other centres along the southern foothills of the Waterberg (Modimolle, Mookgophong, Bela-Bela) and owes its existence to the iron-ore mining operations that developed within and around the town.

The mine at Thabazimbi developed by its owner, the state company Iscor, opened in 1931. The township, established a few years earlier when construction began, was owned and managed by the mine. Located at the foot of the north-facing hill (Sovereign Hill) on which mining operations originally commenced, it accommodated 36 white families with flats for another 29 single miners. Water, electricity and refuse removal were provided at the company's expense.

There was initially a single, company-owned, nonprofit shop and butchery, but before long Mervyn Milner, an enterprising Jewish trader from the Rustenburg area, set up a store on a neighbouring farm, creating rivalry with the company store. The local butcher, "Trader Horn" Botha, would source his fresh meat supplies late each afternoon by making excursions into the neighbouring bush.[2] The first doctor was Dr Durbach, also Jewish. He was followed several years later by another Jew, Dr Shlomo Livni, who was also appointed District Surgeon from 1938–1951,[3] and who was responsible for the civil action that eventually led to the town being separated from the mine's management and proclaimed as a township in 1953. Soon there were 60 white children of schoolgoing age. The company donated land to the government for the construction of a school. Mail was delivered by train ("the Thabazimbi Express"), but because its arrival times were erratic, the local postmaster would put out signs in front of the post office to inform clients: "Post sorted", "Post not yet sorted" or "No train, no post, no anything"![4]

Increased mine production during World War II necessitated more personnel as well as additional water and power supplies. The population of the town during the war grew to over 800 white and over 2500 black people. The white community was mainly Afrikaans-speaking, conservative and opposed to South Africa's participation in the war, which meant that additional security measures had to be put in place to safeguard the operations.[5] Anti-Semitism was rife.[6]

Additional water was sourced from wellpoints and pipelines from the nearby Crocodile River, despite protests from farmers downstream. A planned dam on the river was never built, but in 1943 a road bridge was completed. In February 1944 heavy rain upstream caused the Crocodile River to come down in flood, curtailing mining operations and cutting the town off from its supplies for over a week. The mine dealt with the crisis in characteristically good humour with a bulletin board outside the office building keeping residents informed of developments.

For example:

> 5 FEB – NIGHT REPORT: thousands of disadvantaged frogs, who were unable to learn to swim in the drought of the last few years, have drowned in the deep water; 9 AM: a whole tobacco shed was washed away and passed neatly beneath the bridges; 10 AM: it was not a tobacco shed, but a double storey house with the lights still on upstairs; 11.30 AM: an eye-witness heard the radio playing in the house as it came past; 1 PM: the house that passed here this morning still had a portion of its garden attached to it and the lady of the house was watering the flowers.[7]

A hospital and nurses' home were opened in 1946, as was a clinic to serve black patients. But an attempt to open a bottle store was turned down because such licences were only issued to outlets in urban areas, and Thabazimbi was not yet officially a town! Young mineworkers desperate for "'n dop" learned to discover the whereabouts of illicit distillers of *mampoer* in the Kransberg hills of the south-western Waterberg (see Chapter 4).[8]

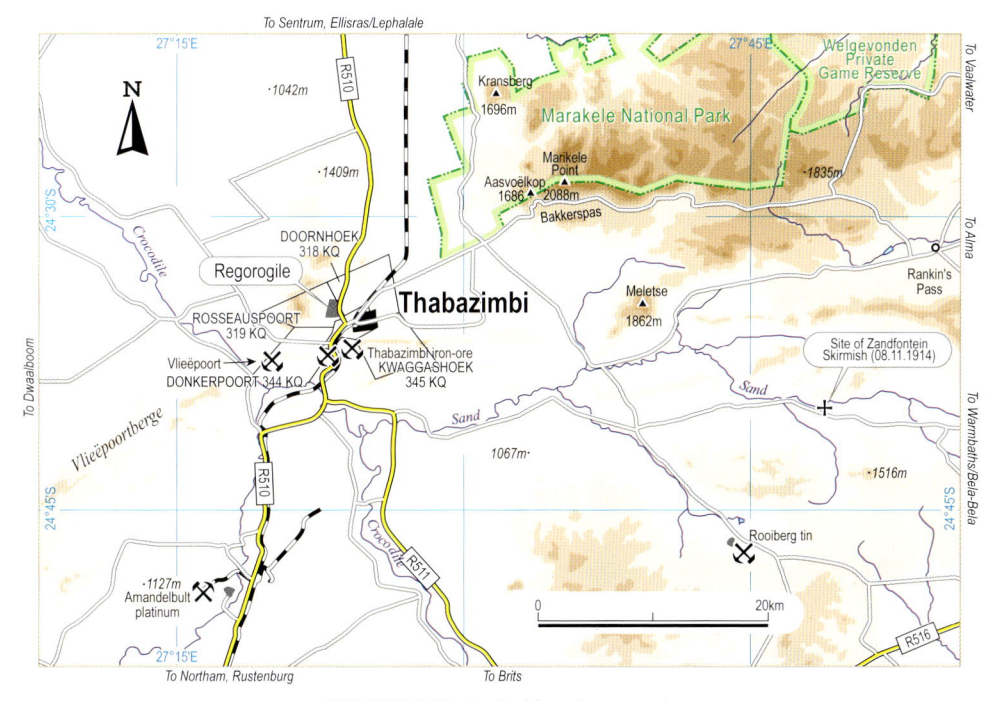

FIGURE 19.1: Thabazimbi and surrounds.

The first Native Commissioner (G. Gibson) was appointed in 1949, together with the first resident magistrate (A.H. de Wet). A school had been operating since 1938, but in the early 1950s departmental approval was granted for the construction of two new (white) schools, a primary and a secondary facility. However, Thabazimbi remained a "private" township: Business and professional people and commercial retailers wishing to serve the community were unable to obtain premises there, just as Mr Milner had discovered in the 1930s.

An application by the District Surgeon, Dr Livni, for a business site on which to open a practice was turned down, precipitating a crisis at the mine. Iscor had been hoping that Thabazimbi would be proclaimed a town by the authorities, as its local authority had been elevated to that of a Health Committee as early as 1947. It submitted an application for township status in October 1949, but this was refused because the existing town failed to comply with a number of criteria. Dr Livni, heading a group of 400 supporters, submitted a petition calling for a new

FIGURE 19.2: Bushveld near Thabazimbi (Pierneef, 1953, oil on canvas). The western Zandriviersberge (earlier known as the Mural Mountains) form the background. The painting is in the collection of the SA Reserve Bank, Pretoria, and is reproduced with its consent.

town, an extension to Thabazimbi, to be formed on its boundary. The current mine manager, Ben Alberts (Sr), recognised the legitimacy of the applicants and after negotiations, the town's Health Committee assumed responsibility for the new extension ("Livnidorp") in addition to the Iscor property. These pressures forced Iscor to upgrade the conditions in the village itself sufficiently to qualify for formal town proclamation, which was eventually granted early in 1953.[9]

The 1950s saw continued expansion of Thabazimbi: The abattoir was modernised, the hospital expanded and sport and recreational facilities upgraded. A new post office, magistrate's office, police station and numerous retail outlets were built. In 1962 the Frikkie Meyer High School was opened with its first principal Alwyn van Zyl and 23 staff, catering for almost 500 pupils. The primary school had over 460 pupils and 14 teachers. The three Afrikaans sister churches and the Apostolic Faith Mission each built a church.

In 1994 SanParks opened the Kransberg National Park, which includes the southwestern corner of the Waterberg plateau as well as the bushveld flats as far as the railway line. Soon renamed Marakele, the park has a common boundary with the Welgevonden and Marakele Private Game Reserves. In due course, perhaps, a mechanism will be found that will enable fences to be dropped between these three scenic, diverse and very large properties, which together comprise an area exceeding 110 000 ha.

Like many rural local municipalities, Thabazimbi has struggled with poor administration and poor to dysfunctional financial management. As a result it has incurred very large debts to parastatal water and power service providers, to the extent that for a period, it came under imposed administration from central government. In the local government elections of August 2016, the ruling ANC lost political control of the local municipality to a coalition of the DA and the EFF, as happened also in the Modimolle-Mookgophong Local Municipality. It remains to be seen whether this would lead to any improvement in the administration of the town.

For many years, it had been realised that the mine's life could not be extended indefinitely, and in 2015 the decision was taken to suspend operations in preparation for its closure. The town of Thabazimbi had a long time to prepare for the closure of the mine which had been responsible for its birth. Fortunately, the sharp growth in hunting and ecotourism operations surrounding the town over the last decade are likely to provide it with a sustainable future.

Along the Crocodile River about 35 km northwest of Thabazimbi is the farm Gana Hoek. This property was apparently first registered (in the name of J.S.

Potgieter) on 12 December 1865, making it one of the earliest formally demarcated farms in the region.[10] The farm is best known today for the large baobab tree (12 metres circumference at chest height) to be found there. According to Clive Walker, who accompanied the author to visit the tree in 2012, it is the most southerly naturally occurring baobab in South Africa (24° 26.27′ South). Evidently, this was a popular hunting area towards the end of the nineteenth century and one of the favourite camping destinations of President Paul Kruger, whose initials are among the many to be found engraved in the bark of the grand old tree.[11]

FIGURE 19.3: The Gana Hoek baobab. Visited on 16 May 2012.

MATLABAS

Altitude: 1000 metres amsl
Population: less than 100 (including Sentrum)

About 40 km north of Thabazimbi and a third of the way to Lephalale (Ellisras) along the arterial route R510 where the road crosses the lovely, slow-moving but perennial Matlabas River, lies the small settlement of the same name (the correct name, Motlhabatsi, means "sandy river" in Setswana). The river drains the highest peaks of the Waterberg plateau in its southwestern corner – the catchment area now protected by the Marakele National Park – and flows northwestwards to join the Limpopo River about 75 km downstream from the confluence of the Marico and Crocodile rivers – below which the river adopts the name Limpopo. (In the recent geological past, before being constrained by crustal warping and faulting, the Limpopo was a much bigger feature, and probably even included the Kavango River amongst its tributaries.)

This is flat, thorny sandveld terrain (formally, "western sandy bushveld"[12]), ranging in altitude between 900 and 1000 metres amsl, with hot summers and cool winters, accompanied by light frost. A 150 years ago, much of it was within the tsetse-fly zone mapped by Jeppe and others, as described in Chapter 12. Today, although the tsetse have long been exterminated, the area is still periodically subject to epidemics of malaria, and the deep still, reed-lined pools of the Matlabas are ideal breeding areas for bilharzia. The soil is reasonable, but rainfall is below 500 mm per annum and unless irrigated, is unable to support crop-based agriculture. These days, as has been the case since the first explorers, hunters and trekboers moved through it, the area is known principally for its hunting potential, with large game ranches and trophy-hunting establishments characterising its occupation.

The settlement of Matlabas dates back to the late 1880s when traders used to come into the area seasonally to collect salt from the numerous pans dotted across the sandy plain. Salt, biltong and hides would have been the principal source of income then. Most early residents hired land from the government and there were few private landowners. A cultural history researcher, the late Wilna Aukema, conducted many of her interviews in the 1980s with longstanding Waterberg farming residents in the vicinity of Matlabas. Although she and her husband, archaeologist Jan Aukema, died in a car accident in 1989 before her work was completed, some of her field notes have subsequently been summarised and published.[13]

The earliest homes were simple, mud-walled, rectangular structures with a gabled thatched roof and comprising three rooms: two bedrooms separated by a living room (*voorkamer*). A stoep (verandah) might later have been added to the front of the house, initially open, later walled in. Cooking facilities were either in a separate structure or absent, with cooking being undertaken on an open fire and in a clay oven. Washing and toilet facilities were also primitive; if there was a formal toilet at all, it would have been a rough seat placed above an excavated hole in the ground. These early homes became widely known throughout the Waterberg as *hartbeesthuisies* (wattle-and-daub homes), although the term is ill-defined and its origin is unknown.[14] Few, if any, still stand in a condition resembling their original form: Either they fell, or were knocked down, or have been incorporated into larger, more robust structures over the years.

Magisterial district borders caused repeated difficulties for stock farmers throughout the Waterberg, but nowhere more so than in the Rankin's Pass – Matlabas area, where restrictions imposed by magistrates on the movement of stock and transport (related to east coast fever and other stock-borne diseases) forced transport riders and cattle migrations into long detours. In the first maps produced by the British administration after the South African War, the boundary between the Waterberg and Rustenburg districts, for example, ran northwards from the position of Rooiberg tin mine (which, having opened in 1908, was just within the

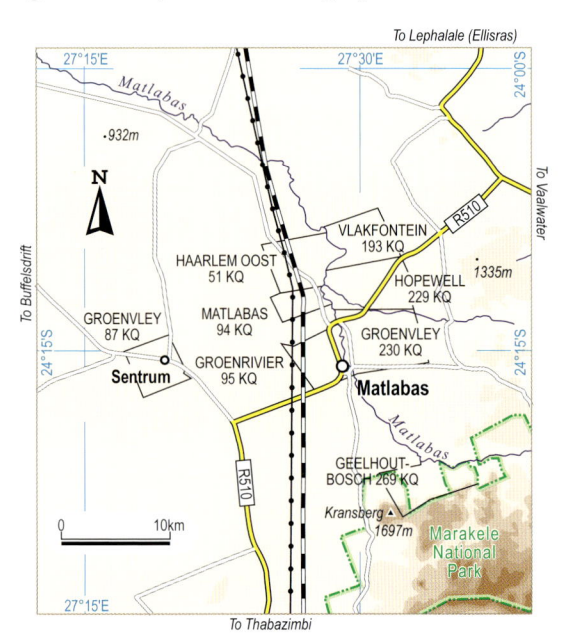

FIGURE 19.4: Matlabas and surrounds.

Waterberg District) up to the peak of Marikele Point and then northwestwards along the course of the Matlabas River to its junction with the Limpopo. This meant that any stock movements from Matlabas in the direction of Warmbaths would have had to be made north and then east from Matlabas (that is, along the road that would pass through Bulgerivier and Vaalwater) to remain within the Waterberg District, whereas a much shorter and better route would have been to go south towards Rooiberg and then directly east to Warmbaths. But this would have meant transgressing the magisterial district boundary, which was not permitted. In December 1904 a large group of farmers in the region (including some from the Zwagershoek) petitioned the office of the resident magistrate in Nylstroom to allow them the use of this southern route. They were told that requests would be dealt with on a case-by-case basis – in other words, their request was not granted.[15] Several similar unsuccessful appeals were made by the residents of Rankin's Pass.

In April 1920 the Matlabas group tried again. A petition, written by P.A. Rankin on behalf of over 30 signatories was received by the Magistrate's Court in Rustenburg, making two requests: 1) that a Justice of the Peace should be appointed at Matlabas, because the nearest existing JoP was at Rooiberg, over 65 miles (104 km) away; and 2) that the area north of the Zandrivier and between the Crocodile and Matlabas rivers be transferred from the Rustenburg to the Waterberg Magisterial District, because the magistrate's office at Nylstroom was considerably closer than that at Rustenburg. Moreover, in the summer months, flooding of the Crocodile often made it impossible to reach Rustenburg. The petitioners also named three preferred candidates for the position of a Matlabas JoP: J.J. Botha of Hoopdaal, J.D. May of Matlabas, or J.D.P. Heystek of Welgevonden.[16] The SA Police supported the recommendations, but they were deferred by the Department of Justice until adequate cells had been built at the Matlabas police post. At length, in 1926, Botha was appointed as a JoP for the area. As for changes to the borders of magisterial districts, the issue was eventually resolved through the creation of the Thabazimbi Magisterial District after the founding of that town in the 1930s.

A police post called Haarlem Oost had been set up after the South African War. However, it was learned subsequently that the post had actually been built on the adjacent farm, Matlabas! In 1918 application was made by the Commissioner of Police for new buildings to be constructed on higher ground which was considered to be healthier (further from the river). This was approved and land for the new station was purchased on Matlabas in 1920 from the owner, John Henry Bailey.[17]

FIGURE 19.5: The original church tree on Geelhoutbosch (visited in December 2018, a hundred years after the congregation was founded beneath it). Visited courtesy of Marataba Safari Lodge, Marakele, 10 December 2018.

The first church (Gereformeerde Kerk) in the Matlabas area was established in 1918 on the farm Geelhoutbosch at the foot of the Waterberg escarpment about 15 km upstream from Matlabas. Services were held under a large tambotie tree (*Spirostachys africana*) – which was still standing a century later.

In 1932 a fine new GK church, still in use, was built at Matlabas (on the farm Groenrivier where a school and new police station were also eventually located). Schooling, which for many years had been locally nonexistent or at best rudimentary, gradually became established in the Matlabas region until by the early 1930s there were three separate farm schools providing classes within a radius of about 30 km. A single centralised school was established on the farm Groenvley about 15 km west of Matlabas.[18] It was called the Groenvlei Sentrale Skool, and the settlement that grew around it became the present hamlet of Sentrum – ultimately at the expense of Matlabas itself which slowly withered and died – apart from the GK church there.

In January 1919 the sub-Native commissioner for Nylstroom received an application from Ehrens Naoa, a chief of the Seleka community, for permission

FIGURE 19.6: The Gereformeerde Kerk at Matlabas with its cornerstone.

to negotiate with the Transvaal Landowners' Association for the purchase of two farms, Vlakfontein and Hopewell, on the east bank of the Matlabas River across from the police post. The Department of Native Affairs rejected the application on the grounds that the properties were too far from the newly designated native areas.[19]

Soon after this application, it seems that Hopewell and several other adjacent farms were acquired by the Oslo Land Company (OLC), a Norwegian-owned company formed in 1918. The property was initially established as a cattle-ranching business and one report said that it had around 10 000 head of cattle on its farms. In October 1919 the company also applied to the Transvaal Land Board for the purchase of ten Crown Land properties (amounting to 10 000 morgen, or 8565 ha) close to the confluence of the Mogalakwena and Limpopo rivers in the Soutpansberg District. It said that it had purchased 3000 head of cattle from Rhodesia to place on its Waterberg property, but owing to restrictions related to east coast fever, it was not allowed to do so and therefore sought alternative land for these cattle. The request was considered and led shortly to a public announcement in the Government Gazette offering the ten farms for sale.[20] It is not known whether OLC bid for the Soutpansberg properties or succeeded in obtaining them.

In the Matlabas area, the company's principal focus was on cattle ranching. Wilna Aukema's research noted that the company provided work for numerous farmers in the region. Their homes were primitive: walls constructed of grass and bark with mud floors.[21] In the meanwhile, the company had begun to grow cotton on at least some of the land it had purchased along the banks of the Matlabas. We know this because of correspondence from the Managing Director of OLC, Volrath Vogt, who was both the founder and chairman of the Matlabas Cotton Planters' Association while resident on Hopewell between 1923 and 1926.[22]

Volrath Vogt was Norwegian. Born in the capital, Kristiania in 1885 (its name changed to Oslo in 1925), he went into banking, but, becoming bored, he migrated to South America in 1905 to work firstly as a *gaucho* (cowhand) and later as an accountant for an Argentine company. He and his wife returned to Norway in 1914 and stayed there for the duration of World War I, during which period it seems he became involved in some confidential strategic chemicals business (possibly related to armaments). The family moved to London for a few years until Vogt accepted the role of manager (actually MD) of the OLC cattle ranch in South Africa in 1922. It is possible that he was also an investor in the company.

It seems that in addition to its cotton-growing operations, OLC needed to supplement its grazing: In May 1924 Vogt wrote twice to the Chief Government

Entomologist, requesting assistance in tackling weevils in the company's grain store on Hopewell and advice for dealing with green caterpillars that had been eating his lucerne. He was advised that he could eliminate the weevils by treating the grain bin with carbon bisulphide and then making it airtight (but to act with caution, as the gas given off was highly flammable and explosive). As for the caterpillars, he was told, nothing could be done: They were the larval stage of the lucerne butterfly, *Colias electo electo*, an insect with "sulphur-yellow wings, bordered with black".[23] The only practical solution was to harvest the lucerne before it flowers.

The Vogts left Hopewell and South Africa at the end of 1926 for New Zealand where they settled. Volrath's son, Herlof Anton (who had been born in Norway and who was sent with his sister to Jeppe boarding school in Johannesburg), noted in his autobiography that at the time they left Matlabas, his parents "were exhausted by malaria", a not uncommon condition in those days. Volrath died in 1937, suffering from complications of a kidney injury sustained while he'd worked as a gaucho many years earlier.

In 1934 when a plague of locusts hit the western Transvaal, OLC was reported to own 31 farms spreading for 60 km along the Matlabas River, with a total area of about 82 000 ha (95 500 morgen). Its own efforts to control the swarms proved insufficient to combat the invasion and the government deployed teams to assist with the spraying of arsenite of soda in both solid and liquid forms. Subsequently, the company claimed that the government teams had been negligent in that during the first three months of 1934, they had sprayed the pesticide in greater concentrations and more densely than was necessary. Furthermore, the teams had discarded used or partially used canisters across the veld. The effects of this overdosing were to poison both the grazing and water for cattle on the properties. The company claimed that as a result 324 cattle had died and a further 3253 animals had to be sold because the soil was no longer safe for them. It had been advised that no stock could be kept on the affected properties for ten years until the level of toxicity had declined.[24]

Accordingly, the OLC lodged a claim against the Department of Agriculture in the amount of £50 000. Government admitted a degree of liability in respect of the cattle lost, but challenged the other claims. In the end, after two years of negotiations and threatened court cases, including an intervention by the Norwegian Consul General (the shareholders of OLC had always been Norwegian) a settlement was reached in terms of which OLC received a payment of £5000.

But that was not the end of it. The manager of OLC at the time of the arsenical spraying programme in 1934 was R.A. Mordt, who lived on one of the properties (called Oslo Ranch in legal papers) with his family. It transpired that his young children became seriously ill as a result of arsenic poisoning, allegedly as a result of drinking milk from cows that had been poisoned. This was confirmed by a government doctor and in 1938 resulted in a claim against the government by Mordt for an amount of £80 000. The government was anxious not to allow the claim to end up in court, because, as the formerly confidential files now lodged with the archives make clear, its attorneys had concluded that the plaintiff's case had a good chance of succeeding. This would have had substantial financial implications for the state, including the possibility of other similar claims. Consequently, an out-of-court settlement was proposed, amounting to an ex gratia payment of £4000 plus costs. Mordt, who described himself as penniless, readily accepted this offer and the matter was finally closed early in 1939.[25]

A year later, OLC was trying to meet its debts through the sale of its remaining farms.[26] It was also having to defend itself against a government policy to seize all property deemed to be associated with the enemy (Nazi Germany) following the forcible occupation of Norway by Germany in April 1940.

Several of the OLC's farms along the foot of the Waterberg escarpment to the north of Hopewell comprised part of what was known as the "Diamond V

Ranch", on account of the shape of the branding iron used on the stock. After the war, six of these properties (Wildebeestfontein, Tiegerfontein, Matlabasfontein, Welgevonden, Rietvly and Rhenosterfontein), amounting to about 19 000 ha, were consolidated into a new state-owned farm, Diamant, which was then subdivided by the government into numerous small sections for settlement.[27]

Today, a traveller along the main road between Thabazimbi and Lephalale traverses an "infrastructure corridor" containing a heavy-haul railway line and high-voltage power lines linked to the coal mines and power stations west of Lephalale. Shortly afterwards, between the corridor and a bridge over the serene Matlabas River, the traveller will encounter an unexpected – and seemingly irrational – series of kinks in the road, originally designed to access the now-faded settlement, but never realigned.

ELLISRAS/LEPHALALE
Altitude: 825 metres amsl
Population (2011) including Marapong: 44 000[28]

Like Thabazimbi, Ellisras owed its establishment to the development of a mine in the region (also by Iscor), followed by Eskom's power station – although a farming community had supported a small settlement in the area for many years prior to these projects. From the late nineteenth century, there had been isolated farmsteads, some occupied by owners, others, like the Peacocks on Cremartardfontein, by managers for foreign-owned land companies. Those who were permanently resident were mainly subsistence farmers at that time, their only sources of monetary income being from the proceeds of hunting or from bounty paid by the government for the scalps of animals considered to be detrimental (e.g. 2/6d a head for baboons and 10/- per tail for wild dog). From about 1910 salt mining (for example at Soutpan) also delivered a meagre, hard-earned income.[29] At Soutpan, about 40 km to the southwest of Ellisras, there was a settlement of up to 13 (white) families even before the South African War, with each family having responsibility for harvesting a section of the pan. Having been scraped off the surface of the pan, the salt was washed through sieves, spread out to dry and then stacked. It was sold for 10/- per *muidsak* (*streepsak* or grain bag), or exchanged for a variety of produce required by the miners.[30]

FIGURE 19.7: The Matlabas River near Matlabas with the Waterberg escarpment behind.

The first white settler in the area now occupied by Lephalale was probably Pieter Ernestus Johannes Kruger, who on 25 January 1870 was granted the farm Waterkloof along the Mokolo River in payment for his services to the government. Although Kruger built a simple home on the farm, it is unknown whether he ever lived there, or merely used it for annual hunting visits.[31]

In 1886 a young man and his family came from the Marico area to hire a portion of the farm from Kruger. They were Johannes Lodewikus (Jan) Lee (1859–1926) and his wife Maria Magdalena Blignaut (1863–1919). Lee was the son of a famous big game hunter and early traveller into the land across the Limpopo River: John (actually also Johannes) Lee of Potchefstroom, also known as John Hunter, who travelled into Matebeleland with Thomas Baines – and was painted by Baines shooting elephant from horseback.[32] Lee was the grandson of an illustrious British Rear-Admiral, Richard Lee, of Londonderry, Ireland, who had been decorated by King George III. Jan Lee (the son) was a member of the Transvaal Republican police (ZARP) before the South African War. During the war, he joined the Waterberg Commando under Cmdt Gen."Groot Freek" Grobler and served with him at Rhodes Drift and Colesberg, before returning to his farm where he joined about 300 other Boer families (including Grobler, Boshoff and Enslin) in crossing the border into Bechuanaland with their stock. This earned them the derogatory epithet of *Gamjanders* by the Boers who continued the guerrilla war.

After the war, Jan Lee and his family (his wife Maria bore him ten children before her death in 1919) returned to Waterkloof, where he first rented and later bought a portion of the farm (apparently from the firm of Farrant & Thompson), together with the farm Rietspruit across the Mokolo River. There he established a small school for the growing community in the area. In 1921 Jan remarried; his second wife, Martha Magdalena Carolina Breedt (née Pelser) (1896–1953), was a young widow with four children from her first marriage. She and Jan had another three children before Jan died in 1926 after injuries sustained in a mule-cart accident. Martha was soon married a third time, to Pieter Erasmus, with whom she had another four children. No wonder a school was needed!

There was now a growing community living on various portions of Waterkloof and this was to form the nucleus of a new town, Ellisras. It effectively commenced its existence in 1927 as a shop, postal agency and bus stop on Waterkloof when Marthinus Frederik Loots (1889–1971) purchased a portion of the farm as a base for the transport business he ran between the village of Vaalwater (the railhead) and Stockpoort, a border post with Bechuanaland at the Limpopo River.

Loots was a remarkable entrepreneur. Born in Pretoria, he went to school in the Marico area where he matriculated before becoming a teacher and part-time farmer. In 1920 he had settled on the farm Graaff-Reinet about 20 km further down the Mokolo River from Waterkloof. There he started the first bus service three years later, having persuaded local farmers to let him take their dairy products to Vaalwater for sale, while returning with their requirements. When he wanted to open a formal postal agency and RMS bus stop on Waterkloof, Loots was requested to submit proposals for the name of the site. The successful name was Ellis Ras, derived from the names of the owners of two other portions of Waterkloof; on 1 August 1930, this name, Ellisras, was formally adopted.[33]

One of the names was that of Patric(k) Henry Ellis (1878–1952), an Irish-born immigrant who, like Loots, had grown up in the Marico and later fought for the Boers. Captured, he had been imprisoned in Ceylon. He and his family moved to the Waterberg in 1925, settling on Ventershoek, an isolated farm in the hilly area about 25 km to the southeast of Waterkloof. He had children of schoolgoing age and to bring them closer to the new school at Hoornbosch, he moved with his family to Waterkloof in 1926 where they stayed with the then owners of one of the portions, the McSeveney family. In 1932 Ellis was able to purchase 65 morgen from Mrs Anne McSeveney – for the price of 12 cows and a bull! He began a market gardening business on the banks of the Mokolo River, an activity for which Lephalale is known to this day.[34]

The other name from which the settlement's name was derived was Pieter

FIGURE 19.8 (LEFT): Marthinus Loots. Photograph courtesy of Johan Beukes, Pioneer Lodge, Lephalale. FIGURE 19.9: Patric(k) Ellis and his wife (CENTRE). Pieter Erasmus (RIGHT). Photographs courtesy of Johan Beukes, Pioneer Lodge, Lephalale.

Barend Elardus Erasmus (1885–1956). Erasmus had been born near Bronkhorstspruit, and as a 14-year-old he had fought in the South African War. In 1928 he too arrived as a settler on the farm Waterkloof where he met the recently widowed Martha Lee, whom he married the same year. Both had been married previously, with the result that they were now responsible for the upbringing of a total of eleven children. In spite of the difficulty of raising this large family during the depression years, Piet Erasmus was closely involved with the community, and instrumental in the establishment of the Gereformeerde Kerk in the area as well as the first school in the new settlement. He was a popular man, known as Klein Pietie or Rooi Pietie, suggesting that he was shortish and red-haired.[35]

Marthinus Phillipus van Staden of the farm Grootfontein, immediately north of Waterkloof, was another founder member of the GK at Ellisras. Van Staden had been an ouderling in the GK at Nylstroom since 1882. During the South African War, he'd served in the Waterberg region as a Commandant. Immediately after the war, he was appointed by the British colonial government to serve on a commission to evaluate claims for reparations. The commission's report written by Major French, its military member, was highly complimentary about Cmdt Van Staden's character:

> He is the most influential and popular man among the Dutch in the district and is personally known to every one of them ... Everywhere we went they were delighted to see him on his arrival. That he deserved his popularity I fully believe, for he was a thoroughly straight man. He did his work on the Commission most honourably and impartially and I never saw him in the least bit influenced by his personal relations or particular friends.[36]

At Grootfontein, after the war, Van Staden, too, established a school on his property.

Just across the Mokolo River from Loots's first property Graaff-Reinet, was another settlement on the farm Oranjefontein, the home of Koos Geldenhuys. A school was built here in 1908 and had a teacher by the unusual name of Daniël (Daan) van der Paardevoort Pronk (1869–1923). He was the son of a Dutch nobleman, Baron Izak Jordaan van der Paardevoort Pronk, who'd lived at Brandfort in the Orange Free State (where Daan was born) and fought in the Basotho wars. The baron, who was English-speaking, had a sister – Daan's aunt – who married the leading Pretoria lawyer Sir Willem van Hulsteyn. Their daughter, Margaretha (Daan's cousin), became an actress with the stage name of Marda Vanne. In 1918 she was married briefly to Advocate J.G. Strijdom (see Chapter 16). Another member of the Pronk family was a teacher in Nylstroom and married Carl Forssman. Daan van der Paardevoort Pronk owned the farm Shotbelt adjacent to Hoornbosch. He later taught at the school at Swartwater, where he died from blackwater fever in 1923.

The old *waenhuis* (wagon shed) on Oranjefontein served as one of the venues for NGK services from around 1915. Others were at Steenbokpan and at Hoornbosch, about halfway between Oranjefontein and Waterkloof. There was also an early school at Hoornbosch, owned by Oom Joël and Tant Sannie van Rooyen. This was chosen as the site for the first formal NG church building (the Mispa Kerk) in that district, consecrated in 1940.[37] In 1947 this Albertyn community purchased 27 morgen on Waterkloof from Marthinus Loots to build a *pastorie* to attract a resident minister. In 1952 more land was purchased, this time from the Waterkloof school (now Laerskool Ellisras), with a view to building a new church to replace the Mispa Kerk on Hoornbosch.

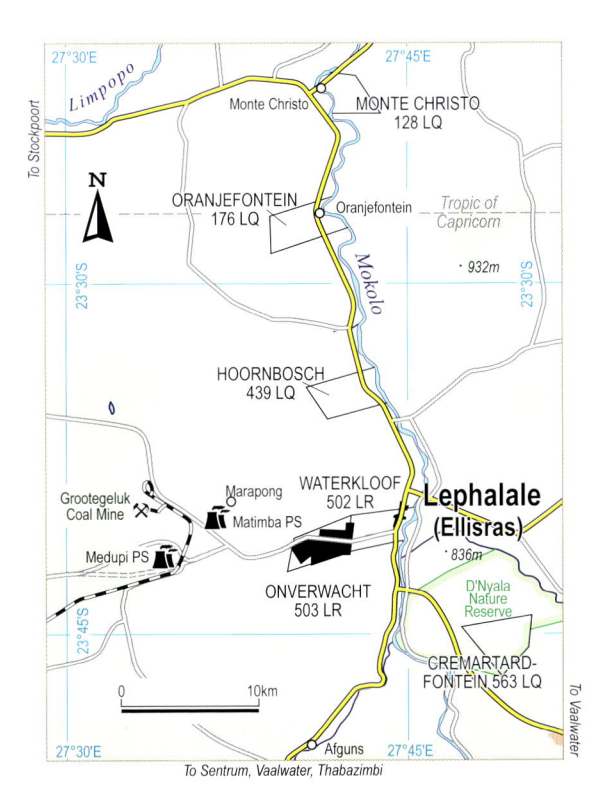

FIGURE 19.10: Ellisras/Lephalale and surrounds.

Ellisras was proclaimed a town on 7 December 1960 but, in common with Vaalwater, languished under the authority of the Transvaal Board for the Development of the Peri Urban (the Peri Urban Board) for several years, before obtaining its first Health Committee in November 1965. From the late-1960s, as work on developing the Grootegeluk coal mine commenced, the township began to grow quickly. It was eventually proclaimed a municipality on 1 July 1986, enabling the local authority to develop roads, schools, housing and a business district. The decision by Iscor to develop a separate township on the farm Onverwacht adjacent to the original town of Ellisras was caused by the inability of the owners of portions of Waterkloof to agree on a single price for plots.[38] This caused considerable initial confusion and rivalry, but the subsequent amalgamation of the two nodes into the single town of Lephalale (the name change occurred in 2002) led to somewhat improved administration (with the constant support of the two major local corporate operators, Eskom and Exxaro). At the time of writing, the town's major problem, apart from its perennial water constraint, concerned serious shortcomings in the management of its sewage disposal system, leading to repeated contamination of the Lephalala River. Today, Lephalale supports more than a dozen schools (including the descendant of the original Waterkloof school), a sophisticated private hospital, a tertiary training college and the administration of the local municipality. An industrial site is under development.

Ten kilometres to the southeast of Lephalale, on the R33 or "new" road to Vaalwater, is the D'Nyala Nature Reserve. Encompassing several former farms, including Cremartardfontein (the property once managed by Arthur Peacock for TCL), the 8000 ha that make up D'Nyala not only boast scenic frontage on the Mokolo River, but also straddle the faulted contact between the 1900 million-year-old Waterberg sandstones and the coal-bearing sediments of the 300–240 million-year-old Karoo sequence. This contact is marked by a steep escarpment down which the road descends via several spectacular turns from over 1100 metres amsl on the plateau to only 800 metres in Lephalale. In the turbulent political times of the 1980s, D'Nyala was used by the National Party government as a discrete venue for cabinet workshops (*bosberade*), for hunting jamborees, and later for some of the first clandestine meetings with senior members of the ANC. Today the reserve is administered by Limpopo Parks and Tourism.[39]

The commissioning of Eskom's Matimba coal-fired power station in 1981 and the more recent construction of the Medupi station, now due for completion in about 2020, increased demand for coal and resulted in the doubling of the production capacity of Exxaro's Grootegeluk mine outside the town. These developments are certain to boost the prosperity and continued growth of Lephalale – although at a slower rate as construction work on the current projects concludes – for many years into the future.

Marken

Altitude: 1000 metres amsl
Population: a few hundred

The residents of the little settlement of Marken on the farm of that name, some 70 km east of Lephalale and 95 km north of Vaalwater, are quick to make it clear that in their view they do not reside in the Waterberg, but rather in the Koedoesrand! This distinction is interesting, because Koedoesrand is strictly the name of an even smaller settlement 5 km to the west, on the farm Murchison, as well as the name given to an area between the village and the N11 arterial road some 30 km to the north, linking Mokopane with Tom Burke. Be that as it may, many Waterbergers see Marken as the northern limit of their plateau as well as the only settlement of any size to the north of Vaalwater.

Marken is located near the track of an old communication trail that ran from

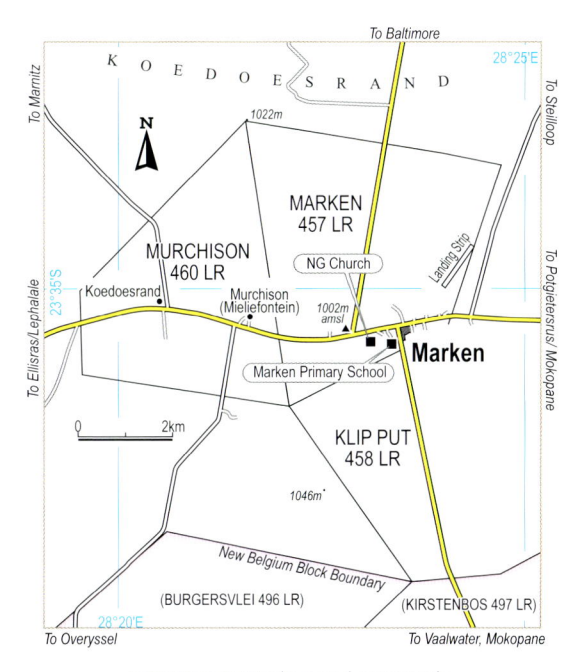

FIGURE 19.11: Marken and surrounds.

the Limpopo River up the Palala River and then south to Zandriviersspoort and which, for example, was used by the British commander Colenbrander and his column during the South African War (Chapter 9).[40] Today, it is characterised by the "almost-intersection" of seven roads – almost, because there are actually five junctions in a 5 km stretch of road. On that basis alone, Marken should be a humming metropolis, but as only one of the seven roads leads to a large town (Lephalale), it has so far been spared this fate.

The story of this quaint settlement and its community, which in the past was notorious for its poaching activities, has been told appropriately by S.W. (Schalk) Breytenbach, who taught at the popular and highly regarded government primary school there for over 40 years and was universally known as "Meester".[41]

The farms Marken and Klipput (strictly, Klip Put) on which the present settlement has developed were originally demarcated in April 1869, and the first inspection report was undertaken, like so many others in the region, by Jan Antonie Smit in February 1870. The farms were awarded by the ZAR – in the person of President M.W. Pretorius himself – to Jan Hendrik Barnard and Johannes Michiel Hattingh Roets respectively. Probably neither of these owners ever saw or occupied the properties, which served a purely speculative purpose.

As with most other Waterberg farms, they were subjected to a second, somewhat more rigorous inspection in 1888 under the supervision of F.A. "Groot Freek" Grobler (the Boer Cmdt Gen.) and were eventually properly surveyed by G.W. Ferguson in 1894. Marken passed through several owners before being sold back to the ZAR government for £20. Klipput similarly went through a rapid sequence of owners before being repurchased by the government for £30.[42]

Adjacent to Marken in the west is the farm Murchison, which had a slightly different and more interesting early ownership history. Having been demarcated and inspected like those above, it was awarded by President Pretorius in January 1870 to Benjamin Beauman Wilson, who immediately sold it for £100 to Christina Johanna Reus, who equally immediately sold it on to Oscar Wilhelm Alaric Forssman – for a mere £35. Forssman, as is described in Chapter 5, was perhaps the country's biggest "collector" of farms, usually in lieu of payment for services rendered to individuals, companies or the government. He was a shrewd and unhurried investor. He kept Murchison until 1889 when he sold it into the land frenzy that followed the discovery of gold on the Witwatersrand. The buyer, Transvaal Land Company, paid him £500. TLC (later to be absorbed into the much bigger land and mining development company, TCL, one of the foundation stones of Rand Mines) kept the property until 1922 when it was sold to Johannes

Nicolaas Janse van Rensburg for £807.7.6d. Van Rensburg immediately on-sold half of the farm to Jan Adriaan de Beer for half the purchase price.[43]

Murchison (also known locally as Mieliefontein) was the site of the original school in the district. In 1930 the school moved to a new site in the far south of the farm Marken, whose new owners were a consortium of Jan Hendrik Joubert, Lourens du Plessis, Daan Rossouw and Kerneels Koekemoer. They had purchased both Marken and Klipput from P.G.W. (Piet) Grobler.[44] Grobler, the MP for Rustenburg, had been appointed Minister of Lands in the first cabinet of General Hertzog in 1924 and would hold this portfolio for the next ten years. A *Bosvelder* and an enthusiastic conservationist, Grobler was responsible for the widely acclaimed promulgation of the National Parks Act (No. 56 of 1926) on 31 May 1926, which formally proclaimed the creation of the Kruger National Park.[45] This was a fitting role for a great-nephew of Paul Kruger, born in Oom Paul's house near Rustenburg in 1873.[46]

The quartet of farmers/developers on Marken and Klipput slowly established a small settlement around the junction of roads. The new school they had built was designed to bring customers to the general store on the lonely road linking Potgietersrust with the settlement at Ellisras. The school opened on 20 January 1930 with 25 pupils in six classes (Standards 1–6) and with Meester Breytenbach as its only teacher. By the time it closed for the Easter holidays, pupil numbers had risen to 43 and Breytenbach had to teach Afrikaans, English, geography, nature study, arithmetic and history. The arrival at the end of April of a second teacher, Mr Goldner, must have been highly appreciated![47]

At that time, there were only five families living in the vicinity of the school on Marken, Klipput and surrounding properties. This was hard country and landowners struggled to make a living. Many took in children from distant farms as boarders and in this way were able to augment their income. In 1931 and again in 1935, successive locust swarms destroyed most of the crops. These were the same locust plagues that led to the demise of the Oslo Land Co (OLC) at Matlabas, described earlier in this chapter. The effects of the global Depression reached even outposts like Marken. Early in 1933 most of the northern part of the Transvaal was placed under quarantine, due to an outbreak of foot and mouth in Rhodesia (it was lifted over the Koedoesrand in May). The years 1934/1935 brought the most severe drought since the one experienced just after the end of the South African War.[48]

The area to the north of Marken (the real Koedoesrand) was still considered prime hunting country and both local and visiting hunting parties would venture into its largely open spaces during the hunting season in search of kudu, impala and wildebeest. Malaria was the principal concern in the wet season, but concerted efforts by the antimalarial teams from Tzaneen, who came into the area to spray during the summer months, gradually took control of the disease.[49] In winter the lower Mogalakwena River was a favourite destination for camping holidays: Eugène Marais accompanied his friends the Prellers there a few times towards the end of his life, and Pretoria socialites like the Bourkes also made elaborate excursions into the area.

In the 1940s, the development – only 20 km to the south of Marken – of the huge New Belgium Block of farms as a settlement area for ex-servicemen and others, brought some prosperity to the hamlet. Although this plan was ultimately a failure (the soil was too poor and the tenant-purchasers too ill-resourced), for several years participants in the scheme sent their children to Marken school and made their purchases locally. In the 1980s most of the old New Belgium farms were consolidated into privately owned game reserves like Lapalala, Touchstone and Keta (see Chapter 9), and the population of the district declined.

Although neglected (like most rural Transvaal communities), Marken benefitted from the efforts of Meester Breytenbach, who, with his university level education (a rarity) and being the only resident government employee, was able to lobby effectively for the provision of services. He obtained a telephone (initially only one in his office, but later a fully automated exchange), managed to have a weekly Road Motor Service bus service introduced and later a post office. Although unsuccessful in his attempts to secure a locally based district surgeon, Breytenbach and his school commission (formed mainly of owners of the property on which the school was built) eventually succeeded in getting the support of the Department of Education to build a centralised school and boarding hostel at Marken, following the closure of other small farm schools. The new school opened in April 1948 with 122 pupils, nearly all of whom were boarders, but had only two classrooms. This was soon rectified and by 1950 there were six teachers, six classrooms and other facilities like a library, sports fields and a pavilion.[50] Breytenbach, by virtue of having headed the previous school, was ineligible for the same position at the new one, but chose to stay on as deputy head rather than accept a headship elsewhere. He had become an institution both at the school and in the neighbourhood and was regarded as the de facto mayor of Marken.

In 1953 the NGK was given a piece of land by its owners on which to build a church, which is still in use today.

The area never had the potential to offer a prosperous lifestyle from conventional farming. Neither the soil nor the rainfall was suitable for intensive stock or crop farming. Towards the end of his life (he died in October 1981), Breytenbach lamented:

> It is ironic that this region, which had been the refuge of those without property in the 1930s, the hideaway for saboteurs and deserters in WW2 and where there were once so many children in families that there was nowhere for all of them to sit, should now become unpopulated. Children grew up, left, studied, took up jobs other than farming and returned to the ancestral farm only when time allowed. So the beloved bushveld, unlike any other part of the country, became empty, without hospitality, soulless. Ordinary farmers could no longer afford the land, which became sold to people who lived in the cities and came out occasionally to their Bushveld farms to hunt.[51]

Today, Marken remains a small unformalised (as opposed to informal) settlement, serving a still small local community. The primary school which now

FIGURE 19.12: The NG Kerk at Marken and its cornerstone.

accommodates over 300 pupils – virtually all as boarders – up to Grade 7, remains renowned for its high standard of education, drawing pupils from places that include Mokopane and Lephalale.[52] To the north, the former Koedoesrand hunting grounds are now large, privately owned game ranches, many, like Shelanti, specialising in the breeding of scarce species like buffalo and sable.

The Jakkalskuilpad and Kloof Pass

Access from Marken southeastwards to Potgietersrust (later Mokopane) was for many years via a circuitous route around the northeastern edge of the Water-

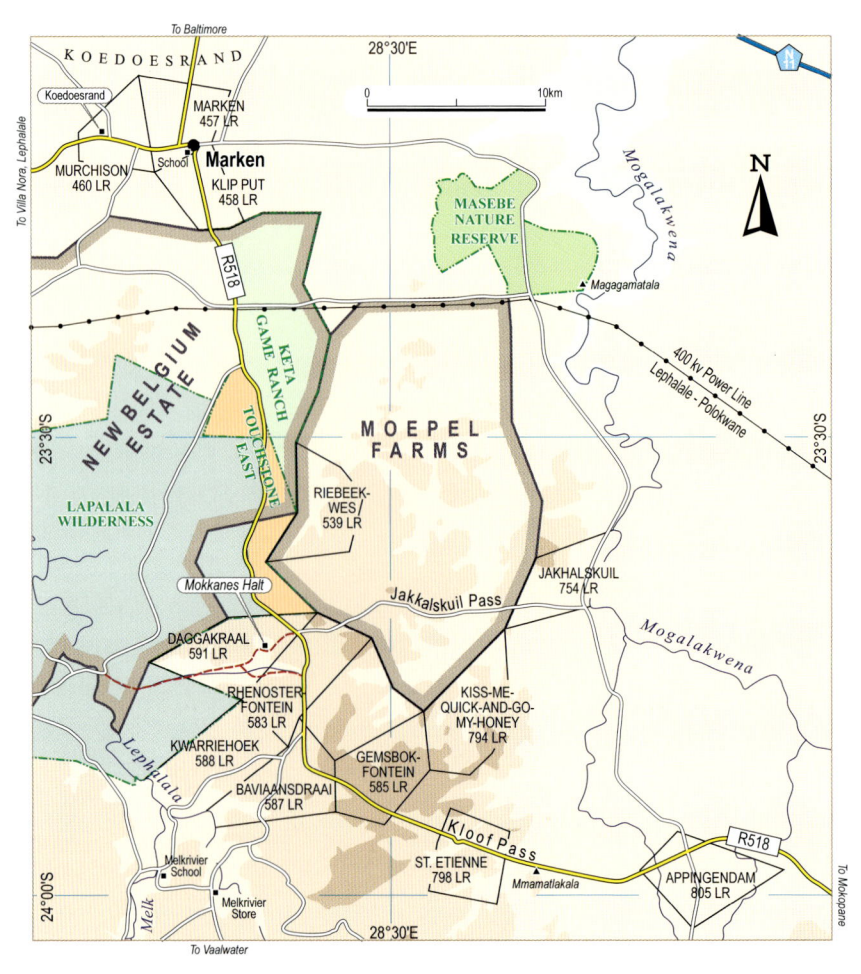

FIGURE 19.13: Marken and Kloof Pass area.

berg escarpment. During the Depression of the 1930s, impoverished whites were employed (as with the Papstraat Pass on Tarentaalstraat) to build a shorter route up the escarpment between the existing road crossing the farm Jakkalskuil below the escarpment and Daggakraal on the plateau, the latter being crossed by the existing road between Vaalwater and Marken. Farmers from various properties on the plateau, including Willem Janse van Rensburg, Johannes Kloppers, Jacobs (Daggakraal), MacDonald, Blignaut and Kiesling (from the farm Riebeek West) were employed in its construction. This route, which shortened the travel time considerably, was known as the Jakkalskuilpad or Jakkalskuil Pass road. Twenty years later, the pass was rebuilt, at which time it was formally named after F.H. Odendaal, who was by then the Administrator of the Transvaal, but it continued to be known by all as the Jakkalskuilpad. It served a large community between Marken and Melkrivier for over 50 years.[53]

Daggakraal is said to have been the first farm in this part of the Waterberg to be permanently settled: Together with the neighbouring properties of Rhenosterfontein and Gemsbokfontein it was settled by Karl Wölff, a surveyor, as early as 1878. Later, as his family multiplied, Daggakraal became divided into numerous smaller portions, each occupied by members of the extended family. As a result, interrelated families like Kloppers, Boshoff and Botha can all trace their ancestry back to this original settler. The neighbouring farm Kwarriehoek was also intensively settled, having been divided up into at least a dozen portions, each with narrow access to the Lephalala River, to accommodate the descendants of a former owner.[54] An irrigation furrow ran along the north bank of the Lephalala River (see Figure 10.5).

In 1950, thanks to Meester Breytenbach's lobbying, Spoornet commenced an RMS bus service from Potgietersrus to Villa Nora via Jakkalskuilpad and Marken. Until then, local mail was delivered to and collected from a postal tree called Groesbeek on a portion of Daggakraal owned by a farmer of that name. From 1950 a rough shelter was built on Daggakraal at a place that came to be known as Mokkanes Halt. This served to accommodate mail and passengers until the Jakkalskuilpad was superseded in 1989 by Kloof Pass. Once, a bus became stuck in the mud on the old pass, closing it to traffic for over a month and creating great disruption, especially for school children going to and from Potgietersrus. The RMS bus driver was often called upon to undertake additional duties. One former resident recalls a driver called Callie Retters, who would receive medicines at his home from the chemist in Potgietersrust the night before his journey and then deliver them safely to patients along the route. The bus would stop at the store on Rhenosterfontein before continuing via Daggakraal to Marken and on to Villa Nora. If all went well, it would return the same afternoon or early the next morning, and repeat the round trip (a total of about 300 km mostly on gravel roads – the stretch as far as Tinmyn was tarred) later in the week. Farmers would sometimes earn a bit of extra income by sending cans of cream back with the bus to "Potties".[55] Oom Hansie Kloppers, who grew up on Daggakraal and went to boarding school in Potgietersrus in the late 1950s, rode by bicycle several times between school and the farm – a long and difficult ride (85 km), especially in the days before the popularisation of derailleur gears.[56] Today, the Jakkalskuilpas road, which is still in use, runs through the southern part of the Moepel Farms, an area set aside for development as a nature reserve by the local community.

Repeated difficulties in negotiating the Jakkalskuilpad in the rainy season prompted several local farmers, led by Kerneels Ackermann and Koos Botha from the Rhenosterfontein agricultural association, to lobby for a new all-weather road to be built up a long valley to the south of Jakkalskuil. In 1978 and again the following year, they appealed to the acting Administrator of the Transvaal (Cruywagen) for support and, eventually, the construction company Basil Read was contracted to build the new pass. The road was built from either end towards the middle: from Baviaansdraai (just south of Kwarriehoek) on the plateau end; and from Appingendam on the plain below the escarpment. (This may explain why bridges on these sections of the road are dated 1981, a full seven years before the road was completed.) The middle section of the new road presented the most formidable set of construction challenges: Not only was extensive excavation needed to provide a long gentle gradient down the steep hillside, together with all the retaining gabions that entailed, but towards the bottom, the only feasible route necessitated a long, curved, high-level bridge across a tributary of the Mmetlane River on the farm St Etienne. This was finally completed in 1988, and the pass was opened on 1 July that year by S.J. Schoeman, Transvaal provincial MEC.

The local farmers lobbied to have the pass named the C.F. Ackermann Pass after the person most responsible for its creation, but the authorities decided to call it Kloof Pass instead. More recently, a graffiti artist used the vertical rock faces in the vicinity of the bridge to name the route the "De La Rey Pas" in memory of the fine Boer leader. Apart from being ideal surfaces for defacement by graffiti vandals, these fresh vertical rock surfaces provide one of the country's finest series of exposures of the processes involved in sedimentary rock formation. Travellers through the pass will occasionally come across geologically minded individuals or groups photographing, measuring or simply staring at the fascinating story to be read and interpreted on the rock faces.

The bridge was the weak point of the whole pass, although the road also suffered from numerous rock-falls during the rainy season. Early in the present century, very heavy rains caused some washaways and threatened to undermine the pillars supporting the bridge, which in any case was proving too narrow for the volume of traffic the road had started to carry. The only practical solution was the complete closure of the pass while repairs were effected and the bridge strengthened and widened. Kloof Pass reopened to traffic in 2008 and has functioned very satisfactorily ever since, although the steep cuttings that were part of its original design will condemn it to experiencing recurring rock-falls into the future.

The pass through which the new road has been built is wild and undeveloped and was always well-known for its snake and baboon population. While geologists pore over the exposures along the road, herpetologists may be found skulking furtively in the valleys below, looking for legless reptiles. Leopard too are known

FIGURE 19.14: View east towards Polokwane from the top of Jakkalskuilpas. Mogoshi peak (1780 metres) is visible on the horizon at left. Part of the road down the pass can just be seen among the trees on the extreme right.

FIGURE 19.15: Kloof Pass, the rugged upper part of the kloof (RIGHT) and the lower end (BELOW) where the road runs out onto the plain below the escarpment with the settlement of Mmamatlakala hidden beneath a blanket of early morning wood-smoke.

to live in the rugged terrain and may occasionally be encountered when traversing the pass at night.

Four kilometres to the north of Kloof Pass, on the particularly rugged terrain between it and the old Jakkalskuilpad, lies the quaintly named Kiss-Me-Quick-And-Go-My-Honey, a farm that still has no formal vehicular access. Many romantic and poignant tales have been told to explain this remarkable and unwieldy name, but all are apocryphal. The property already had its name in 1869 when its details were first entered in the Transvaal Republic Deeds Register for Waterberg: "Kiss me quick and go my Honey, No. 762, Wyk Waterberg, Geinspecteered 25de Maart 1869 (Insp J.A. Smit)". It was granted as a loan farm by President M.W. Pretorius to Johannes Jurgen Cornelis Erasmus on 21 February 1870 upon payment of £1.10.0. He sold the rights four years later to G.J. Verdoorn, the man who was then postmaster at Nylstroom and something of a land speculator.[57]

The most likely origin of the name of this farm, more mundane perhaps but no less romantic, is a song, maybe a favourite of the surveyor, written in 1856 by Silas Sexton Steel, with music by Frederick Buckley. Apparently, it became hugely popular at concerts on Broadway, New York:

> The other night while I was sparking Sweet Turlina Spray,
> The more we whispered our love talking,
> The more we had to say;
> We thought the old folks and the little folks were fast [asleep] in bed –
> We heard a footstep on the stairs,
> And what d'ye think she said?

> Chorus:	O! Kiss me quick and go my honey,
> 	Kiss me quick and go!
> 	To cheat surprise and prying eyes,
> 	Kiss me quick and go![58]

CHAPTER TWENTY
From Waterberg to Waterberg: the plight of the Herero

Search for the word "Waterberg" on the internet and you're as likely to be directed to sites related to the striking plateau and national park of that name in north-central Namibia as you are to the striking plateau and Biosphere Reserve of northern South Africa. Few sites, however, would inform you that the two Waterbergs are linked, not geologically (although they do have much in common) but by a tragic historical thread about which too little has been written in the public domain.

On the morning of 11 August 1904, it is estimated that between 30 000 and 50 000 members of the Herero nation of German South West Africa (SWA, now Namibia) – in other words, between 50% and 80% of the total Herero population – were gathered in the flat basin which lies to the east of the prominent Waterberg plateau, an arcuate feature some 55 km long and 15 km wide standing as much as 400 metres above the surrounding north-central Namibian countryside. The plateau, today a national park, lies some 50 km to the east of the modern town of Otjiwarongo. Its eastern perimeter is defined by steep (often sheer) cliff-faces and scree slopes. The highest point is about 1880 metres amsl; the surrounding plain varies from below 1370 metres amsl along the ephemeral stream beds (omiramba) that drain the eastern face of the plateau to about 1550 metres amsl in the west and north. The surface of the plateau dips gently northwards, eventually diving beneath the surrounding young Kalahari sands.

The weathering-resistant cap of the plateau is formed by a reddish aeolian (fossilised sand dune material) sandstone, called the Etjo Formation, of late Triassic age (210–203 million years ago).[1] It is the western correlate of the Clarens Formation in South Africa, the Karoo-age sequence that forms the cliffs at Mapungubwe in Limpopo and at Golden Gate in the Free State. The superficial appearance of the Namibian Waterberg escarpment is remarkably similar to the eastern escarpment of its South African counterpart, despite comprising rocks a tenth the age of the latter.

The plateau describes a vague arc that protects a drainage basin on its eastern flank: one of the catchment areas of the Omuramba Omatako, an ephemeral watercourse which flows away to the northeast, eventually to join the Okavango River between Rundu and the start of the Caprivi Strip.

In 1873 the Rhenish Mission established a mission station at the foot of the eastern escarpment of the Waterberg plateau at a place called Otjosondjupa ("place of the gourds").

The first missionary there, Heinrich Beiderbecke, described the scene:

> I was surprised by the beauty of the place. It is situated at the foot of a densely wooded mountain range which rises from the plain in terraces. At the top, it ends in a crest which is some 60 feet high and which consists of gigantic blocks of red sandstone which stand vertically next to each other. At the foot of this crest – and that is so remarkable in country so poor in water – there are several springs; the largest of these is situated above Otjosondjupa, where the mountain recedes and forms a semi-circle and thus an enormous amphitheatre.[2]

FIGURE 20.1: Perennial waterhole in the basin below the Waterberg escarpment at the beginning of the twentieth century. National Archives of Namibia Photographic Library, No. 28855.

FIGURE 20.2: The striking red colour of the Etjo Formation fossilised sand dunes that form the cap of the Waterberg plateau.

The planned annihilation of the Herero

The basin supports numerous such perennial waterholes as well as good pasture. For many years, Otjosondjupa had been the home of the important Waterberg Herero clan led by the popular Chief Kambazembi. The surrounding plain had become the last refuge for the Herero people and their remaining herds of cattle – their principal source of wealth apart from land – under the overall leadership of their controversial paramount chief, Samuel Maharero.

Most of the Herero chieftaincies had heeded Samuel's call to join him at the Waterberg. He had at his disposal some 5000 warriors, many armed with rifles, but it seemed that he and his chiefs had no clear strategy with which to withstand and confront the inexorable advance of the German colonial army, the Schutztruppe, under the command of 56-year-old Lieutenant General Lothar von Trotha.

Von Trotha's force that day comprised 96 officers, 1488 men, 30 artillery pieces and 12 machine guns. He had divided his forces into five sections, each under orders to approach the Herero from a different direction, guided by communication via a heliograph (signalling) station he had cleverly arranged to be erected on the edge of the plateau overlooking the battlefield.[3] Contrary to the assertions of some later analysts, including several apologists for the German administration of SWA to play down the tactics and objectives of Von Trotha regarding the Herero, his plan was straightforward and unequivocal.

As he put it in his own memoirs:

> My initial plan for the operation, which I always adhered to, was to encircle the masses of Hereros at Waterberg, and to annihilate these masses with a simultaneous blow, then to establish various stations to hunt down and disarm the splinter groups who escaped, later to lay hands on the captains by putting prize money on their heads and finally to sentence them to death.[4]

What had led to this dire state of affairs, in which a senior German officer was permitted to adopt and implement a policy that was tantamount to genocide? This general was personally selected by the Kaiser and a popular choice amongst the German public within and outside the territory; a career soldier and colonial army veteran, who'd served in German East Africa and China, in the process earning a reputation for his ruthless approach to quelling rebellions. He was a man associated with numerous atrocities including the burning down of populated villages, public hangings and the execution of prisoners of war. How could someone with such a history have been entrusted with overall control of SWA and been allowed such leeway in its military administration? There is no argument that can be made over a hundred years later to justify, let alone condone, the brutal actions of the Schutztruppe under Von Trotha's command in 1904–1905. At the same time, however, it is important to acknowledge that military actions of other colonial governments of the period were no less

FIGURE 20.3: Lieutenant General Lothar von Trotha. Image courtesy of the National Archives of Namibia.

ruthless, for example, the Belgians in the Congo, the Spanish in Cuba, and the British in India and the Sudan (and in South Africa). It is also necessary to place the battle of Waterberg on 11 August 1904 in context, and to do this, we must take a brief tour of the events leading up to it.

Samuel Maharero's ascension to be the first paramount chief of the Herero nation had taken place in August 1891, following the death of his father Kamaharero of Okahandja the previous year. He was neither the customary nor the popular choice of the people, who had favoured one of his uncles. In exchange for concessions (which included permitting the forfeiture of traditional tribal land to the German colonial administrations of Captain Curt von Francois and his successor, Major Theodor Leutwein), Samuel enjoyed their endorsement and connivance in subjugating his opponents. Gradually, in conjunction with Leutwein's cynical carrot-and-stick support, Samuel was able to secure the reluctant allegiance of most of the regional Herero chiefs. Those who rebelled against his authority, like his cousin Nicodemus, could expect to be captured and executed by the Germans on pretexts subtly endorsed by Samuel. In return, Leutwein expected Samuel to assist him in his principal objective, which was to secure land for the settlement of German ranchers. For a while, this arrangement worked to mutual benefit: More land was made available for German settlement and Samuel's wealth and influence grew. However, resentment gradually brewed amongst the Herero over the loss of their traditional grazing lands (cattle constituted by far the most important measure of tribal and individual wealth).

The rinderpest

The arrival of the rinderpest epizootic from the north via the Caprivi Strip early in 1897 had devastating consequences. Within six months at least two-thirds of the Herero herds had been destroyed, with some communities losing virtually all their cattle.[5] By the end of the year, the Herero people had been deprived of much of their principal and traditional measure of wealth, their livelihood, food (especially milk, a staple) and source of revenue. Although the meat of an animal killed by rinderpest could be eaten by humans with no ill effects, the sheer volume of dead beasts soon overwhelmed communities' ability to cook or process meat. Abandoned cadavers began to rot, creating potential for diseases. Where wood was plentiful, corpses were burned and buried, but many were left to decompose in the open. Inevitably, the rinderpest was followed by outbreaks of diseases affecting humans, including typhoid, scurvy and probably anthrax.[6]

For the cynical Leutwein, one benefit arising from the rinderpest epizootic was that transport between Windhoek and the coast was brought almost to a halt owing to the wholesale deaths of trek oxen. This enabled him to motivate successfully for funds to construct a railway line. Labour too, became available and work on the line between Swakopmund and Windhoek commenced in 1897 and was completed in 1902. The rinderpest was also beneficial for the German administration in that, temporarily at least, it not only reduced the need for Herero to have land for their cattle, but it forced them into selling land – their only other asset – in order to live. To try and retain some semblance of their former power and influence, Herero chiefs resorted to selling not only land, but also labour, and to launching raids on each other's territories – for both cattle and labour. Samuel Maharero, who'd inherited almost nothing of value from his father, had already been selling land from within his chieftaincy to cover his expenses, which increasingly included those arising from his growing partiality to,

FIGURE 20.4: Major Theodor Leutwein (LEFT) and Paramount Chief Samuel Maharero (RIGHT). Images courtesy of the National Archives of Namibia.

if not dependence on alcohol.⁷ The northern Herero chief Kambazembi was one of few who had not allowed his people to sell any land to whites.⁸

Although land sales largely ended once the rinderpest epizootic had passed, and as Herero people returned to their pastures and started to rebuild their herds, many had in the meanwhile incurred heavy debts with traders. Some traders required that debts be repaid in cattle; they would then keep the cattle for a while before selling them on at a profit. But in cases where the indebted Herero no longer had cattle to sell (as with Samuel, for example) traders, especially those with aspirations of becoming farmers themselves, began to claim land instead of cattle as debt settlements. The inevitable consequence was the continued impoverishment of the Herero as a people, often by chiefs (especially Samuel) who sold communal land without the authority of their communities to fund their own debts and to dispense patronage to their followers.⁹

The creation of Herero reserves

German Rhenish missionaries, who had been closely associated with Herero society for almost 50 years and who had to a large extent been complicit in Leutwein's policy of facilitating the settlement of German immigrants at the expense of the local people, gradually became concerned that the communities upon which their proselytising work was dependent were in danger of dissolving owing to their loss of land. The missionaries began to lobby for the demarcation and allocation of land (reserves) surrounding their mission stations to which Herero ownership would be inalienable.¹⁰ In 1902 the government, in liaison with the Rhenish Mission, determined that a reserve for the Herero should be established in the Otjimbingwe area. This was eventually proclaimed in November 1903, and work on the definition of similar Herero reserves in the Okahandja and Gobabis areas was in progress.

However, the Herero were unhappy about these developments: With justification, they did not trust the motives of the government, they resented the prospect of being confined in small areas (even if this land could not be sold to settlers or traders) and, of course, they were not consulted about the delineation process. Moreover, they needed to be able to sell land to pay debts and maintain their patronage networks.¹¹ In the same month, the implementation of new credit control legislation resulted in a rush by traders to offset debts of their Herero customers through land seizures, albeit conducted through the courts, before the

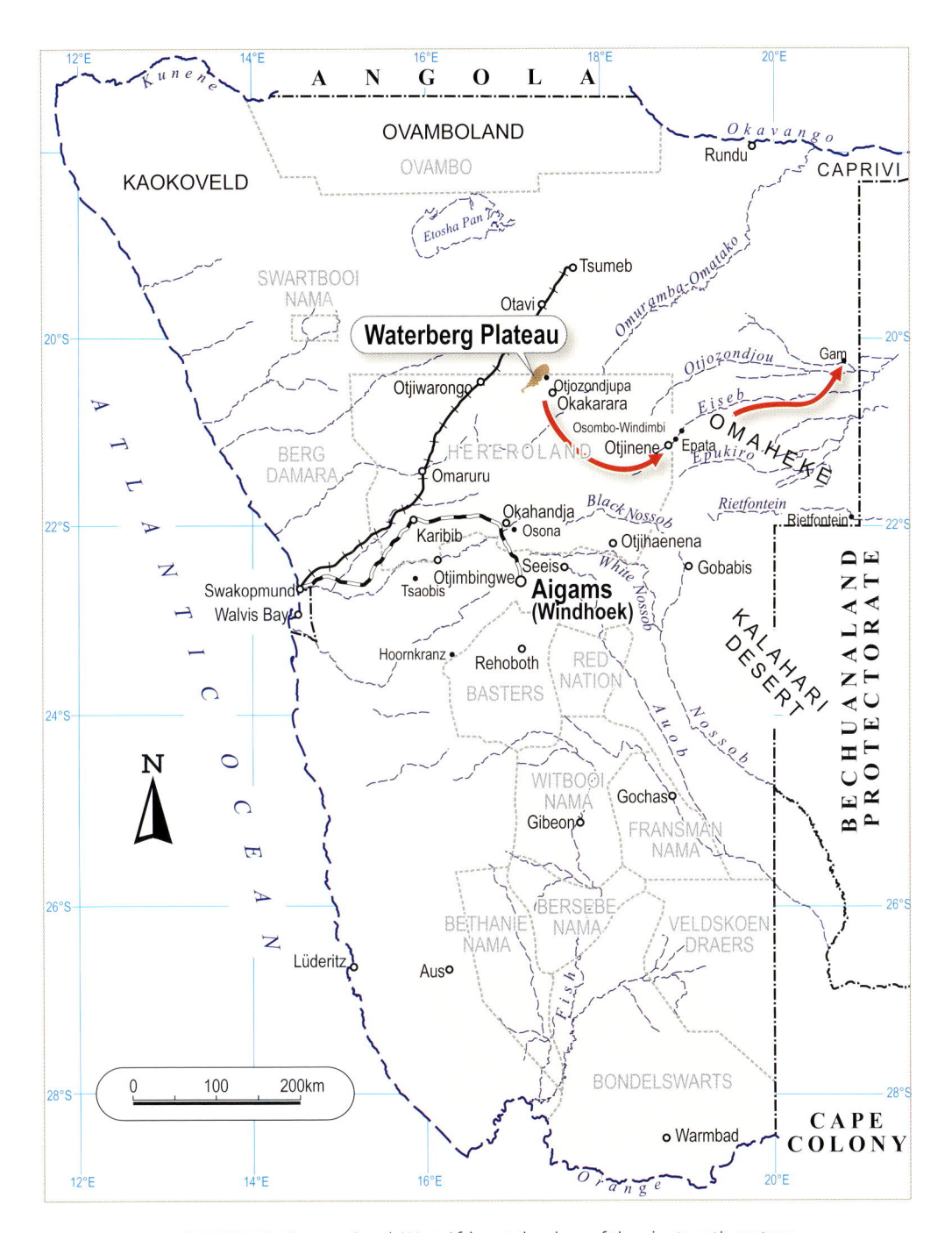

FIGURE 20.5: German South West Africa at the close of the nineteenth century.

new rules came into force at the end of the year. Leutwein was determined, by means of the reserves, to limit the amount of land available for white settlement and to ensure that the various tribes had at least some land of their own. This made him increasingly unpopular with many of the settlers who were of the opinion that it was their right, as colonists, to acquire from the "natives" whatever land they needed.

During November 1903, when an incident at Warmbad in the south of the territory necessitated Leutwein's presence with a military force, the central part of the territory was left militarily depleted. In Okahandja the garrison commander was Lieutenant Zürn, a young, inexperienced and arrogant officer who had already earned a reputation amongst the Herero for his abusive actions against them. Zürn had also been tasked with defining the boundaries of the proposed Herero reserves at Okahandja and Waterberg. He made some pretence of consulting the relevant chiefs about the reserve boundaries, but on 8 December he proclaimed the boundaries of the two new reserves without the approval of the chiefs concerned. Zürn's actions and attitude were to provide the spark that ignited the Herero rebellion against German rule.[12]

Herero anger explodes

On Sunday 10 January 1904, a trader on his way to Okahandja passed a large contingent of armed Herero horsemen also heading for the town. Alarmed, the trader reported the sighting to Zürn, who, being aware of the resentment that had built up against him, immediately – though wrongly – assumed that the incoming horsemen were the vanguard of an imminent Herero attack on the garrison. In fact, Samuel, by now the prosperous and acknowledged paramount chief of the Herero, had no such plans. (On the contrary, he had recently moved to Osona, a fortified hilltop 20 km south of Okahandja because he'd heard a rumour that the Germans, under Zürn's orders, were set to kill *him*.) Zürn instructed all white settlers in the town to evacuate their homes and move into the fort, while his Schutztruppe were placed on 24-hour watch. The Herero observed these actions and concluded that a German attack was imminent – a view reinforced when they learned that Zürn had requested reinforcements from Windhoek. A fuse had been lit.

On Tuesday 12 January, gunfire erupted in the town; the Herero besieged the garrison and killed some Germans they came across who had not heeded Zürn's instruction. They tore up the railway tracks to the south of the town to impede the arrival of reinforcements.[13]

Within the next two days, German garrisons, traders and farmers across Hereroland were attacked, although there was clearly no coordination, consistency or plan behind the assaults. Within the first two weeks, 123 Germans, almost all men, had been killed. (Samuel had given orders that the Herero were not to harm German women and children, missionaries, English settlers, Basters, Bergdamaras, Namas or Boers. He also sent emissaries with letters to Hermanus van Wyk of the Rehoboth Basters and Hendrik Witbooi of the Namas, urging them to join the revolt.[14])

The uprising falters

However, the Herero failed to take advantage of the vulnerability of the Okahandja garrison in the first few days, allowing reinforcements to reach the town. In failing to take advantage of their surprise uprising, the Herero soon began to lose the initiative in the conflict. On 15 January a large German force from the coast managed to relieve the siege on Okahandja. On the same day, Captain Franke from Omaruru, who was on his way south to join Leutwein, turned back at Gibeon, passed through Windhoek, which had not been attacked, and reached Okahandja on 27 January. During the next week, his column relieved the besieged garrisons at Karibib and Omaruru, the latter only after a fierce eight-hour battle. But Samuel and his main force were nowhere to be found.[15] Leutwein hastened back from the south, in time to take charge of 2000 reinforcements who had either arrived at, or were en route to Swakopmund. He also wrote to Samuel in the hope of arranging peace negotiations. Samuel's response was dismissive. He laid the blame for the conflict at the feet of the German authorities, specifically those of Lt Zürn in Okahandja.

By the beginning of April, all the major Herero chiefs had joined Samuel at Ongandjira (Okandjira) east of Okahandja. They included Kanjunga and Salatiel from Waterberg (the good Chief Kambazembi having died the previous August), Zeraua from Otjimbingwe, Tjisiseta of Omaruru and Tjetjo from Gobabis. Perhaps 20 000 Herero had gathered in this area, together with their huge cattle herds, which greatly constrained their mobility. However, their strategy was unclear; it seemed that they were hoping to enter into peace negotiations, although they had also dug defensive trenches in expectation of a German attack.[16] Leutwein too would have preferred a negotiated outcome to the revolt. He was concerned about the loss of Herero labour and cattle, both important components of the territory's economy. However, public opinion had swung against him and he had orders from Berlin not to enter into any negotiations, but rather to ensure that the Herero were defeated decisively on the battlefield.

On 9 April 1904, a German force of 800 under Leutwein's command attacked the Herero encampment at Ongandjira. After spirited resistance, the Herero, whose soldiers numbered around 5000, half of whom were equipped with modern rifles, were eventually forced to retreat and abandon the settlement, mainly because of severe injuries inflicted on women and children by German artillery, which accidentally overshot the Herero positions and landed amongst the families behind.[17] Casualties were some 80 Herero killed, versus five Germans (and another eleven wounded). The Herero force retreated northwards to some waterholes at Oviumbo where the German column caught up with them on 13 April.

This time, the Herero were better prepared and set an ambush for the advancing Schutztruppe in an area unsuitable for the use of artillery. During a battle that ebbed and flowed for over ten hours, the Herero failed repeatedly to take advantage of their strength. Had they done so, they would have inflicted a severe defeat on the Germans, who were both exhausted and almost out of ammunition. Instead, at dusk, Samuel ordered the withdrawal of his men and the whole Herero group moved away to the north towards the Waterberg plateau east of Otjiwarongo where they arrived with their cattle herds three months later.[18]

Leutwein replaced

The Herero were permitted to escape in this way because the inconclusive battles in April made it clear to the Germans that additional reinforcements were needed if they were to maintain the Herero campaign without jeopardising the security of the rest of the territory. In addition, events had undermined Leutwein's authority in the eyes of the German military and public – to the extent that in May he was relieved of his military command (although he remained as civilian Governor until his return to Germany in November), to be succeeded by General Von Trotha. For all his cynical manoeuvrings, Leutwein was the one German official who had developed a good understanding of, and relationships with the Herero, Nama and other groups and he was probably the only person who might still have been able to salvage a stable peace from the rebellion. His ignominious departure sealed the fate of the Herero.[19]

From the time of his arrival in Swakopmund on 11 June 1904, Von Trotha developed plans to remove once and for all the threat posed by the Herero. He refused to allow any negotiations. He declared martial law, thereby neutralising Leutwein's civil authority as Governor and he reorganised his military forces, which grew in strength weekly as more troops arrived from Germany. His instructions to his men were to avoid engagement with the Herero for the time being whilst at the same time encircling the Waterberg area, where the Herero had gathered, to ensure that they were unable to leave it.[20]

The battle of 11 August 1904

So now we return to that fateful dawn of 11 August 1904 when the majority of the Herero people were encamped on the plain below the Waterberg escarpment. As the German advance began, Von Trotha himself led the attack on Samuel's position at Hamakari on the southern edge of the basin. Other columns moved in from the west, east and northeast, one of Von Trotha's principal concerns being to

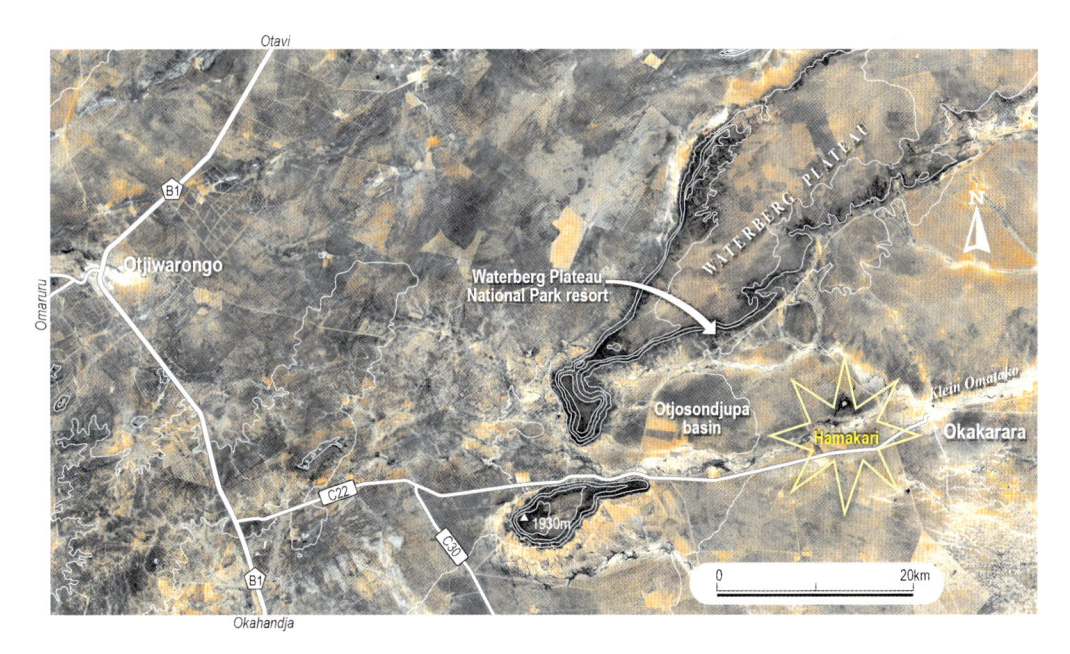

FIGURE 20.6: The Waterberg plateau of north-central Namibia.
(Image modified from Google Earth imagery, 27 July 2016).

prevent the escape of any Herero into the Omaheke region to the east. Despite the presence of the signalling station on top of the escarpment, Von Trotha's tactics were marred by poor communication, especially with the easternmost column, which became lost and failed to close off the eastern escape route.

After a day of fierce fighting in which the German forces gained the upper hand but without achieving Von Trotha's "simultaneous blow", the Germans camped in the hope of concluding the battle the next morning. Instead, they awoke to find that although many Herero had been killed the day before, the vast majority, Samuel Maharero amongst them, had managed to escape towards the southeast with their herds during the night. Von Trotha wanted to pursue the fleeing Herero immediately, but was obliged to allow his troops to rest for a day. This reprieve was enough to allow Samuel and his people to regroup and consolidate their flight.[21]

German pursuit

Over the next several weeks, pursuing German patrols repeatedly caught up with bands of Herero as they fled first towards the southeast, and then upstream along the courses of the Eiseb and Epukiro omiramba which led northeastward through the wastes of the Omaheke. These were the only two possible escape routes to Bechuanaland. After the rains, there would be many waterholes along these watercourses well known to the Herero herders, but in August and September, most of them were dry. Before long, the German pursuers began to encounter abandoned wagons, then abandoned or dead animals, and later, old people and children who were simply unable to keep up with the speed of the retreating Herero mass. Eyewitness reports described how the Schutztruppe under Von Trotha's command bayonetted or shot all those whom they encountered.[22] Different Herero groups followed different routes, though all had the same objective: to reach the perceived safety of Bechuanaland. But the conditions were dire.

Thousands of Herero and their cattle died in the Omaheke in the months following the battle at Hamakari. Some groups managed to flee northwards and to seek sanctuary amongst the Owambo; others crossed from Rietfontein north of Gobabis into the Ghanzi area where their kinsman (mainly Mbandera) had lived long before the colonial political boundaries were established.

FIGURE 20.7: View to the southeast across the flat, waterless plain of the Omaheke from slopes of the Waterberg plateau, August 2017.

Von Trotha was relentless in his pursuit, but the Herero proved elusive and the exercise took its toll on his own troops and their horses. On 2 October 1904, whilst camped at Osombe-Windimbe on the Eiseb omuramba downstream from Otjinene, Von Trotha issued the edict that would seal his notoriety in history:

> I, the Great General of the German soldiers, send this message to the Herero people. The Hereros are no longer German subjects. They have murdered and robbed and have cut off noses, ears and other parts of the bodies of wounded soldiers. Now they are too cowardly to continue fighting. I say to the nation: Every person who delivers one of the captains as a captive to a military post shall receive 1000 Marks. The one who hands over Samuel Maharero shall receive 5000 Marks. All Hereros must leave the country. If they do not do so, I will force them with cannons to do so. Within the German borders, every Herero, with or without weapons, with or without cattle, will be shot. I shall no longer shelter women and children. I shall drive them back to their people, otherwise I shall order shots to be fired at them. These are my words to the Herero nation.
>
> SIGNED, the Great General of the mighty Kaiser, Von Trotha.[23]

The number of Herero who perished from thirst, hunger or exhaustion in their flight from Von Trotha's army has never been, and cannot be, reliably established.[24] But based on the numbers who survived, deaths undoubtedly amounted to several thousand. Many more, possibly the majority, were to die in the concentration camps established to accommodate survivors. In 1905, after Von Trotha's extermination order had been rescinded, as many as 25 000 Herero – perhaps a quarter to a third of the original population – emerged from the Omaheke and other remote areas of the territory, starving, sick and barely able to walk.[25] A recent history of Namibia reports that a census carried out in 1911 recorded a total of 19 423 Herero in the territory and suggests that maybe another 3000 to 10 000 survivors were living in adjacent countries. If the Herero population before the war had been between 60 000 and 100 000, well over half that total had perished in the conflict.[26]

Von Trotha recalled

The German public and government were shocked by the news emanating from South West Africa, in particular the edict issued by Von Trotha. The general, having lost popularity at home as a result of his unduly harsh tactics against the Herero and his failure to bring either them or the Nama to book as he had promised,

FIGURE 20.8: Von Trotha and his men pausing for directions in the Omaheke.
Image courtesy of the National Archives of Namibia.

suffered the additional loss in the summer of 1905 of both his wife (in Germany) and his nephew (some reports say his son) being killed in a battle with the Nama. (His grave may be seen near the southern end of today's Fish River Canyon hiking trail.) In September Von Trotha requested that he be relieved of his command.[27] He left South West Africa in November 1905, having been replaced as Governor by a civilian, Friedrich von Lindequist, who'd previously served in the territory under Leutwein.

Concentration camps

Von Trotha had been responsible, under instructions from Berlin, for the establishment of internment or concentration camps to accommodate male prisoners of war as well as captured or surrendered Herero and Nama women and children. The first camps were primitive structures enclosed in a barbed-wire fence. Several camps were built early in 1905: The largest, with a capacity of 7000, was erected at Windhoek on the slopes below the Alte Feste (its inmates were used to build the Tintenpalast, the present seat of the Namibian government). Others were set up at Okahandja, Karibib, Swakopmund and Lüderitz (the notorious Shark Island camp). Because of the labour shortage in the territory created by the war, the authorities resorted to the deployment of inmates of the camps as forced, unpaid labour, made available for military, government and even private use. Women and girls in the camps were conscripted as prostitutes for the benefit of the German troops. Prisoners were used for the construction of the railway lines from Lüderitz to Aus and from Swakopmund to Otavi in the north. On the Lüderitz–Aus line, over two-thirds (1400) of the workers died during its 18-month construction.[28]

Recent research has revealed much about the conditions of these camps.[29] It is sufficient to requote the description given in December 1905 by Dr Heinrich Vedder, then a young missionary, later a respected historian and governmental advisor:

> At Swakopmund, the Herero "were ... housed in pathetic structures constructed out of simple sacking and planks, in such a manner that in one structure 30–50 people were forced to stay, without distinction to age or sex. From early morning until late at night, on weekends as well as on Sundays and holidays, they had to work

under the clubs of raw overseers until they broke down. Added to this, food was extremely scarce … Like cattle, hundreds were driven to death and like cattle they were buried."[30]

There was also little or no provision for sanitation; illnesses were rife. By the time the camps finally closed in 1908, possibly as many as 4000 inmates had died, half of them at Shark Island and surrounding camps in the Lüderitz area.

The flight of the Herero

News of atrocities committed by German troops filtered through to the fleeing Herero groups, convincing most of them that a return to the territory was not an option, despite the ordeal they faced in the Omaheke and elsewhere. Although many Herero groups took refuge in isolated hilly areas to the north and west of the Waterberg around water sources, the location of which was known to them alone, many others attempted to seek sanctuary in Bechuanaland. The precise route followed by Samuel Maharero and his group after leaving the battlefield at Hamakari on 12 August is not known, although he is reported to have first gone south to Otjinene on the Eiseb drainage. From there he almost certainly followed the omuramba downstream to the northeast, passing the waterholes of Epata and Osombo-Windimbe.[31] This was springtime in the Omaheke: Daily temperatures would have ranged from about 8 °C overnight to as high as 34 °C in the early afternoon. The summer rainfall regions of southern Africa also experienced the worst drought on record during the three seasons from 1903–1906, and it can be assumed that the Omaheke was even drier than usual in the spring of 1904, with little or no rain falling before November.

Arrival in Bechuanaland

We do not know the size of Samuel's group, how many women and children were included, nor how many cattle set off with them on the tortuous journey across the Omaheke. It is believed that the group passed through or close to the village of Gam on the SWA side of the border near the confluence of the Eiseb and Otjozondou omiramba, and that from there they travelled northeast to Nxai Nxai – a total distance of between 550 and 600 km from the Waterberg. There

CHAPTER 20: FROM WATERBERG TO WATERBERG   693

FIGURE 20.9: The Herero concentration camp below the Alte Feste, Windhoek.
Image courtesy of the National Archives of Namibia.

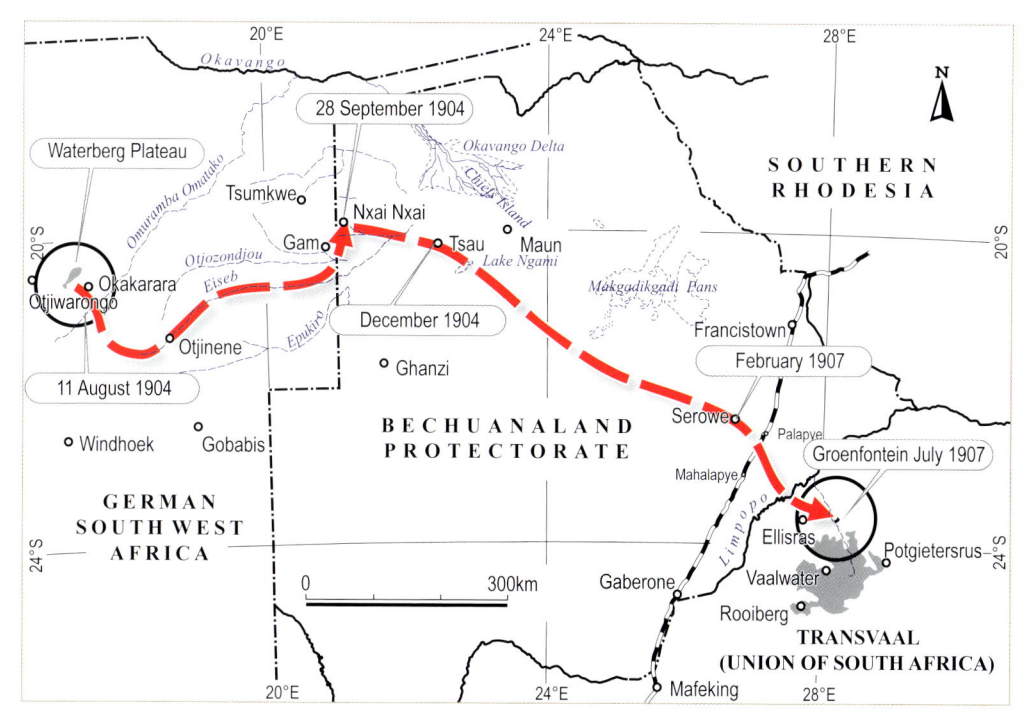

FIGURE 20.10: The route followed by Samuel Maharero and his clan between August 1904 and July 1907. The route is taken from maps included in the magisterial work *The Thirstland Trek*, by Nicol Stassen (2016), published by Protea Book House, Pretoria.

is no record of the number of men, women, children or cattle who fell by the wayside in the course of this epic journey, only that those who survived were wretched, poorly clothed, weak, thirsty and malnourished when they reached their destination. Six weeks after the battle at Hamakari, the *Kgoši* (chief) of the Batawana community in western Bechuanaland received a letter from Samuel, dated 28 September 1904 and giving his location as Nyainyai (Nxai Nxai, a pan near the Aha Hills), that read as follows:

> To the Magistrate,
> I am in the Batawanas country. I am writing to tell you that I have been fighting with the Germans in my country, the Germans were my friends, [but] they have made me suffer so much by the manner in which they troubled me that I fought with them. The beginning of the trouble was that I gave the English some boys to work at Johannesburg, this is the reason that they fought with me. An Englishman called Juda knows this; he was the man who came to get these boys. I have been fighting for eight months and my ammunition is finished. As I come into your country at Mogopa I ask help from Queen Victoria. In the olden times my father was friendly with the English government and on this account I come to the English government for succour and request permission to live in their country. I now ask you to have mercy on me and help me in my heavy trouble, please reply to me. This is my prayer to you that I may follow those of my people who have gone before me till I get there. If you will allow me I will leave here at once. Sir I ask you to answer me as soon as possible. I send my best greetings.
>
> I am the chief of the Damaras, Samuel Maharero.[32]

(Samuel was not to know that Victoria had passed away three years before and had been succeeded by her son, Edward VII.)

The Kgoši of the Batawana people of Ngamiland at the time of Samuel's arrival was Sekgoma Letsholathebe, the most powerful king in the history of the Batawana.[33] Much of Sekgoma's power derived from the patronage he'd acquired through the protection of numerous Herero groups who'd crossed into his territory since he came to power in 1891. Some Herero had started leaving SWA after the succession dispute that followed the death of Kamaharero, Samuel's father, in

1890. In particular, some chiefs or heirs who had aspired to succeed Kamaharero, but were usurped by Samuel, chose to start a new life in Bechuanaland rather than live under the rule of a paramount chief they distrusted or feared – especially after the executions of the Mbanderu leaders Kahimemua and Nicodemus at Samuel's hands in 1896.[34]

Sojourn in Bechuanaland

Sekgoma's capital had been alongside Lake Ngami. However, in 1896 the lake dried up for the first time in living memory, jeopardising the Batawana clan's water supply. At about the same time, the rinderpest epizootic swept through the Protectorate, killing the wild ungulate population of the Okavango delta. This in turn eliminated the infestation of tsetse fly in the region, which had hitherto prevented its development for cattle farming. Sekgoma seized the opportunity to take control of this newly available grazing land. He placed his herds on what is now called Chief's Island in the Okavango Delta and moved his capital from Lake Ngami to the springs at Tsau, about 50 km northwest of the lake.[35] When Samuel arrived at Nxai Nxai in September 1904, he was summonsed to Tsau to be interviewed by Sekgoma, who needed to balance the arrival of this controversial, yet prestigious refugee with the risk he posed for the critical relationships he'd established with Samuel's past enemies.

In the margin of Samuel's letter copied above, Lt Mervyn Williams, the acting magistrate at Tsau, who translated it, wrote the name "Hewitt" against the name "Juda". The historian Jan-Bart Gewald, who in the 1990s undertook detailed ground-breaking research into the story of the Herero (and whose works are extensively referred to in this chapter), has pointed out that Juda refers to an Englishman called Alex Hewitt, who was employed by the Witwatersrand Native Labour Agency (WNLA) to recruit Herero refugees to work on the South African gold mines. Samuel had heard the name pronounced in Afrikaans when it sounded like the biblical Judas.[36]

Samuel's request to be allowed "to follow those who had gone before" was a reference to several other groups of Herero who had already crossed into Bechuanaland since the outbreak of the rebellion nine months earlier. Hewitt and others had been busy since January both at Swakopmund and on the Bechuanaland border, recruiting destitute or displaced refugees fleeing from the conflict – in much the same way that raptors hover around the margins of a veld fire in the hope of snatching fleeing animals. They had already recruited many men to work on the mines (shipping them out via Walvis Bay to Cape Town and thence by train to the Witwatersrand). Their actions enjoyed the support of the British administration in the Protectorate. A letter from the British High Commissioner in South Africa (Lord Milner) to the Resident Commissioner for Bechuanaland as early as March 1904 commented on reports made by Hewitt and advised that Herero refugees should be made welcome in Bechuanaland because their cattle and labour could boost the economy of the Protectorate.

Lt Williams, the magistrate for Ngamiland at Tsau, to whom Samuel's letter had been addressed, had sent a report in April 1904 to his superior, the Resident Commissioner in Mafeking, saying that a large number of Herero men had gathered near Rietfontein (on the SWA side of the border) in preparation to cross the border. At the end of September and early in October, subsequent reports describe numerous small groups of Herero men, women and children in an advanced stage of distress having crossed the border to seek refuge. The reports suggest that in excess of 2000 Herero had managed to cross into Bechuanaland, but that they had left many more compatriots behind dying of thirst and hunger. In January 1905, the Resident Commissioner reported to the British High Commissioner that temporary arrangements had been made to accommodate all the refugees who had entered Bechuanaland, and that they had been disarmed and told that they would not be allowed to return to SWA without permission. Samuel Maharero and his young son Friedrich were among those chiefs named.[37]

In December 1905, the next acting magistrate at Tsau, Lt Merry, submitted a report to the Resident Commissioner in which he listed all the Herero chiefs and headmen who had settled in various camps close to the Bechuanaland border. Samuel and two of his sons were at Tsau, while his oldest son Friedrich had now moved across to Serowe in eastern Bechuanaland. The

Figure 20. 11: Friedrich, the son of Samuel Maharero. Image courtesy of the National Archives of Namibia.

total Herero population in the camps was given as 1175, with 550 cattle.[38]

Throughout this period, recruiting agents from both the WNLA and individual mining companies were active in procuring the services of Herero refugee males to work on the Transvaal gold mines that were suffering from a severe shortage of labour. Recruitment also resumed in SWA once the wars against the Herero and Nama ended, as there were many homeless, displaced people for whom a job underground in a mine was preferable to destitution.

Samuel, who had already been involved in sourcing members of his Herero community in Okahandja for recruiting agents before the outbreak of the Herero rebellion, soon resumed this role, prompted by his almost complete lack of any assets or wealth and by the cool relationship between him, other Herero refugee groups in Bechuanaland (like the Mbanderu) and his Batawana host, Sekgomo. When Samuel's bedraggled group first crossed into Bechuanaland, Sekgomo relieved them of their weapons and effectively banished them to a site on the Boteti River, on the boundary between Ngamiland and territory owned by the Bamangwato under the Tswana-speaking Chief Khama III.[39] Khama did not approve of this arrangement and by mid-1905, Samuel and his people (comprising his son Friedrich, ten other men with their families and about 30 cattle) had moved into the security of the police station at Tsau. As Gewald[40] has summarised, by now Samuel had lost all authority over the Herero people in Ngamiland, most of whom regarded him with fear or even disdain.

In 1906 Kgoši Sekgomo Letsolathebe, on his way home from medical treatment in Mafeking, was deposed by his nephew Mathiba in a coup supported by the British administration. Sekgomo was detained in the fort in Gaborone where he would remain for the next five years.[41] This change in control in Ngamiland had a profound impact on all Herero groups in the region, including the Mbanderu. Samuel Maharero's refugee status also became vulnerable, although he was considered valuable as a conduit for the procurement of mine labour. Indeed, it was this attribute that was to provide a route out of Bechuanaland for Samuel and his group.

Samuel and his group move to the Transvaal Waterberg

Late in 1906 the recruiting agent Hewitt, now working for a loose association of companies that would later become the Corner House Group (Rand Mines), approached Samuel with an idea: Would his people be interested in working on the Witwatersrand gold mines in exchange for being given the right to reside on privately owned land in the Transvaal?[42] No doubt Samuel would also receive payment for each labourer he was able to provide. Samuel, who by now was probably concerned that he and other Herero might be forced to return to German SWA, was receptive – although he was later to write that he'd been tricked by Hewitt.

Accordingly, Anglo-French, a gold mining company which like most of its counterparts owned numerous properties around the Transvaal, offered to make available one of its farms in the Waterberg: Groenfontein on the Lephalala River, about 16 km southeast of the Clermont police station at Villa Nora and a few kilometres beyond the northern boundary of the New Belgium Estate. This arrangement required the approval of the Transvaal government because it would set a precedent for the local settlement of foreign nationals. However, the already acute labour shortage on the Witwatersrand – the importation of indentured Chinese labourers had commenced in June 1904 – was aggravated in June 1907 by a strike by white mine workers which jeopardised production,[43] and government approval was soon forthcoming.

The Anglo-French agreement

On 4 June 1907, a contract was concluded between the Anglo-French Land Company of the Transvaal Limited, represented by its chairman, William Dalrymple, and the Transvaal Government, represented by its Minister of Lands, Johann Rissik.[44]

The agreement provided for the following conditions:
- Anglo-French would lease the surface rights of Groenfontein to the Transvaal Government for a period of 25 years, for the sole purpose of enabling the government to settle Chief Samuel Maharero and his tribe or followers there.
- While the chief and his followers would at any time be free to leave the farm, the government undertook to ensure that every male resident of 18 years or older would work for a period of at least six consecutive months in each year on the mines of the Farrar-Anglo-French Group in return for wages prevailing on the Witwatersrand. A separate contract would be concluded between the company and each of these male residents committing them to these conditions.
- The settlement on Groenfontein would be a Government Native Location;

any male resident failing to comply with the requisite working contract would no longer be eligible to live on the farm and it would be the government's responsibility to enforce this condition. The government would pay to Anglo-French a consideration of 1/- per month for each male over 18 years of age settled on the farm.
- The government would recognise Samuel Maharero as chief or captain of the natives to be settled on the land.

The British Governor of the Transvaal (Lord Selbourne, who'd replaced Milner in 1905) promptly approved the relocation of Samuel and his followers. According

FIGURE 20.12: Sir William Dalrymple (LEFT) and Johann Rissik (RIGHT). Sir William Dalrymple caricature by E.A. Packer (1928). Dalrymple (1864–1941), in addition to being MD of the Anglo-French Exploration Co., was a member of the Transvaal Provincial Legislative Assembly (1907–1910), a member of the Johannesburg Town Council (1901–1904), chairman of the Council of the University of the Witwatersrand and Hon. Col. of the Transvaal Scottish, amongst other roles. Edgar Arnold Packer (1892–1932) was a cartoonist with *The Star*, using the pseudonym "Quip". The caricature was reproduced as No. P40 in the collection *Catalogue of Pictures in the Africana Museum* (Volume 4, p. 53), compiled by R.F. Kennedy (1968), and published by the museum in Johannesburg. Johann Rissik portrait by Dr Irmin Henkel (1966). Johann Friederich Bernhard Rissik (1857–1925) had been Surveyor General of the ZAR. At the time of the Herero agreement, he was Minister for Lands in the Transvaal government; after Union, he became Administrator of the Transvaal. The portrait (in charcoal and light white on a pale brown background) was drawn by Dr Irmin Henkel, a German orthopaedic surgeon (1921–1977) who came to SA in 1951 and continued to practise both medicine and art while living in Pretoria. The work, probably copied from a photograph, was presented to Rissik's son Peter Ulrich in 1967. It was reproduced as No. H235 in the collection *Catalogue of Pictures in the Africana Museum* (Volume 7, p. 4), compiled by R.F. Kennedy (1972), and published by the museum in Johannesburg.

to the historian Gerhard Pool, the group that left Bechuanaland for the Waterberg in June 1907 comprised about 110 men, 140 women and children, 116 head of cattle and 30 head of small stock.[45] It is not stated how the group reached Groenfontein (by road well over 700 km to the southeast of Tsau), but presumably either the mining company or the Transvaal government provided transport. The report from the magistrate at Tsau in December 1905 listed six headmen at Tsau in addition to Samuel (including his two younger sons, Alfred and Johannes), but showed his oldest son Friedrich as being at Serowe in the east of the Protectorate.[46] Since Friedrich is known to have lived with the family at Groenfontein, it is possible that he and his group were picked up at Serowe on the way to the farm.

In his book about Samuel Maharero,[47] Pool described conditions at Groenfontein in some detail, based upon reports from Berlin missionaries who visited the group from Bobididi, a few kilometres downstream on the Lephalala River, and the following comments are summarised from his record. On their first visit, a few months after the Herero had arrived, the missionaries found the people to be in poor health, suffering from malnutrition as well as malaria, which was then

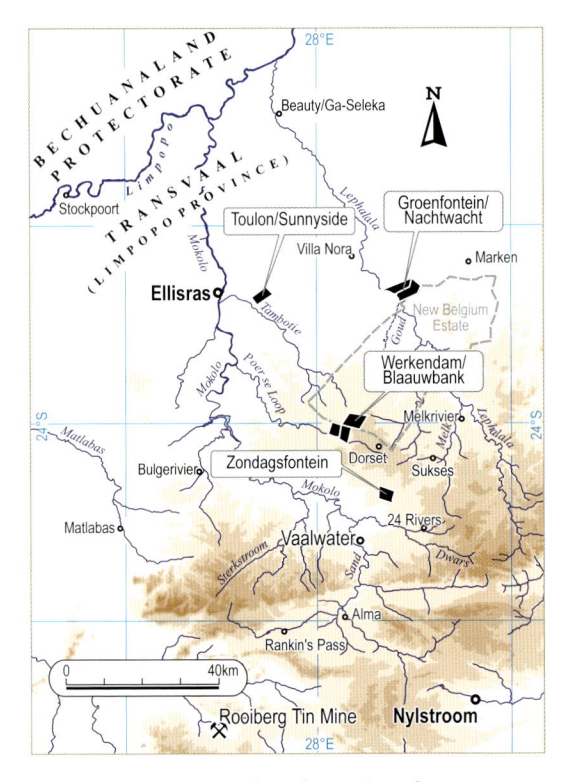

FIGURE 20.13: Location of Waterberg farms to which Herero refugees dispersed, 1907–1923.

endemic along the river. Many men were so weak that they were unable to go to work on the mines despite their contractual obligations, so they were not earning any money. Samuel himself went off to Johannesburg to seek employment, but returned to Groenfontein in about 1908.

It is not clear how many of the Groenfontein Herero eventually did go to work on the gold mines of the Witwatersrand, because in 1908 the Farrar-Anglo-French Group became involved in the development of a tin mine that had begun two years earlier at Rooiberg, just to the south of the Waterberg plateau (see Chapter 18). The company faced the same shortage of labour there as on its gold mines. After using Chinese indentured workers for a while, Anglo-French introduced its new "captive" Herero workers to the mine in 1908.[48] Sir William Dalrymple, who'd concluded the contract to bring the Herero to South Africa, would become chairman of Rooiberg in 1910 and retain that position until his retirement in 1941.[49]

However, the experiment to use the Herero on the mines was not successful for several reasons. Firstly, the Herero, with their proud tradition of pastoralism, struggled to adapt to crop-based agriculture on Groenfontein. Secondly, their customary haughty attitude did not lend itself well to the hierarchical conditions of mine employment – not to mention the working environment underground. Whereas individual Herero men, hired via Samuel and others in Bechuanaland and in SWA may have been forced to adapt to their new conditions, those from Groenfontein failed to do so.

By 1912 Anglo-French had concluded that its plan could not work: It negotiated the cancellation of the lease on Groenfontein with the government (now the Union) and ordered the Herero to vacate the property.[50]

Dispersal of the Herero

This development caused the widespread dispersal of the Groenfontein Herero community. Some moved onto the adjacent farm Nachtwacht either as employees or as tenants; others went to the farms Toulon and Sunnyside, just east of Ellisras, where they worked for Elizabeth Hunter's grandfather, Ted Davidson, and other farmers.[51] Samuel Maharero, now approaching 60, moved with his family to Werkendam in the south of the New Belgium Estate, which they rented from TCL. Apparently, 20 men, 24 women and eleven children lived with Samuel there, together with 260 cattle and 70 small stock.[52] According to records of the Waterberg Native Commissioner's office in Nylstroom in 1914, Samuel's sons

Friedrich and Alfred were amongst this group, while other Herero lived on the adjacent TCL farm Blaaubank and yet others on Zondagsfontein some 15 km north of Vaalwater.[53] At this time, TCL was developing large herds of cattle on its Waterberg farms[54] and it would have made sense for the company to have employed the subcontinent's finest cattlemen to manage them. By October 1917, Herero were living on several farms in the Waterberg district.

It is of interest that the record of the Native Commissioner referred to above also lists a total of 31 Herero men who were said to be proceeding to Cape Town in September 1914. Were they on their way home, or were they part of South Africa's preparations to invade German SWA? Gewald refers to records that Samuel Maharero made available to the Union Defence Force soldiers from amongst his followers for the invasion of SWA, "in the hope that, as his son put it, if the country was taken back from the Germans it would be given back to Samuel and his people".[55]

Since the end of the Herero/Nama-German War, the territory had also experienced a severe labour shortage that had crippled the economy. After the formal closure of the concentration and labour camps in 1908, the German administration had

FIGURE 20.14: Samuel Maharero and family at Groenfontein. At the back (LEFT TO RIGHT): sons Johannes, Alfred and Friedrich. Front (LEFT TO RIGHT): wife Johanna, Samuel, daughter Milka. Baumann, Julius (1968): *Van sending tot kerk: 125 jaar Rynse Sendingarbeid in SWA, 1842–1967*. Karibib, Namibia, p. 81. This image reproduced from Pool (1991), p. 289 with the kind permission of Laurentius Davids on behalf of the publishers, Namib Publishing House (a subsidiary of Macmillan Education Namibia).

repeatedly appealed to members of the surviving Herero and Nama communities to come out of hiding, to return to their lands and to fill employment vacancies. Many Herero (and to a lesser extent Nama) men returned from their places of refuge in Owambo, Angola, Bechuanaland and South Africa. Some sought work whilst others attempted to locate family members and to resettle lands they had vacated. This trend accelerated after the defeat of the Germans in June 1915 and their replacement by a South African administration.

At some point during this period, probably in 1911 or 1912, before the group had dispersed from Groenfontein and while Samuel was still regarded as the paramount chief, Eugène Marais, who had been appointed Resident Justice of the Peace in the Waterberg, made his first visit to the Herero community, presumably in an effort to help resolve their precarious residential status. In an account published almost a decade later, Marais related how he was struck by the physical stature of the people – tall, slim, and elegant and their fluency in High Dutch and German. They were fastidious about personal hygiene and mostly adherents of Christianity. Through interviews with Samuel and a preacher called Julius Kauraisa, Marais learned of their ordeal in the Omaheke and came to understand their current plight as a stateless people driven from their homeland, bereft of their homes and assets (their cattle), and reduced from their proud pastoralist past to working as miners, general workers and servants.[56]

Ted Davidson of 24 Rivers met the Herero during his trading expeditions into the bushveld in the 1900s. His daughter, Elizabeth Clarke, remembered her father's description of the striking appearance of the women:

> [T]all and graceful in build, [they] wore frocks of German print, full, long skirts, billowing out over paddings of numerous petticoats and swept up into a curved waistline, high in front and moulded into a close-fitting bodice. Spotless white aprons were worn over the frocks [and they] ornamented themselves with collarets of lucky beans ... On their heads they wore high turbans, built up from stiff padding of oddments of German prints.[57]

Samuel Maharero applies to return to his homeland

By 1919 Samuel Maharero, now about 64 and in ill health (although he is said to have given up alcohol), was desperate to return home and to have his own land.

He was also being hounded by the South African government which, having passed legislation restricting the rights of black people to live in "white areas", objected to concentrations of Herero on white farms, yet refused to grant them an area (a "location") of their own. In February that year, Samuel followed up a visit to the Bechuanaland Resident Commissioner in Mafeking with a letter, appealing to the British administration to allow him to purchase some property on the western margin of the Protectorate for himself and his people – or alternatively, to enable them to return to their homeland. He pointed out that he was currently living on rented land over which he had no tenure and from which he was liable to eviction at short notice.[58]

Notwithstanding the urgency and sincerity of Samuel's appeal, a full year went by before the office of the British High Commissioner in Cape Town forwarded Samuel's letter to the Governor General in Pretoria. While sympathetic to Samuel's appeal, the High Commissioner felt it

> inadvisable that any steps should be taken with a view to the settlement of [Samuel] Maharero and his followers in the neighbourhood of the conquered [territory] until the future of that territory had been finally determined. The Treaty of Peace with Germany has now been ratified, but before considering the matter further, [the High Commissioner] would be glad to know whether the Union Government or the Protectorate authorities in Windhuk [sic] would see any objection to the settlement of [the Herero] in the South-West district of the Bechuanaland Protectorate, in the event of suitable land being available there.[59]

Within the comparatively short period of two months, the Governor General was advised that neither the Union government nor the administration of the South West African Protectorate had any objections to the settlement of Maharero and his people in western Bechuanaland and this information was relayed speedily to the Resident Commissioner in Mafeking.[60] Time went by once again. In the meanwhile, in July 1920, Friedrich Maharero and eight of his group were given permission to visit SWA ostensibly to raise money to enable them to purchase a farm in the territory for Samuel Maharero. While in the territory, Friedrich met Hosea Kutako, a member of the Okahandja Herero royal family who had been injured in the war with the Germans, had survived and was later appointed as head of the Herero community in Windhoek. Together, they visited the Orumbo reserve northeast of Okahandja. There, a ceremony was held in which Friedrich, on instructions from Samuel, proclaimed Kutako (then about 50 years old) to be his father's representative in the territory.[61]

Repatriation approval granted – and rescinded

At last, in July 1921 – 15 months after approval in principle had been granted for the settlement of Samuel Maharero and his people in western Bechuanaland, there was a communication from the Resident Commissioner's office in Mafeking. Clearly, this was an office where not much happened. The letter, evidently written in response to an enquiry from the Governor General (HRH Prince Arthur of Connaught) as to progress with the matter, reported that in February 1921 a meeting had been held with one of Samuel's sons, Johannes, to discuss possible sites for settlement. The Herero wanted to settle near Kalkfontein, up against the Bechuanaland border with SWA, in the corner formed by the border immediately east of Gobabis. The Commissioner opposed this site (mainly because he was concerned it would offend the Boers living in Gobabis and at Ghanzi, although he also said the area was already occupied by Mbanderu) and instead proposed a remote site about 120 km south of Ghanzi. This site, however, was found to have limited surface and underground water potential. Consequently, the parties could not agree on a possible settlement site and the Commissioner requested that the Governor General advise Samuel accordingly. In September the Commissioner's secretary advised Johannes Maharero of this conclusion directly – after which the Mafeking office no doubt resumed its usual level of somnolent inactivity.[62]

In the meanwhile, it had been decided by the League of Nations (an alliance of countries involved in Germany's defeat) that former German colonies, including SWA, should be administered by "advanced" nations granted mandates to do so. SWA was to be administered as a mandated territory by South Africa. Early in 1920 a commission was established by the interim South African administration in SWA (preparatory to the conclusion of a formal administrative mandate which was only gazetted in June 1921) to investigate the various treaties and agreements that had been concluded during German rule. Samuel Maharero was invited to give evidence before the Commission. He was overjoyed at the prospect of going home and immediately set about making contact with other exiled Herero leaders to accompany him. Unfortunately, in the process of doing so, he missed the deadline to testify.[63] One cannot escape the uneasy conclusion that lack of governmental interest in his case and general bureaucracy on the part of Union government officials was largely to blame for this disappointing outcome.

These setbacks were followed by another one. Early in 1922, Samuel managed to obtain permission – and the relevant documentation – from the Union Native Affairs Department to travel to SWA with eight colleagues. However, this was apparently granted without prior approval of the new SWA administration, which forced the cancellation of the trip.[64]

This succession of frustrations, all of which could be blamed largely on a combination of poor communications and inept (if not malicious) bureaucracy by Union and British petty officialdom, inevitably caused Samuel to become dispirited. Unable to remain any longer in South Africa, he and his remaining followers on Werkendam sought sanctuary with Khama III at Serowe across the border in Bechuanaland. During December 1922 they trekked there on foot with their stock and belongings.[65]

The journey, coming on top of the disappointment of being unable to secure a place of permanent residence either in his homeland or in western Bechuanaland, was too much for the old man, who, it soon transpired, was also terminally ill with cancer. Perhaps a final blow was the death, on 21 February 1923 in Serowe of his only real benefactor, Kgosi Khama III of the Bamangwato, at the grand age of 86.

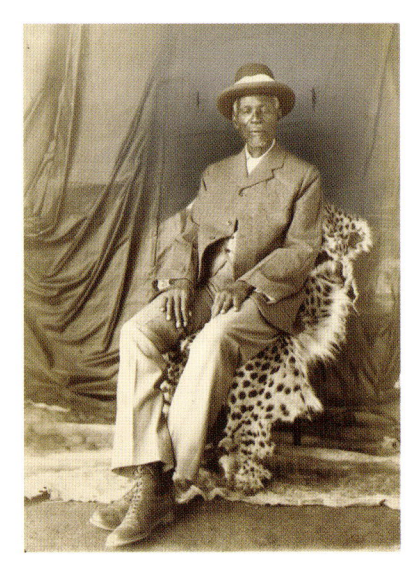

FIGURE 20.15: Samuel Maharero (LEFT) and Bamangwato Kgosi Khama III (RIGHT). Sketch of Samuel Maharero by C. Bleach drawn from images supplied by the National Archives of Namibia; photograph of Kgosi Khama III – source unknown.

Three weeks later, on 14 March 1923, Samuel Maharero died, also in Serowe. He was perhaps 68. The certificate from the district surgeon was insightful:

> Chief Samuel Maharero died this morning. He has been suffering from malignant disease of the stomach – the immediate cause of death being exhaustion and heart failure. He had many other complications but the main cause of death was the malignant growth. Signed, A Worrall, MB ChB., MD, BP Government, Serowe.[66]

Before his death, Samuel had expressed a wish to be buried next to his ancestors in Okahandja. He also appealed to those leaders who survived him to unite in rebuilding the Herero nation, implying that his eldest son Friedrich should be the nucleus of this revival. Indeed, at a meeting of local Herero in Serowe in May, Friedrich was confirmed as his father's heir and in a letter to the local magistrate, Friedrich signed himself as "Chief of the Maharero",[67] although this title had not yet been conferred on him by the Herero communities living in SWA.

Samuel Maharero finally returns to the ancestral Herero homeland

News of the passing of Samuel plunged the Herero community in SWA into mourning – notwithstanding the divisive nature of his rule prior to his exile almost 30 years before. Plans commenced immediately to bring Samuel's body back to Okahandja for burial, although these were met with some apprehension by the administration, which was concerned that they might lead to renewed violence amongst the Herero people. Eventually though, on 19 August 1923, Samuel's remains, placed in a lead-lined coffin, left Serowe by train, accompanied by Friedrich and seven other Herero men. The cortege arrived in Okahandja on Thursday 23 August to be met at the station by a mounted honour guard and 1500 uniformed men from the various Herero regiments. A procession, led by Hosea Kutako (the man whom Friedrich had appointed as Regent in 1920), accompanied by Traugott Maharero, the family's senior surviving representative in Okahandja, and the town's magistrate, together with Friedrich then escorted the coffin to Traugott's house to be laid in state.

The funeral was held in the town's Rhenish church on Sunday 26 August according to a programme determined by the Herero community, with the missionary Heinrich Vedder delivering the sermon. The Herero had arranged a huge ceremony, making good use of their own instinctive military aptitude and their experiences under German rule. Over 2500 uniformed men (wearing a variety of uniforms of German and British origin), 170 riders and a brass band, together with an enormous crowd of men and women attended the event. The crowd was addressed by both the local magistrate and senior authorities, with Hosea Kutako in the role of Master of Ceremonies. After a brief private ceremony, Samuel's remains were laid to rest next to the graves of his father Kamaharero and grandfather Tjamuaha. The worried authorities had to admit being impressed by the dignity and order with which the whole event was managed.[68]

FIGURE 20.16 (TOP): Samuel Maharero's funeral procession in Okahandja. Image courtesy of the National Archives of Namibia.
FIGURE 20.17 (RIGHT): Grave of the Herero kings in Okahandja: Tjamuaha (died 1859), Kamaharero (died 1890) and Samuel Maharero (1856–1923).

Friedrich Maharero, who succeeded his father as paramount chief of the Herero, was not allowed by the administration to remain in the territory, despite the appeals of most Herero leaders on his behalf (the leader of Omaruru being a notable exception). He had apparently been involved in clandestine meetings with members of the Rehoboth Baster community to discuss a black nationalist movement being promoted by Marcus Garvey, a Jamaican pan-Africanist, and was considered a security risk by the authorities. Before leaving for exile in Bechuanaland (he was allowed to return only in 1952, a few months before his death), Friedrich appointed Kutako as paramount chief in his stead.[69] Kutako, who died in 1970, aged 100, would rise to become one of the first nationalist leaders in SWA and a leading player in the process that would ultimately result in the formation of the independent Namibia.

A Waterberg legacy?

What of those Herero people left behind in South Africa – those who had dispersed from Groenfontein following the cancellation of the Anglo-French contract and the promulgation of the 1913 Natives Land Act? There is almost no

FIGURE 20.18: Friedrich Maharero (LEFT) and Hosea Kutako (RIGHT) late in their lives. Images courtesy of the National Archives of Namibia.

documentation of the residual Herero communities or families that remained in South Africa after the 1920s. Elizabeth Hunter's book contains a photograph of her grandfather standing with a Herero housekeeper in the bushveld, evidently taken in the 1950s. And there are some reports of Herero communities related to mineworkers and their families on the East Rand from the 1960s. In 2011 a journalist, her interest piqued by Marais's account of the Waterberg Herero, set out to find any descendants who might still be living in the region. Eventually, she came across a family near Beauty (Ga-Seleka) on the lower Lephalala River whose ancestors were from Omaruru and who had travelled across the Omaheke with Samuel Maharero.[70] But apart from these few isolated known contacts, there will surely be numerous descendants of Samuel Maharero and other Herero clans still resident in the Waterberg and surrounding regions. There is said to be a significant Herero community still living in and around Serowe in Botswana, probably descended from Friedrich, who spent his exile there.

Another Waterberg link

Today, in the beautiful Waterberg Plateau National Park where the formidable red sandstone cliffs still face the flat waterless wastes of the Omaheke and perennial springs still flow onto the plain from the foot of the escarpment, there are few reminders of the battle that took place here over a hundred years ago, or indeed of the prior and subsequent settlements of German missionaries, soldiers, police and farmers. Outside the park amongst the bush lies a forlorn, unkempt walled cemetery housing the graves of the great Herero chief Kambazembi and his descendants. Within the park a neat graveyard contains the remains of the

German soldiers who had died during the conflict. The first casualties were on 14 January 1904 when Kambazembi's men, heeding Samuel's instruction, killed all 17 men posted at the settlement, sparing only the missionary Eich, women and children.[71]

Amongst those who died that day was James Charles Watermeyer, an agricultural and chemical analyst born in Graaff-Reinet in 1864. Watermeyer had worked for the Cape government under the surveyor and geologist Thomas Bain, but in 1899 he had resigned to join the German Imperial Government in South West Africa.[72] Watermeyer was one of 17 children of Gottfried Andreas Watermeyer and Anna Suzanna Ziervogel, a distinguished Eastern Cape family.[73] An elder brother, Frederick Stephanus Watermeyer (Jr), was a leading land surveyor in the colonial government of the Transvaal after the South African War. Coincidently, at the time of James's death, Frederick was engaged in the resurveying of farms in the Waterberg district of the northern Transvaal. His name appears on many of the title deeds records for the area – although not on that of Groenfontein, Samuel Maharero's temporary home.

FIGURE 20.19 (RIGHT): Headstone in the German cemetery at the Waterberg Plateau National Park, including the name of *"techniker"* James Watermeyer, (born in Graaff-Reinet on 8 June 1864 and died in Waterberg on 14 January 1904). FIGURE 20.20 (BOTTOM): The escarpment of the Waterberg plateau (Namibia) today.

CONCLUSION

Echoes from the hills

One would like to hope that, to some extent, the future of a society should depend on the degree to which it has studied, appreciated, absorbed and learned from its past. The relatively peaceful advent of a universal democratic government in South Africa in 1994 – which few thought would be possible – inspired not only South Africans but communities around the world to believe that here at last was a society determined to learn from its past mistakes, to redress historical injustices and to build a new nation based on mutual respect and equality of opportunity, regardless of creed, gender, colour or race. South Africa became a global poster child for the merits of reconciliation and democracy.

A quarter of a century later, the "rainbow nation" has lost most of its hue. Its much-vaunted Constitution and the safeguards thought to have been carefully enshrined within it have been repeatedly challenged, threatened, ignored and disparaged. The belated eviction of Jacob Zuma as president in February 2018 and his replacement (though by the most slender of majorities) by Cyril Ramaphosa may well signal, as the new president has promised, a "new dawn" for our country. But so much has been stolen or forfeited, so much credibility has been lost, so much needs to be rebuilt, that the twilight before sunrise could be long and clouded.

The white population of the country, though now marginalised politically, must share the blame for this tragic decay in what had the potential to become the brightest light on the African continent. Whites have steadfastly remained in denial, privately, if not in public, about the evils that were perpetrated by us, our parents and their ancestors upon the majority of the people of this country, usually in the name of "civilised governance". This book is replete with examples of how we advocated, or at least tolerated, the imposition of gross injustices on the black – or more correctly, nonwhite – population that we would never have countenanced being inflicted upon ourselves or our kin. And yet many amongst the white population still refuse to acknowledge these sins, let alone take steps to atone for them. Instead, we fabricate flimsy arguments to justify – and even to try and resurrect – the bigoted policies our forefathers wrote, and waste our time blaming each other for our own shortcomings.

In the Waterberg many Afrikaners still cannot forgive the English for what they have been taught to believe were cruel, deliberate atrocities committed against their forebears over a hundred years ago. This even despite evidence that in most instances these "crimes" were deliberately exaggerated by Afrikaner republican spin-doctors like Gustav Preller and his ilk, and that many others were the unfortunate but inevitable result of what would nowadays be termed wartime "collateral damage". This is not in any way intended to condone the mandated atrocities that did take place (those of the Breaker Morant gang for example), perpetrated by *both* sides in that conflict, or the draconian farm-burning campaign of Kitchener and his callous officers, or the poorly conceived and appallingly implemented internment camps. But surely enough time should have passed by now for us all to be able to review that history – as I've attempted to do in this work – more dispassionately and objectively rather than with the intention still of identifying points to score against the "other side".

English-speaking residents of the Waterberg (and I am one) are by and large astonishingly ignorant of – and often uncaring about – the injustices, insults and arrogance heaped upon a truculent rural Boer community by our strutting Imperialist forefathers, who were, truth be told, only a few decades into their own post-agrarian economy. Remarkably, English-speakers will often still spring to the defence of rogues like Cecil John Rhodes (I was schooled in the country named after him and indoctrinated to sing his praises); Rhodes, who used the façade (and fraud) of "British Imperialist Good" to advance his own narrow money- and ego-driven goals at the expense of everything and everyone else. In the same way, we are inclined to excuse the excesses of Lord Milner and his coterie of pampered toffs, appropriately called his "Kindergarten". How on earth was Milner able to justify his award of former Boer farms – ravaged and destroyed by his compatriots' army – for the settlement of demobbed British servicemen and then insist on the use of English as the medium of instruction in a territory where this was the language of an invading minority?

The short answer is English arrogance – an arrogance born of the glorious Victorian Imperial era, but which like the Queen herself, had already passed. Milner and his dandies were determined to rub the Boers' noses in the remains of their defeat – which had come about, let's admit, through a vastly superior force of numbers rather than through sharper military acumen. There can be no excuse for the British actions in 1900–1902. They were callous, unjust and unbecoming of a country that considered itself haughtily to be the most civilised in the world.

Several British officers during the war (Colenbrander for example) were little more than mercenaries. To their credit, many Britons, not least Emily Hobhouse, sensed the wrongdoing of the British forces and did what they could to put wrongs to right, as did many English-speakers in the Cape Colony. But it is long past time for the present English-speakers of the rural former Free State and Transvaal, including the Waterberg, to face up to the wrongs perpetrated by our ancestors upon those of our friends and neighbours and to apologise for them.

From the advent of Milner's British colonial government in the Transvaal in 1903 – 45 years before the election victory of Malan's National Party in 1948 – legislation began to be passed that would limit, erode and eventually remove the political rights of people other than whites. It is just too easy for English-speakers (*none* of whom of course *ever* voted for the National Party!) to blame the un-doubted iniquities of apartheid on Afrikaner Nationalists. This is simply not true. The cornerstone Natives Land Act was passed by the Union government in 1913 when Generals Louis Botha and Jan Smuts, leaders of what would eventually morph into the United Party, were at the helm and the Afrikaner Nationalists (under Hertzog) were in opposition! Whilst the forced removal of Indians in Nylstroom and Warmbaths to the artifices of Asalaam and Jinnah Park respectively may only have taken place in 1953 under National Party rule, we cannot ignore that from the days of the British colonial government in the Transvaal, Indians were denied the vote, restricted as to where they could live and trade and even prohibited from using the sidewalks. Mahatma Gandhi's *satyagraha* campaign on behalf of Indian rights took place in 1906 under British colonial administration. The post-flight tribulations of the Herero refugee Samuel Maharero and his clan can be laid squarely at the feet of the somnolent British Resident Commissioner for Bechuanaland, ensconced expediently as he was, in Mafeking, rather than in the country he was supposedly administering. It was the SA government under the premiership of Smuts, not Strijdom (an acknowledged anti-Semite) that banned the further immigration of Jews in 1936.

It's time to face the facts: Save for a tiny, embattled, vilified group of genuine English-speaking liberals (the Liberal Party of Alan Paton in 1953 and a fringe group in the United Party that included Jan Hofmeyr and Helen Suzman) and a handful of Marxists, *virtually all* whites north of the Orange River, Afrikaans- or English-speaking, favoured a policy of white supremacy and the relegation of black people to the status of disenfranchised, subordinate employees, allowed to live only in places demarcated for them by their white masters. The proof, if this statement should be disputed, lies in the voting record of those politicians whose party formed the government from Union until 1924, and again, in coalition from 1933–1948 and the official opposition thereafter: the United Party, its antecedent, the South African Party, and its successor, the New Republican Party. Hypocrisy defined.

At least the Nationalists made no bones about their separate development intentions, odious and unsustainable though these may have been. The "Sappe" on the other hand, practised the artful deceit learned from their perfidious English handlers – *pretending* to be opposed to apartheid but actually endorsing it. Only the methodology and language of its implementation might have differed. (With their questionable integrity and fluid morality as tickets, many members of the UP would not be out of place on the benches of today's ruling party, whose MPs express faux astonishment and feign ignorance about the gross pillaging of national assets brazenly committed by their colleagues.)

Let us acknowledge, it was only in 1948, after the Nationalists came to power, that government (outside of Natal) for the first time claimed responsibility for the education of black children. Of course, they did so with Verwoerd's demeaning "hewers of wood and drawers of water" ideology in mind – but at least they grasped the nettle for so long delegated to missions and other private institutions, and funded by iniquitous poll taxes levied on people who already had so little. At least there began the development of a government-led education policy and programme for the largest population group in the country; a responsibility that ought to have been insisted upon at Union in 1910.

I repeat, my standpoint is not to defend the indefensible. Far from it. The apartheid policy endorsed one way or another by the vast majority of white voters, irrespective of language group or party affiliation for 90 years from 1903 (and covertly advocated by some delusional *bittereinder, swartgevaar* proponents to this day), was a travesty of human rights. A travesty which has bequeathed the nation a suite of deeply entrenched distortions, prejudices, inequalities and skewed demographics that will take decades to resolve, even with a responsible, caring, competent, future-oriented government – the semblance of which has still to make its appearance. We white citizens of South Africa need to accept the blame for that legacy. And having acknowledged our culpability, take active steps, which might well impact negatively on our lifestyles and financial wellbeing, to put in place behaviours, practices and policies that will help rectify these distortions in our national society in ways that will deliver a country where all our children and grandchildren will choose to live. For if we are honest with ourselves, it is difficult to think of somewhere – anywhere – else as beautiful, as vibrant, as engaging, as rich in human potential, faunal and floral diversity with as many challenging, yet attainable possibilities, as this country in which we live.

But the history of the Waterberg, as I've found it to be, is not just a "white blame" story. There are all too few heroes in this history and the historical black population is by no means exempt from criticism, because the record – to the extent that it is documented – shows otherwise. Historians differ on their interpretation of the details of black polities before the tumult of the early nineteenth century, often because of the conflicting and incomplete content of the oral testimonies they tried to record. No doubt some early historians and ethnographers – they were all white – filtered what they heard and recorded in order to paint a particular preconceived picture of how they felt pre-white history *ought* to have been. Many of their modern academic counterparts do the same; but their agenda now is one of unqualified vilification of whites/imperialists/colonialists/capitalists/ liberals (choose any one) irrespective of merit or context. Yet there is little doubt, or contention, that the black population of the Transvaal Highveld in the late eighteenth and early nineteenth centuries was fluid, confused and confusing, as numerous chieftaincies manoeuvred to optimise their location and authority. Factional disputes – often viciously and terminally violent – characterised clan politics. The waves of early Ndebele invasion and settlement on the alluvial plains of the Mogalakwena River from the late seventeenth century onwards were anything but benign. Pre-existing polities were vanquished, eliminated, absorbed into the victors' community or forced to flee, depending on circumstances and personalities. Women and children in particular were mere spoils of war. Tyrannies (for example, those of the Masebe brothers in the late nineteenth century) were common and brutal.

The invasion by Mzilikazi and his Matebele in the 1820s threw what little order prevailed into upheaval. Whole chieftaincies, not just factions, were killed or put to flight (for example, the annihilation of the Hurutshe) and their capitals (like Kaditshwene, the largest human settlement north of the Cape) destroyed by marauding impis. Relatively settled societies were thrown into complete disarray and forced into refuge, often in remote locations inappropriate for the growing of crops or stock herding. Traditional oral historical records (tales) were interrupted or lost, making the reconstruction of many societies' pasts difficult or conjectural. Indeed, so severe was the disruption of settled Sotho-Tswana society on the Transvaal Highveld by Mzilikazi's impis that it would not be out of place – given the modern proclivity for recompense – if the Matebele polity of southern Zimbabwe were to be sued for reparations.

Only the settled Ndebele polities (Langa and Kekana) along the Mogalakwena valley seemed to escape the Matebele predations – perhaps because the latter were unexpectedly interrupted by the arrival of the first Boer trekkers, prompting Mzilikazi to refocus his attention on *their* eviction. To that extent, therefore, the peoples of the Highveld and further north would surely have welcomed the distraction caused by the advancing Boers, for they were undoubtedly spared further oppression from the Matebele. For their part, the Boers took advantage of the dispersed black communities on the Highveld to claim, rather ingenuously, that the land before them was empty and available for occupation (as long as they made sure not to notice the skeletons and ruined villages along the way). In consequence, Sotho-Tswana peoples essentially exchanged Matebele hegemony for that of the Boers – and who's to say, 180 years later, which might have proved the more oppressive?

What has been made clear from both archaeological and historical research, is that for a reason that further work will no doubt reveal, the high ground of the Waterberg plateau was almost unpopulated in the first half of the nineteenth century, as it had also been for a thousand years before the arrival of the first (Iron Age) pastoralists. Boer settlers only moved onto the plateau in the 1860s and found it to be virtually uninhabited – to their dismay, because it meant they struggled to find labourers to undertake, on their behalf, the hard manual work entailed in farming. This fact, which has been supported by several angles of research, gives the lie to some claims lodged by opportunistic "community groups" on the basis that land occupied by them for generations had been seized by white settlers. All the hard evidence gleaned to date would contest this assertion in nearly every case, leaving only questionable, and unverifiable oral testimonies in their favour. For populist political reasons, the present government has been reluctant to reject such evidently spurious claims and has gone out of its way over the last decade to prevent or delay their being challenged in court. Every community claim in the Waterberg that has managed to be dragged to court so far has been rejected, although a number of much smaller individual claims have succeeded.

Another reality which this study of Waterberg history has reaffirmed is the poor agricultural potential across the plateau. For the most part, the soils are derived from coarse, siliceous sandstones, lacking in clay content and poor in moisture retention. Their general acidity leads to the dissolution and removal (leaching) of nutrients, leaving behind a residue that is unable to support successive crop harvests or high stock densities unaided by supplementary nutrients. Even once artificial fertilisers became available, their use could only be sustained under circumstances of subsidies and fixed prices. Although rainfall is above the national average, run-off is high. Underground aquifers are usually small, lenticular chambers (known as fracture-fill aquifers) which are rapidly depleted until refilled during the following rainy season – except in drought when they are not filled at all.

Of course, there are exceptions: areas underlain by intrusive diabase sills or dykes, which produce reddish, clay-rich, alkaline soils; and alluvial plains like that of the Sand River in the Alma valley. There are occasional fractures that do store and yield very large volumes of water. But by and large, the sourveld of the Waterberg plateau has never been able to sustain prosperous, cash-crop agriculture (apart from tobacco) – at least, in the absence of government-supported quotas and price controls. Only high-value products that are able to absorb the cost of continuous artificial fertilisation (or supplementary feed in the case of stud animals), herbicide and pesticide, might have some chance of success. This has never been an area of high potential for conventional agriculture. And the forecast consequences of climate change for the region over the remainder of this century will merely exacerbate this unwelcome fact.

The land claims issue in the Waterberg was born of good legislation enacted in 1994, but it has raised unrealistic expectations amongst claimants and extreme reactionary responses from landowners (most of whom, after all, purchased their properties quite legitimately and who are often in possession of letters from government assuring them that there are no claims on their properties). Politicians of every shade, whom the public should know to be career-driven opportunists and seldom concerned with either the national interest or the facts other than to advance their personal fortunes, keep raising the spectre or promise of land claims depending upon their objectives. The latest of these is the idea of land expropriation without compensation, a populist notion which, if implemented, would prove ruinous for the economy and foreign investment, and for the whole population. For evidence of the consequences, look no further than the north bank of the Limpopo River. But such tub-thumping idiocies attract votes – especially from those who know no better.

They do so at the expense of real but uncomfortable issues about which there is much greater consensus across all sectors of the community. The urgent need for high-quality basic education (from pre-school upwards) at rural state schools is the biggest of these. Rural children are already at an inherent disadvantage compared with their urban peers, by virtue of their upbringing, geography, lack of amenities (often including electricity) and lack of access to, or familiarity with modern technology. They should not have to suffer from inferior teaching either. The plight of rural Afrikaner youth a century ago and the creation of a white "underclass", which the government of the day attempted to accommodate through job reservation and sheltered employment in state institutions and the civil service, is an experience our government should be at pains to avoid, not emulate. Early childhood (pre-school) learning is almost nonexistent in rural areas of the Waterberg, followed by generally appalling primary education. A recent study found that 80% of children in Grade 4 nationally are unable to read comprehensively, even in their mother tongue. Imagine how much worse that number must be for rural schools! Practical assisted access to appropriate tertiary training (every word in that phrase being important) is another vital need, especially for students living far from the relevant institutions. Government today has the resources; what is needed is evidence that it has the will and the ability to learn from the past and that its priorities lie with the interests of its youth, rather than only with those of the unions.

Bilharzia is endemic in Waterberg rivers. A government survey published in 1980, based on comprehensive sampling, indicated a prevalence of the disease among our rural population at between 20% and 70%, concentrated in young children of schoolgoing age. Nothing of consequence has been done by government in the intervening 35 years to address this situation, notwithstanding that it was a signatory in 2001 to a goal by the World Health Organisation to treat at least 75% of all children at risk by 2010. Yet other African countries, much poorer than South Africa, embraced the programme and have made great strides in tackling the disease. This is possible because there is a cheap, readily available, easily administered treatment for it. Our health departments, national and provincial, would rather pursue political dreams like national health insurance and allow already disadvantaged children to languish in the debilitating torpor of bilharzia than do the hard work of implementing treatment programmes in the field.

Unfortunately, politicians and many government bureaucrats still behave much as they have always done. For the same reasons that Vaalwater languished for 40 years in an administrative backwater and the New Belgium settlement project of the 1940s failed because its viability from the outset was never challenged, so today national and provincial bureaucrats find it too troublesome (and politically too risky) to visit rural areas and make a sincere effort to understand local issues and take informed measures to resolve them.

All stories like to have heroes. But one person's hero may be another's villain. And many heroes turn out, once all the facts are eventually tabled, to have had feet of clay. The best of heroes are those whose songs were never sung, who sought no accolades, or who, unchanged by the praise they did attract, merely got on with the task at hand. In my analysis of Waterberg's history, such individuals have been few indeed. But a reading of musty files, journals and minutes does reveal a few personalities, described in these pages, whose selfless efforts over many years did much to enhance the lives of the communities in which they lived and worked. William Caine, I would argue, should count as a hero, notwithstanding his undue sense of British superiority. He shouldered a devastating family history to live alone on New Belgium through the South African War with fortitude and great diligence, faithfully looking after the property of his distant (unappreciative) employers, when most men in the same position would have either exploited the situation or fled. And he left us an invaluable, meticulous, candid record of his experiences, which can be used reliably to corroborate some stories and dispel favourite myths.

School inspectors like Leipoldt and Moerdijk travelled the region tirelessly under difficult conditions, working professionally and diligently for modest remuneration and against considerable resistance from parents and departments to improve the health and learning of pupils in desperate need of both. They sought neither gratitude nor recognition for their services. It is largely thanks to the dedicated, applied research of Botha de Meillon and his team that the Waterberg today (with many other parts of South Africa and the world) is mainly free of malaria – an achievement never adequately acknowledged by his countrymen in his lifetime or since. Ludvig Field, the quiet, unassuming Nylstroom philanthropist, must be remembered for his unqualified generosity towards the community of his second home – and later, towards his home village in Norway. And the heroic service of the Baber brothers, Charles and Colin, cannot pass without further mention. For a period of over 40 years, they toiled, without remuneration, on behalf of local agriculture and the community as a whole, lobbying apolitically for a diversity of services ranging from the provision of a government vet to television reception, from professional assistance with crop selection to the tarring of the roads.

It is a sad fact that the deeds of women in Waterberg's history have gone almost unrecorded (perhaps because it has been written mainly by men) – yet probably amongst them might be found the most heroic personalities of all. One who comes to mind is Judith Susanna "Sannie" Nicholson (née Marais) who, while her husband, Richard Granville, took off to join the Rhodesian Pioneer Column, stayed behind to care for the family farm; who on the outbreak of the South African War joined Cmdt Gen. Grobler's commando to nurse injured burghers; and who, after her husband's imprisonment, continued to nurse the injured from both sides of the conflict. Later, when Nicholson became a Member of Parliament, Sannie remained in the shadows, tending the farm and bringing up her large family. That record is nothing short of heroic! I have no doubt there have been many women like Sannie whose outstanding selflessness and courage have been overshadowed by their noisier and better-known spouses.

The survival and modest development of the Waterberg in years gone by was due largely to these and other unsung efforts of local community members, who persisted with their pleas for goods and services for decades while governments and the transient occupants of ministerial offices came and went. Farmers' organisations and church groups were usually at the forefront of these initiatives, patiently raising funds, lobbying recalcitrant bureaucrats or simply going ahead and building facilities themselves. The commissioning of the Ons Hoop and Ons Toekoms hostels in Nylstroom, the building of schools like Meetsetshehla, Mohlakamotala and the Waterberg Academy, the Farmers' Hall in Vaalwater and the creation of the Waterberg Welfare Society were community responses to the vacuum created by governmental neglect or inadequacy. Similarly, organisations like the Lapalala Wilderness School, Waterberg Nature Conservancy, the Waterberg Biosphere Reserve, Community Policing Forums and, more recently, the Waterberg Security Initiative, were voluntary responses by local citizens to augment or support educational and environmental agencies and the South African Police Services (SAPS) respectively, whose services are at risk of foundering through a lack of governmental funding, capacity or commitment.

What is to be learned from all this? The most obvious lesson must be that communities cannot and must not be lulled into reliance on politicians, political parties or formal governmental departments for the provision of many of the services they need. We must be prepared to roll up our sleeves, find the money, make the time and get down to work ourselves. In the past, communities that worked together to fund, build or manage facilities also built a community spirit, a spirit that occasionally transcended divisions of gender, race or creed. If the Waterberg community is ever to prosper, more opportunities for broad community cooperation must be sought and pursued.

There is so much that needs to be done in the Waterberg. To list a few:

- The education and skilling of our rural children, who shall inherit this land and who vitally need to be equipped to manage it well, or at least, better than their parents
- The environment, always at risk of catastrophic mismanagement by short-term opportunists if they are but allowed to establish a beachhead
- Security, an ever-present and growing concern, compounded by huge formal unemployment, undocumented (that is to say, illegal) foreign immigrants or refugees, and dysfunctional policing

These issues are vital to every resident of the plateau. If they are to be successfully addressed, they will require the cooperative participation and buy-in of *all* the communities of the Waterberg plateau. For too long we have attempted to deal with these matters within our own micro-communities, defined by colour or religious sect, language group, or even occupation. But these issues are much bigger than any one community. They affect, are dependent on and can only be addressed by the whole community.

We must resist the temptation to revert to the familiar but failed ideologies and prejudices of our past. We *have* to acknowledge the realities of the present (what *is*, rather than what we might prefer it to be) and be prepared to break new ground in forging alliances with former adversaries, but current neighbours, to build new community projects together for our mutual benefit. If this wonderful country in which we live is to have the future it deserves, it is time for us all to start thinking of ourselves as South Africans *first*. We must relegate our differing ethnicities, religious and political affiliations, genders and languages to a subordinate status whilst celebrating the variety and depth they add to our national culture.

If we can but learn and heed this one lesson from our tortured past, then Waterberg's history will have served its purpose.

RICHARD WADLEY
February 2019

ENDNOTES

CHAPTER 1: THE CHASING OF TALES, pp.13–18

1 www.localgovernment.co.za/districts/view/30/waterberg-district-municipality. Accessed 10 February 2017.
2 For example: W. Taylor, G. Hinde & D. Holt-Biddle (2003): *The Waterberg*. Struik, Cape Town; Clive Walker & J. du P. Bothma (2005): *The Soul of the Waterberg*. African Sky Publishing, Johannesburg.
3 For example: the collected stories of Cyril Rooke Prance; Elizabeth Clarke (1954): *Waterberg Valley* (unpublished, unfinished); WSG Boshoff (1977): *Sal Ek Ooit Vergeet*. Privately published, Nylstroom; Isabel Hofmeyr (1987): *Turning Region into Narrative: English Storytelling in the Waterberg*. History Workshop: The Making of Class, University of Witwatersrand, Johannesburg; Pat Greer (2004): *The Story of the Waterberg* (unpublished manuscript); Lex Rodger (2010): *Waterberg*. Privately published, Johannesburg; Elizabeth Hunter (2010): *Pioneers of the Waterberg*. Privately published, Lephalale; S.W. Breytenbach (2012): *Koedoesrand-Kronieke*. Privately published; Louis Trichardt (2016): *Swaershoek in die Waterberge*. Privately published.
4 Includes: Roelf Odendaal (1995): *Waterberg op Kommando* and (~2000): *Noordtransvaal op Kommando*, both privately published, Nylstroom; Lieutenant E.I.D. Gordon (Scots Fusiliers) (1901): unpublished diary; G.D. Scholtz (1941): *Generaal Christiaan Frederik Beyers*. Voortrekkerpers, Pretoria; G.D. Scholtz (1942, 2013): *Die Rebellie, 1914–1915*. Protea Boekhuis, Pretoria.
5 Including: Prof. A.D. Pont (1965): *Die Nederduitsch Hervormde Gemeente Waterberg, 1865–1965*. Kerkraad van die NHK Waterberg. Krugersdorp; J.H. Martinson (ed.) (1960): *Eeufeesbrosjure van die Ned. Geref. Gemeente Waterberg, 1860–1960*. Nylstroom; Prof. H. van der Wateren (ed.) (2009): *GKSA – 150. Bosveldfees*. Gereformeerde Kerk Waterberg, Nylstroom; D.W. van der Merwe (1984): Die Geskiedenis van die Berlynse Sending-genootskap in Transvaal, 1860–1900. *Argiefjaarboek vir Suid-Afrikaanse Geskiedenis, Ses-en-Veertigste Jaargang, Deel 1*. State Printer, Pretoria; Louis Nel (2005): Quoted in *Gedenkalbum, NG Gemeente, Vaalwater, 1930–2005*. NG Gemeente Vaalwater; Anon (1987): *A Summarised History of Anglican Worship in the Waterberg Parish of the Diocese of Pretoria for 80 years until 4th October 1987*. Church records, the Church of St Michael and All Angels, Nylstroom (Modimolle); Charlotte Wiener (2016): *The Jewish Country Communities of Limpopo/Northern Transvaal*. Privately published; J.J. Nel and others (Editors) (1966): *Nylstroom 1866–1966*. Published on behalf of the Nylstroom Town Council; Daniel van den Berg (1985): *Naboomspruit – 75 in 1985*. Published by the Naboomspruit Town Council; Delene Snyman (ed) (2008): *NTK 100 Gedenkblad*. A publication to commemorate the centenary of the organisation. NTK, Modimolle; Walter Kolver (1984): *MKTV Kooperasie Beperk: 75 Jaar – van Stryd en Voorspoed*. MKTV, Rustenburg; Ian Finlay (2011): *Welgevonden*. Firefinch Publications, Dublin, Ireland.
6 Gutteridge, Lee (2008): *The South African Bushveld – a field guide from the Waterberg*; Southbound, an imprint of 30° South, Johannesburg.
7 www.waterberg-bioquest.co.za
8 Other works that include a general treatment of the region's natural history include Clive Walker & J. du P. Bothma (2005): *The Soul of the Waterberg*, African Sky Publishing; and Lex Rodger (2010): *Waterberg*, privately published, Johannesburg. *Birds of the Waterberg – a quick guide*, by Philip & Warwick Tarboton (2011) and Retha van der Walt (2009): *Wild Flowers of the Limpopo Valley* provide more detailed coverage. A comprehensive bibliography of books relating to the Waterberg may be found on the website of the Waterberg Nature Conservancy: www.waterbergnatureconservancy.org.za.
9 Caine, W.H.A. (1902): *Rough Diary of Events and my Life during the Transvaal War, 1899–1901*. Noord

Brabant, New Belgium. Unpublished handwritten manuscript. A typed version was prepared by his nephew, G.W.R. Caine of Salisbury Southern Rhodesia in 1958. Copy in Manuscript Library of the University of Cape Town, Accession No. BCS.356; and at National Archives in Pretoria: Accession (*Aanwin*) No. A1443 (1).

10 Conita Walker (with Sally Smith) (2017): *A Rhino in my Garden*. Jacana, Johannesburg.

CHAPTER 2: DIGGING INTO THE PAST, pp.19–44

1 The name *Australopithecus africanus* means "southern ape" and the name was created by Raymond Dart. The type fossil was a skull found in 1924 in the Buxton limestone quarry near Taung (H.J. Deacon & J. Deacon 1999: *Human Beginnings in South Africa: Uncovering the Secrets of the Stone Age*. David Philip, Cape Town). Dart interpreted the specimen as intermediate between humans and living anthropoids. After World War II, Dart supervised the sorting of the Makapan Valley limeworks dumps by James Kitching and Alun Hughes and several specimens of *Australopithecus africanus* were recovered.

2 Huffman, T.N. (2007): *Handbook to the Iron Age: The Archaeology of Pre-Colonial Farming Societies in Southern Africa*. University of KwaZulu-Natal Press.

3 Schoeman, M.H. (2006): *Clouding Power? Rain Control Space, Landscapes and Ideology*. PhD. thesis, University of the Witwatersrand.

4 Sumner, T.A. (2013): A refitting study of late Early to Middle Stone Age lithic assemblages from the site of Kudu Koppie, Limpopo Province, South Africa. *Journal of African Archaeology* 11, pp. 133–153.

5 Mason, R.J. (1962): *Prehistory of the Transvaal*. Johannesburg: Witwatersrand University Press; Mason, R.J. (1988): *Cave of Hearths, Makapansgat, Transvaal*. Archaeological Research Unit Occasional Paper 21. Johannesburg: University of the Witwatersrand; Sampson, G.S. (1974): *The Stone Age Archaeology of Southern Africa*. New York and London: Academic Press; Sinclair, A. (2009): The MSA stone tool assemblage from the Cave of Hearths, Beds 4–9. In McNabb, J. & Sinclair, A. (eds) *The Cave of Hearths: Makapan Middle Pleistocene Research Project: Field Research by Anthony Sinclair and Patrick Quinney, 1996–2001*, pp. 105–137. Oxford: Archaeopress.

6 Herries, A. & Latham, A. (2009): Archaeomagnetic studies at the Cave of Hearths. In McNabb, J. & Sinclair, A. (eds) *The Cave of Hearths: Makapan Middle Pleistocene Research Project: Field Research by Anthony Sinclair and Patrick Quinney, 1996–2001*: pp. 59–64. Oxford: Archaeopress. The Brunhes-Matuyama magnetic reversal boundary at 780 000 years ago was not observed at the site so its occupations must be younger than this (p. 64).

7 Berna, F., Goldberg, P., Horwitz, L.K., Brink, J.S., Holt, S., Bamford, M. & Chazan, M. (2012): Microstratigraphic evidence of in situ fire in the Acheulean strata of Wonderwerk Cave, Northern Cape Province, South Africa. *Proceedings of the National Academy of Sciences of the United States of America* 109, pp. 7593–7594.

8 Mason (1962).

9 Tobias, P.V. (1971): Human skeletal remains from the Cave of Hearths, Makapansgat, northern Transvaal. *American Journal of Physical Anthropology* 34, pp. 335–367.

10 Pike, A.W.G., Eggins, S., Grün, R., & Thackeray, F. (2004): U-series dating of TP1, an almost complete human skeleton from Tuinplaas (Springbok Flats), South Africa. *South African Journal of Science*, 100, pp. 381–383.

11 Ogola, C. (2009): The taphonomy of the Cave of Hearths Acheulean faunal assemblage. In McNabb, J. & Sinclair, A. (eds) *The Cave of Hearths: Makapan Middle Pleistocene Research Project: Field Research by Anthony Sinclair and Patrick Quinney, 1996–2001*: pp 65–74. Oxford: Archaeopress.

12 Milo, R.G. (1998): Evidence for hominid predation at Klasies River Mouth, South Africa, and its

implications for the behaviour of early modern humans. *Journal of Archaeological Science,* 25: pp. 99–133.
13. Cooke, H.B.S. (1962): Notes on the faunal material from the Cave of Hearths and Kalkbank. Appendix 1. In R.J. Mason (1962), pp. 447–453; Cooke, H.B.S. (1963): Pleistocene mammal faunas of Africa, with particular reference to southern Africa. In *African Ecology and Human Evolution,* edited by F. Clark Howell and F. Bourlière, pp. 65–116. Chicago: Aldine Publishing.
14. Lombard, M. (2005): Evidence of hunting and hafting during the Middle Stone Age at Sibudu Cave, KwaZulu-Natal, South Africa: a multi-analytical approach. *Journal of Human Evolution* 48, pp. 279–300.
15. Wadley, L., Hodgskiss, T. & Grant, M. (2009): Implications for complex cognition from the hafting of tools with compound adhesives in the Middle Stone Age, South Africa. *Proceedings of the National Academy of Sciences of the United States of America* 106, pp. 9590–9594.
16. Wadley et al (2009).
17. Tobias, P.V. (1949): The excavation of Mwulu's Cave, Potgietersrust district. *South African Archaeological Bulletin* 4, pp. 2–13; Sampson (1974).
18. Plug, I. (1981): Some research results on the late Pleistocene and early Holocene deposits of Bushman Rock Shelter, eastern Transvaal. *South African Archaeological Bulletin* 36, pp. 14–21; Porraz, G., Val, A., Dayet, L., de la Peña, P., Douze, K., Miller, C.E., Murungi, M.L., Tribolo, C., Schmid, V.C. & Sievers, C. (2015): Bushman Rock Shelter (Limpopo, South Africa): A perspective from the edge of the Highveld. *South African Archaeological Bulletin* 70, pp. 166–179.
19. Mason (1962); Van der Ryst, M.M. (2006): *Seeking Shelter: Later Stone Age Hunters, Gatherers and Fishers of Olieboomspoort in the Western Waterberg, South of the Limpopo.* Unpublished PhD thesis: University of the Witwatersrand.
20. Mason (1962); Van der Ryst (2006).
21. Mason (1962).
22. Mason (1962).
23. Hutson, J.M. & Cain, C.R. (2008): Reanalysis and reinterpetation of the Kalkbank faunal accumulation, Limpopo Province, South Africa. *Journal of Taphonomy* 6, pp. 399–428.
24. Barré, M., Lamothe, M., Backwell, L. & McCarthy, T. (2012): Optical dating of quartz and feldspars: A comparative study from Wonderkrater, a Middle Stone Age site of South Africa. *Quaternary Geochronology* 10, pp. 374–379; Backwell, L., McCarthy, T.S., Wadley, L., Henderson, Z., Steininger, C.M., De Klerk, B., Barré, M., Lamothe, M., Chase,B.M., Woodborne, S., Susino, G.J., Bamford, M.K., Sievers,C., Brink, J.S., Rossouw, L., Pollarolo, L., Trower, G., Scott, L. & d'Errico, F. (2014): Multiproxy record of late Quaternary climate change and Middle Stone Age human occupation at Wonderkrater, South Africa. *Quaternary Science Reviews* 99, pp. 42–59.
25. Backwell et al (2014).
26. Schoonraad, M. & Beaumont, P.B. (1968): The North Brabant Shelter, North Western Transvaal. *South African Journal of Science* 64, pp. 319–331.
27. Van der Ryst, M.M. (1998): *The Waterberg Plateau in the Northern Province, Republic of South Africa, in the Later Stone Age.* BAR International Series 715, pp. 1–158.
28. Van der Ryst, M.M. (1998).
29. Van der Ryst, M.M. (1998).
30. Wadley, L., Murungi, M.L., Witelson, D., Bolhar, R., Bamford, M., Sievers, C., Val, A. & De la Peña, P. (2016): Steenbokfontein 9KR: A Middle Stone Age Spring Site in Limpopo, South Africa. *South African Archaeological Bulletin* 71, pp. 130–145. Steenbokfontein is situated within the Vaalwater Formation which developed in the central part of the main Waterberg Basin. See the chapter on geology for more detail.
31. Huffman (2007): p. 392.
32. Wadley, L. (1987): *Later Stone Age Hunters and Gatherers of the Southern Transvaal.* Oxford: BAR International Series 380, pp. 1–255.

33 Van der Ryst (1998): p. iv.
34 Van der Ryst (1998): p. 43.
35 Van der Ryst (1998).
36 Van der Ryst (1998).
37 Schoonraad & Beaumont (1968).
38 Eastwood, E. & Eastwood, C. (2006). *Capturing the Spoor: An Exploration of Southern African Rock Art.* Claremont: David Philip.
39 Laue, G.B. 2000. *Taking a Stance: Posture and meaning in the Rock Art of the Waterberg, Northern Province, South Africa.* Unpublished MSc thesis, University of the Witwatersrand.
40 Laue (2000): p. 42.
41 Van Schalkwyk, J.A. (2002): Metaphors and meanings: Conceptualising the Schroda clay figurines. In Van Schalkwyk, J.A. & Hanisch, E.O.M. (eds) *Sculptured in Clay: Iron Age Figurines from Schroda, Limpopo Province, South Africa*: pp. 69–79. Pretoria: National Cultural History Museum.
42 Laue (2000): p. 67.
43 Laue (2000).
44 Aukema, J. (1989): Rain-making: a thousand year-old ritual? *South African Archaeological Bulletin* 44, pp. 70–72. Jan Aukema and his wife Wilna (who was a cultural historian attached to the then Cultural History Museum – now Ditsong – also worked in the Waterberg) were killed in a motor accident in 1989 before either of them had completed their research. Jan left field notes at UNISA where he worked and these have been reported on by Van der Ryst (1998) and by Huffman (1990, 2007). Tom Huffman was his supervisor and published his obituary: Huffman, T.N. (1990): Obituary: The Waterberg research of Jan Aukema. *South African Archaeological Bulletin* 45, pp. 117–119. Wilna's field notes are housed with the National Cultural History Museum in Pretoria. Some of her field work data have been published by Mauritz Naudé of Ditsong.
45 Wadley, L. (1996): Changes in the social relations of precolonial hunter gatherers after agro-pastoralist contact – an example from the Magaliesberg, South Africa. *Journal of Anthropological Archaeology* 15, pp. 205–227.
46 Macquarrie, J.W. (ed.) (1962): *The Reminiscences of Sir Walter Stanford.* Volume 2, 1885–1929. Cape Town: Van Riebeeck Society, Cape Town; Jolly, P.A. (1992): Some photographs of late nineteenth century San. *South African Archaeological Bulletin* 47, pp. 89–93.
47 Van der Merwe, D.W. (1984): *Die Geskiedenis van die Berlynse Sendinggenootskap in Transvaal, 1860–1900.* Staatsdrukker, Pretoria, pp. 115–117.
48 Van der Ryst (1998): p. 13.
49 Van der Ryst (1998): p. 13.
50 Van der Ryst (1998).
51 Eastwood, E. & Eastwood, C. (2006).
52 Namona, C. & Eastwood, E.B. (2005): Art, authorship and female issues in a North Sotho rock painting site. *South African Archaeological Society Goodwin Series* 9: 77–85.
53 Huffman (2007).
54 Van der Ryst (1998): p. 111.
55 Huffman (1990): date reference Pta-3616. See also p. 467 for list of dates for Happy Rest, Diamant and Eiland Facies ceramics.
56 Plug, I. (2000): Overview of the Iron Age fauna from the Limpopo Valley. *South African Archaeological Society Goodwin Series* 8: 117–126.
57 Huffman (2007).
58 Miller, D., Boeyens, J. & Küsel, M. (1995): Metallurgical analyses of slags, ores, and metal artefacts from archaeological sites in the North-West Province and Northern Transvaal. *The South African Archaeological Bulletin* 50: 39–54.
59 For a detailed description of the smelting process used in the Iron Age, see Chirikure, S. (2010) *Indigenous mining and metallurgy in Africa.* Cambridge University Press.

60 Naylor, M.L. (2015): *Tracing metals: an archaeo-metallurgical investigation of metal working remains and artefacts from Thaba Nkulu in the Waterberg, South Africa.* MSc dissertation: University of the Witwatersrand.
61 Huffman (1990).
62 Bandama, F., Hall, S. & Chirikure, S. (2015): Eiland crucibles and the earliest relative dating for tin and bronze working in southern Africa. *Journal of Archaeological Science* 62: 82–91.
63 Huffman (1990).
64 Van der Ryst (1998).
65 Aukema (1989).
66 Aukema (1989).
67 A community tourist operation has been established at Telekishi, enabling visitors to see the cave, rock art, walling and an initiation site.
68 Huffman (1990).
69 Chirikure, S., Heimann, R.B. & Killick, D. (2010): The technology of tin smelting in the Rooiberg Valley, Limpopo Province, South Africa, ca. 1650–1850 CE. *Journal of Archaeological Science* 37: 1656–1669.
70 Mason, R.J. (1986): *Origins of Black people of Johannesburg and the Southern Western central Transvaal AD 350–1880.* Johannesburg: University of the Witwatersrand Archaeological Research Unit Occasional paper 16.
71 Van der Ryst (1998).
72 Van der Ryst (1998): p. 24. The date is 1850± 50 BP (Reference Pta-5161; calibrated date AD 144 [233]258.)
73 Wadley (1996): The date is 1840±40 BP (reference is Wits-1398).
74 Huffman (2007).
75 Van der Ryst (1998): p. 24.
76 Huffman (1990, 2007). Site Kb8; Date reference Pta-3612.
77 Boeyens, J. & Van der Ryst, M. (2014): The cultural and symbolic significance of the African rhinoceros: a review of the traditional beliefs, perceptions and practices of agropastoralist societies in southern Africa. *Southern African Humanities* 26, pp. 21–55.
78 Huffman (2007): p. 448.
79 Huffman (1990).
80 Huffman (1990); Boeyens & Van der Ryst (2014): Date reference: Pta-5129.
81 Lombard, M. & Parsons, I. (2003): Ritual practice in a domestic space: evidence from Melora Hilltop, a Late Iron Age stone-walled settlement in the Waterberg, Limpopo Province, South Africa. *South African Archaeological Bulletin* 58, pp. 79–84.
82 Boeyens, J., Van der Ryst, M., Coetzee, F., Steyn, M. & Loots, M. (2009): From uterus to jar: the significance of an infant pot burial from Melora Saddle, an early nineteenth-century African farmer site on the Waterberg plateau. *Southern African Humanities* 21, pp. 213–38.
83 Boeyens *et al.* (2009): Date reference Pta-5139.
84 Hall, S. (1997). Material culture and gender correlations: the view from Mabotse in the late nineteenth century. In Wadley, L. (ed.) *Our Gendered Past: Archaeological Studies of Gender in Southern Africa.* Johannesburg: Witwatersrand University Press.

CHAPTER 3: HOMELAND OR A PLACE OF REFUGE?, *pp.45–84*

1 www.sahistory.org.za/places/Lebowa. Accessed 08 November 2016.
2 *The Star*, 7 May 2012: announcement of death of former Lebowa Chief Minister Nelson Ramodike.

3 SA cadastral maps, 1:250 000 series, also various maps of the so-called independent homelands.
4 Oberholster, J.J. (1972): *The Historical Monuments of South Africa*. Rembrandt van Rijn Foundation for Culture, on behalf of National Monuments Council, Cape Town, p. 318.
5 Baines, Thomas (1877): *The Gold Regions of South Eastern Africa*. Reproduced in a facsimile edition by Books of Rhodesia, Bulawayo, 1968, p. 67.
6 Van Wyk, Braam & Van Wyk, Piet (2013): *Field Guide to Trees of Southern Africa* (2nd Edition). Struik Nature, Cape Town, p. 596.
7 Galpin, Ernest E. (1925): *The Native Timber Trees of the Springbok Flats*. Memoir No. 7 of the Botanical Survey of South Africa, Pretoria, p. 7.
8 Krige, J.D. (1937): Traditional Origins and Tribal Relationships of the Sotho of the Northern Transvaal. *Bantu Studies*, Vol. XI (4), December 1937, p. 327.
9 Setumu, Tlou Erick (2001): *Official Records Pertaining to Blacks in the Transvaal, 1902–1907*. MA thesis, Faculty of Humanities, University of Pretoria, pp. 11–12, 20.
10 Sekikaba Peter Lekgoathi (2009): "Çolonial Experts", Local Interlocutors, Informants and the Making of an Archive on the "Transvaal Ndebele", 1930–1989. *Journal of African History*, 50, pp. 61–80. Lekgoathi compiled a fascinating summary of the work of various ethnologists in an attempt to understand the conclusions they reached and to clarify the relationships between the various Ndebele chiefdoms in the region and their Sotho counterparts.
11 Setumu (2001): p. 19.
12 Jackson, A.O. (1983): *The Ndebele of Langa*. Ethnological Publications No. 54. Dept of Cooperation and Development, Republic of South Africa, Pretoria, pp. 3–5.
13 Hofmeyr, Isabel (1993): *We Spend Our Years as a Tale That is Told: Oral Historical Narrative in a South African Chiefdom*. Witwatersrand University Press, Johannesburg, pp. 257–259.
14 Jackson (1983): pp. 8–12.
15 Jackson (1983): p. 12.
16 Delius, Peter, Maggs, Tim & Schoeman, Alex (2014): *Forgotten Worlds – The Stone-walled Settlements of the Mpumalanga Escarpment*. Wits University Press, Johannesburg, pp. 116–118.
17 Setumu (2001): p. 21.
18 Jackson (1983): pp. 6 & 11.
19 Huffman, Thomas N. (2007): *Handbook to the Iron Age: the Archaeology of Pre-colonial Farming Societies in Southern Africa*. University of KwaZulu-Natal Press, Pietermaritzburg, p. 448.
20 Setumu, Tlou Erick (2009): *Chief Mokopane II (Parts 1 & 2)*. Limpopo Provincial government Heritage Month Souvenir Edition 2009, reproduced on 7 September 2016 by Thobela FM on its website (accessed 14 February 2017): http://www.thobelafm.co.za/sabc/home/thobelafm/kadi fokeng/details?id=0201b317-f515-464a-9b
21 Ysel, Hendrina (compiler) (undated): *History of Langa and Vaaltyn Kekana*. Pamphlet published by the Arend Dieperink Museum, Mokopane (compiled from various sources).
22 Hall, Simon (1995): Archaeological Indicators for Stress in the western Transvaal Region between the seventeenth and nineteenth centuries. In Caroline Hamilton (ed.): *The Mfecane Aftermath: Reconstructive Debates in southern African History*. Jointly published by Witwatersrand University Press and Natal University Press, p. 311. Huffman, T.N. (1990): The Waterberg Research of Jan Aukema, an Obituary. South African Archaeological Bulletin 45, pp. 117–119. Huffman (2007): p. 448.
23 Huffman (2007): p. 444.
24 Parsons, Neil (1995): Prelude to Difaqane in the Interior of south Africa, c.1600–c.1822. In Caroyn Hamilton (ed.): *The Mfecane Aftermath: Reconstructive Debates in southern African History*. Jointly published by Witwatersrand University Press and Natal University Press, p. 328.
25 De Jongh, M. and De Beer, F.C. (1992): A Case of Ambiguous Identity – oral tradition and the Ba ga Seleka of Lephalala. *South African Journal of Ethnology*, 15 (4), pp. 101–107.
26 De Jongh & De Beer (1992): pp. 104–105.

27 De Jongh & De Beer (1992): pp. 106–107.
28 The term difaqane is a southern Sotho term interpreted as "the hammering" that refers to the tumultuous upheaval that took place in the Free State and Transvaal regions of South Africa early in the 19th Century as a result of the development of powerful Nguni kingdoms, which forced weaker and/or disaffected chiefdoms and factions to flee from the wrath of their armies. This term is used here in preference to its Xhosa equivalent, mfecane, ('the crushing') because this book is concerned principally with the region in which Sotho-Tswana (and Ndebele) societies lived, and experienced the consequences of this tumult. The reference for these alternative spellings is the introduction by Carolyn Hamilton to the colloquium proceedings she edited called *The Mfecane Aftermath*, (1995), published jointly by the Witwatersrand University Press and the Natal University Press. The book, a collection of essays by specialists on this aspect of South African history, was the output of a colloquium held to debate the merits of an argument proposed in the late 1980s by a history lecturer at Rhodes University, Julian Cobbing, to the effect that the difaqane was a myth and that the Zulus were not responsible for the widespread destruction of Sotho societies on the Highveld. Rather, this was a consequence of British and Boer incursions, the work of missionaries and the slave trade, especially from Natal and Mozambique. He suggested that the difaqane – and the desolation attributed to it – was used by the Boers as a means of justifying their invasion and settlement of the "empty" Highveld. His beliefs were contentious; later, they were largely and readily discredited in a series of academic debates, in which Cobbing declined to participate. However, historians agreed that his raising of the issue had been useful in that it had forced academics to focus on a topic that had long been taken for granted. One of the colloquium participants summarised Cobbing's contribution as follows: *"Although credit is certainly due to Julian Cobbing for rekindling interest in the early nineteenth century and for postulating a bold new paradigm, it must also be said that his use of evidence is sufficiently reckless to undermine his credibility among all those with the patience to check his footnotes. Even more questionable than his tendency to stretch and distort historical information to suit his own purposes is his determination to ignore all sources which contradict his hypotheses."* – Jeff Peires (1995): in Carolyn Hamilton (ed.) *The Mfecane Aftermath*, p. 238.
29 Edgecombe, Ruth (1986): The Mfecane or Difaqane. *In* Trewhela Cameron (ed.): *An Illustrated History of South Africa*. Jonathan Ball, Johannesburg, pp. 115–126.
30 Edgecombe (1986): pp. 118–119.
31 Edgecombe (1986): p. 124.
32 Schapera, I. (1953): The Tswana. *Ethnographic Survey of Africa: Southern Africa, Part III*. International African Institute, London. (Reprinted 1973), p. 15.
33 Visagie, Jan (2014): Migration and the societies north of the Gariep River. *In* Fransjohan Pretorius (ed.): *A History of South Africa – from the distant past to the present day*. Protea Book House, Pretoria, pp. 105–124.
34 Hammond-Tooke, David (1993): *The Roots of Black South Africa*. Jonathan Ball, Johannesburg, p. 32.
35 Given the enormity of the difaqane and its irrefutably dire consequences for thousands of black South Africans in the period 1790 to 1825, it is frankly astonishing that the Reader's Digest *Illustrated History of South Africa – The Real Story*, published in 1988, with a second edition in 1992, with Christopher Saunders, Associate Professor of History at UCT as Consultant Editor and Colin Bundy, Professor of History at the University of the Western Cape as Historical Advisor, does not even mention the difaqane (or mfecane) by name, referring merely to an *"upheaval in Natal"* (p. 90) and *"Disorder among the Sotho people"* (p. 91). So much for the "Real Story"!
36 Visagie, Jan (2014): The Emigration of the Voortrekkers into the Interior. *In* Fransjohan Pretorius (ed.): *A History of South Africa – from the distant past to the present day*. Protea Book House, Pretoria, pp. 138–142.
37 Giliomee, Hermann (2009): *The Afrikaners: Biography of a People* (2nd Edition). Tafelberg, Cape Town, p. 153.

38 Huffman (2007): pp. 417–425; 453–454.
39 Legassick, Martin (1972): The Sotho-Tswana Peoples before 1800. In Leonard Thompson (ed.): *African Societies in Southern Africa*. Heinemann, London, p. 116.
40 Huffman (2007): p. 429.
41 Pieres, J. (1986): The Emergence of Black Political Communities. *In* Trewhela Cameron (ed.): *An Illustrated History of South Africa*. Jonathan Ball, Johannesburg, pp. 49–50.
42 Schapera (1953): pp. 14–15.
43 Schapera (1953): map.
44 Cadastral Maps, 1:148 700 series, Sheet 6 (Blaauwberg) and Sheet 10 (Nylstroom), produced by the Surveyor-General's Office, Pretoria, April, 1911, reprinted November, 1920. Dordrecht plan produced by Surveyor General's office, Pretoria, April 1895.
45 Benjamin, Chantelle (2015): Mining a threat to holy caves. *Mail & Guardian*, 27 February 2015. http://www.mg.co.za/print/2015-02-27-mining-a-threat-to-holy-caves. Accessed 27 February 2015.
46 Child, Katherine (2016): Holy ground safe from steel miners for now. *Times Live*, 4 April 2016. http://www.timeslive.co.za/thetimes/article1297755.ece. Accessed 14 February 2017.
47 Hall, Simon (1997): Material Culture and Gender Correlations: The view from Mabotse in the late Nineteenth Century. *In* Lyn Wadley (ed.): *Our Gendered Past*. Witwatersrand University Press, pp. 209–219.
48 The actual site where the Van Rensburg group was murdered in July 1836 was first discovered and visited by Louis Trichardt's son Carolus (Karel) two years later. It lies on the west bank of the Limpopo River in Mozambique, just south of where the Tropic of Capricorn crosses the river, about 70km upstream of its confluence with the Olifants. The site was verified in 1959 by a Parks Board expedition led by Dr W.H.J. Punt. Source: W.H.J. Punt (2012): The Fate of Lang Hans van Rensburg. *In* Dr U. de V. Pienaar et al: *A Cameo from the Past: the prehistory and early history of the Kruger National Park*. Protea Book House, Pretoria, pp. 102–114.
49 Fuller, Claude (1932): *Louis Trigardt's Trek across the Drakensberg, 1837–1838*. The Van Riebeeck Society, Series 1 Volume 13. Cape Town. Claude Fuller was an Australian-born entomologist, whose seminal study on the tsetse fly, published in 1923, is still regarded as the definitive work on the early records of this pest (See Chapter 13 of this book). He died in a car accident in Mozambique in 1928, aged 56; his book about Trichardt was published posthumously at the request of his widow.
50 Giliomee (2009): p. 165.
51 Esterhuysen, Amanda (2006): Let the Ancestors Speak: an archaeological excavation and re-evaluation of events prior and pertaining to the 1854 siege of Mugombane, Limpopo Province, South Africa. PhD thesis, University of Witwatersrand, Johannesburg, p. 13.
52 Giliomee (2009): p. 171.
53 Hofmeyr (1993): pp. 109–110.
54 Jackson (1983): P 14; De Waal (2012): The bygone days of the Soutpansberg. *In* Dr U. de V. Pienaar et al: *A Cameo from the Past: the prehistory and early history of the Kruger National Park*. Protea Book House, Pretoria, pp. 175–176; Oberholster (1972): pp. 317–318.
55 Preller, G.S. (1914–1915): Baanbrekers. 'n Hoofstuk uit die Voorgeskiedenis van Transvaal. *Brandwag*, Vol. 5 No. 14/15, December 10, 1914, p. 460. As quoted in AO Jackson (1983): p. 15.
56 Setumu (2009): Mokopane II Part 2.
57 Hofmeyr (1993): pp. 144–147.
58 Giliomee (2009): pp. 184–186; Dougie Oakes (1992): They Called their slaves Inboekselings. *In*: *Illustrated History of South Africa – The Real Story*. Reader's Digest Association, Cape Town, p. 146.
59 Hofmeyr (1993): pp. 125–130.
60 Jackson (1983): pp. 16–17; De Waal (2012): p. 176.
61 Jackson (1983): pp. 17–18. Jackson records that the Voortrekker leader Andries Hendrik Potgieter was best known by his middle name, Hendrik; that this was transformed by the Langa into

Ntereke; and therefore that his younger brother Hermanus, whom they had killed at Fothane, was known as the "little" *Ntereke*, or *Nterekane*.

62 Jackson (1983): p. 18.
63 Esterhuysen (2006): p. 7.
64 De Waal, J.B. (2012): The bygone days of the Soutpansberg. *In* Dr U. de V. Pienaar et al: *A Cameo from the Past: the prehistory and early history of the Kruger National Park*. Protea Book House, Pretoria, pp. 196–199.
65 Anonymous (no date): *Hoe die dorp sy naam gekry het*. Brochure published by the Arend Dieperink Museum, Mokopane.
66 Burke, E.E. (ed.) (1969): *The Journals of Carl Mauch: his travels in the Transvaal and Rhodesia, 1869–1872*. National Archives of Rhodesia, Salisbury, p. 104.
67 Jackson (1983): pp. 24–27.
68 Jackson (1983): pp. 28–34.
69 Jackson (1983): p. 36.
70 Hofmeyr (1993): pp. 46, 68–69.
71 Hofmeyr (1993): pp. 42–43.
72 Jackson (1983): pp. 37–41.
73 Jackson (1983): pp. 60–62.
74 Grobler, J.E.H. (2004): *The War Reporter: The Anglo-Boer War through the eyes of the Burghers*. Jonathan Ball, Cape Town, p. 114; and R.F. Odendaal (c.2000): *Noord-Transvaal op Kommando*. Privately published, Nylstroom, p. 208.
75 Jackson (1983): pp. 43–46; Odendaal (c.2000): pp. 224–234.
76 Odendaal (c.2000): pp. 225–228).
77 Setumu (2001): p. 70.
78 SANA TAB 87 NA2863/02 (Part 1): Native Commissioner, Waterberg: Border dispute between Chiefs Hans and Hendrik Masebe, 1902. Annexure V to Annexure 4 to No. 3: Report by Native Commissioner Waterberg, S.W.J. Scholefield, to Commissioner for Native Affairs, Johannesburg, dated 19 February 1903 at Hartingsburg.
79 SANA TAB 87 NA2863/02 (Part 1).
80 Setumu (2001): pp. 75–77.
81 Massie, H. (Brevet-Major, compiler) (1905): *The Native Tribes of the Transvaal*. Prepared for the General Staff, War Office. His Majesty's Stationery Office, London, pp. 40–41.
82 Jackson (1983): pp. 145–150.
83 Van Warmelo, N.J. (1953): *Die Tlôkwa en Birwa van die Noord Transvaal*. Ethnologiese Reeks No. 29. Departement van Naturellesake, Pretoria, p. 3.
84 Jackson (1983): pp. 63–64.
85 Hofmeyr (1993): p. 82.
86 Hofmeyr (1993): p. 70.
87 Setumu (2009): Mokopane II Part 2.
88 Hofmeyr (1993): pp. 78–83.

CHAPTER 4: THE WILD WEST, *pp.85–118*

1 Burke, E.E. (1969): *The Journals of Carl Mauch: his travels in the Transvaal and Rhodesia, 1869–1872*. National Archives of Rhodesia, p. 32.
2 SANA TAB RAK 3058 Rhenosterpoort 78 (283KQ). Title deeds records.
3 Trichardt, Louis (2016): *Swaershoek in die Waterberge – Afrikaner erfenis of niemandsland?* Privately published, pp. 22–29; 62–64.
4 Boeyens, Jan C.A. & Cole, Desmond T. (1999): Kaditshwene: What's in a Name? In Rosalie Fin-

layson (ed): *African Mosaic – a Festschrift for JA Louw*. Published by UNISA, pp. 250–284. ulr. unisa.ac.za/bitstream/.../Finlayson_R_9781868881383_Section8.pdf. Accessed 30 October 2016. Also: Boeyens, Jan C.A. (2016): A tale of two Tswana towns: in quest of Tswenyane and the twin capitals of the Hurutshe in the Marico. *Southern African Humanities* 28, pp. 1–37.

5 SANA TAB RAK 3058 Rhenosterpoort 78 (283KQ). Title deeds records.

6 Trichardt, Louis (2016): pp. 70–74. Trichardt asserts (p. 70) that Steijn's sister had married Albasini; however, Dr U. de V. Pienaar, in his magisterial book *A Cameo from the Past*, a history of the Kruger Park, states (p. 158) that Albasini married Gertina Petronella Maria van Rensburg at Ohrigstad on 6 March 1850, when they were 36 and 18 years old respectively. They would produce nine children. There is no evidence that João Albasini married a second time.

7 Prance, C.R. (1937): Chapter 19, "Oom Sandy's Earthly Paradise", in *Tante Rebella's Saga – a Backvelder's Scrapbook*. HF & G Witherby, London. In Prance's tale, Sandy McBean (an alias for Harry Bell) and his wife choose to move from a cushy life in Johannesburg, where he was a mine engineer, to a Spartan farming existence on Suikerboswoestyn (Sterkfontein), unluckily after a visit to the property in summer when its two rivers were flowing. It was not long before Sandy heard the sound of blasting: his neighbour, Oom Sarel Suidhuizen (Enslin) had employed a *bywoner* to tap a stream that seeped out of the cliff on their common boundary. One of Sandy's springs promptly lost its flow. A court case ensued, only for Sandy to discover belatedly that his wily neighbour had concluded a water servitude right with Sandy's predecessor. Sandy eventually had to pay Oom Sarel to halt his excavation work – which the latter was by then happy to do, having secured more than enough water for his needs.

8 Trichardt (2016): pp. 85–96.

9 SANA TAB CJC 504 CJC5: John Rankin, of Rhenosterpoort, Swagershoek Ward, Waterberg (1901–1903) – claims for compensation as a Protected Burgher; a file containing numerous claims, affidavits and military correspondence.

10 SANA TAB TAD 59 A328/27: Petition from farmers in the Zwagershoek area, 9 February 1910, and subsequent correspondence regarding east coast fever and the right of the Resident Justice of the Peace, John Rankin, to issue permits allowing movement of cattle.

11 SANA TAB LD 1650 AG2565/08: Correspondence relating to threatened removal of police from quarters at Rankin's Pass, 28 October 1908, in protest over poor accommodation.

12 Trichardt (2016): p. 185.

13 Nattrass, Gail (1983): *The Rooiberg Story*. Published by Rooiberg Tin Ltd to commemorate the 75th anniversary of the mine, p. 27; Trichardt (2016), p. 186.

14 SANA TAB LD 1744 AG1409/09: Appointment of Constable J.W. Chaney in charge of Rhenosterpoort Police Post, by letter from J.C. Smuts, Acting Attorney General. Chaney's appointment was with effect from 15 January 1910. The previous September, he had been appointed as a Public Prosecutor at the court in Nylstroom (SANA TAB LD 1752 AG1645/09).

15 SANA SAB JUS 550 125/30 (1929/1930): Rhenosterfontein (Rankin's Pass) Police Post, Waterberg: dispute over ground enclosed in fenced police property.

16 SANA TAB MHG 0 6159/44: Death Notice of Catherine Rankin, born Bell, parents Edward and Mary Anne Bell, born in February 1885, died 8.12.1944, spouse of late William John Rankin, whom she married on 25.3.1905. And: TAB LD 1250 AG1548/06: Proceedings against W.J. Rankin for the recovery of £11 due to the Immigration Department from his wife Catherine (née Bell), 1906.

17 Grobler, Maritz, Weir, Robbie and others (eds) (1995): *Die Suidelike Springbokvlakte/The Southern Springbok Flats, 1876–1995 – 'n historiese oorsig/an historical overview*. Published by the Berlin Geloftefeeskomitee, Springbok Flats, pp. 135–137.

18 Grobler et al (1995): p. 137.

19 During the period 1980–1986, Wilna Aukema undertook field-based research into vernacular (early settler) homesteads in the Waterberg region under the auspices of the Natural Cultural History Museum (NCHM), as part of her work towards a Masters degree in Cultural History.

Concurrently, her husband, Jan Aukema, was conducting a field survey of archaeological sites in the Waterberg, also towards a Masters degree, but in Archaeology. Tragically, both were killed in a car accident in December 1989, before Wilna had completed her thesis. Wilna's field notes were lodged with the Museum and later formed the basis for a number of papers published by Mauritz Naudé, an employee of the NCHM. Unfortunately, Wilna Aukema's field notes seldom specified the locations of the farms she visited and Naudé too, did not locate the properties he wrote about in his papers. It fell to this author to do so: of the more than 50 farms named in the three Naudé papers examined, 42 – including Kliprivier – have been positively identified and located. Naudé's articles appeared in two issues of the *NCHM Research Journal*: 1998 (Vol.7), pp. 47–91; and 2008 (Vol.3), pp. 35–56; and in *SA Journal of Cultural History* 18 (1), June 2004, pp. 34–62. In summarising and collating Wilna Aukema's research, these papers have provided a wealth of information about rural settler life around the end of the nineteenth century. An obituary for Jan Aukema by Professor T.N. Huffman, supervisor of his MA at the University of the Witwatersrand, appeared in the December 1990 issue of the *SA Archaeological Bulletin* (Vol. XLV, No. 152, pp. 117–119).

20 Naudé, M. (1998): Oral Evidence of Vernacular Buildings and Structures on Farmsteads in the Waterberg (Northern Province). *Research by the National Cultural History Museum*, Vol. 7, pp. 47–91; and Naudé, M. (2004): Oral Evidence on [sic] the Construction of Vernacular Farm Dwellings in the Waterberg (Northern Province). *SA Journal of Cultural History* 18 (1); pp. 34–62. The extract used in the text is an amalgamation of several references to the farm Kliprivier contained in the two papers.
21 Trichardt (2016): pp. 137–147.
22 Trichardt (2016): p. 143.
23 Naudé, M. (2008): Oral Evidence on Aspects of Folk Life during the First Hundred Years (1840–1940) of Frontier Settlement in the Waterberg (Limpopo Province). *NCHM Research Journal*, Vol. 3, pp. 35–56 (p. 51).
24 Anonymous (undated): *Mampoer*. A brochure produced by the Arend Dieperink Museum in Mokopane.
25 Redlist.sanbi.org/species.php?species=3835-110. Visited 3 November 2016. Other plants that are virtually confined to the foothills of the Sandriviersberg include the bulbous species *Pachycymbium carnosum*, *Drimiopsis burkei*, *Orbea carnosa ssp. carnosa* and a cryptic species of *Ledebouria*.
26 SANA TAB MHG 0 0/3172: Jacoba Johanna Carolina Potgieter – Death Notice, Testament and joint Will.
27 De Waal, J.B. (2012): The Bygone days of the Soutpansberg. In U. de V. Pienaar (ed.): *A Cameo from the Past. The prehistory and early history of the Kruger National Park*. Protea Book House, Pretoria, pp. 169–212.
28 Spies, S.B. & Nattrass, Gail (1994): *Jan Smuts: Memoirs of the Boer War*. Jonathan Ball Publishers, Johannesburg, p. 191.
29 Jackson, A.O. (1983): *The Ndebele of Langa*. Ethnological Publications No. 54, Department of Co-operation and Development, Pretoria, p. 41.
30 Engelbrecht, S.P., Chairman of Editorial Board (1955): *Pretoria (1855–1955) – History of the City of Pretoria*. Pretoria City Council, pp. 54–55.
31 Hancock, W.K. & Van der Poel, Jean (eds) (2007) – *Selections from the Smuts Papers, Volume 1: June 1886 – May 1902*. Cambridge University Press. And, Francis Hugh de Souza (2004): *A Question of Treason*. Kiaat Creations, p. 128.
32 SANA TAB MHG 0 84285: Death Notice of William Davidson (15.11.1873–18.12.1933) and Inventory, witnessed by Henry Bell (Sr) and Henry (Harry) Bell.
33 Helme, Nigel (1974): *Major Thomas Cullinan – a biography*. McGraw-Hill.
34 SANA SAB MNW 258 MM3494/14: RE of Sterkfontein 1590, Waterberg District. Allotment of land under Settlement Act of 1912 to Henry Bell.

35 SANA RSA CEN 640 6 4227: Letters to and from H Bell in June 1918 concerning soft scale on leaves.
36 Louis Trichardt (2016) pers. comm.
37 SANA TAB MHG 0 84285: Death Notice for William Davidson; and SANA TAB MHG 0 6960/61: Death Notice for and Will of Margaret Stewart Bell (born Galloway), widow of Henry Bell.
38 The Red Oath, or Africa Oath, was an initiative of Parliament in February 1940 which authorised volunteer soldiers to commit themselves to wear an orange flash on their uniforms, symbolising their willingness to serve South Africa anywhere on the African continent. By 1945, almost ten percent of the country's white male population had enlisted in the army. Opponents of SA's participation in the war called the servicemen *rooiluise* (red lice). Source: Fransjohan Pretorius (ed.) (2014): *A History of South Africa from the Distant Past to the Present Day*. Protea Book House, Pretoria, p. 604.
39 Trichardt (2016): pp. 159–160.
40 Die Boervolk van SA http://www.boervolk.com; http://152.111.1.88/argief/berigte/beeld/2000/03/25/8/12.htm.
41 Odendaal, R.F. (2000): *Noordtransvaal op Kommando*; and *Waterberg op Kommando* (1995), both books privately published. Odendaal, who died in Nylstroom in about 2004, aged 81, had trained and practised as a teacher, while farming part-time near Mabula, west of Warmbad, before moving to Nylstroom. He was active in his church (GK) and was known as an extreme Afrikaner republican, bitterly opposed to the British (even a hundred years after the war). His books recount in great, partisan detail the conditions in the concentration camps, the alleged behaviour of British troops, the Breaker Morant scandal, the heroic exploits of the Boers, especially General Beyers, and his total disdain and revulsion for those Boers like Gen. Grobler, who chose not to fight to the bitter end. Whilst his books contain a vast amount of factual material, this is selective and hard to sift from his forcefully expressed personal opinions. His work is the standard reference for many old-time Afrikaans-speakers in the Waterberg.
42 Grant, M.H. (Capt.) (1903): *History of the War in South Africa, 1899–1902*. Volume IV, Chapter XXV. Produced under the direction of his Majesty's Government; (2)"Events in the Northern Transvaal, April 1901–May 1902", pp. 435–451. Published by Hurst & Blackett Ltd, London, 1910; (3) *Cassell's History of the Boer War 1899–1902*, Volume II. Cassell, London, 1913; (4) *South Africa and the Transvaal War*, Volume VII (The Guerilla War), by Louis Creswicke. Caxton Publishing, London, *circa* 1903.
43 http://www.the peerage.com/p19684.htm: Henry Edmund Buxton *et seq.*; and http://freepages.genealogy.rootsweb.ancestry.com/vivianegan: WIGG Census Extracts: the Suffolk Census of 1891.
44 Dooner, Mildred D. (1903): The Last Post: Roll of officers [of the Norfolk Regiment] who fell in South Africa, 1899–1902. Reprinted by Naval & Military Press; and Royal Norfolk Regimental Museum, Norwich, Norfolk http://www.rnrm.org.uk.
45 http://www.flickr.com/photos/36928008@N08/41228851314/ image of plaque at St Edmund's church submitted by "Claire", 21 November 2009.
46 Prance, C.R. (1923): Dead Man's Corner, in *Under the Blue Roof – Sketches of a Settler's Life in the Transvaal Backveld, 1908–1921*. William Cooper & Nephews Ltd, Berkhamsted, England.
47 Surveyor General's Office, Pretoria (1905): Document SG No. A 1840/03: 333 Form A: Details of farm Langkloof 333 (now registered as 285KQ) as surveyed in October 1904 by F.O. Watermeyer and published in the Government Gazette No. 337 on 21 July 1905. Document provided by Anton Botes, 26.09.2011.
48 Prinsloo, P.W. and others (eds) (2002): *Gedenkalbum van die Doornfontein Geloftefees Waterberg, 16 Desember 1902–16 Desember 2002*. Published by the Doornfontein Geloftefeeskomitee, Nylstroom, p. 45.
49 Marais, Eugène (1972): The World's Most Feared Snake. In *The Road to Waterberg and other essays*.

Human & Rousseau, Cape Town, pp. 70–82. The mamba stories are repeated, with some variation in details, in Leon Rousseau (1982): *The Dark Stream – the Story of Eugène Marais*, Jonathan Ball Publishers, Johannesburg, pp. 231–234; and in V. E. d'Assonville (2008): *Eugène Marais and the Waterberg*, published by Marnix, Pretoria, pp. 35–39.
50 Louis Trichardt (2016): pers. comm.
51 SANA TAB TAD 59 A328/27:09 February 1910: petition from Zwagershoek farmers requesting that the local RJP be authorised to issue permits for the movement of cattle; and subsequent correspondence.
52 SANA SAB MVE 416/16/48: Letter from G van Heerden, Chairman, South African Party, Zwagershoek Branch, to Prime Minister JC Smuts, 28 May 1924.
53 Louis Trichardt (2016): pers. comm.
54 SANA SAB MVE 416/16/48: Documentation and correspondence relating to the upgrading of the siding at Alma to the status of a station, 1925/1926.
55 Trichardt (2016): pp. 165–173.
56 Bester, D.J. (ed.) (1984): NTK 75. *NTK-Nuus*, Vol. 7, No. 11, September/October 1984, Nylstroom. Page 152. A publication to commemorate the 75th anniversary of the founding of the Waterberg Landbouwers Kooperatieve Vereniging (WLKV), the predecessor of the NTK.

CHAPTER 5: BALTIC CONNECTIONS, *pp.119–147*

1 Burke, E.E. (ed.) (1969): *The Journals of Carl Mauch: His travels in the Transvaal and Rhodesia, 1869–1872*. National Archives of Rhodesia. Salisbury, p. 28.
2 Nel, J.J. and others (eds) (1966): *Nylstroom 1866–1966*. Nylstroomse Stadsraad, Nylstroom, p. 23.
3 Van der Merwe, D.W. (1984): *Die Geskiedenis van die Berlynse Sendinggenootskap in Transvaal, 1860–1900*. Argiefjaarboek vir Suid-Afrikaanse Geskiedenis. Staatsdrukker. Pretoria, pp. 46–47.
4 Burke (1969).
5 Burke (1969): p 28.
6 Naudé, M. (1998): Oral Evidence of Vernacular Buildings and Structures on Farmsteads in the Waterberg (Northern Province). *Research by the Cultural History Museum*, 7, pp. 47–91.
7 Jeppe, Friedrich (1899): *Jeppe's Map of the Transvaal or South African Republic and Surrounding Territories (Sheet 2 of 6)*. ZAR, Pretoria.
8 De Bruiyn, H. (1972): *Report on the Nooitgedacht Lead Mine, Northeast of Vaalwater, Northern Transvaal*. Unpublished report, Geological Survey of South Africa. Pretoria.
9 Burke (1969): pp. 28–29.
10 Burke (1969): p. 29.
11 SANA TAB. RAK 3058–3062, Title Deed records of Waterberg farms No. 236 (Oversprong).
12 Bernhard, F.O. (ed.) (1971): *Karl Mauch – African Explorer*. Struik, Cape Town p. 94.
13 Burke (1969): p. 29.
14 Burke (1969): pp. 29–30.
15 Burke (1969): p. 30.
16 Burke (1969): p. 31.
17 Burke (1969): p. 31.
18 South Africa 1:50 000 Sheet 2427BC Kransberg (Second Edition, 1980). Chief Director of Surveys and Mapping, Mowbray, Cape.
19 Burke (1969): pp. 31–32.
20 Burke (1969): p. 32.
21 SANA TAB. RAK 3058–3062, Title Deed records of Waterberg farms No. 78 (Rhenosterpoort).
22 Burke (1969): pp. 32–33.

23 www.genie.com/people/oscar-wilhelm-alrik-forssman.
24 De Villiers, Simon A. (1974): *Otto Landsberg 1803–1905: 19th Century South African Artist*. Struik, Cape Town, p. 109 & pp. 15–24.
25 Schoeman, A.B. (2007): The First Surveyor General of the "Zuid-Afrikaanse Republiek (ZAR) alias Transvaal: Magnus J.F. Forssman (1822–1889); in *SA Deeds Journal* 13 (October 2007), pp. 8–9. Department of Rural Development & Land Reform. Website accessed 26 March 2017. www.ruraldevelopment.gov.za/phocadownload/Deeds-journal/SADJ13.pdf.
26 Jenkins, G. (1971): *A Century of History: the story of Potchefstroom* (2nd Edition), A.A. Balkema, Cape Town. This book by Geoffrey Jenkins, the well-known novelist (e.g. *A Twist of Sand*) was originally published in 1939 from a manuscript written in 1936/37, when the author was a 17 year-old schoolboy at Potchefstroom Boys High School. A copy of this scarce book was loaned from the William Cullen Library at the University of the Witwatersrand (AFRICA DT 2405.P74 JEN). P51.
27 Forssman, O.W.A. (compiler) (1872): *A Guide for Agriculturalists and Capitalists, Speculators Miners Etc, wishing to invest money profitably in The Transvaal Republic, South Africa; containing a description of a number of first class farms situated in different districts of the Republic, and general useful information*. William Foster & Co., Printers, Cape Town. Reprinted by the State Printer, Pretoria in 1984; copy viewed at Brenthurst Library, Johannesburg.
28 Bernhard, F.O. (1971): *Karl Mauch – African Explorer*. Struik, Cape Town, pp. 13, 41, 94.
29 Blomstrand, C. (2008): The Swedish Emigration to Potchefstroom. *Famnea* (Nuusbrief van die Genealogie Genootskap van SA – Noordwes). Februarie 2008 5 (1) pp. 1–5.
30 Jenkins, G. (1971): p. 74.
31 Pienaar, U. de V. (2012): *A Cameo from the Past – the prehistory and early history of the Kruger Park*. Protea Book House, p. 329.
32 Jeppe, F. (1877): *Transvaal Book Almanac and Directory for 1877*. F Davis & sons, Pietermaritzburg, p. 62. Copy viewed at the Library of the Chamber of Mines, Johannesburg.
33 Forssman, O.W.A. (Jr) (1962).
34 Forssman. O.W.A. (Jr) (1962): *OWA Forssman 1822–1889*. Pamphlet biography of his grandfather, SANA TAB Pretoria library, Vol P361 (No. 375).
35 Nel, J.J. et al (eds) (1966): *Nylstroom 1866–1966*. Nylstroomse Stadsraad. P. 66: "Die ou Goewermentskool te Nylstroom na die Engelse Oorlog" – Nylstroom Government School, 5th November 1915.
36 British Concentration Camps of the Anglo-Boer War 1900–1902 http://www2lib.ust.ac.za/mss/bccd/Person/126437/Carl_Forssman/ . Accessed 26 May 2015.
37 SANA TAB LD 752 AG2610/04: Appointment of Carl WG Forssman as Justice of the Peace, Waterberg District, 27 May 1904.
38 Anon (1931): *Nylstroom and Warmbaths and Waterberg District – "where farming, sporting and health-giving opportunities abound"*. 40-page brochure "issued under the joint auspices of the Waterberg and Warmbaths Publicity Association and the South African Railways and Harbours", p. 8. A copy of this brochure was loaned to the author by Mr Pieter Prinsloo of Doornfontein.
39 SANA TAB MHG 0 87497: Death Notice, Carl Wilhelm Gottfried Forssman.
40 Alfred, Mike (no date): The Jeppe Family. In *I Love Kensington, our village in the City*. http://www.kensington.org.za?p=430. Accessed 18 March 2016.
41 Alfred (no date): p. 2.
42 De Villiers (1974): p. 28.
43 Jeppe (1877): p. 29.
44 Engelbrecht, S.P. (Chair of Editorial board) (1955): *Pretoria (1855–1955)*. A history of the town in its centenary year. Pretoria City Council, p. 11.
45 Bernhard (1971): p 40, for example. Also Burke (1969): p. 37.
46 Duminy, A. (2011): *Mapping South Africa – a historical survey of South African Maps and Charts*. Jacana, Johannesburg, pp. 96–98.

47 Engelbrecht (1955): p. 335.
48 Wills, Walter H. (ed.) (1907): *The Anglo-African Who's Who and Biographical Sketch-Book, 1907*. L. Upcott Gill, London. Re-published in 2006 by Jeppestown Press, London. p. 197.
49 Jeppe, C. (1906): *The Kaleidoscopic Transvaal*. Chapman & Hall, London, pp. vii–viii.
50 SANA TAB SS 233/1377 pp. 304–306: Letter from Carl Jeppe, Landdrost Klerk, Waterberg, to Acting Government Secretary, Pretoria, dated 23 April 1877.
51 Wills (1907): p. 197.
52 Jeppe (1906): pp. 15–16.
53 Jeppe, C (1906): p. ix.
54 Glen, H.F. & Germishuizen, G. (2010): *Botanical Exploration of Southern Africa, Edition 2. Strelitzia 26*, SA National Biodiversity Institute, Pretoria, pp. 231–232.
55 SANA TAB CJC 272 CJC899: Ludwig Field, Claim for Compensation file (1902): contains a certificate of Norwegian citizenship issued by the Royal Swedish and Norwegian Consulate in Johannesburg on 5 October 1899.
56 Breedt, J.L. (ed.) (1998): *Nederduitse Gereformeerde Gemeente Waterberg Gedenkblad ter Herdenking van Kerkbou 1898–1998*. NGK Nylstroom, pp. 34–35.
57 Tord Tutturen (2017): pers. comm. Tord, a direct descendant of Ludvig's elder brother and a researcher into the family history, confirmed Ludvig's protestant religion in email correspondence with the author in October 2017. Tord's great-uncle, Lauritz Tutturen, is at the time of writing, a keen genealogist and a source of much information about the family, which has a long and proud Norwegian lineage.
58 Tord Tutturen (2017): pers. comm.
59 SANA TAB CJC 272 CJC899: Ludwig Field, Claim for Compensation file (1902): contains a set of documents and invoices detailing equipment ordered by and delivered to Field in late 1899.
60 SANA TAB CS 161 13758/02: Einar Field – documentation relating to his arrest and detention as a prisoner of war, in 1901.
61 Tord Tutturen (2017): pers. comm.
62 SANA TAB CJC 272 CJC899: Statement from Ludvig Field dated 14 June 1902; statement in support of his claim by neighbour Maurice Freeman (a Russian subject who had been on commando with the Boers but was not arrested) dated 14 June 1902; sundry reports by British officers disputing and finally denying the claims.
63 Nel, J.J. *et al* (eds) (1966): *Nylstroom 1866–1966*. Published by the Nylstroom Town Council; pp. 85–87; 93.
64 SANA TAB TPB 1913 TA2/14903: Letter from Ludvig Field to the Town Clerk, Nylstroom, dated 13 February 1932; letter in reply dated 22 February accepting this condition and another advising the Provincial Secretary of the same.
65 Tord Tutturen (2017): pers. comm. An anthology of newspaper articles was published as *"A Transvaaler's Travels"* by Naspers in Cape Town in 1925.
66 Tord Tutturen (2017): pers. comm.
67 Lewis, Thos. H. (1913): *Women of South Africa*. Le Quesne & Hooton-Smith, Cape Town, p 258.
68 Nel (1966): pp. 91, 93.
69 Information from memorabilia in the possession of Anne Howe, Chaney's granddaughter, viewed on 14 August 2014.
70 Butterfield (1978): pp. 247–248.
71 Hunter, Elizabeth (2010): *Pioneers of the Waterberg – a Photographic Journey*. Privately published, Lephalale, Limpopo, pp 51–53; (1978): Butterfield (1978): pp. 254–257; Brigitte Wongtschowski (1987): *Between Woodbush and Wolkberg – GooGoo Thompson's Story*. Privately published, Pietersburg, p. 43.

CHAPTER 6: THE GATEWAY TOWNS, *pp.148–209*

1. www.statssa.gov.za/?page_id=993&id=modimolle-municipality. Modimolle and Phagameng population as per 2011 national census. Accessed 6 August 2016.
2. Nel, J.J. and others (eds) (1966): *Nylstroom 1866–1966*. Published on behalf of the Nylstroom Town Council in celebration of its centenary. Now long out of print. Although unashamedly concentrating on the Afrikaner (and white) history of the town, including its religious, educational and political perspectives, the book nevertheless contains a wealth of invaluable information that would be difficult if not impossible to source elsewhere. It is sad that the 150th anniversary of the town's founding (in 2016) has passed without any celebration or recognition.
3. Silverman, Melinda (ed.) (2004): *Nylstroom/Modimolle – urban change in a small South African town*. Published by the School of Architecture at the University of the Witwatersrand.
4. Anon (1952): *Oorsig van die Geskiedenis van Nylstroom*. Van Riebeeck Festival brochure. Published by Nylstroom Town Council, 1952. SANA TAB Pamphlet Library No. 4344.
5. Louis Trichardt, Rhenosterpoort, Swaershoek region, pers. comm; 8 March 2016.
6. Buys, J.C.: Die Ontstaan en Ontwikkeling van Plaaslike Bestuur in Nylstroom. In Nel, J.J. et al (1966): *Nylstroom 1866–1966*. Proclamation 86 of the ZAR Volksraad on 15 February 1866.
7. Brits, M.B. (1966): Die Geskiedenis van Waterberg en die Ontstaan van Nylstroom. In Nel, J.J. et al (1966): *Nylstroom 1866–1966*.
8. Brits (1966).
9. De Beer, F.C. (1996): Berge is nie net berge nie: swartmense se persepies oor Modimolle. *South African Journal of Ethnology* 19 (1), pp. 1–6.
10. Wiggett, Matthew & Silverman, Melinda (2004): The history of Nylstroom/Modimolle. In Silverman, Melinda (ed.) (2004): N*ylstroom/Modimolle – urban change in a small South African town*. University of Witwatersrand Architecture Department, pp. 22–24.
11. Anon (1952): *Oorsig van die Geskiedenis van Nylstroom*. Van Riebeeck Festival brochure. Published by Nylstroom Town Council, 1952. SANA TAB Pamphlet Library No. 4344.
12. Mabin, Alan & Conradie, Barbara (eds) (1987): *The Confidence of the Whole Country – Standard Bank Reports on economic conditions in Southern Africa, 1865–1902*. Standard Bank Investment Corporation, Johannesburg, pp. 121–122.
13. Mabin & Conradie (1987): p. 153.
14. Mabin & Conradie (1987): p 165.
15. Buys (1966): p. 82.
16. Distant, W.L. (1892): *A Naturalist in the Transvaal*. RH Porter, London. Chapter V: Through the Waterberg.
17. Dursan, Amraj (2004): the landdrost building. In Silverman, Melinda (ed) (2004): *Nylstroom/Modimolle – urban change in a small South African town*. University of Witwatersrand Architecture Department: pp. 91–98.
18. Dursan (2004).
19. Oberem, Pamela (2009): *The Lazarus Funeral Parlour*. Umuzi (Random House Struik), Cape Town. Dr Pamela Oberem is better known as a veterinarian who, together with her husband Dr Peter Oberem and others, produced books detailing diseases affecting livestock and game animals. The Oberems live west of Rankin's Pass.
20. Anon (1952): *Oorsig van die Geskiedenis van Nylstroom*. Van Riebeeck Festival brochure. Published by Nylstroom Town Council, 1952. SANA TAB Pamphlet Library No. 4344.
21. Buys (1966): p. 83.
22. Boonzaaier, Boon (2008): *Tracks Across the Veld*. Privately published, Bela, p. 297.
23. British Concentration Camps of the Anglo-Boer War 1900–1902: Nylstroom. http://www.lib.uct.ac.za/mss/bccd/Histories/Nylstroom/. Accessed 16 March 2016.
24. Trichardt, Louis (2016): *Swaershoek in die Waterberge*. Privately published, pp. 182–183.

25 British Concentration Camps of the Anglo-Boer War 1900–1902: Nylstroom. http://www.lib.uct.ac.za/mss/bccd/Histories/Nylstroom/ . Accessed 16 March 2016.
26 Odendaal, R.F. (~2000): *Noord Transvaal op Kommando, 1899–1902*. Privately published, Nylstroom. Odendaal's book contains an appendix listing 549 deaths in the Nylstroom concentration camp, together with details of name, sex, age and home address. However, some of these entries appear to be duplicated. (The book contains a similar list of 617 casualties at the Pietersburg Camp.) A memorial plaque in the Nylstroom concentration camp cemetery states: "*Hier rus behalwe 'n paar ander persone 544 Vroue en Kinders wat as slagoffers van die Konsentrasiekamp gedurende die oorlog van 1899 tot 1902 die lewe ingeskiet het en vir wie ook 'n Gedenksteen op die Markplein opgerig is. Onthul, 15 Des. 1942*".
27 Hall, Darrell (1999): *The Hall Handbook of the Anglo Boer War*. University of Natal Press, Pietermaritzburg, pp. 217, 220.
28 Van Heyningen, Elizabeth (2013): *The Concentration Camps of the Anglo-Boer War: a Social History*. Jacana, Johannesburg, pp. 150, 169; and Hall, Darrel (1999): *The Hall Handbook of the Anglo–Boer War*. University of Natal Press, Pietermaritzburg, p. 222.
29 British Concentration Camps of the Anglo-Boer War 1900–1902: Nylstroom. http://www.lib.uct.ac.za/mss/bccd/Histories/Nylstroom/. Accessed 28 March 2016.
30 Anon (1952): *Oorsig van die Geskiedenis van Nylstroom*. Van Riebeeck Festival brochure. Published by Nylstroom Town Council, 1952. SANA TAB Pamphlet Library No. 4344.
31 Praagh, L.V. (1906): Extract from *The Encyclopedic History of the Transvaal*. Praagh & Lloyd, London & Johannesburg.
32 Du Toit, G.J. (1966): Onderwysontwikkeling in Waterberg. In Nel, J.J. et al (1966): *Nylstroom 1866–1966*, pp. 63–77.
33 Du Toit (1966): p 70.
34 Anon (1952): *Oorsig van die Geskiedenis van Nylstroom*. Van Riebeeck Festival brochure. Published by Nylstroom Town Council, 1952. SANA TAB Pamphlet Library No. 4344.
35 Wilks, Brett (2004): Ons Hoop and Ons Toekoms – two school hostels paint an interesting picture of Modimolle's history. In Silverman, Melinda (ed.) (2004): *Nylstroom/Modimolle – urban change in a small South African town*. University of Witwatersrand Architecture Department, pp. 129–137.
36 Ncube, Sabelo (2004): housing in Phagameng: apartheid housing and beyond. In Silverman, Melinda (ed.) (2004): *Nylstroom/Modimolle – urban change in a small South African town*. University of Witwatersrand Architecture Department, pp. 141–149.
37 Sosnovik, Amanda (2004): apartheid in a small town: moving from discrimination to democracy. In Silverman, Melinda (ed.) (2004): *Nylstroom/Modimolle – urban change in a small South African town*. University of Witwatersrand Architecture Department, pp. 33–39.
38 Van Deventer, F.J. (1966): Die Bantoebevolking van Nylstroom. In Nel, J.J. et al (1966): *Nylstroom 1866–1966*, pp. 96–99. Interestingly, this article makes no reference whatsoever to the eviction of the Indian community from the centre of the town and their forced removal to Asalaam. Nor is there an account to be found elsewhere in this official history of the town to 1966.
39 Ncube (2004): pp. 145–147.
40 Sosnovik (2004): pp. 34–35.
41 Tord Tutturen (2018): pers. comm. Tutturen is Field's great nephew and found this information in a copy of a magazine called *Fram*, issued by the Norwegian community in Durban.
42 Vilakati, Sabatha (2004): the land bank building. In Silverman, Melinda (ed.) (2004): *Nylstroom/Modimolle – urban change in a small South African town*. University of Witwatersrand Architecture Department, pp. 119–127.
43 www.statssa.gov.za/?page_id=993&id=mookgopong-municipality. Mookgophong population as per 2011 national census. Accessed 6 August 2016.
44 Erasmus, D. & Du Plessis, J.A. (compilers) (1952)*: NABOOMSPRUIT 1652–1952. Van Riebeeck Fees – Die Dorp Naboomspruit en Omgewing*. Published by the Naboomspruit Town Council to com-

memorate the Festival. Mr Erasmus at the time was the station master at Naboomspruit; and Mr Du Plessis was a teacher in the secondary school.

45 Van den Berg, Daniel (1985): *Naboomspruit – 75 in 1985*. Naboomspruit Town Council. Van den Berg was a teacher at Hans Strijdom High School. The book is long out of print and unavailable. I was lent a scarce numbered copy belonging to Topsy Graham and the late Dr Walter Eschenburg – Topsy grew up outside the town and Walter had a veterinary practice in the area for many years before the couple moved to Vaalwater.

46 Hansen, Mauritz (2010): *Eeu van genade – Naboomspruit 1910–2010*. Privately published, Mookgophong. Mauritz Hansen has spent his whole life in Naboomspruit/Mookgophong. He went to the Hans Strijdom High School in the town and from 1986–1999 was its head. During this period, he organised an annual rugby festival at the school; by 2018, the event, now named in his honour, hosted teams from 82 schools around the country, making it possibly the biggest single such event in the world. Mauritz generously gave me one of the few remaining copies of his coveted book when I visited him in Mookgophong in March 2018.

47 Van Wyk, Braam & Van Wyk, Piet (2013): *Field Guide to Trees of Southern Africa*. Revised Second Edition. Struik Nature, Cape Town, p. 46.

48 See for example Rousseau, Leon (1999): *The Dark Stream: The Story of Eugène Marais*. Jonathan Ball, Johannesburg; D'Assonville, W.E. (2008): *Eugène Marais and the Waterberg*. Marnix, Pretoria; Rousseau, Leon (ed.) (1984): *Eugène Marais: Versamelde Werke*, J.L. van Schaik, Pretoria; Marais, Eugène (1984): *Road to Waterberg and other essays*. Human & Rousseau, Cape Town.

49 Glen, H.F. and Germishuizen, G. (compilers) (2010): Botanical Exploration of Southern Africa, Edition 2. *Strelitzia 26*. South African National Biodiversity Institute, Pretoria, pp. 285–286.

50 Grobbelaar, Nat (2004): *Cycads*. Self-published, Pretoria, p. 107.

51 *The Star*, 22 March 1978.

52 *The Star*, undated article written at the time of E.A. Galpin's death in 1982. See following note.

53 Scholes, R.J. & Walker, B.H. (1993): *An African Savanna: synthesis of the Nylsvley study*. Cambridge, pp. 21–23.

54 Van Den Berg (1985*)*: pp. 100–101.

55 Greenstein, Lisa (ed.) (2002): *Jewish Life in the South African Country Communities, Volume 1: The Northern Great Escarpment, The Lowveld, The Northern Highveld, The Bushveld*. South African Friends of Beth Hatefutsoth, Johannesburg, pp. 137–138.

56 The source of this story was an unpublished manuscript, "*The Story of Naboomspruit*" by E.A. (Jim) Galpin, which was quoted in Van den Berg's (1985) *Naboomspruit 75 in 1985* and extensively, in Hansen's (2010*) Naboomspruit: 1910–2010 – Eeu van genade,* pp. 106–120. Galpin provided a wealth of fascinating anecdotal material in his manuscript.

57 Boonzaaier (2008): The Naboomspruit-Zebediela Line, in *Tracks Across the Veld*. Privately Published, Bela-Bela, p. 301.

58 Naboomspruit Crisis – the Latest News. In The Passing Show, *Sunday Times* of August 15, 1965, p.2. Reproduced in Hansen (2010): p. 91.

59 www.statssa.gov.za/?page_id=993&id=bela-bela-municipality. Bela-Bela population as per 2011 national census. Accessed 6 August 2016.

60 SANA TAB Pamphlet Library, Volume P70, Nos. 4337 & 4338 (1952): Memorandum from Voorsitter, Warmbad Dorpsraad (J.T. Erasmus) dated 21 January 1952; and accompanying article – *Ontstaan en Ontwikkeling van Warmbad*, compiled by J.J. McCord.

61 McCord, J.J. (1952): *Ontstaan en Ontwikkeling van Warmbad*. p 1. "Warmbad is gebore uit die drang na Vryheid, uit die liefde vir 'n land en volk wat die Voortrekkers hierheen gebring het. Dit moet onthou word, Warmbad is 'n mylpaal op die pad van Suid-Afrika. Ook moet onthou word dat Suid-Afrikanisme, Suid-Afrikaanse Nasionalisme, nie 'n saak van ras en taal is nie, maar iets van hart en gees! In my eie geval stam ek af van Skotland en Holland, McCord en Potgieter, maar Suid Afrika is my vaderland. Elke blanke wat hom heelhartig aan Suid-Afrika gee is 'n lid van die Nasie van Suid-Afrika".

62 Lewis, Thos. H. (1913): *Women of South Africa*, Le Quesne & Hooton-Smith, Cape Town, p. 179.
63 Silverman, Melinda & Manoim, Irwin (eds) (2003): *Warmbaths/BelaBela/Settlers: a site investigation by architecture students from the University of the Witwatersrand*. Published by the School of Architecture at the University of the Witwatersrand, Johannesburg.
64 Lichtenstein, Henry (1928): *Travels in South Africa in the years 1803, 1804, 1805 & 1806*. Volume 1. The Van Riebeeck Society, Series 1, Volume 10. Cape Town, p. 261.
65 Lye, William F. (1975): *Andrew Smith's journal of his expedition into the interior of South Africa / 1834–36*. A.A. Balkema, Cape Town, p. 286.
66 Savva, Argvris Clinton Fok & Kinver, Simon (2003): The History of Warmbaths. In Melida Silverman, Melinda & Manoim, Irwin (eds) (2003): p. 15; and J.B. de Vaal (2012): Coenraad de Buys – renegade and pioneer of the Soutpansberg. In U de V. Pienaar (ed.) (2012): *A Cameo from the Past – the prehistory and early history of the Kruger National Park*. Protea Book House, Pretoria, p. 204.
67 Union of South Africa Mines Department (1911): Annual Reports for 1910, Part IV: Geological Survey pp. 19, 106.
68 Rodger, Lex (2010): *Waterberg: Vintage Waterberg and Timeless Waterberg*. Published by the Rodger Family, Johannesburg, p. 20.
69 De Vaal (2012): p. 205.
70 McCord (1952): p. 3.
71 Savva et al (2003): p. 16. Quoting an unattributed, undated article from the files of the Bela-Bela Local Municipality entitled *"Badoorde van die Noorde"*.
72 McCord (1952): p. 4.
73 SANA TAB Maps 3/406: Kaart van Toelichting: Hartingsburg township plan dated 29 April 1895; and 3/408: regional plan of Hartingsburg and Warmbaths with railway line. Undated, probably 1902/3.
74 Savva et al (2003): p. 18.
75 Savva et al (2003): p. 18.
76 Oberholster, J.J. (1972): *The Historical Monuments of South Africa*. The Rembrandt van Rijn Foundation for Culture, on behalf of the National Monuments Council, Cape Town, p. 317. The blockhouse was proclaimed in 1959.
77 McCord (1952): p. 6.
78 McCord (1952): p. 7.
79 Savva et al (2003): p. 21.
80 Warmbaths Government Township – brochure advertising the auction of 650 erven on Wednesday 2 November 1921 by Dely & De Kock on behalf of the owners, the Government. (Poor photocopy in Bela-Bela Publicity Association files).
81 Lacovig, Alessio (2003): The Aventura Resort. In Silverman, Melinda & Manoim, Irwin (eds) (2003): pp. 173–186.
82 Lodge, Kim (2003): Klein Kariba – a community struggling for an identity, an architecture struggling for a language. In Silverman, Melinda & Manoim, Irwin (eds) (2003): pp. 187–91.

CHAPTER 7: VAALWATER, *pp.210–251*

1 www.statssa.gov.za/?page_id=993&id=modimolle-municipality Vaalwater and Leseding population as per 2011 national census. Accessed 6 August 2016.
2 SANA TAB RAK 3058: Waterberg Farms, title deeds register, farm nos. 1–988.
3 SANA TAB RAK 3058, RAK 3061: Waterberg farms title deeds registers, 1st and 2nd series.
4 Kirkman, Peter (2012): *From Manchester to Albany and Beyond: Genealogy and Chronicles of the 1820 Settler John Kirkman and his Descendants*. Published privately, Hilton, KZN, pp. 347–355. Peter's

work was recognised by the Genealogical Society of South Africa as the best study completed in 2013. In 600 pages of detailed family histories, wonderful anecdotes and authoritative historical accounts, profusely illustrated throughout, Peter's account includes references to over 3400 people who have been part of, or are associated with the Kirkman family and their descendants.
5 Howe, Anne (2013): unpublished notes from her aunt, Mary Chaney, about the early days in Vaalwater.
6 SANA TAB TAD 715 G3204/1907: letter dated 3 April 1907 in which the writer also raises, at Kirkman's request, concerns that the irrigation potential of the property was much less than had been described in the Government Gazette; and that a promised pump had not been delivered.
7 Kirkman (2012): p. 355.
8 Kirkman (2012): p. 355.
9 Kirkman (2012): p. 367–370.
10 Zeederberg, Harry (1971): *Veld Express*. Howard Timmins, Cape Town.
11 SANA SAB CDB 2943 & 2944 PB4/2/2/138: Vaalwater township development proposals by I.A. Kirkman, 1923.
12 SANA TAB TRB 2/1/268 61/4/989: Notice 335 (Administrator's) of 1953 – Proclamation of Vaalwater Township by SA Lombard, Deputy Administrator of Transvaal, signed on 18 December 1953 and published in the Provincial Gazette.
13 SANA TAB TRB 2/1/268 61/4/989: Notice 335 (Administrator's) of 1953.
14 SANA TAB TRB 2/1/40 25/1/65: Correspondence with developers relating to Vaalwater township proclamation, including town plan.
15 Minutes of the Waterberg Sentraal Landbou-Unie (predecessor of the Vaalwater District Landbou-Unie), 18 January 1956. Courtesy of VDLU archives, Vaalwater. Accessed December 2015.
16 Carruthers, E.J. (1981): dspace.nwu.ac.za/bitstream/handle/10394/5173/No.10(1981)_carruthers_EJ.pdf. Accessed 5 February, 2016.
17 SANA TPB 3038 TALG3-1-47: Vaalwater – Establishment of Local authorities & correspondence.
18 Minutes of the Waterberg Sentraal Landbou-Unie (predecessor of the Vaalwater District Landbou-Unie), 4 July 1956. Courtesy of VDLU archives, Vaalwater. Accessed December 2015.
19 Minutes of the Waterberg Sentraal Landbou-Unie (predecessor of the Vaalwater District Landbou-Unie), 7 November 1964. Courtesy of VDLU archives, Vaalwater. Accessed December 2015.
20 Minutes of the Waterberg Sentraal Landbou-Unie (predecessor of the Vaalwater District Landbou-Unie), 9 February 1957. Courtesy of VDLU archives, Vaalwater. Accessed December 2015.
21 History of the Hoërskool Ellisras; www.rassies.com (official website). Accessed 13 March 2015.
22 SANA TRB 74/2/989: Proclamation of 15 July 1970 creates Vaalwater Local Area Committee.
23 SANA TAB TRB 25/1/65: Minutes of first Vaalwater LAC meeting, 1 September 1970.
24 SANA TAB TRB 2/4/66 & G31/60/6 (1969–1979): numerous LAC and Peri-Urban board minutes, memoranda and letters, many in connection with the removal of blacks from Vaalwater to Steilloop.
25 Opening date of the Vaalwater-Melkrivier Road: 12 December 1981. Patricia Beith pers. comm. (Her wedding day).
26 Information board at Irish House, the Polokwane Municipal Museum, cnr. Thabo Mbeki and Market Streets, Polokwane.
27 Harlow, Vincent (1957): Sir Frederic Hamilton's Narrative of Events related to the Jameson Raid. *The English Historical Review* 72, p. 279. https://doi.org/10.1093/ehr/LXXII.CCLXXXIII.279
28 *Die Zoutpansberger* (2017): *Vir meer as 130 jaar is koerante deel van Limpopo*. Article under "About Us" tab on newspaper website: https://www.zoutpansberger.co.za/AboutUs. Accessed 17 July 2017.
29 Rosenberg, Valerie (2005): *Herman Charles Bosman: Between the Lines*. Struik, Cape Town, pp. 138–159.
30 Dries van Rooyen (pers. comm). Dries, the founder of the *Mogol Pos* in 1996, had spent a career in the newspaper business, starting off with *Die Transvaler* in Pretoria. Later, he moved to Polo-

kwane to work for *Die Noord Transvaler/Die Môrester*, before retiring to Lephalale to set up *Mogol Pos*. Conversation with him held at *Mogol Pos* offices on 22 September 2017, his 81st birthday.

31. *Die Zoutpansberger* (2017).
32. *Die Zoutpansberger* (2017).
33. Swart, Keina (2017): pers. comm., from the Editor of *Die Pos/The Post* and referral to an article published on the on-line news website Afrikaans.com on 29 May 2017: http://www.afrikaans.com/news-headlines/taalnuus/mediaveertjies-2017
34. See Note 31.
35. Kruger, Leoni (2016, 2017): pers. comm. 17 July 2017 and editorials by the editor of the *Mogol Pos* on 19 February 2016 and 17 February 2017, relating to 20th and 21st anniversaries of founding of the newspaper.
36. SANA TAB RAK 3058 and RAK 3062 pp. 477–478: Original and revised title deeds register for Bulge River 252, now 198KQ.
37. Ben van der Walt, pers. comm., from an interview held with him on his farm Bulge River on 23 May 2017.
38. Frans Faber, pers. comm., from an interview with him at Witfontein on 28 May 2017.
39. Frans Faber pers. comm.
40. SANA SAB URU 2158 912: Issue of Crown Grant to W.R. Howard in respect of farm Rietbokhoek No. 123, Waterberg District, 1944.
41. Minutes of the North Waterberg Farmers' Association and the Vaalwater Distrik se Landbou-Unie. See Chapter 13 for more about these organisations.
42. SANA TAB RAK 3059 and RAK 3064 pp. 905–906: Original and revised title deeds register for farm Zanddrift 1458, then 463, now 94KR.
43. SANA TAB RAK 3064.
44. Pont, Prof. A.D. (1965): *Die Nederduitsch Hervormde Gemeente Waterberg, 1865–1965*. Published by the Kerkraad van die NHK Gemeente Waterberg; pp. 103–105.
45. Keith Dietrich (2017): pers. comm. Dietrich, a Distinguished Professor in the Department of Visual Arts at Stellenbosch University has spent several years with his cousin, Carol Coney, researching the history of his paternal grandfather, Siegsmund Alexander Francis Albrecht Dietrich, who lived on the Waterberg plateau prior to his capture by the British in 1901. This led him to visit the Waterberg a number of times, most recently in August 2017, to search for the sites of his grandfather's stores and the grave of Albrecht's first wife. In homage to the women and children of the Dwars River valley who were captured by the British in 1901 and sent to the Nylstroom and Pietersburg concentration camps, where many died, he created the diptych *Dwarsrivier 1901*, which is included here as Figure 7.14. Visit www.keithdietrich.co.za. (Accessed 27 August 2017.)
46. SANA TAB MHG 13220: Death Notice (1903) of Angelina Cornelia Marian Dietrich (née Meyer).
47. SANA TAB MHG 1009/41: Death Notice of Siegsmund Alexander Francis Albrecht Dietrich; and SANA TAB MHG 1763/48: Death Notice of Susanna Josina Maria Dietrich (née Barnard).
48. SANA TAB RAK 3058 and RAK 3062 pp. 213–214: Original and revised title deeds register for farm Vier-en-Twintig-Rivier 106, now 102KR.
49. SANA TAB RAK 3062.
50. Transvaal Consolidated Land & Exploration Company Limited (TCL),Land Manager's Report for the year ending 31 December 1911: C.A. Madge, 15 March 1912. Johannesburg Public Library, Reference Section. Accessed 7 November 2012.
51. TCL 13th Annual General Meeting, Johannesburg, 3 April 1908, Chairman's Report by H.C. Boyd. Johannesburg Public Library, Reference Section. Accessed 7 November 2012.
52. www.firstworldwar.westminster.org.uk/?p=842: Charles Albert Madge>>Westminster School and World War I. Accessed 31 May 2017. Another relevant site for Madge is: www.1914-1918.invasionzone.com/forums/index.php?/topic/164743-Lt-col-charles-albert-madge-33rd-division. Research undertaken by Alan Tucker, June 2011. Accessed 31 May 2017.

53 Letter dated 9 February 1917, following a meeting of the North Waterberg Farmers Society on 16 January, addressed to Mrs C.A. Madge in Dorking, Surrey, expressing the condolences of the Waterberg farming community on the loss of her husband and their good friend. Among the documents of the late H.C.F. Baber, now in the possession of his widow, Joan Baber, who is thanked for allowing access to them.
54 SANA TAB RAK 3062.
55 Hunter, Elizabeth, compiler (2010): *Pioneers of the Waterberg: a Photographic Journey*. Privately published, Lephalale. Dates of birth and death were obtained from headstones in the graveyard at St John's Church, 24 Rivers.
56 Hunter, Elizabeth (compiler) (2010).
57 www.baberfamilytree.org/worldwide/baber-jones.htm "Descendants of Charles Baber and Phoebe Jones". Accessed 18 February 2017.
58 Members.pcug.org.au/~croe/asb/0z_boer.cgi. Australians in Boer War: Oz-Boer Database Project. Accessed 18 February 2017.
59 Recordsearch.naa.gov.au/SearchNRetrieve/Interface/ViewImage.aspx?B=3043113. Accessed 19 February 2017.
60 As for note 63, Chapter XIV p. 484.
61 www.waterbergcottages.co.za/content/the_baber_family.html. Some historical facts about the farm and the family. Accessed 19 February 2017.
62 Juliet Calcott (2017): pers. comm.: informal notes compiled in response to my questions, followed by much invaluable discussion.
63 Article in *Beeld*, August 1993: "Grond is sy Tweelingbroer": written by Andriette Stofberg; *Beeld*. Article courtesy of Juliet Calcott (née Baber). Translation: "The veld is a living being. I feel in all modesty that we are just temporary caretakers of the land. We on the farm are a biological entity, we all live there together. People say that I am a 'big farmer', one of the 20 percent who provide 80 percent of the country's agricultural production. They forget that there is a whole team on the farm which is responsible for delivery. We do it together".
64 Juliet Calcott (2017).
65 Juliet Calcott (2017).
66 Minutes of the North Waterberg Farmers' Association and of the Vaaalwater Distrik se Landbou-Unie.
67 Joan Baber (2017): pers. comm.

CHAPTER 8: THE GAP THROUGH THE HILLS, *pp.252–277*

1 Burke, E.E. (ed.) (1969): *The Journals of Carl Mauch: His travels in the Transvaal and Rhodesia, 1869–1872*. National Archives of Rhodesia, Salisbury.
2 Nel, J.J. et al (eds) (1966): *Nylstroom 1866–1966*. Nylstroom Stadsraad, Nylstroom.
3 SANA SS 2343 R6807/90: Letter by J.C. de Beer, 22 May 1890, Inspection reports & applications regarding road through Zandrivierspoort.
4 Details of the Forssman family were obtained from www.geni.com/people/Oscar-Wilhelm-Alrik-Forssman.
5 SANA TAB SS 2412 R9003/90: Application by C Forssman dated 21 June 1890 for liquor licence at his Zandrivierspoort store.
6 SANA TAB SS 2892 R7701/91: Application and petition by C Forssman dated 24 June 1891 to open a postal agency at his Zandrivierspoort store.
7 SANA TAB SS 2343 R6807/90.
8 SANA TAB ZAR 141: Annual Reports of the Department of Public Works, 1890–1898.

9. Caine, W.H.A. (1901): The diary of the late WHA Caine written in the Transvaal Republic during the period from 2nd October 1899 to 2nd December 1901. Unpublished manuscript in the possession of his nephew, G.W.R. Caine of Harare, Zimbabwe in 1958; lodged with the Manuscript Library, University of Cape Town (Accession no. BCS.356). p. 1.
10. Caine (1901): p. 4.
11. Caine (1901): p. 11.
12. Caine (1901): p. 15.
13. Hall, Maj. D.H. (1999): *The Hall Handbook of the Anglo-Boer War, 1899–1902*. University of Natal Press, Pietermaritzburg.
14. Scholtz, G.D. (1941): *Generaal Christiaan Frederik Beyers, 1869–1914*. Voortrekkerpers Bpk, Pretoria.
15. Caine (1901): p. 30.
16. Caine (1901): p. 34.
17. Grant, Capt. M.H. (1910): *History of the War in South Africa, 1899–1902* (written by direction of His Majesty's Government). Hurst & Blackwell Ltd, London. Vol IV, Chap XXV: Events in the Northern Transvaal, April 1901–May 1902, p. 439.
18. Caine (1901): p. 34.
19. Odendaal, R.F. (~2000): *Noordtransvaal op Kommando*: "Sandrivierspoort". Privately published, Nylstroom; pp. 237–240.
20. Grant (1910): p. 440.
21. Caine (1901): p. 40.
22. http://www.thepeerage.com/p55272.htm (accessed 01.08.2013).
23. Correspondence from Jo Outhwaite, Temple Reading Room, Rugby School (counter@rugby school.net), dated 10.09.2013.
24. Regimental History of the Kings Own Yorkshire Light Infantry, Vol. 2, pp. 507–508.
25. Regimental History of the Kings Own Yorkshire Light Infantry, Vol. 4 (Register of Officers), p. 180.
26. The coordinates of the grave are: 24° 23' 31.3" South; 28° 08' 02.5" East.
27. *"The Bugle"*, journal of the KOYLI of August 1901, p. 4, courtesy of Steve Tagg, Doncaster Museum & Art Gallery. Steven.tagg@doncaster.gov.uk., dated 18.09.2013.
28. Grant (1910): pp. 442–443.
29. Grant (1910): p. 444.
30. Pakenham, T. (1979): *The Boer War*. Jonathan Ball, Johannesburg.
31. www.rhinoresearchsa.com (accessed 22.05.2014).
32. SANA TAB TA GOV631: French, R., Major (1904): Report on South Africa from impressions formed while serving as Military Member of the Compensation Claims Commission for the Waterberg District, Transvaal during 1903.
33. Van den Bergh, G.N. (1980): Die Polisiediens in die Zuid-Afrikaansche Republiek. Vol. 38, Archival Yearbook of South African History. Government Printer, Pretoria, p. 236.
34. SANA TAB LD 1693 AG116/09: Bredell, H.C. (1909): Report to the Commissioner, Transvaal Police, on Status of Police Posts in the Waterberg District.
35. SANA TAB TP 9 1500: Certificate of appointment of J.W. Chaney as Constable with Transvaal Police, 1 August 1908; letter of extension for two years, record of conduct and service. SANA TAB LIC 8 180: Particulars of Claim to Pension/Gratuity by J.W. Chaney.
36. SANA TAB LD 1752 AG1645/09: Appointment of Const. J.W. Chaney as Acting Public Prosecutor, Court of Sub-Native Commissioner, Nylstroom.
37. SANA TAB LD 1744 AG1409/09.
38. Article and photograph by J.W. Chaney dated 27 December 1912, appearing in the 28 January 1913 issue of the *Lincolnshire Free Press*, published in Spalding, Lincolnshire. Copies available on microfiche at Spalding Library: spalding.library@lincolnshire.gov.uk. This copy obtained directly from the newspaper via rachel.mayfield@jpress.co.uk.

39 SANA SAB LDB 1170 R1478: Resolutions of the North Waterberg Farmers' Association, 1917–1948.
40 SANA SAB URU 1005 2887: Grant to JW Chaney of Portion C of farm Zandriviersspoort 315, Waterberg District.
41 SANA SAB PMG 345 2F1B/21887/64: Permanent closure of Sandriviersspoort Postal Agency.
42 Clarke, Elizabeth Frances (1954): *Waterberg Valleys* – her unfinished, unpublished autobiography. Extracts, including that quoted here, formed the text for the book *"Pioneers of the Waterberg – a Photographic Journey"*, compiled by her niece, Elizabeth Hunter, and published privately in 2010.
43 Nel, Louis (2004) (pers. comm.): *N.G. Gemeente Waterberg-Noord: 75 Jaar*. Unpublished notes written as a contribution to the *Gedenkalbum* published by the NG Gemeente Vaalwater in 2005.
44 SANA TAB RAK 3058-3062: Title Deeds of Waterberg Farms: Vaalwater No.5.
45 SANA SAB LEM 137 925: Department of Agriculture file of transactions with J.W. Chaney of Vaalwater.
46 SANA TAB MHG 0 5890/52: Death Certificate and Estate Dossier for John William Chaney.
47 SANA TAB MHG 6750/76: Death Certificate and Estate Dossier for Winifred Frances Chaney.
48 SANA SAB PMG 345 2F1B/21887/64: Permanent closure of Sandriviersspoort Postal Agency.
49 Lindsay, Bruce (2013): pers. comm., in September 2013, former owner of Mashudu Lodge on Modderspruit.

CHAPTER 9: THE NEW BELGIUM STORY, pp.278–341

1 Government maps 1:250 000 topo-cadastral series Sheet 2328 Polokwane (Third Edition, 2002); and 1:50 000 topographic series, sheet 2328CC Blinkwater (Third Edition, 2008); published by the Chief Directorate, National Geospatial Information, Mowbray, Cape.
2 SANA TAB RAK 2768/130; RAK 3058-3062: Waterberg farms Inspectie Rapports, Deeds Registers.
3 *The Statist*, 3 October 1891, p. 382: Transvaal land companies.
4 New Belgium Land & Exploration Company Limited: 2nd Annual report, for the year ended 30 September 1891, and presented at an Ordinary General meeting of the company's shareholders at Winchester House, London EC, on 31 December 1891. In the files of the Johannesburg Public (City) Library.
5 Briefsofcases.blogspot.com/2014/01/Transvaal-Lands-Co-v-New-Belgium.html. The conflict of interest case between Transvaal Lands Co and New Belgium (Transvaal) Land & Development Co., Ltd. English Court of Appeal, (1914) 2, Ch 488.
6 Extracts from the Annual Reports of the New Belgium (Transvaal) Land & Development Company Limited contained in *The Mining Manual*, published annually from 1887 by Walter R. Skinner in London. This publication would evolve into *Walter R. Skinner's Mining International Year Book*, published by the *Financial Times* until 1976; and to the *Financial Times International Year Book* until at least 1982. Copies of the *Mining Manual* from 1889 until 1914 were located in the National Library, Cape Town, and in the libraries of the Anglo American Corporation and the SA Chamber of Mines, Johannesburg. These extracts were referred to when the actual company reports were not located.
7 Braun, L.F. (2008): p. 258.
8 SANA TAB RAK 2768/130.
9 Jeppe, F. (1879).
10 Jeppe, F. and Merensky, A. (1868): Original Map of the Transvaal or South African Republic, from surveys and observations by Surveyor General M. Forssman, C. Mauch, F. Hammar, Surveyor J. Brooks and other official documents combined with the results of their own explorations. By F. Jeppe and A. Merensky, Potchefstroom and Botsabela. Reconstructed and augmented with data from various exploring travellers … by A. Petermann. Germany. Reproduced on p, 97 of *Mapping*

in South Africa, by Andrew Duminy (2011), Jacana Media, Johannesburg; and also in "Friedrich Jeppe: Mapping the Transvaal, c.1850–1899" by Jane Carruthers (2003), *Journal of Southern African Studies*, 29 (4), pp. 955–976.
11. Merensky, A. (1875): Original Map of the Transvaal or South-African Republic, including the Gold and Diamondfields.
12. Lamb, Doreen (1976): *The Republican Presidents, Vol. 2*. Perskor Publishers, Johannesburg.
13. Kock, Antonius Francois (1897): *Verdragen der Zuid-Afrikaansche Republiek: Compilatie van Tractaten, Conventies enz.*, in behoorlijke volgorde bijeengebracht en gerangschikt. Pretoria, pp. 117–126.
14. SANA SS 993 R4917/84.
15. New Belgium (Transvaal) Land & Development Co. Ltd Annual Report for the year ending 30 September 1891, presented at shareholders' meeting in December 1891. Johannesburg Public (City) Library.
16. New Belgium (Transvaal) Land & Development Co. Ltd Annual Report for the year ending 30 September 1893, presented at shareholders' meeting in December 1893. Johannesburg Public (City) Library.
17. SANA Library: 553.41 G6569GOL: Charles Sydney Goldman (1895–1896): *South African Mines: Their Position, Results & Developments* (3 volumes). Published by Effingham Wilson & Co., London; and Argus Printing & Publishing, Johannesburg.
18. Summarised NBTL annual reports in Walter R. Skinner's *The Mining Manual*, 1889–1914.
19. SANA SS 2892 R7701/91: Application and petition by C. Forssman dated 24 June 1891 to open a postal agency at his Zandrivierspoort store.
20. SANA TAB SS2524 R12844/90 and SS 2567 R14301/90: letters from L.V. Desborough to the State Secretary dated 12 September and 14 October 1890 requesting a permit and elaborating on the description of the weapons to be imported.
21. SANA TAB SS 2680 R864/91 and KG 29 CR131/91: letter from L.V. Desborough to Commandant General, Pretoria dated 21 Jan 1891; and response of same date.
22. SANA TAB PW 4 MR500/91: Correspondence from L.V. Desborough of New Belgium, dated 22 January 1891, seeking support from State Secretary of ZAR for continuing construction of a road from Warmbaths to the confluence of the Lephalala and Limpopo rivers; and the Secretary's negative response.
23. www.vimp.thepiltonstory.org/picture/The-Desborough-Family-at-Broadgate-Villa
24. www.scottishsporthistory.com/sports-history-news-and-blog/the-men-who-wrote-the-laws-of-association-football
25. www.paperspast.natlib.govt.nz/cgi-bin/paperspast?a=d&d=CHP18930102.2.2.2: numerous articles in *The Press* between December 1877 and February 1886.
26. www.tribalpages.com: Laurence Vivian Desborough.
27. www.dia.govt.nz/births-deaths-and-marriages: New Zealand Government Births, marriages and deaths website.
28. https://archive.org/stream/proceedings-of-Royal-Colonial-Institute. Proceedings of 1890, 1891, 1892, 1893: List of Fellows, Resident and non-Resident.
29. www.paperspast.natlib.govt.nz/cgi-bin/paperspast?a=d&d=CHP18930102.2.2.2: *The Press*, Volume L Issue 8370, 2 January 1893, p. 1. Also the British Government site, https://probatesearch.service.gov.uk/Calendar. 1892, p. 279.
30. www.ancestry.co.uk : England & Wales Census of 1871.
31. SANA TAB MKB 80 DRD1576/08; MKB 84 MCD392/09; MKB 88 MCD960/09; MKB 95 MCD1679/09; MKB 100 MCD33/10: Applications for renewals of claims and prospecting licences and the right to dispose of gold: HW Peacock and Atkinson, Kaapschehoop Area, Eastern Transvaal, 1908–1910.
32. SANA TAB MKB 98 MCD1969/09.
33. http://www.umjindi.co.za/pages/geology/geology.htm (accessed on 28 December 2015). This is

the website of the Barberton Museum; the information was discovered and pointed out to me by Marc Cadranel, a Canadian-resident descendant of the Peacock family.
34 Announcement in *The Spectator*, 3 March 1860 (p 10), information kindly provided by Rudi Butt of Hong Kong, 3 October 2013.
35 Foreign Office records from the UK National Archives (FO 228/524; 228/540; 656/25; 656/32; 1092/339); *North-China Daily News* and *North-China Herald*, July 26–August 16 1873 (from the collection of the British Library, London).
36 Various shipping records of the UK National Archives, sourced via the website www.ancestry.co.uk
37 Caine, W.H.A. (1902): *Rough Diary of Events and my Life during the Transvaal War, 1899–1901*. Noord Brabant, New Belgium. Unpublished handwritten manuscript. A typed version was prepared by his nephew, G.W.R. Caine of Salisbury Southern Rhodesia in 1958. G.W.R. Caine received the original manuscript from among the papers of his late maiden aunt, Lilian Caine, who had died in England in 1953. G.W.R. Caine's daughter, Rose Cochrane, later arranged for a copy of the typed version to be deposited with the Manuscript Library of the University of Cape Town, Accession no. BCS.356. A second typed copy is lodged with the National Archives in Pretoria: Accession (Aanwin) No. A1443 (1). Attached to this copy is not only G.W.R. Caine's accompanying letter, but also an undated note from an Archive reviewer, C. Eloff, who was clearly disenchanted with Caine's views, especially about the Boers, and who deemed it an unimportant, parochial and biased document. On the contrary, it is one of very few, meticulously compiled, authoritative, contemporary documents relating to the Waterberg of the time. Moreover, despite the inevitable prejudices of its writer, it provided, as will be seen in this story, numerous examples of evidence to corroborate or refute incidents and events reported in other documents. Thanks to Rose Cochrane of Fish Hoek for allowing me to view and photograph the original manuscript and to quote extensively from its contents.
38 Hunter, Elizabeth (compiler) (2010): *Pioneers of the Waterberg: a Photographic Journey*. Privately published. Lephalale, Limpopo.
39 Topographic Map Series 1:50 000 Sheets 2328CC Blinkwater (2008) and 2428AA Dorset (2005). Chief Directorate: National Geo-spatial Information. Mowbray, Cape.
40 Jeppe, F. (1899): Jeppe's Map of the Transvaal or S.A. Republic and Surrounding Territories; Sheet 2. Edward Stanford, London.
41 Reference details to follow: letter from Commandant HS Lombard, Acting Commandant of the Waterberg Commando Lager and Justice of the Peace, dated 16 October, 1899, recognising Arthur Frederick Peacock as a Burgher of the ZA Republic.
42 SANA TAB SS 8262-8268: Landdrost Waterberg. WH Caine Vraagt permit om hier te Blyven 30.10.1899.
43 Caine (1902): p. 3.
44 Caine (1902): p. 4.
45 Caine (1902): p. 6.
46 Jackson, A.O. (1983): *The Ndebele of Langa*. Ethnological Publications No. 54, Department of Co-operation and Development. Government Printer, Pretoria. p. 39.
47 Jackson (1983): pp. 42–43.
48 Caine (1902): p. 9.
49 Caine (1902): p. 16.
50 Caine (1902): p. 17.
51 Caine (1902): p. 19.
52 Carruthers, E.J. (1995): *Game Protection in the Transvaal 1846–1926*. Fifty Eighth Archives Year Book for South African History. Government Printer, Pretoria. pp. 68–92.
53 Carruthers (1995): p. 75 for example.
54 Caine (1902): p. 25.

55 Caine (1902): p. 26.
56 Caine (1902): p. 28.
57 Caine (1902): p. 33.
58 Jackson (1983): p. 46.
59 Caine (1902): p. 42.
60 Caine (1902): p. 44.
61 *London Gazette*, Issue No. 26343, page 6238, dated 8 November 1892: Lt-Gen William Hull Caine, Royal (late Madras) Artillery, retires on an Indian pension and extra annuity; dated 1 October 1892. https://www.thegazette.co.uk/London/issue/26343/page/6238
62 Caine (1902): pp. 45, 46.
63 Gordon, E.I.D. (1901): The unpublished diaries of Lieutenant Edward I.D. Gordon, No. 4 Company, 12th Mounted Infantry, seconded from 2nd Battalion, Royal Scots Fusiliers. Lt Gordon served with the 12th MI throughout the northern Transvaal during 1901 and was promoted to Captain in October 1901; later, during World War l, he fought with his regiment at the Battle of the Somme on June 25, 1916, ending his career as a Lt Col. His twin sons also served with the RSF; and I am grateful to one of them, Major Antony Gordon, with assistance from his wife Pat, for allowing me to access, quote from and refer to the diaries.
64 Caine (1902): p. 44.
65 Caine (1902): p. 48.
66 Oberem, P. et al (2009): *Diseases and Parasites of Cattle, Sheep and Goats in South Africa*. Afrivet, Johannesburg, p. 48. Contagious bovine pleuropneumonia (CBPP) is a disease caused by the bacterium *Mycoplasma mycoides*. It is considered to be the most serious cattle disease on the African continent. It has been eradicated from South Africa, but is still prevalent in parts of neighbouring countries as well as in Angola and East Africa. According to a 1996 publication by the Department of Agriculture (Directorate of Animal Health), affected cattle develop a fever, are listless, have difficulty in breathing, have moist coughs and a nasal discharge. Vaccinations can be effective. (www.elsenburg.com/info/nda/lunsickness2_cbpp.pdf)
67 Caine (1902): p. 52.
68 Gordon, E.I.D. (1901): diary entries 20 and 21 October 1901. Commandant M.P. van Staden was certainly an important catch: in January 1901, he, together with Cmdt Lodi Krause and their respective commandos had been part of the Boer force under the leadership of Generals Smuts, De la Rey and Beyers that had overwhelmed the army of General Clements at Nooigedacht in the Magaliesberg (see for example J.E.H. Grobler, 2004: *The War Reporter – the Anglo-Boer War through the eyes of the burghers*. Jonathan Ball, Johannesburg, page 101.) C.E. Schutte, a staunch Kruger supporter, had been the Landdrost of Pretoria from 1892–1900 (S.P. Engelbrecht et al., 1955: *Pretoria 1855–1955*. Council of Pretoria, pp. 346, 377).
69 Caine (1902): pp. 54, 55.
70 Caine (1902): pp. 55–56.
71 SANA TAB CJC 9 CJC87: claims submitted for compensation to Military Compensation Board on behalf of the New Belgium Transvaal Land & Development Company Limited; 1902–1903.
72 SANA TAB CJC 129 CJC2513: claim for compensation submitted to Civil Claims Commission on behalf of the New Belgium Transvaal Land & Development Company Limited; 1904.
73 SANA TAB CJC 128 CJC2479: Claims for Compensation: British subjects, Transvaal, Waterberg: £15 for the loss of a Martini-Henry rifle confiscated from its owner, W.H.A. Caine by British forces under Col. Colenbrander, November 1901.
74 SANA TAB CJC 7 CJC49: Arthur Frederick Peacock Claim for compensation; TAB CJC 7 CJC48: Mary Katherine Peacock Claim for compensation; TAB CJC 7 CJC50: Edith Anne Fawssett Claim for compensation; TAB PMO 38 PM2617/01: comments on the characters of Caine and Peacock.
75 SANA TAB GOV631: Report on the Waterberg, by Major Robert French, Military Member of Compensation Claims Commission, Waterberg District. December 1903, p. 3.

76 Report of the Lands Settlement Commission, South Africa, dated 28th November 1900. Part ll. Documents, Evidence. &c. Presented to both Houses of Parliament, June 1901. Printed by HMSO, London, p. 62.
77 Report of Lands Settlement Commission, pp. 257–270.
78 Transvaal Consolidated Land & Exploration Company Limited: Annual report for 1907; Chairman's report, 3 April 1908. Johannesburg City Library records.
79 SANA TAB TAD 997 N354: Letter dated 5 January 1906 from S. Meintjes on behalf of New Belgium (Transvaal) Land & Development Company Limited to Secretary for Agriculture in regard to proposed visit by Departmental representative Mr Altenroxel to New Belgium Block.
80 Information on Heinrich Altenroxel was kindly provided by his great-great niece Bernardine Altenroxel in September 2017, based on descriptions given in A.P. Cartwright (1974): *By the Waters of the Letaba;* and Louis Changuion (ed) (1994): *Tzaneen 75: 1919–1994.*
81 Clarke, Elizabeth (1954): *Waterberg Valley.* Unfinished and unpublished autobiography of Elizabeth Francis Clarke (31.03.1912–15.04.1954), the second daughter of Molly (née Fawssett) and E.A. (Ted) Davidson of the farm 24 Rivers. Mrs Clarke died in a road/railway-crossing accident in the Free State.
82 Walter R. Skinner's *The Mining Manual*, published in London from 1887; and in which New Belgium featured every year from 1889 until 1914. This annual compendium of global mining and exploration companies evolved into the *Mining International Year Book*, published by the *Financial Times* at least until 1982. An almost complete collection is stored in the library of the Chamber of Mines, Johannesburg.
83 Captain Charles Albert Madge (born 26 August 1874), had been with the Royal Warwickshire Regiment in the Anglo-Boer War. In World War I, he was a Lieutenant Colonel with the Divisional Staff of the 33rd Division of the Union Defence Force, when he was killed in action at the Hohenzollern Redoubt on 10 May 1916. A plaque to his memory is on the wall of the St John's Church at 24 Rivers, a church built in part with contributions from his pre-war employer, Transvaal Consolidated Land & Exploration Company Limited.
84 Transvaal Consolidated Land & Exploration Company Limited, Annual Report for 1911: Report by Land Manager, Capt C.A. Madge dated 15 March, 1912. Johannesburg City Library records.
85 Transvaal Consolidated Land & Exploration Company Limited, Annual Report for 1912, dated 24 June 1913. Johannesburg City Library records.
86 www.sahistory.org.za/archive/natives-land-act-no-27-of-1913
87 Transvaal Consolidated Land & Exploration Company Limited, Annual Report for 1912, dated 24 June 1913. Johannesburg City Library records.
88 Transvaal Consolidated Land & Exploration Company Limited, 71st Annual Report, for 1965.
89 SANA TAB RAK 2768/130; RAK 3058-3062: Waterberg farms Inspectie Rapports, Deeds Registers.
90 *London Gazette*, Issue 28794 of 20 January 1914, pp. 532–533: "In the matter of the New Belgium (Transvaal) Land & Development Company Limited. At an Extraordinary General Meeting of the Members of the above-named company, duly convened, and held at the Canon-street hotel on Wednesday the 17th December 1913, the subjoined Special resolution was duly passed; and at a subsequent Extraordinary General meeting of the members of the said Company, also duly convened, and held at the same place, on Wednesday 14 January 1914, the said special resolution was duly confirmed: "That the company be wound up voluntarily". And at such last mentioned Meeting, Mr Henry Samuel, of Eldon-street House, Eldon-street, London EC, was appointed Liquidator for the purposes of such winding-up. – Dated this 19th day of January 1914, Benedict W. Ginsburg, Chairman". https://www.thegazette.co.uk/London/issue/28794/page532
91 Perren, R. (2008): Biographies of William and Edmund Hoyle Vestey. http://www.bluestarline.org/william-vestey.htm
92 www.kessler.co.uk/WP-content/uploads/2013/07/Vestey_Royal_Commission_evidence_and_ensuing_debate.pdf

93 http://www.bluestarline.org/edmund-vestey2.htm Biography of 2nd Edmund Hoyle Vestey. www.wedowproductions.com/vesteyfamily Vestey family trees.
94 Knightley, Phillip (1981, 1993): *The Vestey Affair* and *The Rise and Fall of the House of Vestey*.
95 Obituaries for Edmund Hoyle Vestey, 19 June 1932–23 November 2007. *The Times*, 30 November 2007; *The Daily Telegraph*, 29 November 2007.
96 SANA TAB MHG 4439/54: Death Notice and Estate of John Todd.
97 SANA SAB LDE 1766 32357 (1): Description of the property known as the New Belgium Block, dated 15 January 1937, prepared by the estate manager "in case the property should be sold or otherwise dealt with."
98 John Todd, Nottingham Road, KwaZulu-Natal: notes from conversations with him in April 2015.
99 Clarke, Elizabeth (1954): *Waterberg Valley*.
100 Knightley, Phillip (1981): *The Vestey Affair*. Macdonald. London. In 1993, Knightley's original story was updated and expanded, and published under the title *The Rise and Fall of the House of Vestey: the story of how Britain's richest family beat the taxman – and came to grief* (Warner, London). Knightley (1929–2016) was an Australian-born, UK-resident journalist and author. As a member of the *Sunday Times*'s Insight team, he was credited with exposing the thalidomide scandal in 1971, with interviewing the soviet spy Kim Philby shortly before the latter's death in 1988 and with the exposure of numerous other state, commercial and personal intrigues. He wrote several books before his death in 2016.
101 Oberem, P. et al (2009): *Diseases and Parasites of Cattle, Sheep and Goats in South Africa*. Afrivet, Johannesburg, pp. 172–173.
102 SANA SAB LDE 1767 32357 (4): Letter dated 2 March 1943 from Brigadier WS Maxwell, SAA Transit Camp, Hay Paddock, Pietermaritzburg to the Chief Defence Force Liaison Officer in Johannesburg, contained among documents relating to the New Belgium Land Settlement Scheme. Brig. Maxwell was worried about the consequences for returning ex-servicemen: "*I understand the land is to be parcelled into 1000 morgen plots. If they are to be unirrigated, a capable man may maintain a very low standard of living on the best land, but God help him on the worst.*" The Lands Department responded that the plots were to be at least 2000 morgen in extent; and that those areas where poisonous plants occurred had been excluded from the allotments. (This was not strictly true: the areas excluded from development had been identified by their rugged terrain – the Palala river gorge – rather than by their vegetation.)
103 SANA SAB LDE 1766 32357 (1): Memo dated 3 March 1937 from Secretary of Lands. Followed in subsequent months by exchanges of correspondence between the Secretary, the Land Board and the Minister, leading, in September, to a suggested value of £65 000 (compared with Vestey's initial proposal of £104 000).
104 SANA TAB MHG 94504 Estate of Robert Sharpe.
105 www.ancestry.co.uk : England and Wales Census reports for 1871, 1881, 1891, 1901.
106 www.ancestry.co.uk : Outgoing shipping records from the UK, 1880–1960.
107 SANA TAB MHG 0 4808/65.
108 Supplement to *London Gazette* of 22 June 1918, p. 7388.
109 Supplement to *London Gazette* of 17 July 1919, p. 9125.
110 SANA SAB VWR 98 B645/188/6: Aansoek deur plaaseinaar van Lindley 87, Hendrik Willem Kok vir bywoner W.J. Averre. 1935.
111 SANA SAB LDE 1767 32357 (4): Memo from Minister of Lands following a meeting with Dr W.R. Thompson, Acting Secretary of Lands on 16 May 1942.
112 SANA SAB LDE 1766 32357 (2): Letter to Secretary of Lands from Averre dated 6 July 1939, attaching the death threat, which he believed came from "the crowd of which two members were recently convicted for shooting kudu on their land" (that is Messrs Kruger, Van der Merwe & Bronkhorst of Wynkeldershoek). Both Averre and the Department reported the matter to the police – but the Station Commander at Dorset (Sergeant Rochat) which was adjacent to the farm

in question, was a friend of the three men, and nothing further happened.
113 SANA SAB LDE 1766 32357 (Vols 1-3): Report on the New Belgium Estates project by Dr W.R. Thompson on behalf of the Department of Lands, 24 July 1941.
114 SANA SAB LDE 1774 32357. Memoranda and letters between government departments of finance, land, irrigation and mines regarding the procuring of funds for development of basic infrastructure on New Belgium preparatory to settlement: geological survey, boreholes, houses, dipping tanks, roads.
115 Strauss, C.A. (1942): Geology of a portion of the North East Section of the New Belgium Block of farms, Sheet 31, Potgietersrus District. Unpublished Open file report No. 1942–0052 (dated 3 May 1942) of the Council for Geoscience, Pretoria.
116 SANA SAB LDE 1774 32357. Memoranda and letters between government departments of finance, land, irrigation and mines regarding the procuring of funds for development of basic infrastructure on New Belgium preparatory to settlement: geological survey, boreholes, houses, dipping tanks, roads.
117 SANA TAB MHG O 1452/43.
118 SANA TAB MHG O 1452/43.
119 SANA SAB LDE 1771 32357 (8): Report dated 2 December 1941 from Thompson, now both the Superintendent of the Settlement Project and the Acting Secretary for Lands.
120 SANA SAB LDE 1767 32357 (4): Report from W.R. Thompson for December 1941; note dated 26 March 1942; and other documents in the same file.
121 SANA SAB LDE 1767 32357 (4): Notes of a discussion dated 16 May 1942 between Dr Thompson (Acting Secretary for Lands) and the Minister of Lands.
122 SANA SAB LDE 1767 32357 (4): Reports from Vermeulen to Secretary for Lands dated 13 July 1944; and 10 March, 1945.
123 *Rand Daily Mail*, 25 May 1948.
124 SANA SAB LDE 1768 32357 (6): Memo dated 24 October 1949 from Secretary for Lands to the Minister detailing all the unawarded allotments.
125 SANA SAB LDE 1768 32537, pages 1585/1586: Letter dated 07.12.49 from D.P.J. Botha, Secretary of Lands to Inspector of Lands, Nylstroom.
126 SANA SAB LDE 1768 32537, pages 1667–1699: various reports, minutes and letters within the Department of Lands between January and August 1951.
127 SANA TAB MHG 5512/51.
128 Clive Walker (2016): *Lapalala*. Published by Lapalala Wilderness, produced by Jacana, Johannesburg.
129 Clive Walker (2013): *Baobab Trails – an artist's journey of wilderness and wanderings*. Jacana, Johannesburg, pp. 225–234.
130 Walker (2016).
131 www.lapalala.com. Accessed 16 May 2016.
132 Chunnet, Fourie & Partners (1991): *Lephalala River Catchment Water Resources Development Study*. Interim Report No. 26: Water Resources Studies (November 1991). A report prepared for the RSA/Lebowa Permanent Water Commission.
133 Louw, M.D. (1992): *Executive Summaries of the ROIPs (Relevante Omgewingsinvloedprognose) for the Frischgewaagd, Moerdyk, Alem, Groenfontein and Kaapsvlakte Dams*. A report prepared for the Directorate: Design Services, of the Department of Water Affairs and Forestry, by consultants Chunnett, Fourie & Partners.
134 Bohlweki Environmental (Pty) Ltd (2002): *Environmental Impact Assessment Report for the Proposed Matimba-Witkop No.2 400 kV Transmission Line, Northern Province*. A report prepared for Eskom by the consultants. This report, as well as those above relating to the dam sites on the Lephalala River, were kindly made available by Mr Alby Geerkens, owner of Alkantrant (the consolidation of the farms Uitkomst and Lhea).

135 Land Claims Court of South Africa, Case No. 68/2006; Heard 19 March 2007, Decided 7 May 2007, Justice A Gildenhuys.

CHAPTER 10: SONDAGSLOOP AND TARENTAALSTRAAT, *pp.342–375*

1. Notes of a visit to Tarentaalstraat on 8 August 2005 in the company of Marinus & Hester du Plessis who with their parents were shareholders in the farm Tweebosch 161KR, which straddles the Tarentaalstraat kloof.
2. Pieter Prinsloo (2016): pers. comm.
3. Rodger, Lex (2010): *Waterberg* (Vintage Waterberg & Timeless Waterberg). Privately published by the Rodger family, Johannesburg.
4. Ilva Greer, pers. comm. (2005).
5. The diary of Lt (later Lt Col) E.I.D. Gordon, 12th Mounted Infantry/2nd Bn. Royal Scots Fusiliers. Lt Gordon saw action throughout the war with the Royal Scots Fusiliers from their arrival in South Africa in October 1899. He was seconded to the 12th Mounted Infantry in April 1901 and remained with this unit until the end of that year. The contents of the diary were kindly made available by his son, Maj. Antony Gordon, retired (who also spent his career with the RSF, later renamed the Royal Highland Fusiliers) and his wife Pat. This diary is a valuable archive and is the only known detailed personal record of the activities of the British forces in the Waterberg.
6. Louis Nel (pers. comm., 2005).
7. Louis Nel.
8. Louis Nel.
9. Patricia Beith (pers. comm., 2014).
10. SANA TAB LD 1693 AG116/09: Inspector HC Bredell (1909): Inspection Report of Police Posts in the Waterberg District on behalf of the Commissioner, Transvaal Police, Pretoria.
11. SANA SAB PWD 1160 2815: Dorset Police station – site, repairs, requirements, minor works, 1907–1953.
12. SANA SAB ACT 305 (T) 11332 – Police site on Brakfontein No. 96, Waterberg, 1912–1958, pp. 14–19.
13. SANA SAB ACT 305 (T) 11332 – Police site on Brakfontein No. 96, Waterberg, 1912–1958, pp. 30–33.
14. SANA SAB PWD 1160 2815.
15. SANA KAB MOOC 6/9/22843 4089/54: Estate Papers of Herbert Ernest Tomlinson, who died at Somerset East on 25.07.1954, aged 75.
16. Grobler, L.M. (2005): Uit die Notules van Springbokvlakte Boerevereniging, 1905–2005. Privately published, Pretoria, p. 8.
17. SANA TAB LA28 PET59/08: Petition of H.E. Tomlinson and 53 others of Nylstroom re: the "Asiatic Question".
18. SANA SAB CEN650 E4828: H.E. Tomlinson, of Visgat via Nylstroom.
19. Pienaar, U de V. (2012): *A Cameo from the Past – the prehistory and early history of the Kruger National Park*, Protea Book House, Pretoria, pp. 507–512; 524–526; 532.
20. SANA SAB VWR 145 B645/211/178: Aansoek deur plaaseienaar Dorset. Bywoner Z.H. Swanevelder.
21. Pieter Swanevelder (2017): pers. comm.
22. Carel Dirker (pers. comm., 2015).
23. Carel Dirker (pers. comm., 2015).
24. SANA TAB RAK 2760/75; RAK 2767/169; RAK 3058/802; RAK 3068/260: Inspection reports and title deed registers for the farm Vlugtkraal 448/38KR.

25 Ilva Greer, pers. comm. and quoted in the unpublished manuscript written by her late husband, Pat, entitled *The Story of the Waterberg* (2004); the text was kindly made available by Ilva and her daughter Greta Greer. The manuscript contains a detailed, in parts romanticised, account of the pre-Settler black oral history of the region, as related by Pat's black employees, augmented by the work of ethnologists. It also contains numerous amusing anecdotes of early settler life.
26 H.A. van den Berg (pers. comm., 2005). Mr and Mrs Van den Berg were interviewed when they stayed as guests at Mokabi Lodge on Weltevreden on 12 August 2005.
27 Sam van Coller (pers. comm., March 2017).
28 Ilva Greer, in Pat Greer (2004), p 70.
29 SANA TAB SS849 R4418/83: Inspection Report of 24 farms making up the Devonshire Block, 19 September, 1883. Archival indices also list the following earlier reports which, however, were found to be missing from their files when sought: TAB SS 233 R1381/77; TAB SS 233 R1445/77 and TAB SS 238 R2290/77, all dating from the period April to June 1877 and dealing with aspects of the inspection and surveying of farms in the Devonshire Block.
30 Koos Boshoff, pers. comm. – interviewed on 7 August 2005 on Macouwkuil, where he was visiting the resort with a view to purchasing it.
31 SANA TAB RAK 2760/75; RAK 2767/169; RAK 3058/802; RAK 3068/260: Inspection reports and title deed registers for the farm Weltevreden 802/1469/49KR.
32 Charles Baber (pers. comm, 2007).
33 "Boelie" Pretorius, pers. comm.: I interviewed Boelie at his retirement home in Modimolle on 23 August 2005. He died in 2010.
34 www.archontology.org/nations/south_africa/sa_pres1/naude.php: Biography of Jozua Francois Naudé and Paul H. Butterfield (1978): *Centenary: The first 100 Years of English Freemasonry in the Transvaal, 1878–1978*. Ernest Stanton Publishers, Johannesburg, pp. 247–248.
35 *Cape Jewish Chronicle* (April 2009): Honourable Menschen column, p. 2.
36 SANA TAB TAD24 A72/8: Farmers' Petition to re-open Tarentaalstraat through Donkerhoek; and related correspondence, 27 February–14 March 1911.
37 Prinsloo, P.W., Odendaal, R.F, Dreyer, L.G. and Kriel, M. (eds) (2002): *Gedenkalbum van die Doornfontein Geloftefees Waterberg, 16 Desember 1902–16 Desember 2002*. Published by the Doornfontein Geloftefeeskomitee.
38 SANA TAB RAK 3058, 3059, 3065, 3068: title deeds records of the farms Waterval 576/190KR; Elandsbosch 787/372KR; Doornfontein 1806/839/374KR; and Boekenhoutskloof 1809/842/187KR.
39 Prance, C.R. (1943): The Voortrekkers: their achievement and their aim; in *Dead Yesterdays*, published by the author at Port St Johns, pp. 91–99.
40 This account is summarised from those provided in Lt E.I.D. Gordon's diary for 17/18 August 1901; and *Die Slag van Doornfontein* by P.J.D. Jansen van Vuuren in: Prinsloo, P.W. et al (eds) (2002): *Gedenkalbum van die Doornfontein Geloftefees Waterberg*, pp. 16–19.
41 Kent, Leslie E. (1946): The warm springs at Loubad, near Nylstroom, Transvaal. *Transactions of the Royal Society of SA*, 31 (2), pp. 151–168.
42 Pretorius, P.J. (1948): *Jungle Man: the autobiography of Major P.J. Pretorius, CMG, DSO & Bar*, George Harrap, London. Originally printed in 1947; reprinted twice in 1948; published by Dassie Books, Central News Agency in 1952; by Alexander Books in the USA in 2000; and republished with additional documentation in 2013 by Dream Africa Productions of Bela-Bela (see note 48). In 1976, the novelist Wilbur Smith wrote "*Shout at the Devil*", a novel loosely based on the story of Pretorius and the *Königsberg*. In the same year, the book was turned into a film, starring Lee Marvin and Roger Moore.
43 Pretorius, P.J. (2013): *Jungle Man: Mtanda Batho – Hunter and Adventurer*. Dream Africa Publishing, Bela-Bela, p. 4. This edition of Pretorius's memoirs is far more comprehensive than previous editions. It is described as *"the first and only unabridged, authentic and complete autobiography of Major P.J. Pretorius, printed according to the manuscript he dictated."* The manuscript was made avail-

able for publication by a family member. A note that it had been *dictated* by the author, rather than having been written by him, explains some questions relating to its authenticity. Having had very little formal schooling (there were few competent schools in the Waterberg in the 1880s and in any case, Pretorius left school at age 12), it would be surprising if Pretorius had been able to write well at all, let alone in English. Indeed, a letter written in pencil in Pretorius's hand – in Afrikaans – to his mentor General Smuts in 1943 appealing for a work opportunity is evidence of his difficulty in composing a letter, even at the age of 66 (SANA SAB ARB 3903 1612/1/17/147).

Another source refers to the phonetic and rudimentary spelling of Pretorius's communications in English with a Dr Peringuey in connection with the shooting of the Addo elephants. (See Hoffman, 1993, in note 47 below).

A further example is provided by the letter included in the 2013 version of the memoirs (on p. 634), sent by Pretorius to his daughter while aboard a ship heading for Somaliland in 1941 – it appears to have been dictated to someone else who wrote the letter. In all probability, the memoirs were written by or for L.L. le Sueur on Pretorius's behalf between 1933 and 1939, and the incidents considered most appetising for the market of the immediate post-war period were then selected for publication. World War II intervened and the first edition of the book was only published in 1947, two years after Pretorius's death.

44 Pretorius (1948): p. 52.
45 Pretorius (2013) pp. 297–299.
46 Pretorius (1948): p. 185.
47 Hoffman, M.T. (1993): Major P.J. Pretorius and the decimation of the Addo elephant herd in 1919–1920: important reassessments. *KOEDOE* 36 (2), pp. 23–44.
48 Pretorius (1948): p. 36.
49 SANA TAB MHG 6542/45: Death Notice of Major P.J. Pretorius, with supporting documents – anti-nuptial contract, Will etc.
50 Woods, Gregor (2005): P.J. Pretorius – Jungle Man. *Man MAGNUM* magazine, December 2005, pp. 25–30. Another serialised account of Pretorius's life, paraphrasing his book, was penned by Jim Cornelius in three parts, published online in June 2011, in sections 142, 151 & 170, at: https://frontierpartisans.com/142/cruiser-killer
51 The coordinates of the grave are: 24º 35′ 17.76″ South; 28º 12′ 13.14″ East.
52 Bulloch, John Malcolm (1908): *The Earls of Aboyne*, pp. 38–39. National Library of Scotland. digital.nls.uk/histories-of-scottish-families/archive/95612895?mode=transcription. Accessed 8 December 2017.
53 Details about Lt Besley were obtained from the Graduation List of the Royal Military College, Sandhurst, and http://www.iwm.org.uk/memorials/item/memorial/56282. These and other sites refer incorrectly to the location of Lt Besley's death as being Wedelfontein instead of Middelfontein. Accessed 15 November 2017.
54 *London Gazette*, March 29, 1901, p. 2197. https://www.thegazette.co.uk/London/issue/21300/page/2197/data.pdf. Accessed 9 December 2017.
55 https://archive.org/stream/lastpostrollofal00doonrich/lastpostrollofal00doonrich-djvu.txt
56 Odendaal, R.F. (c.2000): *Noordtransvaal op Kommando*. Privately published, Nylstroom, p. 239.
57 http://www.iwm.org.uk/memorials/item/memorial/57874

CHAPTER 11: MAKING TRACKS TO THE PLATEAU, *pp.376–394*

1 Trapido, Stanley (1978): *Landlord and Tenant in a Colonial Economy: The Transvaal, 1880–1910.* Journal of Southern African Studies. Vol 5 (1), pp. 26–58.
2 SANA SAB/LDE/1157/22057/2: Act No. 30 of 1922 (Railways Construction Act); Third Schedule.

3 SANA SAB/MVE/416/16/48: Petition from Waterberg Farmers to Hon Minister for Railways & Harbours, May 1911, submitted November 1911; response from Minister's office, 14 November 1911.
4 SANA SAB/MVE/416/16/48: Memorandum from the GM, South African Railways & Harbours, 14 November 1919.
5 SANA SAB/MVE/416/16/48.
6 SANA SAB/MVE/416/16/48: Letter from E.A. Davidson, Secretary of North Waterberg Farmers' Association, 26 March 1920.
7 SANA SAB/MVE/416/16/48: Letter from E.A. Davidson, Secretary of North Waterberg Farmers' Association, 9 September 1920.
8 SANA SAB/MVE/416/16/48: Letter from C.E. Read, Secretary of Zwagers Hoek Farmers Association, August 1920.
9 SANA SAB/MVE/416/16/48: Letter to Railways Board from Joint Deputation of Farmers' Association, 26 May 1922.
10 SANA SAB/MVE/416/16/48: Letter from I.B. Grewar, Acting Secretary of North Waterberg Farmers' Association, 10 May 1921.
11 SANA SAB/MVE/416/16/48: Letter from Nylstroom Village Council (undated).
12 SANA SAB/MVE/416/16/48: GM South African Railways & Harbours Report: *Proposed Line – Nylstroom to Vaalwater*, to Chairman of Railways Board, 21 April 1921.
13 SANA SAB/LDE/1157/22057/2: Act No. 30 of 1922: The Railways Construction Act (26 July 1922).
14 Liebenberg, B.J. (1975): The Union of South Africa up to the Statute of Westminster, 1910–1931. In C.F.J. Muller (ed): *500 Years – A History of South Africa*. Academica, Pretoria, p. 395.
15 Liebenberg (1975): p. 398.
16 SANA SAB/MVE/416/16/48: Minute 1639 from Prime Minister's Office to Governor General, 16 May 1924.
17 SANA SAB/MVE/416/16/48: Letter from G. van Heerden, Chairman, South African Party, Zwagershoek Branch, to Prime Minister J.C. Smuts, 28 May 1924.
18 Van Rensburg, R.J. (1966): Die Handel, 1866–1966. In J.J. Nel and others (eds): *Nylstroom 1866–1966*. Nylstroomse Stadsraad, p. 41.
19 *South African Railways & Harbours Magazine*, November 1925, pp. 1082–1084. (Transnet Library, Johannesburg).
20 *South African Railways & Harbours Magazine*, November 1925, pp. 1082–1084. (Transnet Library, Johannesburg).
21 *South African Railways & Harbours Magazine*, November 1925, pp. 1082–1084. (Transnet Library, Johannesburg).
22 Van Rensburg (1966): p. 41.
23 *South African Railways & Harbours Magazine*, November 1925, pp. 1082–1084. (Transnet Library, Johannesburg).
24 SANA SAB/MVE/416/16/48: Minutes of meeting of representatives of the Palala region with Mr C.T.M. Wilcox of the Railway Board, Vaalwater, 1 October 1925.
25 SANA SAB/MVE/416/16/48: Note of a meeting of members of the Mogol Farmers' group with Mr C.T.M. Wilcox on 1 October 1925.
26 SANA SAB/MVE/416/16/48: Letter from Mr R.S. Read, Secretary of Swaershoek Farmers' Association, 21 September 1926 requesting upgrade of Alma siding to a station.
27 Charles Baber, pers. comm. July 2010.
28 SANA SAB/MVE/416/16/48: Letter from E.A. Davidson, Secretary of North Waterberg Farmers' Association, 25 April 1938.
29 Boonzaaier Boon (2008): *Tracks Across the Veld – a Southern African Rail Safari*. Privately published, Bela-Bela. Construction of railway lines to Thabazimbi and Ellisras, p. 299.
30 Van Rensburg (1966): p. 41.

31 Nel, Louis (ed.) (2005): *Gedenkalbum Waterberg Noord, Vaalwater, 1930–2005*. NG Gemeente Vaalwater, p. 14.
32 Boonzaaier (2008): p. 299.
33 Water, rail in place in Waterberg. *Mining Weekly*, 11 December 2009–14 January 2010, p. 24.

CHAPTER 12: PESTS AND PESTILENCE, *pp.395–427*

1 World Health Organisation (WHO) (2011): *African Trypanosomiasis* – Fact Sheet 259, October 2010 http://www.who.int/mediacentre/factsheets/fs259/en/
2 Wikipedia (2011): African Trypanosomiasis. http://en.wikipedia.org/wiki/African_trypano-somiasis
3 Stamp Out Sleeping Sickness (2011) http://www.stampoutsleepingsickness.com/about-sos-/sleeping-sickness-and-nagana/naga
4 Nash, T.A.M. (1969): *Africa's Bane: The Tsetse Fly*. Collins, London.
5 Bengis, Dr Roy (2011): pers. comm. Dr Bengis is the State Veterinarian at Skukuza, Kruger National Park, with a long interest in the tsetse fly.
6 Nash (1969): pp. 46–56.
7 Nash (1969): p. 34.
8 Cumming, R. Gordon (1850): *[Five years of] A Hunter's Life in the far interior of South Africa*, London. Facsimile publication of 1857 edition in 1986 by Galago Publishing, Johannesburg.
9 Cumming (1850): p. 314.
10 Craig, A., and Hummel, C. (1994): *Johan August Wahlberg – Travel Journals And some letters, South Africa and Namibia / Botswana, 1838–1856*. Van Riebeeck Society, 2nd Series, No. 23. Cape Town.
11 Fuller, Claude (1923): *Tsetse in the Transvaal and Surrounding Territories – an historical review*. Entomology Memoirs No. 1. Division of Entomology, Department of Agriculture, Pretoria. Fuller was an Australian who came out to South Africa in 1897 and rose to the position of Chief Entomologist of the Union government in 1927. He became immersed in the history of the Transvaal as a result of his research into the historical distribution of the tsetse fly. This in turn led him to study the remarkable trek made by Louis Trigardt (*sic*) in 1837–1838 across the northern Transvaal, down the precipitous Drakensberg and eventually to Delagoa Bay, where he and most of his group eventually succumbed to malaria. This work was published in 1932, shortly after Fuller's own untimely death in a car accident in Mozambique. (Fuller, Claude (1932): *Louis Trigardt's Trek across The Drakensberg, 1837–1838* (edited by Leo Fouché). Van Riebeeck Society, 1st Series, No.13. Cape Town.
12 Fuller (1923): pp. 13–14.
13 William Cornwallis Harris (1852): *The Wild Sports of Southern Africa*. Republished in Facsimile Edition (1987) by Struik, Cape Town, p. 191.
14 Duminy, A. (2011): *Mapping South Africa: a historical survey of South African maps and charts*. Jacana, Johannesburg, p. 97.
15 Sandeman, E.F. (1975): *Eight Months in an Ox-Waggon: reminiscences of Boer Life*. Facsimile reproduction of the 1880 edition. Africana Reprint Library Vol.1. Africana Book Society, Johannesburg.
16 Duminy (2011): p. 115.
17 Delegorgue, A. (1997): *Travels in Southern Africa, Volume 2*. (A translation by Fleur Webb, introduced by S. Alexander and B. Guest). Killie Campbell Africana Library, University of Natal Press, Durban.
18 Cumming (1850): p. 339.
19 Lye, W.F. (editor) (1975): *Andrew Smith's Journal of his expedition into the interior of South Africa, 1834–1836*. Published for SA Museum by A.A. Balkema, Cape Town.
20 Fuller (1923): p. 18.
21 Nash (1969): p. 23.

22 Fuller (1923): p. 20.
23 Baines, Thomas (1877): *The Gold Regions of South Eastern Africa*. Facsimile edition by Rhodesiana Reprint Library, Volume 1 (1968), Books of Rhodesia, Bulawayo, p. 58.
24 Baines (1877): p. 48.
25 Baines (1877): p. 65.
26 Baines (1877): p 66.
27 Selous, F.C. (1908): *African Nature Notes and Reminiscences*. Facsimile Reprint of 1st Edition, in 1986 by Galago Publishing, Johannesburg.
28 Bengis, Dr Roy (2011): pers. comm.
29 Plug, Ina (1989): Notes on the distribution and relative abundances of some animal species, and on climate in the Kruger National Park during prehistoric times. *Koedoe* 31, pp. 101–120. Pretoria. ISSN 0075-6458.
30 Zeederberg, Harry (1971): *Veld Express*. Howard Timmins, Cape Town.
31 Tamsen, E. (1923): in Fuller (1929), p. 54.
32 Merck Veterinary Manual: Rinderpest (Cattle Plague) Introduction. http://www.merckvetmanual.com/mvm/htm/bc/56300.htm
33 Department of Agriculture (2011): Rinderpest (Cattle Plague). http://www.nda.agric.za/vetweb/History/HDiseases/HAnimalDiseasesin%20SA8.htm
34 *New York Times*, 27 June 2011: Rinderpest, Scourge of Cattle, is Vanquished. http://www.nytimes.com/2011/06/28/health/28rinderpest.html accessed on 28 June 2011.
35 Carruthers, E.J. (1995): *Game Protection in the Transvaal 1846 to 1926*. Archives Year Book for South African History. Government Printer, Pretoria.
36 Buys, J.C. (1966): Die Ontstaan en Ontwikkeling van Plaaslike Bestuur in Nylstroom. In J.J. Nel et al (eds) (1966): *Nylstroom 1866–1966*. Nylstroom Town Council, pp. 81–82. The quotation is from a report by James A. Kay in the *Staats Courant* of the ZAR of 22 June 1880.
37 Wessels, N.J. (1966): 'n Beknopte Oorsig oor die Aardrykskunde van Waterberg. In J.J. Nel et al (eds) (1966): *Nylstroom 1866–1966*. Nylstroom Town Council. P. 14.
38 Leipoldt, C. Louis (1937): *Bushveld Doctor*. Jonathan Cape, London, pp. 88–103.
39 Van Wyk, Ben-Erik & Wink, Michael (2004): *Medicinal Plants of the World*. Briza, Pretoria, p. 102.
40 SANA TAB CS 934 01 19088 (1910): Proceedings of the Imperial Malarial conference, Simla, India, in 1909 reporting on the first scientific use of quinine for the treatment of malaria in 1887.
41 Chanthap Lon and others (2014): Blackwater fever in an uncomplicated *Plasmodium falciparum* patient treated with dihydroartemisinin-piperaquine. *Malaria Journal*, 13:96. Accessed 10 August 2016. https://malariajournal.biomedcentral.com/articles/10.1186/1475-2875-13-96
42 Van Heyningen, Elizabeth (2013): *The Concentration Camps of the Anglo-Boer War: a social history*. Jacana, Johannesburg, p. 148.
43 Coetzee M., Kruger, P., Hunt, R.H., Durrheim, D.N., Urbach, J., Hansford, C.F. (2013): Malaria in South Africa: 110 years of learning to control the disease. *South African Medical Journal*, October 2013. 103 (10), pp. 770–778 (p. 771).
44 SANA TPB 704 01 2580 (1909): Report from Pretoria Municipality about Malaria outbreak, followed by article in *Transvaal Chronicle*, Friday 3 September 1909.
45 SANA SAB GES2650 27/56: Letters from Dr K.J. Dekema, District surgeon in Nylstroom, 15 April and 9 May 1920, reporting outbreak of malaria in Nylstroom district.
46 SANA SAB GES 2650 27/56: Article in *Pretoria News* of 31 August 1921 – Malaria in the North and Land Settlement. Col. Reitz to investigate.
47 SANA SAB GES 2650 27/56: Letter to the Resident Magistrate, Nylstroom, from Dr K.J. Dekema, dated 19 May 1923.
48 SANA SAB GES 2650 27/56: Report to resident magistrate, Nylstroom from Dr K.J. Dekema, District Surgeon, of Special Tour of Malarial Areas, dated 21 June 1926.
49 Ingram, Alexander & Botha de Meillon (1927): A Mosquito Survey of Certain Parts of South

Africa, with Special Reference to the Carriers of Malaria and their Control (Part 1). Publications of the SA Institute for Medical Research, No. XXIII Vol. IV, pp. 1–81.
50 Ingram & De Meillon (1927): p. 14.
51 Coetzee (Prof.) et al (2013): pp. 771–772.
52 SANA SAB GES 2650 27/56: Letter dated 10 June 1935 from E.A. Davidson, Secretary of North Waterberg Farmers' Association to Minister of the Interior requesting declaration of Waterberg as a malarial district.
53 Coetzee et al (2013): p. 773.
54 Simon Hay and co-authors (2010): Developing Global Maps of the Dominant Anopheles Vectors of Human Malaria. PloS Med. 2010 February 7(2) e1000209. http://www..ncbi.nlm.nih.gov/pmc/articles/PMC2817710/?tool=pubmed
55 Maureen (Prof.) Coetzee (2001): These biographical notes are drawn largely from the glowing tribute to Botha de Meillon paid by Coetzee in an Obituary she wrote for Annals of tropical Medicine & Parasitology, Vol. 95 (2), pp. 219–221.
56 Coetzee et al (2013): p. 772.
57 SANA SAB GES 2651 27/56A: Letter from Senior Malaria Officer, Tzaneen, dated 13 February 1948 in response to Farmer's Weekly reader's query about presence of malaria in Nylstroom, Vaalwater and Warmbad districts.
58 Dr Peter Farrant reporting to a meeting in Vaalwater of the North Waterberg Farmers' Association on 18 October 1989. Minutes of the Association.
59 Coetzee et al (2013): pp. 776–777.
60 Mabazo M.L.H., Brian Sharp & Christian Lengeler (2004): Historical review of malarial control in southern Africa with emphasis on the use of indoor residual house-spraying. *Tropical Medicine and International Health*, 9 (8), pp. 846–856 (p. 848).
61 Leipoldt (1937): pp. 124–125.
62 SANA T353 OWM 608: *Laerskool Vaalwater School Journal* (1953–1969): entries for 14 August and 14 October 1953.
63 Griffiths, Charles, Jenny Day & Mike Picker (2015): *Freshwater Life*. Struik Nature, Cape Town, p. 130.
64 Magaisa, K., Myra Taylor, Eyrun F. Kjetland & Panjasaram Naidoo (2015): A review of the control of schistosomiasis in South Africa. *South African Journal of Science*, 111 (11/12), pp. 1–6.
65 Ticks and tick-borne diseases. www.up.ac.zs/animal-zoonotic-diseases-institutional-research-theme/article/276699/ticks-and-tick-borne-diseases-program. Accessed 10 August 2016.
66 Manamela (2001): pp. 11–12.
67 Manamela, David M.S. (2001): *The History of Ticks and Tick-borne Diseases in Cattle in Natal and Zululand (Kwazulu-Natal) from 1896 to the Present.* Masters thesis at University of Natal, Pietermaritzburg. Researchspace.ukzn.ac.za/bitstream/handle/10413/4929/Manamela_DavidM-S_2001.pdf?sequence=1&isallowed=y. Accessed 10 August 2016.
68 www.arc.agric.za/arc-ovi/Pages/tick-borne-disease-vaccine.aspx
69 Bezuidenhout, J.D. (2009): Heartwater: an abridged historical account. *Journal of the South African Veterinary Association*, 80(4), pp. 208–209.
70 Theilerioses (East Coast Fever and Corridor Disease) http://www.nda.agric.za/vetweb/History/H_diseases/H_Animal_Diseases_in%20SA5.htm. Accessed 29 July 2014.
71 Kuttler, K.L. (1973): East Coast Fever (Theilerasis, Theilerosis, Rhodesian Tick Fever, Rhodesian Red Water). Pdf.usaid.gov/pdf_docs/PNAAA681.pdf
72 SANA SAB URU 6 438: Letter from Gen L Botha, Prime Minister, dated 20 July 1910, requesting approval of Governor General Gladstone for the declaration of Palala and Zwagershoek wards of Waterberg District as having prevalence of east coast fever. Approval granted on 28 July.
73 Resource Centre, DAFF (2012): Theilieriosis. A publication of the SA Department of Agriculture, Forestry and Fisheries.

74 Quan, Melvyn (2013): African horse sickness. African Veterinary Information Portal. Accessed 13 August 2016. www.avrivip.org/sites/default/files/african_horse_sickness_complete.pdf
75 McKenna, Thomas St C. (2015): Overview of African Horse Sickness. www.merckvetmanual.com/mvm/generalised_conditions/african_horse_sickness/overview_of_african_horse_sickness.html. Accessed 13 August 2016.

CHAPTER 13: FITTING THE LAND TO ITS PURPOSE, pp.428–467

1 Van der Walt, H. & Van Rooyen, Theo H. (1995): *A Glossary of Soil Science* (2nd Edition). Soil Science Society, Pretoria, p. 61.
2 Low, B. & Rebelo, A.G. (eds) (1998): *Vegetation of South Africa, Lesotho and Swaziland*. Department of Environmental Affairs & Tourism, Pretoria, p. 81.
3 Rainfall data and records from various sources: 1933–1935 from Tafelkop rainfall records and Koedoesrand Kronieke by S.W. Breytenbach (2012); 1962–1964 from Tafelkop records; 1991–1993 from Tafelkop records and local farmers.
4 Mucina, Ladislav & Rutherford, Michael (eds) (2006): *The Vegetation of South Africa, Lesotho and Swaziland*. Strelitzia Series No. 19. SANBI, Pretoria, pp. 468–473.
5 Giliomee, Hermann (2003): *The Afrikaners: Biography of a People*. Tafelberg, Cape Town, p. 181.
6 Delius, Peter (1983): *The Land Belongs to Us*: The Pedi Polity, the Boers and the British in the Nineteenth Century Transvaal. Ravan Press, Johannesburg, p. 127.
7 Braun, L.F. (2008): *The Cadastre and the Colony: Surveying, Territory and Legibility in the Creation of South Africa, c.1860–1913*. PhD dissertation at the Graduate School-New Brunswick Rutgers State University of New Jersey, p. 240.
8 Braun (2008): p. 243.
9 Forssmann, O.W.A. (compiler) (1872): A Guide for Agriculturalists and Capitalists, Speculators, Miners Etc., wishing to invest money profitably in the Transvaal Republic, South Africa containing a description of a number of first-class farms situated in different districts of the Republic; and general useful information. William Foster & Co., Printers, Cape Town. (Copy in the Brenthurst Library, Johannesburg.)
10 Delius (1983): p. 129.
11 Troye, Gustav (1892): Troye's Map of the Transvaal (Sheet 2). Fehr & DuBois, Pretoria.
12 Jeppe, Friedrich & C.W. (1899): *Jeppe's Map of the Transvaal or SA Republic and surrounding territories*, Sheet 2. Edward Stanford, London.
13 Surveyor General's Office (1960): *Alphabetical List of Farms in the Province of Transvaal*. Government Printer, Pretoria.
14 Giliomee (2003): p. 180.
15 Mabin, Alan & Conradie, Barbara (eds) (1987): *The Confidence of the Whole Country: Standard Bank Reports on Economic Conditions in Southern Africa, 1865–1902*. Standard Bank Investment Corporation, Johannesburg, pp. 418–419.
16 Jeppe (1899).
17 Giliomee (2003): p. 322.
18 French, Major Robert (1904): *Report on South Africa from impressions formed while serving as Military Member of the Compensation Claims Committee for the Waterberg District, Transvaal*. Submitted to the Director General of Military Intelligence to the Colonial Office, London on 19 January 1904. SANA TAB GOV631 PS 2/04.
19 Anonymous (2011): Grain SA Fertiliser Report, pp. 39–40. Published by Grain SA www.grainsa.co.za/upload/report_files/Kunsmisverslag-Volledig.pdf. Accessed 9 May 2016.
20 Anonymous (2011): Grain SA Fertiliser Report, p. 40.
21 Agri SA (2015): Agri SA – 110 jaar.

22 Correspondence of the North Waterberg Farmers' Society, among the papers of the late H.C.F. (Colin) Baber, former Secretary of the Society, in the possession of his widow, Joan Baber. Accessed July 2017.
23 Correspondence of the North Waterberg Farmers' Association (HCF Baber papers).
24 Minutes of the North Waterberg Farmers' Association, 1971–2000. In the custody of the District Agricultural Union, Vaalwater (VDLU) Accessed, November 2015.
25 Minutes of the North Waterberg Farmers' Association, 15 October 1985; Peter Kirkman (2012): *From Manchester to Albany and Beyond* – genealogy & Chronicles of the 1820 settler John Kirkman and his descendants. Privately published, p. 370.
26 Minutes of the North Waterberg Farmers' Association, 28 February 1997.
27 Minutes of the North Waterberg Farmers' Association, 1952–1955. Courtesy of Mrs Shelley Zeederberg, December 2015.
28 News clipping from *Rand Daily Mail* of 13 December 1948. (The item was distributed by SAPA, and also appeared in *Die Burger* and *The Cape Times* of the same date.) Item amongst the documents of the late H.C.F. Baber, in the possession of his widow.
29 Bester, D.J. (1984): *NTK 75*. NTK-Nuus, 7, No. 11, Sept/Oct 1984. Nylstroom, pp. 13–14.
30 Bester (1984): pp. 17–26.
31 Bester (1984): pp. 118–119.
32 Bester (1984): p. 124.
33 Bester (1984): p. 127.
34 Bester (1984): p. 61.
35 Snyman, Delene (ed.) (2008): *NTK 100 Gedenkblad*. A publication to commemorate the centenary of the organisation. NTK, Modimolle, p. 7.
36 Bester (1984): p. 132.
37 Bester (1984): pp. 64–68.
38 Aflatoxins are a class of mycotoxins produced by certain fungal species occurring in a number of agricultural products, including peanuts, maize, sorghum, cottonseed and rice. Their development is closely linked to temperature and humidity conditions in the field, during and after harvesting and in storage. They have been identified as being carcinogenic and are associated with the development of liver cancer. They occur naturally in many foodstuffs; countries have differing limits of allowable aflatoxin levels. In Australia the maximum allowable aflatoxin concentration is 15 parts per billion; in the US it is 20ppb. http://poisonousplants.ansci.cornell.edu/toxicagents/aflatoxin/aflatoxin.html; https://www.daf.qld.gov.au/plants/field-crops-and-pastures/broadacre-field-crops/peanuts/growing-peanuts/aflatoxin-in-peanuts. Accessed 29 January 2018.
39 Snyman (2008): p. 8.
40 Snyman (2008): pp. 19–20.
41 Snyman (2008): pp. 9–0.
42 Dr Peter Farrant (2016): pers. comm.
43 Kolver, Walter (1984): *MKTV Koöperasie Beperk: 75 Jaar – van Stryd en Voorspoed*. MKTV, Rustenburg.
44 Tobacco Institute of Southern Africa website, www.tobaccosa.co.za. Accessed 10 May 2016.
45 Anonymous (2006): *Tobacco*. A publication of the Department of Agriculture. www.nda.agric.za/docs/FactSheet/Tobacco06.pdf . Accessed 9 May 2016.
46 Van der Walt, Alita (2007): SA tobacco: necessary monopoly? *Farmers' Weekly*, 29 June 2007. www.farmersweekly.co.za/article.aspx?id=1175&h=SA-tobacco:-necessary-monopoly. Accessed 10 May 2016.
47 Limpopo Tobacco Processors website: www.ltp.co.za . Accessed 10 May 2016.
48 Department of Agriculture, Forestry & Fisheries (DAFF), (2016): Abstract of agricultural statistics, 2016. http://tinyurl.com/hwroscb . Accessed 5 July 2017. Also, Statistics for 2015/2016 season from Tobacco Institute of SA (TISA): www.tobaccosa.co.za/tobacco-farming/

49 Du Preez, Clarina (2017): *Declining Tobacco Production: Analysing key drivers of change*. MCom thesis, University of Pretoria; pp. 17, 21.
50 Anonymous (2011): *A Profile of the South African Tobacco Market Value Chain*. Published by the Directorate of Marketing, Department of Agriculture, Forestry & Fisheries, Pretoria. www.daff.gov.za/docs/AMCP/TOBACCOMVCP2011-12.pdf. Accessed 10 May 2016.
51 Ben Mostert (2017): pers. comm.
52 Minutes of the Vaalwater District Agricultural Union, May–July 1972. In the custody of the District Agricultural Union, (VDLU), Vaalwater. Accessed, January 2016.
53 Oberem, P. (ed.) (2009): *Diseases and Parasites of Cattle, Sheep and Goats in South Africa*. Afrivet, Pretoria, p. 46.
54 Minutes of the North Waterberg Farmers' Association for 29 June 1983.
55 Carel Dirker (pers. comm). Carel Dirker had been born in 1935 on the family farm Uitzoek near Sukses, where he later went to school. The huge maize surplus of 1981/2 was among the many memories he shared when I visited him and his wife Janet at the retirement village near Onderstepoort where they lived in April 2015.
56 Minutes of the North Waterberg Farmers' Association for 17 November 1975.
57 Typical 1980 and 2016 land market values were provided by local landowners; and the former value (of about R150/ha) was escalated using CPI data derived from Stats SA and other government sources.
58 Carson, Annette (compiler) (1995): *Welgevonden – wilderness in the Waterberg*. A high-quality promotional publication. Welgevonden and Think Creative, Johannesburg.
59 www.welgevonden.org/about-us/introduction.html. Accessed 16 May 2016.
60 Minutes of the North Waterberg Farmers' Association, for 25 April 1984.
61 Waterberg Biosphere Reserve Management Plan, 2011, compiled for the Waterberg District Municipality by Contour Project Managers.
62 www.waterbergnatureconservancy.org.za/index.php/conservancy

CHAPTER 14: MATTERS OF FAITH, *pp.468–528*

1 Van Zyl, M.C. (1975): Transition, 1796–1806. In C.F.J. Muller (ed.): *500 Years: A History of South Africa*. Academica Press, Pretoria, pp. 101–116.
2 Kotzé, C.P. (1975): A New Regime, 1806–1834. In C.F.J. Muller (ed.): *500 Years: A History of South Africa*. Academica Press, Pretoria, pp. 117–145.
3 Pont, Prof. A.D. (1965): *Die Nederduitsch Hervormde Gemeente Waterberg, 1865–1965*. Kerkraad van die NHK Waterberg. Krugersdorp, pp. 17–20.
4 Du Plessis, J.D. (1975): The South African Republic. In C.F.J. Muller (ed.): *500 Years: A History of South Africa*. Academica Press, Pretoria, pp. 252–292.
5 Martinson, J.H. (ed.) (1960): *Eeufeesbrosjure van die Ned. Geref. Gemeente Waterberg, 1860–1960*. Nylstroom.
6 Pont (1965): p. 25.
7 www.sahistory.org.za ; www.wikipedia.com.
8 Van der Wateren, Prof. H. (ed.) (2009): *GKSA – 150. Bosveldfees*. Gereformeerde Kerk Waterberg, Nylstroom. This is the only available published history of the GK in the Waterberg. It provides a brief description of the church and its founders. Its text is characterised by the deep, brooding resentment of the author – no doubt reflecting a significant sector of the community – for the ways in which the Afrikaners and their churches had been abused by the British over a hundred years earlier. Another reference source is R.F. Odendaal's *Waterberg Word Gekersten* (undated, but c.2000) – literally translated as "Waterberg becomes Christianised", it describes, in a complex

fashion, the development of the three sister churches in the region. It also mentions the role of the Berlin Mission at Middelfontein, but, consistent with Odendaal's (and van der Wateren's) abhorrence for anything related to the British, fails to include the St Michael's Anglican church in Modimolle. In 1878, Ds. S.J. du Toit had been tasked with the translation of the Authorised Dutch Version of the Bible, dating from 1637 – which was still in general use. His first results were Genesis in 1893 and the gospel of St Matthew in 1895, followed by several other sections. However, they were not well-received by the public. In 1917 S.J. du Toit's son, Ds J.D. du Toit ("Totius"), was instrumental in persuading the Free State Synod of the Dutch Reformed Church to promote a complete new translation. Somewhat ironically, the British and Foreign Bible Society, a charitable organisation originally established in 1804 (by people including the slavery abolitionist William Wilberforce) to translate the Bible into Welsh, offered to pay for the translation and publication of the Bible into Afrikaans. Work commenced in 1923, involving a team of five "end-translators": Totius (GK), Dr H.C.M. Fourie (NHK), Prof. J.D. Kestell, Prof. E.E. van Rooyen (NGK) and Prof. B.B. Keet. In 1929 their publication of *Vier Evangelies en die Boek van die Psalms* (intended as a pilot) was well-received and the entire book was ready for publication in May 1933 when an initial print of 10 000 copies was made in Cape Town. The second, revised edition of the Afrikaans Bible was published in 1953, and a third "modern" edition in 1983.

9 Van Wyk, Daan (2015): *Hervormde Kerk vier 150ste bestaansjaar*. Article in *Die Pos* (Bela-Bela) of 27 November 2015, reporting on the 150th anniversary of the founding of the Modimolle NHK community. Written by Ronél van Jaarsveld, p. 15.
10 Van der Westhuizen, W.J. (1966): Die Kerk in Waterberg in die Afgelope Eeu. In J.J. Nel (1966): *Nylstroom 1866–1966*. Nylstroomse Stadsraad, pp. 48–49.
11 Van der Westhuizen (1966): p. 49.
12 Martinson (ed.) (1960): pp. 8–10.
13 Van der Merwe, D.W. (1984): Die Geskiedenis van die Berlynse Sending-genootskap in Transvaal, 1860–1900. *Argiefjaarboek vir Suid-Afrikaanse Geskiedenis, Ses-en-Veertigste Jaargang, Deel 1*. State Printer, Pretoria, p. 165.
14 Van der Merwe (1984): pp. 65–66.
15 Zollner, L. & Heese, J.A. (1984): The Berlin Missionaries in South Africa/Die Berlynse sendelinge in Suid-Afrika. *Genealogical Publication No. 19*, Human Sciences Research Council, Pretoria; and D.W. van der Merwe (1984): p. 112. Sourced at the Brenthurst Library, Johannesburg.
16 www.sahistory.org.za/topic/berlin-missionary-society. Accessed 30 January 2018.
17 Andrew, Nancy (2006): The dilemmas of apologising for apartheid: South African land restitution and the Modimolle land claim. https://www.mpl.irf.fr/colloque_fancier/Communications/PDF/Andrew.pdf. Accessed 30 January 2018.
18 Van der Wateren (ed.) (2009).
19 Roodt-Coetzee, K. (1949): Die lief en leed van die banneling. In Breytenbach, J.H. (ed.): *Gedenkalbum van die Tweede Vryheidsoorlog*. Nasionale Pers, Kaapstad, pp. 531–532.
20 Reitz, Deneys (1933): *Trekking On*. Faber & Faber, London.
21 De Kock, W.J. (1975): The Anglo-Boer War, 1899–1902. In C.F.J. Muller (ed.): *500 Years: A History of South Africa*. Academica Press, Pretoria, p. 352.
22 Lee, Emanoel (2002): *To the Bitter End – a photographic history of the Boer War, 1899–1902*. Protea Book House, Pretoria, pp. 171–172.
23 Odendaal, R.F. (2000): *Noordtransvaal op Kommando, 1899–1902*. Privately published, Nylstroom, pp. 425–427. The late R.F. Odendaal, a resident of Nylstroom and son of F.H .Odendaal, an administrator of the Transvaal, was himself a town council member and mayor of Nylstroom. In common with his father, Advocate J.G. Strijdom (a local attorney and later prime minister) and other members of the National Party in the region, he formed a particularly conservative, anti-British faction of the party, as reflected in his several books and pamphlets.
24 Van der Westhuizen (1966) p. 52.

25 Spies, S.B. (1986): Unity and Disunity, 1910–1924. In Trewhella Cameron & S.B. Spies (eds): *An Illustrated History of South Africa*. Jonathan Ball, Johannesburg, pp. 238–240. The influenza pandemic of 1918–1920 infected some 500 million people worldwide and killed over 50 million, making it the most lethal event in recorded history. The origin of the virus – which is the ancestor of strains of the H1N1 influenza virus still prevalent today – is not certain, but is thought to have been China, from where it was carried into the susceptible battlefields of Europe by a group of some 94 000 Chinese labourers brought in to work behind the British and French lines. Its attributed Spanish origin was due to deaths in that country (which was neutral during the war) being widely reported by the media there, whereas elsewhere in Europe, censorship prevented the publication of casualties. (Jeffery K. Taubenberger & David M. Morens: 1918 Influenza – The Mother of all Pandemics. *Emerging Infectious Diseases*, Vol 12 (1), January 2006, pp. 15–22. www.cdc.gov/eid; https://virus.stanford.edu/uda/; Dan Vergano, *National Geographic*, January 2014 at news.nationalgeographic.com/news/2014/01/140123-spanish-flu-1918-China-origins-pandemic-science-health/)
26 Martinson (ed.) (1960): p. 15.
27 Martinson (ed.) (1960): p. 19.
28 Pont (1965): p. 122.
29 Spoelstra, B. (2015): Quoted from his work "Ons volkslewe", in *Nagmaal op die Platteland*. Internetmuseum vir Afrikanergeskiedenis, January 22 2015. http://www.afrikanergeskiedenis.co.za/nagmaal-op-die-platteland/. Accessed 5 May 2015.
30 Louis Nel (2005): Quoted in *Gedenkalbum, NG Gemeente, Vaalwater, 1930–2005*. NG Gemeente Vaalwater, pp. 12–13.
31 Breedt, J.L. (ed.) (1998).
32 Du Plessis, R.J. (-): Geskiedenis van die Gereformeerde Kerk, Warmbad Gemeente – Die Tydperk Voor 06 Februarie 1954. www.gkwarmbad.co.za/downloads/Kerk%20Geskiedenis%20Geref%20Kerk%20Warmbad.pdf
33 Van der Wateren (ed.) (2009): pp. 8–9.
34 Du Toit, F.G.M. (ed.) (1972): *Warmbad 1922–1972: Nederduitse Gereformeerde Kerk Warmbad Feesalbum*. NGK Warmbaths, pp. 18–19.
35 Du Plessis (-): p. 4.
36 Van den Berg (1985): p. 120.
37 Erasmus, D. & J.A. du Plessis (1952): *Naboomspruit: Van Riebeeck Fees, 1652–1952*: Die Dorp Naboomspruit en Omgewing: 'n populêr-historiese studie. Naboomspruit Dorpsraad. SANA Pamphlet Library P102 (No. 4295).
38 Van den Berg, Daniel (1985): *Naboomspruit – 75 in 1985*. Published by the Naboomspruit Town Council, p. 123.
39 Notes provided by Ds. Albert Berg of the GK Gemeente Vaalwater, based on extracts from the GK Eeufees Gedenkboek of 1960 and the book by C.T. Harris and others (2010): *Van Seringboom tot Kerkbou – die argitektoniese erfenis van die Gereformeerde Kerk*. Published by the GK Raad.
40 Van Heerden, Marthie (ed.) (1994): *Gemeente Albertyn, Halfeeufees Gedenkblad 1944–1994*. NG Gemeente, Albertyn, Ellisras.
41 Nel, Louis (2005): *NG Gemeente Waterberg-Noord – 75 Jaar*. A comprehensive set of notes compiled by a doyen of the district and son of Jan Nel, first headmaster of the Melkrivier Laerskool. Born in the area, Louis inherited his father's interest in the early history of the Waterberg district, together with a large collection of documents, memorabilia – and a fine memory and writing skill. Louis has long promised to put all his valuable reminiscences and information down on paper; and it is to be hoped that he will do so. In the meanwhile, it is typical of his generosity that he provided me with a copy of the notes he prepared at the request of the NGK to enable it to compile the commemorative volume celebrating the 75th anniversary of the Waterberg-Noord community in 2005.

42. Nel, Louis (ed.) (2005): *Gedenkalbum NG Gemeente Vaalwater 1930–2005*. Kerkraad, NG Gemeente, Vaalwater. Contribution from Louis Nel, pp. 18–20.
43. Nel (ed.) (2005): Louis Nel, pp. 20–23.
44. Nel (ed.) (2005): p. 31.
45. Frans Faber, pers. comm., in an interview on his farm Witfontein 6KR on 28 May 2017.
46. Kloppers, Dr W. (ed.) (2016): 2017 *Gedenkalbum Nederduitsche Hervormde Kerk van Afrika; 500 kerkhervorming, 1517–2017*. Published by SENTIK, Pretoria.
47. Van Wyk, D.J.C. (compiler) (2016): *Nederduitsch Hervormde Gemeente Waterberg 150 (1865–2015)*. Published by the NHK Gemeente Waterberg and SENTIK, Pretoria.
48. www.wikipedia.com: Afrikaanse Protestantse Kerk. Accessed 6 December 2015.
49. Grobler, Maritz et al (eds) (1995): Die Suidelike Springbokvlakte/The Southern Springbok Flats – an historical overview. Published by the Berlin Geloftefeeskomitee, Springbok Flats, p. 252.
50. Nel, Louis (2014): Afrikaanse Protestanse Kerk Gemeente Vaalwater. Published by the church.
51. www.apk.co.za Accessed 6 December 2015.
52. Wolhuter, Ds. W.H.S. (1998): Wit Oeslande in die Waterberg Sendingwerksaamhede 1960–1998. In J.L. Breedt (ed). *NGK Waterberg Gedenkblad 1898–1998*, pp. 23–25.
53. Knipe, Dr W. (2010): Die Geskiedenis van die NG Kerk in Naboomspruit. In Mauritz Hansen (2010): *Naboomspruit 1910–2010 – Eeu van genade*, p. 130.
54. Kane-Berman, J. & Moloi, L. (2016): *South Africa Survey 2016*. Institute of Race Relations, Johannesburg, p. 68. Repeated visits by the author to Moria City during 2017 to obtain updated factual information and imagery about the two churches were met with polite, but firm demurral.
55. Wouters, Jacqueline M.F. (2014): *An Anthropological Study of Healing Practices in African Initiated Churches with Specific Reference to a Zionist Christian Church in Marabastad* [Pretoria]. MA thesis, UNISA, Pretoria, p. 49. Accessed at Easter, 14–17 April 2017. Uir.unisa.ac.za/bitstream/handle/10500/18867/dissertation_wouters_imf.pdf?sequence=1
56. Wouters (2014): pp. 50–54.
57. Wouters (2014): p. 56; and Morton, Barry (undated): *Engenas Lekgonyane and the early ZCC: Oral Texts and Documents* at https://www.academia.edu/14338013/Engenas_Lekganyane_and_the_early_ZCC_oral_texts_and_Documents . Accessed 17 April 2017.
58. Morton (undated): Quoting a letter from Secretary of Native Affairs Commission to the Secretary of Native Affairs, dated 18 September 1925, contained in the National Archives ref: SANA SAB BAO 7264 120/4/68.
59. Lye, William F. (1975): *Andrew Smith's Journal of his expedition into the interior of South Africa / 1834–36*. A.A. Balkema, Cape Town, pp. 59–60.
60. Morton (undated): Quoting from a letter dated 26 October 1942 from the Additional Native Commissioner, Pietersburg to the Secretary for Native Affairs, Pretoria. SANA SAB BAO 7264 20/68.
61. Morton (undated): Quoting from a letter dated 12 February 1943 from Kraut, Hazelhurst, Wagner & Hutchinson to Native Affairs Department, Pretoria. SANA SAB BAO 7264 120/68.
62. Wouters (2014): p. 62, quoting the works of E.K. Lukhaimane (1980): *The Zion Christian Church of Ignatius (Engenas) Lekganyane, 1924 to 1948: An African Experiment with Christianity*. MA dissertation, University of the North, Pietersburg, p. 33; Verwey, E.J. (ed.) (1995): *New Dictionary of South African Biography*, HSRS Publishers, Pretoria, p. 132; and Hanekom, C. (1975): *Krisis en Kultus: Geloofsopvattinge en seremonies binne 'n swart kerk*. Academica, Cape Town, p. 62. Elias Lukhaimane wrote from the vantage point of being a member of the St Engenas ZCC.
63. Wouters (2014): p. 65.
64. Wouters (2014): pp. 128–136.
65. Wouters (2014): p. 66.
66. www.sahistory.org.za/people/joseph-engenas-matlhakanye-lekganyane. Accessed 17 April 2017.
67. Müller, Retief (2015): The Zion Christian Church and Global Christianity: negotiating a tightrope

between localisation and globalisation. *Religion* 45 (2) pp. 174–190 (p. 186).
68 Wouters (2014): p. 125.
69 Müller (2015): pp. 180–181.
70 Allen, Vivien (1971): *Kruger's Pretoria*. AA Balkema, Cape Town, p. 89.
71 Anon (undated): The Catholic Church in Northern Transvaal. Informal notes in the collection of the Church of Our Lady Mediatrix in Bela Bela, September 2015.
72 Anon (undated).
73 Feliciano, R. (2003): The Roman Catholic Church. In: Diane Arvanitakis (ed.) (2003*): Warmbaths/BelaBela/Settlers: Urban Change in a Small South African Town.* A site investigation by Architecture Honours students from the University of Witwatersrand. Published by the University.
74 Anon (undated).
75 Anon (undated).
76 South African Catholic Bishops' Conference, Diocese of Polokwane. Historical record. www.sacbc.org.za/dioceses/Pretoria/Pietersburg/
77 Anon (undated).
78 Anon (undated).
79 Allen (1971) p. 83.
80 Bousfield, H.G. (1886): *Six Years in the Transvaal* (Notes of the founding of the Church there). Society for Promoting Christian Knowledge, London. Sourced at the Bodleian Library, Oxford (Reference 602192276; 1340 f.3).
81 Anon (1987): *A Summarised History of Anglican Worship in the Waterberg Parish of the Diocese of Pretoria for 80 years until 4th October 1987*. Church records, the Church of St Michael and All Angels, Nylstroom (Modimolle).
82 Hunter, Elizabeth (2010): *Pioneers of the Waterberg – a Photographic Journey*. Privately published, Lephalale, Limpopo, pp. 51–53.
83 Anon, (1987).
84 Van der Westhuizen (1966): p. 61.
85 Van der Westhuizen (1966): p. 62.
86 Anon (1987).
87 Jean Kohly (2007): Notes taken of Reminiscences by Mr Freddie Gibb at his hospital bed, 15 April 970. Included among church records, St Michael's Church, Modimolle.
88 Anon (1987): p. 3.
89 Anon (1987).
90 Van der Westhuizen (1966): p. 62.
91 Hunter (2010).
92 Anon (1987): p. 3.
93 Van den Berg (1985): pp. 126–127.
94 Anon (1987): p. 4.
95 Anon (1987): p. 5.
96 Rev. Douglas Kirkpatrick (2014): pers. comm., September 2014. Rev Kirkpatrick was Rector of St Michael's from 2011 to 2014.
97 Rev. J. Whiteside (1906): *History of the Wesleyan Methodist Church of South Africa*. Juta & Co., Cape Town (https://archive.org/stream/historywesleyan00whitgoog_djvu.txt).
98 Mears, W.J.G. (1972): *An Outline of Methodism in the Transvaal*. Privately published, pp. 31–32.
99 Strick, Mary (1995): Methodist Church, Warmbaths/Waterberg Circuit: The Women's Auxiliary, 1945–1995. Private papers of Mr Harold Nicolson.
100 Harold Nicolson, longstanding resident. Notes from a meeting with Mr Nicolson at his home Hawkhill on the farm Tweefontein 463KR between Bela-Bela and Modimolle, on 1 September 2015.
101 Anon (1987, 2007): Methodist Ministers in the Warmbaths-Waterberg Circuit from 1909. Papers of

the Anglican parish of St Michael's, Modimolle.
102 Strick (1995): p. 4.
103 Harold Nicolson (2015).
104 Brian Stone (2015): notes of the reminiscences of Mr Stone and his wife Wendy, collected during a visit to their home at the Renaissance Retirement village in Bela-Bela on 8 September 2015. Brian Stone died on 31 October 2018, aged 98.
105 Law, Anet (2010): The Methodist Church. In: Mauritz Hansen – *Naboomspruit 1910–2010, Eeu van genade*. Privately published, Mookgophong, pp. 147–149.
106 Wiener, Charlotte (2006): *The History of the Pietersburg (Polokwane) Jewish Community*. MA Thesis, UNISA, p. 120. Printed version 2007, UNISA. Available on-line at: www.uir.unisa.ac.za/bitstream/handle/10500/1721/dissertation.pdf.txt?sequence=2.
107 Kollenberg, Adrienne, Rose Norwich, Joan Gentin & Phyllis Jowell (compilers) (2002, reprinted with addenda 2008): *Jewish Life in the South African Country Communities. Volume 1: Northern Transvaal*. South African Friends of Beth Hatefutsoth, Johannesburg, p. 146. This series, which currently runs to five volumes (another three are in compilation) is a marvellous, comprehensive, poignant and unique record of the role played by Jews in the country districts of South Africa over the last 150 years. Professionally collated, fully referenced and attractively published over a period of 15 years, it deserves a much wider, more secular audience than it has enjoyed to date.
108 Wiener (2006): pp. 104–108.
109 Wiener (2006): pp. 70–80.
110 Kollenberg et al (2002, 2008): pp. 140–141.
111 Wiener, Charlotte (2016): The Jewish Country Communities of Limpopo/Northern Transvaal. Privately published, p. 31.
112 Kollenberg et al (2002, 2008): pp. 137–138.
113 Jankelow, Gareth & Penn, Amanda (2003): The Hebrew Communal Hall. In: Melinda Silverman & Irwin Manoim (eds) (2003*): Warmbaths / BelaBela / Settlers: Urban Change in a Small South African Town*. A site investigation by Architecture Honours students from the University of Witwatersrand. Published by the University.
114 Kollenberg et al (2002, 2008): pp. 168–173.
115 Kollenberg et al (2002, 2008): Dr & Mrs Basil Hack, their son and his family were the last Jewish residents of Bela-Bela in 2001; p. 169.
116 Saks, David Y., Associate Director, SA Jewish Board of Deputies (2015): pers. comm. Author of *Boerjode: Jews in the Boer Armed Forces, 1899–1902*. Self-published in 2010.
117 Kollenberg et al (2002, 2008): p. 141.
118 Kollenberg et al (2002, 2008): p. 142.
119 Marquard, Leo (1969): *The Peoples and Policies of South Africa* (4th Edition), Oxford University Press, London, p. 236. This book was first published in 1952.
120 Silberhaft, Rabbi Moshe (2012): *The Travelling Rabbi: My African Tribe*. Jacana, Johannesburg.
121 https://en.wikipedia.org/wiki/Abrahamic_religions: Abrahamic religions are monotheistic religions of West Asian origin, emphasizing and tracing their common origin to Abraham or recognizing a spiritual tradition identified with him. The largest Abrahamic religions in chronological order of founding are Judaism (1st millennium BC), Christianity (1st century AD) and Islam (7th century AD). Accessed 20 September 2015.
122 Rousseau, Leon (1982): *The Dark Stream – The Story of Eugène Marais*. Jonathan Ball, Johannesburg, p. 223.
123 Van Deventer, F.J. (1966): Die Bantoebevolking van Nylstroom. In J.J Nel (1966): *Nylstroom 1866–1966*. Nylstroomse Stadsraad, p. 97.
124 Taraj, Chota (2015): Mr Taraj, whose father Ismail Taraj had been born in Nylstroom in 1917, never lived in the town himself, although he was related by marriage to many members of the Indian community there. Telephonic interview on 10 December 2015.

125 Zunaid Mosam (2015): pers. comm. Zunaid was born and brought up in Nylstroom, where his father was for many years employed by the local Toyota garage. After completing his tertiary studies he became a political activist, which made it necessary for him to leave the town (he spent some time in Botswana) until the change of government in 1994.
126 SANA TAB LA28 PET59/08: Petition of HE Tomlinson and 53 others of Nylstroom re: the Asiatic Question.
127 Zunaid Mosam (2015): pers. comm.; Mr Ravat (2016) – pers. comm.
128 Savva, Argvris & Clinton Fok (2003): Tracing the Origins of the Bela Bela Mosque. In Melinda Silverman & Irwin Manoim (eds) (2003*): Warmbaths/BelaBela/Settlers: Urban Change in a Small South African Town*. A site investigation by Architecture Honours students from the University of Witwatersrand. Published by the University.

CHAPTER 15: A FUTILE BROEDERTWIS, *pp.529–544*

1 Chief Directorate, Surveys & Mapping, Department of Rural Development & Land Reform: 1:50 000 topographical sheet *2427DB Rankin's Pass*, 1st Edition, 1968. Government Printer, Pretoria.
2 Spies, S.B. (1986): Unity and Disunity, 1910–1924. In T. Cameron (Editor): *An Illustrated History of South Africa*. Jonathan Ball, Johannesburg.
3 Oakes, D. et al (eds) (1992): *Illustrated History of South Africa – The Real Story*. Readers Digest Association of South Africa.
4 Spies (1986): p. 237.
5 Cruise, Adam (2015): *Louis Botha's War: The campaign in German South West Africa, 1914–1915*. Zebra Press, Cape Town, pp. 43–48.
6 Scholtz, G.D. (1942, 2013): *Die Rebellie 1914–1915*. Protea Boekhuis, Pretoria, p. 167.
7 Sampson, Philip J. (1915): *The Capture of De Wet: The South African Rebellion, 1914*. Edward Arnold Publisher, London, pp. 187–189.
8 De Wet, J.M. (1940): *Jopie Fourie: 'n Lewensket*s. Voortrekkerpers, Pretoria.
9 Scholtz, G.D. (1942, 2013): p. 146.
10 Australian and New Zealand newspapers referred to Commandant Van Deventer as "Colonel Kirk Van Der Venter"! *Poverty Bay Herald* of 14 November 1914: National Library of New Zealand: www.paperspast.natlib.govt.nz/cgi
11 *Pretoria News*, Tuesday 10 November, 1914.
12 *Pretoria News*, Wednesday 11 November, 1914.
13 Annemarie Folkus, pers. comm. Feb. 2014.
14 Genealogical Society of South Africa: http:/www.eggsa.org/library/main.php?g2_itemId=711095
15 Genealogical Society of SA. Details that have been collected about the ten men buried on Zandfontein are as follows: **Frederik Johannes Botes** – Burgher, Geyser's Commando, Killed in Action, Zandfontein 08.11.1914. Born about 1896 in the Pretoria district; his mother, Catharina Philipina Standers, lived in Nylstroom; **Bartholemew Peter (Piet) Cronning** – Burgher, Geyser's Commando, Killed in Action, Zandfontein, 08.11.1914. Born about 1885 in Zastron, Orange Free State. Widow Anna Maria Johanna du Plessis, died 1957; **H.J. Devanner** – Burgher, Geyser's Commando, Killed in Action, 06.11.1914. **NB**: not killed in the Zandfontein, battle, but two days earlier, presumably nearby; **Albert Cornelis De Wilde** – Burgher, Geyser's Commando, Killed in Action, Zandfontein, 08.11.1914; **F.J.J. du Preez** – Burgher, Geyser's Commando, Killed in Action, Zandfontein, 08.11.1914; **E.R.J. Geldenhuys** – Captain, Geyser's Commando, Killed in Action, Zandfontein, 08.11.1914. Most senior casualty; **Jeremiah Daniel Nel** – Burgher, Geyser's Commando, Killed in Action, Zandfontein, 08.11.1914. Born in Port Elizabeth, married in Nylstroom to Johanna Helena Fourie; seven children, ranging in age at the time from 17 years

to only 4 months; ***J.C. van der Merwe Junior*** – Burgher, Theunis Botha's Commando, Killed in Action, Zandfontein, 08.11.1914; ***Jan Nagel*** – Burgher, T Jones's Pretoria Commando, Killed in Action, Zandfontein, 08.11.1914; ***Albert William Wrighton*** – Lieutenant, Geyser's Commando, Killed in Action, Zandfontein, 08.11.1914. Born in ~1879 at Tuigewich, Buckinghamshire. Husband of Annie Wrighton; parents were from Aynho, Northamtonshire. He was the older brother or cousin of four other Wrighton men, all from Aynho and all of whom were to die later in the War, at Delville Wood (1916) and at Vimy Ridge/Cambrai (1917).

16 Commonwealth War Graves Commission: www.cwgs.org/find-a-cemetery/cemetery/4007210/ZANDFONTEIN%20FARM%20BURIAL%20GROUND

17 ***Daniel Johannes Elardus Opperman*** – Lieutenant, Theunis Botha's Commando; the son of Commandant Opperman, Lt D.J.E. Opperman was given a funeral with full military honours, attended by General Smuts and approximately 800 mounted men of the Botha Commando, at the Dutch Reformed Church in Pretoria on 15 November, followed by burial at Rebecca Street Cemetery. Commandant Opperman gave a stirring, patriotic speech at his son's graveside, reported in full in the *Pretoria News* the next day. ***Herbert Henry Horatio Hitchcock*** – Burgher, Geyser's Commando; born in England in 1879; worked as a miner at Rooiberg tin mine, resided on the neighbouring farm Olievenbosch (15 km SW of Zandfontein) where he was buried on 9 November. His first wife, Sarah Susanna Brink, had died early in 1914; he remarried, but his second wife, Filippina Suzara Helberg (22) had died only two weeks before the battle. ***William Stonewall Johnson*** – Burgher, Geyser's Commando; Johnson was originally buried at Warmbaths, where he had lived. His remains, together with those of two colleagues, Mapham and Dewrance (below), were later mistakenly exhumed and re-interred in the British Garden of Remembrance in the cemetery at Pietersburg (Polokwane).

18 ***George William Mapham*** – Burgher, Geyser's Commando; died 5 December 1914, Warmbaths. He was 34, the husband of Mrs G. Mapham of Warmbaths. His name appears on the Town War Memorial, Bedford, Eastern Cape. ***Herbert John Dewrance*** – Burgher, Geyser's Commando; killed in action near Warmbaths, 6 November 1914.

19 ***Albert S.A. Opperman*** – Burgher, Geyser's Commando; from Potgietersrust; died on 25 November 1914, after an action under Commandant Geyser that resulted in the capture of several rebel leaders and men in the Waterberg. Buried at the Rhenosterpoort Farm Burial Ground just north of Rankin's Pass, about 18 km NE of Zandfontein (see Figure 15.5 and Chapter Four). ***J.J. Scheepers*** – Sergeant, Waterberg Commando; shot by rebels on 20 November 1914 after capture on 11 November; buried at Grootvlei farm cemetery, between Modimolle and Kranskop. This has not been located.

20 Scholz G.D. (1942, 2013).

21 Van den Berg, J.H. (1960): Eensame simbole van een van die Bloedigste Gevegte in Transvaal tydens die Rebellie van 1914. *Historia* 5 (1), pp. 53–55. Translated from the Afrikaans by Dr C. Sievers and R. Wadley.

22 *Pretoria News*, Friday 11 December, 1914: "The Rooiberg Fight".

23 www.weaponslaw.org/instruments/1899-Hague-Declaration. The Hague Declaration of 1899 forbade the use of soft-nosed, expanding bullets (called "dum-dum" after the armaments factory at Dum-Dum outside Calcutta, India, where they were first manufactured). The reason was that such bullets caused extremely severe injuries. This prohibition occurred immediately before the outbreak of the South African War and was largely adhered to by both sides in that conflict.

24 *Rand Daily Mail*, Wednesday 11 November 1914.

25 Spies (1986): p. 238.

26 Fitzpatrick, Sir Henry (1979): *South African Memories*. AD. Donker, Johannesburg, pp. 236–244.

27 Kruger, Rayne (1977): *Good-bye Dolly Gray: The Story of the Boer War*. Pan Books (2nd Ed.), London.

28 Pakenham, Thomas (1979): *The Boer War*. Jonathan Ball, Johannesburg, pp. 201–206.

29 Webb, H.S. (1915): The Causes of the Rebellion. Northern Printing Press Co. Ltd., Pretoria (p. 95).

Sourced at Bodleian Library, (Rhodes House), Oxford University.
30 Scholtz, G.D. (1941): Generaal Christiaan Frederik Beyers, 1869–1914. Voortrekkerpers Beperk. Johannesburg, Pretoria en Potchefstroom, p. 374.
31 *Pretoria News*, 11–14 December 1914.
32 Muller, C.F.J. (ed.) (1977): *Five Hundred Years: A History of South Africa*, pp. 388–390. Academica Press, Pretoria & Cape Town.
33 *The Star*, Monday 21 December 1914.
34 List of Rebel Prisoners released from Gaol. *Pretoriana* 081 (July 1981), pp. 107–19.
35 Spies (1986): p. 238.
36 Scholtz, G.D. (1941): pp. 99–101.
37 https://thegazette.co.uk/London/issue/30857/supplement/9798: Supplement to the London Gazette of 22 August 1918 (Issue 30857), p. 9798. Forsyth, D.R. (1967): Dekoratie voor Trouwe Dienst, 1899-1902. *Journal of South African Military History* Vol.1 No. 1 (December 1967); and No. 2 (June 1968). https://samilitaryhistory.org/journal.html. I am grateful to Peet Geyser, a descendant of Commandant Geyser, for having directed my attention to this source and the one before. Instituted in terms of Government Notice No.2307 of 21 December 1920 and published in the Union of South Africa *Government Gazette* of 24 December 1920, the award of Dekoratie voor Trouwe Dienst (DTD) was made to a total of 655 officers of the former Zuid-Afrikaanse Republiek and the Oranje Vrijstaat who, being citizens of these two republics, had rendered distinguished and especially meritorious service during the South African War of 1899–1902. Richard Granville Nicholson (Chapter 17) was also a recipient of a DTD medal.
38 Sandriviersspoort Hotel register, 1925–1935, loaned by Mrs Anne Howe, 2013.
39 Peet Geyser: Pers. comm. 23 June 2014.

CHAPTER 16: DOMAIN OF THE NORTHERN LIONS, *pp.545–575*

1 Grobler, L.M. (2005): *Uit die Notules van Springbokvlakte Boerevereniging, 1905–2005*. Publiself Uitgewers, Pretoria, p. 4.
2 Lamb, Doreen (1974): *The Republican Presidents, Volume 1*. Perskor Publishers.
3 Grobler, L.M. (2005): p. 4.
4 SANA TAB SS 67 R673/65: (10.06.1865): G.J. Verdoorn, Waterberg. Kennisgewing aan den Gouverments' Secretaris dat de heer H.J. van Staden bij meerderhijd van stemmen is gekozen te Lid van den Volksraad.
5 SANA TAB SS 129 R1699/70: A petition from J. Smit and 24 others requesting that Willem Stellen be appointed as member of the Volksraad for Waterberg. (5.12.1870).
6 SANA TAB SS 199 R2922/75: Landdrost Waterberg. Zendt kiesloten voor den Heer G.J. Verdoorn, Lid Volksraad. (10.12.1875).
7 Lamb, Doreen (1976): *The Republican Presidents, Volume 2*. Perskor Publishers, Pretoria, pp. 58–64.
8 SANA TAB SS 735 R5847/82: Landdrost Waterberg – Zendt in requisitie aan en antwoord van C.S. Potgieter, als Lid Volksraad. (18.10.1882).
9 SANA TAB SS 0 R12741/92: J. du P. de Beer, Waterberg, Zendt in 6 Requisitien en antwoord as kandidaat voor eerste Volksraad. (2.11.1892).
10 Heystek, Jooste (1966): Die Politieke Ontwikkeling in Waterberg. In J.J. Nel et al (eds): *Nylstroom, 1866–1966*. Published by the Nylstroom Town Council.
11 Jeppe, Carl (1906): *The Kaleidoscopic Transvaal*. Chapman & Hall, London. pp. 171–176.
12 Walker, Eric (1963): *Cambridge History of the British Empire, Volume 2*. Cambridge University Press. pp. 395–397.
13 Lamb (1976): p. 75.

14 Rousseau (1999): p. 93.
15 Lamb (1976): p. 83.
16 Swart (2003): p. 83.
17 Carruthers, E.J. (1995): *Game Protection in the Transvaal, 1846–1926.* Archives Year Book for South African history. Government Printer, Pretoria, p. 46.
18 Carruthers (1995): p. 49.
19 Carruthers (1995): pp. 52–54.
20 Bester, D.J. (ed.) (1984): *NTK 75.* A special edition of the magazine *NTK-Nuus* to commemorate the 75th anniversary of the founding of the NTK. Volume 7 No. 11, Sept/Oct 1984. Nylstroom, p. 22.
21 Spies, S.B. (1975): Reconstruction and Unification, 1902–1910. In C.F.J. Muller (ed.): *500 Years – A History of South Africa.* 2nd Edition. Academica Press, Pretoria.
22 Grobler, L.M. (2005): pp. 7–8.
23 Scholtz, G.D. (1941): *Generaal Christiaan Frederik Beyers 1869–1914.* Voortrekkerpers, Johannesburg.
24 Spies (1975): p. 365.
25 Scholtz (1941): p. 165.
26 *The Times* of London, 8 April 1907, p. 3: "*Pietersburg: Mr Granville Nicholson has been selected by Het Volk as its candidate to contest the vacancy in the Waterberg caused by the appointment of General Beyers to the Speakership*". (www.newspapers.com).
27 Simons, P.B. (ed.) (1986): *John Blades Curry 1850–1900,* Brenthurst Press; as quoted in Arne Schaeffer (2008): *Life & Travels in the Northwest, 1850–1899,* Yoshi Publishing, Cape Town, p. 84. Also quoted in Shelagh Nation (2016): *Oupa, OBE,* 30° South Publishers, Pinetown, p. 17.
28 Nation, Shelagh (2016): *Oupa, OBE.* 30° South Publishing, Pinetown, p. 42. Shelagh Nation is the granddaughter of R.G. Nicholson.
29 Nation (2016): p. 43.
30 Nation (2016): pp. 64–72.
31 Spies (1975): pp. 376
32 www.africanelections.tripod.com/za.html Accessed 8 September 2015.
33 Rosenthal, E. (1961): *Encyclopaedia of Southern Africa,* Frederick Warne, London, pp. 435–436. Hancock, Keith & Van der Poel, Jean (2007): *Selections from the Smuts Papers, Volume 3 – June 1910–November 1918,* pp. 321–322: a letter to Smuts dated 25 November 1915 from Ernest Lane, SA Heavy Artillery, Bexhill, Sussex.
34 Hancock, Keith & Van der Poel, Jean (2007): *Selections from the Smuts Papers, Volume 3 – June 1910–November 1918,* pp. 321–322: a letter to Smuts dated 25 November 1915 from Ernest Lane, SA Heavy Artillery, Bexhill, Sussex.
35 Heystek (1966): p. 101.
36 Nation (2016): pp. 75–87.
37 SANA TAB MHG 2189/48: Death Notice, Richard Granville Nicholson.
38 Bester (ed.) (1984): p. 17.
39 Grobler, Maritz et al (eds) (1995): *Die Suidelike Springbokvlakte/The Southern Springbok Flats 1876–1995 – an historical overview.* Published by the Berlin Geloftefeeskomitee, Springbok Flats. p. 219.
40 www.afrikanergeskiedenis.co.za/nasionaliste/biografiese-profiele/jg-strydom/. Accessed 20 March 2017.
41 Lewis, Thos. H (1913): p. 274.
42 Vanne and Ffrangcon-Davies, to their credit, returned to South Africa many times in later years despite a hostile local political climate and pressing career commitments in Britain, and were very active in promoting South African theatre. In the summer of 1936, during one of their visits, they went on a camping trip to the lower Mogalakwena River in the company of Myles Bourke and his wife. Bourke, the son of a wealthy Pretoria family, was a leading South African actor and producer, and founder of the Pretoria Repertory Theatre. During World War II, Bourke was given

the rank of Major in charge of the UDF Entertainment Unit, established to provide light relief to the forces. In the course of the next five years, the unit produced almost 60 shows and toured widely – SA, SWA, Rhodesia, Libya, Morocco, Egypt, Gibraltar, Greece, Sicily and Italy, the latter with the 8th Army. There were several performing troupes (with names like The Amuseliers, Ballyhoos, Whizzbangs and Troopadours) who together gave up to 150 performances a year. They were hugely popular with the forces. Many of the cast members went on to become well known in South African and international theatre: Sid James, for example, moved to England after the war and gained popularity as a member of the cast of the BBC programme *Hancock's Half-Hour*. Marda Vanne died in 1970 at the age of 73 and her partner Dame Ffrangcon-Davies in 1992 (aged 101), both in England.

43 Grime, H. (2015): *Gwen Ffrangcon-Davies, Twentieth Century Actress*. Routledge, London, p. 125; Rose, M. (2003): *Forever Juliet: The Life and Letters of Gwen Ffrangcon-Davies, 1891–1992*. Lark Press, p. 57; Green, M. (2004): *Around and About: Memoirs of a South African Newspaperman*. New Africa Books, pp. 30–31, 48. In 2017 Bookstorm published a biographical novel about Marda Vanne: entitled *The Lion & the Thespian*, it was written by David Bloomberg, a prominent Cape author, actor, lawyer and politician.

44 I thank Johan de Beer of Kokanje, Modimolle, for bringing this information to my attention.

45 Van der Westhuizen, Christi (2007): *White Power & the Rise and Fall of the National Party*. Zebra Press, Cape Town, p. 83.

46 Meiring, Piet (1973): *Ons Eerste Ses Premiers. 'n persoonlike terugblik*. Cape Town. Quoted in Christoph Marx (2008): *Oxwagon Sentinel: Radical Afrikaner Nationalism and the History of the Ossewabrandwag*. UNISA Press, Pretoria, p. 144.

47 Harrison, David (1981): *The White Tribe of Africa: South Africa in Perspective*. Macmillan, Johannesburg, pp. 85–100.

48 Wilkins, Ivor and Strydom, Hans (1978): *The Super-Afrikaners: Inside the Afrikaner Broederbond*. Jonathan Ball, Johannesburg, p. 360.

49 Marx, Christoph (2008): *Oxwagon Sentinel: Radical Afrikaner Nationalism and the History of the Ossewabrandwag*. UNISA Press, Pretoria, pp. 167–168.

50 Grobler (2005): p. 28.

51 Buys, J.G. (1966): Die ontstaan en ontwikkeling van plaaslike bestuur in Nylstroom. In J.J. Nel et al (eds): *Nylstroom 1866–1966*. Nylstroomse Stadsraad.

52 Marx (2008): pp. 251–255.

53 Suzman, Helen (1993): *In No Uncertain Terms – memoirs*. Jonathan Ball, Johannesburg, p. 9.

54 Marx (2008): p. 259.

55 Grobler (2005): p. 30.

56 Marx (2008): pp. 267–282.

57 Delmont, Elizabeth (1992): *The Voortrekker Monument: Monolith to Myth*. University of the Witwatersrand History Workshop "Myths Monuments Museums: New Premises?" 16–18 July 1992, p. 5. At wiredspace.wits.c.za/bitstream/handle/10539/7785/HWS-93.pdf?sequence=1. Accessed on 5 October 2015. Also published in *SA Historical Journal*, 29, pp. 76–101. The Leipzig monument in question is the Völkerschlachtdenkmal, opened in 1913 on the centenary of the victory of European allies over Napoleon at the Battle of the Nations. It had a similar nationalistic objective and was very similar in both exterior and interior design to the Voortrekker Monument. Both structures took over a decade to build. Although Moerdyk did refer to the Völkerschlachtdenkmal as one of several sources of inspiration for the design of the Voortrekker Monument, he denied that it was the model for his design.

58 Marx (2008): pp. 299–303.

59 Harrison (1981): pp. 72–76.

60 Marx (2008): pp. 314–315.

61 Harrison (1981): p. 129.

62 Strydom, Hans (1982): *For Volk and Führer: Robey Leibbrandt & Operation Weissdoorn*. Jonathan Ball, Johannesburg. Within a month of the victory of the Nationalists in 1948, Robey Leibbrandt was released from prison, having served less than five years of his life sentence. He married, had five children, dabbled in far-right-wing politics and died in obscurity of a heart attack in Ladybrand in 1966, aged 53.
63 Marx (2008): pp. 433–437.
64 Harrison (1981): p. 151.
65 Coetzer, P.W. (1986): The Era of Apartheid, 1948–1961. In S.B. Spies, (ed.) (1986): *An Illustrated History of South Africa*. Jonathan Ball, Johannesburg, p. 271.
66 De Villiers, Johan Willem (1992): *Lewe van F.H. Odendaal, 1898–1966*. PhD thesis at UNISA. http://hdl.handle.net/10500/17976. Accessed 21 May 2017.
67 Heystek (1966): p. 101.
68 Horrell, M. (1969): *The African Reserves of South Africa*. SA Institute of Race Relations, Johannesburg, pp. 2, 8, 17.
69 http://www.heijstekfamilie.nl : "Heystekke" in Zuid-Afrika, deel 10. Accessed 5 December 2015.
70 Correspondence from the files of the Het Volk/South African Party and United Party Transvaal Provincial Office, Nos. 1 (1911–1934); 10 (General elections) and 114 (Waterberg Constituency); stored at the UNISA Archives, Pretoria. Accessed on 15 February 2016.
71 Bester (1984): p. 56.
72 Suzman (1993).
73 www.africanelections.tripod.com/za.html. Accessed 8 September 2015.
74 Louis Nel (2015): pers. comm. Louis Nel is a life-long Waterberg resident whose father was a headmaster of Melkrivier school in the 1950s. In 2015 he bred Huguenot cattle on the farm Vrischgewaagd 649LR and maintained a keen interest in the history of the region.
75 B.J. Vorster, speech at Naboomspruit, 17 June 1971. In O. Geyser, editor (1977): *B.J. Vorster: Select Speeches*. Also available at www.sahistory.org.za/archive/extract-speech-during-political-meeting-held-17- june-1971-naboomspruit-waterberg-constituency. Accessed 10 March 2016.
76 Horrell, M., Horner, D. and Kane-Berman, J. (compilers) (1972): White Political Parties: Developments in 1971 – The National Party. In *A Survey of Race Relations in South Africa*. SA Institute of Race Relations, Johannesburg.
77 Giliomee, Hermann (1993): Obituary for Dr Andries Treurnicht, published in the *Independent* newspaper, London, on Tuesday 27 April 1993. http://www.independent.co.uk/news/people/obituary-andries-treurnicht-1457706.htm. Accessed 10 March 2015.
78 Louis Trichardt (2016): interviewed on his farm Rhenosterpoort 283KQ in March 2016.
79 Oakes, D. (1992): *Illustrated History of South Africa* (2nd Edition), Reader's Digest Association, Cape Town, p. 273.
80 Spies, S.B. (1986): Reconstruction and Unification, 1902–1910. In Trewhella Cameron (ed.): *An Illustrated History of South Africa*. Jonathan Ball, Johannesburg, p. 221.
81 Zunaid Mosam, pers. comm., 23 September 2015.

CHAPTER 17: GAINING THE RIGHT TO READ, pp.576–612

1 Visagie, Jan (2014a): Migration and societies north of the Gariep River. In Fransjohan Pretorius (ed.): *A History of South Africa*. Protea Book House, Pretoria, p. 115.
2 Visagie, Jan (2014b): The emigration of the Voortrekkers into the interior. In S.B. Spies (ed.) (1986): *An Illustrated History of South Africa*. Jonathan Ball, Johannesburg, p. 125.
3 Giliomee, Hermann (2003): *The Afrikaners – Biography of a People*. Tafelberg, Cape Town, p. 190.
4 Lamb, Doreen (1976): *The Republican Presidents*, Volume 2. Perskor Publishers, Johannesburg, p. 56.

5. Ploeger, J. (1955): Onderwys in Pretoria. In S.P. Engelbrecht et al (eds): *Pretoria 1855–1955*. Pretoria City Council, p. 203.
6. Du Toit, G.J. (1966): Onderwysontwikkeling in Waterberg. In J.J. Nel et al (eds): *Nylstroom 1866–1966*. Nylstroom Town Council, pp. 65–67.
7. SANA TAB OD 642: Onderwys Departement – Register van Dorp en Wykskole, Waterberg. Also a list in G.J. du Toit (1966), pp. 65–67.
8. Du Toit (1966): pp. 68–69.
9. Van den Berg, D.F. (1985): *Naboomspruit – 75 in 1985*. Naboomspruit Dorpsraad, pp. 128–133.
10. SANA TAB T353 OWM 684–688: Zanddrift Skool No. 463: School journals, 1930–1959.
11. In 1923 Leipoldt left government service to become deputy editor of the newspaper *De Volkstem* in Pretoria at the urging of Jan Smuts. Two years later, he moved to Cape Town to set up a practice as a paediatrician, but continued to write novels, poems and plays in both English and Afrikaans for the rest of his life. Apart from *Bushveld Doctor*, his best-known work in English was *The Ballad of Dick King and Other Poems* (1948) and a novel, *Stormwrack* (1980), which was the middle member of a trilogy that was only published together as *The Valley* in 2001. In 1934 he won the Hertzog Prize for poetry, together with Totius and W.E.G. Louw; a decade later, he won the same prize for drama. He served as Secretary to the SA Medical Association and edited its journal for many years. He established a boarding hostel for boys at his home in Kenilworth and also formally adopted an English boy, Jeff. His home, with its exotic furnishings and extensive library, became a favourite venue for many in Cape society, particularly because, in addition to his other talents, Leipoldt was an outstanding host, gourmet cook and wine connoisseur. He wrote several authoritative books on these hobbies. Leipoldt had suffered from the effects of childhood rheumatic fever all his life and also developed asthma as an adult. He died in April 1947. The notes on Louis Leipoldt were collated from several sources, including C.L. Leipoldt (2001): *The Valley*, a trilogy novel, with biographical note by Etaine Eberhard. Stormberg Publishers, Plumstead, Cape; Pheidippides (the late Dr Louis Leipoldt) (1949): *The Ballad of Dick King and other poems* – with a Memoir and Notes by H.M.L. Bolus. Stewart Publishers, Cape Town; C.L. Leipoldt (1979): *Dear Dr Bolus*. A.A. Balkema, Cape Town; Pieter W. Grobbelaar (1980): *C. Louis Leipoldt*. Skrywers in Beeld (4). Tafelberg, Cape Town; C. Louis Leipoldt (1980): *Bushveld Doctor*. Human & Rousseau, Cape Town.
12. Anon (1937): Bushveld Doctor, by Mr C. Louis Leipoldt, FRCS – book review. *The British Medical Journal*, December 18, 1937, p. 1225.
13. SANA TAB T353 SWG 24: School Journal of Geelhoutkop Government School, Waterberg Middle Block, July 1938–September 1954. (All entries in Afrikaans, translated by author).
14. SANA TAB T353 OWM 448/449: School Journal of Palmietfontein 172 Government School, Waterberg, from January 1917 to January 1970 (except for the period 1938–1944, the journal for which is missing). Initial entries by the Head and Inspector were in English, later in Afrikaans (the latter have been translated by the author).
15. The Palmietfontein school nominal roll at the end of 1927 listed the following pupils: *Standard 1*: Lena de Beer, Sarie de Kock, Maria du Plessis, Jacoba Grobler, James Adlum, Dirk Badenhorst, Jan Roos, Karel van den Berg; *Standard 2*: Katriena Oelofse, Willem de Beer, Hendrik Duvenage, Stoffel Korff; *Standard 3*: Jacomina de Beer, Cornelia Visagie, George Barnes, Johannes Barnard, Andries Oelofse, Gerhardus Visagie; *Standard 4*: Maria de Beer, Anna & Jacoba Duvenage, Lavina Visagie, Gerhardus Barnard, Lukas Korff, Petrus Wakeford, Kemp van der Merwe; *Standard 5*: Jacoba Benade, Archibald de Kock, Gaynor Webster; *Standard 6*: Magdalena & Louisa Adlum, Phyllis Webster, William Wakeford.
16. *Laerskool Melkrivier School Journal, 1950–1991*. This document was made available courtesy of Louis Nel, son of the second headmaster of the school and local resident, after it was salvaged from a rubbish dump on the property while the school was disused in the early 2000s.
17. Ben van der Walt, pers. comm., interviewed on Bulge River 198KQ on 23 May, 2017.

18 SANA TAB T353 OWM 73: School journal of Bulgerivier 252 Government School, Waterberg, from July 1920 to 1949. (Entries in Dutch and Afrikaans, translated by the author.)
19 Ben van der Walt (2017).
20 Pieter Swanevelder: pers. comm. 20 May 2017.
21 SANA TAB T353 OWM 606 (D3/77/391): Witfontein Laerskool 547, 1961–1971. No other journals from the school have been archived.
22 Kirkman, Peter (2012): *From Manchester to Albany and Beyond: Genealogy and Chronicles of the 1820 Settler John Kirkman and his Descendants*. Self-published, Hilton KZN, p. 358.
23 SANA TAB T353: School journal of Hartebeestpoort 1645 Government School, Waterberg, from January 1923 to July 1936. (Entries in English and Afrikaans.)
24 SANA TAB T353 OWM 608 (D3/80/522): Vaalwater School, 1953–1969. The original school journal, maintained by successive heads.
25 Wilkins, Ivor and Strydom, Hans (1978): The Super-Afrikaners: Inside the Afrikaner Broederbond. Jonathan Ball, Johannesburg, p. A106.
26 Giliomee, Hermann 2003: *The Afrikaners: biography of a people*. Tafelberg, Cape Town, p. 468.
27 *Laerskool Vaalwater: School Journal for September 1969 to December 1982*. Archived in the headmaster's office at the school.
28 Pieter Swanevelder: pers. comm. 20 May 2017
29 *Laerskool Vaalwater: School Journal*, May 1972.
30 Wilkins & Strydom (1978): p. A227.
31 www.happywarrior.org/genealogy/lleweni.htm. Accessed 9 September 2015. Indeed, so prominent was one of HTW Salusbury's ancestors, John Salusbury, that when he was knighted by Queen Elizabeth in 1601 for his loyal services as a courtier, a poem, dedicated to him, was written to commemorate the event. In the poem, entitled *Love's Martyr*, the queen was represented by a phoenix and Salusbury by a turtle-dove. When published later that year, it was accompanied by a poetical essay, also dedicated to Salusbury, written by William Shakespeare. The essay is known as "The Phoenix and the Turtle". Bate, Jonathan & Rasmussen, Eric (eds) (2007): *The RSC Shakespeare – William Shakespeare Complete Works*: Poems & Sonnets introduction, p. 2394. Macmillan, London.
32 Weir, Robbie (1995): The Village of Settlers. In Maritz Grobler and others (1995): *Die suidelike Springbokvlakte, 1876-1995 ('n Historiese Oorsig)/The Southern Springbok Flats, 1876–1995 (A Historical Review)*. Published by the Berlin Geloftefeeskomitee, Settlers, pp. 128–129.
33 Weir, Robbie (1995): Knapp. In Maritz Grobler and others (1995): *Die suidelike Springbokvlakte, 1876–1995 ('n Historiese Oorsig)/The Southern Springbok Flats, 1876–1995 (A Historical Review)*. Published by the Berlin Geloftefeeskomitee, Settlers, p. 82.
34 Laurenz, D.G. (ed.) (1982): *Lord Milner School, Settlers: 50 Commemorative Magazine, 1932–1982*, pp. 17–19.
35 Laurenz (1982): p. 27.
36 Stone, Brian (2015): Interview with Mr Stone at the Renaissance Retirement Home in Bela-Bela on 8 September 2015 when he was 95 years old. He died there on 31 October 2018.
37 Van den Berg, Daniel (1985): *Naboomspruit 75 in 1985*. Naboomspruit Town Council, pp. 150–151.
38 SANA NAB MSCE 1/1958: Death Notice of H.T.W. Salusbury, 12 December 1957, Port Shepstone, KZN.
39 Cruse, H.P. (no date): In Hermann Giliomee (2003): *The Afrikaners – Biography of a People*. Tafelberg, Cape Town, p. 109.
40 Seroto, Johannes (2004): *The Impact of South African Legislation (1948–2004) on Black Education in Rural Areas: An Historical Educational Perspective*. PhD thesis at UNISA, p. 70.
41 Doakes, Dougie (ed.) (1989): The birth of "Bantu" education. In *Reader's Digest Illustrated History of South Africa – the Real Story*. Reader's Digest Association, Cape Town, p. 379.
42 Seroto (2004): pp. 86–87.

43 Seroto (2004): pp. 88–89.
44 Jansen, Jonathan D. (1990): Curriculum as a Political Phenomenon: Historical Reflections on Black South African Education, Journal of Negro Education 59 (2), pp. 195–206.
45 Seroto (2004): pp. 79–81.
46 Seroto (2004): pp. 82–83.
47 Seroto (2004): pp. 107–109.
48 Doakes (ed.) (1989): p. 379.
49 Kruger, D.W. (1969): *The Making of a Nation: a history of the Union of South Africa, 1910–1961*, as quoted in Christi van der Westhuizen (2007): *White Power & the Rise and Fall of the National Party*. Zebra Press, Cape Town, p. 71.
50 Harrison, David (1981): *The White tribe of Africa: South Africa in Perspective*. University of California Press, Los Angeles.
51 Seroto (2004): pp. 106–107.
52 Seroto (2004): pp. 120–121.
53 Sam Mafora, pers. comm., interviewed several times, most recently on 11 May 2017.
54 Sam Mafora (2017).
55 Spies, S.B. (1986): Reconstruction and Unification, 1902–1910. In Trewhella Cameron (ed.): *An Illustrated History of South Africa*, Jonathan Ball, Johannesburg, p. 221.
56 Eshak, Y.I. (1982): Teacher Associations in "Indian" Schools in the Transvaal. B Ed dissertation, University of the Witwatersrand. https://v1.sahistory.org.za/pages/library-resources/thesis/teachers-association-yi-eshak/contents.html
57 Information Board at the Irish House Museum, the museum of the Polokwane Municipality, cnr Thabo Mbeki and Market Streets, Polokwane.
58 Marquard, L. (1969): *The Peoples and Policies of South Africa*. Oxford University Press, London, p. 205.
59 Marquard (1969): p. 203.
60 Horrell, M. (1969): *The African Reserves of South Africa*. SA Institute of Race Relations, Johannesburg, p. 122.
61 Hartshorne, K.B. (1985): The State of Education in South Africa: Some Indicators. *South African Journal of Science*, 81, March 1985, pp. 148–151.
62 Hartshorne (1985): pp. 149–150.
63 Seroto (2004): pp. 163–164.
64 Nick van Coppenhagen, pers. comm., May 2017.
65 Minutes of the North Waterberg Farmers' Association, 17 June 1985.
66 Minutes of the North Waterberg Farmers' Association, 21 October 1986.
67 Dr Peter Farrant, pers. comm., interviewed in Vaalwater on 10 May 2016.
68 Mr Harry Malan, pers. comm., interviewed at Mohlakamotala High School on 26 July 2016.
69 *The Economist* of 7 January 2017: https:www.economist.com/news/middle-east-and-africa/21713858-why-it-bottom-class-south-africa-has-one-worlds-worst-education; and www.expatcapetown.com/education-in-south-africa.html. Accessed 22 July 2017.

CHAPTER 18: ROCKS, DYNAMITE AND DREAMS, *pp.613–644*

1 Jansen, H. (1982): The Geology of the Waterberg Basins in the Transvaal, Republic of South Africa. *Memoir 71* of the Geological Survey, Department of Mineral and Energy Affairs, Pretoria.
2 Callaghan, C.C. (1993): The Geology of the Waterberg Group in the southern portion of the Waterberg Basin. *Bulletin 104* of the Geological Survey, Department of Mineral & Energy Affairs, Pretoria.

3 Callaghan (1993).
4 Bulpin, T.V. (1983): *Lost Trails of the Transvaal*. Books of Africa (Pty) Ltd., Cape Town, p. 255. Unfortunately, Bulpin seldom provided verifiable details of the sources of his accounts.
5 Sources examined include H. Kynaston & E.T. Mellor (1912): *The Geology of the Country around Warmbaths and Nylstroom – an explanation of Sheet 10*. Geological Survey, Mines Department, Union of South Africa, Pretoria; H. Jansen (1982); C.C. Callaghan (1993); P.G. Eriksson, B.F.F. Reczko & C.C. Callaghan (1997): The Economic mineral potential of the mid-Proterozoic Waterberg Group, north-western Kaapvaal craton, South Africa. *Mineralium Deposita* 32 pp. 401–409; D.L. Ehlers & M.C. du Toit (2002): *Explanation of the Nylstroom Metallogenic Map, Sheet 2428*. Council for Geoscience, Pretoria.
6 De Bruiyn, H. (1972): *Report on the Nooitgedacht Lead Mine, north east of Vaalwater, Northern Transvaal*. Unpublished report (No. 1972–0041) of the Geological Survey of South Africa. 5 pages.
7 De Bruiyn (1972): p. 2.
8 Wagner, P.A. (1921): Report on the Crocodile River Iron Deposits. *Mem. Geol. Surv. S. Afr.* 17, (65 pages). Summarised and updated in P.A. Wagner (1928): The Iron Deposits of the Union of South Africa. *Mem. Geol. Surv. S. Afr.* 26, pp. 144–176.
9 Barnard, Annette (2007): *Thabazimbi Iron Ore Mine*. Published by the Thabazimbi Iron Ore Mine, a division of Kumba Iron Ore Ltd, pp. 21–28.
10 Barnard (2007): pp. 35–43.
11 Boonzaaier, Boon (2008): *Tracks Across the Veld*. Privately published, Bela-Bela, pp. 342–343.
12 Barnard (2007): p. 58.
13 Wagner, Percy A. &. Tudor, Trevor G (1923): Platinum in the Waterberg District: a description of the recently discovered Transvaal deposits. *The South African Journal of Industries*, December 1923, pp. 577–597.
14 Wagner & Truter (1923): p. 586.
15 Van den Berg, Daniel (1985): *Naboomspruit – 75 in 1985*. Naboomspruit Town Council, pp. 78–79.
16 *The Star*, 6 March 1925.
17 Nattrass, Gail (1983): *The Rooiberg Story* – a publication to mark the 75th anniversary of the Rooiberg Tin Limited. Rooiberg, p. 8.
18 Falcon, L.M. (1985): Tin in South Africa. *Journal of the SA Institute of Mining and Metallurgy*, October 1985, pp. 333–345.
19 Friede, H.M. & Steele, R.H. (1976): Tin mining and smelting in the Transvaal during the Iron Age. *Journal of the SA Institute of Mining and Metallurgy*, July 1976, pp. 461–470.
20 The term *difaqane* is used throughout this book, in preference to *mfecane*. The first is the form of the word associated with Sotho-speakers in the northern part of South Africa, whereas the latter is limited to coastal Nguni-speakers. Source: Carolyn Hamilton (ed.) (1995): *The Mfecane Aftermath*. Witwatersrand University Press, Johannesburg, p. 1.
21 Nattrass, Gail (1989): The tin mines of the Waterberg (Transvaal), 1905–1914. *Contree* 26/1989, pp. 5–12; 7.
22 Du Toit, M.C. & Pringle, I.C. (1998): Tin. In M.G.C. Wilson & C.R. Anhaeusser (eds): *The Mineral Resources of South Africa*. Handbook 16 of the Council for Geoscience, Pretoria, p. 613.
23 Nattrass (1983): p. 54.
24 Du Toit & Pringle (1998): p. 616.
25 Nattrass (1983): pp. 13–33.
26 Godsell, Sarah (2011): Rooiberg: The Little Town that Lived. *South African Historical Journal* 63:1 pp. 61–77. http://dx.doi.org/10.1080/02582473.2011.549374
27 Marais, Eugène (1972): *The Road to Waterberg and other Essays*. Human & Rousseau, Cape Town, p. 9.
28 Van den Berg (1985): pp. 80–86.
29 Van den Berg (1985): pp. 80–86.

30 Du Toit & Pringle (1998): p. 616.
31 Pringle, I.C. (1986): The Union Tin Mine, Naboomspruit District. In C.R. Anhaeusser & S. Maske (eds) (1986): *Mineral Deposits of Southern Africa*, Vol. 2. Geological Society of South Africa, pp. 1301–1305.
32 Crocker, I.T. (1986): The Zaaiplaats Tinfield, Potgietersrus District. In C.R. Anhaeusser & S. Maske (eds) (1986): *Mineral Deposits of Southern Africa*, Vol. 2. Geological Society of South Africa, pp. 1287–1299.
33 Du Toit & Pringle (1998): pp. 613–620.
34 Martini, J.E.J. & Hammerbeck, E.C.I. (1998): Fluorspar. In M.G.C. Wilson & C.R. Anhaeusser (eds): *The Mineral Resources of South Africa*. Handbook 16 of the Council for Geoscience, Pretoria, pp. 269–279.
35 Van den Berg (1985): pp. 87–92.
36 Absolom, S.S. (1986): The Buffalo Fluorspar Deposit, Naboomspruit District. In C.R. Anhaeusser & S. Maske (eds) (1986): *Mineral Deposits of Southern Africa*, Vol. 2. Geological Society of South Africa, pp. 1337–1341.
37 Venter, F.A. (1944), Cillie, J.F. & Visser, H.N. (1945), J.F. Cillie (1951) and J.F. Cillie (1957): Waterberg Coalfield, Records of Boreholes 1–20; 21–40; 41–100; 101–143. *Bulletins 15, 16, 21 & 23* of the Geological Survey Division of the Department of Mines, Pretoria.
38 Ehlers, D.L. & Du Toit, M.C. (2002): *Explanation of the Nylstroom metallogenic map, sheet 2428*. Council for Geoscience, Pretoria, p. 90.
39 Olivier, J., Van Niekerk, H.J. & Van der Walt, I.J. (2008): Physical and chemical characteristics of thermal springs in the Waterberg area in Limpopo Province, South Africa. *Water SA* 34 (2), April 2008, pp. 163–174.
40 Olivier et al (2008).

CHAPTER 19: ON THE PERIPHERY, *pp.645–675*

1 www.statssa.gov.za/?page_id=993&id=thabazimbi-municipality. Thabazimbi population as per 2011 national census. Accessed 6 August 2016.
2 Barnard, Annette (2007): *Thabazimbi Iron Ore Mine 75: 1931–2006*. Published by Thabazimbi Iron Ore Mine (a member of the Kumba Iron Ore Company Ltd.), p. 69.
3 Greenstein, Lisa (ed.) (2002): *Jewish Life in the South African Country Communities: Volume 1 – the northern Great Escarpment, the Lowveld, the northern Highveld, the Bushveld*. South African Friends of Beth Hatefutsoth, Johannesburg, pp. 166–167.
4 Barnard (2007): p. 72.
5 Barnard (2007): pp. 82–83.
6 Greenstein (editor) (2002): p. 167.
7 Barnard (2007): pp. 89–90.
8 Barnard (2007): pp. 107–108.
9 Barnard (2007): pp. 111–116.
10 Naudé, M. (2008): Oral evidence on aspects of folk life during the first hundred years (1840–1940) of frontier settlement in the Waterberg (Limpopo Province). *National Cultural History Museum Research Journal* 3 (2008), pp. 35–56 (p. 42): quoting N.N. Coetzee (1997): *Die Geskiedenis van Rustenburg ongeveer van 1840 tot 1940*. Pretoria: V&R Drukkers, p. 429.
11 Walker, Clive (2013): *Baobab Trails*. Jacana, Johannesburg, p. 266.
12 Mucina, L. & Rutherford, M.C. (eds) (2006): The Vegetation of South Africa, Lesotho and Swaziland. *Strelitzia* 19**.** SA National Biodiversity Institute, Pretoria, pp. 471–472.
13 Naudé, M. (1998): Oral evidence of vernacular buildings and structures on farmsteads in the Waterberg (Northern Province). *Research by the National Cultural History Museum* 7, pp. 47–91;

Naudé, M. (2004): Oral Evidence on the construction of vernacular farm dwellings in the Waterberg (Limpopo Province). *SA Journal of Cultural History* 18 (1), pp. 34–62; Naudé (2008). Mauritz Naudé, a researcher employed by the SA National Cultural History Museum, has written several papers based on the uncollated field notes of Wilna Aukema, which are held in the custody of the museum.

14 Naudé (2004): p. 40.
15 SANA TAB TAD 84 A426/05; TAD 59 A328/28; TAD 22 A55/10: Successive petitions and appeals from farmers in the Zwagershoek – Matlabas area for relaxation of the restrictions on movement of stock and transport wagons; or alternatively, for the re-demarcation of magisterial boundaries.
16 SANA TAB JUS 291 3/396/20: Petition applying for the appointment of a Special Justice of the Peace at Matlabas.
17 SANA TAB JUS 265 3/648/18: Application by Police Commissioner to Secretary for Justice dated 31 July 1918 and subsequent correspondence, culminating in approval in June 1920.
18 Naudé (2008): p. 54, quoting Coetzee (1997): p. 429.
19 SANA SAB NA 255 45/1919/F596: Application from Chief Ehrens Naoa to sub-Native Commissioner Nylstroom, forwarded to Native Commissioner Waterberg (Gilfillan) to purchase the properties Hopewell 54 (229 KQ) and Vlakfontein 56 (193 KQ).
20 SANA SAB LDE 936 18454/17: Application received by the Transvaal Land Board from Oslo Land Company in October 1919 to purchase ten named Crown Land properties in the Zoutpansberg District, on the east bank of the Mogalakwena River close to its confluence with the Limpopo River. The Land Board chose to advertise the properties in the *Government Gazette* and invite bids.
21 Naudé (1998): p. 57.
22 Letter to Mr Volrath Vogt of Hopewell, MD of Oslo Land Company and Chairman of the Matlabas Cotton Planters' Association, dated 1 October 1926, from Mr Walkie Turner, Hon. Secretary of the association, thanking Vogt for his contributions to the association and the district and wishing him and his family well in the next sphere of their lives. This letter was kindly sent to me by Volrath Vogt's great granddaughter, Stephanie Kemp (originally from New Zealand, now living in the US), who is researching the history of her family. Stephanie also provided me with many other details about her great grandfather and his family.
23 SANA SAB CEN 679 E6780; and CEN 679 E6798: Correspondence between V. Vogt of Oslo Land Co., Hopewell, with the Govt Entomologist in May 1924, concerning weevils in a maize bin and caterpillars in a lucerne field respectively.
24 SANA SAB TES 5900 F42/53/8: Compensation claims against the Department of Agriculture by Oslo Land Company and R.A. Mordt in respect of arsenical poisoning of cattle and children respectively. Correspondence between Dept of Agriculture and Legal Advisers, as well as Norwegian Consul in 1937 and 1938.
25 SANA SAB TES 5900 F42/53/8: Compensation claims against the Department of Agriculture by Oslo Land Company and R.A. Mordt in respect of arsenical poisoning of cattle and children respectively. Correspondence and reports involving legal advisers and Dept of Agriculture, 1938.
26 SANA SAB BE 254 N/15/2: Portion of a letter from the Oslo Land Company's agents, Norton & Co., dated 18 August 1940, responding to an SA government demand that properties and businesses owned by the "enemy" (i.e. Germany, or German citizens) be confiscated by the state.
27 Ben van der Walt of Bulgerivier and Neels Riekert of Matlabas, pers. comm., May 2017, in conjunction with a comparison of government 1:250 000 topocadastral sheets from 1938 and 1984.
28 www.statssa.gov.za/?page_id=993&id=lephalale-municipality. Lephalale and Marapong population as per 2011 national census. Accessed 6 August 2016.
29 Van Heerden, Marthie (ed.) (1994): *Albertyn 1944–1994: Halfeeufees Gedenkblad*. A publication by the Albertyn (Ellisras) Gemeente of the Nederduitse Gereformeerde Kerk to celebrate its half-century. Ellisras, p. 48.

30 Naudé, M. (1998): p. 53. According to the *Transvaal Almanac* compiled by Fred Jeppe in 1877 (p. 74), 1 *muid* was a Dutch measure of capacity equivalent to 2.97 Imperial bushels. I Imp. bushel =8 gallons (36.4 litres). Thus a *muid* was equivalent to about 108 litres.
31 Information plaque at Lephalale Municipal Offices, 2014. Repeated in Elizabeth Hunter (2010): *Pioneers of the Waterberg*. Privately published, Lephalale, p. 119.
32 Claris Dreyer (2017): pers. comm. Claris, a long-time resident of Lephalale and Chief Geologist at the Grootegeluk Coal Mine from its inception until his retirement, is a collector of historical and other memorabilia. He kindly collated and lent me a collection of articles and notes about the history of Ellisras/Lephalale and its personalities, including a short report by Hannes Engelbrecht (undated) with the title *Ellisras – my dorp*. Engelbrecht was the proprietor of the Ellisras Nuus Uitgewers. Much of the text about the town's history is drawn from these valuable records.
33 Information plaque at Lephalale Municipal Offices, 2012.
34 Information plaque at Lephalale Municipal Offices, 2012.
35 Information plaque at Lephalale Municipal Offices, 2012.
36 SANA TAB TA GOV631: French, R. Major (1904): Report on South Africa from impressions formed while serving as Military Member of the Compensation Claims Commission for the Waterberg District, Transvaal, during 1903; p. 3.
37 Van Heerden (ed.) (1994): pp. 50–51.
38 Claris Dreyer, pers. comm., February 2018.
39 www.lephalale.gov.za/about/overview.php. Accessed 30 April 2016.
40 The route is clearly shown on Jeppe's 1892 map of the Transvaal, cutting across the farm Murchison before turning south across Klipput.
41 Breytenbach, S.W. (posthumously) (2012): *Koedoesrand – Kronieke*. Edited by Lourie & Makkie du Plessis and privately published by Ms Mart van Heerden, Marken.
42 SANA TAB Waterberg Farms Deeds Transfers; RAK 3058, pp. 561, 570.
43 SANA TAB Waterberg Farms Deeds Transfers; RAK 3064, p. 472.
44 Breytenbach (2012): p. 66 and others.
45 Pienaar, Dr U. de V. (2012): *A Cameo from the Past – the prehistory and early history of the Kruger National Park*. Protea Book House, Pretoria, pp. 465–467.
46 www.gk-Pretoria.weekly.com/piet-grobler.html. Accessed 30 April 2016.
47 Breytenbach (2012): pp. 20–22.
48 Breytenbach (2012): pp. 93–113.
49 Breytenbach (2012): pp. 109–111.
50 Breytenbach (2012): p. 118.
51 Breytenbach (2012): p. 147. Paraphrased free translation from the text.
52 www.markenprimaryschool.co.za/about/historical. Accessed 20 April 2016.
53 Dr Karel Botha and Mr Hansie Kloppers (2017): pers. comm. Interviewed on 22 May 2017.
54 Frans van Staden, pers. comm., 27 May 2017. During the 1960s, his father gradually consolidated most of these portions of Kwarriehoek into a viable farming unit.
55 Dr Karel Botha (2017).
56 Hansie Kloppers (2017).
57 SANA TAB RAK 3058: Kiss me quick and go my Honey, No. 762. Original register of title deeds, Waterberg Wyk, Book 1, 1869.
58 "Kiss Me Quick and Go My Honey" (1856); Words by Silas Sexton Steel, music by Frederick Buckley, published by Firth, Pond & Company, music publishers, 547, Broadway, New York (1851–1863). Library of Congress, Rare Book and Special Collections Division, *America Singing: Nineteenth-Century Song Sheets*. https://www.loc.gov/resource/amss.hc00034a.0/?st=text. Also found at: www.polmusic.org_1800s_56kmqag . Both sites accessed 21 May 2017.

CHAPTER 20: FROM WATERBERG TO WATERBERG, *pp.676–711*

1. Miller, R. McG. (2008): *The Geology of Namibia, Volume 3: Upper Palaeozoic to Cenozoic*. Geological Survey, Windhoek, pp. 16–41 & 16–42; imagery from Google Earth.
2. Mossolow, N. (1993): Waterberg – on the History of the Mission Station Otjosondjupa, the Kambazembi Tribe and Hereroland. Privately published, Windhoek, Namibia, p. 8.
3. Mossolow, (1993: p. 38.
4. Pool, G. (1991): *Samuel Maharero*. Gamsberg Macmillan, Windhoek, p. 251, quoting from Von Trotha's personal archives.
5. Gewald, Jan-Bart (1996): *Towards Redemption: a socio-political history of the Herero of Namibia between 1890 and 1923*. Research School CNWS, Leiden University, Netherlands. pp. 139–141.
6. Gewald (1996): pp. 143–155.
7. Pool (1991): p. 160.
8. Mossolow (1993): p. 24.
9. Gewald (1996): p. 158.
10. Olusoga & Erichsen (2010): pp. 116–120; Gewald (1996): pp. 163, 180.
11. Gewald (1996): pp. 180–181.
12. Gewald (1996): p. 184.
13. Olusoga & Erichsen (2010): p. 125.
14. Pool (1991): pp. 202–203.
15. Pool (1991): pp. 218–219.
16. Pool (1991): pp. 230–235; Gewald (1996): pp. 203–204; Olusoga & Erichsen (2010): pp. 135–136.
17. Olusoga & Erichsen (2010): pp. 136–137.
18. Pool (1991): pp. 235 –241; Gewald (1996): p. 204; Olusoga & Erichsen (2010): p. 137.
19. Goldblatt, I. (1971): *History of South West Africa* (from the beginning of the nineteenth century). Juta & Co., Cape Town, p. 130–131.
20. Gewald (1996): pp. 206.
21. Pool (1991): p. p262–264.
22. Bennett, B. (1939): *Hitler over Africa*. T. Werner Laurie, London. A graphic account of the rise of Nazism in South West Africa in the 1930s. Of relevance here is an appendix to the book that reproduces verbatim extracts from a report published in 1918 and entitled "A Report on the Natives of South-West Africa and their Treatment by Germany", prepared by the Colonial Administrator's office in Windhoek and presented to both houses of the British Parliament. Chapter XV is especially descriptive of the atrocities that are alleged to have taken place. (pp. 185–193 of Bennett's book). There are many other examples, as also numerous denials.
23. Goldblatt (1971): p. 131. Several slightly differing English translations exist. A version of this edict, written in Ovaherero, signed by Von Trotha but undated and quoting the bounty offered in pounds sterling, is in the Botswana National Archives, Ref File RC 11/1.
24. Pool (1991): pp. 278–280.
25. Olusoga & Erichsen (2010): pp. 161–162. Reports of casualties of Von Trotha's offensive vary widely, the numbers given here being conservative. The extreme estimates would have it that of an initial Herero population of around 80 000, less than 20 000 survived by around 1906. N. Mossolow (1993), pp. 40–41, was of the opinion that the pre-war population of the Herero was more like 50 000 and that about half that number survived the conflict, with most deaths occurring in concentration camps rather than in the Omaheke. There is also debate over whether Von Trotha's use of the word "exterminate" in one of his orders was meant in a strictly military way – as in eliminate the threat – or whether he had genocide in mind. His proclamation of 2 October 1904 lent credence to the latter interpretation – in which case he failed in his objective.
26. Wallace, Marion, with John Kinahan (2011): *A History of Namibia*. Jacana, Johannesburg, pp. 177–178.

27 Olusoga & Erichsen (2010): p. 185.
28 Olusoga & Erichsen (2010): p. 205.
29 Gewald, Jan-Bart (1999): The Road of the Man Called Love and the Sack of Sero: The Herero-German War and the Export of Herero Labour to the South African Rand. *Journal of African History*, 40, pp. 21–40; also Gewald (1996); and Olusoga & Erichsen (2010).
30 Gewald (1999): pp. 27–28.
31 Pool (1991): p. 319.
32 Botswana National Archives, Gaborone, Ref File RC 11/1: An English translation of Samuel's letter by Lt Mervyn Williams, the British Acting Magistrate at Tsau in Ngamiland. Williams had previously advised the Resident Commissioner for the Protectorate in Mafeking of numbers of Herero refugees entering the area. (The author visited the Archives on 05 July 2013.)
33 Gewald, Jan-Bart (2002): "I was afraid of Samuel, therefore I came to Sekgoma": Herero refugees and patronage politics in Ngamiland, Bechuanaland Protectorate, 1890–1914. *Journal of African History*, 43, pp. 211–234 (p. 215).
34 Gewald (2002): p. 219.
35 Gewald (2002): p. 221.
36 Gewald (1999): p. 33.
37 Botswana National Archives, Gaborone: Ref File RC 10/18: correspondence between Acting Magistrate at Tsau, Ngamiland and Resident Commissioner, Mafeking, as well as that between the latter and the High Commissioner (Milner), dating from March and April 1904.
38 Botswana National Archives, Gaborone, Ref File RC 11/1: Report by Lt Merry, Acting Magistrate at Tsau, Ngamiland, to Resident Commissioner, Mafikeng, dated 30 December 1905.
39 Pool (1991): p. 282.
40 Gewald (2002): p. 226.
41 Gewald (2002): p. 228.
42 Pool (1991): p. 284.
43 Handley, J.R.F. (2004): *Historic Overview of the Witwatersrand Goldfields*. Privately published, Howick. pp. 54–55.
44 National Archives of Namibia: ZBU D.IV.1.5 Volume 1: Agreement between William Dalrymple and Johann Rissik, dated 4 June 1907, pp. 26–30. The agreement lodged in the Namibian Archives is a copy of the original and was kindly forwarded by its staff. Surprisingly, no original or copy of this agreement could be located in either the National Archives in Pretoria or in those of Botswana. Note that Johann Rissik, the Surveyor General of the former Transvaal Republic, was appointed chairman of the board of The Transvaal Consolidated Land and Exploration Company (TCL), one of Anglo-French's stablemates in the Corner House Group, after the Peace of Vereeniging in 1901. He held that position until 1907. (Source: A.P. Cartwright, 1968: *Golden Age* (the story of the Corner House Group), Purnell & Sons, Johannesburg, p. 270.) In 1910 he was appointed the first Administrator of the Transvaal in General Botha's Union government (see following note).
45 Pool (1991): p. 285.
46 Botswana National Archives, Gaborone, Ref File RC 11/1: Report by Lt Merry, Acting Magistrate at Tsau, Ngamiland, to Resident Commissioner, Mafikeng, dated 30 December 1905.
47 Pool (1991): pp. 287–289.
48 Nattrass, Gail (1983): *The Rooiberg Story*. Published by Rooiberg Tin Limited on the occasion of its 75th Anniversary. p. 23.
49 Nattrass (1983): p. 14.
50 Nattrass (1983): p. 23; also Gail Nattrass (1991): The tin mining industry in the Transvaal, 1905–1914. *SA Journal of Economic History*, **6** (1), pp. 105–106.
51 Hunter, Elizabeth, (compiler) (2010): *Pioneers of the Waterberg*. Privately published, Lephalale. p. 113.

52 Pool (1991): pp. 288–289.
53 SANA TAB GNLB 203 1588/14: Letter dated 15 September 1914 (a week after South Africa's decision to enter World War I in support of Britain and the same day on which Boer hero General De la Rey was accidentally shot dead on his way to a meeting in Potchefstroom) from the Sub-Native Commissioner, Nylstroom to the Native Commissioner, Waterberg and copied to the Secretary for Native Affairs in Pretoria, entitled "List of Damara Natives proceeding to Cape Town". The letter lists followers of Chiefs Samuel Maharero and Motate Humu who were proceeding to Cape Town. Amongst Samuel's followers listed were his sons Friedrich and Alfred from Werkendam, two men from Tomlinson's farm Blaaubank (the TCL property 2 KR), one from Wale's farm Zondagsfontein (67 KR, 15 km north of Vaalwater), 12 men from Witwatersrand Deep gold mine and one from New Kleinfontein mine. Followers of Chief Motate Humu comprised eleven men from Wit Deep and 1 from New Kleinfontein. A total of 31 names appears on the list.
54 Transvaal Consolidated Lands Annual Report for 1912: Land Manager's Report (Captain C.A. Madge) of June 1913, pp. 9–10. Johannesburg City (Public) Library, Reference Section.
55 Gewald (1996): p. 331.
56 Marais (1921), republished in Rousseau (1984): pp. 341–350; D'Assonville (2008): pp. 29–34. Interestingly, Marais names Nagwag (Nachtwacht 492 LR) as the farm on which the Herero were living at the time of his visit(s), whereas in fact they had been placed on the neighbouring farm Groenfontein 494 LR. It is possible that the population (estimated by Marais to be about 400) had spilled over from Groenfontein onto the adjacent property. Marais's account also refers to the Hereros being greeted as they emerged from the Omaheke by members of Khama's Ngwato people, who had been sent to look for them. But in fact, they had arrived in the country of the Tawana, whose king was Sekgoma. Only later were the Herero befriended by Khama. Marais's writings are well-known for their romanticisms, but he was also dependent on the memories and imaginations of his informants, in this case principally Samuel and his confidante, the preacher Julius Kauraisa. His account of the flight of Samuel's people across the Omaheke (which he calls the Kalahari) is nevertheless a vivid and horrifying tale of the privations endured by these and many other (undocumented) Herero groups. It provides an important and unique record of the saga.
57 Hunter (2010): p. 113.
58 SANA GG 1550:50/829: Letter from Samuel Maharero to the Resident Commissioner, Mafeking, dated 25 February 1919, requesting that he be permitted to purchase some land in the Ghanzi area of Bechuanaland with the intention of settling there with as many of his people who would wish to do so. Alternatively, that they be permitted to return to their traditional lands in South West Africa.
59 SANA GG 1550:50/829: Letter from the Imperial Secretary (H. Stanley) at the British High Commissioner's office in Cape Town to the Secretary of the Governor General (Earl Buxton) in Pretoria, dated 19 February 1920, attaching Samuel Maharero's letter of 25 February 1919.
60 SANA GG 1550:50/840: Exchanges of correspondence during February and April 1920 between the office of the Governor General, the Union Department of Native affairs, the Prime Minister's Office and that of the High Commissioner. Delay was caused by a general election in March 1920.
61 Gewald (1996): p. 321.
62 SANA GG 1553:50/955: Correspondence and a report from the government secretary in Mafeking, J. Ellenberger, to the Secretary of the High Commissioner, dated 13 July 1921, requesting that the Governor General's office be advised that it would not be possible to accede to Samuel Maharero's request (dating from February 1919) for a settlement site in western Bechuanaland. In the meanwhile, Earl Buxton had been succeeded as Governor General, in November 1920, by His Royal Highness Prince Arthur of Connaught. As it happened, Prince Arthur visited the Territory of SWA in August 1922 – and the Herero communities of Windhoek and Orumbo

(where Hosea Kutako was now Regent) took the opportunity to assert their growing sense of independence, by submitting a petition for the Governor General's attention.

63 Pool (1991): p. 290.
64 Pool (1991): p. 291.
65 Pool (1991): p. 291.
66 Botswana National Archives S. 601/8: Death Report, Samuel Maharero, Serowe, 14 March 1923.
67 Botswana National Archives S. 601/8: Letter from Friedrich Maharero to the Magistrate at Serowe, Bechuanaland, dated 7 May 1923.
68 Gewald (1996): pp. 330–336; Pool (1991): pp. 298–304.
69 Gewald (1996): pp. 337–338.
70 Beyers, Yvonne (2011): Owakuru, muta wa tere* – Die Laaste Herero's van die Bosveld se stil gebied. *Beeld,* 19 March 2011. Published by Media24.Path:Published/Media_24_Bylae/By/2011/03/19/BY/Texts/19MaartDanaHereros.xml. Accessed 30 March, 2011.
71 Schneider, Ilme (1998): *Waterberg Plateau Park, Namibia*. A Shell Guide, Windhoek; p. 23.
72 James Charles Watermeyer: S2A3 Biographical Database of Southern African Science. http://www.s2a3.org.za/bio/Biograph_final.php?serial=3053 . Accessed 14 August 2017.
73 Family Group Sheet for Gottfried Andreas Watermeyer / Anna Suzanna Ziervogel (F43249). 1820settlers.com/genealogy/familygroup.php?familyID=F43249&tree=master. Accessed 14 August 2017. Surveying was popular in the Watermeyer family: one of James's younger brothers, Godfrey, became, in the 1920s, the first professor of Land Surveying at the University of the Witwatersrand, and an uncle, Charles Philip Watermeyer, was a prominent surveyor in the Eastern Cape, and a Member of the Cape Legislative Assembly in the late nineteenth century. Other family members rose to prominence in the judiciary.

---

* "My Ancestors, I ask for your help"

GLOSSARY AND ABBREVIATIONS

Note: Through the course of the period covered in this history, town names, especially in Limpopo Province, have changed several times from original Dutch, through Afrikaans and/or English to vernacular names. In the text the name used is the one appropriate to the period being discussed. To the extent that rivers and other topographic features have also experienced name and spelling changes, the same convention has been applied.

AB – *Afrikaner Broederbond,* the Afrikaner Brotherhood, an initially secret society of Afrikaners formed in 1918 to protect and promote Afrikanerdom and the Afrikaans language and culture. It recruited members largely from senior civil servants, politicians and teachers and by the early 1980s, membership peaked at around 12 000. It played an important covert role in National Party politics.
amsl – (metres) above mean sea-level.
ANC – African National Congress, a political party, which (in alliance with the Communist Party and a trade union organisation) has formed the national government since 1994.
Anglo-Boer War – the conflict between the British and the Boers between October 1899 and the conclusion of the Treaty of Vereeniging on 31 May 1902. Also known as the South African War (the preferred moniker in current English usage), *die Engelse oorlog* or *die Tweede Vryheidsoorlog.*
apartheid – separateness, the system of racial segregation and discrimination entrenched in South African legislation between 1948 and 1991, and mainly associated with the National Party, but which was embodied in many statutes dating back to the beginning of the twentieth century.
APK – *Afrikaanse Protestantse Kerk,* a break-away faction of the NGK, established in 1987 in response to the NGK's decision to allow black worshippers to attend services at traditionally white churches.
bakkie – pick-up truck.
Bechuanaland – formerly a British Protectorate, independent as the Republic of Botswana since 1966.
Bela-Bela – the town formerly known as Warmbaths/Warmbad.
billion – a thousand million.
bittereinder – die-hard; term used to describe mainly Afrikaners who refused to accept defeat or political change, especially after the fall of Pretoria in June 1900 and who perpetuated the guerrilla war for another two years before reluctantly surrendering on 31 May 1902. Today often used colloquially to describe the last revellers to leave a party.
bosberaad – a strategic or planning meeting held at a venue away from the workplace ("offsite").
bosluis – tick. The Waterberg is notorious for its number and diversity of tick species.
broedertwis – a quarrel between brothers; often applied to factions within Afrikaner political parties.
burgher – *burger*, a citizen.
bywoner – a squatter on a farmer's land, usually with the owner's consent. Some became share-croppers, assisting the landowner with harvests in return for the right to reside and farm on the landowner's property.
CNO – *Christelik-Nasionale Onderwys*, Christian National Education (CNE).
CP – Conservative Party, a political party formed in 1982 by a right-wing faction of the National Party led by Dr Andries Treurnicht, in opposition to the NP's perceived relaxation of the policy of racial segregation.
DA – Democratic Alliance, a political party.
difaqane (or *mfecane*) – the period of tumultuous unrest in the years 1817–1830 associated with the migration of disaffected Zulu clans (including Mzilikazi and the Ndebele) from the Zululand coast onto the Highveld and the destruction, dispersion and flight of Sotho-Tswana communities already resident there.

DLU – *Distriks Landbou-Unie,* district farmers' union.
Dominee (Ds.) – a minister of one of the Dutch Reformed Churches.
dorp – rural village or small town.
Eerste Vryheidsoorlog – the first Transvaal War of Independence, or First Anglo-Boer War, 1880–1881.
Eerw. – *Eerwaarde* or Reverend, a title used in the NGK to address priests serving in its black missionary organisation, the NG Kerk in Afrika (NGKA).
Eeufees – centenary; *Eeufeesgedenkboek* – centennial commemorative volume.
EFF – Economic Freedom Fighters, a political party.
Ellisras – the name, until 2002, of the town of Lephalale.
epizootic – widespread (but temporary) outbreak of a disease, especially in animal population.
erf (pl. *erven*) – stand or plot of land.
ESA – Earlier Stone Age: in southern Africa, *circa* 2.5 million to 300 000 years ago.
field cornet – *veld-cornet*: historically, an unpaid civil officer appointed (sometimes elected) to preserve law and order in outlying areas on behalf of the government. Some were also Justices of the Peace.
Gauteng – one of nine provinces in post-1994 South Africa; formerly part of the Transvaal Province.
gedenk – remember; *gedenkboek* – commemorative volume.
Geloftefees – Festival of the Covenant: a religious and socio-political Afrikaner festival initiated to commemorate the victory of the Boers over the Zulus at the Battle of Blood River on 16 December 1838; now more a celebration of Afrikaner culture. *Geloftefeesdag* – 16 December; *Geloftefeesterrein* – the site in a neighbourhood where celebrants gather and camp to commemorate the event.
gemeente – congregation, parish.
GK – *Gereformeerde Kerk* – one of the Dutch Reformed "sister" churches.
grondboontjie – peanut, groundnut.
hartbeeshuis(ie) – wattle-and-daub hut, often a rural settler's first residence.
Hartingsburg – name of the settlement replaced by Warmbaths, now Bela-Bela.
hen(d)soppers en joiners – the derogatory term used by *bittereinders* to describe Boers who surrendered after the fall of Pretoria in June 1900. Some then joined the British forces (mainly as National Scouts) opposing the Boers during the guerrilla phase of the war.
Het Bad – the original (Dutch) name of Warmbaths/Bela-Bela.
Het Volk – "The People", a political party established in the Transvaal in 1905 under leadership of Generals Botha and Smuts to fight for self-government, which it secured in 1907.
HNP – *Herstigte Nasionale Party,* Reconstituted National Party, a right-wing faction of the ruling National Party that broke away in 1969.
IDC – Industrial Development Corporation, a parastatal development company.
Iscor/Yskor – (South African) Iron & Steel Corporation, a parastatal company established in the 1920s to produce steel (and to mine the requisite raw materials iron ore, coal, lime, zinc and tin). Privatised and listed on the Johannesburg Stock Exchange in 1989, it split in 2001 into Iscor Steel, soon to be purchased by the multinational steel company Arcelor-Mittal, and Kumba Resources, which inherited all the mining assets. In 2006, Kumba's non-iron business was listed separately as Exxaro Resources, the company which currently owns the Grootegeluk coal mine outside Lephalale. Ownership of the now closed Thabazimbi mine passed first to Kumba Iron Ore Ltd and in 2018 to Arcelor-Mittal SA.
JoP – Justice of the Peace (see RJP).
kaptein – captain, or chieftain.
kerkplein – church square.
kerkraad – church council.
Kgoši – chief or king (Sotho/Tswana).
kloof – wooded ravine or valley.
konsulent – relieving minister (usually at rural churches lacking a resident minister).

koshuis – boarding hostel (e.g. at a school).
KP – *Konserwatiewe Party*: see CP.
laager (laer) – encampment.
landdros(t) – magistrate.
Lephalale – the town formerly known as Ellisras.
Limpopo – one of nine provinces in South Africa since 1994, formerly known as Northern Province. Until 1994 a part of the Transvaal, one of four provinces in South Africa before subdivision.
LSA – Later Stone Age: in southern Africa, 30 000 to 100 years ago.
Mabatlane – a name proposed for the village of Vaalwater, but not yet formally tabled or approved; the name adopted by the post office in Vaalwater since about 2005; origin obscure.
mampoer – home-distilled spirit, e.g. peach brandy; moonshine (often illicitly distilled).
Matebele (Matabele) – Ndebele migrants from Zululand associated with Mzilikazi, who marauded across the Transvaal Highveld in the early nineteenth century before being defeated and put to flight (to southern Zimbabwe) by Boers.
mielie – maize, mealie.
minyan – a quorum of 10 adult males required in order to conduct a Judaic religious service.
MKTV – *Magaliesbergse Koöperatiewe Tabakplantersvereniging*: a large tobacco farmers' co-op, based in Rustenburg, but also serving the Waterberg.
Modimolle – the town formerly known as Nylstroom.
Mogol – an earlier spelling of the Mokolo River, previously known as the Sand or the Pongola; the Mokolo Dam on the river was originally called the Strijdom Dam.
Mokolo – the current spelling and pronunciation for Mogol (as in river or dam).
Mokopane – the town formerly known as Potgietersrus(t).
Mookgophong – the town formerly known as Naboomspruit.
morg (pl. *morgen*) – a measure of area; approximately the extent of land that could be ploughed in a morning. 1 *morg* = 600 Cape Square Roods = 0.856532 hectare = 2.1165 acres.
Moria City – the headquarters of both branches of the Zionist Christian Church (Star and St Engena's, or Dove), located on the farm Maclean 1046LS, about 45 km east of Polokwane on the R71 to Tzaneen.
MSA – Middle Stone Age: in southern Africa, 300 000 to 30 000 years ago.
muid – a Dutch measure of capacity, equal to approx. 2,97 Imperial bushels. I Imperial bushel = 8 gallons (36,4 litres). Therefore 1 muid is approx. 108 litres.
Naboomspruit – the name, until 2002, of the town of Mookgophong.
nagana – Trypanosomiasis of cattle, a usually fatal disease transmitted by tsetse flies.
nagmaal – Holy Communion service in the Dutch Reformed Churches.
na(a)rtjie – soft, loose-skinned tangerine.
National Scouts – a unit of burgher scouts (mainly surrendered Boers, or *hensoppers en joiners*) formed after June 1900 to assist the British during the guerrilla phase of the war (a similar unit was formed in the Free State). By the time peace was declared in May 1902, over 5000 Boers had become "joiners".
NBTC – New Belgium (Transvaal) Land & Exploration Company Limited.
NGK – *Nederduitse Gereformeerde Kerk* – the largest of the Dutch Reformed "sister" churches.
NGKA – *NG Kerk in Afrika* – the missionary organisation (for blacks) within the NGK.
NHK – *Nederduitsch Hervormde Kerk*: one of the Dutch Reformed "sister" churches.
Northern Transvaal – that portion of the former Transvaal Province between Pretoria and the Limpopo River, now split between the provinces of North-West, Limpopo and Mpumalanga.
NP – National Party, a political party, which formed the government of the country from 1948 until 1994.
NTK – *Noord-Transvaalse Koöperasie Beperk*, with headquarters in Nylstroom/Modimolle; successor to WKLM and WLKV.

NWFA – North Waterberg Farmers' Association (originally "Society").
Nyl – the name given to the upper reaches of the Mogalakwena River.
Nylstroom – the name, until 2002, of the town of Modimolle.
OB – *Ossewabrandwag*, "Ox-wagon Sentinel", a radical paramilitary Afrikaner Nationalist organisation established during World War II.
Oom – uncle, also used as a term of respect for an older male.
Orange Free State – originally one of the major two Boer republics, subsequently incorporated into the Union of South Africa (a Republic from 1961) as one of its four provinces and changed to Free State after 1994, one of the nine provinces now comprising the Republic of South Africa.
ouderling – a church elder.
Ouri or Oori – a former name for the Limpopo River.
pas – mountain pass, as in *Bakkerspas* or *Jakkalskuilpas*.
pastorie – parsonage or manse of Dutch Reformed Church minister.
Pietersburg – the former name of Polokwane, the capital city of Limpopo Province.
Polokwane – capital city of Limpopo Province, formerly known as Pietersburg.
Pongola – a former name for the Sand or Mokolo (or Mogol) River.
poort – narrow defile or mountain pass.
Potgietersrus(t) – originally Piet Potgietersrust; shortened to Potgietersrust in 1904; in 1954, the last 't' was dropped. The name, until 2002, of the town of Mokopane.
predikant – parson.
quitrent farm – a system of land tenure introduced in the Cape in 1732, in terms of which occupancy of a piece of land was granted for an initial period of fifteen years; renewable. A loan farm, for which annual dues were payable to the State. Eventually converted to permanent tenure.
rant/rand – range of hills, long ridge.
rinderpest – infectious viral disease of ruminants (especially cattle).
RJP – Resident Justice of the Peace (see JoP).
Rooiluise – "Red Lice", a derogatory term used by right-wing Afrikaners to describe men who volunteered to join the UDF units fighting with the Allies overseas during World War II; derived from the red shoulder flashes they wore.
Roomse gevaar – the perceived danger to public morals ascribed to the Roman Catholic church by its Dutch Reformed counterpart.
SAC – South African Constabulary – a force, initially paramilitary, later becoming more typically a police force, created by the British colonial administration to maintain law and order, especially in rural parts of the Transvaal (and Free State) from 1902 until the creation of provincial police forces in 1907/8.
SALU/SAAU – *Suid-Afrikaanse Landbou-Unie*/South African Agricultural Union.
SAP – South African Party, a political party (members were called *Sappe*), which became the United Party.
SAR&H – South African Railways and Harbours, the forerunner of Spoornet and Transnet.
Schutztruppe – troops of German colonial administrations, e.g. in German South West Africa (Namibia).
skriba – church secretary, or parish clerk.
smous – hawker, itinerant trader (usually Jewish in rural Transvaal during early twentieth century).
South African War – the currently preferred name for the conflict between the British and the Boers that took place between October 1899 and the conclusion of the Treaty of Vereeniging on 31 May 1902. Also known as the Anglo-Boer War, *die Engelse Oorlog* or *die Tweede Vryheidsoorlog*.
Staats Courant – ZAR Government Gazette.
stad – town or city.
stamvrug – wild plum, *Englerophytum magalismontanum*.

Stormjaers – storm-troopers; a militia within the *Ossewabrandwag* involved in acts of sabotage in the period 1941–1943.
SWA – South West Africa, a German colony until 1915, now Namibia.
swaer – brother-in-law.
tante or *tannie* – aunt, also used as a term of respect for an older female.
Tarentaalstraat – "guinea fowl street": the name given to all or part of the D579/D710, a road running from west of Modimolle northwards almost to Melkrivier.
TCL – Transvaal Consolidated Land & Exploration Company Limited.
TLC – Transvaal Lands Company Limited.
TLU – *Transvaalse Landbou-Unie*, Transvaal Agricultural Union (TAU).
Transvaal – a former Boer republic (see ZAR), which became one of the four provinces of the Union of South Africa in 1910. After 1994, the Transvaal was split into four new provinces: Gauteng, North-West, Mpumalanga and Limpopo. The Waterberg falls entirely within Limpopo Province.
trekboer – a trek-farmer, a member of the group of early nineteenth century Dutch-speaking farmers who began to migrate out of the Eastern Cape ahead of the Great Trek.
UDF – Union Defence Force; the national armed forces created after Union in 1910.
UP – United Party, a political party.
Volksraad – the parliament of the former Transvaal Republic; later divided into upper (*eerste Volksraad*) and lower (*tweede Volksraad*) houses.
Voortrekker – pioneer (literally, one who goes ahead).
Warmbaths – the name of the town now known as Bela-Bela. Also previously known as Hartingsburg, Het Bad and Warmbad.
WKLM – *Waterbergse Koöperatiewe Landboumaatskappy Beperk*, forerunner of NTK.
WLKV – *Waterberg Landbouwers Koöperatiewe Vereniging*, forerunner of WKLM.
wyk – ward, as in electoral district.
Zandriviersberg/Zandriviersspoort – Sandriviersberg/Sandriviersspoort: the range of hills forming the southern escarpment of the main Waterberg plateau/the gap in this range through which the R33 passes between Modimolle and Vaalwater.
ZAR – *Zuid-Afrikaansche Republiek*, or Transvaal Republic: the independent Boer republic that existed from the conclusion of the Sand River Convention in January 1852 until 1877 and again from 1881–1902.
Zarp – The police force of the ZAR (members were known colloquially as "Zarps").
ZCC – Zionist Christian Church, South Africa's largest religious organisation, divided into two factions, ZCC "Star" and ZCC "Dove" (St Engena's ZCC), both with headquarters at Moria City east of Polokwane. Total combined membership may exceed 15 million.
Zwagershoek(berg) = *Swagershoek* = *Swaershoek* – "Brothers-in-law corner": the name given to at least two separate areas in the southern Waterberg foothills, where closely-related families settled in the mid-nineteenth century. A large part of the Waterberg was included in the Zwagershoek electoral ward of the old *Volksraad*.

SELECTED BIBLIOGRAPHY

From the large number of sources used and cited in the preparation of this history – as referenced in the endnotes to each chapter – the following is a selection of references considered to be most useful to the general reader. Strictly archival, obscure or informal sources are excluded from this selection. The reader is referred to the appropriate chapter endnotes for such sources.

Aukema, J. (1989): Rain-making: a thousand year-old ritual? *South African Archaeological Bulletin* **44**, pp. 70–72.

Baines, Thomas (1877): *The Gold Regions of South Eastern Africa*. Facsimile edition by Rhodesiana Reprint Library Volume 1 (1968), Books of Rhodesia, Bulawayo, Zimbabwe.

Barnard, Annette (2007): *Thabazimbi Iron Ore Mine 75: 1931–2006*. Published by Thabazimbi Iron Ore Mine (a member of the Kumba Iron Ore Company Ltd).

Bester, D.J. (ed.) (1984): *NTK 75*. A special edition of the magazine *NTK-Nuus* to commemorate the 75th anniversary of the founding of the NTK. Volume 7 No. 11, Sept/Oct 1984. Nylstroom.

Boeyens, J. & M. van der Ryst (2014): The cultural and symbolic significance of the African rhinoceros: a review of the traditional beliefs, perceptions and practices of agro-pastoralist societies in southern Africa. *Southern African Humanities* **26**, pp. 21–55.

Boonzaaier, Boon (2008): *Tracks Across the Veld*. Privately published, Bela-Bela.

Breytenbach, S.W. (2012): *Koedoesrand-Kronieke*. Privately published, Marken.

Burke, E.E. (ed.) (1969): *The Journals of Carl Mauch: His travels in the Transvaal and Rhodesia, 1869–1872*. National Archives of Rhodesia, Salisbury.

Butterfield, Paul H. (1978): *Centenary: The first 100 years of English Freemasonry in the Transvaal, 1878–1978*. Ernest Stanton Publishers, Johannesburg.

Caine, W.H.A. (1902): *Rough Diary of Events and my Life during the Transvaal War, 1899–1901*. Noord Brabant, New Belgium. Unpublished handwritten manuscript, lodged with SA National Archives in Pretoria and with the Manuscript Library at the University of Cape Town.

Callaghan, C.C. (1993): The Geology of the Waterberg Group in the southern portion of the Waterberg Basin. *Bulletin 104* of the Geological Survey, Department of Mineral & Energy Affairs, Pretoria.

Cameron, Trewhella (ed.) (1986): *An Illustrated History of South Africa*. Jonathan Ball, Johannesburg.

Carruthers, E.J. (1995): *Game Protection in the Transvaal 1846 to 1926*. Archives Year Book for South African History. Government Printer, Pretoria.

Clarke, Elizabeth (1954): *Waterberg Valley*. Unfinished and unpublished autobiography of Elizabeth Francis Clarke (31.03.1912–15.04.1954).

Coetzee, M., P. Kruger, R.H. Hunt, D.N. Durrheim, J. Urbach, C.F. Hansford (2013): Malaria in South Africa: 110 years of learning to control the disease. *South African Medical Journal,* October 2013, **103** (10), pp. 770–778.

D'Assonville, W.E. (2008): *Eugène Marais and the Waterberg*. Marnix, Pretoria.

Delius, Peter (1983): *The Land Belongs to Us*: The Pedi Polity, the Boers and the British in the Nineteenth Century Transvaal. Ravan Press, Johannesburg.

Delius, Peter, Tim Maggs & Alex Schoeman (2014): *Forgotten Worlds – The Stone-walled Settlements of the Mpumalanga Escarpment*. Wits University Press, Johannesburg.

Duminy, Andrew (2011): *Mapping in South Africa*. Jacana Media, Johannesburg.

Eastwood, E. & C. Eastwood (2006): *Capturing the Spoor: An Exploration of Southern African Rock Art*. David Philip: Claremont.

Engelbrecht, S.P. et al (eds): *Pretoria 1855–1955*. Pretoria City Council, Pretoria.

Erasmus, D. & J.A. du Plessis (compilers) (1952*)*: *Naboomspruit 1652–1952. Van Riebeeck Fees*. Published by the Naboomspruit Town Council.

Forssman, O.W.A. (compiler) (1872): *A Guide for Agriculturalists and Capitalists, Speculators, Miners Etc, wishing to invest money profitably in The Transvaal Republic, South Africa; containing a description of a number of first-class farms situated in different districts of the Republic, and general useful information.* William Foster & Co., Printers, Cape Town. Reprinted by the State Printer, Pretoria in 1984.

Fuller, Claude (1923): *Tsetse in the Transvaal and Surrounding Territories – an historical review.* Entomology Memoirs No. 1. Division of Entomology, Department of Agriculture, Pretoria.

Gewald, Jan-Bart (1996): *Towards Redemption: a socio-political history of the Herero of Namibia between 1890 and 1923.* Research School CNWS, Leiden University, Netherlands.

Gewald, Jan-Bart (1999): The Road of the Man Called Love and the Sack of Sero: The Herero-German War and the Export of Herero Labour to the South African Rand. *Journal of African History*, **40**, pp. 21–40.

Gewald, Jan-Bart (2002): "I was afraid of Samuel, therefore I came to Sekgoma": Herero refugees and patronage politics in Ngamiland, Bechuanaland Protectorate, 1890–1914. *Journal of African History*, **43**, pp. 211–234.

Giliomee, Hermann (2003): *The Afrikaners: Biography of a People.* Tafelberg, Cape Town.

Glen, H.F. & G. Germishuizen (2010): *Botanical Exploration of Southern Africa, Edition 2. Strelitzia* **26**, SA National Biodiversity Institute, Pretoria.

Goldblatt, I. (1971): *History of South West Africa* (from the beginning of the nineteenth century). Juta & Co., Cape Town.

Greenstein, Lisa (ed.) (2002, reprinted with addenda, 2008): *Jewish Life in the South African Country Communities, Volume 1: The Northern Great Escarpment, The Lowveld, The Northern Highveld, The Bushveld.* South African Friends of Beth Hatefutsoth, Johannesburg.

Grobbelaar, Pieter W. (1980): *C Louis Leipoldt.* Skrywers in Beeld (4). Tafelberg, Cape Town.

Grobler, Maritz et al (eds) (1995): *Die suidelike Springbokvlakte, 1876–1995 ('n Historiese Oorsig)/The Southern Springbok Flats, 1876–1995 (A Historical Review).* Published by the Berlin Geloftefeeskomitee, Settlers.

Gutteridge, Lee (2008): *The South African Bushveld – a field guide from the Waterberg.* Southbound, an imprint of 30° South Publishers, Johannesburg.

Hall, Darrell (1999): *The Hall Handbook of the Anglo Boer War.* University of Natal Press, Pietermaritzburg.

Hamilton, Carolyn (ed.) (1995): *The Mfecane Aftermath: Reconstructive Debates in southern African History.* Jointly published by Witwatersrand University Press and Natal University Press.

Hansen, Mauritz (compiler) (2010): *Naboomspruit 1910–2010: Eeu van genade.* Privately published, Mookgophong.

Harrison, David (1981): *The White Tribe of Africa: South Africa in Perspective.* Macmillan South Africa, Johannesburg.

Hofmeyr, Isabel (1987): *Turning Region into Narrative: English Storytelling in the Waterberg.* History Workshop: The Making of Class, University of Witwatersrand, Johannesburg.

Hofmeyr, Isabel (1993): *We Spend Our Years as a Tale That is Told: Oral Historical Narrative in a South African Chiefdom.* Witwatersrand University Press, Johannesburg.

Huffman, T.N. (1990): Obituary: The Waterberg research of Jan Aukema. *South African Archaeological Bulletin* **45**, pp. 117–119.

Huffman, Thomas N. (2007): *Handbook to the Iron Age: the Archaeology of Pre-colonial Farming Societies in Southern Africa.* University of KwaZulu-Natal Press, Pietermaritzburg.

Hunter, Elizabeth (compiler) (2010): *Pioneers of the Waterberg: a Photographic Journey.* Privately published. Lephalale, Limpopo.

Jackson, A.O. (1983): *The Ndebele of Langa.* Ethnological Publications No. 54, Department of Co-operation and Development. Government Printer, Pretoria.

Jansen, H. (1982): The Geology of the Waterberg Basins in the Transvaal Republic of South Africa. *Memoir 71* of the Geological Survey, Department of Mineral and Energy Affairs, Pretoria.

Jeppe, Carl (1906): *The Kaleidoscopic Transvaal*. Chapman & Hall, London.

Jeppe, F. and A. Merensky (1868): *Original Map of the Transvaal or South-African Republic*, from surveys and observations by Surveyor General M. Forssman, C. Mauch, F. Hammar, Surveyor J. Brooks and other official documents combined with the results of their own explorations. By F. Jeppe and A. Merensky, Potchefstroom and Botsabela. Reconstructed and augmented with data from various exploring travellers, by A. Petermann, Germany.

Jeppe, F. (1879): *Map of the Transvaal and the Surrounding Territories*. Published by SW Silver & Co., London.

Jeppe, Friedrich & C.W. Jeppe (1899): *Jeppe's Map of the Transvaal or SA Republic and surrounding territories*, Sheet 2. Edward Stanford, London.

Kirkman, Peter (2012): *From Manchester to Albany and Beyond: Genealogy and Chronicles of the 1820 Settler John Kirkman and his Descendants*. Self-published, Hilton KZN.

Krige, J.D. (1937): Traditional Origins and Tribal Relationships of the Sotho of the Northern Transvaal. *Bantu Studies*, Vol. XI (4), December 1937.

Lamb, Doreen (1976): *The Republican Presidents, Vols 1 & 2*. Perskor Publishers, Johannesburg.

Laue, G.B. 2000. *Taking a Stance: Posture and meaning in the Rock Art of the Waterberg, Northern Province, South Africa*. Unpublished MSc. thesis, University of the Witwatersrand.

Leipoldt, C. Louis (1937): *Bushveld Doctor*. Jonathan Cape, London.

Marais, Eugène (1972): *The Road to Waterberg and other Essays*. Human & Rousseau, Cape Town.

Martinson, J.H. (ed.) (1960): *Eeufeesbrosjure van die Ned. Geref. Gemeente Waterberg, 1860–1960*. Nylstroom.

Marx, Christoph (2008): *Oxwagon Sentinel: Radical Afrikaner Nationalism and the History of the Ossewabrandwag*. UNISA Press, Pretoria.

Mason, R.J. (1962): *Prehistory of the Transvaal*. Witwatersrand University Press, Johannesburg.

Mossolow, N. (1993): *Waterberg: On the history of the mission station Otjozondjupa, the Kambazembi tribe and Hereroland*. Privately published, Windhoek, Namibia.

Mucina, Ladislav & Michael Rutherford (eds) (2006): *The Vegetation of South Africa, Lesotho and Swaziland*. Strelitzia Series No. 19. SANBI, Pretoria.

Muller, C.F.J. (ed.) (1975): *500 Years – A History of South Africa*. 2nd Edition. Academica Press, Pretoria.

Nattrass, Gail (1983): *The Rooiberg Story* – a publication to mark the 75th anniversary of the Rooiberg Tin Limited. Rooiberg.

Naudé, M. (1998): Oral Evidence of Vernacular Buildings and Structures on Farmsteads in the Waterberg (Northern Province). *Research by the Cultural History Museum*, 7, pp. 47–91.

Naudé, M. (2004): Oral Evidence of the Construction of Vernacular Farm Dwellings in the Waterberg (Limpopo Province). *SA Journal of Cultural History* 18 (1), pp. 34–62.

Naudé, M. (2008): Oral Evidence on Aspects of Folk Life during the first hundred years (1840–1940) of Frontier settlement in the Waterberg (Limpopo Province). *National Cultural History Museum Research Journal* 3 (2008), pp. 35–56.

Nel, J.J. et al (eds) (1966): *Nylstroom 1866–1966*. Nylstroomse Stadsraad.

Nel, Louis (ed.) (2005): *Gedenkalbum, NG Gemeente, Waterberg Noord, 1930–2005*. NG Gemeente, Vaalwater.

Oakes, D. et al (eds) (1992): *Illustrated History of South Africa – The Real Story*. Readers Digest Association of South Africa.

Oberholster, J.J. (1972): *The Historical Monuments of South Africa*. Rembrandt van Rijn Foundation for Culture, on behalf of National Monuments Council, Cape Town.

Odendaal, R.F. (Roelf) (1995): *Waterberg op Kommando*. Privately published, Nylstroom.

Odendaal, R.F. (Roelf) (c.2000): *Noordtransvaal op Kommando*, Privately published, Nylstroom.

Odendaal, R.F. (Roelf) (c.2000): *Waterberg word Gekersten*. Privately published, Nylstroom.

Olusoga, D. & C.W. Erichsen (2010): *The Kaiser's Holocaust – Germany's forgotten genocide and the colonial roots of Nazism*. Faber & Faber, London.

Pienaar, U. de V. (ed.) (2012): *A Cameo from the Past – the prehistory and early history of the Kruger National Park*. Protea Book House, Pretoria.

Pont, Prof. A.D. (1965): *Die Nederduitsch Hervormde Gemeente Waterberg, 1865–1965*. Kerkraad van die NHK Waterberg, Krugersdorp.

Pool, G. (1991): *Samuel Maharero*. Gamsberg Macmillan, Windhoek, Namibia.

Prance, C.R. (1923): *Under the Blue Roof – Sketches of a Settler's Life in the Transvaal Backveld, 1908–1921*. William Cooper & Nephews Ltd., Berkhamsted, England.

Prance, C.R. (1937): *Tante Rebella's Saga: a backvelder's Scrapbook*. H.F. & G. Witherby, London.

Prance, C.R. (1941): *The Riddle of the Veld: Sketches of Life, Character and history in the Backveld of South Africa*. The Knox Publishing, Durban.

Prance, C.R. (1943): *Dead Yesterdays*, privately published by the author at Port St Johns.

Pretorius, Fransjohan (ed.) (2014): *A History of South Africa – from the distant past to the present day*. Protea Book House, Pretoria.

Rodger, Lex (2010): *Waterberg (Vintage Waterberg & Timeless Waterberg)*. Privately published by the Rodger family, Johannesburg.

Rousseau, Leon (ed.) (1984): *Die Versamelde Werke van Eugène Marais*. JL van Schaik, Pretoria.

Rousseau, Leon (1999): *The Dark Stream: The Story of Eugène Marais*. Jonathan Ball, Johannesburg.

Sampson, G.S. (1974): *The Stone Age Archaeology of Southern Africa*. New York and London: Academic Press.

Scholtz, G.D. (1941): *Generaal Christiaan Frederik Beyers 1869–1914*. Voortrekkerpers, Johannesburg.

Scholtz, G.D. (1942, 2013): *Die Rebellie 1914–1915*. Protea Boekhuis, Pretoria.

Schoonraad, M. & P.B. Beaumont (1968): The North Brabant Shelter, North Western Transvaal. *South African Journal of Science* **64**, pp. 319–331.

Sekikaba, Peter Lekgoathi (2009): "Colonial Experts", Local Interlocutors, Informants and the Making of an Archive on the 'Transvaal Ndebele', 1930–1989. *Journal of African History*, **50.**

Seroto, Johannes (2004): *The Impact of South African Legislation (1948–2004) on Black Education in Rural Areas: An Historical Educational Perspective*. PhD thesis at UNISA.

Setumu, Tlou Erick (2001): *Official Records Pertaining to Blacks in the Transvaal, 1902–1907*. MA thesis, Faculty of Humanities, University of Pretoria.

Setumu, Tlou Erick (2009): *Chief Mokopane II (Parts 1 & 2)*. Limpopo Provincial Government Heritage Month Souvenir Edition 2009.

Silverman, Melinda (ed.) (2004*): Nylstroom/Modimolle – Urban Change in a Small South African Town*. Published by the School of Architecture at the University of the Witwatersrand.

Silverman, Melinda & Irwin Manoim (eds) (2003): *Warmbaths/Bela-Bela/Settlers: a site investigation by architecture students from the University of the Watersrand*. Published by the School of Architecture at the University of the Witwatersrand, Johannesburg.

Strydom, Hans (1982): *For Volk and Führer*. Jonathan Ball, Johannesburg.

Surveyor General's Office (1960): *Alphabetical List of Farms in the Province of Transvaal*. Government Printer, Pretoria.

Taylor, W., G. Hinde & D. Holt-Biddle (2003): *The Waterberg*. Struik, Cape Town.

Trapido, Stanley (1978): Landlord and Tenant in a Colonial Economy: The Transvaal, 1880–1910. *Journal of Southern African Studies* **5** (1).

Trichardt, Louis (2016): *Swaershoek in die Waterberge – Afrikaner erfenis of niemandsland?* Privately published.

Troye, Gustav (1892): *Troye's Map of the Transvaal* (Sheet 2). Fehr & DuBois, Pretoria.

Van den Berg, Daniel (1985): *Naboomspruit – 75 in 1985*. Naboomspruit Town Council.

Van der Merwe, D.W. (1984): Die Geskiedenis van die Berlynse Sending-genootskap in Transvaal, 1860–1900. *Argiefjaarboek vir Suid-Afrikaanse Geskiedenis, Ses-en-Veertigste Jaargang, Deel 1*. State Printer, Pretoria.

Van der Ryst, M.M. (1998): *The Waterberg Plateau in the Northern Province, Republic of South Africa, in the Later Stone Age*. BAR International Series **715**, pp. 1–158.

Van der Ryst, M.M. (2006): *Seeking Shelter: Later Stone Age Hunters, Gatherers and Fishers of Oliebooms- poort in the Western Waterberg, South of the Limpopo*. Unpublished PhD thesis: University of the Witwatersrand.

Van der Wateren, Prof. H. (ed.) (2009*): GKSA – 150. Bosveldfees*. Gereformeerde Kerk Waterberg, Nylstroom.

Van Heyningen, Elizabeth (2013): *The Concentration Camps of the Anglo-Boer War: a social history*. Jacana, Johannesburg.

Wadley, L. (1987): *Later Stone Age Hunters and Gatherers of the Southern Transvaal.* Oxford: BAR International Series **380**, pp. 1–255.

Wadley, L., M.L. Murungi, D. Witelson, R. Bolhar, M. Bamford, C. Sievers, A. Val. & P. de la Peña. (2016): Steenbokfontein 9KR: A Middle Stone Age Spring Site in Limpopo, South Africa**.** *South African Archaeological Bulletin* **71**, pp. 130–145.

Walker, Clive (2013): *Baobab Trails – an artist's journey of wilderness and wanderings*. Jacana, Johannesburg.

Walker, Clive (2016): *Lapalala*. Published by Lapalala Wilderness, produced by Jacana, Johannesburg.

Walker, Clive & J du P. Bothma (2005): *The Soul of the Waterberg*. African Sky Publishing, Johannesburg.

Wallace, Marion, with John Kinahan (2011): *A History of Namibia*. Jacana, Johannesburg.

Wilkins, Ivor and Hans Strydom (1978): *The Super-Afrikaners: Inside the Afrikaner Broederbond*. Jonathan Ball, Johannesburg.

Wouters, Jacqueline M.F. (2014): *An Anthropological Study of Healing Practices in African Initiated Churches with Specific Reference to a Zionist Christian Church in Marabastad* [Pretoria]. MA thesis, UNISA, Pretoria.

ACKNOWLEDGEMENTS

Interviews, discussions & information

The following people should be credited as being the sources of much of the material collated in this book. Between them, over the last decade and longer, they have provided information, given or loaned me books, albums and reports, supplied or allowed me to copy photographs and paintings, arranged for, or given me access to properties, accompanied me on visits, subjected themselves to interviews (sometimes repeatedly), listened politely to my research findings, and often offered valuable contributions on topics about which I was completely ignorant. Without their knowledge, involvement and interest, the result would have been much drier and less complete. To you all (and surely others whose names I've inadvertently omitted), may I say, "Thank you – were it not for your contributions, this book would not have been worth producing."

Mariana & Johan Adams, the late Nina & Charles Baber, Joan Baber, Gerrit Bakker, Prof. Terri Bakker, Hayley Barnes, Patricia & Peter Beith, Dr Roy Bengis, Ds. Albert Berg, Zane Bera, Johan Beukes, Susanne Blendulf, Jan Blignaut, Prof. Jan Boeyens, the late Boon Boonzaaier, Rev. Terry Booysen, the late Koos Boshoff, Anton Botes, Dr Karel Botha, Prof. Lindsay Braun, Anna-Mari Bronkhorst, André Burger, Prof. Bruce Cairncross, Juliet Calcott, Malesela Chokwe, Prof. Maureen Coetzee, Mary Connor, Laurentius Davids, Christo Davidson, Johan de Beer, Chris Derksen, Pierre de Wet, Erna Duvenhage, Carel & Janet Dirker, Claris Dreyer, Neels van Emmenis, Topsy Eschenburg, Frans and Hennie Faber, Dr Peter Farrant, Roger Fisher, Len Fletcher, Annamarie Folkus, David Gardener, Frank Gaylard, Tokkie, Sylvie and the late Alby Geerkens, Dr Jan-Bart Gewald, Stephen Gomersall, Major & Mrs Antony Gordon, Ilva Greer, Jannie Grové, Carol Guest, Mauritz Hansen, Heilet Hattingh, Nikki Haw, the late Henry Hayward, Reinhard & Caroline Heuser, Zandra Houston & her brother John Burnell, Anne Howe, Prof. Tom Huffman, Elizabeth Hunter, Chris Janisch, Stephanie Kemp, Ros King, Peter Kirkman, Rev. Douglas Kirkpatrick, Hansie Kloppers, Sas Kloppers, Leoni Kruger, Siegwalt Küsel, Val Landau, Ina Lerchbaumer, Jeran le Roux, Jan Lewis, Henk Lingenfelder, Fatima Lorgat, Sam Mafora, Mashudu Makhokha, Harry Malan, Rob Manson, Richard Marais, William Masalesa, Lesiba Masibe, Prof. Revil Mason, Ken Maud, Jan Middeljans, John Miller, Zunaid Mosam, Ben Mostert, Koos Mostert, Ivan Muller, Colbert Munarini, Morkel Munnik, Shelagh Nation, Gail Nattrass, Mauritz Naudé, Louis Nel, Richard Nicholson, Harold Nicolson, Harry & Peggy Parham, Barend Pretorius, the late "Boelie" Pretorius, Corrie Pretorius, Prof. Fransjohan Pretorius, Ian Pringle, Pieter Prinsloo, Linda Raath, Neels Riekert, the late Lex Rodger, Alison Rose, Karina Sevenhuysen, Patrick Shika, Chrissie Sievers, Rabbi Moshe Silberhaft, Delene Snyman, Fanie Stander, Billy Steenkamp, the late Brian Stone, Fred Stowe, Magda Streicher and her brother, Frikkie Brits, Keina Swart, Prof. Johan Swart, Pieter Swanevelder, Dr Warwick & Michèle Tarboton, Andrew Taylor, John Todd, Louis Trichardt, Lauritz & Tord Tutturen, Dr André Uys, Ian Valentine, Sam van Coller, H.A. van den Berg, Schalk & Grieta van der Merwe, Dr Maria van der Ryst, Ben van der Walt, Jaco van der Walt, Prof. Henk van der Wateren, Dries & Elsa van Rooyen, Frans van Staden, the late Harry van Staden, Dr Mich Veldman, Henning Viljoen, Allan Vlag, Leonard Vorster, Clive & Anton Walker, Coen Weilbach, Robbie Weir and Shelley Zeederberg.

Archival sources

Many formal archival and reference sources were accessed and consulted in the compilation of this work. The staff of the National Archives in Pretoria in particular, had to put up with my presence for several hundred hours in the course of many visits and to deal with over a thousand requests for information under their curation. Without their tolerance, patience and helpful cooperation, this project would simply not have been possible. Thanks in particular to: Tshepo Malemela, Mavis Xaba, Granny Makhubela, Ntombi Godwa Gumede, Oupa Mahlori, Mologadi Mmamabolo, Jonathan Mufumade and, especially, Mary Masango.

In addition, thanks are due to staff of the following institutions for their assistance with various aspects of the research, including the provision of copies and imagery:
- Arend Dieperink Museum, Mokopane: Hendrina Ysel
- Botswana National Archives, Gaborone: Linda Magula, John Modisane, Moses Mafathle
- Brenthurst Library, Johannesburg: Jennifer Kimble, Inarié de Vaal
- British National Archives at Kew; and British Library, Paddington
- Dramatic, Artistic & Literary Rights Organisation (DALRO): Ilyana van Tonder
- Ditsong: National Museum of Cultural History, Pretoria: Jan Middeljans, Mauritz Naudé, Linda Raath
- Hugh Exton Photographic Museum and the Irish House Museum, both operated by the Municipality of Polokwane: Paulina Mapheto
- Johannesburg City Library (formerly the JPL) – Strange Library of Africana
- Library of the University of the Witwatersrand, Johannesburg
- Merensky Library of the University of Pretoria
- Museum Africa, Johannesburg: Kenneth Hlungwani
- National Archives of South Africa, Cape Town Depository: Erika le Roux, Hendrick April
- National Archives of Namibia: Martha Nakangula, Albertina Nekongo, Hertha Ipinge, Ralph Ntelamo
- National Library of South Africa, Pretoria Campus: Mafadi Bapela, Tshepo Mohlala
- Reserve Bank of South Africa: Sheenagh Reynolds (Secretary), Prof. Rory Bester (Art Consultant)
- The Technical Library (Chamber of Mines building, Johannesburg): Katinka van Straaten
- The Chief Directorate, National Geospatial Information, Dept. of Rural Development & Land Reform, Mowbray, Cape, especially Messrs Rabelane & Ambesuwa, for finding, reproducing and printing early editions of many relevant maps of the Waterberg region
- Oxford University, UK: Rhodes Library and Bodleian Library
- UNISA Library, Pretoria

For material relating to British officers who died during the South African War, I am indebted to:
- The Kings Own Yorkshire Light Infantry Museum, Doncaster (Steve Tagg of the Doncaster Museum and Art Gallery, 2013) relating to Lt Charles Powell
- The Royal Norfolk Regimental Museum, Norwich (Kate Thaxton, Curator, 2011) relating to Lt Ronald Buxton
- The National Army Museum/Royal Fusiliers Regimental Museum, London (Alastair Massie, 2017; and Stephanie Killingbeck-Turner, Mick Garrard & David Carter, 2018); and Caroline Jones & Jeanette Lawrence of Wellington College, Berkshire, relating to 2nd Lt Arthur Besley
- I would also like to thank in particular Rosemary Cochrane, who allowed me to look through the original handwritten diary of her great-uncle, William Caine; and Major and Mrs Antony Gordon, who very kindly sent me copied extracts from the handwritten diaries of Lt (later Lt Col) E.I.D. Gordon of the Royal Scots Fusiliers. Typed copies of Caine's diary are lodged with the Manuscript Library of the University of Cape Town and the National Archives in Pretoria. Copies of Lt Gordon's four diaries are lodged with the National Army Museum, London. Both these valuable reference sources warrant publication in their own right.

Editing & reading

I am deeply indebted to Emeritus Professor Jane Carruthers, the eminent and widely published historian with a long career at UNISA, who very bravely took on the task of critically reviewing the entire content of the draft manuscript and made numerous valuable observations and recommendations, most of which have been incorporated into the text.

My wife Lyn not only wrote the chapter on the archaeology and pre-history of the Waterberg (which as a professor of archaeology with the Evolutionary Studies Institute at the University of the Witwatersrand, she is superbly qualified to do), but also read critically the remainder of the manuscript. Our friend and colleague Dr Maria van der Ryst also took on the job of reviewing the whole work and I thank her for the many valuable and insightful comments and suggestions she made.

I am grateful to the following persons for having accepted the task of reading and commenting on various chapters and sections of the text with a view to improving its veracity: Isabel Hofmeyr, John Miller, Wendy Adams, Jan Boeyens, Ds. Albert Berg and Sam van Coller. Nearly all the changes they recommended were incorporated, but of course, responsibility for flaws in the content and expressions of opinion remains mine alone. John Miller was also involved very early in this project when he and I began scouting the western Waterberg for historical information as the basis for a tourist brochure envisaged by the Waterberg Biosphere Reserve. The arrangement was that he would provide the company if I provided the vehicle, fuel, maps – and lunch. It was an enjoyable partnership that contributed much of the material making up Chapter 4.

Illustrations & maps

I was determined that this work should be accompanied by professionally drawn, accurate, yet attractive maps depicting the places described in the text. My Namibian colleague and friend, Colin Bleach, who has illustrated numerous books of a technical nature with maps and figures, was persuaded to draft the maps that appear in the text, working from my rough, frequently inaccurate sketches and often near-illegible captions e-mailed to him in Swakopmund. He also transformed several poor photographs of various subjects into the delightful tinted pen-and-ink sketches that appear throughout the book. In addition, he undertook to liaise on my behalf with the National Archives of Namibia in Windhoek to source documents and photographs. Later, I was privileged to accompany Colin and his wife Maeve on a visit to the Herero graves in Okahandja and to the Waterberg Plateau Park, the site of the final battle, on 11 August 1904, between the Herero and the Germans, as described in Chapter 20. I'm sure there were occasions when Colin was about ready to throw in the towel and abandon this difficult, indecisive, yet demanding client; but fortunately, he persevered and his work greatly enhances the final product. I'd like to think that our association during its production has served to renew and strengthen our longstanding friendship. *Prost!*

Photographs

All photographs used in the book other than those taken by me are acknowledged individually. Many are drawn from private collections and I thank those who made them available for my use; but the majority are from the extensive collections of the National Archives in Pretoria or Cape Town, the Ditsong National Museum of Cultural History (Pretoria) or Museum Africa (Johannesburg). The images selected were painstakingly scanned or photographed at high resolution for me by archival staff. In only a few cases did it prove impossible to identify the source of an image used.

Several images were selected from the collected works of the outstanding late 19th century photographer H. Ferdinand Gros, a Swiss national who came to South Africa in 1867 and had a studio on the diamond fields before establishing one on the corner of Church and Van der Walt Streets in Pretoria in 1877. Gros took many studio photographs as well as images of Pretoria and the goldfields. Numerous photographs were published as albums. The album relevant to this work was entitled *Pictorial Description of the Transvaal*, a copy of which is with the Ditsong National Museum of Cultural History. Most of the photographs contained therein were taken during a tour that Gros made of the Transvaal in 1888. He returned to Europe in 1895.

Twenty years after Gros toured the Transvaal, Emil Tamsen of Nylstroom recorded numerous scenes in and around Nylstroom, including some at the newly opened Rooiberg tin mine. Several of Tamsen's photographs are stored at the National Archives in Pretoria. Others exist as postcards in the collection of Museum Africa.

Photographs of trains on the Vaalwater–Nylstroom line in Chapter 12 were obtained for me by

Chris Janisch from fellow members of Friends of the Rail – Nathan Berelowitz, Dick Manton, Dennis Mitchell – and I thank Chris and his colleagues for permission to reproduce their work.

Liz Hunter and Juliet Calcott generously shared several photographs from the Davidson and Baber family collections respectively and I thank them for permission to reproduce these.

Artwork

In addition to the pen and ink sketches produced by Colin Bleach, I have been able to include reproductions of some other artistic works:

- *Figure 7.10:* "Donkey cart with firewood" by Elizabeth Hunter (2008) – watercolour; in a private collection, reproduced with kind permission of the artist. Elizabeth Hunter spent much of her childhood at the family farm of 24 Rivers in the Waterberg and continued to visit frequently as an adult. She still lives in the area with her husband, Ken Maud, along the northern Waterberg escarpment between Lephalale and Marken. She paints actively and has held several successful exhibitions of her work in Vaalwater and elsewhere in the region.
- *Figure 7.15:* "Dwarsrivier 1901" by Keith Dietrich (2015). Keith is a Distinguished Professor in the Department of Visual Arts at Stellenbosch University and an accomplished artist with works in many galleries and private collections around the world. This particular work is especially poignant and appropriate to this book: It records the names of members of families who lived on several farms along or close to the Dwars River, (which runs from east to west across the centre of the Waterberg plateau) and who either survived or perished in one of the concentration camps of the South African War. The 91 names listed in the two panels include several members of Keith's ancestral family who lived on Zanddrift and Vier-en-Twintig-Rivier. The work is one in a series entitled *Between the Folds*.
- *Figure 10.15:* Detail from "Heuningfontein Store" by Lyn Wadley (2006), watercolour. Reproduced with her permission. A painter, mainly in oils, from an early age in Zimbabwe, Lyn held numerous exhibitions there. Later, she successfully exhibited works, principally Namibian landscapes, reflecting her life in remote corners of that country as the wife of an exploration geologist. This is one of all too few watercolours that Lyn has found the time to complete during our time in the Waterberg.
- *Figure 11.6:* "C.T.M. Wilcox – Caricature (1930)" by "Nemo", pseudonym of Daniel Cornelis Boonzaier (1865–1950). Boonzaier was a professional – and prolific – cartoonist based in the Cape. He worked for *Die Burger* for 25 years, and also contributed caricatures, including this one, to *The Sjambok*. At the time (1930), Wilcox was serving as Administrator of the Orange Free State. Image reproduced courtesy of Museum Africa, Johannesburg; Accession No. 65/3322.
- *Figure 13.3:* "Rain Storm near Hanglip, Waterberg" by Wenselois J. Coetzer (1993), oil on board; in the collection of the artist. Wens Coetzer (born 1947) trained and worked as an aeronautics engineer, whilst at the same time developing his skills as an artist in oils and pastels. Having been introduced to painting by his father, the well-known artist W.H. (Harry) Coetzer (1900–1983), Wens went on to participate in many local exhibitions. For over 30 years, he has owned a small farm on the eastern side of the Waterberg plateau where he now lives with his wife Gill.
- *Figure 13.10:* "Tabakskuur by Vier-en-twintig-rivieren" by Karel Bakker (1988), pencil and watercolour. Reproduced by kind permission of his widow, Prof. Terri Bakker, of Prof. Johan Swart, the current head of the Department of Architecture at Pretoria University, and of the web platform UPSpace, where I first saw the painting. Karel Bakker was a much-loved professor and head of architecture at the University of Pretoria. He died in 2014. The work is included in the art collection of the University's Architecture Department.
- *Figure 14.10:* "Mission station, Waterberg, Transvaal, 19 March 1885" a pencil sketch by Hermann Wangemann (from the collection of the Ditsong National Museum of Cultural History, Acquisition No. HG 6089). Dr Wangemann was Director of the Berlin Missionary Society from 1865 until his death in 1894 at the age of 76. He made two expeditions to visit the Society's

missions in South Africa – in 1884/5 and 1886/7. On each occasion, he completed numerous representations of scenes along his journey, as watercolours or as pencil sketches. The Ditsong Museum was fortunate to purchase 106 of these drawings (for R130) from a Berlin art dealer in 1954. Wangemann used engravings of some of the works from his first journey to illustrate a book describing his experiences. He may have visited the newly established mission at Modimulle whilst en route to Potgietersrust in mid-1867, but he certainly spent a week there in March 1885, during which time this sketch was completed. Details were obtained from a catalogue of Wangemann's works compiled for the Ditsong Museum in 1992 using the notes of Rita Botha.

- *Figure 14.17:* "Nagmaal by die Witkerk, Nylstroom" by Erich Mayer (1941), watercolour. Ernst Karl Erich Mayer was born in Karlsruhe, Germany, in 1876 and studied architecture in Berlin. He came to SA in 1898 and fought with the Boers in the South African War. He was captured at Mafeking in 1900 and sent to St Helena, being repatriated to Germany after the war. His poor health benefitted from the southern African climate, but he was not permitted by the British to return to SA, so went instead to German SWA where he lived from 1904–1911. After Union, he was allowed to return to SA. Later, he married and travelled around the rural parts of the country by ox-wagon with his wife, camping and painting rural scenes. He died in 1960, leaving no family. His work is regarded as an important reflection of early twentieth century rural life; his architectural training resulted in numerous accurate and reliable depictions of rural homesteads from that time. A large collection of Mayer's work (over 900 pieces) is housed at the Ditsong National Museum of Cultural History in Pretoria. The work illustrated has Acquisition No. HG 7014 404 and is one of at least four paintings (and numerous sketches) by Mayer over a 20-year period, depicting *nagmaal* on the *kerkplein* at the NGK's *Witkerk* in Modimolle.
- *Figures 15.8 & 16.3:* "General de la Rey" and "General Louis Botha", charcoal sketches made in London in 1902 by the Dutch artist Antoon van Welie (1866–1957) during a fund-raising visit to Europe by the three Boer generals De la Rey, De Wet and Botha following the Peace of Vereeniging on 31 May 1902. Van Welie was best known for his portraits of European celebrities of the time, including Popes Pius X, Benedict XV and Pius XI, Sarah Bernhardt, Isidora Duncan and Benito Mussolini. The two sketches included here – of Botha (Acquisition No. HG 9374, dated 20 October 1902) and De la Rey (Acquisition No. HG 9361, dated 17 November 1902), form part of a collection of Van Welie's works acquired by the Ditsong National Museum of Cultural History, Pretoria, in 1965. Details about Van Welie were obtained from a note compiled by Jeanne Joubert for the Pretoria Art Museum.
- *Figure 18.9:* "Reduction Works at Union Tin Mine" by Roy Gardener (1985), pastel. Roy Gardener, who died in 2017, had painted essentially full time since 1969. He was what is known as a "traditional realist" and specialised in land- and seascapes. His attention to detail and accurate depiction led to his receiving commissions from most of the old South African mining houses for renditions of their operations. This work was commissioned by the then owners of Union Tin, Goldfields of SA. The chief geologist at Union Tin, Ian Pringle, purchased the painting from the company upon his retirement after the mine closed. Ian now lives in Bela-Bela and invited me to use the painting if I could obtain the requisite permission to do so. This was granted by Roy's son, David Gardener, in July 2017.
- *Figure 19.2:* "Bushveld near Thabazimbi" by J.H. Pierneef (1953) – oil on canvas; from the collection of the South African Reserve Bank, Pretoria (Accession No. 35548). The painting depicts the western Zandriviersberge (the Waterberg escarpment) in the background – the range known to nineteenth century travellers as the Mural Mountains. Jacob Hendrik Pierneef (1886–1957) was born in Pretoria, the son of a Dutch master builder. His artistic talent was apparent from an early age, both in Pretoria and in Holland, where the family lived during the South African War. Although his best-known works are in oil on canvas or board, he also

experimented with etchings and linocuts, as well as with renditions of San art, of which he was an early, enthusiastic exponent. He travelled to Europe and East Africa in the 1920s. On his return, he began to receive commissions for landscape panels, perhaps the most famous of which were the 32 mural panels he painted for the Johannesburg Railway Station (Park Station), commencing in 1929. He painted numerous landscapes of the Cape, but very few of the northern Transvaal.

- *Figure 20.12:* "Sir William Dalrymple – Caricature (1928)" by Edgar Arnold Packer (1892–1932). A London-trained cartoonist, Packer became a cartoonist for the Johannesburg *Star*, using the pseudonym "Quip"; he also contributed to *The Sjambok*. This image is reproduced by courtesy of Museum Africa, Johannesburg; Accession No. 50/753.
- *Figure 20.12:* "Johann Rissik – Portrait, 1966" by Dr Irmin Henkel (1921–1977); charcoal and Chinese white on a light brown background; presumably drawn from a photograph of Rissik, who had died in 1925. Henkel was a German orthopaedic surgeon and a self-taught artist who came to South Africa in 1951 and became known after completing a portrait of the then Prime Minister Dr D.F. Malan. This image is reproduced by courtesy of Museum Africa, Johannesburg; Accession No. 70/1089.

Publication

Several of the themes covered in this work have been presented in an informal visual format to audiences of the Waterberg Nature Conservancy over the years. I am grateful to my fellow members of this organisation for the opportunities to air the results of my research as it progressed and for their encouragement and support in urging completion of the project.

An early version of Chapter 15 about the skirmish on Zandfontein during the 1914 Rebellion was originally published to commemorate the centenary of that event, in the *Military History Journal* of December 2014, **19** (4), pp. 143–149. Ditsong Museums of South Africa, the publishers of that journal, edited by Susanne Blendulf, are thanked for their permission to republish the story here.

Several topics in the book were the subjects of a series of heritage sites commemorated during Heritage Month, September 2017, on the Facebook page of Marlene van Staden, the Mayor of Modimolle-Mookgophong Local Municipality, whose brainchild it was to increase public awareness of the diversity of such sites within the municipal area.

I was thrilled when Nicol Stassen of Protea Book House agreed to take on the publication of this work, appointing one of his senior editors, Danél Hanekom, to oversee the editing process. For an aspiring, first-time writer, the editor looms large as a formidable, intimidating cloud on the horizon. I was lucky to find in Danél someone who sensed from the beginning what sort of outcome she wanted to achieve and who gently, but persistently coached and coaxed me into complying with her objectives, while still allowing me enough latitude to appease my fragile ego! Danél's team included Hanli Deysel with her outstanding design and layout skills and proofreader Carmen Hansen-Kruger, whose meticulous attention to detail ferreted out several lingering gremlins. Martie Eloff provided logistical support and motherly care. And I must not omit reference to the magic of Frans Petersen, who managed to restore many faded or indistinct photographs to almost their original quality. Between them, the team transformed my rough, inconsistent material into the fine product you are holding. I could not have wished for a more constructive, supportive introduction to the publishing world and thank them for all they have done.

To my family, and especially to Lyn, my long-suffering wife, life-partner and closest friend, thank you for your innumerable contributions, debates and gentle critiques. Thank you for having tolerated so patiently and so encouragingly my interminable ramblings, repetitions and growing obsession with this project over the last decade and more. Its publication will have been accompanied by a collective sigh of relief. But wait! There are other stories to tell …

INDEX

1820 Settlers 211, 356, 602
1914 Rebellion *see* Rebellion of 1914
24 Rivers (Twenty-Four Rivers) *see also* Vier-en-Twintig-Rivier 148, 212, 237, 238, 240, **242**, 243, 274, 347, **350**, 378, **453**, 614; 24 Rivers (E.A. Davidson) school 245, **578**, 595, 608; Davidson/Peacock/Baber families 244, 245, 247, 248, 274, 309, 321, 335, 358, 440, 514, 517, 703; St John's church 243, **246**, 247, 314, 335, 468, 514, 515–**516**, 517–518

Aari River *see* Limpopo River
Aasvoëlkop **95**, 125, 129, 428, **647**
AB *see* Afrikaner Broederbond
Abbotspoort 201LR 82
Aberdeen-Angus cattle breed 316, 320
Abraham Kriel Kinderhuis (Children's Home) 485
Abrahamic faiths 526
Acacia galpinii 184
Acacia spp. 429
Ackermann, C.F. (Kerneels) 672
Ackhurst, C.H., principal **166**
Acts and Ordinances — Asiatic Amendment Ordinance, No. 29 of 1906, 574; Bantu Authorities Act, No. 68 of 1951, 54, 83, 568; Bantu Education Act, No. 47 of 1953, 168, 604; Co-operative Agricultural Societies Act, No. 17 of 1908, 446; Game Theft Act, No. 105 of 1991, 461; Group Areas Act, No. 41 of 1950, 168, 169, 216, 528, 567, 568, 606; Immorality Amendment Act, No. 21 of 1950, 567; Iron & Steel Industry Act, No. 11 of 1928, 621; Land Settlement Act, No. 12 of 1912, 101, 323; National Parks Act, No. 56 of 1926, 667; Natives Land Act, No. 27 of 1913, 314, 709, 714; Peri-Urban Areas Health Board Ordinance No. 20 of 1943, 219; Population Registration Act, No. 30 of 1950, 567; Probationary Lessee Scheme Act, No. 38 of 1924, 328; Prohibition of Mixed Marriages Act, No. 55 of 1949, 567; Promotion of Bantu Self-Government Act, No. 46 of 1959, 568; Railways Construction Act, No. 30 of 1922, 381; Reservation of Separate Amenities Act, No. 49 of 1953, 528, 567; Restitution of Land Rights Act, No. 22 of 1994, 340–341; Settlers' (Land Settlement) Ordinance, No. 45 of 1902, 211, 311, 552; South Africa Act of 20 September 1909, 556

Adam Matsega's Kraal 61
Adansonia digitata see baobab tree
Adlum, James 775
Adlum, Louisa 775
Adlum, Magdalena 586, 775
Adlum, R.J. 585
Afguns **663**
Africander cattle 356
African Explosives & Chemical Industries Ltd (AECI) 439
African horse sickness (AHS) 426–427
African Independent/Initiated Churches (AIC) 499–500
African National Congress (ANC) 230, 528, 575, 649, 664
African Parks Foundation 466
Afrikaanse Protestanse Kerk (APK) 498–499
Afrikaanse Taal- en Kultuurvereniging (ATKV) 208
Afrikander Proprietary Gold Mines 618
Afrikaner Broederbond (AB) 363, 368, 526, 561–563, 572, 596, 604
Afrikaner nationalism 208, 484, 557, 563, 621
Afrikaner Party 567
Agri SA 459–460
Agricultural Journalists Association 249
Agriculture, Department of 101, 274, 440, 657
Aha Hills 694
Aigams (Windhoek) **683**
Albasini, João 64, 70, 88
Alberts, Ben Snr 649
Albertyn, Ds. J.R. 493
Alem dam site **338**, 340
Alheit, Ds. 488, 492, 493
Alkantrant 535LR **25**, **36**, **338** *see also* Uitkomst 507LR
Alldays 619
Alma **95**, 365, 369, 370, **371**, 375, **386**, 446, 451, 497, 515, 569, 614, 619; Central School **578**, 595; rail siding **117**, 385, **386**, 390, 394, 446; Alma valley **95**, 343, **364**, 429, 718 *see also* Koppie Alleen
Altefeste, Windhoek, Namibia 691, **693**
Altenroxel, Heinrich 312–313
AmaNdebele 199
Amandebult platinum mine **647**
American Pentecostal Church 500
anaboom, ana tree 47, **48** *see also* Livingstone's trees
Anderson Hall 188, 491

Anderson, David 188
Andrews, Rev Stanley 515
Anglican Church — Anglican Diocese of St Mark the Evangelist 518; Anglican Ladies Guild 516; Pretoria Diocese **511;** St John's 24 Rivers church **242**, 243, **246**, 335, **350**, 468, **516**, 517, 518; St Mary's, Warmbaths 516, 518; St Michael's, Modimolle 512, **513**, 514
Anglo-Boer War *see* South African War
Anglo-French Exploration Company 630
Anglo-French Land Company of the Transvaal Ltd **435**
Anglo-Rand Mining Finance Corporation 634
Annecke, Dr Siegfried 416
Anopheles funestus 417, 420
Anopheles spp. 412–414, 417
apartheid 45, 49, 168–169, 207, 216, 227, 314, 450, 498, 498, 526, 560, 567, 572, 602–604, 714–715
Apie (peanut butter brand) 170, **450**
Apostolic Faith Mission (AFM) 500, 649
Appingendam 805LR **670**, 673
Areas Reservation, Immigration and Registration Bill of 1925, 574
Armistice 133, 246, 377, 382
Armsorgraad (Poor Relief Board) 493
Armstrong, Edgar 186, 189
Armstrong, Robert MacLachlan 211
Artemisia annua 413
Artemisinin 413
Asalaam *see under* Nylstroom
Athlone, 1st Earl of **383**, 384
Aukema, Jan 19, 35, 37, 39, 40, 41, 651, 727, 729, 734
Aukema, Wilna 650, 656, 727, 733, 734, 780
Australopithecus africanus 19–20
Averre, Eleanor (Annie) (née White) 324, 325, 335
Averre, Kate Elizabeth Maria (née Overton) 324
Averre, Major William James 324, 325, 326, 331, 332, 333, 334, 335, 339, 351
Averre, Sidney H. 324
Averre, William Frederick 324
Avondale 665KR 347

Baber, Alfred Henry 245, **246**, 247
Baber, Anthony 249
Baber, C.E. (Charles) 16, 227, 245, 247, 248, **248**, 249, 250, 251, 390, 442, 443, 453, 456, 517, 720
Baber, Charles Edward, lawyer 245, 247
Baber, David 248
Baber Eulalia 247
Baber, Dr Rupert 249, 467
Baber, Dr Tanya 228

Baber, H.C.F. (Colin) 16, 227, 245, 247, 248, **248**, 250, 251, 442, 456, 460, 517, 596, 720
Baber, Jennifer (m. Pullinger) 245, 247, **248**
Baber, Joan (née Fraser) 244, 248, 250
Baber, Juliet (m. Calcott) 249
Baber, Lois (née Davidson) 245, **246**, 247, 517
Baber, Nina (née Wynne) 249
Baber, Ruth 246
Bad se Loop (Badzynloop) 174, **176**, **185**
Badenhorst, Commandant C. 108, 266
Baden-Powell, Colonel Robert 267
Bafokeng 53
Bailey, Abe 180, 553
Bailey, John Henry 655
Bain, Thomas 711
Baines, Thomas 48, 404–405, 660; travel route 1871 **403**
Bakeberg Masibi scheduled area 314
Bakenberg (Bakenberg Langa) 32, **67**, 74, 77, 78, 80, 82, 298
Baker, Sir Herbert 468, 515
Bakgaga 49
Bakker, Anton, chemist 194
Bakker, Gerrit II 88, 163, **171**, 194, 561, 566
Bakker, Prof. Karel **453**
Bakker's Pass (Bakkerspas) 88, **95**, 97, 111, 127, **647**
Bakone 49
Bakwena 49, 53, **60**
Balkis Land Company 311
Ballindalloch Farms **320**
Bamangwato 54, 59, **60**, 697, 706
Bangor 664KR 347
Banning (Bannink) family **241**
Bantu Administration & Education, Minister of 572
Bantu Affairs, Department of 222, 605
baobab tree (*Adansonia digitata*) **38**, **650**
Bapedi 49, 503, 505
Bar's Kuil 124, **128**
Barnard, J. 585–586
Barnard, Jan Hendrik 666
Barnard, Martin & Anna 586
Barnes se Winkel **357**
Barnes, Basil **358**, **359**, 360
Barnes, George & Mary **357**, **358**
Barnes, Lennox 357
Barnes, Magdalena Petronella ("Baby") (née Joubert) 356, **359**, 360, 361
Barrow, Rev. James 521
Basters 199, 685
Basters Nek 141
Basutoland (Lesotho) 409, 501, 502

Batawana community 694–695, 697
Batswana 49, 80
Bauhinia galpinii 184
Baumann, Max 633
Baverkuil 78KR 233, **234**
Baviaansdraai 587LR **670**, 672
Baxter, L.M. 569
Beauty 56LQ 54, **65**, 80, 216, 399, 597; Beauty Road 232, 236, 597, 606 *see also* Ga-Seleka
Bechuanaland Protectorate 215, 219, 239, 261, 264, 301, 302, 303, 390, 392, 404, 408, 438, 502, 660, 688, 692, **693**, 694, 695, 696, 697, 700, 701, 703, 704, 705, 706, 709, 714 *see also* Botswana
Beckett, T.W. 519
Beechwood 398KR 381
Beiderbecke, Heinrich 677
Beit, Alfred 347
Beith, Patricia & Peter 346, **456**, 457
Bekker, J.P. 491, 595
Bela-Bela *see also* Warmbaths 13, 107, **148**, 149, 155, 196, 199, 203, **207**, 209, 518, 613, 628, 636, 643; APK 499; local municipality **573**, 632; hot springs 155, 545, 643; town 519, 520, 528
Belgium, Treaty with 284
Belhar Confession 498
Bell, Catherine 94
Bell, Elizabeth (Gladys)
Bell, Henry 101–103; grave site **95**, 103
Bell, Henry Jr (Harry) 102
Bell, Margaret (Mary) Stewart (née Galloway) 101
Bell, Margaret (Peggy) 101
Bellevue 74LQ 402
Bellevue 98KR & 99KR **242**, 247, 248
Bellevue Kosher Hotel 524
Benade (Benadi/Bernardie) family **241**
Benedictine Order of Monks 510
Berg, Ds. Albert 492
Bergdamaras 685
Bergh se Bad 195
Berlin Missionary Society (BMS) 32, 51, 70, 72, 74, 75, 76, 81, 137, 173, 477, **478**, 479, 603, 604, 700 *see also* Modimulle Berlin Mission
Berrange, Toby 190
Besley, 2nd Lt Arthur Charles Gordon 371, **374**, **375**
Best, Lt Alexander Archie Dunlop 179
Beyer, Rev. Emil 478
Beyers, General Christiaan Frederik 79, 104, 108, 179, **261**, 262, 266, 299, 305, 349, 363, 367, 375, 481, 483, 540, **541**, 543–544, **552**, 553, 556, 558
Bezuidenhout, Waterberg UP candidate 566
Bham's Store 208

Bididi *see* Bobididi
Bilharz, Dr Theodor 421
bilharzia (also schistosomiasis) 351, 421, 423, 587, 590, 651, 719; distribution map **422**
bittereinders 79, 105, 482
Blaauwbank 2KR **349**
Blaauwbank 3KR 242, 243, **286**, 294, 296, **296**, 297, 304, 314, 347, **349**, **700**
Blaauwbank 515KQ **20**, 24
black education 168, 602, 604, 605, 607, 608
Black Hand, the (a gang) 326
black labour 310, 330, 382, 386, 430, 633
blackwater fever 413, 419, 663; distribution map **418**
Blair, Major Hunter 306
Blignaut, Maria Magdalena (m. Lee) 360, 660
Blinkwater 177KQ **234**, 235
Blinkwater 820LR 82
Blokland, Beelaerts van – *see* Van Blokland, Beelaerts
Bloklandspruit (Kgokong River) 285, **286**, **338**
Blood River, Battle of 63, 365, 367, 562–563
Blouberg 121, 624
Blue Star Line 316–317
BMS *see* Berlin Missionary Society
Bobididi 51, **65**, 82, 700
Bochum 47
Boekenhoutskloof 187KR 262, **350**, 365, 367, 369, 370, 610
Boeyens, Jan 19, 39–42
Böhmer, W.H. 595, 598
Bok, F.J. Eduard, State Secretary 356
Bokpoort Pass 79, 174, 177, 182, 183, 237, **350**, 614
Bolus, Dr Harry 583
Bondelswarts 382
Bonsmara, a breed of cattle 16, 17, **456**, 457
Boonzaaier, Boon 158
Boonzaier, Daniel Cornelis 388
Bornman, W.R. 598
Bosch, P.B. 593
Boschdraai 60KR 132, **242**, 247, 248, 251, 619
Boschhoek 579
Boschoek school **578**
Boshoek 131KR 241
Boshoek 621
Boshoff, Carel Frederik 356, 379
Boshoff, Carel Jr 368
Boshoff, Carl 303
Boshoff, Christiaan Hendrik 356
Boshoff, F. 475
Boshoff, Frederik (Groot Freek) 367, 446, 551
Boshoff, Jacobus (Koos) 356

Boshoff, Willie 367, 580
Bosman, Ds. H.S. 412, 474, **475**
Bosman, Herman Charles 224–225
Botes, Anton 114
Botes, Burgher Frederik Johannes 531, 533, **534**, 769
Botes, Petrus 231
Botes, P.W. 193
Botes, Pieter Willem (Oom Willie) **114**
Boteti River 697
Botha, Acting Commandant Theunis 531
Botha, Ds. J.C. 493
Botha, General Louis 100, 104, 139, 243, 299, 382, 425, 484, 529, 540, 542, 548, 551, **552**, 553, 556–557, 559–560, 714
Botha, J.J. 653
Botha, J.L. 585
Botha, Johannes 586
Botha, Koos 672
Botha, President P.W. **506**, 507, 572
Botha, "Trader Horn" 646
Botha's Commando 531–532, 537–538
Botswana 13, 59, 60, 230, 252, 401, 411, 638, 642, **696**, 710 *see also* Bechuanaland
Bousfield, Bishop Henry Brougham 510, **511**, 512
Boyd, H.C. 311, 314
Boyne **501**, 502
Brabant's Horse 111, **112**
Braemar Castle 324
Brakfontein 16KR 294, 346, 347, 349, **349**; Laager 258, 294, 295; police post 269, 270, **286**, 325, 332, 346, 347, **441** *see also* Dorset
Brandt, Ds. L.E. 483, 490
Braun, Prof. L. 431
"Breaker" Morant 713
Bredell, Inspector H.C. 269–270
Breedt, Martha Magdalena Carolina (née Pelser) 660
Brereton, Major 91
Bresler, Ross 601
Breytenbach, S.W. (Schalk) (*Meester*) 666–669
Brink brothers, Guillaume & Cornelis 85–86, 127
British South Africa Company (BSA Co) 371; Sleeping Sickness Commission 402
Brits, Frikkie **359**, **494**, **592**
Brits, Gert Pieter 356
Brits, H. 589
Brooks and Froude, surveyors 121
Brown, Edward 329
Brown, Rev. H. Milton 519
Brown, William 224
Brown's Cutting **329**, 351

Brown's Hotel, Warmbaths **206**
Brucoporko, Marcel 448
Buchanan, David Dale 210
Buckley, Frederick 675
Budler, B.D. 569
buffalo 23, 85, 122, 339, 406, 408, 410, 424, 425, 463, 670
Buffalo Fluorspar **185**, 613, **627**, 636, 637, **637**, 638
Buffalo River 50, 63
Buffelfontein 65KR **242**, 248
Buffelsfontein 237KQ **25**, 39, 40, **41**, 53
Buffelsfontein 347KR 636
Buffelshoek 351KQ 621
Buffelshoek 41KR 356, 389
Buffelshoek 446KQ 61
Buffelshoek 546LQ 131
Buffelshoek spring 644
Buffelspoort 280KQ 108, 111, 127
Bulawayo Road 529
Bulge River 198KQ 230, **231**
Bulge River 53, 131, 220, 230, 231, **231**, 232, 591, 614
Bulgerivier area 35, 40, **41**, 132, 148, 219, 230, 231, **231**, 232, 274, 277, 429, 495, 544, 653; agricultural union 443; school **578**, 591, 594
Bulhoek (farm) 382
Buller, General Redvers 260
Bulpin, T.V. 617–618
Burgers, President Thomas François 47, 137, 165, 201, 284, 547, **577**
Burgersdorp 472
Burkea africana 354, 429
Bushman Rock Shelter **20**, 23
Bushmen *see* San
Bushveld Carbineers 245
Bushveld Igneous Complex 625, **627**, 634–636
Bushveld, the 36, 278, 391, 412, 419, 492, **437**, 648
Butcher, Burgher A. 532
Button, Edward 72
Buxton, Lt Ronald Henry 108–110; family members 110; grave site **95**, **109**
Buys, Coenraad de (also De Buys) 197–199
Buys, Deon 609
Buys, Gabriel 199
Buysdorp 199
Buyskop *see* Warmbaths
bywoners 159, 232, 238, 351, 434, 551, 576

Caine, William Henry Atwood — diary 134, 257–264, **293**; family background 291; in East Africa 310; New Belgium sojourn 291–295, **296**, **297**, 298–307

Cambridge, Major General Alexander, 4th Governor General of SA 384
Campher, Hans 570–571
Canvas Town 294–295, **286,** 299, 301, 302, 326
Capricorn High School, Pietersburg 321, 595
Cardwell, Norman, principal 165
Carey, Rev. C.S. 512
Carmel Hotel, Warmbaths 524
Carnegie Commission 564
Carter, Bishop William 512
Carwell, N. 580
Castera, Captain 307
cattle ranching 313, 318, 455, **456,** 458, 656
Cave of Hearths **20,** 21, 23
Celliers, Charl Andries 241
Celliers, Jan 138
Central Sandy Bushveld 429
Chain, Burgher J. 532
Chaney, Frances Mary
Chaney, John & Winifred
Chaney, John Everard
Chaney, John William 92, 146, 215, 270–275, 277, **442**; Chaney's hotel & store **272**, **273**, **275**, **276**; constable at Rhenosterpoort police post **271**; Vaalwater mill & store **215**
Chaney, Mary 211
Chaney, Winifred 276–277
Charles Hope 260KR 79
Chazdan, Rabbi 525
Chief's Island, Botswana **693,** 695
Choate, Rev. Denis 521
Christelik-Nasionale Onderwys (CNO) *see* Christian National Education
Christian Catholic Apostolic Church (CCAC) 500
Christian National Education (CNE) 165, 238, 560, 580, 596, 605
Church of Our Lady of Mediatrix 510
Church Ordinance of 1804 469
Cilliers, Sarel 63
Cinchona pubescens 413
Clanwilliam 73MR 405
Clarens Formation (sandstone) 199, 677
Clarke, Elizabeth (née Davidson) 274, 294, 321, 703
Clarke, Rev. Frank 514
Clements, General R.A.P. 543
Clermont police post 346, 347, 698 *see also* Villa Nora
Clover, Newlyn 442
Cochrane, Rose **293**
Coetzee, Constable, SAP 189
Coetzee, Dr M.J. 168
Coetzee, Johannes Cornelius 221
Coetzer, Wenselois **461**
Coggin, Rev. Kenneth 521
Cohen & Widder, prospectors 617
Cohen, Mickey 524, 601
Cohen, Philip 186, **187**, 188, 193, 633
Colenbrander, Lt Col Johan W. **306**
Colley, General George 547
Collins, Ernest Olferman (& Emma) 121, 151, **151,** 152, 153, 473, 474, 485, 546; owner of Elandsbosch 372KR 366; owner of Waterval 190KR 365
Collinson, Roger 338
Combretum spp. 429
Commando Path 402–404
Compensation Claims Commission 267, 310, 436
concentration (internment) camps 376, 414, 540, 690–691 — Howick 556, Irene 361, Nylstroom 159–161, 168, 239, 241, 361, 481, Pietersburg 168, 241
Connaught, HRH Prince Arthur of 705
Conradie, Ds. 497
Consolidated Lead & Copper Company Ltd 618
Cooke, Henry, Nylstroom Camp Superintendent 159
Cooper, Harry 225
copper 36, 42, 618, 628
corn 446, 451
Corner House Group 697
Cornwallis Harris *see* Harris
cotton 140, 243, 249, 311–314, 352, 433, 451, 656
Courtney-Acutt, Nancy (née Lovelock) 177
Cousyn brothers 189
Crassula cymbiformis Toelken 98
Cremartardfontein 563LQ 256, 659, **663**, 664
Creswell, F.H.P. 557
Creswell, Walter 469
Crocodile River 57, 399, **403**, 529, 620, 621, 642, 646, **647**, 649 *see also* Limpopo River
Cronje, Willem 536
Cronning, Burgher Bartholemew Piet 531, 533, **534**, 769
Crook's Corner 642
Cruywagen, W.A., Deputy Minister for Bantu Affairs 222, 672
Culicoides imicola 427
Cullinan area 448, 569
Cullinan, Sir Thomas 101
Culmpine Estate 212, **214**
Cumming, Gordon 398–399, 401; travel route 1846, **403**
Curlewis Mrs A.H. 569

Currie, Richard 219, 380
Cuyler, Eben 228
Cyferskuil (Radium) 222

Daggakraal 591LR **670**, 671, 672
Dalrymple, Sir William 630, 698, **699**, 701
Daly, Jean, attorney 157
Damant's Horse 211
Davidson, Douglas Haig
Davidson, Edward Alexander (Ted) 17, 101, 215, 227, 244–245, **244**, 247, 248, 274, 321, 335, 358, 378–379, 389–391, 416, 440–441, **442**, 512, 514, 703; E.A. Davidson Primary School, 24 Rivers **242**, 245, 608
Davidson, Elizabeth (m. Clarke) 514
Davidson, Jane (m. Gaylard) 514
Davidson, Lois (m. Baber) 245, **246**
Davidson, Molly (née Fawssett) 242, 244–246, 274, 300, 302, 512, 514, 515–517
Davidson, William 101; grave site **95**
Davis, R., horticulturist 212
Dawkins, Lt Col J.W.G. 108, 266, 305
DDT *see* dichlorodiphenyltrichloroethane
De (Die) Volkstem, newspaper 137
De Bad 1198 200–201, purchased by the State **202**
De Beer, Abraham 88
De Beer, General I. du Plessis 254
De Beer, J.C., road inspector 254
De Beer, Jacomina Elizabeth 85
De Beer, Jan Adriaan 667
De Beer, Jan du Plessis, LV for Waterberg 300, **550**, 547
De Beer, Johanna Petronella 86, 97
De Beer, Stephanus Arnoldus 86
De Beer, Zacharias, Landdrost 165
De Buys, Coenraad *see* Buys
De Clerck, F.C. 210
De Jager, Naomi 225
De Klerk, Dr Annemie 467
De Klerk, Frederik Willem (F.W.), President 507, 572
De Klerk, J.C. 103, 237
De Klerk, Susan (m. Strijdom) 560
De Klerk, Willem A. 180, 186
De Kock, Michiel 586
De la Rey, General Jacobus Herculas (Koos) 104, 139, 482–483, 540–543, **541**, 548, 551, **552**, 558
De Lange Commission 607
De Lange, J.L., attorney 157
De Loskop 205LS 24
De Meillon, Dr Schalk Jacobus Botha 416–420, **417**

De Mist, Governor 469
De Naauwte 393KR **371**, 374
De Nyl Zyn Oog 423KR 86, **148**, 150, **162**
De Ridder, Ds. J.A. 474, 489
De Villiers, R.J. 101
De Wet, A.H., magistrate 648
De Wet, General Christiaan 483, 529–530, **541**, 542–544, 551, **552**
De Wilde, Burgher Albert Cornelis 531, 533, **534**, 769
Dekema, Dr K.J. (Dykema, Dekeman), Nylstroom District Surgeon 163, 415
Delegorgue, Adulphe 400–401; travel route 1844 **403**
Delfos, Cornelis Frederik 620–621
Den Boerenvriend, a newspaper 158
Dennison, Jerome 179, 190
Depression, Great, 1929–1933, 95, 195, 224, 323, 345, 355, 390, 434, 485, 512, 564, 601, 630, 635, 662, 667, 671
Dercksen, Hennie 337
Derdekraal 1138, New Belgium **286**, 319, 324, 326, 332, 334, 335, 339
Desborough, Lawrence Vivian 287–290, **288**; family members 287, 289
Devanner, Burgher H.J. 533, **534**, 769
Devonport, H. **186**
Devonshire Block **350**, 356
Dewrance, Burgher Herbert John 535, 770
Dewsnap, Lance Corporal 266
diabase 36, 428, 618, 718 *see also* dolerite
Diamant 228KQ **25**, 35, 443, 659
Diamond V Ranch 658–659
diamonds 140, 280, 619
Dichepetalum cymosum, gifblaar 323, 334
dichlorodiphenyltrichloroethane (DDT) 416, 420, 425
Dichrostachys cinerea (sickle bush) 430, 463–464
Die Hoofstad, a newspaper 572
Die Monitor, a newspaper 225
Die Moot 367, 369
Die Morester, a newspaper **225**, 444
Die Noord-Transvaler, a newspaper 224–225
Die Oog 195, 643
Die Pos/The Post, a newspaper **226**, 227
Die Warmbad Nylstroom Pos, a newspaper 226
Die Zoutpansberg Wachter, a newspaper 223
Die Zoutpansberger, a newspaper 224, 225
Diemont, Nick 464
Diener, Eugene & Nicola 226
Diephiring School 606
Dietrich, Albert 238, 239

Dietrich, Alberta (Cornelis Elizabeth) 239
Dietrich, Captain Moritz 239
Dietrich, Engela 239
Dietrich, Franziska Carolina (née Jeppe) 239
Dietrich, Keith, Prof. 239, **241**
Dietrich, Susanna Josina (née Barnard) 239
difaqane (mfecane) 42, 44, 55–59, 62, 86, 630
Dingane 55, 57, 63
Dingiswayo, Mthethwa chief 55
Diplorhyncus condylocarpon 429
Dirker, Anna 495
Dirker, Carel 352
Dirker, Samuel 353, 494
Distant, W.L. 155
Distinguished Service Order (DSO) 373, 544
D'Nyala Nature Reserve 232, 256, 614, **663**, 664
Dönges, Dr Eben, State President 363
Donkerpoort 344KQ 623, **647**
Donkerpoort 406KR 141
Donkerpoort dam 170
Donnisthorpe, settler family 179
Doorndraai 282KR **185**, 599
Doorndraai dam 179, **185**, 429
Doornfontein 374KR **350**, 365, **366**, 367, 368, 369, **371**, 551; Geloftefeesterrein 114, 367; school 365, 368, **578**, 580, 595; skirmish 1901 **371**
Doornhoek 318KQ **647**
Doornhoek 342KR 177, 180, 182, **185**, 188, 633
Doornhoek Platinum Mines Ltd 625, 626
Doornkom 376KR 259, **364**
Dordrecht 578LR 61
Dorset 26, 61, **234**, 236, **248**, 268, 270, 332, 343, 346–356, 494, 618; police station 123, **234**, 258, 294, 325, 332, **338**, 343, **349**, 592 *see also* Brakfontein police post
Double R Ranch 335–336
Dove, Rev. Nathan 523
Doyle, settler family 179
Drenthe 778LR 82
Dreyer, Ds. J.A. 495
Dreyer, Magdalena Adriana (Lenie) 239
Driefontein 164KR 361, 583, 584
Driefontein 317KR 24, **25**, 196, 599, 601
Drummond, Willy 175
Drummondlea sliding 175
Du Plessis, Commandant Jan Harm (Oom Rooi Jan) 535–537
Du Plessis, Dr Felix 221
Du Plessis, J.A. 581
Du Plessis, Johann 225
Du Plessis, Lourens 667
Du Plessis, Manie 536

Du Preez, Burgher F.J.J. 531, 533, **534**, 769
Du Preez, Ds. J.L. 476, 481, 483
Du Toit, Ds. Dr Jacob Daniel (Totius) 472, 483, 577
Du Toit, Ds. S.J. 472, 577
Du Toit, Francois 368
Du Toit, Frans Petrus Johannes 237
Du Toit, G.J. 598
Du Toit, Percy 191, 637
Du Toit, Thomas 366–367
Dubbelwater, New Belgium 335
Duncan, Patrick, Minister of Education, Governor General **383**
Durbach, Dr 646
Dutch Reformed Church in Africa 499
Dutton, Major Frank 190
Dwaalhoek 185KQ **25**
Dwars River 119, **120**, 123, **128**, **214**, 237–238, **242**, 256, 294, 361, 378–379, 581
Dykema (Dekeman), Dr *see* Dekema

Earlier Stone Age (ESA) 21, 23–24
East Coast Fever (ECF) 425–426
Eckstein, Hermann Ludwig 241, 354
ecotourism 410, 428, 649
Eensaamheid fault 642
Eersbewoond 62, 86, 88, **148**, 149, 150, 153, **207**, 365, 499, 545
Eiseb omuramba 688–689, 692, **693**
Eiselen, Dr W.W.M. 604; Eiselen Commission 604
Elandsbosch 122KR 582; Elandsbosch School **578**, 582, 583, 584, 597
Elandsbosch 372KR **350**, 365, 366, 367, 369, **371**
Elandshoek 263KQ (Welgevonden Game Reserve) **95**, 100, **100**, 101
Elandspoort 151
Elaret 222
Ellesmere Domain 288
Ellis, Patric(k) Henry **661**
Ellisras *see also* Lephalale 149, 216, 219, 220–223, 227, 248, 363, 443, 444, 561, 645, 659–665, **663**, 701; churches 489, 492–493, 495, 517, 522; coal 638, 639; Marapong 639, 659; *Mogol Pos* 226–227; Onverwacht 227, 639, 664; railway 390–392, 623; roads to 232–233, 651, 667; schools 221, 571, 590, 663
Eloff, Emma Petronella 91
Eloff, F.C. 175
Eloff, Mrs E.P. 579
Eloff, Philippus A. 86, 88, 89–90
Emmanuel church 510
Emslie, Bea 226
Encephalartos eugene-maraisii 183

Engelbrecht, Koot 178, 193, 580
Engelmohr, Susanna **543**
Enslin, Anna Susanna (Sannie) (m. Trichardt) 88–89
Enslin, Catharina Maria (m. de Beer) 88
Enslin, Catharina Maria (née Grobler) 88, **89**, 97
Enslin, Christoffel Bernardus 85
Enslin, J.J. 446
Enslin, Jacomina (née de Beer) 85, 86, 87, 149
Enslin, Johan Adam I 16, 85, 150
Enslin, Johan Adam II 85, 86, 88, **89**, 91, 92, 96, 97, 105, 660
Enslin, Johan Adam III 88
Enslin, Johanna Petronella (m. Swanepoel) 88
Enslin, Jurgina (m. Eloff) 88
Epukiro omuramba **693**
Erasmus, (Rooi Pietie) Piet 622
Erasmus, Adolph (Dolf) 145, 180, 181, **186**, 191, 517, 625, **626**, 628, 633
Erasmus, D.E., principal 194
Erasmus, Danie 186
Erasmus, Johannes Jurgen Cornelis 675
Erasmus, Martha Magdalena (née Lee) 660
Erasmus, Pieter Barend Elardus 660, 661–662, **661**
Erasmus, Rooi Daniel 345
Erikssonia edgei, a butterfly 107
Erleigh, Norbert Stephen 329
ESA *see* Earlier Stone Age
Eskom 10, 172, 192, 216, 223, 340, 391, 631, 638–639, 641, 664
Etjo Formation 677–678
Euphorbia ingens, naboom, mookgopho 173
Exxaro Resources Limited 394, 623, 664

F.H. Odendaal Hospital 172
Faber, Anna Margaretha (née Nel) 233
Faber, Frans & Hennie 235
Faber, Petrus Frans 233
Faber, Rudolph & Isabella 236
Fadogia homblei, gousiektebossie 323
Faidherbia albida, ana tree *see* Livingstone's trees 47–48
Fancy 556LQ **22**, 23
Farrant & Thompson 660
Farrant, Dr Peter & Janet 17, 213, 227, 228, 244, 441, 459, 460
Farrant, Jack (John) 212, **442**, 452, 594
Farrant, Kathleen 593
Farrant, Major Rupert, DFC with Bar 227, 442, 456, 595–596
Farrant, Margaret (Garry) 213

Farrant, Rhoda (née Kirkman) 212, 593
Farrar-Anglo-French Group 698, 701
Farrar, George 553
Faul 582
Faurea saligna 429
Fawssett, Edith Anne 242, 244, **296**, **297**, 302, 309, 511
Fawssett, Katherine (m. Peacock) *see* Peacock, Katherine 293, 511, 515
Fawssett, Mary (Molly) 244, 300, 512
Ferguson, E.N., surveyor 281
Ferguson, G.W. 666
Ferreira, Rob 580
Ffrangcon-Davies, Gwen 560
Field Arme Fonds 143, 145
Field, Einar 140–141, **142**, 143
Field, Ludvig (Ludwig) Johannesen 140–145, **141**, **142**; donor of town hall **171**
Finch-Dawson, settler family 179
First Anglo-Boer War 132–134, 145, 165, 356, 507
First South African War *see* First Anglo-Boer War
Fischer, Abraham 551, 553
Fisher, Bishop Edward 515
Fisher, John William 210
Fitzpatrick, Percy 553
flue-cured Virginia tobacco (FCV) 452–453, 455
Fokeng **51**, 53, 59, **60**
Fold, The (an orphanage) 232
Ford, Ernest 633
Ford, L.P. 138
Forssman, Carl Wilhelm Gottfried 134, **135**; Forssman's hotel and store **257**, **264**
Forssman, Emelie 130, 134, 136
Forssman, Louis 134
Forssman, Magnus, Surveyor General 130, 132, **133**, 135, 137, 431
Forssman, Oscar Wilhelm Alric (OWA) 124, 129–137; Portuguese Consul General, Chevalier **133**
Forssman, Oscar William Alric Jr 132, 140, 666
Forssman, Siemsen & Zoon 156
Forssman's Pass 134, **135**, 177
Foster Gang 540
Fothane 52, **67**, 69, 77 *see also* Moordkoppie
Fouché, Dr Jim, State President 363
Fouche, P.R. 584
Fourie, Captain Josef (Jopie) 529–530, **543**, 558, 566
Francistown, Botswana **693**
Frank Martin & Co. (Pty) Ltd 637
Franks, Dr Kendall 159
Fraser, Joan (m. Baber) 148

Freeman, S. **186**
freemasonry **44**, 146–147
French, Major Robert 268, 310, 438
Frikkie Geyser dam *see under* Naboomspruit
Frikkie Meyer High School 649
Frischgewaagd 649/652LR dam site **338**, 340
Furman, Alex 213, 221, 227, 442, 444, 456
Furstenburg, Sarah, midwife 353

Gaberone, Botswana **693**
Galloway, Margaret (Mary) Stewart (m. Bell) 101
Galpin, E.A. (Jim) 180, 184, 189
Galpin, Ernest 184, 517
Galpin, Marie Elizabeth (née de Jongh) 184
Gam, Namibia **683**, 692, **693**
GaMahlanhle 500
Gamlanders (Gamjanders) 79, 105
Gana Hoek 111KQ 649, **650**
Gandhi, M.K. (Mohandas, or Mahatma) 574
Gansieskraal **33**, 35, **36**, **38** *see also* Thaba Nkulu
Garaty, H.S., Detective Inspector 79
Gardener, Roy **634**
Gardiner, Rev. John 521–522
Garvey, Marcus 709
Ga-Seleka 60, **65**, 236, 287, 301, 351, 399, **403**, 499, 710 *see also* Beauty
Gatkop caves 61
Gaylard, Jane (née Davidson) 514
Gaylard, Rev. Horace 514
Gaza kingdom 58, **60,** 62
Gedenk (Witfontein) 232, 235, 443
Geelhoutbosch 269KQ **652**, **654**, 655
Geelhoutkop 342, **342**, 345, **350**, 359, 360, 361, **362**; agricultural union 443; police post 269, 346, 425, **441;** school 578, 583, 584, 586, 587, 589, 594; Gen. Beyers's camp 108, 266, **350**
Geldenhuys, Captain E.R.J. 531, 533, **534**, 769
Geldenhuys, Koos 493, 662
Gemsbokfontein 132KR 241, 243, 314, 606
Gemsbokfontein 585LR **650**, 671
Gerber, Ben 636
Gereformeerde Kerk in Suid-Afrika (GK) — formation of 471–472, 482; Ellisras 489, 662; Matlabas 489, **654, 655;** Naboomspruit 491; Nylstroom **472**, **473**, 474, 476, 481, 483, 485, **486;** Vaalwater 491, **492;** Warmbad 489
Gerlachshoop 477
Gesuiwerde Nasionale Party 561
Gewald, Jan-Bart 695, 702
Geyser, Frikkie and family 178, 191, 192, 193
Geyser, Hendrik 180, 193
Geyser, Lt Col/Commandant Andries Hermanus 275, 531, 544; Geyser's Waterberg Commando 90, 531–532, 535
Gibb, Freddie 514
Gibeon, Namibia **683**
Gibson, G., Native Commissioner 648
Gibson, Rev. John 522
gifblaar 323, 334
Gilfillan, John 446
Gilfillan, W.H., surveyor 205
Gillbanks, Edgar 600
Gladstone, Lord, British Prime Minister 134, 556
Gladwin 636
Glen Alpine dam 569
Glossina spp. *see* malaria
Gloucester Regiment, 2nd 310, 438
Gobabis, Namibia 682, **683**, 688, 705
Goddefroy, Ds. M.J. **475**, 476, 481, 483, 490
Godfrey, Rev. 512
Goedehoop 83KR 277, 453
Goedgedacht 130KR 241
Goergap 113KR **25**, 26, 28, 29
Goetz, Andreas Marthinus 134
Gojela I, a Kekana regent 81
gold 72, 75, 83, 132, 138, 140, 182, 278–280, **290**, 291, 294, 351, 357, 382, 390, 420, 434, 439, 617–618, 666, 695, 697, 698, 701
Gold Fields of South Africa 630–632
Gondwana 638
Goodenough, Rev. Dudley 522
Goodwin, Rev. Walter 521
Goossens, Father Emmanuel 510
Gordon Highlanders 179, 307
Gordon, Lt Edward I.D. 304–305, 307, 345, 369–370
Gordon-Moore, Emily 375
Gorkum (Gorcum) 577LR (New Belgium) **282**, 333, 495
Goud River (Goudrivier) 24, 294, 304, **329**, **338**, 351
Graaff-Reinet 360, 471, 711
Graaff-Reinet 179LQ 661, 662, **663**
Graham, settler family 179
grain silos **117**, 216, 352, 394, 451, 458
Grass Valley chrome mine 624
Great Trek 365, 469, 471, 491, 562, 576, 602
Green, Dr Pierce 159
Green, M.G. 594
Greer, Ilva (née Webster) 454–355, 357, 359
Greer, Patrick 354
Grenfell, Lt Col H.M. 262–263, 266, 369
Grewar, a settler family 211
Grewar, Bertie 211
Grewar, I.B. 380

Grewar, R.A. 441, 444
Grewar, T.B. 441, **442**
Griffiths, H.B., Vaalwater postmaster 216
Grigson, settler family 179
Griqua 56, 57, **60**
Grobler, Aletta (née Potgieter) 489
Grobler, Andries Nikolaas (Juttie) 97
Grobler, Catharina Maria (m. Enslin) 88, **89**
Grobler, Frederik Albertus (Groot Freek), Commandant General 78, 103, **105**, **106**, 114, 198–199, 258–259, 261, 280, 555, 660; grave site **95**, **105**
Grobler, Frederik A. (Jr) **106**
Grobler, G.F. 580
Grobler, Jan Adriaan Pieter 200
Grobler, Johannes Hendrick 544
Grobler, Johannes Hendrik (Hansie) 97
Grobler, Nicolaas Johannes 86–87, 97
Grobler, P.G.W. (Piet), MP for Rustenburg 667
Grobler, Zacharias Christiaan (Grysman) 97
Groblersdal 57, 477, 522
Groenfontein 141KR (smallholdings) **214**, 223
Groenfontein 227KR 635,
Groenfontein 458KQ 87
Groenfontein 494LR 183, **693**, 698, 700, **700**, 701, **702**, 703, 709, 711
Groenfontein dam site **338**, 340
Groenfontein Tin Mine 635
Groenrivier 95KQ **652**, 655
Groenvlei Sentrale Skool 655
Groenvley 87KQ **652**, 655
Groenvley 230KQ **652**
Groesbeek (postal tree) 671
Grootegeluk 459LQ **663**
Grootegeluk coal mine 236, 390, 391, 392, 394, 623, **627**, 638, 639, **640**, **641**, 642, **663**, 664
Grootfontein 392KR **181**
Grootfontein 501LQ 662, **663**
Groothoek 278KQ 87, **95**, 97, 98, 544
Gros, H.F., photographer **66**, **76**, **150**, **174**, **200**, **437**, **479**
groundnuts 170, 249, 352, 428, 440, 447–452
Grundlingh, T.A., teacher 193
Gubbins, J.G. 636
Gujarat area/state 527, 574
Gunfontein 71KR 123, 349
Gwaša Cave **20**, 66

Haakdoorn 177, 179
Haakdoringdraai 220KQ 126
Haarlem Oost 51KQ **652**; police post 653 *see also* Matlabas

Haenertsburg 49, 145, 250
Hamakari, Waterberg, Namibia **687**, 688, 692, 694
Hamburg 644KR 95
Hamilton, Frederic 223
Hammer, Father Stephen 508
Hanglip **181**, **461**
Hans Strijdom dam 569, 639 *see also* Mokolo dam
Hans Strijdom Technical High School (Hansies) 193–194, 611
Hansen, Mauritz 194
Hardwick, Rev. James 520
Harley, Dr J. 421
Harmony Block 333
Harris, Captain William Cornwallis 399; travel route 1835, **403**
Harry Smith 200LR 54
Hartbeestfontein 511KQ 630
Hartbeestlaagte 525KR 258
Hartebeesdrift police station **231**, 232
Hartebeespoort 84KR 17, **214**, 216, 221, 452, 453; school **578**, 593
Harting, Prof. Pieter 201
Hartingsburg 155, 158, 201–202, **202**, **207**, 287, 508, 518, 579 *see also* Warmbaths
Harvey, director of TLC 280
Hawley, Edward 279
Hayes, Hugh 224
Hayward, F.E., Secretary for Lands 282, 328
Healey, Dorril 359
heartwater (disease) 424
Hebrew Community Hall, Warmbaths **524** *see also* Judaism
Helberg, H.L.B. 591
Henderson, Manie 123
Henkel, Dr Irmin, surgeon & artist **699**
Henning, T.J. 325, 331–332
Henry, Sister, matron 511
hensoppers en joiners 106, 483
Herbst, J.B.H. **186**
hereford, a breed of cattle 243, 456
Herero, the — concentration camps **693**, 702; flight to Bechuanaland/Botswana 690, 692, **693**; Hereroland 685; people 183, 676–682, 684, 689, 709–710; planned annihilation of 678–681; refugees 183, 632, 695–696; residence in Waterberg **700**, **702**; return to Namibia 702–709
Hermanusdoorns 233, 274, 495
Herstigte Nasionale Party (HNP) 570–571
Hertzog, Albert 570
Hertzog, General J.B.M. 83, 362, 382, 384, 385, 541–542, 544, 551, 553, 557, 561–562, 570, 621, 667, 714

Het (De) Bad 1198/465KR 155, 200–201, 203, 208, 489, 508, 545 *see also* Warmbaths
Het Volk Party 551
Heuningfontein 364; school **350**, **578**; store **350**, **364**
Heuningneskloof 174
Heuser, Dieter 339
Hewitt, Alex 695–698
Hewson, chairman of North Waterberg Farmers' Association 389
Hewson, Vaalwater residents 219
Heystek, J.D.P. 653
Heystek, Jooste 168, 195, 446, 567, 568, **568**, 569, 571, 580, 590
Hindon, Captain Jack 179, 266
Hitchcock, Burgher Henry Horatio 531, 534, 770
Hlubi region of Nguniland 50, 51; Hlubi people 56
Hobhouse, Emily 482, 714
Hoffman, Dr J.J. 156
Hoffman, S.J. 581
Hoffman, S.S.B., principal, Laerskool Vaalwater 219, 422, 581, 595, **596**, 598
Hoffmann, field cornet 532
Hofmeyr, Eerw. Stephanus J.G. 499
Hofmeyr, Jan 714
Hofmeyr, Jannie, Administrator of the Transvaal 88
Holloway, E.C. 274
Holtzhausen, C.A. 593
Holtzhausen, I.C.H.O. 215, 219, 222
Hoopdaal 96KQ 653
Hoornbosch 439LQ 493, 663, **663**; school 661, 663
Hopewell 229KQ **652**, 656, 657, 658
Horak, Ds. 494, 595
Horn, Johannes 79
Hotine, John 636
Howard, W.R. 236
Howe, Ann 146, 272, 274
Huffman, Prof. Tom 35, 37, 40
Hughes, George F & Company 262
huguenot, a breed of cattle 16
Hulpmekaar Drift 621
Hume, David 399
Humfrey, Lt, S.A.C., Nylstroom 161
Humphris, Alma Evadne (m. Todd) 319
Huneberg, Major Joseph 524
Hunter, Elizabeth **147**, **228**, 244, 294, 516
Hunter, John 660
hunter-gatherers 27, 32–33, 40, 42–43, 56
Hurter, W.T. 594, 598

Hurutshe 51, 53, 57, 59, **60**, 82, 716
Hwaduba **60**

Immerpan 320, 365
Impey, Maria 165
In Den Berg, New Belgium 335
Indian families, early 527
influenza, "Spanish" flu epidemic 166, 194, 484
Ingram, Dr Alex 416–417
internment (concentration) camp 482, 691, 713 *see also* contration camps
Iron Age (pottery, smelting, walling) 20, 27–44, **36**, **41**, 406, 617, 619, **629**, 630, 717
iron ore 36, 61, 391, 613, 620–625, 638, 646
Iscor 220, 390–391, 621–623, 638–639, 645–646, 648–649, 664
Islam 526–528 — Muslim families 527, 528; Nylstroom/Modimolle community 169, **527**; Warmbaths/Bela-Bela community 528

Jackson, Eric 517
Jackson, Murray 179
Jacobs, J.P., teacher at Vischgat/Zandfontein 325, 331, 334, 349, 351, **592**
Jacobs, Jan 227
Jaffet, Z. **186**
Jagger, J.W., Minister for Railways 380, 389
Jakhalskuil 754LR **650**, 671
Jakkalskuil Pass/road **650**, 671, **672**
Jameson Raid 138
Jameson, Dr Leander Starr 550
Jan Trichardt's Pass 88, 89, 92, **95**, 106, 114, 304, 617
Jan's Nek 123, 124, **128**
Janse (Jansen), Abraham Christoffel 79
Janse, Johannes Hermanus 79
Janse van Rensburg, Johannes Nicolaas 667
Janse van Rensburg, Willem 671
Jehovah's Witness sect 359, 584, 589
Jensen, principal 600
Jensen, Rev. Rasmus **479**, 480
Jeppe, Barbara (née Brereton) 140
Jeppe, Carl 138, **139**, 165, 548
Jeppe, Carl Wilhelm Friedrich 136
Jeppe, Franziska Carolina (m. Dietrich) 239
Jeppe, Friedrich Heinrich (Fred) 136, **137**, **435**, **403**
Jeppe, Herman Otto, first Postmaster General of ZAR 136
Jeppe, Julius 136, 138
Jeppe, Julius Jr 138
Jeppestown 138

Jerusalemgangers 86, 98, 149, 150, 152, 201
Jesuits *see* Roman Catholic Church
Jewish community 363, 523–526, anti-Semitism 525, 562, 601, 646; cemetery, Pietersburg 523; immigration 562, 714;
J.H. Dewhurst butchery chain 316–317
Jinnah Park *see under* Warmbaths
John Chaney's general store 272, 274
Johnson, Burgher William Stonewall 531, 534, 535, 770
Johnstone, John 240
Jones, Commandant 531–532
Jooste, Ds. Willie 498
Jordaan, H.S.E. 569
Jordan, Trevor 464
Joubert, Aletta (m. van Staden) 360
Joubert, Christina Johanna (née Blignaut) 361
Joubert, General Piet 54, 139, 154, 360, 476, 480, 547–550
Joubert, Eerw. P.J. 499
Joubert, Hannetjie 353
Joubert, Jan Hendrik 360–361, 667
Joubert, Magdalena Petronella (m. Barnes) **358**, 359, 361
Joubert, Pieter Gabriel 361
Jubilee Shelter, Magaliesberg 40
Judaism *also see* Hebrew, Jewish Community 523, **524**, 526
Jurrius, J.P. 579

K2–Mapungubwe 37
Kaalvallei 163KR **345**
Kaapsche Hoop 291
Kaapsvlakte dam site **338**, 340
Kaditshwene (*also* Tswenyane, Karechuena, Kurrichane) 57, 85, 716
Kafferskraal 168KR 582
Kafferskraal 55LQ 54
Kafue River 372
Kahimemua 695
Kaiser 679, 689
Kalahari 543, 558, 614, 677, **683**
Kalkbank **20**, 24
Kalkfontein 705
Kamaharero, a Herero chief 680, 694, 695, **708**
Kambazembi, a Herero chief 678, 682, 685, 710
Kameelfontein 108KR 345, 361, 585
Kameelfontein 4LR 405
Kanjunga, a Herero chief 685
Kareefontein 377KR 238
Karibib, Namibia **683**, 685, 691
Kariefontein 579

Karoo Supergroup **627**, 638
Kasteelberg 98
Kaufmann, storekeeper 175, 177
Kauraisa, Julius 703
Keen, Sergeant 189
Kekana 41, 49, 50, **51**, 52–53, 60, 64–66, 68–76, 78, 83, 717
Kemp, General Jan 483, 529, 543–544, 563
Kennedy, Dr Joel 194
Keta Game Ranch 299, **338**, 339, 668
Keyter, Theunis Cornelius 221
Kgaba 52, 64
Kgatabedi II (Bernard), a Kekana chief 83
Kgatla **51**, 59, **60**, 61–62, 68, 82
Khama III, Kgoši (chief) 254, 697, **706**
Khoekhoe herders 34–35
Khuba 53
Khumalo, a Zulu clan 55
King Edward VII 556
King George III 660
King George V 317, 325, 544, 558, 594
King, Captain, Waterberg Native Commissioner 73
Kings Own Yorkshire Light Infantry 265
Kirkman family 211–215, 274, 276, **442**
Kirkman & Armstrong 213, **215**, 274, 276
Kirkman, Eustace 444
Kirkman, Gilbert Rufus 276, **442**
Kirkman, J.B. 276
Kirkman, Walter (Harry) 348
Kirkman, William Rufus 211, 219, 440, **442**, 446, 593
Kirstenbos 497LR (New Belgium) **25**, 37, 39, 328, **329**, 330, **331**, 332, 339, 340, **665**
Kirstenbosch, New Belgium **282**, **286**, 299, 304, 325, 328
Kiss-me-quick-and-go-my-honey 794LR **670**, 675
Kitchener, Lord 376, 482, 483, 713
Klaas Zyn Nek 7KR 124, **128**
Klein Kariba resort 208
Klein-Nyl River 151, 169, 170
Klein Sand River 122, 252, 254, 270, 277, 345, **350**, 363
Klipfontein 157KR 241
Klipfontein 53KR **349**, **350**, 352
Klipheuwel 40KR 354, 356
Klipplaat 34KR **25**, 30, **31**, 132
Klipput 458LR **665**, 666, 667, **670**
Kliprivier 464KQ 95, **95**, 96
Klipspringer-Marsfontein diamond field 619
Klipspruit school 443, **578**
Kloof Pass 45, **46**, 47, 69, 343, 614, 670–675

Kloppers, Johannes 671
Kloppers, Oom Hansie 672
Kloppers, T.J. 491
Knapp 651KR 600
Knapp-Fisher, Rt Rev. Bishop Edward 517
Knightley, Phillip **316**, 322
Knobel, Rankin's Pass 92–94
Knopfontein 184KR (Alma) 116, 118, 446
Kobe, son of Chief Seleka I 54
Koboldt, Rev. Heinrich B.M.S. 121, 477–478
Koch, Ds. J.A. 493
Koch, Harold, immunologist 409
Koedoesrand 335, 361, 561, 665, **665**, 667, 670, **670**
Koekemoer, Kerneels 669
Koenap, New Belgium 239, **282**, **286**, 299, 306, 495
Koevoet 632
Kololo 56
Königsberg, SMS, German cruiser 372, **373**
Koperfontein 37KR 356
Koppie Alleen 359KR (Alma) 304, **386**; railway siding 116, 385, **386**
Korana 56–57, **60**, 198
Kotzé, Ds. C.R. 563
Kotze, L.C. 589
Kotzé, Sir John, Chief Justice 137
Kotzee, Dr A.L. 597
Krabbefontein 313
Kralingen **371**, 381
Kransberg 88, 98, 125, 428, 465, 647, 649, **652**
Kransberg agricultural union 443
Kransberg Nature Reserve/National Park 465, 649
Kranskop 86, **150**, **152**, 158, 473, 477, 528, 563
Krause, Rev. Oswald 478, **479**, 480, 603
Kriel, teacher **166**
Kroep, H.J., magistrate 158
Krogh, J.C. 580
Kromkloof 203KR **148**, 173, **181**, 183, 193, 259; school 193, **578**, 580, 581,
Kruger National Park 113, 348, 406, 411, 667
Kruger, Leoni 227
Kruger, Pieter Ernestus Johannes 660
Kruger, President Stephanus Johannes Paulus (Oom Paul) 90, 99, 139, 158, 203, 221, 474, 478, 523, 545–548, **549**, 577, 650, 667
Kuduveld trading store **350**, 357, **357**
Kühl, Berlin missionary 70
Kumba Resources Limited 623
Kuschke Game Reserve 72
Kutako, Hosea 704–705, 708, **709**
Kwaggashoek 345KQ 391, 620, 621, **647**
Kwalata Wilderness Nature Reserve 278, 320, **338**, 339, 351, **466**

Kwa-Makapane **25**, 44, 61
Kwarriehoek 584LR **670**, 671, 672
Kwarriehoek 588LR **350**, 352, 590; school **578**, 589
Kwena **51**, 59

Laas, Col. J.C.C. 563
Labour Party 553, 557
Lacanti family 601
Laerskool Eenheid 159, 168
Laerskool Ellisras 663
Laerskool Eugène N. Marais 193
Laerskool Vaalwater 216, 250, 421, 571, 581, 595–598, 606
Laing's Nek 547
Lake Ngami **693**, 695
Lamprecht, Albertus 180
Lamprecht, Anna Jacoba 180
Lamprecht, A.S. **186**
Lamprecht, Cornelius Johannes 180
Lancers, 17th 244
Lancers, 19th 110
Land Bank, Modimolle **172**, 190, 247, 248, 447, 457, 570
Landmans Lust 595LR, New Belgium **282**, **329**, 336
Landsberg & Forssman 157
Landsberg, Anna Maria (Mimi) 136
Landsberg, Carl & Catharina 130
Landsberg, Emelie Henrietta Amalia (m. OWA Forssman) 130
Landsberg, Otto 130, 136
Langa Ndebele 32, 33, 41, 44, 50, **51**, 52, 54, 64–65, 69–70, 73–74, 76, 77
Langa, Hendrik Madikwe, a Mapela chief 82
Langkloof 285KQ **95**, 106, 114, **114**, 115
Langkloof, Cape 197
Langschmidt, Eduard 193
Lanslots, Dom Ildefons 508
Lapalala Wilderness Private Nature Reserve 17, 33, 40, 61, 278, 336, **337**, **338**, 339–340, 464–467, 614, 618, 668
Lapalala Wilderness School 10, 17, 337, 721
Later Stone Age (LSA) 23, 26, **28**, 29
Laveran, Dr Charles 412
Law, Willy 522
Le Feuvre, Rt. Rev. Bishop Philip 518
Le Roux, Aletta (née van Heerden) 115, 116
Le Roux, Koos 102
Le Roy, Rt. Rev. Bishop Fulgence 510
Lebowa 45–50, 54, 78, 222, 340, 502, 568 *see also* Northern Sotho
Lebowakgomo 48, 502

Ledger, Lt Col 275
Ledwaba 49
Lee, Johannes Lodewikus (Jan) 660
Lee, John (Johannes) *also* John Hunter 660
Lee, Maria Magdalena (née Blignaut) 660
Lee, Martha Magdalena Carolina (née Pelser) 662
Lee, Rear-Admiral Richard 660
Lee, Rev. 522
Leeuwdrift 88KR **214**
Leeuwarden 633KR 558
Leeuwpoort 554KQ 631
Leibbrandt, Robey 565, **566**
Leipoldt, Dr C. Louis 412, 421, **582**, 583, 720
Lekalakala 48, 64, **67**
Lekganyane, Bishop Barnabas Ramarumo 505, **506**, 507
Lekganyane, Bishop Edward 503, 505
Lekganyane, Bishop Joseph 504, 505
Lekganyane, Bishop Joseph Engenas **506**
Lekganyane, Bishop Ignatius Barnabus (Engenas) 500, **501**
Lemmer, General 104
Lensly, F.D. & N.E. **442**, 446
Leopard Rock 345
Lephalale *see also* Ellisras 13, 23, 37, 117, 123, 149, 173, 227, 230, 232, 233, 242, 256, 356, 445, 452, 453, 645, 659–665, **663**, 670; churches 489, 493, 499, 522; coal and power stations 236, 340, 394, 613, 638, 639, 642, 659; local municipality **573**; Marapong 639, **641**, 659, **663**; roads and rail (R33 and others) 252, 351, 390–391, 394, 399, 476, 592, 623, 625, 651; schools 664
Lephalala (Palala) River 108, 119, 123, 124, 266, **286**, 303, **327**, **329**, 330, 342, 346, 347, **350**, 352, 377, 379, 389, 419, 495, 582, 590, 664, **650**, 666; archaeological sites 26, 27, 28, 35, 53; Lapalala Wilderness 336–340, **338**; malaria 351, 420; New Belgium 278, 283, 287, 294, 299, 304, **327**, 328, **329**, 333, 335;
    reservoir sites **338**, 340; settlements along 33, 49, 51, 53, 54, 60, 80, 82, 301, 352, 399, 499, 671, 698, 700, 710
Leseding *see under* Vaalwater
Lesiba Kekana 75, 83
Letsholathebe, Sekgoma 694–695
Letty, Cythna Lindenberg (m. Forssman) 135, 140
Leutwein, Major Theodor **680**, 681, 684–686, 691
Levine, Rev. J.I. 523, 525
Lewis, Rev. Hubert Collin 514, 516–517
Leyds, Dr W.J. 549
Leydsdorp 50
Lhea 534LR **286**, **329**, **338** *see also* Alkantrant

Libanon 653LR 132, **350**, 352, 354, 358, 488; school 350, 354, 355, **578**, 589
Liberal Party 714
Lichtenstein, Henry 197
Liebenberg, Christiaan 536
Liebenberg, Jan Johannes 544
Limburg 47
Limpopo Province 13, 19, 20, **20**, 21, 23, 24, 26, 35, 43, 225, 406, 407, 423, 424, 457, 458, 499, 500, 523, 525, 526, 575, 664, 677; Limpopo Provincial Department of Economic Development, Environment and Tourism (LEDET) 467, 664
Limpopo River 20, 23, 48, 62, 64, 71, 73, 104, 131, 198, 219, 264, 287, 295, 301, 302, 306, 311, 342, 379, 389, 390, 403, **404**, 410, 411, 434, 620, 638, 651, 653, 656, 660, 666, 718; archaeological sites: 20, **20**, 27, 30, 34, 35, 37; early travellers/settlers along river 53, 54, 58, 60, 399, 400, 401, 402, **403**, 405; diseases of 285, 396, 398, 400, 406, 407, 408, 409, 412, 419, 423
Limpopo Tobacco Processors 455
Limpopodraai agricultural union 443
Lincolnshire Free Press, a newspaper 272
Lindani Game Reserve 10, **25**, 354–355, 356, 389
Lindley, Rev. Daniel 469
Linsley, Burgher D. 532
Little, Mary Ann (m. Desborough) 289
Little Wonder brick-making machine 444, **445**
Livingstone's trees **48**
Livni, Dr Shlomo 646, 648
Livnidorp 649
Lleweni School 177, **181**, **578**, 597–598, **599**, 600–601
Lobengula, Matebele chief **403**, 405
Lodewyk brothers (Lodewyk & Erasmus) 249, 548, 592
Lombard, H.S., Commandant **296**, 297
Lombard, S.A. 215
London Missionary Society 469, 477
Loots, Marthinus Frederik 660, **661**, 663
Loram, Charles 603
Lorentz, Burgher H. 532
Loskop Dam 333
Lottering, Gert, Frans & Cornelis 121, 477, 480
Lotz, J.W. 598
Loubad 140, **350**, 369, 370, 371, **371**, 373, 375, 381, **391**, 643, 644
Louis Trichardt (Makhado), town 63, 191, 225, 452, 455, 523, 525
Lourenço Marques 62, 141, 143, 415 *see also* Maputo
Lourens, J.C.F. (Jan) 227, 442

Louw, Eric 562
Louw, Willem Petrus 235
Louw, W.J. 581
Loveday, R.K., LV for Barberton 548
Lowveld 47, 117, 219–222, 242, 248, 342, 347, 248, 351, 319, 412, 414, 425, 443, 445, 453, 455, 492–493
LSA *see* Later Stone Age
Luderitz, Namibia 691–692, **683**; Luderitz-Aus railway line 691
Luna Road, Warmbaths 519, 520, **524**
Lutheran Church 517
Lydenburg 63, 64, 133, 280, 511, 546; Lydenburg Commando 554; Lydenburg Gold Exploration 280; NGK parish 470–471, 473–474

Maas, Mike 495
Mabatlane *see* Vaalwater
Mabalingwe golf estate 209
Mabotse 44, 61
Mabula Game Reserve 209
Macabe, Joseph 405
Maclean 1046LS **501**, 502
Macouwkuil 45KR 356
Maddocks, settler family 179
Madge, Lt Col Charles Alfred **243**, 341, 347, 442
Madikana, wife of Chief Mokopane II 71
Madimatle mountain 61
Madlala Pan 405
Mafora, Sam 16, 598, 606, 608
Magagamatala **67**, **69**, 70
Magalakynsoog 201KR 367
Magaul area *see* Mokolo area
Magersfontein, Battle of 542
Maggs, C. 580, 635
magnetite 36, 619, 624, 625
Magope Hill 77, 81
Mahalapye (also Magalapye), Botswana 690, **693**
Maharero, Johanna **702**
Maharero, Johannes **702**
Maharero, Milka **702**
Maharero, Samuel, a Herero chief 679, 671, **680**, 688, 689, 692, **693**, 694, **696**, 698–708, 710, 714
Maherero, Alfred **702**
Maherero, Friedrich **696**, **702**, **709**
Maherero, Samuel, Herero paramount chief **680**, **702**, **706**, **708**
Maherero, Traugott 707
Mahmoud, Ahmed (Taraj) 527
Mahwelereng *see under* Potgietersrus
maize 16, 92, 127, 249, 299, 313, 332, 352, 394, 428, 443, 446, 447, 450–451, 458

Majebas (Matjibas) Kraal 1005LS 554, 556, 558
Majuba Hill 547, 554
Makabeng **20**, **65**
Makapan Valley 19, 20, **20**, 23; Cave of Hearths **20**, 21; Gwasa cave **20**; Limeworks **20**; Mwulu **20**, 23
Makapansgat 66–68, **67**, 70, 71 *see also* Gwaša cave
Makapanspoort 71
Makhado *see* Louis Trichardt
Makhado, chief 99
Makwela, principal 606
Malan, C., Minister of Railways 117, 386
Malan, D.F. 334, 561, 564, 566, 567, 574, 596, 604
Malan, Ds. Dirk 498
Malan, Harry 610
malaria 62, 71, 73, 121, 155, 160, 174, 184, 194, 198–199, 351, 355, 410, 411–413, **414**, 415–420, 423, 427, 478, 583, 651, 657, 668, 700, 720 ; distribution map **418**; *Glossina morsitans morsitans* 398, 402; *Glossina pallidipes* 398; *Glossina* spp. distribution map **407;** Imperial Malaria Conference 413; *malaria* outbreaks 71, 414, 420, 427
Malebana community 82
Malherbe, Dr Ernest 564
Malmanies (Mamane) River 131, 230, **231**, 232
Malokong mission 32
Mamiaanshoek 297KQ **95**, 111
Manala 49
Mandela, President Nelson 153, 465, 507, 560
Mankopane 52, 64, 70, 72, 73
MaNthatisi, a Tlokwa regent 56
Manukozi Soshangane 62
Mapela (Mapela Langa) **67**
Mapham, Burgher George William 535, 770
Mapungubwe **20**, 27, 35, 37, 43, 411, 677
Mara 199; Mara Research Station 451, 455
Marais, Bennie 193, 580
Marais, Eugène 115, 174, 180, **181**, 182–183, 549, 668, 703; hypnotism 182; morphine addiction 182, 527; Resident Justice of the Peace 703; the scientist 194; Marais Tin Syndicate 181, 633
Marais, Jaap 570
Marais, Jan, Sergeant, SAP 189
Marais, Judith Johanna (Sannie) (m. Nicholson) 554, **555**, 721
Marais, Pieter Johannes 240
Marais, Soloman 554
Marakele National Park **95**, 111, 125, 127, 400, 433, 465, 614, 645, **647**, 649, 651, **652**
Marapong *see under* Lephalale
Marble Hall 448, 528

Marcus Masibi scheduled area 314
Marcus, Solly 224–225
Marè, Ds. Roelof 498
Maria 564LR **29**, **34** *see also* Swebeswebe
Marico district 42, 53, 57–58, 62, 82, 85, 86, 88, 399, 660, 661
Marikele Point 87, 125, 126, **128**, **647**, 653
Maritz Rebellion 529
Maritz, Gert 58
Maritz, Lt Col Manie 483, 529–530, **532**, 543–544
Marken 37, 47, 222, 237, 335, 343, 361, 389, 561, 614, **617**, 619, 625, 635, 645, 665, **665**, 666, 668, **670**, 671–672; Marken 457LR **665**, 666, 667, **670**; NG Church 493, 495, **665**, **669**; school **578**, 590, **665**, 667–669
Marks, Sammy 158
Marloth, Dr Rudolf, botanist 182–183
Marnitz, Jan 278
Marquard, Leo 526
Marrieres Wood, Battle of 484
Martinique 171LR 82
Martjie se Nek 123, **242**, **349**
Marx, J.M., magistrate 190
Marx, Mienie, nursing sister 194
Masebe I 32, 50
Masebe II 50,
Masebe III 52, 73–75, 77, 83, 298
Masebe Provincial Nature Reserve **25**, 30, **67**, 69, 340, **670**
Masebe, Bakenberg 298
Masebe, Cornelius 77
Masebe, Hans, a Langa chief 79, 81, 258, 298
Masebe, Marcus 79, 82, 314
Mashishi 52
Mashonaland 372, 554, 556; road to 285, 287, 294
Mashudu Lodge 122, 262
Masibe, Lesiba 230, 467
Mason, Prof. Revil 19, 21, **629**
Massey, Gabrielle Elizabeth 111
Matabeleland 372
Matebele 41–42, 52, 56–59, **57**, 62, 630, 716–717
Mathews, Charles 520
Mathewson, A.J. (Arthur), Justice of the Peace 347, 440, **442**
Mathiba 697
Mathombeni 52
Matimba power station 232, 236, 340, 391, 394, 638–639, **640**, **641**, 663
Matjesgoedfontein 57KR **349**, **350**, 353
Matlabas 35, 394, 451, 614, 645, 651, **652**, 653, 657, 667; Matlabas 94KQ 216, 232, **652**, 653, 655; Matlabas Cotton Planters' Association 656; Gereformeerde Kerk 489, **654**, 655, **655;** Matlabasfontein 229 (now Diamant 228KQ) 659
Matlabas River (Mothlabatsi River) 37, 39, 126, 127, 256, 311, 400, 415, 419, 465, 651, **652**, 653, 656, 657, **658**, 659
Matlala 50, 73
Matlasedi Primary School 609; Secondary School 609
Matlou, chief 50
Mauch, Carl (Karl) 72–73, 85, 96, **119**, 121, 123, 124, **126**, 127–132, 137, 253, 352, 361, 478
Maun, Botswana **693**
Maxwell, Brigadier W.S. 110
May, J.D. 653
Mayer, Erich **487**
Mbandera 688
Mbeki, President Thabo 229
McCallum, William 630
McCord, J.J. 146, 196, 197, 199, 201, 203, 569
McCord, Johanna (née Potgieter) 197, 489
McCorkindale, Alexander 210
McKerral, David 618
McMicking, Major H. 91, 141, 262, 263, 265
McPhearson, Dr 194
McSeveney, Mrs Anne 661
Meads, George 186
Mecklenburg, Duke of 136
Medupi power station 236, 340, 394, **616**, 638, **641**, **663**, 664
Meetsetshehla Secondary School *see under* Vaalwater
Meintjes, Stephanus, agent 303
Meletse Game Reserve 534
Meletse Mountain **20**, **25**, 61, 623–624, **647**
Melinda Fault zone **626**, 642
Melkboschkraal 431LQ 445
Melk River (-rivier) 16, 17, **31**, 124, 228, 233, 237, 238, **242**, 276, 342, 343, 346, **350**, 352, 354, 358, 379, **670**; NG Church *see* NGK Gemeente Melkrivier; school 294, **350**, 567, 571, **578**, 589–591 595, **670;** store **350**, **352**, **670**
Melora Hill, Lapalala **25**, 39–41, **42**, **53**, **336**, 618
Menees, Rev. Richard 518
Mentz, Col 389
Merensky, Alexander 137
Merensky, Dr Hans 83, 313
Merriespruit bridge **234**, 236
Merry, Lt 696
Merz, Brian & Val 610
Messines Ridge, Battle of 484
Methodist Church 517, 519–520; Warmbaths/Bela-Bela congregation **519**, **520**, 524

Methodist Jubilee Hall 520–521
Methuen, Brigadier General 541–542
Methuen, David 337
Metorex 632
Meyer, headmaster 580, 600
Meyer, Lukas 139, 548
mfecane *see* difaqane
Michau, Gerald 355
Michau, Martha Sophia (m. Webster) 354
Middelfontein 391KR 370, **371**, 375, 480 *see also* Loubad
Middelfontein 564KR **65**, 68, 99, 121, 168, 173, 175, 177, **185**, 477, 481, 603 *see also* Modimulle mission
Middle Stone Age (MSA) 21, **22**, 24–26, 43
Middleton, A.H. 586–587, 589
Middleton, Stephen 359
Mieliefontein **665**, 667
Miller, John 467
Milner, Lord Alfred 161, 179, 267, 436, 559, 601, 696, 713; Lord Milner Farm School 600; Milner's Kindergarten 713
Milner, Mervyn 646
mineral springs 195–196; at Warmbaths **200**
Minnaar, Cornelis Jacobus 200
Minnaar, Johannes Christoffel 241
Mispa (Mizpah) Kerk 493, 663
mispel 97, 123
Mmamabolo, principal 606
Mmamabolo Pedi chieftaincy 500, 502
Mmamatlakala 45, **46**, 50, **67**, 69, **670**, **674**
Mmatshwaana 54, **65**
Mmetlane River 45, 673
Moddernek 114, **128**, 141, **350**, **371**
Modderspruit 150KR 122, 262
Modimolle *see also* Nylstroom 13, 62, 121, 149, 152, **153**, **162**, 170, **172**, 227, 235, 252, 343, 365, 367, **371**, 374, 376, 394, 474, 476, 477, 482, **484**, **486**, 497, 499, **527**, 528; town hall **171**
Modimolle-Mookgophong local municipality **171**, 228, 230, 575, 649
Modimulle Berlin Mission 68, 99, 121, 168, 173, 177, **185**, 477, **478**, **479**, **480**, 481, 603
Moepel Farms 614, **670**, 672
Moerdijk, J.L. (Jan) 563, 582, 583–585, 591, 593, 594, 598, 720; family **166**,
Moerdyk, Gerard 485, **486**, 560, 563, 583
Moerdyk, New Belgium **282**, **329**, 330, **338**, 340
Moffat, John Abraham, architect 205
Moffat, Robert 56
Mogalakwena Formation **615**
Mogalakwena Local Municipality **573**

Mogalakwena Platinum Mine 52, 624, 625, **627**
Mogalakwena River 41, 45, 47, 50, 52, 64, **66**, **67**, 69, 71, 75, 222, 335, 405, 419, 625, 656, 668, **670**, 716, 717 *see also* Nyl River
Mogemi (a Kekana regent) 72
Mogol Pos/Post, a newspaper **226**, 227
Mogol River *see* Mokolo River
Mogoshi peak **672**
Mohlakamotala Secondary School **578**, 610–611, 721
Moima, Godfrey 606
Mokamole River 45
Mokerong 46, **67**, 78
mokhukhu 504
Mokkanes Halt **670**, 671
Mokolo area 389–390
Mokolo (formerly Hans Strijdom) Dam 221, **231**, 232, 569, 591, 639
Mokolo (Mogol, Pongola) River **214**, 221, 223, 377, 378, 379, 389, 400, 402, **403**, 416, 419, 433, **663**
Mokopane *see also* Potgietersrus(t) 9, 13, 19, **20**, 47, 50, 52, 64, **67**, 71, 73, 243, 340, 343, 452, 455, 522, 613, 619, 620, 624, 625, 645, 665, 670
Mokopane II (Chief Klaas Mokopane) 68–75
Mokopane, chief 20, 64, 66, 68, 71, **76**
Molekoa 50
Molepolole 53
Moll, C., Nylstroom magistrate 138
Molokomme community 82
Monarch Cinnabar (Pty) Ltd 618
Montanha, Joaquim de Santa Rita, Catholic priest 99
Monte Christo 128LQ **663**
Monyebodi, a Sotho chief 73
Mooihoek 226KR 135
Mookgophong *see also* Naboomspruit 13, 24, 121, **135**, **148**, 149, 173–196, 179, **185**, **192**, 230, 297, 371, 452, 476, 499, 575, 611, 613, 628, 636, 646, 649
Moorddrift 64, **66**, **67**, 177
Moordkoppie 52, 77
Moore, a prospector 635
Moorrees, Dr (NGK) 491
Mordt, R.A. 658
Morgenrood, Louisa Mary (m. Hendrik Read) 94
Moria/Moriah/Morija 500, **501**, 502–507
Morris, Sergeant M. 532
Mortimer, Jack 212
Mortimer, John 244, 274, **442**
Mosam, Zunaid 575
Mosdene **148**, 180, **185**, 517
Moshoeshoe, Sotho chief 55

Mostert, Ben 227, 442, 459
Mostert, postmaster 346
Motaung, Edward 503
Motlhabatsi River *see* Matlabas River
Mounted Infantry units 108, 120, 260–264, 304, 375; 4th Battalion 375; 12th Battalion 304, 345, 369; 20th Battalion 110, 260, 264, 265, 305
Mpande, Zulu chief 63
Mpangazitha, a Hlubi chief 56
Mpatamatcha 346
MSA *see* Middle Stone Age
Mt Hampden 554
Mthethwa chiefdom 55, **57**
Mugombane 64
Muisvogelkraal 654LR 294, 590
Muller, Ivan **506**
Müller, Retief 505
Munnik, Morkel 497
Munro & Maggs 635
Munro, William 240
Mural Mountains 400, **403**, **648**
Murchison 460LR 665, **665**, 666, 667, **670**
Murray, Ds. Andrew Jr 200, 470
Murray, Eerw. John N. 499
Murray, Rev. Andrew Snr 469
Musi (Msi) chiefdom **51**
Muslim families *see under* Islam
Mwulu's Cave **20**, 23
Mzilikazi, Matebele chief 41, 51, 52, 55–58, 62, 64, 197, 405, 716, 717 *see also* difaqane

Naauwpoort 518KR **135**
Naauwpoort School 353, **578**
Naboomfontein 320KR 175
Naboomspruit *see also* Mookgophong 112, 122, 135, 173–184, **185**, **187**, 188–196, 319, 320, 436, 444, 448, 545, 561, 569, 571; churches **490**, 498–499, 515, 516, 517, 518, 522, 524; founding 175, 180; development 186–188, 191; Frikkie Geyser dam **185**, 191, 192, 637; healthcare 194; mineral springs 195–196, 643–644; mining 625–626, 628, 631, 633–638; policing 188–189; roads 177, 259, 365, 561; SA War 178–179, 262, 297; schools 93, 194, 580, 595, 598–600; siding/station 177, **178**, 190, 191, 194, 490, 633, 636; town centenary **192**; Town Council 186, **186**, 192, 637; Zebediela branch line 179, 190, 320
Naboomspruit 348KR 177, 178, 180
Nachtwacht 492LR 183, **700**, 701
nagana *see* tsetse fly
Nagel, Burgher Jan 531, 534, **534**, 770
Namas 382, 685, 686, 690–691, 703

Naoa, Ehrens, chief 655
Nash, Anne 211
Nathan Sacks Chemist, Warmbaths **205**
Nathanson, a Jewish trader 92
National Bank of South Africa, 161, **162**
National Party (NP) 45, 64, 88, 115, 192, 225, 325, 334, 362–363, 368, 382, 384, 446, 521, 553, 557–558, 561–564, 566–572, 596, 604–608, 621, 664, 714
National Scouts 482
Native Affairs Commission 81, 603–604
Native Affairs, Department of 605, 656; Minister of 604
Natorp, Carl, trader 145, 153, 175
Nature Conservation, Department of 465
Naudé, Beyers 363
Naudé, Jozua Francois (Tom), Senator 146, 362; house on Vrymansrus road **350**
Ndebele 39, 41, 49, 50–59, **60**, 62, 64, 66, 70–77, 82, 83, 430, 716–717 *see also* Kekana *and* Langa Ndebele
Ndwandwe chiefdom **57**
Ndzundza 49, 52
Neckar 183LR 82
Nederduitsche Hervormde Kerk (NHK) 151, 470–473 — Lydenburg 470; Naboomspruit/Mookgophong **490**; Nylstroom/Modimolle 474–475, 481, 483, **484**, 485; Rustenburg 470; Soutpansberg 470; Vaalwater **496**, 497; Waterberg 497
Nederduitse Gereformeerde Kerk (NGK) 150, 468–471, 473–476, 481, 483, 485, 488, 492, 498, 522, 572 — Albertyn 493; Lydenburg 270, 473, 474; Marken **669**; Melkrivier (Waterberg Noord) 238, **350**, **494**, 495; Naboomspruit 491; NG Kerk in Afrika (NGKA) 499; Nylstroom/Modimolle **476**, 481, 483, 485, **487**, 488; Steenbokpan 493, 663; Vaalwater **496**, 497; Warmbad-Wes 489; Warmbaths 489–490
Nederduitse Hervormde of Gereformeerde Kerk (NH of GK) *see* NGK
Neethling, Ds. J.H. 470
nege maande arrangement 330, 332
Nek, the 254, **264**, 275, 441
Nel, Burgher Jeremiah Daniel 531, 532, 533, **534**, 769
Nel, Christoffel Bouwer 221
Nel, Jan Christoffel 589
Nel, Susara Hendrika 374
Nelson, C.W. 165
New Belgium 608LR 278, **329**
New Belgium Block/Estate 15, 239, 274, 278–280,

281, **283**, 284–341, 343, 347, **349**, 351, 356, 443, 514, 598, **665**, **670**, 698, **700**, 701; archaeological sites 24, 33, 35, **36**, **38**, 40–42; conservation investments 335–341, 464; farms list 280, **282**, **286**; John Todd 318–320, **319**, **320**, 565; New Vestey ownership 315–323, **316**, **317**, 381; origin of name 283–285; Union Goldfields Limited 328–329
New Belgium Settlement Scheme (1940s) 324–335, **330**, 353, 495, 668, 719; William Averre 324–331, 335, 351
New Belgium (Transvaal) Land & Development Company Limited (NBTL) 256, 279–280, 281, 283, 285, 291, 294, 297, 303, 307, 310–313, 315, 434, **435**; Caine era 14, 134, 257, 259, 260, 261, 291–310, **297**, 367, 435, 720; Desborough era 287–289, **288**; O'Brien era 312–313; Peacock era 290–291, **290**
New Ellisras Road R33 232–233
newspapers 223–227
Ngamiland, Botswana **693**, 694, 696–697
Nguni, Nguni-speakers 39–41, **51**, 52, 53, 55–56, 59, 456
Ngwape 54
Nichol, Dr W. 219
Nicholson, Judith Johanna (Sannie) (née Marais) **555**, 721
Nicholson, Richard Granville OBE, Major, first MP for Waterberg 553–554, **555**
nickel 624
Nicodemus, a Herero chief 680, 695
Nicolson, Harold 519, 521
Niemandt, Koos, field cornet 530, 535
Nkgalabe Johannes 82
Nood Hulp (Noodhulp) 492KR 98, **148**, 201, **207**
Nooitgedacht 136KR **214**
Nooitgedacht 167KR 174
Nooitgedacht 50KR 493, **494**, **495**
Nooitgedacht 92KR 618
Nooitgedacht lead mine 123, 618, **627**
Nooitgedacht, Magaliesberg 543
Noord (North) Brabant 608LR **25**, 257, 258, 259, **282**, **286**, 292, 294, 297, 298, 299, 304, 307, 309, 312, 320, 339, **349**
Noord-Transvaalse Koöperasie Beperk (NTK) 117–118, 170, 172, 442, **449**, 450, 451, 455, 457–459
Norman, Sergeant, Brakfontein police post 346
North Waterberg Farmers Society (Association) *see under* Vaalwater
Northam 391, 621, 624
Northern Educational Trust (NET) 609–610

Northern Review, a newspaper 225
Northern Sotho 45, 49, 52, 507, 510, 568 *see also* Lebowa & Sotho-Tswana speakers
Northern Transvaal Lands Company 311
Northern Transvaal Nut Growers Association 447
Notwane 401
Nterekane, the War of 70
Nutfield 448, 515
Nxai-Nxai (Nyainyai) **693**, 694
Nyathi estate 353
Nyl (peanut butter brand) 170, 448
Nyl River 174, 473, 477 *see also* Mogalakwena River
Nylstroom *see also* Modimolle 13, 68, 86, 88, 97, 100, 111–114, 122, 129, 131, 134, 135, 149–156, **162**, **164**, 171, 172, 177, 181, 183, 191, 215, 220, 277, 310, 312, 313, 325, 334, 372, 374, 405, 407, 431, 485, 544, 545, 563, 633, 655, 701, 720;
  black community (incl. Phagameng, Sikoti Pola) 149, 161, **162**, 168–170, 602; churches 113, 121, 155, 163, 236, 238, 254, 358, 392, 471–476, 481, 483–490, 492, 493, 494, 497, 498, 511, 512, **513**, 515, 516, 517, 518, 522, 523, 525–528, 595, 662; commerce 88, **142**, 143, 145, **146**, 153, 161, 163, 188, 226, 444, 475, 558; concentration camp and cemetery 134, 143, 159–161, **160**, 239, **240–241**, 310, 361, 367; farmers' co-operatives 443, 446–448, **449**, 457; founding & development of 72, 121, 136, 140, 151, 153, 154, 158, 170, 201, 254, 365, 546; hospital 567; Grape Festival 367, 394; Health Committee, Village & Town Council 143, 145, 153, 163, **171**, 380, 562, 569; Indian community (incl. Asalaam) 169, 170, 183, 208, 347, **527**, 528, 568, 575, 606, 714; magisterial district, court & post office 79, 81, 103, 136, 138, 143, 153, 155, **157**, 219, 271, 287, 295, 365, 438, 546, 547, 653, 675; malaria 156, 160, 411–415, 419; Masonic Lodge **144**, 145, 146, 272; Nylstroom–Vaalwater railway line 115, 116, 213, 217, 227, 239, 275, 376–394, 441; police 161, 189, 268, 270; politics 88, 220, 228, 325, 347, 525, 546–548, 551, 553, 556, 558, 560, 561, 565, 567, 568, 569, 575, 649; Pretoria–Pietersburg railway line 141, 158, 177, 302, 490; SA War 108, 143, 159–161, 258–263, 266, 295, 297, 298, 301, 304, 307, 369, 481, 512; roads 254, 256, 268, 276, 297, 351, 355, 358; schools 134, 163, 165, **166**, 167, 168, 170, 238, 568, 576, 577, 580, 587, 594, 595, 602, 606, 610; town hall 143, **171**
Nylsvley 560KR 177, 180, 185, **185**, 336, 569
Nyl-Zyn-Oog *see* De Nyl Zyn Oog

O'Brien, W.F. (Mike) 312–313
Oceana Transvaal Land Company 256, 278, **435**, 630
ochre **29**, 39
Odendaal, F.H. (Fox) 144, 172, 220, 252, 256, 561, **567**, 590, 671
Ohrigstad **20**, 23, 63, 64, 150, 200, 365
Okahandja, Namibia 680, 682, **683**, 684–685, 691, 697, 704, 705, 707–708
Okakarara, Namibia **687**, **693**
Okavango River 677, **683**, **693**, 695
Olckers, Martiens 180
old coach road 135, 175, 545
Olieboomspoort **20**, **22**, 23, **25**, 34, 219
Olievenbosch 506KQ 630
Olievenhoutfontein 74KR 606
Olifantsdrift 172LR 54, 82
Omaheke, Namibia **683**, 688, **689**, 690, 692, 703, 710
Omaruru, Namibia 683, **685**, 709, 710
omiramba 677, 688, 692
Omsingel 262, **264**
Omuramba Omatako, Namibia 677, **683**, **687**
Ongandjira (Okandjira), Namibia **683**, 685
Ongegund, New Belgium 335
Ongelukskraal 48KR **25**, 39, 40, 433
Ons Hoop, hostel, Nylstroom **167**, 168, 568, 587
Ons Toekoms, hostel, Nylstroom **167**, 587, 721
Onverwacht 503LQ **663** see Ellisras
Oosthuizen, Gerhardus (Mossie) 180, 360, 361
Oosthuizen, Gert Hendrik 221
Oosthuizen, John R. (*Meester*) 233, 238
Oosthuizen, Magdalena (Maggie), (née Willemse) 233
Oosthuizen, Magdalena Petronella (née Joubert/ Blignaut) 359, **538**, 360
Oosthuizen, Rev. Brian 522, 592–593, 597
Oosthuizen, Sampie 234
Opperman, Burgher Albert S.A. 90, **95**, 531, 534, **534**, 535, 770
Opperman, Commandant 770
Opperman, Lt Daniel Johannes Elardus (Rooi Daniel) 531, 534, 537, 538, 770
Oranjefontein 176LQ 493, 662, 663, **663**
Oranjia Unie party 551, 553, 557
Orsmond, Arnold 146, 163, 438
Oslo Land Company (OLC) 656, 667
Oslo Ranch 658
Osombe-Windembe, Namibia **683**, 689
Ossewabrandwag (OB) 526, 563–565, 566
ostriches 131, 234, 243, 558; eggshell beads **28**
Ostwald process 439

Oswell, Cotton 401–402; travel route with Warden 1846 **403**
Otavi, Namibia **683**, 691
Otjimbingwe, Namibia 682, **683**
Otjinene, Namibia **683**, 689, 692, **693**
Otjiwarongo, Namibia 677, **683**, 686, **687**, **693**
Otjosondjupa, Namibia 677–678, **687**
Otjozondou omiramba, Namibia **683**, 692
Ou Voortrekkerpad 175
Ouri River *see* Limpopo River
Outspan (oranges brand) 190
Oversprong 650LR 124, **128**, 132, **350**, 352
Overton, Kate Elizabeth Maria (m. Averre) 324
Overysel 815LR **65**, 82, 83
Oviumbo 686
Oxwagon Sentinel *see* Ossewabrandwag

Paardeberg, Battle of 110
Pachystigma pymaeum, gousiektebossie 323
Pachystigma triflorum, Waterberg medlar 97
Packer, Edgar Arnold, cartoonist **699**
Palala region, or ward 79, 174, 245, 287, 377, 378, 425, **441**
Palala River *see* Lephalala River
Palapye, Botswana 91, 301, 303, 305, 307, **693**
Palmietfontein School **350**, 359, **578**, 585–587, **588**, 589
Papstraat **344**, 345, **350**, 361, 671
Parham, Harry 452
Paris Evangelical Mission Church (PEMS) 502
Parish of St Andrew 518, 520
Parish of Waterberg 518
Parker, Bishop 515
Parker, Dale 17, 336–337, 464
Parker, Duncan 339
Parker, Rev. Tom 522
Parr's Halt 219
Parrott, Lt Col 311
Pasch, Ds. G.B.S. 491
Pasteur, Henry 278, 280
Phatsane 45, **46**
Pathudi, Dr Cedric 45, **47**
Paton, Alan 714
Peacock, Arthur F. 242, 244, 256, **290**, 293, 294–295, **296**, 297, 301, 307, 309, 347, 435, 511, 664
Peacock, Harry Wilkinson 285, 287, 289, **290**, 291; Peacock gold nugget **290**
Peacock, Katherine (née Fawssett) 242, 244, 256, **290**, **296**, **297**, **308**, 515, 517
peanut butter 170, 448, **449**, **450** *see also* Apie *and* Nyl
Pearson Nature Reserve 336

Pedi 41, 50, **51**, **57**, 62, 64, 500
Pegasus Petroleum 191
Percy Fyfe Reserve 336
Perdekop **95**, 106, **107**, 426
Phagameng *see under* Nylstroom
Phalaborwa 36, 50, 439
Phalane 50
Phatsane 45, **46**
Philips, Lionel, a Randlord 313
Pie van Teneriffe scheduled area 314
Pienaar, Major Jack 530, 531, 543
Pienaars River 201, 451, 643
Pierneef, J.H. **648**
Piet Potgietersrust *see* Potgietersrus
Pietersburg *see also* Polokwane 23–24, 26, 49, 79, 146, 170, 173, 208, 213, 223–225, **224**, 267, 298, 313, 321, 362–363, 448, 553–554, 556, 558, 561, 595, 607; concentration camp 239, **240**–**241**, 310; Pretoria–Pietersburg railway line 158, 175, 261, 266, 302, 311, 377, 490; religions 483, 500, 502, 508–510, 514, 515, 518, 522, 523–525
Pietersburg Horse (Bushveld Carbineers) 245
Pilanesberg 50, 127, 129
Pilgrim's Rest 407, 511
Pioneer Column, Rhodesian 554, **555**, 721
Pitout, Arthur 448
Plange, Konrad 313
Plasmodium falciparum 412–414
Plasmodium vivax 412, 414
Plat River Dam 207
Platbank 243KQ 126
platinum 52, 83, 181, 194, 613, 624–625, **626**, 628, 633, 643; platinum group metals (PGMs) 624, 643
Platrivier 201
Plumer, Col H.C.O. 79, 104, 143; Plumer's Column 244
poaching 372, 463, 666
Pock, P. 580
Podocarpus latifolius 342, 343, 584
Poer se Loop 124, **234**, 236, 294, 305, 314, 347
Pole Evans, Dr I.B. 184
police posts *see under individual stations:* Brakfontein/Dorset; Clermont/Villa Nora; Rhenosterpoort (fontein)/Rankin's Pass; Geelhoutkop; Zandriviersspoort
Polokwane *see also* Pietersburg 24, 50, 72, 340, 362, 500, 521, 522, 523, 535, **673**
Polokwane Observer, a newspaper 225
Pongola River *see* Mokolo River
Pongolapoort 252, **386**
Pool, Gerhard 700

Posselt, Rev. Otto 478
Postma, Ds. Dirk 85, **471**, 472, 474
Postma, Ds. M. 483
Postma, Maria (m. Du Toit) 472
Potgieter, Catharina Aletta Magdalena (née Maré) 100
Potgieter, Catherina Frederica 98, 489
Potgieter, Commandant General Andries Hendrik 58, 66, 69, 71, 85, 99; Potgieter Trek 63–64, 85, 149–150, 200, 365, 576
Potgieter, Carel S., LV for Waterberg 280, 547
Potgieter, Elsie-Maria (widow of P.J.; m. Stephanus Schoeman) 69, 99
Potgieter, Hermanus 64–66, 69, 70
Potgieter, Jacoba Johanna Carolina (née van Heerden) 98, 99, 100–101, 102; memorial site **95**, **100**
Potgieter, Johanna Hendrina (m. J.J. McCord) 197, 489
Potgieter, J.S. 649–650
Potgieter, Paul Maré 100
Potgieter, Commandant General Pieter Johannes 64, 69, 71, 99, 173, 197, 241, 546; Makapansgat 66, 68, 99; Piet Potgietersrust 71, 99; Potgieter Street, Nylstroom 153, 154
Potgieter, Pieter Johannes Jr., Native Commissioner 77–78, 98, 99, **100**; mayor of Pretoria 100
Potgietersrus(t) *see also* Mokopane 47, 53, 68, 71–72, 75, 77, 81, 83, 84, 99, 135, 151, 153, 177, 194, 195, 225, 263, 267, 275, 306, 311, 325, 416, 448, 451, 452, 483, 490, 491, 515, 522, 523, 525, 544, 553, 595, 631, 635, 667, 670, 671; Mahwelereng 48, 52, **65**, 84; Pretoria–Pietersburg railway line 158, 177
Powell, Lt Charles Folliott Borradaile **264**, **265**, 266
Powell, Sir Richard Douglas, MD 265
Prance, C.R. (Cyril Rooke) 93, 98, 111, **112**, 113–114, 367, 514
Prance, Edith Lena **113**, 514
Prance, Rev. Ernest Reginald 113, **514**, 515
Preller, Gustav 65, 68, 713
Pretoria Group of rocks **627**, 630
Pretorius, Barend 584
Pretorius, "Boelie" 361
Pretorius, B.P. 361
Pretorius, Commandant General Andries 63, 371, 546
Pretorius, Major P.J. (Phillipus Jacobus), ("Jungle Man") 371–374, **372**, **374**, 375
Pretorius, President Marthinus W. 66, 68, 69, 124, 163, 210, 231, 247, 352, 358, 366, 470, 546, 547, 548, 549, 666, 675,

Prinsloo, Burgher P.G. 532
Prinsloo, J.J., Magistrate in Nylstroom 115, 153, 546
Prinsloo, Pieter 368
Prinsloo, S.J., principal 589
Prinsloo, Willem (Moorddrift) 64
Progressive Party 138–139, 553, 570
Pronk, Baron Izak Jordaan van der Paardevoort 662
Pronk, Daniël (Daan) van der Paardevoort 662, 663
Pronk, Dr Izaac J. 134, 559
Pronk, Margaretha (m. van Hulsteyn) 559
Pronk, Maria Elizabeth (m. Forssman) 134, 165
Pronk, Ms A., teacher 134, **166**
Protea caffra 429
Protectorate of Bechuanaland *see* Bechuanaland
Pruissen 52, **67**
Public Health, Department of 194, 418
Pullinger, settler family 179, 248
Purekrans 271KR 79, 174, **185**, 582
Purves, S.O. 580
Pusompe Mission 20, **25**, 32–33
Pyramids Masonic Lodge **144**, 145–146, 272

quarantine 183, 268, 425, 426, 590, 667
quinine 194, 413, 415, 419
Quirk, settler family 179

R33 road 122, 123, **214,** 230, 232, 236, 237, 252, 277, 351, 367, 376, 592, 610, 639, 642, 664
Radikwekwe 643
Radium 222, 515, 518
railway lines — Nylstroom-Vaalwater *see under* Nylstroom; Pretoria-Pietersburg 158, 313, 377
Railway Mission 515
rain-making rituals/sites 20, 30, 32–34, **36**, 37, **38**, **39**, **42**, 43, 73
Ramakoko, village 129
Ramaphosa, President Cyril 507, 712
Ramodike, Nelson 45
Rankin, Catharine (née Bell) 94
Rankin, Gerty 94
Rankin, Hester Carolina (née Eloff) **90**, 91, 94, 95
Rankin, Jacoba Wilhelmina (née Swanepoel) 90
Rankin, James Creighton (Jimmy) 90, 94
Rankin John, Resident Justice of the Peace 90–92, **94**, 95, 115, 116, 274, 446, 579, 580
Rankin, Nicolina Jacoba Dorothea (née Strydom)
Rankin, P.A. 653
Rankin, Phillip 90
Rankin, William 94
Rankin's Pass 86, 88, 90–92, **93**, **95**, 102, 112, 115, 125, 148, 266, 275, 304, 369, 381, 390, 429, 497, 529, 550, 579, 595, 614, 617, 652, 653; police station **93**, 268, 269, 270, 271, 347 *see also* Rhenosterpoort/fontein police post; Rankin's Post (Pos) 88–89, 91, **441**
Ransome, Simms & Jeffries 335, 363
Ratelhoek 158KR 241
Ravat, Ebrahim, trader 183, 527–528
Ravazotti, Gianni 339
Read, C.E., Secretary of Zwagers Hoek Farmers' Association (1919/1920) 379
Read R.S., Secretary of Swaershoek Farmers' Society/Association (1926) 117, 390
Read, Hendrik Petrus Johannes 94
Read, Henry Rankin 95
Read, Hester (née Rankin) 90
Read, James Oliver 94
Read, John Charles Edward 94, 95
Read, Raymond Sydney 94, 95
Rebecca Street Cemetery, Pretoria
Rebellion of 1914, 90, 146, 183, 372, 483–484, 529–531, **532**, 533–544, 558, 566; Zandfontein (8 November 1914) 529–532, **533**, **534**, **536**, 538, **539**
Reconstruction and Development Programme (RDP) 216, 228
red beds 616–617
Red Oath (Africa Oath, or Rooi-eed) 102
redwater 268, 323, 421, 424, 425, 438; Rhodesian redwater 323
Rehoboth Basters 685, 709
Reichman, Barney 188
Reitz, Col Deneys 415, 482
Rens, Douggie 521
Responsible (National) Party 553, 557
Retief, Piet 58, 63, 368, 505
Retters, Callie 671
Reus, Christina Johanna 666
Reynolds, J.R. 158
Rhenish Mission 677, 682, 707
Rhenosterfontein 286 (now Diamant 228KQ) 659
Rhenosterfontein 284KQ 125
Rhenosterfontein 289KQ 125
Rhenosterfontein 465KQ 89, 90, 91, 92, **95**, 115; Rhenosterfontein/poort police post 90, 92, **93**, 268, 269, 271, 346, **441** *see also* Rankin's Pass police station
Rhenosterfontein 583LR **650**, 671, 672
Rhenosterfontein school committee 90
Rhenosterkloof 483KQ **25**
Rhenosterpoort 283KQ 85, 86, 87, 88, 89, 90, 91,

92, **95**, 96, 97, 102, 104, 106, 114, 125, 127, 268, 269, 271, 274, 381, **539**; Geloftefeesterrein **89**
Rhenosterpoort/fontein police post *see* Rankin's Pass police station**;** school **578**, 579, **579**
Rhenosterpoort 402KR 371, **371**, 373, 374
Rhinosterhoek Mountains 124, **125**, 126, **128**
Rhodes Drift 104, 556, 660
Rhodes, Cecil John 550, 554, 713
Rhodesian Pioneer Column 554, 555, 721
Richey, J.A. 580
Richter, Carla Pauline (m. Emil Tamsen) 145
rickettsial parasites 424
Riebeek-West 539LR **670**, 671
Rietbokhoek 4KR **234**, 236
Rietbokspruit 302KR 179, **185**, 517
Rietfontein, Namibia 688, 696
Rietfontein 191KR 259, 266, 270, **350**, 360, 363, 365, 377; school 365
Rietfontein 345KR 625, 626, **626**
Rietfontein 389KR 140
Rietfontein 460KQ 531, **539**
Rietfontein 45LQ 54
Rietfontein 513KR (Marais's home) **148**, **181**, 182, 195; police station 183, 189, **441**
Rietpoort 390KR **371**
Riet River **22**
Rietspruit 23, 219, 416
Rietspruit 527LQ 219, **403**, 660, **663**
Rietvallei 151
Rietvlei (Rietvly) 287KQ **95**, 99, 102, 115, 116
Rietvly 285 (now Diamant 228KQ) 659
Rimington's Guides 211
rinderpest 242, 285, 291, 294, 300, 307, 309, 310, 322, 357, 376, **408**, 409, 424, 434, 681, 682, 695
Rissik, Johann, Surveyor General, First Administrator of Transvaal 356, 389, 432, 557, 698, **699**
Road 600 **175**, 177
Road Motor Service (RMS) 215, 391, 392, 668
Roberts Heights 563
Roberts, Burgher G.H. 532
Roberts, Lord 100, 104, 110, 259, 267
Robertson, settler family 179
Robinson, Mary Dorril (m. Barnes) 357
Robinson, Miss 321–322
Rochat, Sergeant, Dorset police station 326, 332
rock art 29–30, **31**, 33, **34**, 35, **42**, 43
Roedtan 153, 156, 186, 189, 251, 319, 320, 565, 624
Roentgen, N.J. 347
Roets, Frans 102
Roets, Johannes Michiel Hattingh 666
Rolong **51**, 59, 60
Roman Catholic Church 99, 208, 507–510, 518, 554; Polokwane parish 508; Warmbaths/Bela-Bela parish 168, 208, **509**, 510, 603
Rondalia Group 195
Rondeboschje 295KR 243
Roodekuil 496KR 451
Roodepoort 222KR 635
Roodepoort 467KR **148**, 200, 201, **202**, **207**, 518
Roodepoort 557KR 184
Rooiberg 24, 37, 39, 92, 129, 515, 518, 524, 529, 537, **539**, 653; group of rocks **627**, 634; Rooiberg Tin **627**, 628–635, **629**, **631**, **647**, 653, **700**, 701; Rooiberg Minerals Development Company Limited 630
Rooi-eed *see* Red Oath
Roos, Ds. G.J. van Staden 492
Roos, Frans 536
Roos, Mr & Mrs 588
Roos, Tielman 542
Rootshain, Rev. 524
Rosenburg, Marcus 523
Ross, Mrs M.M. 594
Ross, Sir Ronald 412
Rosseauspoort 319KQ **647**
Rossouw, Daan 667
Royal Colonial Institute, Proceedings 289
Royal Fusiliers, 2nd Battalion 375
Royal Military College, Sandhurst 265, 375
Royal Norfolk Regiment, 2nd Battalion 108, **109**
Royal Scots Fusiliers 345
Royal Warwickshire Regiment, 6th Battalion 243
Rundgren, Eric 335; Double R Ranch (Rundgren & Roberts) 335, 336
Rustenburg Layered Suite 624–625, **627**

Sacke, A.V. 524
SA Iron & Steel Corporation (Iscor/Yskor) 220, 390, 391, 621–623, 638–639, 645–649, 659, 664
Salatiel, a Herero chief 685
salt 298, 333, 422, 430, 651, 659
Salusbury, Aileen 598
Salusbury, Gerald 598, **599**
Salusbury, Geraldine de Lyon Wentworth (Mop), (née Dillon) **599**, 601
Salusbury, Henry (Harry) Thelwall Wingfield 598–599, 601
Samuel, director of TLC 280
San, the 27, 29–35, 37, 39, 42; *see also* rock art
Sand River 72, 86, 116, 252, 263–264, 266, 304, 369, 370, **371**, 377, 385, 529, **533**, 580, **647**, 718
Sand River Convention 70, 530
Sandrivierberge 119, **122**, **253**, 343, **350**
Sandrivierspoort *see* Zandrivierspoort

SanParks 465, 649
Santa Gertrudis, a breed of cattle 456
satyagraha 574, 714
Schaaf, Harald 459
Scheepers, Jacobus J. 86–87, 104
Scheepers, Sergeant F.J.J. 535, 770
Schikfontein 115KR 356
Schimmel Paard Pan 405
Schistosoma haematobium 422
Schistosoma mansoni 422
schistosomiasis (bilharzia) 421–423; distribution map **422**
Schlömann, Hermann 32
Schmedeskamp, mill owner 188
Schmidt, Rev. Theo 517
Schoeman, S.J. 673
Schoeman, Stephanus 69–70, 99, 546
Schoemansdal **65**, 69, 71–72, 75, 99, 150, 151, 153, 200
Scholefield, S.W.J., Native Commissioner for Waterberg 80
Schoongelegen 200KQ 231, **231**, 232, 591
Schoongelegen Safaris **338**
Schrikfontein 715LR 39
Schroda 46MS **20**, 30, **31**, 35
Schüling, Father Laurence 508
Schurfpoort 112KR 24, **25**, 29, 356
Schutte, C.E., Pretoria magistrate 305
Schutztruppe 678, 679, 684, 686, 688
scorched earth policy 376, 434
Sebetwane, Kololo chief 56
Second Frontier War 197
Second Volksraad 548, 551
Sedibu, Alfred 82
Seegers, C.F. 446
Seegers, C.P. 584
Sefakaola Hill 52, **67**, 71, 75
Sekete IV, Fokeng chief 59
Sekhukhune, chief 74, 547
Sekhukhuneland 49, 51, 62
Sekungwe *see* Tafelkop
Sekwati, Pedi chief 64
Seleka, the 33, 49, **51**, 53–54, **60**, 78, 80; chiefs 54, 301, 399, 499, 655; Ga-Seleka 60, **65**, 236, 287, 301, 306, 351, 399, **403**, 499, 710; Seleka Tribal Authority 54 *see also* Beauty
Selous, Frederick Courtenay 374, 402, 406, 407, 409, 554, **555**
Sentrum 651, **652**, 655
Seripheum plumosum 429, 464
Serowe, Botswana **693**, 696, 700, 706, 707, 710
Seshego 47, **67**

Setateng 51, **65** *see also* Shongoane, Villa Nora
Sethethong sa Modiki 500, **501**
Settlers (village) 95, 179, **207**, 314, 347, 377, 451, 515, 518, 521, 522, 552, 558, 568, 597, 599–601, 610
Seven Sisters hills (Sandrivierberge) **122**, 252, **253**, 274
Shaka, Zulu chief 55, 58, 62, 63
Shambala Game Reserve 465
Shandoss, Joe, Public Prosecutor 189, 192
Shark Island 691–692
Sharpe, Ellen Nichols (née Wright) 324
Sharpe, Robert 318, 324
Shelanti game ranch 670
Shepstone, Sir Theophilus 547
Shimmin, Rev. Isaac 518, 520, 603
Shingwedzi Reserve 348
Shongoane (Setateng) 51, 82
Shorthorn, a breed of cattle 456
Shotbelt 438LQ **663**
SHV Holdings 466
Silberhaft, Rabbi Moshe 526
Siloquana Ridge (tsetse type site) 402, **403**
Singlewood 190, 599
Skandinawiedrif 130
Skeurkrans 39
Skirvings 180
Skolsky's shop 636
Skrikkloof 346, **350**
Slagtersnek 368
slavery, abolition of 57, 68, 70, 469
sleeping sickness *see* tsetse fly
Slypsteendrift 91KR 123
Smetrijns, Father Humbertus 508, **509**
Smit, C.A. 231
Smit, Gert 186, 189
Smit, Jan Antonie 237, 431, 666
Smit, J. de V. of the Waterberg 549
Smit, Kerneels 96
Smith, Andrew, Dr 401–402
Smith, Hendrik Johannes 358
Smith, Ola 456
Smits, Ds. G.W. 474
SMS *Königsberg*, German cruiser 372–373
Smuts, General Jan Christiaan 92, 100, 115–116, 271–272, 373, 382, 384, 385, 481, 482, 529, **532**, 540, 542, 543, 551, 553, 560, 561–566, 574, 621, 623
Smuts, Jacob 636, 714
Smuts, M. 165
Snyman, D.B., Commandant 73
Snyman, Ds. J.T.C. 236, 491–492

Snyman, settler family 180
Soloman Maraba scheduled area 314
Soloman, Sir Richard, Attorney General of Transvaal 553
Somme, Battle of 484
Sondagsloop 123, 258, 342–343, **350**, 352–353, 443, **456**, 457, 587
Sondagsloop River **350**
sorghum 35, 446, 451
Sotho-Tswana (Setswana) speakers 33, 39, 40, 42, 53, 56, 59–62, 82, 398, 430, 651, 716, 717 *see also* Tswana, Lebowa & Northern Sotho
sourveld (*suurveld*) 27, 37, 323, 334, 351, 428, 455, 718
South African Agricultural Union (SAAU) 440, 459
South African Army Service Corps 324
South African Constabulary (SAC) 111, 112, 161, 188–189, 267, 268, 270
South African Democratic Teachers' Union (SADTU) 598
South African Heritage Resources Agency (SAHRA) 61
South African Institute for Medical Research (SAIMR) 416–417
South African Native Affairs Commission 81, 603, 604
South African Party (SAP) 224, 557, 715
South African Railways (SAR) 213, 274, 376
South African Railways & Harbours Administration (SAR&H) 376
South African Tin Mines Company Ltd 633
South African War (Anglo-Boer War) 14, 81, 89, 92, 101, 104, 111, 134, 140, 145, 155, 158, 159–161, 169, 174, 177, 178, 182, 188, 203, **204**, 211, 212, 224, 238, 243, 244, 245, 256, 267, 270, 281, 282, 300, 310, 323, 345, 346, 351, 360, 367, 372, 377, 409, 424, 434, 435, 436, 476, 477, 480–482, 491, 499, 500, 507, 512, **532**, 540, 541, 544, 548, 551, 555, 559, 566, 583, 598, 600, 653, 659, 660, 662, 666, 667, 711, 720, 721; and black communities 54, 78, 602; education after war 579–580; Jews with Boers 525; malaria 411, 413, **414**; Waterberg engagements 108–110, 178–179, 259–267, 345, 368–370, 375
South African War (First) *see* First Anglo–Boer War
Soutpan (Zoutpan) 62, **65**, 659
Soutpan 981LQ 659
Soutpansberg/Zoutpansberg 33, 58, 62, 64, 69, 71, 72, 73, 104, 121, 129, 198, 199, 200, 259, 452, 470, 614, 656

Sovereign Hill *see* Thabazimbi Iron Ore Mine
Soweto riots 572, 607
Spanish flu epidemic *see* influenza
Speciale Commissie 280
Spink, Rev. Dorothy 521
Spink, Rev. Reginald 521
St Engenas ZCC *see* Zion Christian Church
St Etienne 798LR **670**, 673
St Vincent's Private Hospital *see* Warmbaths
Staats Courant, a newspaper 157
stagecoach services 133, 135, 156, 158, **174**, 175, 178, 213, 407, 524, 545
Stallard, Col 522
Standard & Diggers News, a newspaper 295
Stander, Willem 368
Steel, Silas Sexton 675
Steenbokfontein 9KR **25**, 26, **348**, 349, **349**, 592
Steenbokpan 295LQ 493, **616**, 663, **663**
Steenbokpan agricultural union 443
Steijn, Hans Jurgens 87, 107
Steilloop 47, 222, **403**; Steilloopbrug **65**, 405, 605
Stellen, Willem, field cornet 547
Sterk River 79, 429, 599
Sterkfontein on Rietspruit **403**
Sterkfontein 282KQ 88, 94, **95**, **103**
Sterkrivier settlement 452
Sterkstroom River 126, 127, 416, 433, 440
Stevenson, Mary 228
Stevenson-Hamilton, Col James 348
Steyn, Douw 465
Steyn, Dr Colin 569
Steyn, H.J.T. 589
Steyn, Oom Kassie **488**
Steyn, President Marthinus Theunis 105, 542
Steynberg, P.J. 580
Stilling-Anderson, J.M.B., Registrar 446
Stockpoort 215, 216, 391, 392, 401, 660, **663**
Stoebe vulgaris 429, 464
Stoltz, Danie & Isabel 226
Stone Age — Earlier (ESA) 21; Later (LSA) 23, 26–27, **28**; Middle (MSA) 21, **22**, 24–26, 43
Stone, Brian 521–522, 599–601
Stork, F.W. 479
Stormjaers 565–566
Stratford, Donny 522
Strauss, J.G., Postmaster General 221
Strauss, J.H. 581
Streicher, Magda 359
Strijdom Dam *see* Mokolo Dam
Strijdom, Johannes Gerhardus 144, 167, 558–560, 562, 592, 596, 663; Lion of the North 325, 560; MP for Waterberg 191, 325, 334, 525, **559**, 561,

564, 566, 569, 572; Minister of Lands 567; Prime Minister 560, 567, 590, 714; Strijdom Hoërskool ("Hansies") 193, 611
Strijdom, Susan (née de Klerk) 193, 560
Stronach-Dutton road-rail system 190
Strydpoortberge 502
Sukses 237, 294, 295, 343, 346, **349**, **350**, 351, 495; Boerevereniging 395, 443; postal agency 353 *see also* Tretsom
Sunnyside 532LQ 248, **700**, 701
Sunward Park 497
Suzman, Helen 570, 714
Swaershoek/Swagershoek/Zwagershoek 17, 87, 88, **95**, 112, 114, 115, 116–117, 150, 182, 256, 261, 304, 377–378, 384–385, 426, 446, 550, 653; Farmers' Association 88, 95, 117, 379, 390; Geloftefeesterrein **89**, 90; Pos 91; ward 278, 425, **441**, 578–579
Swaershoekberge 304, **350**, 363, **533**
Swakopmund, Namibia 681, **683**, 685, 687, 689, 691, 695
Swanepoel, A.P. 500
Swanepoel, Christoffel Bernardus 446
Swanepoel, Sarel 88, 92
Swanepoel, S.J.M., principal, Laerskool Vaalwater 86, **597**, 598
Swanepoel, Stephanus J.M. 446
Swanepoelsrus 86
Swanevelder, Erasmus Albertus (Bert) 349
Swanevelder, Lodewyk Willem (Lood) 592
Swanevelder, Zacharias & Helena 592
Swart, Ds. J.J. 363, 497,
Swart, G.J. 584
Swart, Keina 226
Swart, Nico 118
Swartwater 399, 405, 663
Swebeswebe 870LR **25**, **29**, **34**
sweetveld (*soetveld*) 35, 37, 39, 327, 455, 465
Swift, Captain, SAC, Nylstroom 161
Syferkuil 15JR 518
Syzigium cordatum 343

Taaiboschspruit 126
Tafelkop (Sekungwe) 26, 39, 343, 346, **350**, 356
Tafelkop 46KR 356
Tambotie River 124, 236, **329**, 347
Tambotierand 370, **371**
Tamsen, Adolph 145
Tamsen, Carla Pauline (née Richter) **147**
Tamsen, Emil 145–146, **147**, 156–158, 161, **162**, 163, 166, 172, 183, 313, 407, 475, 511–512, 553, 580; Freemason 145–146, 272; mayor of Nylstroom 145, 163; Tamsen's Store 145, **146**, 153, 154, 163
Tarentaalstraat **242**, 256, 343–346, **344**, **350**, 353, 357, 358, 360, 361, 363–365, 377, 425, 488, 614, 671
Tati goldfields 131, 404
Telekishi **25**, 39, 41, 42
Terminalia sericea 429
Thaba Nkulu **25**, **33**, 35, **36**, 37, **38**, 39, **42** *see also* Gansieskraal
Thabakgone (now GaMahlanhle) 500, **501**
Thabazimbi 13, 61,97, 149, 230, 304, 391, 399, 400, 420, 465, 498–499, 545, 561, 638, 642, 645–650, **647**, 649, 651, 659; establishment 88, 391, 632; Express 646; Iron Ore Mine 613, 620–621, **622**, 623–624, **627**, 638, 642, 645–646, **647**, 649; local municipality 230, **573**, 575, 632, 649; magisterial district 653; railway line 220, 390, 391, 394, 623, 639, 642; Regorogile township **647**;Thabazimbi-Murchison lineament **627**, 643
Theiler, Sir Arnold 425
therianthropes 32
thermal springs 98, 196, 197, 208, 375, 643–644
Thomas, C. Fred 633
Thompson, Dr W.R. 327, 330, 331, 335
Thorland, Jack 634
Thothwane 54
Thulare, chief 51
Thutlane hill **67**, 70
tick-borne (disease) 323, 423–427
Tiegerfontein 230 (now Diamant 228KQ) 659

tin 37, 177–178, 180–182, 613, 619, 625, 628–635, 643; for individual deposits and production *see* Rooiberg Tin, Union Tin & Zaaiplaats Tin Mining Company; Zwartkloof tin **627**
Tinmyn 135, 635, 672
titaniferous magnetite 625
Tjamuaha, a Herero chief **708**
Tjetjo, a Herero chief 685
Tjisesta, a Herero chief 685
Tlokwa, a Sotho chiefdom 56, 82
tobacco 16, 140, 228, 243, 249, 299, 301, 313, 314, 352, 356, 357, 359, 392, 428, 452–455; industry collapse 393, 458; production **453**, **454**; Tobacco Institute of Southern Africa (TISA) 453; Tobacco Producer Development Company (TPDC) 455; United Tobacco 444
Tobias-Zyn-Loop 179
Todd, Alma Evadne **319**
Todd, David 320
Todd, George **186**, 192
Todd, John 318, **319**, **320**, 321–323

Todd, John II 319
Todd, Mary 318
Todd, settler family 179
Tokodi 73, 74
Tom Burke 47, 405, 665
Tomlinson, Herbert Ernest (Bert) 347, **348**, **442**
Tomlinson, Martha (née Amm) 348
Tomlinson's Hotel 347, **349**
Tompi Zachariah, regent 54
Totius 472, 483, 577
Touchstone Reserve 337, 340, 467, 668
Toulon 495LQ 248, **700**, 701
Toulon, in Sabie Lowveld 348
Toutswe 37
Towoomba Pasture Research Station **207**, 245
Tracson *see also* Tretsom 346
Transvaal and Goldfields Extension Transport Company 133
Transvaal Agricultural Union (TAU) 440, 459–460
Transvaal Consolidated Land & Exploration Company Ltd (TCL) 111, 241–244, 247, 290, 294, 311, 314–315, 347–349, 354, 432, 434, **435**, 440, 516, 557, 598, 592, 618, 635, 664, 666, 701–702
Transvaal Estates & Development Company 311
Transvaal Land Board 656
Transvaal Landowners' Association 656
Transvaal Lands Company Limited (TLC) **132**, 133, 280, 666
Transvaal Lewendehawe Koöperasie (TLK) 456–457
Transvaal Mortgage, Loan & Finance Company 303
Transvaal Platinum Ltd 626, 628
Transvaal Provincial Council 172, 219
Transvaal Registrar of Deeds 434
Transvaal Republican police (ZARP) 89, 660
Transvaal Supergroup 624, **627**, 628, 630
Transvaal, British Annexation, 1877–1881, 73, 137, 153, 201, 284, 507, 547, 577
trekboers **437**, 576, 651
Tretsom *see* Sukses
Treurnicht, Dr Andries 570, **571**, 572–573, 607
Trichardt, Jan Christiaan 88–89, 92, 107; *see also* Jan Trichardt's Pass
Trichardt (Tregardt), Louis, Voortrekker leader 62, 63, 85, 195, 199, 412, 424, 576
Trichardt, Louis, Swaershoek farmer 16, 85, 86, 88, 91, **95**, 102, 116, 227, 442, 459, 568, 573
Triegaardt (Tregardt, Trichardt), Susanna Sophia 85
Trouw Zyn Nek 21KR 123, 346, **349**, 350

Troye, Gustav 137
trypanosomiasis *see* tsetse fly
Tsau, Botswana 695, **696**, 697, 700
tsetse fly 37, 63, 86, **283**, 396–411, 413, 427, 620, 651, 695; infected area **403**, **407**; *nagana* 396, 398, 401, 410, 427; sleeping sickness (trypanosomiasis) 396, **397**, 398, 402; type locality, Siloquana ridge **403**
Tshipise 642
Tsonga (Shangaan) **60**, 639
Tsumkwe, Namibia **693**
Tswana (Batswana) 80, 86, 200, 697 *see also* Sotho-Tswana speakers
tuberculosis 357, 395
Tucker, Lt Gen 108
Tudor, Trevor 625
Tuinplaats (Tuinplaas) 678KR 498, 515; archaeological site **20**, 21
Tuli Block 390, 411
Turbult 494KR **148**, 200, 201, **207**
Turf Farmers Association 170
Turfloop *see* University of Limpopo
Turner, settler family 179
Tweebosch 161KR **344**
Tweebosch, Western Transvaal 541
Tweefontein 463KR 86, **148**, 149, **207**, 313, 365, 545, 547
Tweestroom siding **264**, 274, **390**
Twenty-Four Rivers *see* 24 Rivers
Tzaneen 313, 416, 419, 420, 500, 511, 614, 668

Uitkomst 507LR **36**, 278, 279, 281, **329**, 334, 340 *see also* Alkantrant
uitlanders 548, 550
Uitzoek (Uitsoek) 63KR **349**, **350**, 352, 353
Uitzoek, New Belgium 335
UNESCO-accredited Waterberg Biosphere Reserve 14, 278, 339, 464, 467
Union Cold Storage Company 316–317
Union Defence Force (UDF) 529, 530, 531, 534, 535, 543, 544
Union Tin Mines Ltd 181, **185**, 188, 580, 625, **627**, 631, 633–635, **634** *see also* tin
Unionist Party 557
United Party (UP) 220, 362, 448, 561, 562, 566, 569–570, 714, 715
University College of the North, Turfloop *see* University of Limpopo
University of Limpopo **501**, 607
University of the North *see* University of Limpopo
Uys, Piet 58

Vaalbank 222
Vaal-Harts settlement scheme 327
Vaalkop 405KR 140
Vaalkop 819LR 82
Vaal River 58, 130, 389, 542, 546, 622
Vaalwater No.5/137KR 210–211, 212, 213, **214**, 276, 311
Vaalwater 9, 13, 16, 17, 95, 102, 122, 123, 127, 146, 148, 149, 173, 210–213, **214**, 215–223, 227, 250, 343, 443, 495, 645, 719;
    APK Gemeente 498–499; Bantu public school 606; Bosveld Land Board 228; Bosveld Regional Services Council 227–228, 459; Farmers' Hall 216, 440, 444, 721; GK Gemeente 491–492, **492**; grain silos **215**, 352, 394, 451, 458; Health Committee application 219; Laerskool 216, 221, 240, 421–422, 571, **578**, 581, 595–596, **597**, 598; Leseding 210, **214**, 216, **218**, 222, 228, **229**, 230; Local Area Committee (LAC) 221–222; Local Management Authority application 220; Mabatlane 210; malaria 416, 419, 420; Meetsetshehla Secondary School **214**, 216, **578**, 584, 601, 609, 721; mill 92, 212, **215**, 254, 276; mineral discoveries 613; Modimolle Local Municipality incorporation 228; NG Gemeente 353, 392, 495, **496**;
    NHK Gemeente 236, 238, **496**, 497; North Waterberg Farmers Society (Association) 212, 223, 243, 245, 249, 274, 379, 380, 389, 391, 416, 440, **442**, 443, 459–460; NTK silos 458; old cemetery 229, 234, 235, **496**; post office 102, 213, 216, 221, **441**; private township 215–216, **217**, **218**; R33 and other roads 220, 222, 223, 230, 232, 236, 237, 252, 268, 276, 346, 639; railway line, service & station 95, 102, 115, 116, 118, 122, 213, **214**, 220, 239, 252, 275, 376, **378**, **384**, **386**, **387**, 390–391, **392**, **393**, 394, 441, 452, 457; Road Motor Service (RMS) 215, 217, 219, 661; Vaalwater Development Company (Pty) Ltd 215; Vaalwater District Agricultural/Landbou Union (VDLU) 249, 443, 456, 458, 459; Zeederberg complex 213, 216
Vacuum Oil Company (Voco) 353
Valtyn Lekgobo 75
Valtyn Makapan scheduled area 80, 81, 314
Valtyn Ndebele 52–53
Valtyn, Kekana chief **67**, 71, 75–78
Van Alphen, Isaac 157
Van Blokland, Beelaerts **284**, 285
Van Brink 582
Van Coller, Captain 531
Van Coller, Sam 354, 356

Van de Velden, nursing sister 194
Van den Berg, H.A. 354–356
Van der Bijl, Dr Hendrik 621
Van der Hoff, Ds. Dirk 470, **471**
Van der Merwe, Burgher J.C. Jr 531, 533, **534**, 770
Van der Merwe, Constable, Brakfontein police post 347
Van der Merwe, Janneman 453, 459, 601
Van der Merwe, Louwrens Petrus (Louw) 365
Van der Merwe, Lucas Cornelius (Kerneels) 365
Van der Merwe, Ouderling F. 597
Van der Merwe, Petrus **366**
Van der Merwe, T.F. (Tillie) 452, 453
Van der Merwe, Z. 227
Van der Ryst, Dr Maria 19, 23, 24, 26, 34
Van der Walt, Burgher J.C. 532
Van der Walt, Ds. P.G.L. 492
Van der Walt, Gertbreggie Andriesina (m. Pretorius) 371
Van der Walt, J.S. (Faan) 591
Van der Walt, Lenie & Ben 232
Van der Walt, N.G.S. 563
Van der Westhuizen, A.G. 584
Van der Westhuizen, P.B.E., principal, Laerskool Vaalwater **597**
Van Deventer, E.P. 569
Van Deventer, Frederik 368
Van Deventer, Lt Col Dirk 531, **532**, 537, 538
Van Dyk, F.J. 595
Van Emmenis, A.S. (Nols) 230–231, **442**
Van Heerden, Andries 115, 182
Van Heerden, C. 318
Van Heerden, Carel & Willem 115
Van Heerden, Carl Sebastiaan, field cornet 98, 99
Van Heerden, Catharina Frederika (née Swanepoel) 98
Van Heerden, C.C., LV for Waterberg 550
Van Heerden, Charles 115
Van Heerden, Elsie-Maria Aletta (m. Pieter Potgieter) 99
Van Heerden, Fanie 115
Van Heerden, Gert, Chairman of SA Party, Zwagers Hoek 115, 384, 446
Van Heerden, Jacoba Johanna Carolina 98, 101 *see also under* Potgieter
Van Heerden, J.C. 115
Van Heerden, Nicolaas & Dorothea 173
Van Heerden, S.J.N. 579
Van Heerden, Willem, chemist 194, 195, 274
Van Heerden, W.S.J. Ms 115
Van Heyningen, Ds. (NGK) 471
Van Hoeck, Father Clemens 509

Van Hulsteyn, Margaretha (née Pronk)
Van Hulsteyn, Margaretha (Scrappy) 559 *see also* Vanne, Marda
Van Hulsteyn, Sir Willem 559, 662
Van Loggerenberg, F.J.L. 347
Van Niekerk, Pieter (Oom Piet) Wynand le Roux, MP for Waterberg 558, **559**
Van Niekerk, S.G.J., Administrator of Transvaal 221
Van Nispen, R.A., magistrate 73
Van Pittius, Mrs M.M. 595
Van Ravenswaay, station master 178
Van Rensburg, Siener *541*
Van Rensburg trek 62
Van Rensburg, C.J. 546
Van Rensburg, Daniel Janse 149
Van Rensburg, D.R.J. 569
Van Rensburg, Eerw. G.H.J. 499
Van Rensburg, P.J. **186**
Van Riebeeck Festival (*Fees*) 195
Van Rooy, Ds. J.A. 472, 476, 483
Van Rooyen, D.P. 580
Van Rooyen, Dries & Elsa 277
Van Rooyen, Oom Gysbert & Tant Maria 182
Van Rooyen, Hans "Purekrans" 79, 582
Van Rooyen, Oom Joël & Tant Sannie 663
Van Rooyen, Petrus Jacobus 174
Van Ryneveld, Roelof 447
Van Ryneveld, Sir Pierre 447
Van Schalkwijk, Burgher J. 532
Van Staden, A.S. 581
Van Staden, Ben 360, 361, 365
Van Staden, Christiaan 633
Van Staden, Commandant Marthinus Phillipus 662
Van Staden, Jacobus & Michiel 584
Van Staden, J.F. 318
Van Staden, Johannes Jurie (Lang Hans) 365
Van Staden, Marlene, mayor 230
Van Staden, H.J., field cornet 201, 546
Van Tonder, "Buks" 498
Van Tonder, prospector 633
Van Vlissingen, Paul Fentener 465, **466**
Van Welie, Antoon 541, **552**
Van Wyk, Ds. L.A.S. 492
Van Wyk, G.J. 581
Van Wyk, Hermanus 685
Van Wyk, J.H. (Oom Jan) 535, **536**, 537
Van Wyngaard, J.J. 370
Van Zyl, Alwyn 649
Van Zyl, Dr 228
Van Zyl, Fanie 228
Van Zyl, Jan Kasper 173
vanadium 624, 625, 643
Vangueria sp. 97
Vanne, Marda 559
Vardon, Frank 401–402; travel route with Oswell, 1846, **403**
Vause Raw, Wyatt 448, 569
Vedder, Dr Heinrich 691, 708
Vegkop, Battle of **57**, 58
Venda 40, 49, **60**, 507; people 49, 62, 71
Venter, Ds. H.J. 486, 490, 595
Venter, Johannes Hendrik 247
Venter, Maarten 191; trailers 191
Ventersdorp 568
Ventershoek 579LQ 661
Verdoorn, Clare Inez, botanist 183
Verdoorn, Goosen Johannes, Justice of the Peace, postmaster 122, **128**, 131, 153, 165, 278, 547, 675
Vereeniging, Treaty of 106, 267, 482, 512, 551, 704
Verenigde Kerk, Die (United Church) 474–475, 485
Vermaak, P.S. 491
Vermeulen, C.W.P. 332–334
Vermeulen, Jan 236
Verster, Ds. J.P. 497
Verwoerd, Anna (m. Boshoff) 368
Verwoerd, Dr H.F. 562, 596, 604
Vestey brothers 315, **316**, 317–318
Vestey, Edmund Hoyle 315, **316**, 317, 329
Vestey, Edmund Hoyle II **317**, 318
Vestey, George Ellis 329
Vestey Group 316, 317
Vestey, Ronald 317
Vestey, Samuel & Hannah 315
Vestey, Samuel II, 2nd Baron of Kingswood 329
Vestey, Samuel III, 3nd Baron of Kingswood 318
Vestey, William, 1st Baron of Kingswood **316**, 317, 328
Victor, Aletta 586
Vier-en-Twintig-Rivier 102KR 239, 240, 241, **242** *see also* 24 Rivers
Viljoen, Commandant Johannes Stephanus 201
Viljoen, Ds. D.J. 475
Viljoen, J.H., Waterberg Commandant 546
Viljoen, O.J. 195, 196
Viljoen, P., Vaalwater postmaster
Villa Nora 51, 78, 82, 420, 625, **627**, 671, 672, *see also* Clermont, Setateng
Vischgat 11KR 18, 325, 331, 349, **349**, 353, 592
Vischgat 520KR 173, **185**
Visgat 123, 216, **234**, 236, **286**, 343, 351
Viviers, Ds. 595

Vlakfontein 177KR **148**, 183
Vlakfontein 193KQ **652**, 656
Vlakfontein 270KR 582
Vlakfontein 522KR **148**, 175, 178, **185**, 490
Vleissentraal 456
Vlieëpoort **403**, **647**
Vlieëpoortberge **403**, **647**
Vlugtkraal 38KR **350**, 354, 355, 356, 586
Voco *see* Vacuum Oil Company
Vogelstruisfontein 69KR 581
Vogt, Herlof Anton 657
Vogt, Volrath 656–657
Volspruit 326KR 624
Von Francois, Captain Curt 680
Von Lettow-Vorbeck, Col 373
Von Lindequist, Friederich 691
Von Trotha, General Lothar **679**, **690**; Von Trotha's edict 689–690
Voortrekker wagon **87**
Vorster, Ben 227, 235, 442, 456
Vorster, B.J., Prime Minister 570–571, 593
Vorster, Commandant Barend 70, 259
Vos, J. 247
Vredenburg 71
Vrischgewaagd 649LR 595
Vroom, A. 175
Vroom's store **176**, **185**
Vrymansrus Road 243, 305, 362, 606
Vrystaat Koöperasie Beperk (VKB) 458

Wachteenbietjiesdraai 350KQ 621
Wagner, Dr Percy 620, 625
Wahlberg, Johan August 399–400, 401; travel route 1844, **403**
Wakeford, William 586
Wakkerstroom 100, 511
Wale, Horace & "Pard" **442**
Walker, Clive 17, 336, 337, 464, 467, 650
Walker, Conita 17, 337
Wallmansthal 74
Wangemann, Dr Hermann **478**
Warmbad *see* Warmbaths
Warmbad, German SWA/Namibia 530, **683**, 684
Warmbaths *see also* Bela-Bela, Hartingsburg, Het Bad 86, 98, **148**, 149, 172, 182, 191, 196–209, **205**, **206**, 419, 451, 485, 545, 547, 561, 569, 601, 628, 632, 653; Anglican church 515, 516, 518; Belabela township 203, 207, 509, 510, 518, 568; Buyskop **198**, 199, 201; *Die Pos* 226–227; GK 489–490; hot springs 155, **200**, 201, **202**, 203, 205, 208, 371, 470, 525, 643, 644; Jewish community 523–525, **524**; Jinnah Park 207–208, 528, 714; magisterial district 81, 207, 539; Methodist Church 518–519, **519**, **520**, 521–522, 524, 603; NGK 489; railway line & station 158, **204**; Rebellion of 1914, 531, 535, 537; roads 156, 208, 268, 287, 476, 529, 545, 561; Roman Catholic Church 168, 508, 510, 603; SA War 108, **204** (blockhouse), 242, 244, 259, 266, 298, 301, 304, 306–307; St Vincent's hospital 208, 508, 509–510; town proclaimed & plans 155, 203, 205, **207**, 539; Towoomba Agricultural Research Station 451
Warmburg 376KS 502
Waterberg — Academy **214**, 216, **578**, 584, 594, 601–602, 721; Agricultural Union 443; Biosphere Reserve (WBR) 14, 17, 278, 339, 466–467, **573**, 721; coalfield 13, 613, 638–641; Commandant 104, 201, 295, **296**, 299, 302, 546, 662; Commando 90, 104, 281, 531, 535, 544, 660; copper, a butterfly 107; District Municipality 13, 467, 528, **573**, 613, 645; Game Park, timeshare resort 358, 488; Ko-operatiewe Landboumaatskappy Beperk (WKLM) 170, 449–451; Landbouwers Koöperatieve Vereniging (WLKV) 117, 170, 446–451, 551, 558, 569; Land Company 180; magisterial district (or *wyk*) 13, 97, 131, 136, 138, 153, 157, 210, 245, 268–269, 271, 278, 311, 314, 377, 411, 416, 425, 431, 546, 549, 579–580, 652–653; medlar 97, 123; Mountain Bushveld 429; Native Commissioner 73, 80, 99, 701; Nature Conservancy (WNC) 17, 467, 721; Parliamentary/Volksraad representatives 17, 168, 191, 195, 300, 303, 325, 367, 388, 446, 547, **550**, 551, 553, 556–561, 566–573, 590; Posture 30, **31**, 32; Provincial Council Representatives 88, 172, 186, 192, 220, 561, **567**, 571, 590; Road Board 440; Rock Group **627**; sandstones 36, 45, 152, 252, 342, 343, 349, 614–619, **615**, 624, 625, 633, 638, 642–643, 664, 717; red coloration of sediments **617**; depositional model **616**; School Committee/Board 165, 580; Waterberg-Noord NG gemeente 238, 353, 392, 492–497; Waves radio station 230; Welfare Society 228, 229, 611, 721
Waterberg plateau, Namibia 677–689, **683**, **687**; National Park **687**, 710–711
water-berry tree 343
Waterkloof 502LQ 221, 493, 660, 661, 662, 663, **663**, 664
Watermeyer, Frederick Stephanus Jr., surveyor 280, 711
Watermeyer, F.V., surveyor 114, 203
Watermeyer, Gottfried Andreas 711

Watermeyer, James Charles **711**
Waterval 237
Waterval 190KR **350**, 365, 367, 369, **371**
Watkins, Rev. Owen 518
Watson, Rev. Roland 521
Webb, H.S. 542
Webster, Ilva (m. Greer) 354
Webster, Martha Sophia (née Michau) 354, 586–587
Webster, Wallace 355
Wedelfontein *see* Middelfontein 391KR
Weilbach, Coen 90
Weir, settler family 436
Welgevonden 16KQ 653
Welgevonden 234KQ 464
Welgevonden 284 (now Diamant 228KQ) 659
Welgevonden 343KR **148**, 180, **185**, 189, 194, 625, 626, 628, 637
Welgevonden Private Game Reserve **95**, 99, 125, 126, **231**, 433, 464, 465, 466, 614, 649
Welsh Commission 603–604
Weltevreden 469KQ 86
Weltevreden 478KQ 531, 536, **539**
Weltevreden 49KR/848KR **29**, **350**, 354–361
Weltevreden 523KR 180
Weltevreden 508LR, New Belgium **282**, 333
Wentworth concentration camp 556
Wentzel 342LQ archaeological site **20**, 37
Werkendam 628LR 351, **700**, 701, 706
Westfalia 313
Westwood, Prof. J.O., Oxford University 402
Whitbread, Jane 228
White, Eleanor (Annie) (m. Averre) 324, 325, 335
White, William, principal 165
Whitehouse, engineer-in-chief, SAR&H 377, 378
Whitehouse, George 185
whooping cough epidemic 480, 586
Wiener, Charlotte 523
Wilcox, C.T.M., Railways Board member 386, **388**
Wildebeestfontein 297 (now Diamant 228KQ) 659
Willard, Charlie 188
Willemse, Hermanus, teacher/farmer 227, 233, 236–238, **442**, 444, 497
Willet, Eric **442**
Willets, Lt, SAC, Nylstroom 161
Williams, A.J. 600
Williams, J.H. 620–621
Williams, Lt Mervyn 695–696
Williamson, H. 581
Willson, G.V. 380
Wilson, Benjamin Beauman 666
Wilson, Lt Col Henry Hughes 262
Windhoek, Namibia 681, **683**, 684, 691, **693**, 704
Wit Kerk, Die **476**, 487, 488
Witbooi, Hendrik 685
Witfontein 6KR **234**
Witfontein 86LQ 54
Witfontein 232, 233, **234**, 235, 236, **349**, 497; family cemetery **235**, 236; postal agency (Gedenk) & store **233**, 235; school 233, **234**, 238, 497, **578**, 592, 593, 597
Witfonteinrant 620
Witklip 100KR 237, **241**
Witkop ridge (Siloquana ridge) **403**
Witwatersrand Native Labour Agency (WNLA) 695
Wölff, Karl, surveyor 671
Wolvenfontein 149KR **214**, 305
Wonderkrater 25 *see also* Driefontein 317KR
Woodbush area 145, 313, 511
Woolonsky, Burgher D. 532
World Alliance of Reformed Churches 498
World Health Organisation (WHO) 396, 417, 420, 423
World Heritage site 20, 43
World War, First — after 179, 190, 349, 373, 377, 434, 525, 580, 620; before 92, 346; consequences 434, 484, 512, 557; during 146, 275, 316, 362, 434, 480, 620, 656; outbreak 183, 243, 372, 377, 516, 557, 558; service in 110, **113**, 184, **243**, **246**, 262, 310, 372–373, 484, **514**, 558, 583
World War, Second — after 170, 191, 225, 333, 419, 455, 480, 497, 510; before 170, 177, 425, 485; consequences 480, 485, 512, 604, 622; during 191, 328, 416, 455, 480, 490, 514, 525, 526, 564, 622, 636, 646; outbreak 168, 277, 328, 362, 510; service in 102, 184, 212
Worrall, Dr A., Serowe District Surgeon 707
Wrighton, Lt Albert William 531, 534, **534**, 770
Würth, Theodor 447
Wynkeldershoek 608LR, New Belgium **282**, **286**, 294, 326, 351
Wynne, Nina (m. Baber) 249

yellowwood 342, 345, 584
Young, Frederick 201
Youyou, Dr Tu 413
Ypres siding 179
Ypres, Battle of 484

Zaaiplaats 223KR 635
Zaaiplaats Tin Mining Company Limited 613, **627**, 628, 631, 635; *see also* tin

Zanddrift 94KR 123, 148, 233, 236, 237, 238, 239, 240, 241, **242**, 256, 497; maternity home **242**, 353; school **242**, 353, 492, 493, **578**, 581, **581**, 589, 596; settlers 237

Zandfontein 20KR **349**; Zandfontein-Vogel School **349**, 351, 353, **578**, 589, 592, **592**

Zandfontein 476KQ (1914 Rebellion) 529, 530, 531, 532, **533**, 534, 535, 539, **539**, 542, 544, **647**

Zandfontein 566KR 480, 569

Zandrivier 138KR 211, **214**

Zandriviersberge *see* Sandriviersberge

Zandriviersspoort/Sandriviersspoort **128**, 148, 220, 252, **253**, 256, 259, 261, 262, **264**, 363, 433, 569, 599

Zandriviersspoort 146KR 274; hotel, postal agency & store 92, 134, 146, 215, **257**, **264**, 270, **272**, **273**, 274, **275**, **276**, 277, 286, 293, 318, 436, **441**, 544; police post **264**, 268, 269, **269**, 270, 346, 425, **441**, rail line **264**, 377, 379, 385; road through poort 122, 123, **255**, 256, **264**, 294, 407, 441, 527, 580, 666; war skirmishes 108, 257, 259, 261, 262, **262**, 263, **263**, 264, **264**, 265, **265**, 266, 302, 305, 307, 369

ZCC *see* Zion Christian Church

Zebediela scheduled area 52, 53, 62, 80, 179, 190, 314, 320, 491, 515, 619

Zeederberg, Arthur & Shelley 213

Zeederberg, Dolf 213

Zeederberg, Harvey 213

Zeederberg, James 213

Zeederberg, Kathleen (née Farrant) 213

Zeederburg Coach Service 175, 407, 524

Zeerust 59, 85, 511

Zeraua, a Herero chief 685

Zerbst, J.B.H. 192–193

Ziervogel, Anna Suzanna (m. Watermeyer) 711

Zimmerman & Jaffet, attorneys 189

Zion Africa Church (ZAC) 501

Zion Christian Church (ZCC) 499–500, **501**, 502–507; Dove **501**, **504**, 505–506; factions 499, 503; *na leebana* 504; Star (of David) **501**, **503**; Zion City Moria 499, **501**, 503, **506**

Ziziphus spp. 429

Zondagsfontein 67KR **700**, 702

Zondagsloop 56KR 295, 343, **349**

Zoutpan 256, 390

Zoutpansberg Commando 104, 555

Zoutpansberg Hebrew Congregation 423

Zoutpansberg Review & Mining Journal, a newspaper 145, 276, 223, **224**, 225

Zoutpansbergdorp 63, 64, 99, 150, 200

Zuid-Afrikaansche Republiek (ZAR) 62, 72, 73, 104, 121, 124, 130–138, 151, 199, 203, 210, 211, 254, 280, 284, 295, 298, 303, 311, 431, 433, 469–470, 477, 548, 577, 666

Zulu, the 49, 55, 57, 63, 469; Zulu War 290

Zürn, Lt 684, 685

Zwagershoek *see* Swaershoek

Zwagers Hoek Farmers' Society *see* Swaershoek

Zwartfontein 818LR 82

Zwartkloof Fluorspar **627**, 636

Zwide, Ndwandwe chief 55, 62

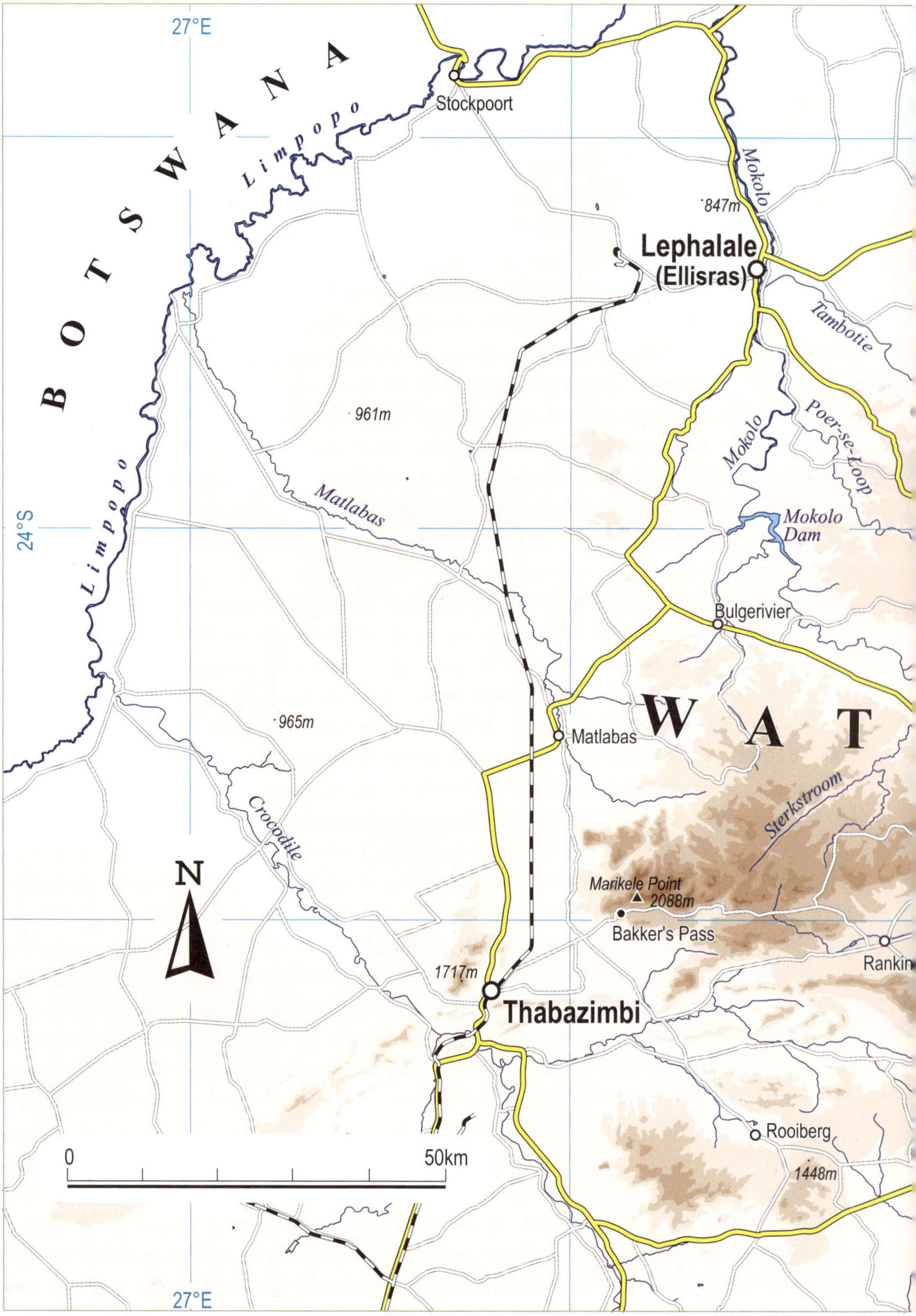

27°E
BOTSWANA
Limpopo
Stockpoort
Mokolo
847m
Lephalale
(Ellisras)
Tambotie
961m
Poer-se-Loop
Mokolo
Matlabas
24°S
Mokolo
Dam
Limpopo
Bulgerivier
965m
Matlabas
W A T
Sterkstroom
Crocodile
N
Marikele Point
2088m
Bakker's Pass
Rankin
1717m
Thabazimbi
Rooiberg
0
50km
1448m
27°E